T0191534

Lecture Notes in Computer Science 13315

More information about this series at https://link.springer.com/bookseries/558

Gabriele Meiselwitz (Ed.)

Social Computing and Social Media

Design, User Experience and Impact

14th International Conference, SCSM 2022
Held as Part of the 24th HCI International Conference, HCII 2022
Virtual Event, June 26 – July 1, 2022
Proceedings, Part I

 Springer

Editor
Gabriele Meiselwitz
Towson University
Towson, MD, USA

ISSN 0302-9743 ISSN 1611-3349 (electronic)
Lecture Notes in Computer Science
ISBN 978-3-031-05060-2 ISBN 978-3-031-05061-9 (eBook)
https://doi.org/10.1007/978-3-031-05061-9

This Springer imprint is published by the registered company Springer Nature Switzerland AG
The registered company address is: Gewerbestrasse 11, 6330 Cham, Switzerland

Foreword

Human-computer interaction (HCI) is acquiring an ever-increasing scientific and industrial importance, as well as having more impact on people's everyday life, as an ever-growing number of human activities are progressively moving from the physical to the digital world. This process, which has been ongoing for some time now, has been dramatically accelerated by the COVID-19 pandemic. The HCI International (HCII) conference series, held yearly, aims to respond to the compelling need to advance the exchange of knowledge and research and development efforts on the human aspects of design and use of computing systems.

The 24th International Conference on Human-Computer Interaction, HCI International 2022 (HCII 2022), was planned to be held at the Gothia Towers Hotel and Swedish Exhibition & Congress Centre, Göteborg, Sweden, during June 26 to July 1, 2022. Due to the COVID-19 pandemic and with everyone's health and safety in mind, HCII 2022 was organized and run as a virtual conference. It incorporated the 21 thematic areas and affiliated conferences listed on the following page.

A total of 5583 individuals from academia, research institutes, industry, and governmental agencies from 88 countries submitted contributions, and 1276 papers and 275 posters were included in the proceedings to appear just before the start of the conference. The contributions thoroughly cover the entire field of human-computer interaction, addressing major advances in knowledge and effective use of computers in a variety of application areas. These papers provide academics, researchers, engineers, scientists, practitioners, and students with state-of-the-art information on the most recent advances in HCI. The volumes constituting the set of proceedings to appear before the start of the conference are listed in the following pages.

The HCI International (HCII) conference also offers the option of 'Late Breaking Work' which applies both for papers and posters, and the corresponding volume(s) of the proceedings will appear after the conference. Full papers will be included in the 'HCII 2022 - Late Breaking Papers' volumes of the proceedings to be published in the Springer LNCS series, while 'Poster Extended Abstracts' will be included as short research papers in the 'HCII 2022 - Late Breaking Posters' volumes to be published in the Springer CCIS series.

I would like to thank the Program Board Chairs and the members of the Program Boards of all thematic areas and affiliated conferences for their contribution and support towards the highest scientific quality and overall success of the HCI International 2022 conference; they have helped in so many ways, including session organization, paper reviewing (single-blind review process, with a minimum of two reviews per submission) and, more generally, acting as goodwill ambassadors for the HCII conference.

This conference would not have been possible without the continuous and unwavering support and advice of Gavriel Salvendy, founder, General Chair Emeritus, and Scientific Advisor. For his outstanding efforts, I would like to express my appreciation to Abbas Moallem, Communications Chair and Editor of HCI International News.

June 2022   Constantine Stephanidis

HCI International 2022 Thematic Areas and Affiliated Conferences

Thematic Areas

- HCI: Human-Computer Interaction
- HIMI: Human Interface and the Management of Information

Affiliated Conferences

- EPCE: 19th International Conference on Engineering Psychology and Cognitive Ergonomics
- AC: 16th International Conference on Augmented Cognition
- UAHCI: 16th International Conference on Universal Access in Human-Computer Interaction
- CCD: 14th International Conference on Cross-Cultural Design
- SCSM: 14th International Conference on Social Computing and Social Media
- VAMR: 14th International Conference on Virtual, Augmented and Mixed Reality
- DHM: 13th International Conference on Digital Human Modeling and Applications in Health, Safety, Ergonomics and Risk Management
- DUXU: 11th International Conference on Design, User Experience and Usability
- C&C: 10th International Conference on Culture and Computing
- DAPI: 10th International Conference on Distributed, Ambient and Pervasive Interactions
- HCIBGO: 9th International Conference on HCI in Business, Government and Organizations
- LCT: 9th International Conference on Learning and Collaboration Technologies
- ITAP: 8th International Conference on Human Aspects of IT for the Aged Population
- AIS: 4th International Conference on Adaptive Instructional Systems
- HCI-CPT: 4th International Conference on HCI for Cybersecurity, Privacy and Trust
- HCI-Games: 4th International Conference on HCI in Games
- MobiTAS: 4th International Conference on HCI in Mobility, Transport and Automotive Systems
- AI-HCI: 3rd International Conference on Artificial Intelligence in HCI
- MOBILE: 3rd International Conference on Design, Operation and Evaluation of Mobile Communications

List of Conference Proceedings Volumes Appearing Before the Conference

1. LNCS 13302, Human-Computer Interaction: Theoretical Approaches and Design Methods (Part I), edited by Masaaki Kurosu
2. LNCS 13303, Human-Computer Interaction: Technological Innovation (Part II), edited by Masaaki Kurosu
3. LNCS 13304, Human-Computer Interaction: User Experience and Behavior (Part III), edited by Masaaki Kurosu
4. LNCS 13305, Human Interface and the Management of Information: Visual and Information Design (Part I), edited by Sakae Yamamoto and Hirohiko Mori
5. LNCS 13306, Human Interface and the Management of Information: Applications in Complex Technological Environments (Part II), edited by Sakae Yamamoto and Hirohiko Mori
6. LNAI 13307, Engineering Psychology and Cognitive Ergonomics, edited by Don Harris and Wen-Chin Li
7. LNCS 13308, Universal Access in Human-Computer Interaction: Novel Design Approaches and Technologies (Part I), edited by Margherita Antona and Constantine Stephanidis
8. LNCS 13309, Universal Access in Human-Computer Interaction: User and Context Diversity (Part II), edited by Margherita Antona and Constantine Stephanidis
9. LNAI 13310, Augmented Cognition, edited by Dylan D. Schmorrow and Cali M. Fidopiastis
10. LNCS 13311, Cross-Cultural Design: Interaction Design Across Cultures (Part I), edited by Pei-Luen Patrick Rau
11. LNCS 13312, Cross-Cultural Design: Applications in Learning, Arts, Cultural Heritage, Creative Industries, and Virtual Reality (Part II), edited by Pei-Luen Patrick Rau
12. LNCS 13313, Cross-Cultural Design: Applications in Business, Communication, Health, Well-being, and Inclusiveness (Part III), edited by Pei-Luen Patrick Rau
13. LNCS 13314, Cross-Cultural Design: Product and Service Design, Mobility and Automotive Design, Cities, Urban Areas, and Intelligent Environments Design (Part IV), edited by Pei-Luen Patrick Rau
14. LNCS 13315, Social Computing and Social Media: Design, User Experience and Impact (Part I), edited by Gabriele Meiselwitz
15. LNCS 13316, Social Computing and Social Media: Applications in Education and Commerce (Part II), edited by Gabriele Meiselwitz
16. LNCS 13317, Virtual, Augmented and Mixed Reality: Design and Development (Part I), edited by Jessie Y. C. Chen and Gino Fragomeni
17. LNCS 13318, Virtual, Augmented and Mixed Reality: Applications in Education, Aviation and Industry (Part II), edited by Jessie Y. C. Chen and Gino Fragomeni

18. LNCS 13319, Digital Human Modeling and Applications in Health, Safety, Ergonomics and Risk Management: Anthropometry, Human Behavior, and Communication (Part I), edited by Vincent G. Duffy

19. LNCS 13320, Digital Human Modeling and Applications in Health, Safety, Ergonomics and Risk Management: Health, Operations Management, and Design (Part II), edited by Vincent G. Duffy

20. LNCS 13321, Design, User Experience, and Usability: UX Research, Design, and Assessment (Part I), edited by Marcelo M. Soares, Elizabeth Rosenzweig and Aaron Marcus

21. LNCS 13322, Design, User Experience, and Usability: Design for Emotion, Well-being and Health, Learning, and Culture (Part II), edited by Marcelo M. Soares, Elizabeth Rosenzweig and Aaron Marcus

22. LNCS 13323, Design, User Experience, and Usability: Design Thinking and Practice in Contemporary and Emerging Technologies (Part III), edited by Marcelo M. Soares, Elizabeth Rosenzweig and Aaron Marcus

23. LNCS 13324, Culture and Computing, edited by Matthias Rauterberg

24. LNCS 13325, Distributed, Ambient and Pervasive Interactions: Smart Environments, Ecosystems, and Cities (Part I), edited by Norbert A. Streitz and Shin'ichi Konomi

25. LNCS 13326, Distributed, Ambient and Pervasive Interactions: Smart Living, Learning, Well-being and Health, Art and Creativity (Part II), edited by Norbert A. Streitz and Shin'ichi Konomi

26. LNCS 13327, HCI in Business, Government and Organizations, edited by Fiona Fui-Hoon Nah and Keng Siau

27. LNCS 13328, Learning and Collaboration Technologies: Designing the Learner and Teacher Experience (Part I), edited by Panayiotis Zaphiris and Andri Ioannou

28. LNCS 13329, Learning and Collaboration Technologies: Novel Technological Environments (Part II), edited by Panayiotis Zaphiris and Andri Ioannou

29. LNCS 13330, Human Aspects of IT for the Aged Population: Design, Interaction and Technology Acceptance (Part I), edited by Qin Gao and Jia Zhou

30. LNCS 13331, Human Aspects of IT for the Aged Population: Technology in Everyday Living (Part II), edited by Qin Gao and Jia Zhou

31. LNCS 13332, Adaptive Instructional Systems, edited by Robert A. Sottilare and Jessica Schwarz

32. LNCS 13333, HCI for Cybersecurity, Privacy and Trust, edited by Abbas Moallem

33. LNCS 13334, HCI in Games, edited by Xiaowen Fang

34. LNCS 13335, HCI in Mobility, Transport and Automotive Systems, edited by Heidi Krömker

35. LNAI 13336, Artificial Intelligence in HCI, edited by Helmut Degen and Stavroula Ntoa

36. LNCS 13337, Design, Operation and Evaluation of Mobile Communications, edited by Gavriel Salvendy and June Wei

37. CCIS 1580, HCI International 2022 Posters - Part I, edited by Constantine Stephanidis, Margherita Antona and Stavroula Ntoa

38. CCIS 1581, HCI International 2022 Posters - Part II, edited by Constantine Stephanidis, Margherita Antona and Stavroula Ntoa

39. CCIS 1582, HCI International 2022 Posters - Part III, edited by Constantine Stephanidis, Margherita Antona and Stavroula Ntoa
40. CCIS 1583, HCI International 2022 Posters - Part IV, edited by Constantine Stephanidis, Margherita Antona and Stavroula Ntoa

http://2022.hci.international/proceedings

Preface

The 14th International Conference on Social Computing and Social Media (SCSM 2022) was an affiliated conference of the HCI International (HCII) conference. The conference provided an established international forum for the exchange and dissemination of scientific information related to social computing and social media, addressing a broad spectrum of issues expanding our understanding of current and future issues in these areas. The conference welcomed qualitative and quantitative research papers on a diverse range of topics related to the design, development, assessment, use, and impact of social media.

The importance of social computing and social media in today's society has dramatically increased during the COVID-19 pandemic, as people worldwide have been forced to communicate, work, study, shop, and spend their free time online. This has brought about a renewed interest in renovating the design and user experience of online environments, as well as the analysis of their impact on society in general and on critical application domains such as education and commerce more specifically.

Two volumes of the HCII 2022 proceedings are dedicated to this year's edition of the SCSM conference, entitled Social Computing and Social Media: Design, User Experience and Impact (Part I) and Social Computing and Social Media: Applications in Education and Commerce (Part II). The first focuses on topics related to novel approaches to design and user experience in social media and social live streaming, text analysis and AI in social media, and social media impact on society and business, while the second focuses on topics related to social media in education as well as customer experience and consumer behavior.

Papers of these volumes are included for publication after a minimum of two single-blind reviews from the members of the SCSM Program Board or, in some cases, from members of the Program Boards of other affiliated conferences. I would like to thank all of them for their invaluable contribution, support, and efforts.

June 2022   Gabriele Meiselwitz

14th International Conference on Social Computing and Social Media (SCSM 2022)

Program Board Chair: **Gabriele Meiselwitz,** Towson University, USA

- Rocio Abascal Mena, Universidad Autónoma Metropolitana-Cuajimalpa, Mexico
- Francisco Alvarez Rodríguez, Universidad Autónoma de Aguascalientes, Mexico
- Andria Andriuzzi, Université Jean Monnet, France
- Karine Berthelot-Guiet, Sorbonne University, France
- James Braman, Community College of Baltimore County, USA
- Adheesh Budree, University of Cape Town, South Africa
- Adela Coman, University of Bucharest, Romania
- Tina Gruber-Muecke, Anton Bruckner Private University, Austria
- Hung-Hsuan Huang, University of Fukuchiyama, Japan
- Ajrina Hysaj, University of Wollongong in Dubai, United Arab Emirates
- Aylin Imeri, Heinrich Heine University Düsseldorf, Germany
- Ayaka Ito, Reitaku University, Japan
- Carsten Kleiner, University of Applied Sciences and Arts Hannover, Germany
- Jeannie Lee, Singapore Institute of Technology, Singapore
- Ana Isabel Molina, University of Castilla-La Mancha, Spain
- Takashi Namatame, Chuo University, Japan
- Hoang D. Nguyen, University of Glasgow, Singapore
- Kohei Otake, Tokai University, Japan
- Oronzo Parlangeli, University of Siena, Italy
- Daniela Quiñones, Pontificia Universidad Católica de Valparaíso, Chile
- Cristian Rusu, Pontificia Universidad Católica de Valparaíso, Chile
- Virginica Rusu, Universidad de Playa Ancha, Chile
- Christian W. Scheiner, Universität zu Lübeck, Germany
- Tomislav Stipancic, University of Zagreb, Croatia
- Simona Vasilache, University of Tsukuba, Japan
- Yuanqiong Wang, Towson University, USA
- Brian Wentz, Shippensburg University, USA

The full list with the Program Board Chairs and the members of the Program Boards of all thematic areas and affiliated conferences is available online at

http://www.hci.international/board-members-2022.php

HCI International 2023

The 25th International Conference on Human-Computer Interaction, HCI International 2023, will be held jointly with the affiliated conferences at the AC Bella Sky Hotel and Bella Center, Copenhagen, Denmark, 23–28 July 2023. It will cover a broad spectrum of themes related to human-computer interaction, including theoretical issues, methods, tools, processes, and case studies in HCI design, as well as novel interaction techniques, interfaces, and applications. The proceedings will be published by Springer. More information will be available on the conference website: http://2023.hci.international/.

General Chair
Constantine Stephanidis
University of Crete and ICS-FORTH
Heraklion, Crete, Greece
Email: general_chair@hcii2023.org

http://2023.hci.international/

Contents – Part I

Design and User Experience in Social Media and Social Live Streaming

Communication and Information About Breast Cancer: A Comparative
Study Between a Physical and an Online Environment 3
Leticia Barbosa and André Pereira Neto

User Experience Evaluation of a Computational Thinking-Enhanced
Problem-Solving Tool: Findings and Next Steps 13
Juan Felipe Calderon, Luis A. Rojas, Katrina Sorbello, and Nibaldo Acero

Automated Tools for Usability Evaluation: A Systematic Mapping Study 28
John W. Castro, Ignacio Garnica, and Luis A. Rojas

Exploring Links Between the Interaction with Social Media and Subjective
Well-Being: An Exploratory Study 47
Beatriz de Paulo and Manuela Quaresma

Improving EEG-based Motor Execution Classification for Robot Control 65
*Sumeyra U. Demir Kanik, Wenjie Yin, Arzu Guneysu Ozgur,
Ali Ghadirzadeh, Mårten Björkman, and Danica Kragic*

User Experience in Mobile Social TV: Understanding Requirements
and Presenting Design Guidelines 83
Taume Dery and Yavuz Inal

Perspectives for Using Smart Augmented Reality for the Future in Social
Computing and Collaborative Assistance 97
Ralf Doerner

Analysis of the Relationship Between Food and the Writer's Emotions
Using a Meal Diary 110
Shoki Eto, Kohei Otake, and Takashi Namatame

Temporal and Geographic Oriented Event Retrieval for Historical Analogy 123
Kengo Fushimi and Yasunobu Sumikawa

Eye Tracking to Evaluate the User eXperience (UX): Literature Review 134
Matías García and Sandra Cano

Psychological Characteristics Estimation from On-Road Driving Data 146
  Ryusei Kimura, Takahiro Tanaka, Yuki Yoshihara, Kazuhiro Fujikake,
  Hitoshi Kanamori, and Shogo Okada

FLAS: A Platform for Studying Attacks on Federated Learning 160
  Yuanchao Loh, Zichen Chen, Yansong Zhao, and Han Yu

A Study on the Influencing Factors of User Interaction Mode Selection
in the Short Video Industry: A Case Study of TikTok 170
  Haoxuan Peng, Xuanwu Zhang, and Cong Cao

User Preferences for Organizing Social Media Feeds 185
  Kristine M. Rogers

A Property Checklist to Evaluate the User Experience for People
with Autism Spectrum Disorder 205
  Katherine Valencia, Federico Botella, and Cristian Rusu

Operation Strategies for the PUGC Social Media Video Platform Based
on the Value-Chain Perspective—A Chinese Case 'Bilibili' 217
  Mu Zhang and Han Han

Behind the Scenes: Exploring Context and Audience Engagement
Behaviors in YouTube Vlogs 227
  Hantian Zhang

Text Analysis and AI in Social Media

Development of Bilingual Sentiment and Emotion Text Classification
Models from COVID-19 Vaccination Tweets in the Philippines 247
  Nicole Allison Co, Maria Regina Justina Estuar, Hans Calvin Tan,
  Austin Sebastien Tan, Roland Abao, and Jelly Aureus

Fine Grained Categorization of Drug Usage Tweets 267
  Priyanka Dey and ChengXiang Zhai

Text Mining for Patterns and Associations on Functions, Relationships
and Prioritization in Services Reflected in National Health Insurance
Programs 281
  Maria Regina Justina Estuar, Maria Cristina Bautista,
  Christian Pulmano, Paulyn Jean Acacio-Claro, Quirino Sugon,
  Dennis Andrew Villamor, and Madeleine Valera

Analyzing Change on Emotion Scores of Tweets Before and After
Machine Translation 294
   Karin Fukuda and Qun Jin

Automatic Meme Generation with an Autoregressive Transformer 309
   Denis Gordeev and Vsevolod Potapov

Multimodal Emotion Analysis Based on Visual, Acoustic and Linguistic
Features 318
   Leon Koren, Tomislav Stipancic, Andrija Ricko, and Luka Orsag

Linguistic and Contextual Analysis of SNS Posts for Approval Desire 332
   Erina Murata, Kiichi Tago, and Qun Jin

Social Media Engagement Anxiety: Triggers in News Agenda 345
   Kamilla Nigmatullina and Nikolay Rodossky

14 Days Later: Temporal Topical Shifts in Covid-19 Related Tweets After
Pandemic Declaration 358
   Hamzah Osop, Basem Suleiman, and Abdallah Lakhdari

Development of a Text Classification Model to Detect Disinformation
About COVID-19 in Social Media: Understanding the Features
and Narratives of Disinformation in the Philippines 370
   Hans Calvin Tan, Maria Regina Justina Estuar, Nicole Allison Co,
   Austin Sebastien Tan, Roland Abao, and Jelly Aureus

A Comparison of Web Services for Sentiment Analysis in Digital Mental
Health Interventions 389
   Toh Hsiang Benny Tan, Sufang Lim, Yang Qiu, and Chunyan Miao

An Extendable Sentiment Monitoring Model for SNS Considering
Environmental Factors 408
   Yenjou Wang, Neil Yen, and Qun Jin

Empirical Evaluation of Machine Learning Ensembles for Rumor Detection ... 422
   Andrés Zapata, Eliana Providel, and Marcelo Mendoza

A Methodological Framework for Facilitating Explainable AI Design 437
   Jiehuang Zhang and Han Yu

Social Media Impact on Society and Business

The Role of Moral Receptors and Moral Disengagement in the Conduct
of Unethical Behaviors Against Whistleblowers on Social Media   449
  Stefan Becker and Christian W. Scheiner

Dynamics of Distrust, Aggression, and Conspiracy Thinking
in the Anti-vaccination Discourse on Russian Telegram   468
  Svetlana S. Bodrunova and Dmitry Nepiyuschikh

Gender-Sensitive Materials and Tools: The Development
of a Gender-Sensitive Toolbox Through National Stakeholder
Consultations   485
  Eirini Christou, Antigoni Parmaxi, Maria Perifanou,
  and Anastasios A. Economides

A Multidisciplinary Approach to Leadership During the COVID 19 Era.
The Case of Romania   503
  Adela Coman, Mihaela Cornelia Sandu, Valentin Mihai Leoveanu,
  and Ana-Maria Grigore

Twitter and the Dissemination of Information Related to the Access
to Credit for Cancer Survivors: The Case of the "Right to Be Forgotten"
in France   517
  Renaud Debailly, Hugo Jeaningros, and Gaël Lejeune

Patient-Led Medicalisation and Demedicalisation Processes Through
Social Media - An Interdisciplinary Approach   529
  Juliette Froger-Lefebvre and Julia Tinland

Social Intelligence Design for Social Computing   545
  Renate Fruchter, Toyoaki Nishida, and Duska Rosenberg

Influence Vaccination Policy, Through Social Media Promotion
(Study: West Java, East Java, and Central Java)   559
  Ekklesia Hulahi, Achmad Nurmandi, Isnaini Muallidin,
  Mohammad Jafar Loilatu, and Danang Kurniawan

The Role of the Financial Services Authority (OJK) in Preventing Illegal
Fintech Landing in the COVID-19 Pandemic in Indonesia   568
  Bella Kharisma, Achmad Nurmandi, Isnaini Muallidin,
  Danang Kurniawan, and Mohammad Jafar Loilatu

Engagement as Leadership-Practice for Today's Global Wicked Problems:
Leadership Learning for Artificial Intelligence   578
  Wanda Krause and Alexandru Balasescu

Open Innovation within Life Sciences: Industry-Specific Challenges
and How to Improve Interaction with External Ecosystems 588
Niclas Kröger, Maximilian Rapp, and Christoph Janach

Moderation of Deliberation: How Volunteer Moderators Shape Political
Discussion in Facebook Groups? 602
Sanna Malinen

The Platform-of-Platforms Business Model: Conceptualizing a Way
to Maximize Valuable User Interactions on Social Media Platforms 617
Jürgen Rösch and Christian V. Baccarella

The Impact of Tweets, Mandates, Hesitancy and Partisanship
on Vaccination Rates 631
Cheng Lock Donny Soh and Indriyati Atmosukarto

Exploring Public Trust on State Initiatives During the COVID-19 Pandemic ... 643
*Austin Sebastien Tan, Maria Regina Justina Estuar, Nicole Allison Co,
Hans Calvin Tan, Roland Abao, and Jelly Aureus*

Twitter as a Communication Tools for Vaccine Policy in Indonesia:
An Analysis 661
Iradhad Taqwa Sihidi, Salahudin, Ali Roziqin, and Danang Kurniawan

Author Index 673

Contents – Part II

Social Media in Education

Design and Evaluation of a Programming Tutor Based on an Instant
Messaging Interface 3
  Claudio Alvarez, Luis A. Rojas, and Juan de Dios Valenzuela

Embedding Social, Community Projects Within Contemporary Curricular 21
  Pranit Anand

Emergency Remote Teaching in the University Context: Responding
to Social and Emotional Needs During a Sudden Transition Online 30
  Magdalena Brzezinska and Edward Cromarty

Development of an Explicit Agent-Based Simulation Toolkit for Opening
of Schools: An Implementation of COMOKIT for Universities
in the Philippines 48
  Maria Regina Justina Estuar, Roland Abao, Jelly Aureus,
  Zachary Pangan, Lenard Paulo Tamayo, Elvira de Lara-Tuprio,
  Timothy Robin Teng, and Rey Rodrigueza

Exploring Faculty Members Perception of Utilizing Technology
to Enhance Student Engagement in the United Arab Emirates: Technology
and the ICAP Modes of Engagement 67
  Georgina Farouqa and Ajrina Hysaj

Active Learning in the Lenses of Faculty: A Qualitative Study
in Universities in the United Arab Emirates 77
  Georgina Farouqa and Ajrina Hysaj

Food Sector Entrepreneurship: Designing an Inclusive Module Adaptable
to Both Online and Blended Learning Environments in Higher Education 91
  Marco Garcia-Vaquero

Dimensions of Formative Feedback During the COVID-19 Pandemic:
Evaluating the Perceptions of Undergraduates in Multicultural EAP
Classrooms 103
  Ajrina Hysaj and Doaa Hamam

Online Formative Assessment and Feedback: A Focus Group Discussion
Among Language Teachers 115
  Ajrina Hysaj and Harshita Aini Haroon

Analyzing the Impact of Culture on Students: Towards a Student
eXperience Holistic Model 127
  Nicolás Matus, Ayaka Ito, and Cristian Rusu

Enhancing Concept Anchoring Through Social Aspects and Gaming
Activities 136
  Marie J. Myers

Work Organization and Effects of Isolation on the Perception
of Misconduct in Italian Universities During Covid-19 Pandemic 147
  Oronzo Parlangeli, Margherita Bracci, Stefano Guidi,
  Enrica Marchigiani, and Paola Palmitesta

Re-imagining the Distributed Nature of Learner Engagement
in Computer-Supported Collaborative Learning Contexts
in the Post-pandemic Era 161
  Andriani Piki

Experiential Learning Through Virtual Tours in Times of COVID-19 180
  Roxana Sandu

Undergraduate Emirati Students' Challenges of Language Barrier
in Meeting Expectations of English Medium University in the UAE 199
  Sara Suleymanova and Ajrina Hysaj

Building an Educational Social Media Application for Higher Education 210
  Felix Weber, Niklas Dettmer, Katharina Schurz, and Tobias Thelen

Customer Experience and Consumer Behavior

Augmented Reality Filters and the Faces as Brands: Personal Identities
and Marketing Strategies in the Age of Algorithmic Images 223
  Ruggero Eugeni

Analysis of the Behavior of E-Sports and Streaming Consumers in Latin
America 235
  Cristobal Fernandez-Robin, Diego Yañez, Scott McCoy, and Pablo Flores

Information Consumer eXperience: A Chilean Case Study 248
  María Paz Godoy, Cristian Rusu, and Jonathan Ugalde

The Biodigital Rises: A New Digital Brand Challenge 268
  Marie-Nathalie Jauffret and Frédéric Aubrun

Yellow or Blue Dress: How a Product Page Can Impact the Customer
Experience 278
    Catalina Montecinos and Camila Bascur

Evaluating Store Features Using Consumer Reviews in Beauty Salons 292
    Ryo Morooka, Takashi Namatame, and Kohei Otake

The Internet-of-Things and AI and Their Use for Marketers 308
    Marc Oliver Opresnik

Analyzing Methods, Instruments, and Tools for Evaluating the Customer
eXperience 317
    Luis Rojas and Daniela Quiñones

Evaluating the Post-pandemic Tourist Experience: A Scale for Tourist
Experience in Valparaíso, Chile 331
    Virginica Rusu, Leslie Márquez, Patricia González, and Cristian Rusu

Purchasing Behavior Analysis Model that Considers the Relationship
Between Topic Hierarchy and Item Categories 344
    Yuta Sakai, Yui Matsuoka, and Masayuki Goto

Resale Price Prediction Model for Used Electrical Products Using Deep
Neural Network 359
    Shinnosuke Terasawa, Kohei Otake, and Takashi Namatame

An Indicator to Measure the Relationship Between Firms and Consumers
Based on the Subjective Well-Being of Consumers: Promoting Corporate
Social Contribution Activities to Maintain Socially Sustainable
Development 375
    Masao Ueda

Evaluation of Analysis Model for Products with Coefficients of Binary
Classifiers and Consideration of Way to Improve 388
    Ayako Yamagiwa and Masayuki Goto

Clustering and Feature Analysis of Shoes Brands Using Questionnaire
Data and Word-of-Mouth Review Data 403
    Haruki Yamaguchi, Kohei Otake, and Takashi Namatame

Corner-Shopping: Studying Attitudes and Consumer Behavior
on the Cornershop App 422
    Diego Yáñez, Cristóbal Fernández-Robin, and Florencia Bohle

Author Index 437

Design and User Experience in Social Media and Social Live Streaming

Communication and Information About Breast Cancer: A Comparative Study Between a Physical and an Online Environment

Leticia Barbosa and André Pereira Neto(⊠)

Oswaldo Cruz Foundation, Rio de Janeiro, Brazil
leticiatbs@gmail.com, andrepereiraneto@gmail.com

Abstract. Breast cancer is a severe health problem, implying countless, diverse impacts and transformations in the patient's life. In this context, increasingly more breast cancer patients turn to the Internet, especially online health communities, to address the experience of illness obtaining information and social support. Against this backdrop, this study aimed to investigate the following question: how is the expert patient phenomenon established among women with suspected or diagnosed breast cancer in a physical and an online space? Thus, a comparative ethnographic study was conducted in two different environments: the waiting room of a public university hospital and an organized Facebook online group. We identified that the expert patient phenomenon was present among the participants of the investigated online community but not among those who waited in the hospital waiting room. We could observe that the establishment of the expert patient is not a linear process and, among other factors, is associated with the individual's socioeconomic context.

Keywords: Internet · Expert patient · Breast cancer

1 Introduction

In the current epidemiological setting, breast cancer is one of the main types of cancer that stand out in the morbimortality trends in Brazil and globally. Except for non-melanoma skin cancer, it is the second most frequent type among the general Brazilian population and most frequent among women [1, 2].

Breast cancer implies numerous physical, emotional, and social impacts and challenges in patients' lives and can reduce their well-being, produce a traumatic event, and engender long-term distress. The type and intensity of such impacts may vary with women's disease stages, that is, when developing the pathology [3, 4].

In this context, increasingly more breast cancer patients turn to the Internet to address the illness experience. The speed and convenience of online access, associated with the practicality of the research, the possibility of remote communication, and the feeling of anonymity of digital networks, contribute to the search for health information on the Internet being an increasingly common habit among women with breast cancer.

G. Meiselwitz (Ed.): HCII 2022, LNCS 13315, pp. 3–12, 2022.
https://doi.org/10.1007/978-3-031-05061-9_1

The Internet facilitates their access and sharing of information about the neoplasm, knowing more about their clinical condition, diagnostic tests, available treatments, and their respective side effects, prognoses, relapses, and metastases, among other aspects and matters related to the experience of having breast cancer [5].

Besides searching for information, another daily use of the Internet among breast cancer patients refers to communicating with other women who have also been diagnosed with the neoplasm. Heaton [6] pointed out that "the Internet is not just a repository of information. It has also given rise to new forms of interaction and new groupings, or communities, that can bring together people across geographical space and time in all spheres of life". Faced with suspicion or diagnosis of breast cancer, women can turn to online health communities to discuss cancer.

The growing use of online health communities can be associated with the numerous advantages of this type of environment, including the transposition of space-time barriers, the home access convenience, continuous availability, and, in some instances, the feeling of anonymity [7]. These are easily accessible and readily available environments that provide patients with different forms of social support, such as informational support, tangible help, and emotional support [6].

Horrell et al. [8] affirm that "patients are motivated to join these communities to access support, advice, and accountability in reaching health goals". As a strategic space for accessing information and exchanging social support, online health communities can operate as an "extension of the health system", especially for patients with chronic diseases or conditions [9].

The growing participation in online communities has benefited the health of breast cancer patients and contributed to reconfiguring the patients' relationship with their condition and the medical professional. Among them, we highlight the establishment of the expert patient and the empowerment process [10, 11].

The term expert patient has been used to refer to individuals who practically become experts in their disease or health condition by carrying out an extensive search, mainly accessing and exchanging information on the Internet. These patients would be better able to manage their illness, engaging in the self-management of their health [12]. Therefore, they would enjoy a greater possibility of establishing a more collaborative and participatory relationship with the doctor regarding decisions regarding their clinical condition [9, 13].

Expert patients could increase their empowerment by finding out more about their health conditions. Although its definition may vary, empowerment refers to the transformation of the situation of vulnerability, inequality, or powerlessness of an individual or group from the use of different resources, tools, and competencies, encompassing the individual and collective dimension [14].

Online communities can be strategic social networks to receive and offer social support for women with breast cancer. By participating in an environment that gathers patients with the same neoplasm, women talk to each other and still feel the experience of living with this disease. In doing so, they can get support to address stressful situations, emotional crises, or problems arising from their health condition. The social support exchanged in online communities can also help them deal with insensitivity or misunderstanding regarding their disease expressed by health professionals and family and friends [8, 10, 11]. Besides obtaining social support, online communities can be an essential source of information for women with breast cancer to address self-care, self-management, and decision-making about their clinical condition.

In this context, this study aimed to investigate the following question: how is the expert patient phenomenon established among women with suspected or diagnosed breast cancer who are in a physical space and an online space? Thus, we conducted a multi-situated comparative study, adopting the methodological framework of ethnography. Next, we shall present the methodological procedures and discuss the results.

2 Methods

The study was carried out in two different environments that were important for the research subjects throughout their experience with the disease: the waiting room of the mastology outpatient clinic of the Gynecology Institute of the Federal University of Rio de Janeiro and an online health community about breast cancer. Such environments have specificities regarding their materiality, establishment process, and role in the disease experience.

The Gynecology Institute is a teaching, research, and extension unit part of the Federal University of Rio de Janeiro (UFRJ) – one of the oldest and most prestigious Brazilian universities. It provides medium-complexity care and is a reference specialized care establishment in the region. While it is not a high-complexity cancer care specialized unit, the Institute becomes an essential space for diagnosing and treating breast cancer among primary care users in the municipality and, ultimately, the entire state of Rio de Janeiro. The unit offers appointments in the mastology specialty, ultrasound and mammography tests, and breast cancer-related surgeries, thus becoming an essential environment for diagnosing and treating the neoplasm in the city's healthcare network.

Fieldwork was carried out in the hospital's waiting room, where women with suspected or diagnosed breast cancer were waiting for care. Thus, we aimed to explore a space and a specific stage of the patient's daily life: waiting for an appointment with a specialist doctor in a waiting room of a public hospital.

Another field of investigation was selected besides the Gynecology Institute: the online group "Câncer de Mama" organized on Facebook. It is an open group for sharing breast cancer information, especially among patients, an online environment with public access. We selected this community because it has more than 17,000 members. Its primary target audience is Brazilian women with suspected or diagnosed breast cancer, and it has a regular flow of activities: members posted content and informative links on the topic, shared personal experiences and concerns regarding different aspects of the disease, and commented on posts made by other people daily.

We chose this group because of the role that online health communities have assumed in the routine of cancer patients. While not mandatory for the diagnosis or treatment of the disease, like the hospital, participating in an online environment like this has been increasingly common among women with breast cancer to connect with others undergoing a similar experience and obtain information about the disease.

After selecting the "*Câncer de Mama*" online community and the waiting room of the mastology outpatient clinic of the Institute as research fields, we employed ethnography as a methodological reference, considering the specificities of each location.

The ethnographic approach has been widely used in the field of Anthropology to understand and analyze the ways of life and the beliefs and values system typical of human groups. Thus, in general terms, we understand ethnography as the science and art of describing a human group and their respective customs, institutions, practices, systems, and interpersonal relationships [15]. From the late 20th century to the early 21st century, with the spread of the Internet and other digital technologies, ethnography also was also employed to explore and analyze the socio-cultural aspects imbued in the practices, social relationships, and the uses associated with NICTs [16].

Direct observation was carried out in each selected environment to identify the flows of information and communication about breast cancer in the spaces and the dynamics of interaction between the patients considering the premises of the ethnographic approach. Besides observation, a questionnaire and a semi-structured interview were also used for data collection, guided by a roadmap with open-ended questions. The application of these tools facilitated the identification of four aspects regarding women with breast cancer: the socioeconomic profile, the current stage of their breast cancer course, their experience with the disease, and the use of the Internet in this process.

3 Results and Discussion

We considered two aspects to analyze the expert patient phenomenon among the research participants: the socioeconomic profile and the use of the Internet for health purposes. Thus, we intended to investigate whether women in the waiting room or the online community used the Internet to learn more about their health situation, contributing to their training as expert patients, and whether this had any relationship with their socioeconomic conditions.

3.1 Socioeconomic Profile, Internet Use, and Online Information Search on Breast Cancer

Regarding the socioeconomic profile, most of the Gynecology Institute and the online group participants were between 40 and 69 years old (n = 29/40). However, we identified more younger women in the community: 9 of the 24 participants in the community were 39 years old or younger, while only one of the 16 participants was in this age group at the Institute.

Participants at the Institute had more unfavorable socioeconomic conditions than those in the online community. Most of the hospital patients had an education level equal to or less than high school education and had low monthly household income, with an amount equal to or less than approximately US$749. Regarding employment status, they were unemployed or out of the workforce, and some were the leading providers of the monthly household income, which differs in part from the situation identified in the online community. More than half of the participants had completed high school or higher education (complete or incomplete). Fourteen out of 24 had an income of up to three minimum wages; eight had from three to five minimum wages; and three had above five minimum wages, and less than half of the total were responsible for the household income. Most were employed, and 13 had a formal job.

We emphasize that these results do not have statistical significance or are not intended to represent all the patients treated by the hospital or members of the *"Câncer de Mama"* community. However, they help us contextualize internet access, consumption and sharing of health information on the Internet, and the expert patient phenomenon.

Regarding internet access, most participants in the hospital and the online community use the technology regularly, mainly through cell phones. However, we observed that frequent access was not related to the use of the Internet in the breast cancer experience, especially concerning online information research activities at different stages of the disease.

The search for online information about breast cancer was investigated in the following stages: before the suspicion; the first appointment with a specialist doctor; performing diagnostic tests; receiving the results of diagnostic tests; confirmation of diagnosis; and intervals between diagnosis and treatment. We considered that such stages could be associated with a greater or lesser search for information about breast cancer due to the potential information needs among the patients. We mapped whether and how often the participant sought information online regarding each stage.

Only one patient of the Gynecology Institute patients sought information about breast cancer before the disease was suspected, with a frequency of at least once a month. Only three of the participants in the online community said they searched for breast cancer on the Internet before they suspected the disease, with two doing so less than once a month and only one searching daily or almost every day.

Few hospital patients sought information before or after the first appointment with the specialist doctor and the diagnostic tests. The opposite situation was observed among the online community participants: most searched for information online, before or after the appointment and diagnostic tests, and spent an hour or more a day searching.

Regarding the stage after receiving the test results, most hospital patients reported not seeking information online then, while most women in the online community indicated doing so, spending up to two hours daily in the search process. Regarding the diagnosis, four of the 11 participants from the Gynecology Institute who had received the diagnostic confirmation had searched for information about the disease on the Internet. In contrast, most of the participants from the online community carried out online search at this stage of the breast cancer course. Just before the onset of treatment, only three patients from the Institute stated that they had searched for information online about the disease. In the online community, 18 out of 24 participants reported seeking information.

Among hospital patients who reported having done online research about breast cancer at some point in the disease course, the time spent in the search process was, in general, less than an hour, while the time spent in the search process had a mean of one to two hours in the online community.

It is worth noting the research practices identified from the results obtained. We observed that the search for online information about breast cancer was found throughout the experience with the disease among most of the participants in the group. However, this practice was performed by a few participants in the online community sample before the course of the disease; that is, before the suspicion of neoplasm was installed. However, after the suspicion was initiated, almost all participants reported researching the disease online: 23 out of 24 indicated that they had searched for breast cancer on the Internet at least one of the stages related to the investigation, diagnosis, and treatment of breast cancer. The data obtained in the online community sample differ from that observed among the Institute's patients, who did not use the Internet to address breast cancer, despite regularly accessing the network.

The stage when more participants in the online community indicated searching for information about breast cancer online was after receiving the results of diagnostic tests (n = 22/24), followed by having the tests performed (n = 21/24); after confirmation of the diagnosis (n = 20/24); the first appointment with the specialist doctor (n = 19/24) and before the onset of treatment (n = 18/24), which differs from the situation observed at the Institute, where the stage in which more hospital participants searched for information online was the performance of tests (n = 6/16).

The responses obtained indicate that the hospital patients who searched for information about breast cancer on the Internet did not do so at all times or situations in the investigated disease course. In the case of community participants, while resorting to searching online information about breast cancer at different stages of the disease course, the search process was not linear: that is, the patients did not necessarily do so in all situations, sometimes alternating between one stage and another.

It should be noted that the moment of diagnosis does not seem to have been sufficient to encourage the search for online information among participants in the physical and online environments.

The suspicion or diagnosis of the disease implies different changes in women's routine, besides psychological impacts [17]. These changes can produce countless concerns for the patient, shaping information needs on addressing the disease. From this perspective, we consider that the confirmation of the diagnosis could be a central stage to encourage patients to search for breast cancer information on the Internet since it is associated with several transformations and impacts on the patient's life and, thus, can produce information needs. However, this was not observed among most Institute and online community participants. In the case of Institute patients, only one patient who did not seek information about the disease in pre-diagnosis situations reported doing so after confirmation. Regarding online community participants, those who sought information upon diagnostic confirmation had already carried out an online search previously.

3.2 The Expert Patient Phenomenon in the Analyzed Online and Physical Environments

Overall, we identified a residual use of the Internet in the breast cancer experience among Institute patients who participated in the research. Participants did not use digital networks to learn more about their clinical condition or obtain information that they could discuss with the medical professional. In this sense, we could consider that they were not expert patients since they did not carry out extensive online research on their disease to enhance their decision-making process or their self-care.

The low use of NICTs in the disease experience observed among hospital patients may be partly associated with the socioeconomic profile: most Institute patients had an income of up to three minimum wages and a schooling level equal to or less than high school. In Brazil, while women seek more information about health or health services, the percentage of users who perform this type of search declines with decreasing schooling levels [18]. In the case of Institute patients, although it was not investigated, the level of health literacy may have been low, considering their socioeconomic conditions, which could partly explain the low search for breast cancer information online: despite accessing the Internet, patients may not have developed the necessary skills to access, understand, and use online health information.

Community participants, in turn, showed more significant use of technology to address the disease. They searched for breast cancer information at different stages in the disease course and spent more time searching. To a greater or lesser extent, they played the figure of the expert patient: that is, they used the Internet to research and exchange information about their disease, getting to know their clinical condition better. Researching and sharing content about their disease may have contributed to individual empowerment: by knowing more about the different breast cancer dimensions, these patients may have developed a greater sense of control and autonomy regarding their health situation, enhancing their self-care and self-management [11]. However, we could not observe that searching for information online implied the refutation of knowledge or the medical professional.

The research participants engaged in the disease course as prescribed by the biomedical order. That is, they went to the professional to investigate the suspicion of breast cancer and performed diagnostic tests. Therapeutic procedures prescribed by a medical professional were initiated after confirmation. A doctor followed them up to check their clinical condition during treatment and post-treatment. As discussed in the literature, these results reiterate the idea that expert patients and their empowerment process, while possibly implying transformations and reconfigurations in the traditional ways of their relationship with the medical professional, do not necessarily break with the biomedical order. On the contrary, they dialogue with it and, ultimately, reiterate it [11–13].

Considering what was identified among the Institute and online community participants, we should make an observation about the figure of the expert patient. Especially among the online group participants, researching breast cancer on the Internet and knowing more about their disease were standard practices, classifying them as expert patients. However, we could observe, especially when compared to the Institute's results, that this expert patient condition does not occur in a monolithic or linear way. The current spread of ICTs does not mean that individuals with a disease or health condition research extensively about their condition, possibly becoming an expert in the matter. On the contrary, it is a complex process, which is linked, in other factors, to people's socioeconomic conditions.

We should consider that it is necessary to have a device with access to the network and a means to make such a connection to use the Internet. We should know how to handle the device and the resources available on the Internet. Another requirement is knowing how to retrieve the information required and recognize that the content accessed is of good quality. Finally, we should transform such use of the Internet into positive results for our health condition. To a greater or lesser extent, all these requirements are related to the patient's socioeconomic conditions. Higher income and education levels are associated with greater access and diversified use of the Internet and younger age groups and higher health literacy levels [18, 22].

Although they cannot be generalized because they have no statistical relevance, the results obtained in this study reiterate this relationship. Internet access and the use of technology for health purposes were lower at the Gynecology Institute, where more participants in more unfavorable socioeconomic conditions were identified. In the online community sample, the participants had a more favorable socioeconomic profile and showed greater Internet use during the disease. While carrying out this comparative study, we could observe that the expert patient phenomenon occurs in a diversified backdrop. Thus, when designing and analyzing it, we should combine patients' dimensions, including their uses, practices, and motivations, with their life context since socioeconomic conditions can be facilitators or barriers for people to use digital technologies as a means to address their illness or health condition.

4 Conclusion

Breast cancer is a severe health problem with different physical and psychological impacts. Faced with a suspected or confirmed diagnosis, women can turn to the Internet to learn more about the disease or interact with other people who share a similar health situation. In this setting, online health communities have emerged as a space that increasingly more breast cancer patients use to obtain and share information and emotional support.

Considering this context, we aimed to analyze how the expert patient phenomenon is identified among women with suspected or diagnosed breast cancer in a physical environment and an online environment. Thus, we conducted a study guided by the methodological framework of ethnography in two different environments: the waiting room of a public university hospital and an online group organized on Facebook.

From the results obtained, we identified that the patients of the investigated hospital accessed the Internet regularly; however, they do not use such technology in the breast cancer experience. Few searched online for information about the disease and, when they did, it was specific, which differs from what is seen in the online group. Community participants indicated frequent access to the Internet, using this technology as a resource to address the disease. In general, they searched for information on breast cancer at different stages of the disease course and spent more time on this activity.

In this sense, we could consider that the expert patient phenomenon is found among the participants of the investigated online community but not among the patients waiting in the hospital's waiting room. To a certain extent, the absence of the expert patient in the physical environment can be explained by the unfavorable socioeconomic conditions of the Institute's participants. In general, they had low schooling, low income, and were older. Such characteristics can be a barrier to health literacy: that is, to the development of skills related to access and use of health information. Thus, the socioeconomic profile of the hospital participants may have acted as an obstacle to searching online health information and, consequently, establishing the expert patient.

We can also observe that the establishment of the expert patient is not a linear process. The dissemination of NICTs does not necessarily imply using digital technologies for health purposes, primarily to address a disease or health condition. Regular access and use of the Internet are not enough for individuals to develop the habit of extensively researching their health situation: other factors and motivations are involved, including the socioeconomic context in which they find themselves.

Another point to be highlighted is the relationship between the expert patient and biomedicine. As observed in other studies, the practice of researching and knowing more about their disease or health condition did not imply breaking with biomedical knowledge, diagnostic and therapeutic procedures recommended by medicine, or the health professional. Thus, we noticed that the figure of the expert patient observed mainly among the online community participants ultimately displayed a posture of acceptance and submission to the biomedical order.

This study has limitations. Our analysis focused on breast cancer patients in two different environments: an online community and a Brazilian university hospital. The results could have been different if the research focused on another disease or health condition, in other types of environments, or other geographical contexts. We also had a small sample of participants. Studies with more extensive and diversified samples may obtain different findings than ours.

Increasingly more people use the Internet, particularly online health communities, as a resource for dealing with the experience of an illness or health condition. Thus, more studies are required to analyze how the expert patient phenomenon is established in this backdrop and what specificities it can assume.

References

1. Global Cancer Observatory, Cancer Fact Sheets – Breast. Accessed 26 Nov 2020
2. Instituto Nacional de Câncer José Alencar Gomes da Silva. A situação do câncer de mama no Brasil: síntese de dados dos sistemas de informação. Rio de Janeiro, INCA (2019)
3. Lovelace, D.L., McDaniel, L.R., Golden, D.: Long-term effects of breast cancer surgery, treatment, and survivor care. J. Midwifery Women's Health **64**(6), 713–724 (2019)
4. Smit, A., et al.: Women's stories of living with breast cancer: a systematic review and meta-synthesis of qualitative evidence. Soc. Sci. Med. **222**, 231–245 (2019)
5. Perrault, E., et al.: Online information seeking behaviors of breast cancer patients before and after diagnosis: from website discovery to improving website information. Cancer Treatment Res. Commun. **23**, 1–5 (2020)
6. Heaton, L.: Internet and health communication. In: Consalvo, M., Ess, C. (eds.) The Handbook of Internet Studies, pp. 212–231. Blackwell Publishing, Cingapura (2011)
7. Van Uden-Kraan, C.F., et al.: Empowering processes and outcomes of participation in online support groups for patients with breast cancer, arthritis, or fibromyalgia. Qual. Health Res. **18**(3), 405–417 (2008)
8. Horrell, L.N., et al.: Attracting users to online health communities: analysis of Lung-cancer.net's Facebook advertisement campaign data. J. Med. Internet Res. **21**(11), e14421 (2019)
9. Willis, E.: The making of expert patients: the role of online health communities in arthritis self-management. J. Health Psychol. **19**(12), 1613–1625 (2014)
10. Falisi, A.L., Wiseman, K.P., Gaysynsky, A., Scheideler, J.K., Ramin, D.A., Chou, W.-Y.: Social media for breast cancer survivors: a literature review. J. Cancer Surviv. **11**(6), 808–821 (2017). https://doi.org/10.1007/s11764-017-0620-5

11. Pereira Neto, A., Lima, J., Barbosa, L., Schwartz, E.: Internet, Expert patient, and empowerment: activity profiles in virtual communities of chronic kidney patients. In: Pereira Neto, André, Flynn, Matthew B. (eds.) The Internet and health in Brazil, pp. 87–111. Springer, Cham (2019). https://doi.org/10.1007/978-3-319-99289-1_6

12. Fox, N., Ward, K.J., O'Rourke, A.J.: The "expert patient": empowerment or medical dominance? The case of weight loss, pharmaceutical drugs, and the Internet. Soc. Sci. Med. **60**, 1299–1309 (2005)

13. Song, F., et al.: Women, pregnancy, and health information online: the making of informed patients and ideal mothers. Gend. Soc. **26**(5), 773–798 (2012)

14. Zimmerman, M.A.: Psychological empowerment: issues and illustrations. Am. J. Community Psychol. **23**(5), 581–599 (1995)

15. Angrosino, M.: Doing Ethnographic and Observational Research. SAGE Publications, Thousand Oaks (2007)

16. Ardévol, E., Gómez-Cruz, E.: Digital ethnography and media practices. In: Valdivia, A.N. (ed.) The International Encyclopedia of Media Studies, pp. 1–21. Wiley, New Jersey (2013)

17. Coetzee, B., et al.: Exploring breast cancer diagnosis and treatment experience among a sample of South African women who access primary health care. South African J. Psychol. **50**(2), 195–206 (2020)

18. Brazilian Internet Steering Committee. Survey on the use of information and communication technologies in Brazilian households: ICT Households 2019. São Paulo, Brazilian Internet Steering Committee (2020)

19. Batterham, R., et al.: Health literacy: applying current concepts to improve health services and reduce health inequalities. Public Health **132**, 3–12 (2016)

20. Sørensen, K., et al.: Health literacy in Europe: comparative results of the European health literacy survey (HLS-EU). Eur. J. Pub. Health **25**(6), 1053–1058 (2015)

21. Schaeffer, D., Berens, E., Vogt, D.: Health literacy in the German population. Deutsches Arzteblatt Int. **114**(4), 53–60 (2017)

22. Chen, X., et al.: Health literacy and use and trust in health information. J. Health Commun. **23**(8), 724–734 (2018)

User Experience Evaluation of a Computational Thinking-Enhanced Problem-Solving Tool: Findings and Next Steps

Juan Felipe Calderon[1] ⑩, Luis A. Rojas[2](✉) ⑩, Katrina Sorbello[3] ⑩, and Nibaldo Acero[4] ⑩

[1] Facultad de Ingeniería, Universidad Andres Bello, Quillota 980, Viña del Mar, Chile
juan.calderon@unab.cl

[2] Facultad de Ciencias Empresariales, Departamento de Ciencias de la Computación y Tecnologías de la Información, Universidad del Bío-Bío, Chillán, Chile
lrojas.larp@gmail.com

[3] The Stella Way, 17 Enford Street, Hillcrest, QLD, Australia
kat@thestellaway.com

[4] Dirección de Investigación y Postgrado, Universidad de Aconcagua, Santiago, Chile
nibaldo.caceres@uac.cl

Abstract. IT-related works are highly demanded. However, this discipline does not have sufficient and accurate dissemination. In this context, it has missed the application of computational thinking abilities, such as the development of logical reasoning, algorithmic thinking, and problem-solving with creativity. A common issue is the capability of these tools to provide a design environment, where teachers and instructors can design and deploy more contextualized and curriculum-aligned activities. Problock is proposed as a tool to develop computational thinking, in the context of learning based on problem-solving. The resolution of these exercises can be carried out using the basic concepts related to computer science such as abstraction, division of a problem into sub-problems, algorithmic schematization, and programming as a tool to concretize everything in the final solution. An evaluation of the Problock was carried out with 15 high school students to measure the degree of usability. User experiences indicate positive and acceptable evaluations in terms of usefulness, ease of use, ease of learning, and satisfaction.

1 Introduction

IT-related works are highly demanded. However, this discipline does not have sufficient and accurate dissemination, especially in elementary and secondary schools. In this context, it has overlooked the application of computational thinking abilities, such as the development of logical reasoning, algorithmic thinking and problem solving with creativity [1]. In the academic and educational technology market there are tools that are accessible by self-study students to be used as a complement to other activities [2]. A common issue is the capability of these tools to provide a design environment, where teachers and instructors can design and deploy more contextualized and curriculum-aligned activities [3].

Problock is proposed as a tool to firstly develop computational thinking, in a context of learning based on problem-solving [4]. Secondly, it disseminates and motivates students for future professional development in higher education in computer science and related disciplines. Problock considers three (3) user profiles. The first profile is the Researcher, who can extend the capabilities of the platform when designing new activities, that is, the elements that can be used by teachers when formulating and personalizing the problems that students must carry out. The second profile is the Teacher, who must have the power to design problems related to their area of study. They make use of the available elements or activities integrated into the platform by researchers, following a learning pattern based on problem solving. The third profile is the Student who can be enrolled in a virtual course created by a teacher, where they will find the problems designed and published by the teacher.

These exercises can be solved by using the basic concepts related to computer science such as abstraction, division of a problem into sub-problems, algorithmic schematization, and programming as a tool to concretize everything in the final solution. The resolution of these problems must be done through the graphical interface implemented in the platform, where a customized version of programming using blocks will be used [5, 6].

The aim of this work is to study the usability of Problock by Chilean high school students. An evaluation of the Problock was carried out with 15 high school students to measure the degree of usability. To effectively evaluate the usability a questionnaire was designed and launched. The questionnaire is based mainly on the USE questionnaire [7], with some variations provided by Davis's Perception of Utility and Ease of Use questionnaire [8], as well as the Purdue Usability Questionnaire [9]. User experiences indicate positive and acceptable evaluations in terms of usefulness, ease of use, ease of learning and satisfaction.

This paper is organized as follows. Section 1 introduced the motivation of our work through the problem identified and the proposed solution. Section 2 presents related work, which provides an overview of the research field. Section 3 describes the details of our proposed solution. Section 4 describes the experiment accomplished to implement and evaluate the proposed solution in a real environment. Section 5 presents conclusions and future work.

2   Related Works

Currently, computer science has been strongly integrated in various knowledge areas. Being prepared for careers and job opportunities, and being competitive against peers, requires having a clear understanding of the principles and practices of computer science as a means of solving problems, regardless of the field of study or work. It is estimated that many of the future jobs that today's students will have in 10 to 20 years have not yet been invented and will be strongly related to computing [1]. Therefore, there is an urgent need to improve the level of public understanding and discussion of computer science and its related disciplines as an academic and professional field. In primary education, schools have a unique opportunity in their role as educators of new generations of students.

Computer science students learn logical reasoning, algorithmic thinking, and problem-solving concepts and skills that go beyond the study of computing itself and that apply to other sciences and areas of modern life, in the most diverse contexts. For this reason, learning the fundamentals of computer science should not be exclusive to those who are dedicated to studying a higher education career in the field, but rather should be skills developed by all students [2].

Students need to stop simply being consumers of technology and become creators of new and innovative solutions to problems related to their areas of interest. Despite the growing job opportunity and high employability rates for first year computer science graduates, they are not as popular careers compared to other engineering or health-related careers. One of the main factors of this low popularity is the ignorance of computer science at a global level and the illusion that has been generated around it.

Although the Chilean Ministry of Education, with the Enlaces project, has indeed made great progress at the implementation level, there are still limitations. These efforts made to incorporate the use of technology in the student's education have been limited to only the consumption of these technology, through the means of research and collaboration with their peers. However, the basic concepts of computational thinking have not been instilled, nor have the benefits that these can bring to its development been clarified. This generates a type of "ignorance" regarding computer science as an academic and professional field.

At the end of 2014, the results of the Evaluation of ICT Skills for Learning (former SIMCE TIC) were presented. The results of the second national application of SIMCE TIC show that 46.9% of the students are at an initial level. While 51.3% of the students are at an Intermediate level and 1.8% at an Advanced level. In general terms, the results indicate that the students have achieved the necessary skills to communicate with their peers and search for information in digital media. However, the more complex cognitive skills that involve the processing and generation of information are achieved by a much lower percentage of students (1.8%), even considering that they are used by them daily.

When reviewing the report on the development of digital skills for the 21st century in Chile [10], it is concluded that the skills measured by SIMCE TIC (12 skills, organized in four areas) are not directly related to computational thinking skills (at least not the most complex). It must be remembered that the SIMCE is directly aligned with the skills standards, achievement levels, and learning objectives defined by the national curriculum, with which we can affirm that there is an absence of computational thinking skills in the official curricular documents.

This means that from the central level there is also no curricular proposal that builds awareness of the importance of these skills. Therefore, we cannot expect educational centers, let alone teachers, to take charge of teaching or transmitting content and a set of skills that they probably do not even know. From this perspective, Problock appears as a truly innovative proposal that brings into the pedagogical discussion the issues associated with computational thinking.

In 2016, the report Guidance on digital citizenship for citizen training [11] was published, in which the MINEDUC proposes the concept of digital citizenship, which implies the development of digital skills that will allow students to manage new ways of learning and participating when using technologies. This view, in addition to the variable of technical handling of ICTs, introduces an ethical perspective and co-responsibility in the use of technologies. The indication is that educational communities implement pedagogical activities so that students develop ICT Skills for Learning, which have been defined by the Ministry of Education as "The ability to solve problems of information, communication and knowledge, as well as legal, social and ethical dilemmas in the digital environment" and that are included in the curriculum.

In conclusion, neither in the most important standardized test in the country, SIMCE, nor in the ministerial curricular documents are the proposals for specific computational thinking skills presented or discussed.

3   A Computational Thinking-Enhanced Problem-Solving Tool

The alternatives that can be proposed to solve this problem, start from the awareness, first, of those who generate educational policies, and then mobilize this awareness in schools and colleges. This awareness lies in understanding the nature of computing and its impact on the life of the school community. Bearing this in mind, training, and updating in the new techniques and teaching methods, related to the subject must be trained and updated, which today are gaining more and more strength and presence.

In this sense, Problock is consolidated as part of a pedagogical avant-garde that faces the problem, addressing more emphatically those who build the curriculum, motivating a change in the appreciation that is still held about technologies in the school system. The server will contain the business logic and information storage. On the other hand, the client will be a graphical interface that runs on the web browser and allow the user to interact with the platform and receive real-time feedback on actions or events. Among the main advantages that a web application presents, in contrast to more traditional desktop applications, are Multiplatform compatibility, ease of updating, immediate access and centralized information. Multiplatform compatibility permits schools to use the platform regardless of the brand or version of the computers they have as the use of a modern web browsers is sufficient. Ease of updating reduces technical maintenance costs by the school since updates in terms of bug fixes or improvements to the platform are automatically available to users. Immediate access, also contributes to reducing costs associated with technical support, since web-based applications do not need to be installed and configured in the work environment. The platform centralizes the information, thus it is available from any device that has internet access, even facilitating remote work or work outside the educational institution. A screenshot of the teacher's view is in Fig. 1.

Fig. 1. User interface of teacher's view of Problock.

The interface presented to the user is consistent both for the teacher who designs problems and for the student who answers them. As can be seen in Fig. 1, on the left side is the pedagogical sequence of problems that the student must solve. In the case of the teacher, this is shown as editable, enabling the customization of the flow and quantity. In the center, there is a free space as a central canvas, where the student can build the solution to the proposed problem. Problock posits that each problem and subproblem must be capable of being solved by proposing an algorithm built by blocks. For this reason, in the left grey panel of the central canvas, there are drop-down menus with all the available blocks that the student can use to solve the current problem. This canvas has a tab where the student can check the equivalence of their solution's built-in JavaScript code. In the lower-left sector, the student can review the screen output of their respective solution. The teacher's interface is very similar, also having tabs for defining the instructions to be shown to the student and the educational objectives. The objectives can be defined by the teacher as the use of certain structures and variables, together with the definition of a solution as a guideline or template. The system will contrast what the student builds against these guidelines or templates and evaluate its solution through similarity and completeness.

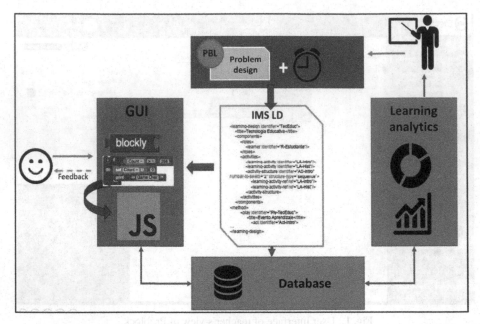

Fig. 2. Problock architecture [4].

Regarding the technological architecture (see Fig. 2), Problock is constituted as a typical web application, based on the model-view-controller (MVC) pattern. It has the incorporation of the Google Blockly engine [12], which allows validation at the syntactic level of the construction of solutions carried out by the students, both visually and at an algorithmic level. On the other hand, the representation of the sequence of problems presented to the student is built internally according to the IMS-LD specification [13]. This allows for a standard internal specification in the field of learning design, which allows interoperability with other applications that implement the LTI interoperability standard, but under a known formal specification (i.e., in [14]).

4   User Experience Evaluation

This section describes the experimentation carried out to measure the degree of usability achieved in the Problock tool.

4.1   Experimental Framework

The main objective of the experimentation was to study and analyze the degree of usability of Problock. Since a random selection of the study groups was not carried out, a quasi-experimental design has been carried out, that is, the study group has been chosen within the environment close to the researcher.

This experiment has been designed with the objective of meeting the three usability conditions (effectiveness, efficiency and satisfaction), described in the ISO-9241–11 standard [15].

Variables and Research Questions. According to the main objective of the experimentation, the variables and questions of the investigation have been defined. The independent variables were the number of tasks performed by the users, the total time used by the users to complete the tasks, the traits of the participants (age and gender), and the responses of the users to the questions of a usability questionnaire.

On the other hand, the dependent variables are relative to the three usability conditions, indicated in the ISO-9241–11 standard: effectiveness, efficiency, and satisfaction. Effectiveness corresponds to the degree of success that users achieve in the execution of tasks, which is measured as the percentage of achievement of tasks by users. Efficiency is the time it takes users to complete each task and is measured in seconds. Finally, satisfaction is the subjective perception of usability that is achieved with the Problock tool, and which is measured through the average value of the users' responses to the questions in the usability questionnaire. As stated earlier, the questionnaire used is based mainly on the USE questionnaire [7], with some variations provided by Davis's Perception of Utility and Ease of Use questionnaire [8], as well as the Purdue Usability Questionnaire [9].

The research questions have been defined to answer the three dependent variables, which correspond to:

RQ1: What is the effectiveness achieved by users in carrying out problem-solving tasks using Problock?
RQ2: What is the efficiency achieved by users in carrying out problem-solving tasks using Problock?
RQ3: What is the user perception of usability regarding the Problock tool?

Participants. 15 high school students participated in the evaluation of the Problock tool. 14 men and one woman participated in the evaluation, aged between 14 and 17 years (M = 15.47, SD = 1.19).

Tasks. To carry out the experiment, each user was asked to perform five (5) problem-solving tasks using the Problock tool. Specifically, the five tasks performed by the users correspond to:

T1: Enter authentication credentials. This task consists of entering the Problock application with the account provided for authentication and authorization.
T2: Select a problem. A problem to be solved must be selected from a list of available problems.
T3: Take the welcome tour. On the main screen choose to take the welcome tour.
T4: Use blocks to display messages on the screen. The purpose of this task for the student to complete an activity to display a message on the screen.
T5: Use blocks to request data from the user. The objective of this task is for students to complete an activity that asks for their name and displays it on the screen.

4.2 Execution of Experimentation

The experimentation has been carried out remotely. In this way, a video conference was held using Skype or Google Meet. In both applications, users shared their screens to check the steps taken. The following steps were performed:

First Stage. Each participant in the evaluation was introduced the general objective of the evaluation and the concepts of the Problock tool. This introduction had an average duration of one minute and 43 s (SD = 24 s).

Second Stage. Each of the participants was required to perform the five above tasks, using the Problock tool. Likewise, users were asked to indicate, aloud, the thoughts they had while using the tool to apply the Thinking Aloud protocol.

Third Stage. Once the session with the users was over, each of the participants was asked to fill out a questionnaire to evaluate the usability of the Problock tool.

4.3 Results Obtained

In this section, the results obtained from the experimental session with the 15 users (First and Second Stages) and from the usability questionnaire (Third Stage) are detailed and analyzed.

First Stage. This stage had an average duration of one minute and 43 s (SD = 24 s), with a total duration (with the 15 users) 17 min and 3 s. The maximum time required by one of the users was one minute and 54 s, while the minimum time was 43 s. The differences between the maximum and minimum times, indicated above, are mainly because there were users who knew the nomenclature normally used in Problock, and who asked minimal questions. Meanwhile, in other cases there were users who were unaware of some pillar concepts of Problock (e.g., variables, cycles, logic, among others), which required extending the time of this stage to clarify these concepts.

Second Stage. This stage consisted of providing users with a statement containing the description of five problem-solving tasks, which had to be solved using Problock. The first point measured was the effectiveness obtained in carrying out the different tasks, calculated through the degree of success in completing the tasks of the experiment by the users. The results show that in all cases the users completed the tasks successfully. Therefore, it can be indicated that 100% effectiveness has been reached.

The next point measured corresponds to the efficiency achieved in the elaboration of the different tasks. This point was measured as the time required by users to carry out the tasks. In Table 1, the time statistics (hh:mm:ss) used by the users in the execution of the tasks in the experiment are presented. Likewise, Fig. 3 has been prepared to present, graphically, the detail of the average times (hh:mm:ss), used in each of the five tasks, with error bars with a confidence level of 95%.

Table 1. Statistics of the time (hh:mm:ss) of execution of the tasks in the experiment.

Descriptive statistics	Tasks				
	1	2	3	4	5
Average	0:00:27	0:00:09	0:02:00	0:00:54	0:01:04
Standard deviation	0:00:06	0:00:03	0:00:58	0:00:19	0:00:13
Typical error	0:00:02	0:00:01	0:00:15	0:00:05	0:00:03
Median	0:00:29	0:00:10	0:02:03	0:00:51	0:01:02
Mode	0:00:22	0:00:10	#N/D	0:00:40	0:00:54
Range	0:00:21	0:00:13	0:03:26	0:01:22	0:00:43
Minimum	0:00:15	0:00:03	0:00:53	0:00:30	0:00:43
Maximum	0:00:36	0:00:16	0:04:19	0:01:52	0:01:26

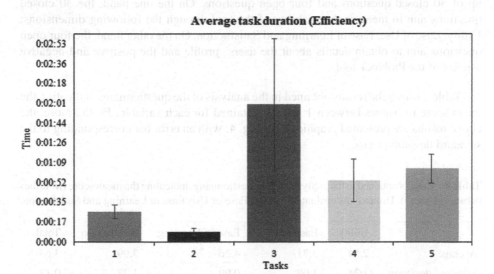

Fig. 3. Time (hh:mm:ss) average duration of the experimentation tasks (efficiency).

As can be seen, users took longer in tasks T3, T5 and T4, which had an average duration of 2 min (SD = 58 s), 1 min and 4 s (SD = 13 s), and 54 s (SD = 19 s), respectively. These tasks have in common the fact that they required the use of different Problock functionalities, which resulted in users spending more time to work.

In contrast, users spent less time on tasks T2 and T1, which had an average duration of 9 s (SD = 3 s) and 27 s (SD = 436 s), respectively. In these tasks, users only had to perform actions related to authentication and review certain functionalities of the tool, and that did not require more effort, generating, as a result, that users spend less time.

The complete performance of the tasks using the Problock tool had an average duration of 8 min and 3 s (SD = 1 min and 6 s). The maximum time required by one of the users was 10 min and 19 s, while the minimum time was 6 min and 34 s.

In general, the differences in the times used by the users are mainly due to the different skills they had to manipulate and perform the different tasks proposed.

Finally, the use of the Thinking Aloud protocol has facilitated obtaining qualitative data from users, regarding the design and functionalities of the tool. For example, three users expected to find the functionality to run tasks outside of the main environment. Also, four users tried to use the tool's blocks by double-clicking and not dragging the block with the cursor to the area. This and the other data, obtained with the Thinking Aloud protocol, will be used to incorporate different improvements in the tool, so that users can meet their goals efficiently.

Third Stage. At the end of the experimental session, users were asked to complete a questionnaire to measure the usability of the Problock tool. This questionnaire is made up of 30 closed questions and four open questions. On the one hand, the 30 closed questions aim to measure the usability of the tool through the following dimensions: Utility, Ease of Use, Ease of Learning and Satisfaction. On the other hand, the four open questions aim to obtain details about the users' profile and the positive and negative aspects of the Problock tool.

Table 2 shows the results obtained in the analysis of the questionnaire, indicating the mean score (in values between 1 and 5), obtained for each variable. Furthermore, the above results are presented graphically in Fig. 4, with an error bar corresponding to the standard deviation ($\pm\sigma$).

Table 2. Results obtained in the analysis of the questionnaire, indicating the mean score (in values between 1 and 5). Dimensions evaluated: Utility, Ease of Use, Ease of Learning and Satisfaction.

	Utility	Ease of Use	Ease of Learning	Satisfaction	Total
Average	2.54	3.71	4.26	3.99	3.63
Standard deviation	1.24	1.08	0.99	1.27	0.13

As shown in Fig. 4, the Ease of Learning dimension is the one that obtains the highest average score of the entire sample, with a mean of 4.26 (SD = 0.99), which also corresponds to the lowest dispersion. between the measured variables. The Satisfaction dimension obtains the second-highest average of the measured scores, with an average of 3.99 (SD = 1.27), which also corresponds to the highest dispersion between the measured variables. It is followed by the Ease of Use variable, which obtains an average score of 3.71 (SD = 1.08). Finally, the Utility variable was the one that obtained the lowest mean score, 2.54 (SD = 1.24).

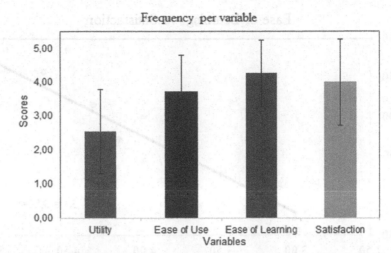

Fig. 4. Problock usability. Mean score for each dimension that is measured in values between 1 and 5, including error bars (± σ).

In general terms, it can be noted that the average of the scores of the entire sample is 3.63 (SD = 0.13). The fact that all the dimensions obtain scores higher than 2.5 stands out. Therefore, this measure can be considered as a good indicator of the general usability of the Problock tool, based on the perception of the users.

4.4 Multivariate Analysis

Since the results of the means of the four dimensions are close to each other, the possible correlation of such values was studied. Thus, a calculation was made to measure the correlation coefficient between the dimensions evaluated. This analysis aims to measure the degree of linear intensity of linkage of the dimensions. To carry out this calculation, a matrix was first created to study the correlation of all the dimensions and thus be able to select the ones that were most related (with a correlation coefficient threshold greater than 70%). Next, scatter plots were made of the average scores of each of the selected dimensions, comparing some scores with others and calculating the linear correlation coefficient (R^2). This coefficient is between 0 and 1, with a greater linear relationship between the dimensions when approaching 1.

The highest correlation found was between the Ease of Learning and Satisfaction dimensions, with a positive linear correlation coefficient of $R^2 = 0.80$. As can be seen in Fig. 5, only two users have given a neutral score regarding the conformity of these two dimensions. However, as shown in the graph, most users expressed high levels of agreement regarding the Ease of Learning qualities of the Problock tool, which is increased to the extent that the tool offers a greater satisfaction in its use.

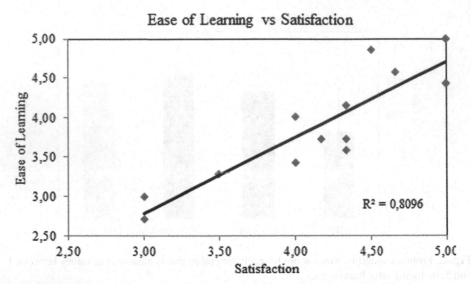

Fig. 5. Correlation between the dimensions Ease of Learning and Satisfaction.

Figure 6 shows the second-highest correlation found between the Ease of Use and Ease of Learning dimensions, obtaining a linear correlation coefficient of $R2 = 0.70$, which shows that there is a positive correlation between these dimensions. This correlation allows indicating that users show greater Ease of Use in the use of the tool, which is increased to the extent that the tool offers greater facilities in learning its use.

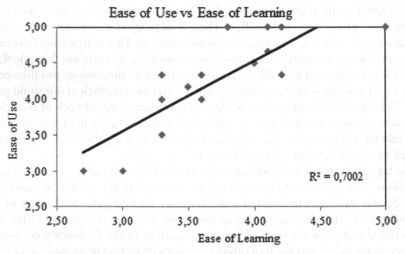

Fig. 6. Correlation between the dimensions Ease of Use and Ease of Learning.

4.5   Discussion of the Results

In this section, a discussion of the results is made, obtained in the experimental session with the 15 users and in the usability questionnaire.

Regarding the degree of effectiveness of the tasks of the experiment, the results show that in all cases the users completed the tasks correctly, reaching 100% effectiveness.

Regarding efficiency, when examining the average times used in each task, it can be indicated that they are reasonable. The tasks T3, T4, and T5 required an understanding of the problem of the task and to fully use the functionalities of the tool, that is why their times are higher.

It should be noted that even though the T3 task required a greater amount of time from the users, with an average duration of 2 min (SD = 58 s), this task was carried out without the users requiring assistance. Likewise, the tasks that required the most time had in common the fact that they required understanding the problem of the task. These results indicate that the Problock tool did not present architectural or design difficulties that would influence the time to complete the tasks, since it is subject to the amount of information that users needed to enter in each case.

Regarding satisfaction, the results of the usability questionnaire show high values for the Ease of Learning (4.26) and Satisfaction (3.99) dimensions. These results of the dimensions, together with the degree of effectiveness and efficiency achieved in the tasks of the experiment, confirm that the Problock tool provides qualities and an architectural model that is easy to handle, and that has generated satisfaction among users.

Likewise, the results obtained provide sufficient evidence to answer the research questions RQ1, RQ2 and RQ3, indicated above:

- RQ1: What is the effectiveness achieved by users in carrying out problem-solving tasks using Problock?

  The results show that in all cases the users completed the tasks correctly, therefore, it can be indicated that 100% effectiveness has been achieved.

- RQ2: What is the efficiency achieved by users in carrying out problem-solving tasks using Problock?

  The complete performance of the tasks had an average duration of 8 min and 3 s (SD = 1 min and 6 s). The average time used in the problem-solving tasks was around a minute and a half. These results are considered acceptable for the operations requested from the users and provide an advantage to support problem resolution.

- RQ3: What is the user perception of usability regarding the Problock tool?

  The usability perception of the 15 high school students regarding the Problock tool is positive in all cases. The results of the usability questionnaire obtained an average of 3.63 (SD = 0.13) for the entire sample, being a good indicator of the users' perception of usability with respect to the proposed tool.

5 Conclusions and Future Work

In this work, Problock was presented as a tool that is designed for the development of computational thinking, which uses problem-based learning as pedagogical support. One of its main features is the possibility of personalized design of the sequence of problems that students must solve, by the teacher. In its technological architecture, the capabilities of Google Blockly are used to support the visual programming of the algorithms that students must build as a solution to the problems posed. In addition, the sequence of problems is formally specified internally using the IMS-LD standard, which allows interoperability with other learning tools from the design.

Finally, regarding the results of the evaluation, a quasi-experimental design was carried out with 15 high school students to evaluate the usability of the Problock tool. The results obtained showed positive and acceptable opinions about the user's perception in terms of Utility, Ease of Use, Ease of Learning, and Satisfaction, achieving scores above 2.5 in all cases (M = 3.63, SD = 0.13). The results of this evaluation provide an affirmative answer to the research questions RQ1, RQ2, and RQ3. Therefore, it is possible to affirm that the Problock tool is easy to use and learn by its potential users, and it allows developing computational thinking, in a context of learning based on problem-solving.

As for future work, we expect to add more functionality to Problock to provide more alternatives in building blocks. We also want to incorporate intelligent help support, which allows users to be guided in solving problems. In addition, we also expect to conduct an in-depth, qualitative evaluation of Problock's usability, in order to obtain more details on opportunities for improvement [16]. Finally, we also want to incorporate a mechanism to qualitatively prioritize the different options to solve problems, according to the main priorities and guidelines of the problem statements [17].

References

1. Kong, S.-C., Lai, M.: A proposed computational thinking teacher development framework for K-12 guided by the TPACK model. J. Comput. Educ. 1–24 (2021)
2. Moreno-León, J., Román-González, M., Robles, G.: On computational thinking as a universal skill: A review of the latest research on this ability. In: 2018 IEEE Global Engineering Education Conference (EDUCON), pp. 1684–1689 (2018)
3. Angeli, C., Giannakos, M.: Computational thinking education: Issues and challenges. Comput. Hum. Behav. **105**, 106185 (2020)
4. Calderón, J.F., Ebers, J.: Problock: a tool for computational thinking development using problem-based learning. In: 2017 36th International Conference of the Chilean Computer Science Society (SCCC), pp. 1–5 (2017)
5. Adi, P.D.P., Kitagawa, A.: A review of the Blockly programming on M5Stack board and MQTT based for programming education. In: 2019 IEEE 11th International Conference on Engineering Education (ICEED), pp. 102–107 (2019)
6. Weintrop, D.: Block-based programming in computer science education. Commun. ACM **62**(8), 22–25 (2019)
7. Lund, A.M.: Measuring usability with the use questionnaire12. Usabil. Interf. **8**(2), 3–6 (2001)
8. Davis, F.D.: Perceived usefulness, perceived ease of use, and user acceptance of information technology. MIS Q. **13**(3), 319–340 (1989). https://doi.org/10.2307/249008

9. Lin, H.X., Choong, Y.-Y., Salvendy, G.: A proposed index of usability: a method for comparing the relative usability of different software systems. Behav. Inf. Technol. **16**(4–5), 267–277 (1997)
10. Jara, I.: Desarrollo de habilidades digitales para el siglo XXI:¿ Qué nos dice el SIMCE TIC. Santiago Chile LOM (2013)
11. Peña, P.: Orientaciones de ciudadanía digital para la formación ciudadana. Ministerio de Educación, República de Chile (2016)
12. Fraser, N.: Ten things we've learned from Blockly. In: 2015 IEEE Blocks and Beyond Workshop (Blocks and Beyond), pp. 49–50 (2015)
13. Koper, R., Miao, Y.: Using the IMS LD standard to describe learning designs. In: Handbook of Research on Learning Design and Learning Objects: Issues, Applications, and Technologies, pp. 41–86. IGI Global (2009)
14. Wague, A., Bousso, M., Capus, L.: Valorization of non-formal learning situations using IMS-LD specification. J. High. Educ. Theory Pract. **21**(3), 123–130 (2021)
15. Stewart, T.: Ergonomic requirements for office work with visual display terminals (VDTs): Part 11: Guidance on usability. International Organization for Standardization ISO, 9241, pp. 89–122 (1998)
16. Rojas P., L.A., Truyol, M.E., Calderon Maureira, J.F., Orellana Quiñones, M., Puente, A.: Qualitative evaluation of the usability of a web-based survey tool to assess reading comprehension and metacognitive strategies of university students. In: Meiselwitz, G. (ed.) HCII 2020. LNCS, vol. 12194, pp. 110–129. Springer, Cham (2020). https://doi.org/10.1007/978-3-030-49570-1_9
17. Rojas, L.A., Macías, J.A.: Toward collisions produced in requirements rankings: A qualitative approach and experimental study. J. Syst. Softw. **158**, 110417 (2019)

Automated Tools for Usability Evaluation: A Systematic Mapping Study

John W. Castro[1(✉)], Ignacio Garnica[1], and Luis A. Rojas[2]

[1] Departamento de Ingeniería Informática y Ciencias de la Computación,
Universidad de Atacama, Copiapó, Chile
john.castro@uda.cl, ignacio.garnica.14@alumnos.uda.cl
[2] Facultad de Ciencias Empresariales, Departamento de Ciencias de la Computación y
Tecnologías de la Información, Universidad del Bío-Bío, Chillán, Chile
lrojas.larp@gmail.com

Abstract. Usability is one of the most critical indicators in determining the quality of a software product. It corresponds to how users can use a software system to achieve specific objectives with effectiveness, efficiency, and satisfaction. A usability evaluation is necessary to ensure that the software system is usable, but this has certain disadvantages (e.g., a high cost of time and budget for the evaluation to be implemented). While these disadvantages can be a bit daunting despite the benefits they provide, some tools can automatically generate and support usability testing. We conducted a systematic mapping study to identify the tools that support automatic usability evaluation. We identified a total of 15 primary studies. In addition, we classify the tools into four categories: measure usability, support usability evaluation, detect usability problems, and correct usability problems. We identified that the automatic evaluation of the usability of web platforms and mobile devices is the most interesting.

Keywords: Usability · Evaluation · Tool · Automation

1 Introduction

Currently, there is a growth of developed software systems, causing an increased demand for higher quality systems, which can be ensured with specific standardized measures and methods through different activities and techniques. One of the essential measures when developing a software system is usability [1]. Usability is the extent to which users use a system, product, or service effectively and with satisfaction, given a context of use [2]. Usability is also related to the acceptability, by users, of a specific system, considering that it is good enough to meet the needs of users [1]. To ensure that these requirements are met, the developed systems must undergo a usability evaluation [3, 4].

Despite the importance of usability evaluation for any software system, it has certain disadvantages, such as a high cost of time and budget given its characteristics. Additionally, some techniques related to usability evaluation need at least one usability expert to be implemented [3, 4]. Guidelines, metrics, and heuristics can guide this expert to

© The Author(s), under exclusive license to Springer Nature Switzerland AG 2022
G. Meiselwitz (Ed.): HCII 2022, LNCS 13315, pp. 28–46, 2022.
https://doi.org/10.1007/978-3-031-05061-9_3

support the work of usability evaluation. However, this expert evaluator will always provide a certain level of subjectivity in their analysis [4, 5]. Although these disadvantages can be discouraging, despite the benefits they provide considering the finished software product, they can be mitigated by implementing usability evaluation tools [5–8].

Usability evaluation tools are systems that support this task. Many tools directly benefit usability evaluation activities in an automated way, allowing, for example, during a usability test to store user registration data such as (i) keystrokes, (ii) clicks made with the mouse, and (iii) the distances traveled by the mouse pointer, among others. These tools allow, in some cases, to analyze the data collected to provide feedback to developers and usability experts, providing information on usability errors and, depending on the tool, automatically correcting them [5–9].

Currently, there is a wide variety of these tools. However, the related literature is composed of a set of independent publications. To the best of our knowledge, no study has comprehensively focused on this literature nor reports on the current state of automated tools for usability evaluation. This research seeks to generate a body of knowledge to classify the tools that support the automatic evaluation of usability. For this, we conducted a systematic mapping study (SMS).

This paper is organized as follows. In Sect. 2, we present the related work. In Sect. 3, we describe the research method of the SMS. In Sect. 4, we discuss the results of the SMS. Section 5 presents possible threats to validity, and finally, the conclusions are presented in Sect. 6.

2 Related Work

From our pilot search, we found that there were only four [4, 10–12] literature reviews related to our research. The first paper by Ivory and Hearst [4] reported the state-of-the-art usability evaluation methods, organized according to a taxonomy that emphasizes the role of automation. Ivory and Hearst [4] focused their efforts on identifying aspects of usability evaluation automation that are useful in future research and suggested new ways to expand existing approaches to better support usability evaluation. This study is interpreted as a precursor to automated approaches that, over time, became the development of tools that allow automatic evaluation of usability. Throughout his study, several tools are named, although not as sophisticated as those that currently exist, considering the year of publication of this study.

The second paper [10] reported *widgets* to help testers in the early evaluation of user interfaces. The authors explain that these *widgets* can detect certain ergonomic inconsistencies in the design of user interfaces. This study does not perform an SMS, and it focuses on exposing the *widgets* that were known. The authors explain the *widgets* in terms of functionality and application and show their experimental phase where they are tested. This study shows the *widgets* in a period before the one we consider (i.e., between 2016 and 2021), so the study is not considered in our research work.

The third paper by Bakaev et al. [11] provided an overview of the methods and tools of traditional, semi-automated, and automated approaches to website usability evaluation. The main difference from our research work, apart from the fact that the authors do not perform an SMS, is that Bakaev et al. [11] focused only on tools that support automated usability evaluation of web user interfaces. In contrast, we focus our efforts on knowing the current panorama of these tools, whether they are focused on the web and desktop applications and mobile devices. Furthermore, the work of Bakaev et al. [11] only briefly describe the tools.

Finally, Khasnis et al. [12] presented a series of tools that support the usability evaluation in their research work, briefly explaining their operation, advantages, and disadvantages. It is important to note that the authors do not perform an SMS, as in this study. Furthermore, Khasnis et al. [12] focused on relating automatic usability evaluation tools with usability evaluation methods. Our approach focuses on relating the reported tools to the catalog of usability evaluation techniques proposed by [13, 14].

After analyzing these papers, we find that the SMS reported in this paper differs from the above reviews in that it aims not only to identify the automated tools to support the usability evaluation but also to (i) identify the techniques related to evaluation that benefit from these tools, (ii) determine the existing problems and challenges of using automated tools for usability evaluation and (iii) classify these tools. None of the reviews in the literature address this issue. Therefore, it is necessary to investigate the current state of automated tools for usability evaluation.

3   Research Method

The secondary study presented in this paper has been developed following the guidelines established by Kitchenham et al. [15] for conducting an SMS. Following these guidelines, the activities we carried out were: (i) formulating the research questions, (ii) defining the search strategies, (iii) selecting the primary studies, (iv) extracting the data, and (v) synthesizing the extracted data.

3.1   Research Questions

The information extracted from the primary studies aims to answer the following research questions: (RQ1) What are the automated tools that support the usability evaluation? (RQ2) Which usability evaluation-related techniques benefit from automated tools? (RQ3) What are the existing problems and challenges of using automated tools for usability evaluation? (RQ4) How can automated tools for usability evaluation be classified?

3.2   Define the Search Strategy

The SMS begins with identifying the keywords, for which it is necessary to find an initial set of articles that answer the research questions. This set is known as the Control Group (CG). The CG is a set of research papers representing, as accurately as possible, the set of primary studies that answers the research questions of the SMS [16]. Furthermore, the CG serves as a source of training samples for refining search strings and determining the sensitivity of the search strategy defined for the SMS. Keep in mind that a highly sensitive search strategy will retrieve many results. However, many of these may be unwanted articles, and a more precise search strategy will retrieve a few articles. However, it may miss many studies that may be useful for research. Therefore, the formation of a CG must have a balance between these two factors [16]. To form the CG, a traditional search for studies related to the research context and, according to the previous explanation, that responds to the research questions was carried out. As a result of this search, six studies were identified [5, 7, 8, 17–19]. Before building the search string, it is verified if the CG studies are found in the Scopus database since it is the one that hosts the most studies. Within Scopus, there are five of the six that belong to the CG; they are [5, 7, 8, 18, 19]. Therefore, we can ensure that Scopus is the best option for research.

To obtain the keywords, a table was generated with the frequency of all the words and combinations of words that appeared in the CG articles, with the help of the Atlas.ti 9 software [20]. We selected only those words directly related to the research questions and that were present in a significant percentage of the CG articles. Subsequently, each one of the words obtained was assigned a value from 0 to 1, determined by its frequency of use, so that the word most frequently repeated in the various CG articles had the value 1. Table 1 shows a fragment of the list of words obtained as a result of this selection process. It shows the words, the percentage of CG studies it appeared in (coverage), the frequency of its appearance throughout the CG studies, and its assigned weight, based on the two previous columns. The weight is calculated based on the percentage of appearance and the frequency as follows (see Eq. 1):

$$((Word\ coverage)/(Maximum\ coverage)$$
$$+ (Word\ frequency)/(Maximum\ frequency))/2 \tag{1}$$

Table 1. Fragment of the list of words obtained from the selection process.

Words	Coverage (%)	Frequency	Weight
Usability	100	1156	1
Evaluation	100	577	0.7496
User	100	388	0.6678
Tool	100	240	0.6038
Interface	100	147	0.5636

3.3 Formation of the Search String

Once the keywords were identified, several search strings were constructed. For constructing the strings, four components are considered that correspond to a classification of the words considered. To define the components, the context of this research was considered, that is, knowing the current panorama of automatic tools that allow the usability evaluation to be supported. The defined components were the following: (i) tools, (ii) automation, (iii) evaluation, and (iv) usability. The logical operator AND was used to join each of these components, while the logical operator OR was used to include synonyms of words from the same component. A total of four search strings were constructed, as shown in Table 2. We used these strings to search for CG studies within the Scopus database. It is important to remember that five of the six CG studies are in the Scopus database.

Table 2. Search strings.

ID	Search string	Studies found	GC found	Ratio X	Ratio Y	Average
1	(usability OR "user experience") AND (evaluation OR testing OR measure OR evaluating OR study OR evaluate OR tests OR assess) AND (tool OR systems OR applications OR tools OR software OR system OR application OR product) AND (automated OR automatic OR automatically OR automating)	2620	5	0.8333	0.0019	0.4176
2	(usability) AND (evaluation OR testing OR measure) AND (tool OR systems OR applications) AND (automated OR automatic OR automatically)	1004	5	0.8333	0.0049	0.4191
3	(usability OR "user experience") AND (evaluation OR testing) AND (tool OR tools OR software OR systems) AND (automated OR automatic)	912	5	0.8333	0.0054	0.4194
4	usability AND (evaluation OR testing OR evaluate OR study) AND (tool OR software OR systems) AND (automated OR automatic)	1304	5	0.8333	0.0038	0.4185

Table 2 shows the number of studies found and the number of CG articles found for each search string tested. All search strings find all five GC studies. Because of this, it was necessary to use additional indicators. These indicators are the X ratio (see Eq. 2), the Y ratio (see Eq. 3), and the average between both (see Eq. 4).

$$XRatio = (No.\ of\ articles\ found\ in\ the\ control\ group)$$
$$/(Total\ of\ articles\ in\ the\ control\ group) \tag{2}$$

$$YRatio = (No.\ of\ articles\ found\ from\ the\ control\ group)$$
$$/(Total\ of\ articles\ found\ per\ search\ string) \tag{3}$$

$$Average = (XRatio + YRatio)/2 \tag{4}$$

As shown in Table 2, the X ratio remains the same for all search strings. This is because, with all strings tested in the Scopus database, the same number of articles belonging to the CG was found. However, the Y ratio shows specific differences since it is based on calculating the proportion of the CG articles found in the total of the results obtained by each string. The string with the highest Y ratio is string 3. To ensure that the selected string is the ideal one for our investigation, the average between the X ratio and the Y ratio is calculated. According to Table 2, string 3 has the highest average, so it is selected as the best search string. The structure of the final search string is shown in Table 3.

Table 3. Final search string.

Keywords						
usability OR "user experience"	AND	evaluation OR testing	AND	tool OR tools OR software OR system	AND	automated OR automatic

Although the search string tests were performed in Scopus, the largest database of peer-reviewed literature [21], the searches were also performed in the IEEE Xplore and Web of Science (WoS) in order to acquire more results. In the search, only studies from 2016 to September 2021 are considered. The databases were analyzed sequentially, using the search fields shown in Table 4. The search fields used were determined by the options provided by each database, due to the different query syntaxes [22–24]. If a duplicate appeared, the first result was kept.

Table 4. Search field per database.

Database	Search fields	Number of results
Scopus	"Title OR Abstract OR Keywords"	904
IEEE Xplore	"Abstract"	162
Web of Science	"Title OR Abstract OR Keywords"	191

3.4   Inclusion and Exclusion Criteria

The inclusion criteria used to select the primary studies are summarized below:

- The article describes one or several tools that support the evaluation of usability or user experience, explaining in detail its operation (e.g., implemented algorithms, architecture, methodologies, theory involved).
- The article reports a testing phase in actual use cases where the tools are tested, and conclusive results are reported, demonstrating that the described tool meets the objective of supporting the evaluation of usability.

It is essential to mention that selecting a study must meet both inclusion criteria. In contrast to this, the exclusion criteria are as follows:

- The tools reported in the study do not perform or support automatic usability evaluation.
- The article does not explain the operation of the presented tools in detail.
- The article does not report a testing phase of the tools.
- The testing phase reported in the article does not deliver conclusive results that answer the research questions.
- The results of the testing phase reported in the article do not show that the tools described meet the objective of supporting usability evaluation automatically.
- The tools described in the article deliver only raw data without any analysis or critique.
- The tool presented in the article is a framework.
- The article is written in a language other than English.

Note that it is enough for a study to meet one of the exclusion criteria to be discarded.

3.5   Select the Studies

A total of 1811 papers were found in the different databases. After excluding duplicate articles, the number was reduced to 1257. Subsequently, a selection of studies was made by applying the inclusion and exclusion criteria to the title and abstract of each of these 1257 studies. The selected articles were validated during a consensus meeting, in which we analyzed the abstracts of articles with conflicting decisions, thus reducing the total to 133 pre-selected articles. After the meeting, the selection criteria were again applied to the full text of the remaining articles. Figure 1 shows the entire filtering and analysis process with the inclusion and exclusion criteria used to select 15 papers. A complete list of the primary studies can be found in Appendix A. The results of applying the different filters during the selection process for each database can be seen in Table 5.

Table 5. Number of remaining studies after filtering the database results.

Database	Studies found	Duplicate-free	Pre-selected studies	Primary studies
Scopus	912	904	110	13
IEEE Xplore	306	162	16	2
Web of Science	593	191	7	0
TOTAL	**1811**	**1257**	**133**	**15**

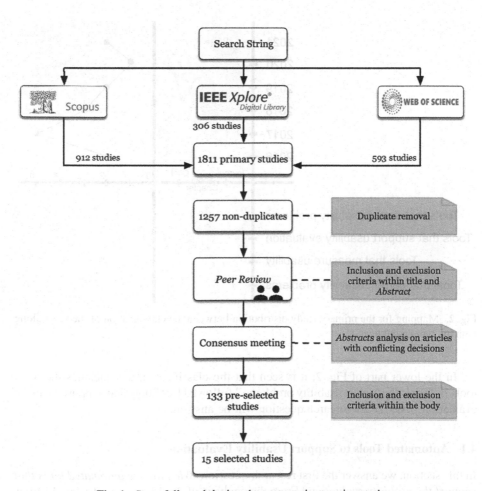

Fig. 1. Steps followed during the systematic mapping study.

4   Results and Discussion

Figure 2 synthesizes the results using two bubble scatter plots. The upper graph represents the number of articles published per year, according to publication type (journal, book chapter, or conference). Similarly, the lower graph plots the publication type against the classification tools (see Sect. 4.4). Thus, the bubbles are located at the intersections between the two axes and their size is proportional to the number of publications for each combination of values.

As can be seen in the upper part of Fig. 2, in 2016, only two studies were found. An excellent interest in tools that support the usability evaluation can be seen in 2017, where five of the 15 primary studies are concentrated. This interest progressively declines, finding three studies in 2018, two in 2019, and only one in 2020. Interest in this area of research recovers a little in 2021, with two studies.

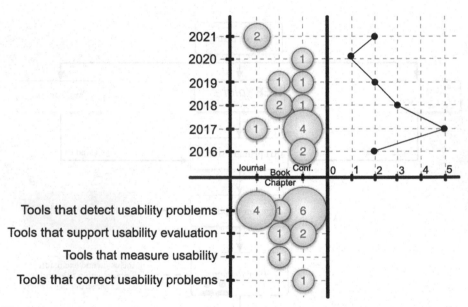

Fig. 2. Mapping for the primary study distribution between the classification of the tools along with the type of publication.

In the lower part of Fig. 2, it is seen that the classification that includes the most tools is "Tools that detect usability problems," followed by "Tools that support usability evaluation." Next, each research question will be answered.

4.1   Automated Tools to Support Usability Evaluation

In this section, we answer the first research question: *What are the automated tools that support the usability evaluation?* From the analysis of the 15 primary studies, 14 tools are obtained, which will be described below.

- **MOBILICS** [PS1] is an extension of USABILICS, so it inherits its methodology. This extension arises from the need to evaluate the usability of web pages in mobile environments considering the touch elements of these devices. MOBILICS performs the usability evaluation by comparing the actual interaction of a user performing a usability test with the interaction predefined by the evaluator who designs the test.
- **Environment for Supporting Interactive Systems Evaluation** [PS2] is a tool that automatically supports the usability evaluation of desktop web user interfaces. This tool performs usability evaluation by detecting usability problems through indicators, using usability data obtained from objective (e.g., an ergonomic guideline inspector) and subjective (through questionnaires) methods.
- **USF (Usability Smell Finder)** [PS3] is a tool that automatically supports usability evaluation of web user interfaces in desktop environments, operating as Software-as-a-Service (SaaS). This tool performs usability evaluation focusing on detecting usability smells, which serve as clues that point to possible usability problems.
- **MUSE (Mobile Usability Smell Evaluation)** [PS4] supports automatic usability evaluation of web user interfaces in desktop and mobile environments. MUSE records user interaction in usability testing sessions. Thanks to its proxy server approach, it can inject JavaScript code to the website to be evaluated without the need for the owner to do so manually.
- **Kobold** [PS5] supports automatic usability testing of web user interfaces in desktop environments, running as SaaS. This tool performs a usability evaluation focusing on detecting usability smells, providing refactorings that can be implemented manually, semi-automatically, or automatically to correct usability problems. Kobold is built on USF, so it uses a similar strategy when detecting usability smells.
- **Plain** [PS6] supports automatic usability evaluation of mobile applications. Plain is an Eclipse Plug-in that allows predicting the usability of a user interface by comparing usability metrics (e.g., composition, symmetry) with the properties of the elements that make up the mobile user interface to be evaluated.
- **UTAssistant** [PS7–PS9] allows supporting the usability evaluation automatically of user interfaces with a web focus. UTAssistant is a web platform that supports usability evaluation work by collecting mouse and keyboard log data during usability testing and allowing audio and video recording (both screen and user face).
- **Guideliner** [PS10] supports automatic usability evaluation of web user interfaces, both desktop, and mobile environments. Guideliner comprises several Java modules and uses Selenium WebDriver as its usability evaluation engine, allowing to search and analyze user interface elements and their features, comparing their values to guidelines to determine usability issues.
- **I2Evaluator** [PS11] supports automatic usability evaluation of web user interfaces, both mobile and desktop environments. i2evaluator seeks to measure user interfaces using aesthetic metrics (e.g., the balance of user interface objects) by incorporating an image decomposition algorithm that helps detect user interface elements to perform metric calculations.

- **PlatoS** [PS12] supports automatic usability evaluation of user interfaces in mobile application environments. The evaluator must create the tasks to perform in the usability tests and simulate the ideal interaction with the user interface. Using predefined usability metrics, PlatoS performs a statistical analysis of the times and actions performed by the evaluator and users to detect usability problems.
- **OwlEye** [PS13] supports automatic usability evaluation of mobile application user interfaces. OwlEye implements a CNN (Convolutional Neural Network) model for usability problem detection. With a set of 66,000 screenshots of more than 9,300 Android applications and using the CNN model, OwlEye can detect problems in user interfaces with a high level of efficiency.
- **ADUE (Automatic Domain Usability Evaluation)** [PS14] automatically allows usability evaluation of desktop applications. ADUE detects domain usability issues based on the domain usability approach. This approach covers aspects related more to the content of the elements that make up the user interface than to their characteristics. ADUE shows the tester the errors and associated components and provides recommendations to correct these problems.
- **GTmetrix** [PS15] supports automatic usability testing of web pages in desktop environments, detecting performance issues by comparing page metrics against 23 rules (related to performance aspects) provided by Yahoo. These values are compared with those detected on the page, and with this the problems to be solved are determined.
- **Dareboost** [PS15] supports the automatic usability evaluation of web pages in mobile environments, using performance metrics (e.g., load times) and comparing them with the metrics obtained from the elements of the web page to be analyzed. The tool provides reports showing the general score of the page, the number of problems, and the improvements recommended for their respective corrections.

4.2 Usability Evaluation Techniques Benefited by Automated Tools

This section answers the second research question: *Which usability evaluation-related techniques benefit from automated tools?* It is essential to mention that the tools have different functionalities and cover usability evaluation differently; therefore, the techniques benefited using these vary according to each case. The techniques used are described below.

- **Interaction Logging** is a technique that records the complete interaction of a user testing a system in such a way that it can be fully reproduced in real-time [25]. The tools that support this technique are Environment for Supporting Interactive Systems Evaluation [PS2], USF [PS3], MUSE [PS4], Kobold [PS5], UTAssistant [PS7–PS9], and PlatoS [PS12].
- **Standards Conformance Inspection** is an inspection method where technology specialists inspect the system determining whether it meets the previously proposed standards [25]. Tools that support this technique are Plain [PS6], I2Evaluator [PS11], PlatoS [PS12], and GTmetrix [PS15].
- **Questionnaires** are an indirect method for studying the user interface that allows knowing the user's opinions about the use of the interface but not giving direct information about it [1]. This technique is supported by two tools: Environment for Supporting Interactive Systems Evaluation [PS2] and UTAssistant [PS7–PS9].

- **Consistency Inspection** is a technique in which a team of designers inspects a set of interfaces for a family of products [25]. This technique is supported by two tools: Environment for Supporting Interactive Systems Evaluation [PS2] and ADUE [PS14].
- **Guidelines review** is a technique in which experts check the conformity of the user interface with the organizational guidelines document or with other guidelines [26]. This technique is supported by Guideliner [PS10] and GTmetrix [PS15].
- **Continuous Recording of User Performance** is a technique that emerges from the evaluation during the active use of the software that is intended to be evaluated [26]. The software architecture should make it easy for system administrators to collect data about system usage patterns, user performance speed, error rate, or frequency of online help replays. This technique is supported by the Environment for Supporting Interactive Systems Evaluation [PS2] and MUSE [PS4] tools.
- **Usage Logging** is a technique that seeks to record the user's actual usage in their interaction with a system, which implies having the computer automatically collect statistics about the detailed usage of the system [1]. This technique is supported by a single tool: MOBILICS [PS1].
- **Video/Audio Recording**, s its name suggests, seeks to generate audiovisual records of user interaction with systems in usability tests [27]. This technique is supported by a single tool: UTAssistant [PS7–PS9].
- **Time Keystroke Logging** is a technique that seeks to generate a record of each keystroke pressed by a user testing a system [25]. Each of these keystrokes is stored along with the exact time the event occurred. This technique is supported by a single tool: PlatoS [PS12].
- **Performance Metrics** is a technique in which essential aspects of the actual use of the software system to be evaluated are quantified, either in a controlled laboratory environment or in the usual work environment [28]. This technique is supported by a single tool: Dareboost [PS15].
- **Heuristic Evaluation** is a technique in which a usability expert observes an interface and tries to obtain an opinion on its good and bad characteristics [1]. This heuristic evaluation technique is supported by a single tool: OwlEye [PS13].

4.3 Problems and Challenges of Using Automated Tools

This section answers the third research question: *What are the existing problems and challenges of using automated tools for usability evaluation?* The main problems and challenges identified in the primary studies are described below.

- **Event Detection in Software Systems** is a technical problem based on the difficulties reported by the authors to identify when events occur in the systems to be evaluated. Goncalves et al. [PS1] reported that considering the MOBILICS tool, the main challenge was detecting events related to touch screens (i.e., *touchstar*, *touchmove* and *touchend*).
- **Detection of Indicators and Thresholds** can be considered an implementation difficulty when determining the metrics to use and how to capture the data that will make the corresponding comparisons with the user interface elements to be evaluated. Assilla et al. [PS2] referred to this problem in the context of the presentation of the Environment for Supporting Interactive Systems Evaluation tool.

- **Validation of Metrics** corresponds to the difficulty of choosing the metrics and quality standards so that the results of detecting usability problems are accurate. Generally, the tools that are based on these indicators must consider analysis models that allow detecting aspects of the user interface and translating them into values that can be interpreted and compared with these metrics. This is precisely the problem presented by the I2Evaluator tool [PS11].
- **A usability expert is still needed in some cases.** This is the main problem that tools seek to automate the usability evaluation. During their development, it must be determined how these tools will guarantee results that allow delivering a synthesis that defines the usability problems of a user interface. Avoiding depending on usability experts is one of the general problems [PS3].
- **General limitations and tool performance improvement.** It corresponds to the challenges reported in the primary studies [PS5, PS6, PS10]. Grigera et al. [PS5] considered improving the accuracy in detecting usability issues to automate new refactorings and select the most suitable one. Soui et al. [PS6] stated that some issues related to quality defect detection need to be investigated. In this way, the authors plan some refactoring operations (e.g., reorganization of the content of the mobile user interface). Marenkov et al. [PS10] specified the tool's limitations considering that it focuses on web environments, indicating that web pages that use Flash or Java Applets are not considered for the use of Guideliner.
- **Tools cannot completely replace manual evaluation.** The use of automated tools has several advantages, such as suitability for large-scale evaluation and less effort in terms of time. However, it is considered essential that these tools cannot completely replace manual tests since usability problems can be found but not how serious the problem is. Therefore, it is necessary to have a specific criterion that the evaluator must exercise based on the interpretation he wants to give to the information provided by the tool [PS15].

4.4 Classification of Automated Tools

In this section, the last research question is answered: *How can automated tools for usability evaluation be classified?* After analyzing the primary studies and the functionalities of each tool reported in each study, a total of four categories were identified, which will be described below.

- **Tools that measure usability.** The tools that belong to this category perform analysis of software systems and deliver an indicator (e.g., percentage of usability, a rating from 1 to 10) that describes the system's usability. These tools do not provide a very detailed analysis, nor do they provide feedback on the specific errors of the analyzed system in terms of usability. In this category, there is only the I2Evaluator tool [PS11].
- **Tools that support usability evaluation.** Tools belonging to this category perform analysis of software systems, provide an indicator that describes the usability of a system, and provide valuable functions that support the usability evaluation. Among these additional functions are (i) automated data capture (e.g., event log, log files), (ii) generation of forms for usability surveys, and (iii) timelines that support evaluation traceability usability, among others. In this category, there is only UTAssistant [PS7–PS9].

- **Tools that detect usability problems.** The tools that belong to this category perform an analysis of software systems and provide feedback on the specific errors found related to usability. The tool displays these errors, for example, in the form of alerts, reports, warnings. To this category belong the tools MOBILICS [PS1], Environment for Supporting Interactive Systems Evaluation [PS2], USF [PS3], MUSE [PS4], Plain [PS6], Guideliner [PS10], PlatoS [PS12], OwlEye [PS13], ADUE [PS14], GTmetrix [PS15] and Dareboost [PS15].
- **Tools that correct usability problems.** The tools that belong to this category analyze software systems and, in addition to providing feedback on the specific errors found in terms of usability, are given the option of correcting them automatically. Tools that perform error correction in a fully automated manner (without prior validation by the tool's user) or semi-automatically (the user authorizes the corresponding automatic correction) are considered in this category. In this category, there is only the Kobold tool [PS5].

5 Validity Threats

The first threat to validity is bias in the article selection process. The articles found with the search string used were evaluated according to the defined inclusion and exclusion criteria. Other researchers may have evaluated the publications differently. To corroborate the concordance in the selection of studies, meetings were held between the researchers to check the discarded preselected articles. Another aspect related to the selection of primary studies is the declared scope of our research since we only consider works published between 2016 and 2021. We may have missed some articles directly related to our research by only considering this period. We only consider the Scopus, IEEE Xplore, and WoS databases for the SMS performed. Although we found many results, more tools could have been reported in other databases. Another point regarding the scope of our research is that we do not consider the grey literature, which will most likely include results that are in line with the objective of this work.

6 Conclusions

A conclusion will be delivered according to each research question.

RQ1: What are the automated tools that support the usability evaluation?

According to the SMS carried out, it was possible to know the general panorama of the tools that support usability evaluation automatically reported in the literature. Between 2016 and 2021, 15 studies were found, of which 14 tools were identified. The reported tools show different approaches to support the usability evaluation. Note that these are presented with different methodologies and ways to support the evaluation of automated usability. The variety of reported tools spans desktop, mobile, and web applications (focused on desktop and mobile) that can be evaluated.

RQ2: Which usability evaluation-related techniques benefit from automated tools?

The most used technique is interaction recording, which makes sense since one of the most used approaches in tools is to perform usability tests to record the interaction of users with the evaluated interfaces. It can be noted that the tools focus on the methodology they use according to the usability evaluation techniques. Some techniques reported [13, 14] were widely used (e.g., standards conformance inspection, consistency inspection), as well as techniques that were not addressed (e.g., collaborative usability inspection [28], pluralistic walkthrough [29]). This highlights that there is still work to be done to develop tools that support automatic usability evaluation. Covering the techniques that have not yet been addressed is a reason to encourage their development.

RQ3: What are the existing problems and challenges of using automated tools for usability evaluation?

According to what was identified in the primary studies, specific challenges, limitations, and problems can be highlighted when implementing the tools. One of these challenges is user interface event detection. This makes sense because it is the most important part of usability testing. Problems in detecting events in usability tests can cause erroneous results, which translates into poor usability for the evaluated interface. A similar aspect is that of the detection of indicators and thresholds. These must be defined and validated so that the tools, which focus on these aspects, can deliver a correct evaluation of usability. Although the tools greatly help the work of implementing an automated usability evaluation, usability experts are still needed, in some cases, to review the results [PS3, PS15].

RQ4: How can automated tools for usability evaluation be classified?

According to the analysis carried out on the primary studies identified in the SMS, the tools can be classified according to their approach and the functionalities that support automated usability evaluation. The classification of the tools is: (i) measure usability, (ii) support usability evaluation, (iii) detect usability problems, and (iv) correct usability problems.

The classification that includes the most tools correspond to those that detect usability problems. This is because it is a broader scope when facing a usability evaluation. These tools provide recommendations that guide the developers of the evaluated applications to correct the usability errors detected. A broader scope is that the tool automatically detects usability problems; only one tool belongs to this category. Kobold [PS5] is presented as one of the most exciting tools because it integrates automatic and semi-automatic refactoring of web user interface elements.

As future works, we will consider more databases (e.g., ACM Digital Library, SpringerLink). Additionally, study and consider the grey literature to expand the results when looking for tools that support the usability evaluation automatically. We want to explore tools that perform qualitative usability evaluations [30]. Finally, we expect to study the usability evaluations results to prioritize and recommend the most relevant aspects [31].

Acknowledgment. This work was supported by the Chilean Ministry of Education and the University of Atacama (ATA1899 project).

Appendix A: Primary Studies

This appendix lists the references of the primary studies used for the mapping study described in this paper.

[PS1] Gonçalves, L. F., Vasconcelos, L. G., Munson, E. V., Baldochi, L. A.: Supporting adaptation of web applications to the mobile environment with automated usability evaluation. In: 31st Annual ACM Symposium on Applied Computing (SAC'16), ACM, Pisa, Italy, pp. 787–794 (2016). https://doi.org/10.1145/2851613.2851863.

[PS2] Assila, A., de Oliveira, K. M., Ezzedine, H.: An environment for integrating subjective and objective usability findings based on measures. In: 2016 IEEE Tenth International Conference on Research Challenges in Information Science (RCIS'16), IEEE, Grenoble, France, pp. 1–12 (2016). https://doi.org/10.1109/RCIS.2016.7549320.

[PS3] Grigera, J., Garrido, A., Rivero, J. M., Rossi, G.: Automatic detection of usability smells in web applications. International Journal of Human-Computer Studies 97, 129–148 (2017a). https://doi.org/10.1016/j.ijhcs.2016.09.009.

[PS4] Paternò, F., Schiavone, A. G., Conti, A.: Customizable automatic detection of bad usability smells in mobile accessed web applications. In: 19th International Conference on Human-Computer Interaction with Mobile Devices and Services (Mobile-HCI'17), ACM, Vienna, Austria, article 42, pp. 1–11 (2017). https://doi.org/10.1145/3098279.3098558.

[PS5] Grigera, J., Garrido, A., Rossi, G.: Kobold: web usability as a service. In: 2017 32nd IEEE/ACM International Conference on Automated Software Engineering (ASE'17), Urbana, IL, USA, pp. 990–995 (2017). https://doi.org/10.1109/ASE.2017.8115717.

[PS6] Soui, M., Chouchane, M., Gasmi, I., Mkaouer, M. W.: PLAIN: PLugin for predicting the usability of mobile user interface. In: 12th International Joint Conference on Computer Vision, Imaging and Computer Graphics Theory and Applications (VISIGRAPP'17) - Vol. 1: GRAPP, Porto, Portugal, pp. 127–136 (2017). https://doi.org/10.5220/0006171201270136.

[PS7] Desolda, G., Gaudino, G., Lanzilotti, R., Federici, S., Cocco, A.: UTAssistant: A web platform supporting usability testing in italian public administrations. In: 12th Biannual Conference of the Italian SIGCHI Chapter (CHItaly'17), Cagliari, Italy, pp. 138–142 (2017).

[PS8] Federici, S., Mele, M. L., Lanzilotti, R., Desolda, G., Bracalenti, M., Meloni, F., Gaudino, G., Cocco, A., Amendola, M.: UX evaluation design of UTAssistant: A new usability testing support tool for italian public administrations. In: Kurosu M. (ed.) Human-Computer Interaction. Theories, Methods, and Human Issues. HCI 2018 (55–67). Lecture Notes in Computer Science, vol 10901. Springer, Cham (2018). https://doi.org/10.1007/978-3-319-91238-7_5.

[PS9] Federici, S., Mele, M. L., Bracalenti, M., Buttafuoco, A., Lanzilotti, R., Desolda, G.: Bio-behavioral and self-report user experience evaluation of a usability assessment platform (UTAssistant). In: 14th International Joint Conference on Computer Vision, Imaging and Computer Graphics Theory and Applications (VISIGRAPP'19) - Vol. 2: HUCAPP, Prague, CZ, pp. 19–27 (2019).

[PS10] Marenkov, J., Robal, T., Kalja, A.: Guideliner: A tool to improve web UI development for better usability. In: 8th International Confe- rence on Web Intelligence, Mining and Semantics (WIMS'18), ACM, Novi Sad, Serbia, article 17, pp. 1–9 (2018). https://doi.org/10.1145/3227609.3227667.

[PS11] Chettaoui, N. Bouhlel, M. S.: I2Evaluator: An aesthetic metric-tool for evaluating the usability of adaptive user interfaces. In: Hassanien, A. E., Shaalan, K., Gaber, T., and Tolba, M. F. (eds.) Proceedings of the International Conference on Advanced Intelligent Systems and Informatics. AISI 2017 (374–383). Advances in Intelligent Systems and Computing, vol 639. Springer, Cham (2017). https://doi.org/10.1007/978-3-319-64861-3_35.

[PS12] Barra, S., Francese, R., Risi, M.: Automating mockup-based usability testing on the mobile device. In: Miani, R., Camargos, L., Zarpelão, B., Rosas, E., and Pasquini, R. (eds.) Green, Pervasive, and Cloud Computing. GPC 2019 (128–143). Lecture Notes in Computer Science, vol 11484. Springer, Cham (2019). https://doi.org/10.1007/978-3-030-19223-5_10.

[PS13] Liu, Z., Chen, C., Wang, J., Huang, Y., Hu, J., Wang, Q.: Owl eyes: Spotting UI display issues via visual understanding. In: 35th IEEE/ACM International Conference on Automated Software Engineering (ASE'20), ACM, Virtual Event Australia, pp. 398–409 (2020). https://doi.org/10.1145/3324884.3416547.

[PS14] Bacíková, M., Porubän, J., Sulír, M., Chodarev, S., Steingartner, W., Madeja, M.: Domain usability evaluation. Electronics 10(16), 1–28, article 1963, (2021). https://doi.org/10.3390/electronics10161963.

[PS15] Al-Sakran, H. O. Alsudairi, M. A.: Usability and accessibility assessment of saudi arabia mobile E-government websites. IEEE Access 9, 48254–48275 (2021). https://doi.org/10.1109/ACCESS.2021.3068917.

References

1. Nielsen, J.: Usability engineering. Morgan Kaufmann Publishers Inc., San Francisco (1994).ISBN: 978-0080520292
2. ISO 9241–11:2018. Ergonomics of human-system interaction–part 11: Usability: Definitions and concepts, ISO (2018)
3. Ferré, X.: Marco de integración de la usabilidad en el proceso de desarrollo software. Facultad de Informática, Universidad Politécnica de Madrid, Madrid, Spain, Tesis doctoral (2005)
4. Ivory, M.Y., Hearst, M.A.: The state of the art in automating usability evaluation of user interfaces. ACM Comput. Surv. 33(4), 470–516 (2001). https://doi.org/10.1145/503112.503114
5. Marenkov, J., Robal, T., Kalja, A.: Guideliner: a tool to improve web UI development for better usability. In: 8th International Conference on Web Intelligence, Mining and Semantics (WIMS 2018), ACM, Novi Sad, Serbia, article 17, pp. 1–9 (2018). https://doi.org/10.1145/3227609.3227667
6. Fabo, P., Durikovic, R.: Automated usability measurement of arbitrary desktop application with eyetracking. In: 2012 16th International Conference on Information Visualisation, IEEE, Montpellier, France, pp. 625–629 (2012). https://doi.org/10.1109/IV.2012.105
7. Federici, S., et al.: UX evaluation design of UTAssistant: a new usability testing support tool for Italian public administrations. In: Kurosu, M. (ed.) HCI 2018. LNCS, vol. 10901, pp. 55–67. Springer, Cham (2018). https://doi.org/10.1007/978-3-319-91238-7_5
8. Grigera, J., Garrido, A., Rossi, G.: Kobold: web usability as a service. In: 2017 32nd IEEE/ACM International Conference on Automated Software Engineering (ASE 2017), IEEE, Urbana, IL, USA, pp. 990–995 (2017). https://doi.org/10.1109/ASE.2017.8115717

9. Liyanage, N. L., Vidanage, K.: Site-ability: a website usability measurement tool. In: 2016 Sixteenth International Conference on Advances in ICT for Emerging Regions (ICTer'16), IEEE, Negombo, Sri Lanka, pp. 257–265 (2016). Doi: https://doi.org/10.1109/ICTER.2016.7829929

10. Charfi, S., Trabelsi, A., Ezzedine, H., Kolski, C.: Widgets dedicated to user interface evaluation. Int. J. Hum.-Comput. Interact. **30**(5), 408–421 (2014). https://doi.org/10.1080/10447318.2013.873280

11. Bakaev, M., Mamysheva, T., Gaedke, M.: Current trends in automating usability evaluation of websites: can you manage what you can't measure? In: 2016 11th International Forum on Strategic Technology (IFOST 2016), Novosibirsk, Russia, pp. 510–514. IEEE (2016). https://doi.org/10.1109/IFOST.2016.7884307

12. Khasnis, S. S., Raghuram, S., Aditi, A., Samrakshini, R. S., Namratha, M.: Analysis of automation in the field of usability evaluation. In: 2019 1st International Conference on Advanced Technologies in Intelligent Control, Environment, Computing & Communication Engineering (ICATIECE 2019), Bangalore, India, pp. 85–91. IEEE (2019). https://doi.org/10.1109/ICATIECE45860.2019.9063859

13. Ferré, X., Juristo, N., Moreno, A.M.: Deliverable D.5.1. selection of the software process and the usability techniques for consideration. STATUS Project (code IST-2001–32298) financed by the European Commission from December of 2001 to December of 2004 (2002). http://is.ls.fi.upm.es/status/results/STATUSD5.1v1.0.pdf

14. Ferré, X., Juristo, N., Moreno, A. M.: Deliverable D.5.2. specification of the software process with integrated usability techniques. STATUS Project (code IST-2001–32298) financed by the European Commission from December of 2001 to December of 2004 (2002). http://is.ls.fi.upm.es/status/results/STATUSD5.2v1.0.pdf

15. Kitchenham, B.A., Budgen, D., Brereton, O.P.: Using mapping studies as the basis for further research–a participant-observer case study. Inf. Softw. Technol. **53**(6), 638–651 (2011). https://doi.org/10.1016/j.infsof.2010.12.011

16. Zhang, H., Babar, M.A., Tell, P.: Identifying relevant studies in software engineering. Inf. Softw. Technol. **53**(6), 625–637 (2011). https://doi.org/10.1016/j.infsof.2010.12.010

17. Assila, A., de Oliveira, K. M., Ezzedine, H.: An environment for integrating subjective and objective usability findings based on measures. In: 2016 IEEE Tenth International Conference on Research Challenges in Information Science (RCIS 2016), Grenoble, France, pp. 1–12. IEEE (2016). https://doi.org/10.1109/RCIS.2016.7549320

18. Barra, S., Francese, R., Risi, M.: Automating Mockup-based usability testing on the mobile device. In: Miani, R., Camargos, L., Zarpelão, B., Rosas, E., Pasquini, R. (eds.) GPC 2019. LNCS, vol. 11484, pp. 128–143. Springer, Cham (2019). https://doi.org/10.1007/978-3-030-19223-5_10

19. Paternò, F., Schiavone, A. G., Conti, A.: Customizable automatic detection of bad usability smells in mobile accessed web applications. In: 19th International Conference on Human-Computer Interaction with Mobile Devices and Services (Mo-bileHCI 2017), Vienna, Austria, article 42, pp. 1–11. ACM (2017). https://doi.org/10.1145/3098279.3098558

20. Atlas.ti9 Atlas.ti 9 desktop trial (windows) (2021). https://atlasti.com/

21. Scopus.com: An eye on global research: 5000 Publishers. Over 71 M records and 23,700 titles 2020. https://www.scopus.com/freelookup/form/author.uri. Accessed 16 Sept 21

22. Castro, J. W., Acuña, S. T.: Comparativa de selección de estudios primarios en una revisión sistemática. In: XVI Jornadas de Ingeniería del Software y Bases de Datos (JISBD 2011), A Coruña, España, pp. 319–332 (2011). http://hdl.handle.net/10486/665299. Accessed 16 Sept 21

23. Magües, D., Castro, J.W., Acuña, S.T.: Usability in agile development: a systematic mapping study. In: XLII Conferencia Latinoamericana de Informática (CLEI 2016), Valparaíso, Chile, pp. 677–684. IEEE (2016). https://doi.org/10.1109/CLEI.2016.7833347

24. Ren, R., Castro, J.W., Acuña, S.T., De Lara, J.: Evaluation techniques for chatbot usability: a systematic mapping study. Int. J. Software Eng. Knowl. Eng. **29**(11n12), 1673–1702 (2019). https://doi.org/10.1142/S0218194019400163
25. Preece, J., Rogers, Y., Sharp, H., Benyon, D., Holland, S., Carey, T.: Human-Computer Interaction. Concepts and Design. Addison-Wesley, Harlow (1994). ISBN: 978–0201627695
26. Shneiderman, B.: Designing the User Interface: Strategies for Effective Human-Computer. Pearson, Boston (1998). ISBN: 978–0201694970
27. Hix, D., Hartson, H.R.: Developing User Interfaces: Ensuring Usability Through Product & Process. Wiley, New York (1993).ISBN: 978-0471578130
28. Constantine, L.L., Lockwood, L.A.: Software for use: A Practical Guide to the Models and Methods of Usage-Centered Design. Addison-Wesley Professional, New York (1999).ISBN: 978-0321773722
29. Nielsen, J.: Usability inspection methods. In: Conference Companion on Human Factors in Computing Systems (CHI 1994), Boston, Massachusetts, USA, pp. 413–414. ACM (1994). https://doi.org/10.1145/259963.260531
30. Rojas P., L.A., Truyol, M.E., Calderon Maureira, J.F., Orellana Quiñones, M., Puente, A.: Qualitative evaluation of the usability of a web-based survey tool to assess reading comprehension and metacognitive strategies of university students. In: Meiselwitz, G. (ed.) HCII 2020. LNCS, vol. 12194, pp. 110–129. Springer, Cham (2020). https://doi.org/10.1007/978-3-030-49570-1_9
31. Rojas, L.A., Macías, J.A.: Toward collisions produced in requirements rankings: a qualitative approach and experimental study. J. Syst. Softw. **158**, 110417 (2019). Article 42. https://doi.org/10.1016/j.jss.2019.110417

Exploring Links Between the Interaction with Social Media and Subjective Well-Being: An Exploratory Study

Beatriz de Paulo[✉] [iD] and Manuela Quaresma [iD]

LEUI | Laboratory of Ergodesign and Usability of Interfaces, Department of Arts and Design, PUC-Rio University, Rio de Janeiro, Brazil
beatrizsandp@gmail.com

Abstract. During the past decade, social apps like Facebook and Instagram have gained relevance in our lives. Implicitly or explicitly, designers are in part responsible for the relationship between product experience and well-being. Past studies have tackled the relationship between social app interaction and well-being, stating that it remains ambiguous. The present work analyzes the diversity of emotional experiences and well-being impacts related to social media apps. Our study presents a users' point of view of how social apps relate to their well-being, adding up to objective studies about the same phenomenon. We have carried out a Collage study with 16 participants, analyzing their reports on emotional experiences and stimulus of psychological needs in the interaction with social apps. Our results describe a duality between positive feelings and concerns about use time, security, and anxiety stemming from the experience with social apps. Based on our findings we present several recommendations for designing social apps that foster healthier and positive experiences for well-being.

Keywords: Social media · Wellbeing · Positive design · Collage study

1 Introduction

In the last decade, the use of social networks has become a significant part of people's daily lives. According to the annual marketing report "The State of Mobile", Brazilians spend an average of 15.6 h per month using Facebook, and 14.5 h per month using Instagram [1].

Social networking applications are defined as "network-based services that allow individuals to (1) build public profiles in a pre-defined system, (2) articulate a list of other users with whom they wish to establish a connection, (3) view and change your list of connections made by others in the system" [2]. As the market changes, popular social apps change as well to permeate more areas of users' lives, moving from messengers and feeds to video, marketing, and business management hubs. Consequentially, social apps gain relevance as interactive products and garner an active user base of billions. Facebook and Instagram, respectively, had 2.3 and 1 billion monthly active users between 2018 and 2019 [3].

G. Meiselwitz (Ed.): HCII 2022, LNCS 13315, pp. 47–64, 2022.
https://doi.org/10.1007/978-3-031-05061-9_4

Like products with a massive audience, there is a diversity and a deep subjectivity of people's experiences with social apps. Indirectly or directly, products can influence the well-being of those who interact with them [4]. Thus, the subjective study of the interaction-experience-well-being relationship becomes relevant.

The Design field begins in the early 2010s to develop tools for assessing the stimulation of well-being through products [5]. On that subject, our work hopes to apply these tools and knowledge to understand how users perceive their relationship with social apps and well-being, theorizing how design can help make this relationship more positive. The present work focuses on understanding how emotions, affectivity, and psychological needs in product experiences relate to the subjective well-being of social app users. The goal is to establish recommendations for designing social apps that promote and preserve the well-being of their users. It hopes to contribute to the field by providing a subjective look at users' perception of how social apps affect their well-being, juxtaposed with the existing theory about the same phenomenon. According to [6], in order to reach scientific truths on any topic, it is necessary to carry out successive research on it, applying different methods and approaches. Collectively with existing objective and quantitative research, it hopes to advance knowledge about user-social apps interaction.

Deriving from this goal, we carried out four Collage [7, 8] sessions with 16 participants aiming to collect deeper insights about their emotional experiences with social apps and their perception of psychological needs. Our general perception is that although there is a predominance of positive perceptions about the use of social networks, a share of users intensely perceives the harmful effects of their continued use. However, most of them cannot control their use time as they wish, recurring to other coping mechanisms to make their experience more positive.

2 The Role of Wellbeing, Emotions, and Psychological Needs in Experiences with Products

Well-being, consciously or not, is one of the central goals of every individual. Objects surrounding us act as platforms for practices and effects that contribute to greater well-being, as commented by Hassenzahl [9] and Pohlmeyer [10]. Interactions with products create contexts to fulfill psychological needs and evoke emotions, and for social apps, it is no different.

Although the concept of well-being was first discussed in Psychology, it also appears in several areas, such as public administration, social sciences, marketing, and design. Definition and understanding of well-being have grown in interest for several reasons, pointed out by Petermans and Cain [5]. First, there is a growing societal interest in seeking practices that lead to a greater state of well-being, such as ampler work-life balance and preventive health. In societies where basic needs are already met, individuals are more concerned about how material resources are exploited rather than their volume. It has also been found that happier individuals are potentially more productive, as greater well-being increases problem-solving and creativity [11, 12]. Well-being has also gained space in quality of life indicators of entire populations, as pointed out by the Organization for Economic Co-operation and Development (OECD) [13]. The fourth motivator, closer to the scope of the present work, points out that objects and services we interact with daily are perceived as "silent companions" that, directly or indirectly, influence how we evaluate our lives [4].

2.1 Objective and Subjective Wellbeing

The conceptualization of well-being involves the distinction between objective and subjective well-being, which can be seen as co-dependent. Objective well-being focuses on external and collective factors, such as the green area in a city or environment luminosity [14]. On the other hand, subjective well-being deals with autonomous reports of how a group of people evaluates their own lives [15]. Subjective well-being is achieved through a high level of life satisfaction, emotional and psychological well-being, in conjunction with personal growth, purpose, and positive relationships with others [16–19].

Despite common beliefs, pleasure and predominant positive affectivity (positive emotions) alone are not enough to achieve a complete state of well-being. An example can be seen in social media feeds, where passive and prolonged consumption can boost momentary pleasure, but not necessarily contribute to well-being. Another example is the consumption of alcoholic beverages and specific sorts of drugs: although considered pleasurable activities, they negatively stimulate other components of well-being, such as physical health or psychological needs.

It is essential to note that the concept of well-being lacks a consensus on its definition, as researchers have predominantly focused on what constitutes well-being and how to measure it [20]. In a review of well-being from a Design point of view, Petermans and Cain [5] reiterate this fact.

2.2 Tools for Assessing Wellbeing in Experiences with Products

Understanding subjective well-being is the first step towards the practice of design for well-being. Diefenbach [21] discusses the growing interest of the Human-Computer Interaction (HCI) field in providing technologies that stimulate subjective well-being through interactions. This interest is enabled by the convergence of two movements: first, the recognition of human and ethical values in the design and development of interactive systems, paired with the growing relevance of the User Experience (UX) field in HCI. The second movement is the rise of positive psychology as the study of human prosperity, that is, the constant evolution and improvement of a human being [22]. The confluence of these movements in design and psychology has led, over the last two decades, to proposals for characterizing design processes and models for well-being.

Based on the work of Sheldon [23], Hassenzahl [24] analyzed the relationship between fundamental psychological needs and positive/negative affectivity in technology experiences. Results indicate that experiences can be differentiated according to the needs met, as well as an "evident link" between psychological needs and positive affect [25]. The model is based on the understanding that well-being does not only stem from activities that have it as a goal, such as taking care of one's health, exercising, or keeping in touch with loved ones. Well-being (subjective or objective) can also be stimulated by mundane day-to-day activities, such as making coffee at a machine, listening to music, driving, or using an app. In this way, the role of designers is to create the circumstances for the stimulation of psychological needs through experiences of all kinds.

The need for **Autonomy** is achieved when an individual is in control of their actions, in contrast to feeling that forces or pressures are the cause of them.

The need for **Competence** is stimulated by creating a feeling that the individual is capable and succeeds in what they are set to do, in contrast to feeling incompetent or inadequate.

The need for **Stimulation** is associated with pleasure and sensory stimulation and is evoked by the feeling of enjoyment, in contrast to feeling bored.

The need for **Security** is to feel safe and in control of your life and actions, as opposed to uncertainty and threats from external circumstances.

The need for **Popularity** is the need to feel respected and influential over others, being relevant to a community. The contrary of this need is the feeling of being disregarded, left out, or uninteresting. As well as stimulation, it is at the heart of social apps to create a sense of constant popularity. New users are continually encouraged to grow their network and create content.

Ultimately, the need for **Relatedness** is to feel that you are in regular contact with dear ones, and not feeling lonely and left behind. Along with Stimulation and Popularity, Relatedness is also a relevant psychological need on social apps. Significantly during the pandemic, social media has become one of the only ways to connect with acquaintances.

In any experience, all six needs do not need to be stimulated completely and simultaneously, but in an ideal balance. By achieving this balance of stimuli, the experience promotes meaning [9] and positive emotions, crucial aspects of product interaction [26] and well-being. Considering the consonance between psychological needs and positive emotions (positive affectivity) in experiences, we aim to assess these aspects in user-social app interactions. The following section will describe our methodology for reaching this goal, based on the theory discussed.

3   Methodology

When a research topic is considered recent, the use of exploratory research methods is recommended [6]. Hence, we selected a predominantly exploratory method for the present work, highlighting the subjective particularity of the Collage [7, 8].

3.1   Collage Board

A Collage session is a design research technique that enables the visual expression of thoughts, feelings, desires, and other aspects that would be challenging to extract by traditional means. Collage sessions are conducted in small groups of 4 to 6 people, where participants start from a pre-defined kit of images, words, or shapes to create individual panels on a theme [7, 8]. We provided a word kit for participants to report their emotions, seeking to provide a certain degree of objectivity in the following analysis.

The panels were divided into four scenarios, based on the previous work by De Paulo and Quaresma [27] on social media use and well-being: (1) General use of social network applications (2) Use of social networks to connect with others (3) Use of social networks to seek information (4) Use of social networks for personal expressions and to see expressions of others. For each scenario, a set of suggested emotions was provided, based on Plutchik's emotion wheel [28]. An empty panel example can be seen on Fig. 2 (Fig. 1).

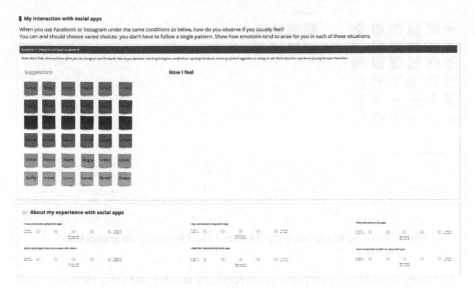

Fig. 1. Empty collage panel for Scenario 1 with suggested words on the left side, and psychological needs' statements at the bottom section.

Participants were also encouraged to make drawings, write new words, and make other interventions in their panels. Regarding the psychological needs, we provided a set of six positive statements for each scenario (for example, "I feel safe using social apps to express myself"). Participants were instructed to mark how much they agreed with each statement on a five-point scale, considering their experiences of using social apps. Figure 2 depicts an example of a finished panel by one of the participants.

All Collage sessions were carried out via *Zoom* through remote board tools. Groups of participants were invited to join the video call at a pre-set timeslot and access their collage kits as pre-built boards. Each session had four participants, totaling 16 participants for the whole study. We did not work with any criteria to assemble the groups, besides time and date availability. Pre-requirements for participation were age (18 and up) and being a regular social app user (at least three use sessions per week). Participant ages ranged between 21 and 33 years old.

Participants had a timeslot of 20 (twenty) to 25 (twenty-five) minutes to fill in their panels for each scenario, considering the emotions most present in their experiences with social apps. We did not ask for participants to consider a specific social app, though the majority mentioned mainly Instagram or Facebook. Participants were instructed to consider the social app they use the most. After the panels were assembled, participants narrated the choices behind their panels and the situations they had in mind. Their panel presentations were the source for associating positive or negative affectivity to each emotion.

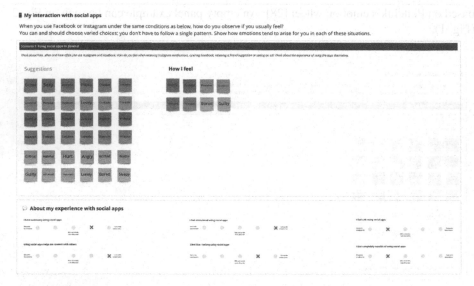

Fig. 2. Example of a finished Word Collage panel by one of the participants.

For analysis purposes, each participant received a number and letter identifier, determined by the order in which the sessions took place (1, 2, 3, or 4) and by the order in which they presented (A, B, C, or D). To determine the psychological need stimulation, we have calculated the medians of the scales of agreement with each statement. The positive stimulation measure is how close the median is to 5. The panel presentations were also considered in combination with the numeric data.

We aimed for a balance between the subjectivity of the reports and the objectivity of the scale questions, providing greater support for the writing of recommendations. The emotions' affectivities were analyzed on a table, providing the percentages of positive and negative affectivity for each scenario.

4   Results Discussion

Based on the Collage study analysis, we will discuss its main findings according to the research questions previously established. Our main goal is to establish connections between the emotions reported, the stimulation of psychological needs, and implied consequences on users' wellbeing. Therefore, we will discuss each aspect separately and draw conclusions based on the results presented.

4.1   Scenario 1: Interacting with Social Media Apps in General

Scenario 1 was described to participants as follows: "Think about how, when, and how often you use Instagram and Facebook. How do you feel when receiving Instagram notifications, opening Facebook, receiving a friend suggestion, or seeing an ad? Think about the experience of using the apps themselves". The general themes evoked in this scenario, given its generalist nature, functioned as a forecast of the others. Participants reported a total of 112 emotions (36 unique), half with positive affectivity and half with negative affectivity.

Psychological Needs in Scenario 1. The Relatedness need tends to be positively stimulated (Md = 4.5), while Security (Md = 2.0) and Popularity (Md = 3.0) are the ones with the lowest positive stimuli. The other three psychological needs tended to be positively stimulated (Md = 4.0). A radar chart demonstrates all tendencies on Fig. 3.

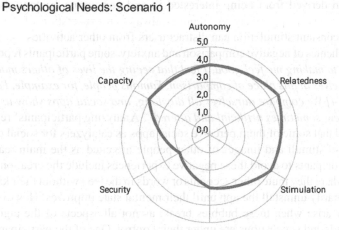

Fig. 3. Radar chart with median values for all psychological needs in Scenario 1.

Although the social media apps work as a way of connecting with loved ones, thus stimulating the "Relatedness" need, this does not necessarily convert into a sense of belonging. Some participants report using social media apps to keep in touch with loved ones due to a lack of options, citing social distancing due to the pandemic, or a lack of time and/or intimacy for face-to-face meetings. This group of participants, although able to maintain contact through the apps, rarely feels stimulated as a result.

Emotions and Affectivity in Scenario 1. Scenario 1 bears the highest percentage of reported emotions of negative affectivity. Due to the general approach to the scenario, reports focused on various aspects of the interaction with social media apps. There is a tie between the emotions "Anxious" and "Interested", followed by "Bored". This result is aligned with the fact that part of the participants expects to feel stimulated every time they open the apps when these expectations are often not met. Notifications also appeared as anxiety triggers, causing polarized expectations that a positive (a message from a friend, a long-awaited reply) or an unpleasant occurrence (a work message, unexpected bad news) is on the way. When the anticipation is considered good, "Anxious" appears with positive affectivity.

Participants report feeling excited to connect with others through social media apps, and stimulated to consume and generate creative content through them. *"[D4] Social networks brought me a job and financial return for many areas, I feel confident"*, *"[A4] Instagram for me is a showcase, of mine and for my work."* This theme connects to the **Popularity**, **Stimulation**, and **Relatedness** needs. We have observed a positive stimulus coming from creative expression and consumption of creative content through social apps. This observation is closely related to the feeling of belonging to a community of common interests or origins. **Relatedness** is also positively stimulated when the social media app acts as an aggregator of the lives of loved ones, making it possible to get in touch without the burden of starting conversations or the need for an in-person meeting. **Stimulation** derived from being interested in and consuming content also emerged, although not always positively. As there is always an expectation of "getting out of boredom", constant stimulation can distract users from other activities.

On the themes of negative comparison and anxiety, some participants reported: *"[B3] in addition to making me feel good, I feel that seeing the lives of others makes me sick, we put ourselves in this place of comparison. Famous people, for example, I don't follow them"*. *"[C4] We compare ourselves all day long, and social apps show us many more points of view, sometimes artificial and altered"*. Analyzing participants' reports, it can be observed that some of them perceive social apps as catalyzers for social comparison. The influx of stimuli and images of other people are cited as the main reasons. Tools used by participants to avoid these negative experiences include the creation of "content bubbles", where they mute other accounts or words in the feed without their knowledge or even temporarily uninstall the app until their mental state improves. However, irritation and anxiety arise when these bubbles break, as not all aspects of the algorithms that regulate feeds and suggestions are under their control. One of the participants reported: *"[A2] sometimes the social network just irritates me when I am avoiding someone or a subject"*, *[C2] I end up creating bubbles. I don't like it when something breaks my bubble, and I get anxious about all the bad news"*.

On the topic of **Security** and **Autonomy**, relevant sub-themes included insecurity in relation to algorithms and data collection and concern about who is able to see your posts. One participant cited: *"[A1] I don't like to post where I'm at, I don't feel safe"*. Participants who report this concern seek to filter as much as possible who can see their posts, although it is necessary to do this for each unwanted person, or completely close the profile to new followers, as can be done on Instagram. However, these measures require great effort on the part of the user, and that they watch at all times who is viewing their content. The sheer need to be vigilant can be seen as a negative stimulus to the need for **Security**. The need for **Autonomy** is also apparently impaired: assuming that all feed content is selected by algorithms allegedly aimed to increase use time, the user feels out of control of their experience.

Control of use time was one of the sub-themes with the highest number of reports. Although users find positive experiences within the applications, there is a perceivable concern with high usage time and addiction. Participants report the desire of spending less time on the platforms or even doing other activities in episodes of boredom, but not feeling able to do so. There are expectations that social apps will mitigate feelings of boredom or even loneliness, and when this expectation is broken (e.g., when content starts to repeat itself) negative stimuli arise, occasionally followed by feelings of guilt. Strategies used by participants include setting time limiters on the apps or forcing themselves to do activities without their phones. This theme brings a direct negative stimulus to **Autonomy** and indirectly to **Stimulation,** since there is a desire to have a more rational use of social apps that cannot be put into practice given the way applications are designed.

4.2   Scenario 2: Using Social Media Apps to Connect with Others

Scenario 2 was described to participants as follows: *"Imagine when you exchange messages with someone, such a friend or relative. Or when you interact with the Instagram Story of a person you follow or leave a comment on their post. In general, occasions when you use social media to communicate with people, close or not."* The participants' understanding aligned well with the scenario proposal, addressing conversations with loved ones, flirting, or simply getting in touch. Participants reported a total of 97 emotions (26 unique), predominantly positive (68.04%).

Psychological Needs in Scenario 2. Addressing the stimulation of the six psychological needs, the **Popularity** and **Security** needs (Md = 3.0) were least stimulated, while **Relatedness** (Md = 4.5) was the most stimulated. The other six needs tended to be positively stimulated (Md = 4.0). A radar chart demonstrates all tendencies on Fig. 4.

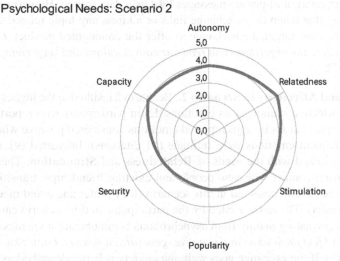

Fig. 4. Radar chart with median values for all psychological needs in Scenario 2.

Alongside the reports, mirroring from Scenario 1, participants report feeling connected with others through social apps, although feelings of belonging (thus stimulating the **Popularity** need) still substantially originate from other forms of contact. These observations are aligned with the existence of a "social deficit" as theorized by Clark et al. [29] where while social networks are used to connect with others, the stimulus of Relatedness and Popularity does not follow. This "deficit" is not always perceived by the user, creating a break in expectations. One participant reported feeling "taken hostage" in some moments of social isolation, for not having other ways to seek social connection other than social networks (*"[C3] because of the pandemic it became a way of communicating, but I felt taken hostage"*).

The other group, who reported enjoying maintaining relationships via social media apps, felt stimulated and connected when doing so. Some reports encapsulate these feelings: *"[D2] in general I feel good because my loved ones are there"*. *"[A2] There is anxiety, but a good kind of anxiety, like wow, this person sent me an incredible message, how am I going to respond to that? I get embarrassed, but it's positive"*. *"[B2] I don't post much but I like to follow what's happening, people's new hobbies, weddings"*. Instagram's "Close Friends" feature was frequently brought up, where participants feel happy, intimate, and safer when posting only for selected people or being in someone else's selected group. Being able to control exactly who sees your posts was reported as a positive stimulus for the **Autonomy** need. The predominance of positive affectivity in this scenario and the tendency of positive stimulation of the need for **Relatedness** permeate both groups, indicating that this is a predominantly positive scenario for users' well-being. On the other hand, the **Popularity** need is more volatile and depends on reciprocity.

The negative stimulus of **Security** in this scenario is similar to Scenario 1, evoked by the uncertainty of how algorithms treat pieces of data shared in conversations. Recent publications on how Instagram algorithms work, for example, do not cite the reading and interpretation of private messages [30] as a data source. However, participants report feeling that when they exchange links or address any topic related to purchases and products, their targeted ads change to offer the commented product. (*"[C4] I feel insecure because the algorithms can read our conversations and keep giving us targeted advertising"*).

Emotions and Affectivity in Scenario 2. Scenario 2 resulted in the highest percentage of positive affectivity amongst its emotions. Even participants who reported avoiding extended contact via social apps reported emotions considered positive when doing so. The most frequent emotions were Intimate (11 citations), Interested (9), and Excited (9), strongly linked with the needs of **Relatedness** and **Stimulation**. These emotions both arise from contact with new people and existing friendships, transiting between two distinct behaviors observed in this scenario (they prefer and avoid meeting people via social media). The anxiety cited by the participants in this scenario can be positive or negative, emanating mainly from asynchronous communication via messaging functions (*"[C2] I feel anxious as in a negative, generalized anxiety, I don't know what they will answer"*). If the exchange goes well, the anxiety is better described as excitement. However, it can quickly turn into negative feelings of anxiety and shame.

In general, the balance of Scenario 2 in emotional experience and psychological needs tends to be positive, promoting feelings of **Relatedness, Autonomy, and Stimulation** by enabling people to stay in touch with loved ones. Social apps were reported to promote an extension of in-person communication, which requires greater physical presence and availability. However, feelings of insecurity due to the environment controlled by algorithms, false or malicious accounts, and negative anxiety due to asynchronous and multimedia communication on the networks are also very present, which can cause negative stimuli and impair the Relatedness and Connection needs.

4.3 Scenario 3: Interacting with Social Media Apps to Seek Information

Scenario 3 was described to participants as follows: *"Think about when you read the news, text posts, or search for an account that posts about art or any topic that interests you"*. Remember when you consume information in general through social media, of any kind, not just journalistic." Participants understood the scenario in a close link with news and journalistic content, although other types of information were also mentioned. Participants reported a total of 131 emotions (32 unique), with a slight predominance of positive affectivity (57.25%).

Psychological Needs in Scenario 3. Needs in Scenario 3 show an overall positive stimulus, with only Security showing a stimulus trend below 3 (Md = 2.5). Other needs tend to be positively stimulated in the participants' perception (Md = 4.0). A radar chart demonstrates all tendencies in Fig. 5.

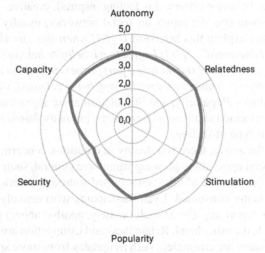

Psychological Needs: Scenario 3

Fig. 5. Radar chart with median values for all psychological needs in Scenario 3.

Mirroring previous scenarios, a negative stimulus to Security was the awareness of an algorithm regulating what appears in the feed based on personal data (*"[C2] I think about the algorithms that recommend things to us"*). Another significant negative stimulus is the concern with fake news or biased headlines (*"[D4] I don't feel that social networks are safe because everyone can manipulate information"*, *"[D3] I feel that I always make a filter of what I'm going to consume, and I think that's not a lot of autonomy"*). Although it has a positive stimulus trend, Autonomy also appears in pair with negative Security stimuli.

Emotional Experiences and Affectivity in Scenario 3. The emotions mentioned in Scenario 3, although mostly positive, are not as predominant as in other scenarios (42.75% of the emotions' affectivity are negative). There was an unanimity of themes and causes in this scenario for negative emotions. Even participants who actively use social networks to seek journalistic information and content report feeling anxious (9 citations) and irritated (7) due to the content of the news. These participants reported feelings of powerlessness in the face of negative or fake news. The Thoughtful emotion (10 citations), third-most cited, is cited in a positive to neutral affectivity, depending on the situation. Interested (13 citations) and excited (11 citations), the most cited emotions, usually refer to an initial feeling of searching and locating relevant content.

In general, in the reports, information actively sought by the participant brings more emotions of positive affect, and emotions of negative affectivity are associated with news that the feed delivers or are found passively in other ways (Stories, sent via direct message). However, there are exceptions. Some quotes embody this feeling: *"[A2] News bring me this moment of feeling depressed and irritated with the type of information that people are sharing", "[C1] we end up using the bubble, filtering to see only what we like it, but it does not always work"*. The strategy of creating "content bubbles" can be observed again to compensate negative stimuli to the **Autonomy, Stimulation** and **Security** needs.

Another behavior observed is feeling inspired, creative, and belonging when being informed about specific topics on social networks, usually not linked to news. Some reports better explain this behavior: (*"[B4] when they are things that interest me, I get excited and interested", "[D3] I agree that it helps to belong because I know things about my community", "[A2] But there is a part further away from the news, news about what I like, art content... it makes me excited, stuff being released, videogame news"*). The need for Relatedness, Popularity, and Stimulation arose significantly in these reports, given that being informed about your community, interests shared with others, or hobbies tend to bring this type of feeling.

Given the above, there is a duality in emotions concerning consuming information through social apps: although some stimuli are general, such as concerns of security and autonomy (occurrence of fake news, biased sources), emotions depend on what type of content is being consumed. Even participants who actively seek information through social apps report negative stimuli (anxiety, apathy, anger) by the content of the news they find. On the other hand, Relatedness and Connection are positively stimulated other types of content are consumed, such as updates from more specific contexts (sports, art, culture) or information about hobbies. However, it is a shared responsibility to create conditions for these stimuli to happen, and users do not feel able to trust the social app to ensure the experience is positive.

4.4 Scenario 4: Interacting with Social Apps for Personal Expression

Scenario 4 was described to the participants as follows: "Think about when you upload a photo of yourself, or a text you wrote, observe photos and publications of friends, even without interacting. Imagine how you express yourself in social apps. Likewise, consider how you view other peoples' expressions. Think not only about images or aesthetics but also about ideas and opinions". Despite being the most subjective scenario, the participants' understanding aligned with our expectations. Reports focused on photos, with citations to other forms of personal expression. Participants reported 130 emotions (31 unique), with 66.9% of positive affectivity.

Scenario 4, despite its majority of positive affectivities, has its two most reported emotions being predominantly negative. Anxiety (9 negative citations and two positive citations), followed by "Insecure" (9 negative citations) were the most present emotions throughout the panels.

"Excited", "Creative" and "Interested" also appeared in half of the panels, being cited eight times for this scenario. The forenamed emotions tend to vary in temporality. A share of the participants is positively excited first, feeling anxious only after expressing themselves. In this case, the anxiety connects with the number of likes and comments arriving via notifications. Another group feels anxious before, during, and after, wondering before posting what the impressions, comments, and number of likes will be.

In the case of the first group, anxiety emerges positively, encouraging users to express themselves and follow how their posts perform. Nevertheless, the general perception is that even positive emotions, in this case, can quickly gain negative light depending on this performance. Positive emotions connect with social apps working as personal records, similar photo albums. Creative and artistic expressions also appear associated with positive emotions. According to the participants, Relatedness with others is also strengthened by observing their expressions via social apps.

The stimulus of psychological needs, like most reports in this scenario, are linked with **Popularity** and **Relatedness**. Regarding psychological needs, all except **Security** (Md = 2.0), are positively stimulated, with medians between 3.5 and 4.0. Security stimulus is the lowest of the four scenarios (Md = 2.5). This low stimulus for Security connects with two factors. Firstly, participants reported concern about who sees their content. In addition, they have worries about the perceived expectations of the network members. A radar chart demonstrates all tendencies in Fig. 6.

Regarding Scenario 4, participants present into two dominant tendencies: (1) individuals who only express themselves (in any way) when they already feel safe and positively stimulated to do so (2) individuals who express themselves desiring to feel safe and positively stimulated. Behaviors related to anxiety about likes, or safety concerns, permeate both groups.

The first group is positively stimulated before making a post with any type of personal expression. Participants report: *"[B3] I already have pretty good self-esteem, for me to post I got to have my self-esteem on a high". "[D4] I feel proud of what I showed to the world". "[B4] When I post anything it is because I already knew it was cool, so I feel confident and important"*. Stimulus to Popularity and Relatedness is also present as a way of participating in collective occasions *("[D2] the stimulus exists to also participate in collective occasions, such as New Year's Eve", [D3] Social apps help me to connect by establishing a proximity with people I like")*. In these cases, the expressions go beyond self-portraits, with emotions tending to be stabler, although anxiety about likes persists. There is also satisfaction in keeping the social network as a personal chronological record. Participants say: *"[D1] I use Instagram as a record because I think it will never*

Psychological Needs: Scenario 4

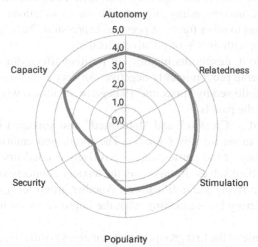

Fig. 6. Radar chart with median values for all psychological needs in Scenario 4.

get lost, [A1] I turn my social networks into a notebook, a record because later I like to see it. I feel lively, creative, playful, sexy."

Another observed behavior associates more negative affectivity with the likes and with the whole experience of expressing oneself through the networks. This behavior represents a negative Popularity stimulus, associated with fear and anxiety about content reception. Participants say *"[D2] There is an expectation of what we will see and generate for others on social networks. [A2] Sometimes I need to calm down before checking likes, messages, replies. It's a difficult cycle to get out of so I avoid getting into it". [B1] I feel like I expose myself a lot. I feel like I wasn't supposed to post that, and I worry about what people will think".*

Following the work by Verduyn et al., [31], negative comparison social comparison, can function as a mediator in the negative relationship between SMA use and well-being. A portion of the participants appears to be aware of that relationship, desiring to shorten use time because of social comparison tendencies. Possibly because of its focus on graphic images, Instagram was nearly the only social network mentioned for this theme.

Concerning **Security** and **Autonomy** needs, negative stimuli stems from places similar to other scenarios. Concerns about having the account cloned or stolen for illicit purposes are present. There is also uncertainty about how IA and ML algorithms employ these expressions, and how content is ranked by them as interesting or not. Network size on Instagram is cited in this case as a negative stimulus for Autonomy, creating patterns of content creation that are prioritized. Some strategies employed by participants to regain Autonomy include signing up for secondary accounts, with fewer followers and a brand-new algorithm.

In Scenario 4, the needs for **Popularity**, **Relatedness**, and **Security** are most stimulated, both positively and negatively. Catalyzed by a need to measure or reinforce their popularity, participants tend to feel anxious about likes and other reactions their posts garner. When personal expressions are well accepted, especially creative expressions, they might bring feelings that stimulate **Relatedness** and **Popularity**. However, there is a mutable character in the experience affectivity. A reception considered insufficient quickly negatively provoke psychological needs. Participants employ a range of resources to create safe spaces, such as avoiding personal expressions altogether or creating closed selections of accounts that may or may not see their content. **Security** appears to be impaired, given the uncertainty about the algorithms that analyze data they make available on the apps, or about having the content misinterpreted, misused, or even stolen for illicit purposes.

4.5 Synthesis and Recommendations

In general, there is a duality in how users report their experience with using social apps. Despite the predominance of positive affectivity and benefits stemming from the interaction, there is a poignant desire to use social apps less.

As interactive products, social apps that stimulate content creation and active participation in social circles can become tools for the needs of Relatedness, Popularity, and Autonomy, providing an outlet for personal expression in new ways. There is a link between creative expression and consumption of creative content considered fun and engaging by users, which relate to the feeling of belonging to a community of shared interests and experiences. The discovery of new content, avoiding the excessive repetition of themes and sources, can also be perceived in the results as a source of positive influence on well-being. In this way, the AI and ML algorithms that curate content on social apps can help discover content that is less linked to complete similarity and more to associations, avoiding the feeling of stagnation and repetitiveness. Heterogeneous and organic feeds can also help users become less fixated on content that may generate negative comparisons. The move towards deprioritizing likes and focusing on the content also aligns with this goal. Other non-numerical forms of feedback for content can also be considered.

Relatedness is also positively stimulated when the social app acts as an aggregator of contact with loved ones, bringing positive stimuli to well-being. The use of social apps when there is enough Autonomy to select who can interact with you, combined with interaction outside the apps, can positively stimulate Relatedness, although not always the need for Popularity. Using social networks to extend and enhance relationships that are strengthened in other ways can be positive for well-being. AI and ML algorithms can also help with this relationship within the platform, making their connections considered "close" more customizable for the user. This can provide more flexible options beyond the exclusion or total blocking of an account from the network. Closer, bilateral forms of interaction (such as the "Close Friends" feature on Instagram, often cited by participants) can help bring positive stimulus of Popularity.

Considering users with a tendency to negative comparison, the constant exposure to images of others without distinction can exacerbate this tendency, harming the Relatedness with others through networks and bringing negative Stimulation. Greater diversification of the content displayed in users' feeds can mitigate or reduce the intensity of this exposure since content that is considered captivating by the platform can become repetitive or even harmful to well-being. Encouraging a more diversified use of the platform, balancing connection with other people, and consumption of diverse content can reduce this catalysis of negative comparison, reinforcing the positive stimulus to Relatedness and Popularity.

5 Conclusions and Future Research

The experience of using social networks, focusing on the emotional experience, proved to be volatile and ambiguous with a set of common variants. Confirming claims made by other researchers who have observed the same phenomenon, it is delicate to establish universal truths about the experience with social apps and well-being. Positive emotions and affectivity were predominant. However, when negative episodes were mentioned, they came with an unfulfilled desire to stop social app use altogether. We have observed tension and, in the words of the participants, a relationship of submission. Social apps are necessary to communicate with physically distant acquaintances, meet new people, and find information about diverse subjects. However, its use carries anxiety and concern about assumed expectations from followers. It also brings anxiety about the frequency and nature of the content that feeds deliver. As a user's audience (their following) grows and more expectations are created on both sides, anxiety increases as well. However, constant invites for returning to the social app, whether for reasons relevant to the in-person social life or out of curiosity, puts a barrier on critical and conscious use.

The present research sought to juxtapose numerous existing studies (mainly in the psychology field) on the long-term effects of using social apps. A subjective analysis from the users' point of view brings valuable discoveries to the subject, glimpsing on how people perceive how social app use affects their lives. Collectively with more objective studies, it can provide important tools for a complete understanding of this phenomenon. Limitations include the subjective nature of the research object and the methods chosen. A method like the Collage session relies on the accuracy and veracity of the reports provided by the participants.

Future research directions include validating the recommendations, structuring them in applicability and relevance. Testing them in real scenarios and verifying their impact on subjective well-being is also desired. Additionally, the observation of this phenomenon can benefit from the combination with objective methods.

Acknowledgements. This study was financed in part by the Coordenação de Aperfeiçoamento de Pessoal de Nível Superior - Brasil (CAPES) - Finance Code 001.

References

1. App Annie. The State of Mobile in 2021 (2021)
2. Boyd, D.M., Ellison, N.B.: Social network sites: definition, history, and scholarship. J. Comput.-Mediat. Commun. **13**(00), 210–230 (2007)
3. Ortiz-Ospina, E.: The Rise of social media (2019). https://ourworldindata.org/rise-of-social-media. Accessed 25 Jan 2022

4. Dorrestjin, S., Verbeek, P.P.: Technology, wellbeing, and freedom: the legacy of utopian design. Int. J. Des. **7**(3), 45–56 (2013)
5. Petermans, A., Cain, R. (eds.): Design for Wellbeing: An Applied Approach. Routledge (2020)
6. Lazar, J., Feng, J.H., Hochheiser, H.: Research Methods in Human-Computer Interaction, pp. 1–560 (2017). https://doi.org/10.1016/b978-0-444-70536-5.50047-6
7. Sanders, E.B.-N., Stappers, P.J.: Co-creation and the new landscapes of design. CoDesign **4**(1), 5–18 (2008). https://doi.org/10.1080/15710880701875068
8. Martin, B., Hanington, B.: Universal Methods of Design: 100 Ways to Research Complex Problems, Develop Innovative Ideas, and Design Effective Solutions. Rockport Publishers (2012)
9. Hassenzahl, M., Eckoldt, K., Diefenbach, S., Laschke, M., Lenz, E., Kim, J.: Designing moments of meaning and pleasure. Experience design and happiness. Int. J. Des. **7**(3), 21–31 (2013)
10. Pohlmeyer, A.E.: Enjoying joy. In: Proceedings of the 8th Nordic Conference on Human-Computer Interaction: Fun, Fast, Foundational, pp. 871–876 (2014). https://doi.org/10.1145/2639189.2670182
11. Desmet, P., Hassenzahl, M.: Towards happiness: possibility-driven design. In: Zacarias, M., de Oliveira, J.V. (eds.) Human-Computer Interaction: The Agency Perspective. SCI, vol. 396, pp. 3–27. Springer, Heidelberg (2012). https://doi.org/10.1007/978-3-642-25691-2_1
12. Desmet, P., Pohlmeyer, A.E.: Positive design: an introduction to design for subjective well-being. Int. J. Des. **7**(3), 5–19 (2013)
13. OECD: OECD Guidelines on Measuring Subjective Well-being. OECD Publishing (2013). https://www.oecd-ilibrary.org/economics/oecd-guidelines-on-measuring-subjective-well-being_9789264191655-en
14. Constanza, R., et al.: Quality of life: An approach integrating opportunities, human needs, and subjective wellbeing. Ecol. Econ. **61**(2–3), 267–276 (2007)
15. Veenhoven, R., et al.: Het rendement van geluk. Inzichten uit wetenschap en praktijk. Stichting Maatschappij en Onderneming, Den Haag (2014)
16. Diener, E.: Subjective wellbeing: the science of happiness and a proposal for a national index. Am. Psychol. **55**(1), 56–67 (2000)
17. Ryff, C.D.: Happiness is everything, or is it? Explorations on the meaning of psychological well-being. J. Pers. Soc. Psychol. **57**, 1069–1081 (1989)
18. Seligman, M.E.P.: Flourish: A Visionary New Understanding of Happiness and Wellbeing. Free Press, New York (2011)
19. Seligman, M.E.P.: Authentic Happiness: Using the New Positive Psychology to Realize your Potential for Lasting Fulfillment. Free Press, New York (2002)
20. Dodge, R., Daly, A., Huyton, J., Sanders, L.: The challenge of defining wellbeing. Int. J. Wellbeing **2**(3), 222–235 (2012)
21. Diefenbach, S., Hassenzahl, M., Eckoldt, K., Hartung, L., Lenz, E., Laschke, M.: Designing for well-being: a case study of keeping small secrets. J. Posit. Psychol. **12**, 151–158 (2017)
22. Botella, C., Riva, G., Gaggioli, A., Wiederhold, B.K., Alcaniz, M., Baños, R.M.: The present and future of positive technologies. Cyberpsychol. Behav. Soc. Netw. **15**(2), 78–84 (2012). https://doi.org/10.1089/cyber.2011.0140
23. Sheldon, K.M., Elliot, A.J., Kim, Y., Kasser, T.: What is satisfying about satisfying events? Testing 10 candidate psychological needs. J. Pers. Soc. Psychol. **80**(2), 325–339 (2001). https://doi.org/10.1037//0022-3514.80.2.325
24. Hassenzahl, M.: User experience (UX): Towards an experiential perspective on product quality. In: Bastien, J.M.C., Carbonell, N. (eds.) Proceedings of the 20th International Conference of the Association Francophone d'Interaction Homme-Machine, pp. 11–15. ACM Press, New York (2008)

25. Hassenzahl, M.: Experience Design: Technology for All the Right Reasons, p. 354. Morgan & Claypool, San Rafael (2010)
26. Desmet, P., Hekkert, P.: Framework of product experience. Int. J. Des. **1**(1), 57–66 (2007)
27. de Paulo, B., Quaresma, M.: Outlining experience and well-being in the interaction with social Media Apps. In: Black, N.L., Neumann, W.P., Noy, I. (eds.) IEA 2021. LNNS, vol. 223, pp. 12–19. Springer, Cham (2022). https://doi.org/10.1007/978-3-030-74614-8_2
28. Karimova, H.: The Emotion Wheel (2021). PositivePsychology.Com. https://positivepsychology.com/emotion-wheel/
29. Clark, J.L., Algoe, S.B., Green, M.C.: Social network sites and well-being: the role of social connection. Curr. Dir. Psychol. Sci. **27**(1), 32–37 (2018). https://doi.org/10.1177/0963721417730833
30. Mosseri, A.: Shedding More Light on How Instagram Works (2021). https://about.instagram.com/blog/announcements/shedding-more-light-on-how-instagram-works
31. Verduyn, P., Ybarra, O., Résibois, M., Jonides, J., Kross, E.: Do social network sites enhance or undermine subjective well-being? A critical review. Soc. Issues Policy Rev. **11**, 274–302 (2017). https://doi.org/10.1111/sipr.12033

Improving EEG-based Motor Execution Classification for Robot Control

Sumeyra U. Demir Kanik, Wenjie Yin, Arzu Guneysu Ozgur,
Ali Ghadirzadeh(✉), Mårten Björkman, and Danica Kragic

Division of Robotics, Perception and Learning (RPL),
KTH Royal Institute of Technology, Stockholm, Sweden
{sumeyra,yinw,arzuo,algh,celle,dani}@kth.se

Abstract. Brain Computer Interface (BCI) systems have the potential
to provide a communication tool using non-invasive signals which can
be applied to various fields including neuro-rehabilitation and entertainment. Interpreting multi-class movement intentions in a real time setting
to control external devices such as robotic arms remains to be one of the
main challenges in the BCI field. We propose a learning framework to
decode upper limb movement intentions before and during the movement execution (ME) with the inclusion of motor imagery (MI) trials.
The design of the framework allows the system to evaluate the uncertainty of the classification output and respond accordingly. The EEG
signals collected during MI and ME trials are fed into a hybrid architecture consisting of Convolutional Neural Networks (CNN) and Long
Short-Term Memory (LSTM) with limited pre-processing. Outcome of
the proposed approach shows the potential to anticipate the intended
movement direction before the onset of the movement, while waiting to
reach a certainty level by potentially observing more EEG data from the
beginning of the actual movement before sending control commands to
the robot to avoid undesired outcomes. Presented results indicate that
both the accuracy and the confidence level of the model improves with
the introduction of MI trials right before the movement execution. Our
results confirm the possibility of the proposed model to contribute to
real-time and continuous decoding of movement directions for robotic
applications.

Keywords: BCI · Deep learning · Human-robot interaction

1 Introduction

Brain Computer Interface (BCI) systems have been shown to effectively translate
the intentions of humans into a series of commands. BCIs enable people to control
external devices such as robotic arms, wheelchairs, computer cursors or even

This work was supported in part by the ERC (European Research Council), Swedish
Research Council and EnTimeMent (EU Horizon 2020 FET PROACTIVE project).

G. Meiselwitz (Ed.): HCII 2022, LNCS 13315, pp. 65–82, 2022.
https://doi.org/10.1007/978-3-031-05061-9_5

drones using only brain activity [4,25]. The fields of application for such systems range across neurorehabilitation, neuro-marketing, gaming and education [16].

One of the main applications of BCI is decoding movement intentions and the kinematics of related movements from electroencephalography (EEG) signals. It is well known that the brain activity during the execution and imagination of the same movement share similar characteristics, and the same areas of the brain are activated [23]. As a consequence, motor imagery (MI) BCI systems have become quite popular in the BCI field, especially in terms of motor learning and neurorehabilitation of stroke patients. MI BCI systems, in which participants are asked to imagine the movement instead of moving, are proven to be successful in separating the intention of moving right hand from left hand, especially since contra-lateral hemispheres oversee the left and the right-side body movements [6]. However, distinguishing between more complex motions of the same limb, and decoding the trajectory of the imagined movement remain as main challenges in the motor-BCI field due to the low spatial resolution of EEG signals and low signal-to-noise ratio [6,14].

The execution of the movement (ME) might include more artefacts originating from muscles compared to MI. On the other hand ME tasks are cognitively less demanding than the MI tasks, due to the higher mental load of MI sessions. Researchers have observed stronger brain activities during ME sessions as opposed to the MI sessions of the same tasks, making them easier to be detected by BCI systems, despite the muscle artifacts [17]. The advancement of deep learning techniques has allowed BCI systems to cope with muscle artefacts along with the ability of moving with the introduction of wireless EEG systems. Regardless of the success of ME based BCI systems in recent years, the delay introduced between the movement intention and control command, especially for avatar/robot controlling applications reveals a need to focus on the detection of motor planning rather than motor execution [12,24].

In the light of previous research, our contribution in this work can be summarized under three items:

- We study a setting in which motor imagery trials are added before motor execution trials to allow the decoding of motor-related EEG even before the onset of the movement to facilitate real-time control of external devices.
- The model recurrently process MI and ME data until it is certain about the output. The decoding starts with available MI data, and we evaluate the confidence level of the model before generating a control command. That way, if the model is not confident of the output, it waits for more data (the motor execution EEG) avoiding unwanted/uncertain feedback.
- We claim that our proposed approach is not only able to decode the direction of intentional movement earlier than conventional approaches, but also does it with higher accuracy.

2   Background

The motor correlates of the EEG signals originates from the motor cortex which is responsible for planning, control, and execution of voluntary movements. Two main oscillations produced by the movement intention are called SMRs (sensory motor rhythms) and MRCPs (movement related cortical potentials) [1,22]. SMRs can be described as the change in mu (8–12 Hz), beta (13–30 Hz) and gamma (>30 Hz) frequency band powers starting from 2 s before the onset, and they are often referred as event related synchronization/desynchronization (ERS/ERD) [21]. MRCPs are the low frequency variations in the EEG signals starting before the onset of voluntary movements or imagination of such movements over the motor cortex [24]. The changes in the EEG signals occur before the onset of the movement such as the Bereitschaftspotential, readiness potential in other words (a negative potential seen 1.5 s before the movement) or ERDs [2].

In a recent study by Mammone et al. [14], researchers used a publicly available 61-channel EEG data set [19] consisting of 15 subjects to differentiate between 6 movements of the same upper limb (elbow flexion/extension, forearm pronation/supination, hand open/close). They claimed it is the first study in the EEG-based BCI field to classify motor planning via deep architectures. Reconstruction of source signals was performed using a beamforming approach, and the signals from the premotor and primary motor cortex are selected to generate time-frequency maps via Continuous Wavelet Transform. A 3-D input consisting of time-frequency maps of the reconstructed sources were fed into a 3-layer CNN architecture for 21 binary classifications of 1 s epochs preceding different movements and rest epochs. They reported an average accuracy of 90.30% for classifying rest vs 6 movements, and 62.47% to differentiate between the movements.

Valenti et al. [24] used the same dataset and represented EEG channels in a 2-D rectangular space resulting a 3-D input. They claimed that a combination of 3-D CNNs with LSTM yields higher classification accuracy with respect to Mammone et al. [14] with an average accuracy of 50.79% for multi-class classification. Mammone et al. [15] later removed the source reconstruction approach and reduced the complexity of their CNN architecture to show that they can achieve comparable results with hand open/close vs rest classification. Compared to the rest class, they reported 90.77% accuracy for Hand Close class, and 92.48% accuracy for Hand Open class from 1 s pre-movement EEG trials.

The main goal in this line of work is to detect and decode the movement the subject is planning to execute as soon and accurate as possible, which would enable almost real-time control of external devices, and would enhance the performance of both assistive and restorative BCI systems [14,24]. The detection of movement intention successfully before or at the onset of movement would make BCI-controlled systems applicable in real-time settings without the need to wait for the processing of the whole epoch. Even though mentioned research focuses on hand and wrist related tasks instead of arm reaching, they show the potential to anticipate the direction of goal oriented upper limb movements as well.

3   Related Work

Due to the challenging nature of MI-EEG signal decoding, previous research on MI-EEG based external device control systems involve artificial matching between the control command and the intention when multi-directional movement is included. Researchers often associate imagining the movement of different parts of the body such as feet or tongue for issuing various directional control commands [13]. Most of the work in this field is built to distinguish among the limbs subjected to the MI such as left hand, right hand, both hands or feet [18]. This is due to the fact that distinguishing the activated areas for different body parts is a less challenging problem than identifying the direction of the movement for the same limb in an EEG based BCI setting. This approach not only limits the number of classes, but also results in an unnatural command system such as a hand-open command corresponding to the imagination of moving both feet. Resulting systems correspond to exhaustive training sessions along with a significant gap between the intention and the issued control command [20].

A research group, Jeong et al. [9] has studied the intentional arm and hand related movements both with MI and ME with healthy participants with the aim of controlling a robotic device using EEG signals. In a recent paper, they proposed a two layer CNN combined with bidirectional LSTM architecture to estimate velocity profiles of 3D arm movements in 6 different directions (left, right, up, down, backward, forward) for 15 subjects. They have showed the application of their approach via a MI-BCI controlled robotic arm. For the online robotic experiments, they placed items in 3 or 4 different directions and claimed that this was the first attempt to decode motor imagery in a 3D multi-direction setting (left, right, forward, backward, up, and down).

In another study, Lee et al. [11] have transformed the 2D EEG signals (channel × time) to 3D by mapping the channel locations to 2D. They proposed that the inclusion of ME would improve the decoding performance of MI. They have reported their results from 9 subjects in 4-class classification setting with either horizontal (left, right, forward, backward) or vertical (left, right, up, down) plane.

Although previous research shows the potential of using EEG based decoding for both MI and ME, decoding the direction of intentional movement as early as the motor planning phase remains unexplored. As stated in Dose et al. [3], another open research question in motor related BCI systems is including a control mechanism to stop the model from providing an undesired feedback when it is uncertain. Our main motivation is combining MI and ME EEG to start decoding as early as possible while evaluating the confidence level of the BCI controlled system for the actions performed by the robot before producing a control command in an almost real-time setting.

4   Method

4.1   Experimental Paradigm

In the proposed experimental set-up, participants with an EEG cap sit on a chair with their dominant arm resting on a desk while following the experimental flow through visual stimuli displayed on a screen. In order to have consistent movement output among participants, a rectangular layout is designed and put on a desk to guide the participants towards 4 target motion directions, and to indicate the resting place in the middle. The task layout with four targets placed in the four corners of a rectangle can be seen in Fig. 1b. Each experimental trial is composed of Rest, MIP, MI, MEP and ME phases for one specific direction, and each phase of the trial is guided through a visual stimulus (Please see Fig. 1c for visual stimuli.). One trial lasts for 13 s as can be seen in Fig. 1a.

Before the experiment, the participants were informed about the nature of the experiment, and they provided their written consents. The experiments started with a practice session of 6–10 trials to familiarize the participants with the stimuli and the tasks. Each trial started with a rest period of 3 s where a plus sign was displayed on the screen, followed by a 3 s motor imagery preparation (MIP) period. During both preparation phases (MIP and MEP), the yellow arrow on the screen shows the direction of the upcoming direction. To differentiate the imagery from the execution sub-trials, a cloud around the arrow was presented as visual stimuli. During the preparation periods, the target was displayed as a red circle in one of the four corners. MIP was followed by a 3 s MI period where the target circle was turned green and a hand image was displayed instead of the yellow arrow. After the completion of MI part, there was a 1 s period of preparation for the actual movement (MEP) followed by 3 s of movement execution (ME). One session of recording consisted of 8–10 runs where each has 15 trials with random directions. A 4–5 min break was introduced between runs and a longer break in the middle of the experiment to avoid exhaustion. The whole recording session including the setup lasted for around 2 h per participant.

4.2 Data Acquisition

The experiments were conducted with four participants with a mean age of 25.25 (±1.71). Two of the participants identified themselves as female, two were right-handed and only one of the participants was BCI naive. The experiments were conducted in RPL lab at KTH using BrainAmp devices from BrainProduct GmbH, Germany. EEG signals were recorded from 32 channels located according to the international 10–20 electrode position system. The ground electrode was placed on the FPz channel and the FCz channel was selected as the reference electrode. Since the electrodes were active EEG electrodes, keeping the impedance level for all electrodes below 20 kΩ provided sufficient signal quality. The data was sampled at the rate 500 Hz.

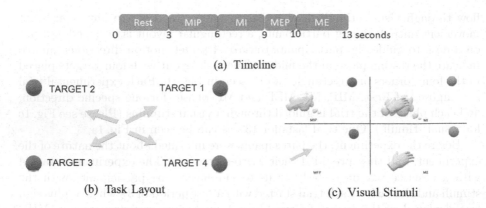

(a) Timeline

(b) Task Layout   (c) Visual Stimuli

Fig. 1. The experimental paradigm for the proposed approach. (a) The timeline of one trial shows the order and duration of the rest, motor imagery preparation (MIP), motor imagery (MI), motor execution preparation (MEP) and motor execution (ME) periods. (b) The task layout is designed by placing the four targets in four corners of a rectangle with the designated resting place in the middle. (c) The visual stimuli for MIP, MI, MEP and ME. The yellow arrows and the red targets stand for the preparation phases. The green target and the hand image means it is time for either MI or ME. The cloud around the arrow and the hand represents the imagery phases (MI or MIP). (Color figure online)

4.3   Pre-processing

Recorded EEG signals were band-pass filtered to the range of 0.5–70 Hz with a Hamming-windowed finite impulse response filter (FIR). Even though there were remaining artefacts caused by muscle movements including eye blinks, no further artifact removal was applied. The data was re-referenced to the two mastoid electrodes TP9 and TP10 to enhance the signal at central electrodes above the motor cortex and then downsampled 125 Hz to enable faster computation. Following this, the continuous signal was extracted into epochs with respect to the trigger signals corresponding to the visual stimuli. The epochs were started with the onset of a MI period and ended with the completion of a ME period, lasting 7 s in total. There were in average 200 trials per participant except one participant (P07) having 120 trials due to a technical issue.

4.4   Proposed Architecture

In order to decode the direction of intentional arm reaching movements, we utilized the EEGNet architecture proposed by Lawhern et al. [7]. Lawhern et al. developed a compact CNN structure named EEGNet utilizing depth.wise and spatial convolutions for feature extraction, and reported the classification performance with different BCI datasets including MI. It has shown that for a 4-class MI dataset, BCI competition IV-2a, the average classification accuracy rate was approximately 69% where the 4 tasks were moving left hand, right hand, both hands or both feet. The performance of EEGNet was similar to the traditional filter bank-common spatial pattern (FB-CSP) method specifically designed for extracting ERD-ERS related features, and they were able to show that the feature representation from the EEGnet aligned with the FB-CSP approach. Since we are interested in decoding the movement intention in a timely manner without a need to wait for the movement to be completed, we added an LSTM layer following the convolutional layers of EEGNet. The diagram of the proposed model is presented in Fig. 2.

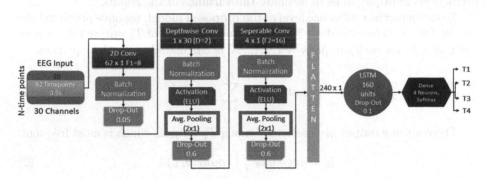

Fig. 2. Schematic view of the EEGNet+LSTM architecture. EEGNet processes each 0.5 s input separately before passing them to the LSTM layer.

The input of our proposed system is a NxCh pre-processed EEG signal where N represents the time-points and Ch is the number of EEG electrodes. A 0.5 s window is applied to the signal without overlap. The original EEGNet architecture starts with a temporal convolution layer with 8 filters, where the temporal kernel is selected as half the sampling rate of the EEG signals. It is followed by a Depthwise convolution layer of size (Ch,1) for spatial convolution followed by a Separable convolution layer, a dense layer and softmax activation. We have added a dropout layer following the temporal convolution to be able to apply the Monte-Carlo Dropout approach for certainty approximation [5]. The dropout rate at this level is kept low with 0.05. The two dropout layers following the second and third convolutions are operated with a rate of 0.6. We also modified the size of average pooling and the size of separable convolution. The LSTM layer is added following a flatten layer with 160 units.

The data for each subject is splitted into training, validation and test groups with a percentage of %64, %16, and %20. The model is trained for each subject individually with q categorical cross-entropy loss. Training is done using the Adam optimizer for a maximum of 350 epochs. To avoid over-fitting, an early stopping strategy is used based on the loss of the validation set. A batch size of 32 provided the optimum performance. The pre-processing steps of the raw EEG signals are performed on MATLAB 2021a and the model is trained on Python using the Keras library.

4.5   Confidence Level of the Model

In order to calculate the confidence level of the model at each time step, we have utilized dropout as a Bayesian Approximation and applied a simple probabilistic approach. Dropout as a Bayesian Approximation was introduced by Gal et al. [5] to evaluate the uncertainty level of a deep learning model. They showed that a dropout layer placed for each weight layer is computationally equivalent to a Bayesian approximation of a Gaussian process. Since in the regular approach, drop-out is applied only for the training phase, they suggested repeating predictions multiple times to estimate the variance of the results.

To estimate the confidence level of the proposed model, we have predicted the output for the same test data T_x times. The frequency of T_x outputs $\theta_{i,k}$ where $i = 1, 2, ...T_x$ for each sample $k = 1, 2, ..., N$ is calculated using the equation:

$$count(\theta_{i,k}) = \sum_{x=1}^{T_x} 1_{x=\theta_{i,k}} \tag{1}$$

The resulting output $\hat{y}_k$ is selected among T_x outputs which is most frequent:

$$\hat{y}_k = argmax_{\theta_{i,k}} \mid count(\theta_{i,k}) \mid \tag{2}$$

Finally we evaluated the confidence level of the output $\hat{y}_k$ based on the probability of occurrence:

$$Confidence_k = \frac{max(count(\theta_{i,k}))}{T_x} * 100 \tag{3}$$

The output is produced at each 0.5 s time-window with the data accumulated up until that data point, and the confidence level is calculated. The confidence level is subjected to a pre-defined threshold Th_c and if it fails to pass the threshold, no output is produced. Applying such an adaptive approach means that there might be time-points during the experiment where the model does not produce any output. Instead, adaptive to each trial and each subject, it will let the trial to continue where the subject starts moving their arm, and the model will evaluate its output and threshold level for the next time-point until it reaches a minimum acceptable certainty level.

5   Results

The conventional motor execution decoding approach does not include MI data for training or testing but only ME data. In order to assess the impact of the proposed method with initial MI phase to decrease the uncertainty of the intended motion before the motor execution, we compared it with the conventional approach.

As the conventional approach, we trained and tested the model with 0.5 s windowed ME data (accumulated at each time-point) including a 1 s preparation period right before the onset. The results of this approach are called 'conventional ME only' for the rest of the paper. The addition of confidence level evaluation before producing an output is evaluated for the 'conventional ME only' approach. For this category, the model is again trained and tested with ME data but provided output only if it reaches the determined confidence threshold level. Otherwise, it waits for the subsequent windowed data. This approach is called 'Confidence Check ME only'.

For our proposed approach, we trained the model with MI and ME trials together in the third category, 'w/MI trials' where the model provides an output for 0.5 s windowed (accumulated) ME data without certainty check. Finally, the results from our proposed approach are presented as 'Confidence Check w/MI trials' where MI trials are included in training and testing. In this option, the output of the model is subject to certainty evaluation.

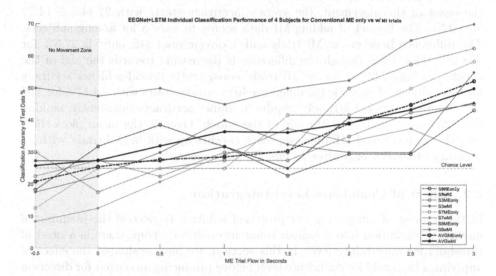

Fig. 3. Classification performances of 'Conventional ME only' (dashed line) with 'w/MI trials' (straight line) where no confidence evaluation is applied. The x-axis represents the time point at which the model produces an output with the data accumulated up until that point. Individual results are presented using the same color for both approaches. The grand average performance are visualized using black lines. (Color figure online)

5.1   The Effect of MI Sub-trials

To understand the effect of including MI sub-trials for training the model, we first compared the results of 'conventional ME only' to 'w/MI trials'. We predicted the intended movement direction 0.5 s before the onset of the movement for both approaches. Cumulatively, we added the upcoming 0.5 s data corresponding to the onset of the movement so that the epoch is now 1 s. The results reported in this section show the outputs of the accumulated data where the incremental step is 0.5 s. In the final step, the whole 3 s ME is used to make a prediction. The percentage of classification accuracy for all 4 subjects is presented in Fig. 3 where the x-axis represents the time-point at which a prediction is made using the data up until that point.

The average classification performances for 4 subjects are shown in colors for 'conventional ME only' (dashed lines) and 'w/MI trials' (straight lines). Their grand averaged results are presented with black line plots.

The results show that 0.5 s preceding the movement where only 0.5 s motor execution preparation data is available for 'conventional ME only', the accuracy lies around chance level with an average of 20.80% ± 6.01. As more data becomes available with the motor execution, the prediction accuracy increases (with an average of 52.26% ± 9.61 at the end of 3 s epoch) as expected for both approaches. With our proposed 'w/MI trials' approach, where MI data is available before the onset of the movement, the average accuracy starts with 27.44% ± 14.72 at −0.5 s. The impact of adding MI data seems to vary a lot among subjects. The difference between 'w/MI trials and 'Conventional ME only' is 27.5% for S08 at −0.5 s. Even though the difference is decreasing towards the end of the epoch, reaching 7.5% at 3 s, 'w/MI trials' consistently provides higher accuracy for this subject. Meanwhile the difference in accuracy starts with −4.17% for S07 at 0.5 s, 'Conventional ME only' results in higher accuracy consistently, making the difference −29.17% at the end of the epoch. Overall, the mean plots show that the 'conventional ME only' approach catches up with 'w/MI trials' with an accuracy of 39.05% only after 2 s into the movement.

5.2 Effect of Confidence Level Integration

In the interest of integrating our proposed solution to decode the planning of movement execution into a human robot interaction set-up, there is a need of threshold for uncertainty level. In this section, we have evaluated the effect of applying a threshold for confidence level before producing an output for direction decision. Even though the analyses were carried offline, we applied a confidence threshold at each time step to decide whether the system is going to produce an output or wait for more data to mimic an online experiment. If the model has a confidence level above $Th_c = 75$ for the 0.5 s window output, the model makes a decision. Otherwise it includes the next 0.5 s window until it reaches the confidence threshold level. The results from this approach are compared with the conventional ME approach, where MI trials are not included in training, but confidence levels are evaluated in the same manner. For the 4 subjects that were collected in the lab, Table 1 provides the classification accuracy, the average confidence levels at that point for the outputs it provides, and the average of the percentage of samples it is confident about. According to the Table 1, the 'ME w/Confidence Check' method is 85.97% confident about 52.27% of the samples with an accuracy of 25.55% 1 s preceding the onset (1 s MEP). Meanwhile, 'w/MI Confidence Check' approach provides in average 37.68% accuracy with a confidence level 87.68% for 74.58% of the samples.

Table 1. Classification Accuracy of MIME data with 4 subjects for the proposed approach (w/MI) vs. conventional approach (ME only)

Metric %	Subject	Type	1sMEP	0.5sME	1sME	1.5sME	2sME	2.5sME	3sME
Accuracy	S03	MEonly	18.8	12.5	12.5	9.4	7.5	7.5	10.4
		w/MI	41.5	47.8	47.8	48.3	40.2	40.2	41.6
	S07	MEonly	15.0	43.3	43.3	45.0	56.0	56.0	59.2
		w/MI	35.42	35.42	39.06	51.25	51.25	51.25	42.71
	S08	MEonly	15.8	18.9	14.1	11.3	26.1	36.7	44.6
		w/MI	39.8	39.8	51.9	43.2	51.3	51.3	49.1
	S09	MEonly	50.0	33.3	31.3	30.0	25.0	31.0	34.2
		w/MI	34.0	50.5	40.4	50.3	50.3	44.0	44.7
	Avg	MEonly	24.88	27.00	25.31	23.92	28.65	32.78	37.09
		w/MI	**37.68**	**43.39**	**44.89**	**46.25**	**48.27**	**46.70**	**44.52**
Confidence	**Avg**	MEonly	85.97	86.61	87.00	86.16	86.00	85.96	**83.23**
		w/MI	**87.68**	**86.63**	**87.38**	**87.09**	**87.08**	**86.74**	82.54
Sample	**Avg**	MEonly	52.27	64.39	67.29	74.15	78.67	81.00	100.00
		w/MI	**74.58**	**80.10**	**84.62**	**88.73**	**91.74**	**92.31**	100.00

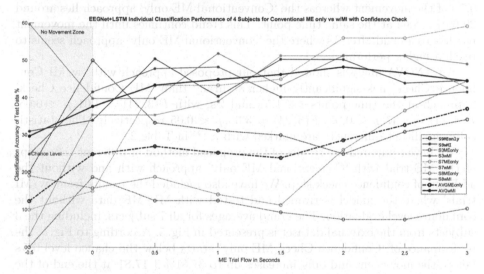

Fig. 4. The classification accuracies for 4 subjects (colored) and their grand averages (black) are depicted for 'ME w/Confidence Check' (dashed lines) and 'w/MI trials and Confidence Check' methods. The same colors represent the same subjects. (Color figure online)

To visualize the results, Fig. 4 is presented where the colored lines correspond to the individual performance and the black lines correspond to the grand averages. The results from 'ME w/Confidence Check', depicted in dashed lines, start from below the chance level and increase in time, but does not reach the performance of 'w/MI Confidence Check' even at the end of the movement.

5.3   Validation with an External Dataset

In order to validate our proposed approach with an external data set we use the data from [8] due to the similarity of their experimental setting to ours where they have recorded MI and ME sessions separately with arm reaching tasks in 3D. This dataset was used by Jeong et al. [9], and [11] and was made publicly available in [8]. They recorded 60 channel EEG in addition to Electrooculogram (EOG) and Electromyography (EMG) signals for 22 subjects, 3 sessions, 11 tasks with 300 trials per task for each subject. We concatenated the MI and ME trials of the same tasks for 3 subjects to simulate having MI sub-trials right before the ME. The data from these subjects are marked as 'J20, J21 and J22' in the Table 2 where the performance of all 7 subjects for all 4 approaches are listed. The highest numbers for each subject among various approaches are highlighted in bold. We see that for most of the subjects in this experiment, the proposed approach results in higher performance in terms of classification accuracy (an average of $30.44\% \pm 17.06$) before the beginning of the movement. The grand average for all 7 subjects for 'w/MI Confidence Check' method reaches $39.70\% \pm 9.85$ into the 0.5 s of the movement whereas the 'Conventional ME only' approach lies around $29.69\% \pm 8.64$ at the same time point. This trend continues until the movement reaches to an end after 3 s where the 'conventional ME only' approach seems to reach a higher performance.

An ANOVA analysis showed that our proposed approach with 'w/MI Confidence Check' has significantly higher accuracy than 'ME Confidence Check' approach at the time points 1 s, 1.5 s and 2 s, with $F(3, 24) = 4.01, p < 0.05$; $F(3, 24) = 4.47, p < 0.05$, $F(3, 24) = 3.36, p < 0.05$ respectively. The statistically significant differences are marked with a '*' in Table 2.

We have compared the results of the proposed method at different time steps of the ME trial with 'Conventional ME only' approach with and without the addition of confidence check step. We have also included the results from 'w/MI trials' where the model is trained and tested with MI+ME data without the confidence level evaluation. The grand averages for all 7 subjects, including the 3 subjects from the external dataset is presented in Fig. 5. According to Fig. 5, the performance of 'Confidence Check ME only' starts below the chance level 0.5 s before the movement and only increases up to $37.81\% \pm 17.81$ at the end of the epoch, less than 'Conventional ME only' approach. The proposed method 'w/MI Confidence Check' starts with a relatively higher accuracy at 0.5 s preceding the movement with $30.44\% \pm 17.46$ and increases when there is more data available. We would like to highlight that at the onset of the movement, this method yields to an average accuracy of $37.83\% \pm 5.96$ whereas all other methods fail to rise above the chance level of 25% for 4 classes.

Finally, Table 3 reports the results of MI decoding when the model is trained with MI+ME data. In this part, the predictions are made before the start of the MEP period. We see a huge variation among participants when it comes to MI classification accuracy. For the subject with lowest amount of training data (S07), the accuracy is below the chance level, 12.5%. Whereas for S08, the accuracy goes as high as 45.0%.

Table 2. Classification Accuracy of MIME data with 7 subjects for the conventional approach (ME only), ME data with confidence check, w/MI subtrials, and finally the proposed approach w/MI subtrials and confidence check

Subject	Type	0.5sMEP	1sMEP	0.5sME	1sME	1.5sME	2sME	2.5sME	3sME
S03	MEonly	**30.0**	17.5	25.0	27.5	27.5	35.0	**42.5**	45.0
	ME Conf	0	18.8	12.5	12.5	9.4	7.5	7.5	10.4
	w/MI	17.5	22.5	30.0	40.0	32.5	30.0	30.0	**55.0**
	w/MI Conf	24.8	**41.5**	**47.8**	**47.8**	**48.3**	**40.2**	40.2	41.6
S07	MEonly	12.5	25.00	25.00	25.00	41.67	41.67	50.00	58.33
	ME Conf	10.0	15.0	**43.3**	**43.3**	45.0	**56.0**	**56.0**	**59.2**
	w/MI	8.3	12.5	20.8	29.2	37.5	33.3	37.5	29.2
	w/MI Conf	3.1	**35.4**	35.4	39.1	**51.3**	51.3	51.3	42.7
S08	MEonly	22.5	27.5	22.5	30.0	30.0	50.0	57.5	62.5
	ME Conf	19.1	15.8	18.9	14.1	11.3	26.1	6.7	44.6
	w/MI	50.0	**47.5**	**50.0**	45.0	**50.0**	**52.5**	**65.0**	**70.0**
	w/MI Conf	**59.7**	39.78	39.8	**51.9**	43.2	51.3	51.3	49.1
S09	MEonly	18.2	31.8	38.6	31.8	22.7	29.6	29.6	43.2
	ME Conf	20.0	**50.0**	33.3	31.3	30.0	25.0	31.0	34.2
	w/MI	27.3	27.3	27.3	31.8	25.0	41.0	41.0	**45.5**
	w/MI Conf	**31.0**	34.0	**50.5**	40.4	**50.3**	**50.3**	**44.0**	44.7
J20	MEonly	15.0	17.5	20.8	30.8	32.5	**36.7**	**36.7**	35.8
	ME Conf	16.7	19.4	17.4	23.1	18.4	21.0	25.1	34.4
	w/MI	21.7	25.8	25.0	25.0	28.3	33.3	35.0	38.3
	w/MI Conf	**36.6**	**31.0**	**29.5**	**29.5**	**43.6**	36.4	36.4	**45.4**
J21	MEonly	28.3	37.5	32.5	40.0	44.2	46.7	47.5	48.3
	ME Conf	**38.0**	28.5	26.4	22.3	22.3	22.3	22.3	37.9
	w/MI	28.3	30.0	35.0	44.2	46.7	45.8	45.8	53.3
	w/MI Conf	23.0	**48.7**	**49.0**	**59.2**	**59.2**	**59.2**	**59.2**	**59.2**
J22	MEonly	26.7	30.0	**43.3**	49.2	53.3	55.8	54.2	57.5
	ME Conf	26.0	31.4	34.3	38.2	44.6	48.2	48.5	44.0
	w/MI	30.0	22.5	21.7	26.7	41.7	**55.8**	53.3	55.0
	w/MI Conf	**35.0**	**34.4**	25.8	40.7	40.7	50.6	50.6	50.6
Avg	MEonly	21.9 ± 7	26.7 ± 7	29.7 ± 9	33.5 ± 8	36.0 ± 11	42.2 ± 9	45.4 ± 10	$\mathbf{50.1 \pm 10}$
	ME Conf	18.5 ± 12	25.6 ± 13	26.6 ± 11	26.4 ± 12	25.9 ± 15	29.4 ± 17	32.4 ± 16	37.8 ± 15
	w/MI	26.2 ± 13	26.9 ± 10	30.0 ± 10	34.6 ± 8.4	37.4 ± 9	41.7 ± 10	44.0 ± 12	49.5 ± 13
	w/MI Conf	$\mathbf{30.4 \pm 17}$	$\mathbf{37.8 \pm 6}$	$\mathbf{39.7 \pm 10}$	$\mathbf{44.1^* \pm 9}$	$\mathbf{47.0^* \pm 7}$	$\mathbf{48.5^* \pm 7}$	$\mathbf{47.6 \pm 7}$	47.6 ± 6

Table 3. Classification Accuracy for 3s MI trials where MI+ME is used for training and MI for prediction with EEGNet+LSTM

Subject	S03	S07	S08	S09	J20	J21	J22	AVG
	25.0	12.5	45.0	20.5	21.7	24.2	26.7	25.07 ± 9.93

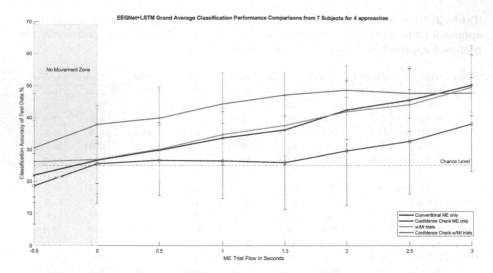

Fig. 5. The grand average performances of all 4 approaches are depicted at different time steps of the ME trial. The bars around the lines present the standard deviation of the classification accuracy at that time point. (Color figure online)

6 Discussion

In this paper, we have proposed a learning framework to decode the movement execution from the EEG signals as early as possible with the inclusion of MI trials. We have shown that the classical EEGNet architecture combined with an LSTM layer successfully predicted the upcoming movement direction at the onset of the movement not only for the data we have collected in our lab, but also for the EEG data from an external dataset. Additionally, we have implemented a confidence level evaluation for the model before producing an output to avoid undesired control commands for external devices. As stated by Dose et al. [3], it is crucial for EEG-based BCI control systems to have a mechanism to assess their confidence level. Our results show that adding a confidence check step for our 'with MI' approach results in better performance while the 'Conventional ME only' approach results in a lower performance even if the whole epoch is available.

The results presented in Fig. 5 and Table 2 suggest that including MI trials ahead of ME, and applying a confidence level evaluation before producing an output would enable decoding of movement direction as early as the onset of the movement with significantly higher accuracy for motion prediction compared to conventional approaches using only motor execution data. This promising approach has a potential to increase the certainty of the motion prediction as early as possible, and might lead to real-time human robot interaction scenarios where the intended motion is predicted before the execution to control the robots without a delay. Minimizing the uncertainty before the execution might lead to more robust human machine control interfaces where the user receives real time output from the machine. In future studies, our approach will be tested with a robot integration in such scenarios.

The integration of confidence level evaluation to the 'conventional ME only' approach resulted in poorer decoding performance for almost all the subjects except 'S07' as it can be observed in Table 2. Since there was less amount of data for this subject, checking the certainty of the model before producing an output actually improved the accuracy in this case. On the other hand, the addition of confidence level evaluation improved the performance for 'w/MI trials' where we have introduced MI trials accompanying ME trials. Since the data becomes more diverse in this case, using a preventive mechanism to assure the certainty of the model results in a boost in terms of decoding accuracy.

Our analysis showed that the optimum threshold for the confidence level might depend on the variations in the EEG data. An empirical search revealed that the optimum threshold is 95% for 3 out of 4 subjects in our dataset where the recording is completed in one sitting. Whereas for the external dataset with recordings from 3 sessions in different weeks, the optimum threshold level drops to 60%.

Another possible future direction of this approach might be the HCI systems designed for neuro-rehabilitation of people with physical impairments who can not execute motor functions. According to Pereira et al. [20], reducing the time delay between the movement intention and the feedback would substantially activate neural plasticity, and boost the recovery at the motor cortex. They highlight the value of such improvement especially for the robot-assisted therapy applications for stroke patients.

One rather surprising result seen in Fig. 5 is that there seems to be no difference between the performance of 'w/MI trials' (green line) and 'Conventional ME only' (blue line). Since 'w/MI trials' starts decoding earlier with available MI data to begin with, we would expect relatively higher results especially before and around the onset of the movement. One explanation might lie under the classification performances of MI data presented in Table 3. The vast disparity among participants is actually in line with the literature, where the MI classification accuracy relies on the concentration of the participants for the duration of the trial consistently. Another challenge in the MI-BCI field is the need of large training data. Since one of the subjects had less amount of data, the inclusion of MI trials did not add much value. We can say that suggested approach would work best when there is a reasonable amount of training data.

One of the main limitations of this study is the number of the subjects. We are planning to collect more data to validate our results in a larger population. We also believe that the results would benefit from modifying the architecture instead of the default EEGNet model. Since EEGNet is specifically designed to mimic FB-CSP features, it would be beneficial to broaden the model to consider ME related features as well. The end goal of the proposed approach is achieving real-time BCI-based control of the robotic devices. In that aspect, we are planning to extend our work to integrate this approach in a robotic experiment. We will design an online system to decode the direction of the movement using EEG signals and provide feedback through robot. It would enable us to observe the impact in terms of early detection and the success rate of the control commands.

Finally, the confidence level evaluation approach adopted in this study is rather simplistic, which is subject to underestimating the uncertainty [10]. In the future work, we are planning to exploit ensemble learning methods.

7   Conclusion

We have presented a novel learning framework to decode the intentional movement direction from EEG signals to generate control commands for external devices such as robotic devices and computer cursors. The setup allows the model to start making predictions 1 s spreceding the movement onset with the inclusion of MI trials. The addition of confidence level evaluation assures that the model is certain about the output before generating a control command. We have shown that our approach achieved decoding the intended movement as early as 0.5 s ahead of the movement with higher accuracy than the conventional approach.

References

1. Barsotti, M., Leonardis, D., Vanello, N., Bergamasco, M., Frisoli, A.: Effects of continuous kinaesthetic feedback based on tendon vibration on motor imagery BCI performance. IEEE Trans. Neural Syst. Rehabil. Eng. **26**(1), 105–114 (2018). https://doi.org/10.1109/TNSRE.2017.2739244
2. Buerkle, A., Eaton, W., Lohse, N., Bamber, T., Ferreira, P.: EEG based arm movement intention recognition towards enhanced safety in symbiotic human-robot collaboration. Robot. Comput.-Integr. Manuf. **70**, 102137 (2021). https://doi.org/10.1016/j.rcim.2021.102137, https://www.sciencedirect.com/science/article/pii/S0736584521000223
3. Dose, H., Møller, J.S., Iversen, H.K., Puthusserypady, S.: An end-to-end deep learning approach to MI-EEG signal classification for BCIS. Expert Syst. Appl. **114**, 532–542 (2018). https://doi.org/10.1016/j.eswa.2018.08.031
4. Edelman, B.J., et al.: Noninvasive neuroimaging enhances continuous neural tracking for robotic device control. Sci. Robot. **4**(31), eaaw6844 (2019). https://doi.org/10.1126/scirobotics.aaw6844
5. Gal, Y., Ghahramani, Z.: Dropout as a Bayesian approximation: representing model uncertainty in deep learning. In: International Conference on Machine Learning, pp. 1050–1059. PMLR (2016)
6. Ieracitano, C., Mammone, N., Hussain, A.E.A.: A novel explainable machine learning approach for EEG-based brain-computer interface systems. Neural Comput. Appl. (2021). https://doi.org/10.1007/s00521-020-05624-w
7. Lawhern, V.J., Solon, A.J., Waytowich, N.R., Gordon, S.M., Hung, C.P., Lance, B.J.: EEGNET: a compact convolutional neural network for EEG-based brain-computer interfaces. J. Neural Eng. **15**(5), 056013 (2018). https://doi.org/10.1088/1741-2552/aace8c

8. Jeong, J.H., et al.: Multimodal signal dataset for 11 intuitive movement tasks from single upper extremity during multiple recording sessions. GigaScience **9**(10) (2020). https://doi.org/10.1093/gigascience/giaa098
9. Jeong, J.H., Shim, K.H., Kim, D.J., Lee, S.W.: Brain-controlled robotic arm system based on multi-directional CNN-BILSTM network using EEG signals. IEEE Trans. Neural Syst. Rehabil. Eng. **28**(5), 1226–1238 (2020). https://doi.org/10.1109/TNSRE.2020.2981659
10. Kahn, G., Villaflor, A., Pong, V., Abbeel, P., Levine, S.: Uncertainty-aware reinforcement learning for collision avoidance. arXiv preprint arXiv:1702.01182 (2017)
11. Lee, D.Y., Jeong, J.H., Shim, K.H., Lee, S.W.: Decoding movement imagination and execution from EEG signals using BCI-transfer learning method based on relation network. In: ICASSP 2020–2020 IEEE International Conference on Acoustics, Speech and Signal Processing (ICASSP), pp. 1354–1358 (2020). https://doi.org/10.1109/ICASSP40776.2020.9052997
12. Livezey, J.A., Glaser, J.I.: Deep learning approaches for neural decoding: from CNNs to lSTMS and spikes to fMRI. arXiv (2020). https://doi.org/10.48550/ARXIV.2005.09687
13. Lotte, F.E.A.: A review of classification algorithms for EEG-based brain-computer interfaces: a 10 year update. J. Neural Eng. **15**, 031005 (2018). https://doi.org/10.1088/1741-2552/aab2f2
14. Mammone, N., Ieracitano, C., Morabito, F.C.: A deep CNN approach to decode motor preparation of upper limbs from time–frequency maps of EEG signals at source level. Neural Networks **124**, 357–372 (2020). https://doi.org/10.1016/j.neunet.2020.01.027, https://www.sciencedirect.com/science/article/pii/S089360802030037X
15. Mammone, N., Ieracitano, C., Morabito, F.C.: MPnnet: a motion planning decoding convolutional neural network for EEG-based brain computer interfaces. In: 2021 International Joint Conference on Neural Networks (IJCNN), pp. 1–8 (2021). https://doi.org/10.1109/IJCNN52387.2021.9534028
16. Mercado, L., Quiroz-Compean, G., Azorín, J.M.: Analyzing the performance of segmented trajectory reconstruction of lower limb movements from EEG signals with combinations of electrodes, gaps, and delays. Biomed. Sig. Process. Control **68**, 102783 (2021). https://doi.org/10.1016/j.bspc.2021.102783, https://www.sciencedirect.com/science/article/pii/S1746809421003803
17. Miller, K.J., Schalk, G., Fetz, E.E., den Nijs, M., Ojemann, J.G., Rao, R.P.N.: Cortical activity during motor execution, motor imagery, and imagery-based online feedback. Proc. Natl. Acad. Sci. **107**(9), 4430–4435 (2010). https://doi.org/10.1073/pnas.0913697107, https://www.pnas.org/content/107/9/4430
18. Müller-Putz, G., Schwarz, A., Pereira, J., Ofner, P.: Chapter 2 - from classic motor imagery to complex movement intention decoding: the noninvasive GRAZ-BCI approach. In: Coyle, D. (ed.) Brain-Computer Interfaces: Lab Experiments to Real-World Applications, Progress in Brain Research, vol. 228, pp. 39–70. Elsevier (2016). https://doi.org/10.1016/bs.pbr.2016.04.017, https://www.sciencedirect.com/science/article/pii/S0079612316300437
19. Ofner, P., Schwarz, A., Pereira, J., Müller-Putz, G.R.: Upper limb movements can be decoded from the time-domain of low-frequency EEG. PLoS ONE **12**(8), 1–24 (2017). https://doi.org/10.1371/journal.pone.0182578
20. Pereira, J., Ofner, P., Schwarz, A., Sburlea, A.I., Müller-Putz, G.R.: EEG neural correlates of goal-directed movement intention. NeuroImage **149**, 129–140 (2017). https://doi.org/10.1016/j.neuroimage.2017.01.030, https://www.sciencedirect.com/science/article/pii/S1053811917300368

21. Pfurtscheller, G., Brunner, C., Schlögl, A., Lopes da Silva, F.: Mu rhythm (de)synchronization and eeg single-trial classification of different motor imagery tasks. NeuroImage **31**(1), 153–159 (2006). https://doi.org/10.1016/j.neuroimage.2005.12.003, https://www.sciencedirect.com/science/article/pii/S1053811905025140

22. Pfurtscheller, G., Neuper, C.: Motor imagery activates primary sensorimotor area in humans. Neurosci. Lett. **239**(2), 65–68 (1997). https://doi.org/10.1016/S0304-3940(97)00889-6, https://www.sciencedirect.com/science/article/pii/S0304394097008896

23. Sobierajewicz, J., Szarkiewicz, S., Przekoracka-Krawczyk, A., Jaśkowski, W., van der Lubbe, R.: To what extent can motor imagery replace motor execution while learning a fine motor skill? Adv. Cogn. Psychol. **12**(4), 179–192 (2016). https://doi.org/10.5709/acp-0197-1

24. Valenti, A., Barsotti, M., Bacciu, D., Ascari, L.: A deep classifier for upper-limbs motor anticipation tasks in an online BCI setting. Bioengineering **8**(2) (2021). https://doi.org/10.3390/bioengineering8020021, https://www.mdpi.com/2306-5354/8/2/21

25. Wang, H., et al.: Brain-controlled wheelchair review: From wet electrode to dry electrode, from single modal to hybrid modal, from synchronous to asynchronous. IEEE Access **9**, 55920–55938 (2021). https://doi.org/10.1109/ACCESS.2021.3071599

User Experience in Mobile Social TV: Understanding Requirements and Presenting Design Guidelines

Taume Dery[1] and Yavuz Inal[2]([⊠]) [iD]

[1] Department of Information Science and Media Studies, Faculty of Social Sciences, University of Bergen, Bergen, Norway
[2] Department of Design, Faculty of Architecture and Design, Norwegian University of Science and Technology, Gjøvik, Norway
yavuz.inal@ntnu.no

Abstract. Social TV has received a lot of attention recently as a concept of enabling viewers to have the feeling of togetherness over distance. Social TV services are not necessarily limited to conventional television platforms. With the widespread use of portable devices, mobile social TV has the potential to enrich the traditional viewing experience. However, there still exists limited research on how to design a usable mobile social TV application to create a decent user experience. In this study, we, therefore, aim to understand user requirements through conducting a three-week field study and to present design guidelines for mobile social TV applications. Results of the study show that mobile social TV appear to be an intriguing concept that enables viewers to socialize with others over distance. The participants, in the study, cherish the opportunity to engage with their friends or family members via mobile social TV. Factors such as genre, platform, feeling of co-presence, and participants' personal and social predispositions play a profound role in user experience.

Keywords: User experience · Mobile social TV · Usability · Visual design · Co-presence · Togetherness · Sociability

1 Introduction

In the past few years, a new type of interactive television application, social TV, has received much attention as a concept of intriguing viewers' participation and enabling viewers to have the feeling of togetherness over distance because watching a program on television with others is found to be more enjoyable than watching alone [6]. Turning isolated television viewing into socially dynamic engagement [2, 13, 22], social TV provides a unique viewing experience [11] and makes it possible to interact with remote viewers online on a single platform such as a television set, personal computer, tablet device, or mobile phone [4]. It enables viewers to engage with their friends, family members as well as other unknown viewers with whom they share a common interest. They talk remotely by utilizing a combination of audio, video, or text chat functions while watching programs, send invitations to their friends, create their buddy lists, share video clips, and join other viewer groups.

Previous research shows that social TV viewers enjoy sharing viewing experiences with others over distance instead of streaming video content solitarily [e.g., 5, 8]. The main motives behind their social TV usage are convenience [9], excitement [9], gathering information [9, 12], entertainment [12], relaxation [10], sharing their favorite programs [10], and communicating with others while watching video content [3, 10–12]. User data on co-viewing experience [10] shows that most viewers engage in co-viewing activity while watching video content on the big screen. About half of viewers prefer watching with others using their mobile phones. Among co-viewers on the platform, spouse/partner is more pronounced, followed by children, parents, and siblings. Most viewers talk with their companions about what they are watching when they watch video content together.

Social TV is a growing area of interest. It has seen a surge in popularity during the coronavirus pandemic as social distancing keeps people in their homes. However, the television industry encounters difficulties in designing and developing social TV applications that fulfill users' requirements of being "useful, sociable, enjoyable, and most importantly, user-centered" [20, p. 940]. The design of social TV applications requires to be usable so that users with a wide range of demographics can easily adapt [7]. Viewers' social TV experience is positively linked to their perceived usability and sociability [21]. Further, when it includes more usable features and sociable functions, viewers are more likely to engage with functions of the platform comfortably which in turn influences a positive viewing experience [21].

Social TV services are not necessarily limited to conventional television platforms. With the widespread use of portable devices, mobile social TV has the potential to enrich the traditional viewing experience and allow viewers to have more customized, flexible, and authentic forms of media consumption. However, there still exists limited research on how to design a usable mobile social TV application to create a decent user experience. Taken together, our aim is to understand the user requirements of mobile social TV. To this end, we present low-fidelity prototypes together with design guidelines for mobile social TV applications.

2   Methodology

We applied the purposive sampling method to recruit participants through a demographic questionnaire. We received 440 inquiries in total and ended up having 15 participants narrowed down based on age, gender, and technical competency. We also chose people who were familiar with YouTube as one of their preferred streaming platforms since mobile social TV applications in this study provided YouTube as their central service video platform. After the recruitment process, we conducted a field study with 15 participants over a three-week period. The participants were divided into three groups of three females and two males in each group. Three participants were between 20–30 ages, one between 30–35 ages, and one between 18–20 ages in each group. We asked the participants to experience a mobile social TV application and provide feedback accordingly. Rave [19], Airtime [1], and PlayNow [18] were selected as free mobile social TV applications available on iOS and Android operating systems. These mobile applications were chosen based on the differences in user interfaces they had and the interaction modes they offered to viewers.

We used multiple methods for data collection including weekly diaries, session diaries, and semi-structured interviews. The participants were assigned to one of the chosen mobile social TV applications and the data was collected as regards their usage for three weeks. The participants were asked to use an application for at least three days a week during the study period. After each daily session, they all made an entry to the session diary that enabled us to collect usage data and understand their approach to the mobile social TV concept. The participants were prompted to input the genre of videos they watched together with others, the number of distant viewers they engaged with, their preferences regarding communication modalities, and any problems or difficulties they encountered. In addition, the participants were asked to make another entry, a weekly diary, at the end of the first and second week to report their experiences in a more detailed manner. They wrote a brief description of their experience with mobile social TV applications, reported interaction methods they preferred, additional features they would add to make the experience more useful, and any features they would like to remove to make it less frustrating. At the end of the study, we administered a semi-structured interview with all participants to obtain data about their expectations from and acceptance of the mobile social TV concept. In the interviews, the participants provided more detailed explanations, elaborated their experiences regarding the assigned mobile social TV applications, and clarified unclear entries in the session and weekly diaries.

3 Results

Here, we report the findings from the study. The participants assigned to Airtime are referred to as Airtime participants (AP). Accordingly, the participants assigned to PlayNow and Rave are referred to as (PP) and (RP), respectively.

3.1 Mobile Social TV Concept

Time Slot. As seen in Fig. 1, time slots between 18 and 21 accounted for 35% of the sessions. The participants explained that evenings were when they usually had the time to have a session in their busy schedule. Many participants found the act of having to find time in their schedule became a hustle in their busy daily lives. *"It became a chore because we felt like this was something, we had to find time to do. It was a bit stressful since we are all busy. However, once we got started, we enjoyed it"* (AP1). This sentiment was expressed by a large portion of participants. AP4 further presented a specific situation that *"I have a little son who goes to bed at 7 o'clock so we can usually do it after he has gone to bed. I have also noticed that it was bizarre to plan to see things together. However, it became a great way of doing something together"*.

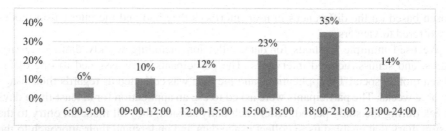

Fig. 1. Time slots of registered sessions.

Platform. Most participants expressed that using the screen in a mobile platform became too small and that it played a massive role in their satisfaction with the experience. When asked what platforms would be ideal for such a concept, seven participants answered PC or tablet, with one explaining that *"I think it would have worked best on a tablet since it is a bigger screen that makes it easier to watch. I also think it would have been good to have it on PC. This would have made it easier to chat."* (AP2).

The fact that the participants were bound to a mobile platform greatly affected their experience. All three applications extend this issue by adding social elements in the viewing window, further minimizing the video content. Three participants noted that they would have enjoyed the experience better on another platform. The limited screen size was listed as one of the main reasons among the participants as to why they would not be using their assigned application after the study.

Second Screen vs. Mobile Social TV. Second screen is a concept of using multiple devices to accomplish the goal of interaction while streaming video. The participants were asked how they felt about the approach of the second screen as compared to social TV. There was a mixture of responses regardless of the assigned applications. Five participants were in support of the idea, with one explaining that *"I think that for this specific situation, it would have been tidier. If I were to repeat this experience, I would have chosen to watch YouTube on the computer and separately texted with them. Doing it all in the same application was a bit too much"* (AP4). However, three participants felt the exact opposite, claiming that the social TV approach is better organized. *"With multiple screens, you have to focus on two things. You are much more independent when you only have one platform. I would rather have it all on one platform instead of using two"* (RP5).

Despite there being two opposing views to what approach is best, participants in both camps noted screen size as the most important factor, with one saying that *"It depends on the screen size. I wouldn't mind just watching videos with my friends on the same screen if I could have everything on that screen"* (AP3). As seen, there was no agreement among the participants as to which approach is superior.

The Concept. The participants addressed their thoughts on mobile social TV as a concept. In order to direct attention away from individual experiences with a given application, the scenario was given; *"Provided that all your feedback was taken into account* *and there was a perfect mobile social TV application created. Does the concept of mobile social TV have potential?"*. Patterns in responses can be separated into three sections: positive, case-specific, and negative.

Four participants were very positive to the concept, with one stating that *"Yes! If everything was fixed, I think I would use the application a lot more. I think it is a good concept. Watching with others makes you feel that you are not watching something alone"* (AP4). RP5 further added that *"It's the feeling of knowing that someone else is watching the same thing as you. It is not the chatting that makes a good experience. It is just the feeling of knowing that there is someone else there"*.

Most participants were positive to the concept but specified that they only felt that it could be viable in specific use cases when people are not physically able to be together. *"Yes, it has potential, but for people who have moved away and have already started a show together. For example, if your friends are traveling abroad and still want to stay together. I think that has to be a unique situation"* (AP4). Other participants cited different factors like content, age range, and sociability as conditions that would affect their decision on potential. These participants were, in general, more skeptical of the idea. They gave the impression that in a specific context, the concept might be viable. However, their experience did not convince them of a need for such a service as an addition to their daily lives.

Four participants were dismissive of the concept. These participants expressed that the concept in any form could not replace the feeling of togetherness that physical presence provides. Out of the participants negative to the concept, three were over the age of 30, with two noting that they felt too old for the concept.

3.2   Mobile Social TV Experience

Genre. Most participants reported that genre affected their viewing experience. Communication was more challenging regarding specific genres. *"Some genres are less suitable for talking. It disturbs the viewing experience a bit if a lot is going on in the video"* (PP2). RP5 further added that *"Some videos require more concentration, and it can be distracting to chat"*. The participants were required to register the genre of each video they watched. Comedy, sports, reality TV, and live events were the most registered genres, respectively. The categories such as tutorial, technology, education, and dance were among the genres which the participants registered the least.

Comedy was cited as a suitable genre by six participants. *"Watching a video that everyone thinks is fun/interesting will make it easier to laugh or talk about the content of the video"* (AP3). AP2 further pointed to a specific situation that *"I think it was fun at our last session watching a "carpool karaoke" video. It automatically became more fun because we could respond and laugh at it together and experience the video together. In such situations, it was cozy"*. The participants seemed to indicate that viewing humorous content with others made their experience better. The apparent need for affirmation from others when watching funny content seems to be met in social TV.

Sports was mentioned by five participants. *"Watching American football games, when there was a touchdown or something cool in the game, I could see the reaction of the others. It's one of those kinds of videos where I think it's better with others than to watch alone"* (AP4). All five participants that listed sports as a suitable category were male and were split across platforms. No females mentioned sports, signifying a connection between gender and interaction methods.

Several participants pointed out "Reality" and "Live Events" as categories that they found appropriate for the concept. Videos in these categories are not prevalent on YouTube. *"I often watch Netflix and YouTube alone. Reality shows like "Skal vi danse," "Farmen", "Paradise Hotel". Such content is much more appealing to watch with others and do not require the same type of focus"* (RP2). In addition to content, this feedback further iterates the importance of offering a service that is facilitated around users existing social habits and not the other way around.

However, some participants specified genres that they felt became distracting. *"If you see a documentary, there is a lot of information and facts being presented, and if you talk at the same time, it might be a bit difficult to concentrate on both"* (AP3). There was a consensus that content that required a lot of concentration became distracting.

Room Capacity. The average registered number of co-viewers in a session was 1,7. Most of the participants independent of mobile platforms cited that their maximum number would be 3–5, with one explaining that *"If there are more than two participants, it becomes a little harder to give attention to all the parties that were involved. It often turned into users interrupting each other and focusing on the video because a lot was being said"* (AP3). As noted by the participants, the number of viewers in a mobile social TV experience highly affects how distracting the viewing experience was. Perfect balance is achieved when the number of viewers is an amount that provides engagement and connectedness without becoming chaotic.

3.3 Interaction Method

Several participants expressed that interaction while streaming sometimes became distracting. *"If we are talking about the video, then people I am interacting with become a distraction, and when we are talking about things irrelevant to the video, then the video becomes a distraction"* (AP2). PP3 stated that it just did not feel natural for her to chat and watch videos at the same time. Factors like genre played a significant role in participants' opinions on how distracting their experience was.

The participants also noted that interaction in the applications made it easier to share content. For example, AP3 explained that *"You had to somewhat share the experience live, which made it easier to share content. I didn't have to go into Facebook and post the link and hit send. We could talk about it live, which was a little cool"*. RP3 further iterated this notion that *"It is gratifying discussing videos with my friends. It was fun to talk with my friends and say what I wanted to say right away and know that they had seen the same thing as me"*.

Text Chat. All participants in Rave and PlayNow registered using text chat where text chat is used as the main interaction method. In Rave, the participants were pleased with how the text chat function was integrated. *"We are really pleased with the chat feature.*

It has been very easy to use" (RP3). However, five participants in PlayNow and Rave expressed that they wanted more alternatives of interaction. For example, PP3 stated that *"This has not been my ideal way of communication in the way of chatting and watching movies together. I would rather video call or hear their voices".* These responses showed that even though the participants in PlayNow and Rave were satisfied with chat function as an interaction method, they felt like it did not adequately replicate the feeling of connectedness with other viewers.

Two participants in Airtime expressed that they felt using text chat was redundant due to the fact they were already communicating through video chat. *"I don't think it added any value since we were already talking and seeing each other, we didn't need to write to each other in addition"* (AP1). AP5 added that he would rather send a regular text message instead of using the inbuilt chat function. As Airtime offers video chat as the main interaction method, which arguably adds a greater feeling of co-presence, feedback, therefore, indicates that text chat is less appealing as it does not enhance the experience. However, as the application focused on video chat as its main interaction method, the lack of text chat as a prominent focus of the user interface may have deterred the participants from using the interaction method.

Video Chat. Despite technical difficulties, most participants in Airtime were reasonably happy with the interaction method. *"The way we feel is that you express more with audio and video. A picture says more than 1000 words. It allows you to look at facial expressions etc."* (AP3). However, one participant instead preferred a less distracting interaction method. Given the fact that video chat was the most prominent interaction method in the application, the interaction became a more substantial distraction compared to the other two applications. *"Seeing the viewing screen is the most important part. What you are watching is not necessarily the faces of other participants. If I had had the option, I would have chosen to use an audio chat function"* (AP4).

As video chat was the most used interaction method with several participants specifically praising the connectedness of the method, it can be argued that the feeling of co-presence plays a vital role in participants' perception of a mobile social TV experience. However, taking into account that a minority of participants would trade the feeling of connectedness for a less distracting interaction method, alternative options should also be available to suit different users and situations.

Audio Chat. All participants in Rave were discouraged from using the audio chat feature, with many saying that it was distracting and had multiple technical issues. RP5 noted that she would have used the function if it had worked as intended. Moreover, two participants in Rave expressed that they wanted a video chat function instead of an audio chat, with one saying that *"Ideally, I would rather use a video feature as compared to an audio feature, but the chat is good"* (RP1).

Emojis. Three Airtime participants felt that the emojis in combination with sound effects became a bit childish, while others found it fun. *"We used them when something happened. For example, if something scary is happening, then you can use a «shock» emoji face. It was not distracting"* (AP5). However, two other participants felt that they were too old for the feature. *"We believe that the audio sounds and emojis target the age group 10–15. For those of us who are a bit 20 older, we are used to things being a little more basic. It is satisfying enough to see and talk to each other"* (AP3). The participants that found the function childish were in their mid 20–30, while the one who enjoyed the feature was the only participant under the age of 20. Nevertheless, using emojis could minimize the distraction of texting while viewing as expressed by a Rave participant. *"Quick response emojis would have been good for my dad. He spends a lot of time writing a text. Here, you should be able to respond immediately while something in a video is happening"* (RP5). Hence, regardless of the disruption, emojis might cause to some participants, it could still be an effective way of limiting both the distraction of text chat and increasing interaction.

3.4 Design Guidelines

Here, we present low-fidelity prototypes with design guidelines for mobile social TV applications (see Figs. 2 and 3).

Notification and Awareness (1). Most participants were favorable to the idea of getting notified when other viewers were using the application, given the condition that it had limitations and did not become an irritation. The solution presented is a less destructive approach to notifying users. Firstly, only seeing what friends are watching when they have entered the application instead of getting a notification. Secondly, explicitly only signaling and notifying viewers within a certain group.

Search in the Platform (2). Several participants were dissatisfied with the method in which the video search engine was implemented. Two participants pointed to the fact that they instead used YouTube's search engine to find videos before returning to their assigned application. The suggested solution is, therefore, to redirect the user to the search engine in the specified streaming service when finding new videos. By allowing users to search in the streaming service, they can find videos in a manner that they are accustomed to.

Streaming Services (3). Most participants felt that the service would have been more appealing if they supported other streaming services that were more in line with their social groups streaming habits. A mobile social TV application, therefore, should support a variety of streaming services, or at the minimum, offer a streaming service that reflects the target audience's viewing preference.

Recommended Videos (4). Two factors determined the kinds of videos recommended in a group profile according to knowledge gained in the study. First, it is the genre suitable for mobile social TV applications. There was, to a certain degree, a consensus on what type of genres the participants felt were most appropriate. A feature recommending videos should, therefore, be a reflection of this data. Second, videos should also reflect the group's interests.

Group Members (5). Most participants felt that the number of viewers in a mobile social TV experience should be 3–5. An application could, therefore, recommend a maximum of five viewers in a group to have the best experience. However, the amount should not be fixed, given the fact that certain social groups might be bigger, and setting limitations on room capacity might discourage certain social groups from using the service.

Community Events (6). The participants deemed the community rooms to be messy and unorganized in PlayNow and Rave applications. The functionality is similarly implemented in both applications, whereby there are few constraints and limitations. Several participants were favorable to the idea, but, with better execution. The solution presented is, therefore, to organize the feature as community events whereby users can add and join events at a specified time and invite other users. The events can also be limited by country or city for participants to communicate in the same language or encourage a more personal connection. An additional way of keeping the feature orderly is adding a filter allowing users to determine a variety of factors. Having these features and limitations addresses the users' complaints of disorganization and simultaneously enables them to engage in a community experience on individual terms.

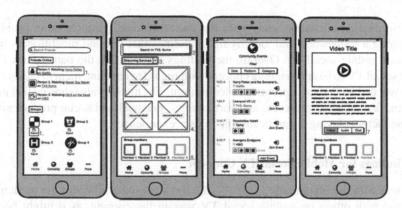

Fig. 2. Low-fidelity prototypes addressing design guidelines (1).

Interaction Method (7). Most participants favored video function. The proposed design suggests video as the default interaction method. There was a minority of participants that also expressed interest in text chat and audio chat as viable interaction methods. An ideal mobile social TV application should, therefore, be flexible enough to offer these options to best suit the users' needs and requirements.

Video (8), Audio (9), and Text (10) Chat Functions. The suggested solution proposes webcam windows that are visible enough for users to develop the feeling of co-presence, but not so large that it serves as a distraction or annoyance. To allow flexibility, already established video chat features like muting sound/video are also advised. As sound is not a visual medium, the feature does not require as much visibility. Nevertheless, as a feedback feature, it could be wise to have a visual cue to indicate when

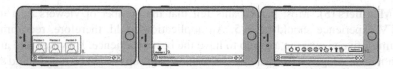

Fig. 3. Low-fidelity prototypes addressing design guidelines (2).

users are speaking. This could be specifically useful if there are many users in a group or multiple people talking at the same time.

The participants were mostly pleased with how the text chat was integrated. They expressed wanting more choices of interaction. The proposed solution, therefore, adds emojis as a form of communication. The idea also presents a solution that addressed both the limited screen space and users missing out of a conversation. However, the option to have a regular keyboard should always be available for users to write and express long-form sentiments.

4   Discussion

We performed an elicitation study with the intent of identifying common problems, establishing a framework, and identifying guidelines that will aid practitioners in the context of mobile social TV. Overall, the findings showed that the participants found mobile social TV to be an intriguing concept with regard to enabling them to engage with others over distance when consuming video content. The findings also showed that genre, platform, feeling of co-presence, and participants' personal and social predispositions were the factors playing a significant role in user experience.

4.1   Findings for Mobile Social TV Concept

Consistent with the literature [e.g., 3, 10], the most common time slot the participants interacted with others on mobile social TV was in the evening, as it might become hard to find an available time for watching together as a leisure time activity during the daytime. Limitations of the mobile social TV applications played a critical role in participants' satisfaction. A bigger screen such as a tablet or PC was reported to be ideal to have a better viewing experience. This is in line with a previous report [10] that an overwhelming majority of people watch video content with others on a television screen, followed by a computer screen, tablet device, and mobile phone, respectively. Our study showed that small screen size distracted the participants when engaging with other viewers.

People usually engage in a second screen such as a mobile phone or computer to communicate with others while watching a television program [10, 14, 17, 23, 24]. The most common activity of social TV is using a mobile phone and text messaging [12]. This indicates that second screen viewing is a popular practice when consuming video content. Similarly, in our study, we found that there was no agreement among the participants on which approach was superior in terms of providing a better viewing experience. Further, one-third of the participants believed that dealing with all on the same platform might become overwhelming, so they preferred watching on a device when interacting with others on another one.

Multiple studies underlined the feeling of togetherness, social connectedness, and co-presence as the main assets of social TV [e.g., 15, 20]. Our study garnered similar results that most participants were positive to the concept. They liked the act of simultaneously watching video content on mobile social TV. In line with previous research [e.g., 3, 11, 12], in our study, communication was a significant motive that the participants valued of social TV in specific conditions such as when people were physically at different locations and wanted to engage with others when watching video on the same platform. Overall, the participants believed that the concept provided advantages of feeling co-presence and togetherness to distant viewers which outweighed the limitations of mobile social TV applications mentioned previously. However, there were also some participants who expressed that the concept cannot replicate the feelings that a physical presence provides.

Most participants found interaction with others to be distracting when watching some types of video content. Genre played a profound role in their opinions on how their experience was distracted. This finding is consistent with extant literature [e.g., 6, 14] that viewers prefer watching certain types of TV programs in groups as they encourage social interaction more than others. In our study, comedy, sports, reality TV, and live events were most suitable for watching together on a social TV platform. Watching funny content together seemed to be suitable. Only male participants found sports to be a suitable genre which signified a connection between gender and interaction method. However, interacting with others when watching specific genres was more challenging to the participants. Documentary, tutorial, technology, and education were found to be unsuitable as engaging with others might be disturbing when they needed more concentration to follow the streaming and understand the content. Similar findings were reported by [10] that people mostly watched movies together followed by sports, news, sitcoms, and reality TV. Further, Baillie et al. [3] showed that the participants in their study would like to engage on social TV when especially watching video content such as sports TV shows.

In line with previous research [10], the participants usually interacted with one or two co-viewers when watching video content on social TV. Engaging with maximum of 3–5 co-viewers was generally reported to be manageable as a high number of people in a platform might distract the viewing experience regardless of video content.

Baillie et al. [3] showed that voice chat was a very useful interaction method to communicate with other viewers over distance when watching television. Similarly, Harboe et al. [8] found that people usually started sending messages or emojis when viewing together and this sometimes ended up with a phone call as an ambient component which provided rich conversation and thereby full involvement. However, the findings of the current study diverge slightly from previous research. We found that video chat was the superior interaction method to other chat options such as text chat, audio chat, and emojis. It provided a greater feeling of co-presence when compared to other interaction methods. Video chat was also found to be the most prominent method offering a less distracting way of communication when watching video content on mobile social TV. Conversely, text chat, audio chat, and emojis were not evaluated as adequate interaction methods that satisfactorily replicate the feeling of togetherness with other viewers.

4.2 Findings for Design Guidelines

Based on participants' feedback, we provided a list of design guidelines for practitioners. We found that several factors such as age, gender, social dynamics, personal interests, platform, video genre affected how the participants adapted mobile social TV and thereby played a profound role in user experience.

The design guidelines, explained in the results section, corroborate, and contribute to previous research [e.g., 7, 16]. For instance, Geerts and de Grooff [7] presented twelve sociability heuristics targeting social TV and suggested that social TV services should provide notification and awareness features when someone from a buddy list starts watching video content or a viewer is available to interact with, etc. Social TV requires to enable different communication channels such as voice chat and text chat for users with different experiences and abilities. Besides, we found that the search function in a social TV application was demanded feature that increased user experience. The application should enable users to search for video content using the application's internal search engine. The recommendation feature was found to be useful functionality. Recommended videos in genres were appreciated by the participants so that they could customize their viewing experience based on their interests and preferences. Similar findings were reported by [16] where the context-specific or personalized recommendation feature was reported as an important function of a social TV application. The number of people affects the viewing experience as a high number might distract the viewers when watching together, thus, up to five viewers are mainly preferable. A social network feature that enables viewers to create viewing networks based on their interests and availability is an important asset.

Furthermore, usability is an important factor affecting viewers' motivation and engagement in a social TV application [21]. After following these guidelines to develop a mobile social TV application, performing a usability evaluation is, therefore, necessary to ensure that the design fulfills user requirements that influence user experience.

5   Conclusion and Future Research

We conducted this study in an attempt to understand user requirements and present low-fidelity prototypes together with design guidelines for mobile social TV applications. Primarily, we found that the participants valued the mobile social TV concept that has the potential to provide new manifold opportunities to engage with their friends, family members, or unknown others with whom they share a common interest while consuming media content. Platform and screen distribution were significant considerations in the participant's approach to the concept. Video chat was the superior interaction method. The concept of social TV seems to gain more popularity in the future as socializing with others when consuming media content and communicating on a single platform simultaneously is considered to be more fun than watching alone. Based on the current study's findings, we suggest more empirical research to further understand how to design more usable mobile social TV applications to be able to create a decent user experience.

Acknowledgments. This paper was originally prepared as an MA thesis by Taume Dery under the supervision of Dr. Yavuz Inal.

References

1. Airtime Media homepage. https://www.airtime.com/. Accessed 31 Jan 2022
2. Auverset, L.A., Billings, A.C.: Relationships between social TV and enjoyment: a content analysis of the Walking Dead's story sync experience. Social Media+ Soc. **2**(3), 1–12 (2016). https://doi.org/10.1177/2056305116662170
3. Baillie, L., Frohlich, P., Schatz, R.: Exploring social TV. In: Proceedings of the 29th International Conference on Information Technology Interfaces, IEEE, Cavtat, Croatia, pp. 215–220 (2007). https://doi.org/10.1109/ITI.2007.4283773
4. Cesar, P., Geerts, D.: Past, present, and future of social TV: a categorization. In: Proceedings of the IEEE Consumer Communications and Networking Conference (CCNC), IEEE, Las Vegas, NV, USA, pp. 347–351 (2011). https://doi.org/10.1109/CCNC.2011.5766487
5. Doughty, M., Rowland, D.A., Lawson, S.: Who is on your sofa?: TV audience communities and second screening social networks. In: Proceedings of the 10th European Conference on Interactive TV and Video. ACM, Berlin, Germany, pp. 79–86 (2012). https://doi.org/10.1145/2325616.2325635
6. Ducheneaut, N., Moore, R.J., Oehlberg, L., Thornton, J.D., Nickell, E.: Social TV: Designing for distributed, sociable television viewing. Int. J. Hum.-Comput. Interact. **24**(2), 136–154 (2008). https://doi.org/10.1080/10447310701821426
7. Geerts, D., De Grooff, D.: Supporting the social uses of television: Sociability heuristics for social TV. In: Proceedings of the SIGCHI Conference on Human Factors in Computing Systems, pp. 595–604. ACM, Boston (2009). https://doi.org/10.1145/1518701.1518793
8. Harboe, G., Metcalf, C.J., Bentley, F., Tullio, J., Massey, N., Romano, G.: Ambient social TV: Drawing people into a shared experience. In: Proceedings of the SIGCHI Conference on Human Factors in Computing Systems, pp. 1–10. ACM, New York (2008). https://doi.org/10.1145/1357054.1357056
9. Hwang, Y., Lim, J.S.: The impact of engagement motives for social TV on social presence and sports channel commitment. Telematics Inform. **32**(4), 755–765 (2015). https://doi.org/10.1016/j.tele.2015.03.006
10. Interactive Advertising Bureau: The OTT co-viewing experience (2017). https://www.iab.com/wp-content/uploads/2017/11/The-Co-Viewing-Experience-2017_IAB__.pdf. Accessed 2 Feb 2021
11. Kim, J., Merrill, K., Jr., Yang, H.: Why we make the choices we do: Social TV viewing experiences and the mediating role of social presence. Telematics Inform. **45**, 101281 (2019). https://doi.org/10.1016/j.tele.2019.101281
12. Krämer, N.C., Winter, S., Benninghoff, B., Gallus, C.: How "social" is social TV? the influence of social motives and expected outcomes on the usage of social TV applications. Comput. Hum. Behav. **51**, 255–262 (2015). https://doi.org/10.1016/j.chb.2015.05.005
13. Mantzari, E., Lekakos, G., Vrechopoulos, A.: Social TV: introducing virtual socialization in the TV experience. In: Proceedings of the 1st International Conference on Designing Interactive User Experiences for TV and video. ACM, California, USA, pp. 81–84 (2008). https://doi.org/10.1145/1453805.1453823

14. Marinelli, A., Andò, R.: Multiscreening and social TV: The changing landscape of TV consumption in Italy. VIEW J. Europ. Television History Culture **3**(6), 24–36 (2014). https://doi.org/10.25969/mediarep/14105

15. Metcalf, C., Harboe, G., Tullio, J., Massey, N., Romano, G., Huang, E.M., Bentley, F.: Examining presence and lightweight messaging in a social television experience. ACM Trans. Multimed. Comput. Commun. Appl., Article No: 27 (2008). https://doi.org/10.1145/1412196.1412200

16. Mitchell, K., Jones, A., Ishmael, J., J.P. Race N.J.P.: Social TV: Toward content navigation using social awareness. In: Proceedings of the 8th European Conference on Interactive TV and Video. ACM, New York, United States, pp. 283–292 (2010). https://doi.org/10.1145/1809777.1809833

17. Nielsen. Action figures: How second screens are transforming TV viewing. https://www.nielsen.com/us/en/insights/article/2013/action-figures-how-second-screens-are-transforming-tv-viewing/. Accessed 31 Jan 2022

18. Playnow homepage. https://playtotvnow.com/. Accessed 31 Jan 2022

19. Rave Inc homepage. https://rave.io/. Accessed 31 Jna 2022

20. Shin, D.H.: Defining sociability and social presence in social TV. Comput. Hum. Behav. **29**(3), 939–947 (2013). https://doi.org/10.1016/j.chb.2012.07.006

21. Shin, D.H.: Do users experience real sociability through social TV? Analyzing parasocial behavior in relation to social TV. J. Broadcasting Electron. Media **60**(1), 140–159 (2016). https://doi.org/10.1080/08838151.2015.1127247

22. Van Es, K.: Social TV and the participation dilemma in NBC's the voice. Telev. New Media **17**(2), 108–123 (2016). https://doi.org/10.1177/1527476415616191

23. Weisz, J.D., Kiesler, S., Zhang, H., Ren, Y., Kraut, R.E., Konstan, J.A.: Watching together: Integrating text chat with video. In: Proceedings of the SIGCHI Conference on Human Factors in Computing Systems, pp. 877–886. ACM, San Jose (2007). https://doi.org/10.1145/1240624.1240756

24. Winter, S., Krämer, N.C., Benninghoff, B., Gallus, C.: Shared entertainment, shared opinions: the influence of social TV comments on the evaluation of talent shows. J. Broadcast. Electron. Media **62**(1), 21–37 (2018). https://doi.org/10.1080/08838151.2017.1402903

Perspectives for Using Smart Augmented Reality for the Future in Social Computing and Collaborative Assistance

Ralf Doerner[✉] [ID]

Hochschule RheinMain, Unter den Eichen 5, 65195 Wiesbaden, Germany
ralf.doerner@hs-rm.de
http://www.cs.hs-rm.de/~doerner

Abstract. Overall, the paper explores perspectives for future developments, identifies challenges, and potential advantages for using smart augmented reality for the future in social computing and collaborative assistance. The paper identifies seven major challenges when using computer-mediated communication today. It provides perspectives and visions on how future computer-mediated communication might be improved when Smart AR is used, especially when smart assistants are employed. Moreover, a computer system architecture has been conceived that provides direction on how the approaches suggested in the paper might be realized.

Keywords: Computer-mediated communication and collaboration · User interfaces in social computing · Augmented reality · Smart assistants · Software agents

1 Introduction

Catalyzed by the Corona pandemic and followed by the increase in the amount of homeschooling or work from home offices, the usage of web conferencing tools such as Zoom or WebEx has become more commonplace. In contrast to the well-established voice-based communication via telephone, communication partners can be seen, and computer screens can be shared. But the potential for Social Computing in this trend has not been fully exploited yet. Camera images from participants that are now becoming available can be processed with AI-based computer vision and 3D reconstruction. As a result, information about the context of the communication situation and about the communication partners themselves (e.g., their emotional state) can be gathered, for instance, using techniques from Affective Computing. An open question is still how this information can be used sensibly.

One possible answer is to introduce additional, communication and collaboration supporting visual information into the view of the user employing methodologies from Augmented Reality (AR). Utilizing Artificial Intelligence (AI), AR can become Smart AR, i.e. the augmentation process is performed more intelligently and the augmentations are chosen and adapted supported by sophisticated

G. Meiselwitz (Ed.): HCII 2022, LNCS 13315, pp. 97–109, 2022.
https://doi.org/10.1007/978-3-031-05061-9_7

software systems. Smart AR-based communication support is particularly valuable if collaboration partners are not limited to persons, but smart assistants that act as agents for software applications are also taken into consideration. Many questions arise about the topic of how these smart assistants present themselves to the users, how users become aware of them and how users discern potential affordances offered by these assistants.

In this paper, we identify current challenges for improving computer-mediated communication that can potentially be met with Smart AR. We then share our ideas and perspectives we see to tackle these challenges. Accordingly, the contributions of this paper are (1) a checklist of challenges when using computer-mediated communication today, (2) perspectives and visions on how future computer-mediated communication look like when Smart AR is used, (3) ideas on how smart assistants can benefit from Smart AR and how they can improve collaboration in computer-mediated communication, and (4) approaches how computer system architectures can be conceived in order to accommodate the new ideas. For this, the paper is structured as follows. In the next section, we briefly review fundamentals and provide pointers to literature for further reading in relevant areas such as AR, AI, or affective computing. In Sect. 3, we list challenges that we identified based on existing literature and our own research experience. Sections 4 and 5 contain ideas on how Smart AR can be employed to enhance communication or collaboration assistance respectively. Approaches and ideas related to system architectures are presented in Sect. 6. Section 7 sketches visions for future, improved computer-mediated communication by extrapolating our ideas and approaches. And finally, a conclusion can be found in Sect. 8.

2   Fundamentals

2.1   Computer-Mediated Communication/CSCW

Computer-mediated communication simply denotes human communication that is supported by electronic devices capable of computing such as smartphones or video conferencing systems. A special case exists where only one human is involved in the communication and the communication partner is realized by a software system, for instance, an intelligent assistant such as Amazon's Alexa or Apple's Siri. In this sense, computer-mediated communication can constitute a form of human-computer interaction that strives for being natural, easy-to-learn, and intuitive as human-human communication serves as a metaphor for human-computer interaction. In this paper, we focus on synchronous communication although computer-mediated communication has also the potential to support asynchronous communication, for example using email or a messaging system such as Whatsapp. We also assume that the human communication partners are not co-located.

There exists a bulk of literature on the various aspects of computer-mediated communication. A good introduction can be found in [6]. In the literature, various drawbacks of computer-mediated communication have been identified such as (1) the loss of information especially in non-verbal communication (e.g., the complete body language of communication cannot be perceived when only images of the face are transmitted), (2) the requirement to change communication (e.g., natural ways of negotiating turn-taking may not work and novel forms such as explicit speaker lists are started to be maintained), (3) an increased mental load and fatigue [19], (4) social consequences such as social isolation and detachment [16]. However, there exist also advantages of computer-mediated communication that are self-evident, e.g., when there is no alternative to face-to-face communication as, for example, the communication partners are at different locations or personal contacts need to be avoided. In the literature, additional advantages of computer-mediated communication have been described such as in the hyperpersonal model of interpersonal communication [29] where benefits such as the more fine-grained control over one's self-representation are named. A subset of advantages that are characterized by the support of collaboration is a topic in the field of CSCW (computer-supported cooperative work) where computer-mediated communication often serves as an integral part of groupware or decision support systems. Here, a good starting point into literature is the proceedings of the yearly ACM conference on CSCW and social computing, e.g., [2].

2.2 Affective Computing

Affective computing is concerned with the ability of a computing device to recognize, interpret and simulate human affective states such as emotion, mood or personality [25]. These states can be represented employing discrete models (e.g., using labels such as "joy" or "sadness" [15]) or dimensional models (e.g., the Big Five personality traits [7]). The recognition of affective states is based on sensor data. While there exists a whole range of sensors such as sensors for measuring skin conductance or electrodes for measuring brain waves, non-obtrusive sensors are usually preferred. Here, cameras play an important role as they work contactless and the stream of images they provide about a person can be analyzed. For example, facial expressions and the way a person walks can be recognized which serves as a basis for identifying affective states. However, this interpretation is challenging and often not robust especially because there is much variation not only between human beings but also within an individuum and signals which makes it hard to map a person's body signals to an affective state. For example, a person's pulse depends on age and physical shape and also on the current stress level of a situation and emotional experiences. This makes it difficult to normalize the sensor data to allow for a general interpretation, particularly across the boundaries of different contexts and cultures. In addition, for research in this area, it is difficult to establish a ground truth because a test subject's statements about their feelings are not always in alignment with their actual feelings [24]. A good introduction to affective computing can be found in [11] and [9]

2.3 Artificial Intelligence

Machine learning, intelligent behavior of computing machinery, and artificial life are topics of Artificial Intelligence (AI). In AI, a broad spectrum of methodologies have been proposed ranging from rule-based systems over decision trees to neural networks. A good introduction is available in [27]. With the advent of GPUs (graphical processing units) and their computing power, computation-intensive methodologies such as deep learning have become more widespread in their use [14]. These methodologies have been successfully applied to various application areas, for example in the AI-based rendering of virtual imagery. Another example is 3D scene understanding where a model of 3D space is constructed from several images, objects are segmented and categorized [20]. AI has also been applied to computer-mediated communication where a specific area named AI-mediated communication is forming [12].

2.4 Internet of Things

The Internet of Things (IoT) is a network of connected devices that are (1) linked to the Internet, (2) able to collect data through embedded sensors, and (3) remotely monitored and controlled [21]. While an IoT ecosystem is evolving and commercial products exist (e.g., IBM Cloud) to realize it, there are still barriers [13]. Among these barriers is fragmentation that ranges from lack of interoperability at the network layer to a multitude of different, inconsistently designed user interfaces for each service provided. IoT is of relevance for computer-mediated communication as it is able to provide a more ubiquitous and rich infrastructure and has according effects on their users [28]. This is not only true for public places such as workplaces or conference rooms but also for private homes where the IoT plays a fundamental role in smart homes in general [10]and specific areas such as ambient assisted [1] living in particular that are all concerned with sophisticated computer-mediated communication.

2.5 Software Agents and Virtual Assistants

Software agents (also called bots) are a specific subset of software programs that exhibit specific characteristics such as proactive and reactive behavior, persistence, and autonomy [4]. These characteristics allow software agents to act on behalf of users. Of particular interest in the context of computer-mediated communication is the ability of software agents to exhibit social behavior (e.g., including the understanding of natural speech or even the maintenance of an anthropomorphic form). Here, intelligent virtual assistants [23] are specific software agents that provide assistance in computer-mediated communication and can act as an interface to computer software in general. Examples of intelligent virtual assistants are Bixby, Siri, or Alexa which have achieved a wide coverage, especially as they are available per default on many smart devices or part of an operating system. Still, these systems are not per se integrated into web-conferencing systems that are independent products, e.g., zoom, Webex, or Microsoft teams.

2.6   Augmented Reality

Augmented Reality (AR) strives to alter the immediate perception of the real world by enriching it with virtual content in real-time. This virtual content must be seamlessly integrated (e.g., regarding its appearance and behavior) [3] and ideally becomes indistinguishable from reality. AR can be considered a part of the Mixed Reality (MR) continuum between Virtual Reality (VR) and reality [17]. There exist several alternatives to how AR can be realized. In video see-through AR a stream of video images is taken from reality, augmented, and then played back to the user. For instance, the user can hold a tablet, point its video camera in reality, and see an augmented image of the reality on the tablet's display which acts as a magic lens that introduces AR. In optical see-through AR, the user perceives the real environment directly and the augmented parts are superimposed optically, e.g., by using semi-transparent displays that users wear similar to glasses or that are mounted to their heads. In projection-based AR, the augmentation with virtual content is directly achieved in the real environment by projecting the virtual imagery on real-world objects or directly on the users' retinas (e.g., employing small laser projectors that are mounted to glasses the users are wearing). A good introduction to AR can be found in [8] and [5]. In the literature, there exists work that highlights the usefulness of AR for intelligent agents (and computer-mediated communication) [30] and identifies a convergence of AR with IoT and software agents [22]. This is an aspect that will be also explored further in this paper.

3   Challenges

In this section, we identify seven challenges that pose obstacles for future progress in social computing and especially for improved computer-mediated communication and collaboration.

3.1   Information Gap Challenge

Compensating the loss of information inherent in computer-mediated communication when compared to face-to-face communication is a challenge. Various types of information are affected such as (1) digitized information by a loss in resolution and inadequate reconstruction (e.g., the extent to which tiny changes in facial expressions can be perceived), (2) spatial information (e.g., missing information about the spatial relationship between communication partners and objects which makes for example deictic gestures difficult to interpret), (3) multisensory information (e.g., only video and audio information is available but not olfactory information or tactile information), (4) information from probing (e.g., a communication partner cannot check objects at a remote site out and build a mental model how they behave), (5) concurrency information (e.g., due to delays in processing data or transporting data via computer networks), (6) 360° information on the environment (e.g., technical restrictions in the field of the view and inability to change viewing direction at the remote site), (7) information on fidelity (e.g., the volume of speech might not be reproduced correctly at a remote site).

3.2   Presence Challenge

Achieving a believable feeling that the communication partner is present in a common communication context is hard to achieve. This includes the notion of common ground (e.g., due to spatial fragmentation), social presence (e.g., due to missing contact opportunities such as handshakes), and telepresence (e.g., due to the experience of a barrier between local and remote sites in the communication).

3.3   Remote Action Challenge

The ability to act at a remote site in computer-mediated communication is often severely limited. For instance, in web-conferencing, it is usually impossible to freely choose the viewing parameters of the remote camera (e.g., to orient oneself or to look in a certain direction).

3.4   Data Fusion Challenge

IT-systems in the environment of the user (e.g., an app of the user's smartwatch that monitors the user's heart rate, software in a home appliance the user is currently using, and a software system recording and archiving the input of a security camera that films the user) are often separated or even isolated making it difficult to collect and relate all the data that software systems possess about the user. Moreover, data may not be stored and data from different points in time may not be available. This limits the potential for the fusion of data from different sources and, as a consequence, reduces the basis for intelligent support for computer-mediated communication.

3.5   Exploiting Semantics and Context Challenge

While it is a challenge in itself to compute information about the context where computer-mediated communication takes place and try to interpret data in order to draw conclusions about semantics, it is also a challenge to exploit this information sensibly. For instance, even if affective computing is able to provide information on the emotional state of the user, ideas are missing what to do with this information or even to formulate requirements how fine-grained and reliable this information needs to be in order to be valuable.

3.6   Representation Challenge

The human communication partners are represented differently in computer-mediated communication, e.g., a person might be represented as an entry in the list of participants, as a photo on a screen, as part of a live video stream, or as an avatar that has a virtual geometry and appearance. The form of the representation affects the quality of communication and collaboration and may even have effects on the user itself, for instance, an avatar might influence a person's self-perception [26]. It is a challenge to find a fitting representation that best supports a high quality in computer-mediated communication and collaboration. This is particularly true for the representation of software agents or virtual assistants that possess no natural appearance in contrast to human communication partners and where a wide choice of representations exists (e.g., anthropomorphic visual representation that comes as close as possible to a real person, a representation as a cartoon-like character, or a representation as an abstract entity).

3.7   IT-System Configuration and Maintenance Challenge

Computer-mediated communication requires respective software and hardware. With this, the need arises to monitor, repair, replace, update, connect, integrate, configure, and adapt software and hardware about the users' needs devising dedicated strategies and policies (e.g., concerning privacy). With the rising complexity and sophistication of hardware and software, it is a challenge to not overburden the user with these IT-system maintenance and configuration tasks. The challenge is usually more severe in private environments where specialists or dedicated personnel are not as readily available as in business environments where companies usually can afford some form of IT support.

4   Smart AR and Information Gaps

In this section, we present some approaches and ideas on how Smart AR can be used to tackle challenges identified in Sect. 3, particularly the information gap challenge.

4.1   Visual Augmentations to Outweigh Loss of Information

One seldomly pursued approach is to alter the video imagery that is shared between communication participants. Here, Augmented Reality (AR) methodologies can make a crucial contribution as they allow the integration of virtual content in the perception of the real world in a seamless and context-sensitive way. Thus, AR can be used to enrich the video images conveyed via computer-mediated communication. For example, additional artificial and exaggerating visual cues can be inserted in images that counter the loss of cues in natural communication. The level of exaggeration could be chosen according to the affective state or stress levels of the users. A simple example would be a virtual arrow that clarifies or amplifies deictic gestures. A more sophisticated example would be a semantic pointer that highlights the fitting volumes, areas, points, objects, or object parts that a deictic gesture refers to (e.g., by including a glow or changing the appearance). Visual markers can be introduced that help to better understand the spatial situation at a remote site or enable the user to tell objects apart that are similar-looking under reduced image resolution. The image perceived from a real object can be replaced by a magnified image that supports a better perception of details. Objects can be introduced to visually represent system functions and help the user realize affordances. Annotations can be integrated to relieve the user's memory. Virtual content can be inserted in images with the aim to hide real objects and thus remove distracting visual information allowing the user to better focus on relevant items. These augmentations could be either manually inserted or automatically added with the use of Artificial Intelligence.

4.2 Reuse of AR Information

In order to perform the augmentation in AR, a whole range of information needs to be gathered, e.g., the 3D pose of one or more cameras whose image streams are available to AR, the 3D pose of real objects, and their spatial relation to each other, material properties of real objects (in order to insert fitting virtual objects), lighting conditions, the size of areas in reality (in order to see whether virtual objects fit), semantic segmentation of objects, environmental conditions such as fog or smoke, information about movements of real objects (in order to extrapolate them and keep the virtual augmentations at the correct position in real-time). This information is usually only used within the AR system, however, it also has value for other IT systems such as virtual assistants that are part of computer-mediated communication. For instance, the information can be used to detect the body posture or gestures of persons which in turn can be used for affective computing or semantic interpretation of gestures (e.g., identifying the target of a deictic gesture).

5 Smart AR and Smart Assistants

This section focuses on the representation challenge identified in Sect. 3 and suggests ideas and approaches to mitigate it.

5.1 Representation of Smart Assistants

AR provides the opportunity to visualize smart assistants by giving them a virtual visual representation and integrating them into the view of reality. But not only the smart assistant can be made visible but also abstract concepts such as their actions or the features and support functions they offer. For example, a virtual assistant could place virtual numbered cards next to communication participants to manage turn-taking.

One idea for the AR-based visual representation of a smart assistant is not to strive to emulate a real person but to use a magical creature such as a house-elf that has been made well-known by Harry Potter books and movies. Thus, the elf as a magical creature is used as a metaphor for an assistive IT system. This metaphor is an original approach to provide an interface that is neither mimicking a human nor completely technical. It has the advantage that users have no fine-tuned expectations of how house-elves behave, communicate, or look like. Novel rules of communication can be introduced to the communication between humans and elf using sophisticated UI technologies such as AR or IoT sensors that will be employed and combined for communication support. This way, the house-elf can serve as one point of contact to system functionality and counter the fragmentation of IT systems, especially in IoT. Moreover, the metaphor makes it credible that the house-elf as a ghost-like creature cannot be seen as brightly as a real object (due to display limitations), can teleport, or magically possess hardware and robots in particular. This way, a house-elf can be embodied in a robot and novel forms of human-computer interaction/human-robot interaction can be created for virtual assistants.

5.2   Pro-active Shaping of Communication Environments

An approach seldomly pursued is that smart assistants proactively change the environment in order to better support communication. For instance, smart assistants could change the lighting situation by turning lights on or closing blinds (which could also lead to an improvement in the quality of AR), turning off noisy appliances, or installing cameras (e.g., using cameras mounted on drones that can be made to stick to walls or ceilings with specific adhesive materials [18]).

6   Smart AR and System Architectures

Future developments in social computing require not only novel concepts in human-computer interaction and the formation of new communication cultures but also suitable architectures of the IT systems that support them. Here, a promising idea is to use biology-inspired approaches in order to strengthen the self-organizational capabilities of the smart assistants, e.g., for making them more flexible and resilient. Currently, mobile devices such as smartphones or tablets possess computing power, high-resolution cameras, displays, and support natural interactions such as multi-touch gestures making them suitable platforms for smart AR. In the future, the advent of large-scale wall-mounted displays, lightweight devices that can be worn similar to sunglasses as well as personal drones can significantly add to the way, smart AI can be realized. For example, remote communication partners can be shown as life-size holograms conveying not only their facial expressions but also their whole-body posture, the viewpoint in a remote location could be chosen by remotely controlling a drone, virtual information items or virtual representations of abstract concepts can be spatially arranged and accessed in the real-world to foster visual thinking and collaboration that deals with abstract subjects.

Figure 1 shows a suggestion for the system architecture of a virtual assistant that can serve as a central building block for a system supporting computer-mediated communication. A smart layer is foreseen that ties together low-level

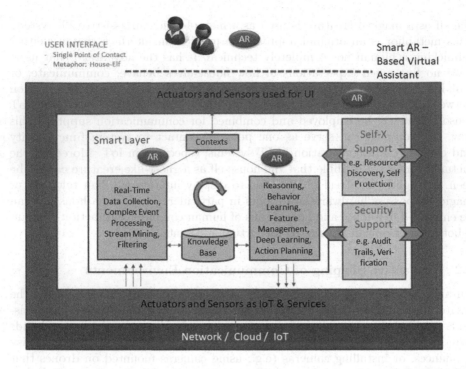

Fig. 1. Integrating AR in the system architecture of a smart AR-based virtual assistant that enables new quality levels in computer-mediated communication

hardware-related aspects as well as user-related aspects. AR can be integrated easily as the necessary procurement of information for augmentations can be seamlessly included in the smart layer where also capabilities for AI exist. The smart layer has access to sensors such as video cameras that are also relevant for realizing AR. All information is collected in a knowledge base where the AR in the user interface part can access output devices for displaying the AR content using AR hardware.

7   Future Potential

Possible future developments are (1) a personal assistant that is available cross-application independent of the user's current location in the real world and the information space and that is able to learn to adapt to the user's preferences over a long period, (2) a virtual agent that executes actions on behalf of the user and represents the user in case of personal unavailability, (3) a dedicated guard that shields the user's privacy and negotiates the exchange of information, (4) a communication coach and translator who points out potential cultural misunderstandings and provides advice how to improve the quality of communication, (5) a historian who can act as an advisor by accessing, filtering and displaying information from the past or even making predictions based on experiences from historical data.

This smart AR-based assistant is not limited to assisting the elderly (as is the focus in ambient assisted living) but offers

- comfort functions (e.g., reminders, entertainment offers, reporting, home automation features, concierge services, wearable assistance),
- coordination with other smart assistants, IT systems, or human communication partners.
- advice on how to best facilitate and support a communication taking context and IT infrastructure into account,
- protection against real and virtual breaches of privacy or security.

8   Conclusion

In this invited paper, we shared our thoughts on the future of social computing, specifically computer-mediated communication and collaboration, where we identified seven major challenges. We see smart assistants as a key element of future communication supporting IT systems. These software agents can benefit from smart AR, as AR is capable of mitigating the information gap challenge. Here, we proposed visual augmentation to counteract a loss of information as well as the reuse of information typically gathered by an AR system for more general purposes. For addressing the representation challenge, we suggest using the metaphor of a house-elf. For addressing other challenges such as the data fusion challenge, we presented a possible system architecture for our proposed smart-AR-based virtual assistant that ties reasoning together in a smart layer and that is firmly built on an IoT ecosystem employing biology-inspired self-X methodologies to organize itself. We show that all parts of a smart AR system can be integrated seamlessly into this system architecture. As a result, a whole range of functions can be added and better integrated to achieve a novel level of computer-mediated communication, e.g., assistants that maintain themselves, act as consultants for optimizing communication to other persons or software entities, or proactively prepare the environment for good and fruitful communication.

References

1. Abtoy, A., Touhafi, A., Tahiri, A., et al.: Ambient assisted living system's models and architectures: a survey of the state of the art. J. King Saud Univ. Comput. Inf. Sci. **32**(1), 1–10 (2020)
2. ACM (ed.): CSCW '21: Companion Publication of the 2021 Conference on Computer Supported Cooperative Work and Social Computing. Association for Computing Machinery, New York, NY, USA (2021)
3. Azuma, R.T.: A survey of augmented reality. Presence: Teleoperators Virtual Environ. **6**(4), 355–385 (1997)
4. Bradshaw, J.M.: Software Agents. MIT Press (1997)

5. Qu, N., van der Meer, W.J.: Augmented reality. In: Aps, J.K.M., Boxum, S.C., De Bruyne, M., Jacobs, R., van der Meer, W.J., Nienhuijs, M.E.L. (eds.) Het tandheelkundig jaar 2017, pp. 183–195. Bohn Stafleu van Loghum, Houten (2016). https://doi.org/10.1007/978-90-368-1030-2_14

6. Carr, C.T.: Computer-mediated Communication: A Theoretical and Practical Introduction to Online Human Communication. Rowman & Littlefield (2021)

7. Cobb-Clark, D.A., Schurer, S.: The stability of big-five personality traits. Econ. Lett. 115(1), 11–15 (2012)

8. Doerner, R., Broll, W., Jung, B., Grimm, P., Göbel, M., Kruse, R.: Introduction to virtual and augmented reality. In: Doerner, R., Broll, W., Grimm, P., Jung, B. (eds.) Virtual and Augmented Reality (VR/AR): Foundations and Methods of Extended Realities (XR), pp. 1–37. Springer, Cham (2022). https://doi.org/10.1007/978-3-030-79062-2_1

9. Fernández-Caballero, A., González, P., Lopez, M.T., Navarro, E.: Socio-Cognitive and Affective Computing. MDPI (Multidisciplinary Digital Publishing Institute) (2018)

10. Goyal, P., Sahoo, A.K., Sharma, T.K., Singh, P.K.: Internet of things: applications, security and privacy: a survey. Mater. Today: Proc. 34, 752–759 (2021)

11. Guthier, B., Dörner, R., Martinez, H.P.: Affective computing in games. In: Dörner, R., Göbel, S., Kickmeier-Rust, M., Masuch, M., Zweig, K. (eds.) Entertainment Computing and Serious Games. LNCS, vol. 9970, pp. 402–441. Springer, Cham (2016). https://doi.org/10.1007/978-3-319-46152-6_16

12. Hancock, J.T., Naaman, M., Levy, K.: Ai-mediated communication: definition, research agenda, and ethical considerations. J. Comput.-Mediat. Commun. 25(1), 89–100 (2020)

13. Hansen, E.B., Bøgh, S.: Artificial intelligence and internet of things in small and medium-sized enterprises: a survey. J. Manuf. Syst. 58, 362–372 (2021)

14. Kelleher, J.D.: Deep Learning. MIT Press (2019)

15. Kleinginna, P.R., Jr., Kleinginna, A.M.: A categorized list of emotion definitions, with suggestions for a consensual definition. Motiv. Emot. 5(4), 345–379 (1981)

16. Meier, A., Reinecke, L.: Computer-mediated communication, social media, and mental health: a conceptual and empirical meta-review. Commun. Res. 48(8), 1182–1209 (2021)

17. Milgram, P., Takemura, H., Utsumi, A., Kishino, F.: Augmented reality: a class of displays on the reality-virtuality continuum. In: Telemanipulator and Telepresence Technologies, vol. 2351, pp. 282–292. International Society for Optics and Photonics (1995)

18. Myeong, W.C., Jung, K.Y., Jung, S.W., Jung, Y., Myung, H.: Development of a drone-type wall-sticking and climbing robot. In: 2015 12th International Conference on Ubiquitous Robots and Ambient Intelligence (URAI), pp. 386–389. IEEE (2015)

19. Nadler, R.: Understanding "zoom fatigue": theorizing spatial dynamics as third skins in computer-mediated communication. Comput. Compos. 58, 102613 (2020)

20. Naseer, M., Khan, S., Porikli, F.: Indoor scene understanding in 2.5/3d for autonomous agents: a survey. IEEE Access 7, 1859–1887 (2018)

21. Nord, J.H., Koohang, A., Paliszkiewicz, J.: The internet of things: review and theoretical framework. Expert Syst. Appl. 133, 97–108 (2019)

22. Norouzi, N., Bruder, G., Belna, B., Mutter, S., Turgut, D., Welch, G.: A systematic review of the convergence of augmented reality, intelligent virtual agents, and the internet of things. Artificial intelligence in IoT, pp. 1–24 (2019)

23. Norouzi, N., Kim, K., Hochreiter, J., Lee, M., Daher, S., Bruder, G., Welch, G.: A systematic survey of 15 years of user studies published in the intelligent virtual agent's conference. In: Proceedings of the 18th International Conference on Intelligent Virtual Agents, pp. 17–22 (2018)
24. Peter, C., Herbon, A.: Emotion representation and physiology assignments in digital systems. Interact. Comput. **18**(2), 139–170 (2006)
25. Picard, R.W.: Affective Computing. MIT Press, Cambridge (1997)
26. Ratan, R.A., Dawson, M.: When mii is me: a psychophysiological examination of avatar self-relevance. Commun. Res. **43**(8), 1065–1093 (2016)
27. Russell, S., Norvig, P.: Artificial intelligence: a modern approach, global edition 4th. Foundations (2020)
28. Van Der Zeeuw, A., Van Deursen, A.J., Jansen, G.: Inequalities in the social use of the internet of things: a capital and skills perspective. New Media Soc. **21**(6), 1344–1361 (2019)
29. Walther, J.B.: Group and interpersonal effects in international computer-mediated collaboration. Hum. Commun. Res. **23**(3), 342–369 (1997)
30. Wang, I., Smith, J., Ruiz, J.: Exploring virtual agents for augmented reality. In: Proceedings of the 2019 CHI Conference on Human Factors in Computing Systems, pp. 1–12 (2019)

Analysis of the Relationship Between Food and the Writer's Emotions Using a Meal Diary

Shoki Eto[1($\boxtimes$)], Kohei Otake[2], and Takashi Namatame[3]

[1] Graduate School of Science and Engineering, Chuo University, 1-13-27, Kasuga, Bunkyo-ku, Tokyo, Japan
`etonasri1010@gmail.com`

[2] School of Information and Telecommunication Engineering, Tokai University, 1-13-27, Takanawa, Minato-ku, Tokyo, Japan
`otake@tsc.u-tokai.ac.jp`

[3] Faculty of Science and Engineering, Chuo University, 1-13-27, Kasuga, Bunkyo-ku, Tokyo, Japan
`nama@indsys.chuo-u.ac.jp`

Abstract. The purpose of this study is to gain useful marketing insights by estimating the writer's emotions from the text and clarifying the relationship between the text's characteristics and food. First, each day's emotional evaluation, taste evaluation, and the presence or absence of an event are clarified from the text, and each text is evaluated in three items. Next, the overall score is calculated from the average value of the three items, and the characteristics of sentences with high and low scores are clarified. The result of our analysis, the characteristics of the text reveal the relationship between emotions of the day and food choices.

Keywords: Natural language processing · Emotion estimation · Word-emotion dictionary

1 Introduction

In recent years, with the progress of digitalization, customer value has diversified, and appropriate instore promotion at retail stores is required. Recently, using natural language processing technology, which is one of the AI technologies, it has become possible to discover the hidden needs of products and services by grasping the characteristics of customers from the sentences written by customers. We think that it is possible to correctly grasp the customer's emotions [1] and obtain useful knowledge for marketing by utilizing the emotion estimation that clarifies the inner emotions.

2 Research Purpose

The purpose of this study is to gain marketing-helpful knowledge by clarifying how emotions in writing texts affect food choices from diary data of consumer. Specifically,

G. Meiselwitz (Ed.): HCII 2022, LNCS 13315, pp. 110–122, 2022.
https://doi.org/10.1007/978-3-031-05061-9_8

partial speech decomposition by morphological analysis is performed in the life behavior diary, and each word is given emotion score from -1 to 1 and, taste score from -1 to 1. Emotion score indicates whether the emotion is positive or negative, and taste score indicates whether the meal was delicious or not delicious.

Then determine if the text contains words that represent the event and give it event score. The main purpose of this study is to gain useful marketing insights by comprehensively assessing sentences from these three scores and clarifying the relationship between emotions and food choices.

3 Analysis Method

In this study, we will use the data of the housewife's life behavior diary provided by a survey company of Japan for analysis. The data period is from 1990 to 2014, and the number of records is 35,373. The column names are customer ID, date of entry, weather, four seasons, behavior data, background data, food data, and meal scene. Behavior data and background data are text data, behavior data describes the behavior and meal content of the day, and background data describes the background leading to the behavior. Morphological analysis of behavior data and background data, which are text data, is performed, and part-of-speech decomposition is performed. We select nouns, verbs, adverbs, and adjectives by part of speech decomposition, and each word is scored. At this point, we use a tool called fasttext [2], an extension of Word2vec [3–5], to create the dictionary needed to determine if the content of the text is positive or negative, and whether it tastes good or bad.

First step is to arbitrarily specify multiple words that are considered very negative and multiple words that are considered very positive. Then we use the fasttext model to learn all Japanese Wikipedia and measure the average value of the very positive and very negative similarities of the morphemes of each sentence. Using the polarities with high mean similarity to determine the negative and positive values of the morpheme. If it is positive, the same value is used as it is, and if it is negative, the value with a minus is used. We convert the score scale of all words from -1 to 1 to complete the polarity dictionary. In this way, we create a dictionary to get the emotion score to judge whether the emotion of each word is positive or negative, and a dictionary to get the taste score to judge whether the taste of each word is good or bad. These dictionaries are used to set a score for each word, calculate the mean for each sentence, and give each sentence the emotion score and the taste score.

Next, we get the event score for each sentence. The event score is obtained by determining if each sentence contains a word for the event. For sentences include positive events such as Christmas and birthdays, the event score was taken from the average interquartile range of the emotion score and the taste score. For sentences containing words of negative events such as colds and illnesses, the event scores were obtained by adding a minus to the average interquartile range of the emotion score and the taste score. Also, for statements that do not contain events, the event score was given 0. Each sentence is evaluated by calculating the overall score, which is the emotion score plus the taste score plus the event score divided by two. We clarify what characteristics they have, and which foods are selected in sentences with high and low overall scores (Fig. 1).

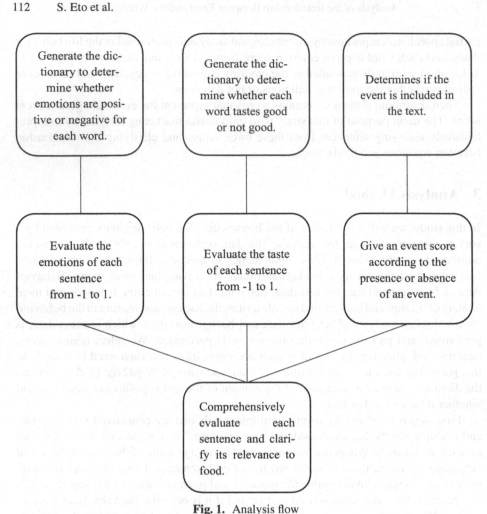

Fig. 1. Analysis flow

4 Result

4.1 Results of Emotion Score and Taste Score

Using the created dictionary, emotion scores of the behavior data and background data for each word are shown in Figs. 2 and 3. The average value of the emotion score of each word in the behavior data was 0.04, the maximum value was 1, and the minimum value was −1. The average value of emotion scores of each word in the background data was 0,041, the maximum value was 1, and the minimum value was −1. In addition, Figs. 4 and 5 show the emotion score of the behavioral and background data for each sentence. The average value of the emotion score of each sentence in the behavior data was 0.071, the maximum value was 0.458, and the minimum value was −0.327. The average value of emotion scores of each sentence in the background data was 0,114, the maximum value was 0.493, and the minimum value was −0.349. The average value of the behavior data and the average value of the background data were both positive values. Table 1 shows one example sentence with a high emotion score, and Table 2 shows one example sentence with a low emotion score. Sentences with high emotion scores contained many positive words such as "fun" and "happy". Overall, there were a lot of positive sentences. Sentences with a low emotion score included negative words such as "boring", "dislike", and "difficulty", and many sentences had a low score even if the content was positive overall.

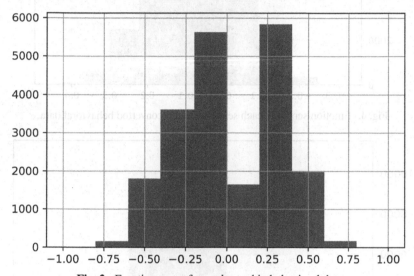

Fig. 2. Emotion score for each word in behavioral data

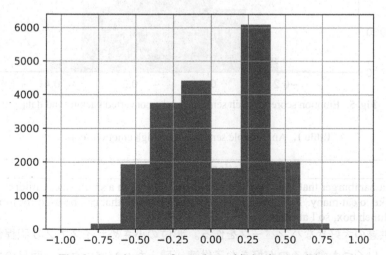

Fig. 3. Emotion score for each word in background data

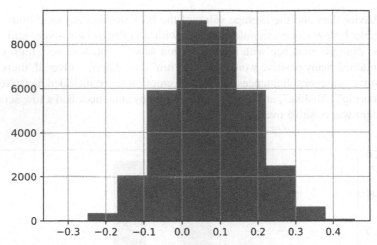

Fig. 4. Emotion score for each sentence in the converted behavioral data.

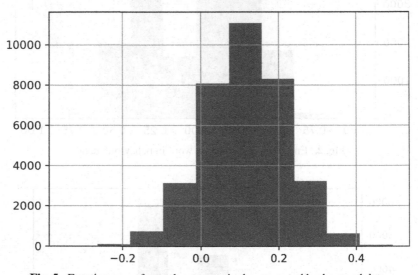

Fig. 5. Emotion score for each sentence in the converted background data.

Table 1. An example sentence with high emotion score

Emotion score: 0.422
Make a hamburger that kids love. I made about 20 eggs in a small size. Children who like to eat many. The remaining hamburger is a size that can be used for tomorrow's lunch box, so I am happy.   （子供達の大好きなハンバーグを作る。卵位の小さめサイズで２０個程作った。いくつも食べるのが好きな子供達。残ったハンバーグは、明日のお弁当に使えるサイズなので実は私にもとてもうれしい。）

Table 2. An example sentence with low emotion score

Emotion score: -0.327
"I'm dying. Please do something about this smell. I smell milk,"shouts the daughter who opened the refrigerator. Tired of mother and daughter smelling radish kimchi. （たまらないよ。この臭いどうにかしてよ。牛乳まで臭うよ」と冷蔵庫を開けた娘が叫ぶ。大根のキムチの臭いに母娘してうんざり。）

Similarly, using the created dictionary, taste scores of the behavior data and background data of each word are shown in Figs. 6 and 7. The average value of the taste score of each word in the behavior data was 0.017, the maximum value was 1, and the minimum value was −1. The average value of taste scores of each word in the background data was 0,014, the maximum value was 1, and the minimum value was −1. In addition, Figs. 8 and 9 show the taste score of the behavioral and background data for each sentence. The average the taste score for each sentence in the behavioral data was 0.204, the maximum was 0.508, and the minimum was −0.251. The average the taste score for each sentence in the background data was 0.149, the maximum was 0.516, and the minimum was −0.251. The average value of the taste score was high in both the behavior data and the background data. This is probably because the target data was a food diary, so the frequency of words related to food was high. Table 3 shows one example sentence with a high taste score, and Table 4 shows one example sentence with a low taste score. Sentences containing many words such as "delicious" and "good" that indicate good taste and words related to ingredients had a high score, sentences containing words such as "bad" and "odor" that indicate bad taste had a low score.

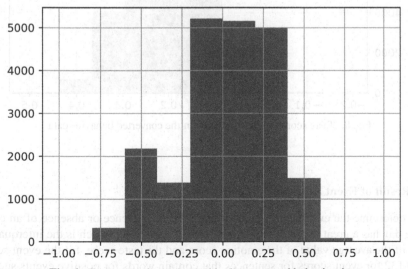

Fig. 6. Taste score for each word in the converted behavior data.

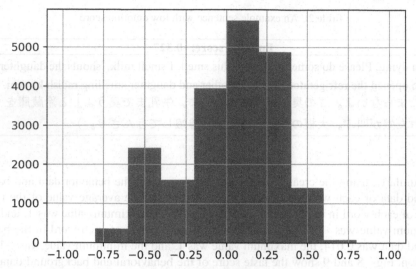

Fig. 7. Taste score for each word in the converted background data.

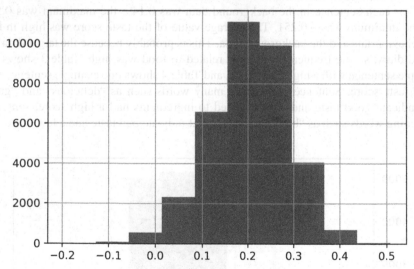

Fig. 8. Taste score for each sentence in the converted behavior data.

4.2   Result of Event Score and Overall Score

Next, determine the event score. Considering that the presence or absence of an event on the day has a great effect on emotions, we adopted 0.132, which is the interquartile range of the average value of the emotion score and the taste score, for the event score. Use 0.132 for event scores for sentences that contain words for positive events such as Christmas and birthdays, and −0.132 for event scores for sentences that contain words for

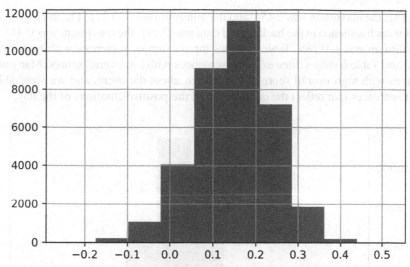

Fig. 9. Taste score for each sentence in the converted background data.

Table 3. An example sentence with high taste score

Taste score: 0.508
I made simmered Japanese mustard spinach and fried tofu. After frying komatsuna and fried tofu in sesame oil, boiled in a sweet and spicy soup stock, the fried tofu tasted good and was delicious. 小松菜と油揚げの煮物を作った。小松菜と油揚げをゴマ油で炒めてから、甘辛いだしで煮たら、油揚げに味がよくしみておいしくできた。

Table 4. An example sentence with low taste score

Taste score: -0.198
I made my favorite pork cutlet for my husband who came home after a long time, but he says, "It smells bad."I'm disappointed even though I made it. 久しぶりに帰宅した夫のために好物のトンカツを作ったのに「臭いがあってまずい。」と夫が言う。せっかく作ったのにガッカリだ。

negative events such as colds and illnesses. Use 0 for event scores in sentences that do not contain a word that means the event name. Overall score is calculated from the emotion score plus the taste score plus the event score divided by two. The overall score is a score that comprehensively evaluates three items: emotion, taste, and the presence or absence of an event. Figures 10 and 11 show the overall score of the behavior and background data for each sentence. The average overall score for each sentence of behavioral data was 0.137, the maximum was 0.454, and the minimum was −0.221. The average overall score for each sentence in the background data was 0.131, the maximum was 0.445, and the minimum was −0.285. Table 5 shows three example sentences with high overall scores, and Table 6 shows three example sentences with low overall scores. Many of the sentences with high overall scores were written about the event, and we were able to extract sentences that reflect the good taste and the positive emotions of the day.

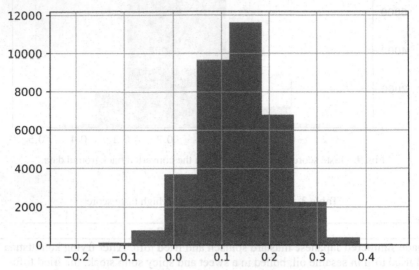

Fig. 10. Total score for each sentence in the converted behavior data.

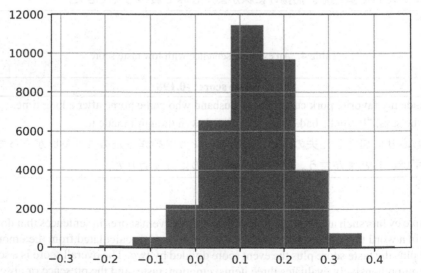

Fig. 11. Total score for each sentence in the converted background data.

Table 5. The example sentences with high overall scores

Behavior data (Overall score -0.041)
Liver is slimy and unpleasant to prepare at home, so I decided to eat liver ginger boiled at a side dish shop.   （レバーは家で下ごしらえをするのは、ぬるっとしていて気持ち悪いので、惣菜屋で買ったレバーのしょうが煮を食べることにした。）
Background data (Overall score -0.241)
My stiff shoulders are so tired that I go to the doctor and it is said that I should eat liver because I am anemic.   （あまりに肩こりがひどく疲れ気味なので医者に行くと、貧血気味なのでレバーなど食べるとよいと言われる。）
Food
Boiled liver ginger
Behavior data (Overall score -0.218)
In case my sick daughter became dehydrated, I gave her a well absorbed Pocari Sweat. Apparently, it fits in my stomach, and I was able to rehydrate.   （体調の悪い娘が脱水症状になるといけないので、吸収の良いポカリスエットを飲ませた。どうやら胃に収まり水分補給ができたようだ。）
Background data (Overall score -0.115)
My daughter has a bad stomach and will get back no matter what she eats. It seems better not to feed anything until the nausea stops.   （娘は胃の調子が悪く、何を食べてももどしてしまう。吐き気が止まるまでは、何も食べさせないようにした方が良さそうだ。）
Food
Pocari Sweat
Behavior data (Overall score -0.135)
Boil iron-rich dried radish and garlic sprouts to help absorb iron with book dashi and basic seasonings. It is a dish that seems to be effective for anemia.   （鉄分豊富な切干し大根と鉄分の吸収を助けるニンニクの芽をほんだしと基礎調味料とで煮る。これは貧血に効果ありそうな一品ね。）
Background data (Overall score -0.19)
I was fluttering when I stood up suddenly. It seems to be anemic. But I don't like livers that are effective against anemia, so I wonder if some other food can be used as a substitute.   （急に立ち上がった時にフラフラした。貧血気味らしい。でも貧血に効くレバー類は嫌いだし何かほかの食品で代用できないかしら。）
Food
Kiriboshi daikon, garlic sprouts, soy sauce, sugar, salt, cooking liquor, Hon-Dashi

Table 6. The example sentences with low overall scores

Behavior data (Overall score 0.451)
I made a chocolate cake called carat and gave it as a gift. The children were pleased that"the handmade cake is delicious after all".
（私は、カラットというチョコレートケーキを作りプレゼントした。子供達は「やっぱり手作りケーキはおいしい」と喜んでくれた。）
Background data (Overall score 0.195)
Today is Valentine's Day. My sons have been excited since the morning. They don't feel comfortable wondering how many chocolates they can get.
（今日は、バレンタイン。我が家の息子達は、朝から、ウキウキソワソワ。いくつチョコレートもらえるかなぁと、落ち着かない。）
Food
Eggs, Flour, Sugar, Cocoa, Chocolate, Butter, Cream
Behavior data (Overall score 0.421)
Go to Chinatown for a meal. It's fun to be able to eat various dishes while turning them on the round table. The food was delicious, and it was a happy birthday feast.
（中華街へ食事に行く。色々な料理を円卓で回しながら食べられるのは楽しいもの。料理もおいしく嬉しい誕生日のごちそうだった。）
Background data (Overall score 0.274)
Today is my birthday. When I thought it would be strange to make something for myself, my husband said, "Let's go eat delicious food together."lucky.
（今日は私の誕生日。自分のために何かを作るのも変だしと考えていたら、「皆でうまい物を食べにいこう」と主人。ラッキー。）
Food
Stir-fried crab egg whites, beef lettuce wrap, small shrimp tempura
Behavior data (Overall score 0.44)
I think I will eat cakes at every house at Christmas, so I decided to serve ice cream cakes and snacks so that they would not overlap.
（クリスマスにはどこの家でもケーキを食べると思うので、重ならない様にアイスクリームケーキとスナック菓子などを出そうと決める。）
Background data (Overall score 0.117)
A daughter who wants to invite her classmates to a Christmas party at Christmas. You need to make a reservation for the cake early and talk with your daughter about what to serve.
（クリスマスにクラスの友だちを呼んでクリスマス会をやりたいと言う娘。ケーキは早めに予約が必要だし、何を出すかなど娘と相談する。）
Food
Ice cream cake, snacks

Even on days when there are no events, there are many sentences that do special things such as eating out, and even if you look at the words in the sentences, you can see that you are uplifting. As for the carefully selected dishes, we found many dishes prepared at the event (cakes, sushi, pizza, etc.). Many of the sentences with low overall scores had many negative words about poor physical condition, cooking failure, and food taste and quality. Negative word explanations about negative events and situations may have lowered the score. In addition, the overall score of sentences with many negative words was low, even if the content was not negative as a whole. Most of the selected foods were foods that relieve poor physical condition (liver, udon, Pokari Sweat, etc.) and nutritious foods.

5 Discussion

From the written text by the customer, we clarified the food when emotions are positive and negative. When the emotion is positive, there was plenty of text about events such as birthdays, Christmas, and celebrations, and many of the foods chosen were prepared for special events such as cakes and sushi. In addition, overall score of sentences containing many positive words related to food such as delicious, feast, and favorite was high. In addition, the overall score was low in sentences explaining the cooking failure situation and in sentences with many negative words related to food such as bad, fat, and rot. This time, we evaluated emotions and identified food, but we would like to propose promotion as a future task.

6 Conclusion

In this study, we performed part-of-speech decomposition by morphological analysis of daily life diary and gave emotion score and taste score to each word. Next, we calculated the average score of the whole sentence from the score given to the word and obtained the emotion score and taste score of each sentence. An event score was given based on whether the sentence contained a word meaning an event, and the overall score was calculated from the emotion score, the taste score, and the event score. From the overall score, it was judged whether the emotion was positive or negative, and the characteristics of the text and the selected foods were considered, and the analysis showed that foods prepared for the event and premium foods were selected when emotions were positive, and nutritious foods and foods for the sick were selected when emotions were negative.

For our future work, some written menu name in the food section of our daily diary, while others wrote the detailed material. If this food data can be integrated into the menu name, it will be possible to clarify in more detail the influence of the emotions of the day on food selection. Also, I evaluated emotions as positive and negative, but I would like to consider more emotions and evaluate the writer's emotions in detail [6].

Acknowledgement. This work was supported by JSPS KAKENHI Grant Number 19K01945, 21H04600 and 21K13385.

References

1. Kobayashi, N., Inui, K., Matsumoto, Y., Tateishi, K., Fukushima, T.: Collecting evaluative expressions for opinion extraction. J. Natural Lang. Process. **12**(3), 203–222 (2005). (in Japanese)
2. Joulin, A., Grave, E., Bojanowski, P., Mikolov, T.: Bag of Tricks for Efficient Text Classification: Facebook AI Research (2016)
3. Mikolov, T., Yih, W., Zweig, G.: "Linguistic Regularities in Continuous Space, Word Representations": Microsoft Research (2013)
4. Mikolov, T., Chen, K., Corrado, G., Dean, J.: Efficient Estimation of Word Representations in Vector Space. Google Inc. (2013)
5. Mikolov, T., Sutskever, I., Chen, K., Corrado, G., Dean, J.: Distributed Representations of Words and Phrases and, their Compositionality. Google Inc. (2013)
6. Averill, J.: On the Paucity of Positive Emotions: Blankstein, K., Pliner, P., Polivy, J.. (eds.) Assessment and Modification of Emotional Behavior, Chapter 2, pp. 7–45 (1980)

Temporal and Geographic Oriented Event Retrieval for Historical Analogy

Kengo Fushimi and Yasunobu Sumikawa[✉]

Department of Computer Science, Takushoku University, 815-1 Tatemachi,
Hachioji-shi, Tokyo, Japan
r88464@st.takushoku-u.ac.jp, ysumikaw@cs.takushoku-u.ac.jp

Abstract. There are many benefits to studying history. Recently, the
study of support for learning history has emphasized the development of
the ability to use knowledge of the past in an analogous manner when
considering solutions to problems that arise in the present day, a process
called historical analogy. Although previous studies have developed the
ability of historical analogy using card games or datasets prepared in
advance by experts, the situation is not ready for anyone to learn about
a subject that is in line with their interests. In this study, we propose a
Twitter chatbot that presents past events recorded on Wikipedia. When
this chatbot receives a reply from a Twitter user, it analyzes the user's
past tweets to collect geographic and temporal information of interest
and returns past events that are close to those. We conducted experiments to confirm the effectiveness of our algorithm, and confirmed that
the accuracy of our algorithm was approximately 30% points higher than
that of the methods used in previous studies.

Keywords: History · Twitter · Wikipedia · Analogy

1 Introduction

The importance of learning history has been widely recognized. In fact, studies
have been conducted on how to enhance historical analogy, which is the ability
to use knowledge regarding historical events as an analogy for thinking about
solutions to contemporary problems [15], to propose a search engine designed
for history learning [11], and to implement a chatbot to support communication
between people and history [19]. These systems that mediate between people and
history are useful when searching for history that users want to know. However,
they do not allow users to search for histories that can be used for historical
analogies or for other similar histories that can be used as alternatives when the
history required by the users is not available.

In this study, we propose a Twitter bot that outputs a historical event to
enhance the historical analogy. The key idea of this bot is that the outputs are
generated with calculations to satisfy the condition that the analogy works well.
According to [10], there are two important conditions for effectively utilizing

G. Meiselwitz (Ed.): HCII 2022, LNCS 13315, pp. 123–133, 2022.
https://doi.org/10.1007/978-3-031-05061-9_9

analogies. The first is the incompleteness. This indicates that, when generating a plausible inference from a source to a target, they cannot be exactly the same and cannot be completely different. The second is the explicit thinking ability pertaining to higher-order relations that exist between the source and the target.

To satisfy the above two conditions, the bot uses temporal and location information to create common aspects and higher-order relations between user input texts and historical events. The bot first takes the user text as a reply to the bot on Twitter. It then retrieves similar historical events that are close in time and location to the input text of the user. This bot infers the location from the input text and the user's past tweets if there is no explicit information in the text. It also obtains time information from the input timestamp. The bot calculates the ranking score for each historical event using these three pieces of information to measure the similarity between the input text and historical events. It then outputs the history with the highest score.

The remainder of this paper is organized as follows. Section 2 describes the related work. Section 3 details how this study collected data to output historical events using our bot and estimated the location. In Sect. 4, we describe the proposed algorithm. The results of the experimental evaluation are presented in Sect. 5 presents. The final section concludes the study and describes future work.

2   Related Works

Activities to create opportunities for the public to easily learn about history had already been carried out by Italian historian Guiccardini [14] in the 15th century. While historians' activities to encourage dialogue between non-researchers and history have been performed for a long time, research to determine the expertise of history that would influence the relationship between historians and the public began at the end of the 19th century [5]. In the early twentieth century, Rebecca Conard identified the use of history in activities other than teaching and proposed the value of making history relevant to the present [6]. Furthermore, history workshops have been held to help workshop organizers promote new practices in local and oral history [17].

The purpose of this research was to encourage dialogue between history and Twitter users through chatbots. A chatbot is a program that can communicate with people using natural language processing techniques. It is defined by [4] as "an artificial construct designed to converse with human beings using natural language as input and output." Classic chatbots were designed to entertain people; however, in recent years, attempts have been made to improve the service quality [1,7,9]. Chatbots are also beginning to be used in education. There have been studies using chatbots for foreign language learning [12,13]; however, in recent years, chatbots have been used not only for the study of other subjects (e.g., computer science [3]), but also for educational activities in general [2,16].

A previous study used Twitter chatbot to encourage interaction with history [19]. This chatbot was designed to disseminate past events recorded on Wikipedia. The purpose of this research is to make people more familiar with

Fig. 1. Examples of descriptions of events in Wikipedia.

the history of Twitter by automatically tweeting about past events that occurred in the same month and day or by spreading past events tweeted by other Twitter users by retweeting them. When another user replied to the chatbot in this previous study, it presented past events that included words in the sentence. At this point, if there are no appropriate past events in the DB, the chatbot simply replies "No relevant history." While the above chatbot was aimed at disseminating history, the chatbot in this study presented history with the aim of facilitating historical analogies. This difference in purpose makes it possible to present other histories that may be of interest to the user, even if there is no appropriate history in response to the user's reply.

3   Data Collection

In this section, we describe the data collected to implement our chatbot in detail.

3.1   History Event Data

Twitter allows users to post sentences described in 280 characters at a time; thus, we store the events described in short sentences in the DB. In addition, as Twitter is used worldwide, it is desirable to store the history of many countries and regions in a DB.

To meet the above criteria, in this study, we used Wikipedia. Wikipedia records past events in the year pages[1] from year 1 to the current year and the day pages[2] from January 1 to December 31. Figure 1 shows an example of Wikipedia's events described in short texts. These events are listed in the "Events" section. We collected these data by using BeautifulSoup[3]. We collected past events and stored not only their description but also the date and related location information in our DB. The collected data covered events from all years spanning from 1 AD to 2019 AD. In total, the dataset contained the descriptions of 71,374 events. In Wikipedia, when an article describes each word in the text in detail, the word has a link to that article. To determine the location of each event, we retrieved the infobox of the linked article and name of the country or region in the Wikipedia category.

---

[1] E.g., https://en.wikipedia.org/wiki/1945.

[2] E.g., https://en.wikipedia.org/wiki/March_21.

[3] https://www.crummy.com/software/BeautifulSoup/.

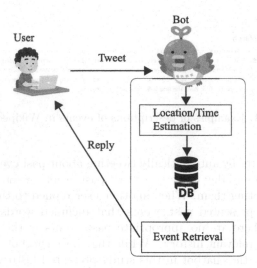

Fig. 2. System overview

3.2 Location Data

We used the latitude and longitude to determine geographic proximity. The collected texts of past events often use the names of the countries. To compare the geographic proximity between countries, we used the latitude and longitude of their capital cities. We collected the latitude and longitude of the capitals of each country from the Web[4].

4 Proposed Algorithm

In this section, we describe our algorithm for presenting history in order to enhance historical analogies. Figure 2 presents an overview of the proposed algorithm. Our chatbot first takes reply texts from users. To collect the reply texts, we used the standard search API provided by Twitter as an official API[5]. It then extracts temporal and geographic words from the text to estimate the geographic and temporal interests of the user from the user's tweets. If these two pieces of information exist in the text, the bot uses them to retrieve historical events; otherwise, it infers information that is not in the text from the metadata and contextual information of the tweet. After retrieving the history close to this estimated geographic and time of interest from the DB, our chatbot replies to the user with the event that it thinks is the most appropriate.

In the remainder of this section, we describe algorithms for estimating and retrieving the geographic and temporal aspects of the events of interest.

---

[4] https://amano-tec.com/data/world.html.
[5] https://developer.twitter.com/en/docs/twitter-api/v1/tweets/search/api-reference/get-search-tweets.

Fig. 3. Example of estimation of geographic interests

Fig. 4. Estimating geographic interests from past tweets

4.1   Estimation of Geographic and Temporal Interests

To present an appropriate past event in response to a reply from a user, the chatbot estimates the geographic and temporal interests of the user.

Geographic Interest Estimation. We perform the geographic estimation in the following two steps. 1) The analysis of the reply text if it clearly includes the geographic words the user wants to know about. 2) The analysis of the user's timeline to infer the location of a particular interest from past tweets if it is not specified.

Figures 3 and 4 show how our chatbot performs the above two analyses. Figure 3 shows an example in which the Twitter user's replies explicitly contain the name of the country (Japan). In contrast, as shown at the top of Fig. 4, if there is no country name in the reply text, the chatbot collects past tweets of that user and analyzes the countries the user is interested in. In this example, as past tweets contain the words "Japan" and "Asia," this chatbot assumes that this user is interested in Japan.

To reveal the specified geographic information from the reply texts, we perform TAGME [8], a tool for named-entity detection, to extract country names. TAGME is a word sense disambiguation tools that links the extracted entities to Wikipedia articles. We extract the geographic information described in the infoboxes and Wikipedia categories of Wikipedia articles that are the result of TAGME. If the names of countries and capitals collected in Sec. 3.2 exist, they will be used as estimated results.

If there is no information regarding the location in the input text, we infer the location by applying TAGME to tweet texts posted by the user. If there are several location names in the TAGME results, we use the most frequent names.

Temporal Interest Estimation. Temporal interest estimation analyzes which year the Twitter user has been interested in from the reply text as well as the geography. If there is no explicit year information in the reply text, we use the year with the smallest difference between the date of the event and that of the reply. This is because many Twitter users tend to be more interested in events close to the present day [18].

4.2　Event Retrieval

By applying the above estimation algorithms, we obtained geographic and temporal information in which Twitter users who replied may be interested. We use these pieces of information to retrieve the history from our DB. To determine the top 1 from the retrieval results, we also use importance score among the results. In the remainder of this section, we describe how to determine the importance of a history and the algorithm to determine the top 1 output by our chatbot.

Event Importance. We determine the importance of an event by the number of references for it on Wikipedia as well as the chatbot spreading history [19]. We collected historical text from Wikipedia as described in Sect. 3.1. Wikipedia text may contain links to Wikipedia articles for the words in that text. We regard the sum of the number of references to these links as the importance of history. The formal definition is as follows:

$$Impr(evt) = \frac{\sum_{ett \in Entity(evt)} Link(ett)}{Max(\{\sum_{ett \in Entity(evt')} Link(ett)) \mid evt' \in E\})}$$

where evt denotes an event, ett denotes an entity, $Entity(evt)$ denotes the set of entities used in the description of argument evt, $Link(ett)$ denotes the function to calculate the number of references of ett given by the argument, and Max is a function that returns the maximum value from its argument.

Score Integration. We integrate all scores obtained by applying the above algorithms to each history to output a history by our chatbot. To perform this integration, we calculate the ranking score for each history by using the following formula:

$$score(evt, t, Loc) = \alpha \, Temp(evt, t) + \beta Spatial(geo(evt), Loc) + \gamma imprt(evt) \quad (1)$$

where t is the result of temporal interest estimation, Loc is the result of geographic interest estimation, $Temp$ is a function that calculates the difference between the time when this algorithm is applied and the year when the event evt occurred, $Spatial$ is a function that calculates the distance between the regions where the event evt occurred and the user's interest, geo provides the geographic information of the given event, and $imprt$ is a function that calculates the importance of the event. α, β, and γ are the hyper-parameters. We set their values such that their sum is 1.

Algorithm 1 shows the replying algorithm for the chatbot. Lines 1~8 and 9~10 estimate geographic and temporal information, respectively. The first step

---

Algorithm 1. Replying algorithm

    Input: A replying tweet *reply*
    Output: A past event
 1: **Function** *Replying(reply)*
 2:   // Geographic and temporal information collection
 3:   $Loc = ExtractGeoInfo(reply.text)$
 4:   //If the reply does not contain any country names
 5:   **if** $(Loc = \emptyset)$
 6:     // Geographic information estimation from past tweets
 7:     $Loc = GeoInference(reply.userID)$
 8:   **end if**
 9:   // Temporal interest estimation
10:   $t = TempInference(reply)$
11:   // Loading history data
12:   $HD = HistoryData(Loc)$
13:   $vals = []$
14:   **for** $e \in HD$
15:     $val = score(e)$// Applying Eq. 1
16:     $vals$.append(val)
17:   idx=argmax($vals$)
18:   **return** $HD[idx]$

---

in estimating geographic information is to extract only named entities from the reply text, and determine if they are names of countries or regions (3rd line). If we miss obtaining these names (5th line), we use Twitter's official API to retrieve the past tweets of the Twitter user who replied. Next, we estimate the temporal interests of the user from the replies using the same analysis as for the estimation of geographic information (lines 9~10). After completing the above estimation analyses, we load the relevant historical data from the DB using geographic information. We then apply the Eq. 1 for each loaded history; we calculate its ranking score using temporal, geographic, and importance scores. Finally, we return the history with the highest ranking score so that our chatbot replies with a single history to the user.

5 Experimental Evaluations

5.1 Experimental Setting

Dataset. Because our algorithm uses geographic information, the dataset must include tweets from several regions of the world for evaluation. In addition, because we will also use chronological information, we require a dataset that includes tweets posted at various times for the evaluation. However, there was no ground-truth dataset that met these conditions; thus, the authors created one. To collect tweet data, the tweets to be collected were made in English so that tweets from many countries could be collected. In addition, to collect tweets from a wide variety of periods, we limited the collection of tweets to official accounts of the following news organizations.

1. **TheHinduScience**: This is an Indian daily newspaper published in English by the Hindu Group. The headquarters are located in Chennai, Tamil Nadu, India.[6]
2. **BBCSport**: BBC Sport is the sports programming division of the British Broadcasting Corporation. The main office is located at Media City, UK, in Salford.[7]
3. **CBCPolitics**: CBC is the public broadcaster of Canada. It operates television and radio services collectively, forming a nationwide public broadcasting network.[8]
4. **CNNBusiness**: CNN is a U.S. cable and satellite news channel owned by CNN Worldwide, a division of WarnerMedia News Channel & Sports. CNNBusiness is a financial news and information website operated by CNN.[9]

We retrieved 25 tweets from each account, and collected 100 tweets. We checked whether the events output by this bot were the latest news articles in regions reported by each account or if they were the latest events that occurred close to the location of the account's company. Two people manually checked all the outputs of the bot to ensure that they were correct. One of the workers is a Ph.D. researcher specializing in machine learning.

Baselines. We compared our algorithm with the following three algorithms.

– HistoChatbot: This calculates the similarity of text and returns one history [19]
– Temp: This uses only temporal interest information proposed in this study
– Imprt: This uses only the event importance proposed in this study

Evaluation Criteria. We assessed whether the sentences resulting from the application of the baselines and the proposed algorithm were past events that were close to the region of each account. In addition, because our algorithm is triggered by replies from Twitter users, we confirmed that the past events output by our algorithm occurred even before each tweet in the dataset was posted.

5.2   Results

Table 1 shows all the results of the evaluation. At the beginning, the sixth column (All) that represents the accuracy for all results shows that the proposed method performed the best.

---

[6] https://mobile.twitter.com/TheHinduScience.
[7] https://mobile.twitter.com/BBCSport.
[8] https://mobile.twitter.com/CBCPolitics.
[9] https://mobile.twitter.com/CNNBusiness.

Table 1. Accuracies

	TheHinduScience	BBCSport	CBCPolitics	CNNBusiness	All
HistoChatbot	56%	52%	50%	60%	54%
Temp	0%	0%	20%	24%	11%
Imprt	16%	64%	72%	56%	52%
Proposed	**80%**	**92%**	**88%**	**76%**	**84%**

Next, we check the results for each Twitter account used in the evaluation. The proposed algorithm achieved the best accuracy for all four accounts. In particular, for the results of TheHinduScience and BBCSport, the accuracy was about 25–30% higher than that of the baselines.

Because our algorithm does not use text similarity, next, we analyzed the results of HistoChatbot, which uses text similarity. For this analysis, we used a tweet about the Zika virus tweeted by TheHinduScience[10]. When the text of the tweet was replied to HistoChatbot, we got "1/28/2016. January 28 - The World Health Organization announces an outbreak of the Zika virus." as the output result. In contrast, when the same text was input into our algorithm, we got "November 27 - Indian Prime Minister Jawaharlal Nehru appeals to the United States and the Soviet Union to end nuclear testing and to start nuclear disarmament, stating that such an action would "save humanity from the ultimate disaster." as a result. The output result of HistoChatbot was the same as the input text; the output event was related to the Zika virus. However, the output results of our algorithm returned an event in which the Prime Minister of India claimed to be working to protect the health of humanity. Thus, we can see that HistoChatbot gave better output results when we wanted to know the past events related to a specific topic. However, our algorithm gave better results when we wanted to know the events in a specific region.

6   Conclusions

In this study, we have proposed an algorithm for retrieving past events that are close to the user's geography and time of interest. This algorithm is applied to users who reply to the Twitter chatbot created in this study. This algorithm first analyzes the location and time the user is interested in from the text of replies and the user's past tweets. Using the geographic and temporal analysis results, the algorithm calculates a ranking score for each past event collected from Wikipedia according to the importance of the event in addition to its geographical and temporal proximity. Finally, the algorithm outputs the event with the highest ranking score.

---

[10] https://mobile.twitter.com/TheHinduScience/status/1459566755718877190.

Future Work. Future work will *identify how our bot is effective for inquiry-based history-learning.* By using this bot, it is possible to determine history that the user did not expect. By analyzing how this affects inquiry-based learning of history, it becomes possible to examine the effectiveness of bots as a new history learning environment.

Acknowledgments. This work was supported in part by MEXT Grant-in-Aids (#19K20631).

References

1. Agus Santoso, H., et al.: Dinus intelligent assistance (dina) chatbot for university admission services. In: 2018 International Seminar on Application for Technology of Information and Communication, pp. 417–423 (2018)
2. Auccahuasi, W., Santiago, G.B., Núñez, E.O., Sernaque, F.: Interactive online tool as an instrument for learning mathematics through programming techniques, aimed at high school students. In: ICIT 2018, pp. 70–76. Association for Computing Machinery, New York (2018)
3. Benotti, L., Martnez, M.C., Schapachnik, F.: A tool for introducing computer science with automatic formative assessment. IEEE Trans. Learn. Technol. **11**(2), 179–192 (2018)
4. Brennan, K.: The managed teacher: Emotional labour, education, and technology. Educ. Insights **10**(2), 55–65 (2006)
5. Cauvin, T.: The rise of public history: An international perspective. Historia Crítica No.40 68, 3–26, April 2018
6. Conard, R.: The pragmatic roots of public history education in the united states. Public Hist. **37**(1), 105–120 (2015)
7. Dibitonto, M., Leszczynska, K., Tazzi, F., Medaglia, C.M.: Chatbot in a campus environment: design of LiSA, a virtual assistant to help students in their university life. In: Kurosu, M. (ed.) HCI 2018. LNCS, vol. 10903, pp. 103–116. Springer, Cham (2018). https://doi.org/10.1007/978-3-319-91250-9_9
8. Ferragina, P., Scaiella, U.: Tagme: On-the-fly annotation of short text fragments (by wikipedia entities). In: Proceedings of the 19th ACM International Conference on Information and Knowledge Management, CIKM 2010, pp. 1625–1628. Association for Computing Machinery, New York (2010)
9. Griol, D., Molina, J.M., Callejas, Z.: Incorporating android conversational agents in m-learning apps. Expert. Syst. **34**(4), e12156 (2017)
10. Holyoak, K.J., Thagard, P.: Mental Leaps: Analogy in Creative Thought. MIT Press (1980)
11. Ikejiri, R., Sumikawa, Y.: Developing world history lessons to foster authentic social participation by searching for historical causation in relation to current issues dominating the news. J. Educ. Res. Soc. Stud. **84**, 37–48 (2016). (in Japanese)
12. Jia, J.: Csiec: a computer assisted English learning chatbot based on textual knowledge and reasoning. Knowl.-Based Syst. **22**(4), 249–255 (2009)
13. Katchapakirin, K., Anutariya, C.: An architectural design of scratchthai: a conversational agent for computational thinking development using scratch. In: IAIT 2018. Association for Computing Machinery, New York (2018)
14. Knevel, P.: Public history: the European reception of an American idea? Levend Erfgoed **6**, 4–8 (2009)

15. Lee, P.: Historical literacy: theory and research. Int. J. Historical Learn. Teach. Res. **5**(1), 25–40 (2005)
16. Pérez, J.Q., Daradoumis, T., Puig, J.M.M.: Rediscovering the use of chatbots in education: A systematic literature review. Computer Applications in Engineering Education (2020)
17. Samuel, R.: 'editorial introduction', in samuel, ed., history workshop: A collectanea 1967–1991 (oxford: Hw 25, 1991), iii; colin ward, 'fringe benefits', new society (November 1989)
18. Sumikawa, Y., Jatowt, A.: Analyzing history-related posts in twitter. Int. J. Digit. Libr. **22**(1), 105–134 (2020). https://doi.org/10.1007/s00799-020-00296-2
19. Sumikawa, Y., Jatowt, A.: Designing chatbot systems for disseminating history-focused content in online social networks. IEICE Trans. Inf. Syst. (Japanese Edition) J104-D(5), 486–497 (2021). (in Japanese)

Eye Tracking to Evaluate the User eXperience (UX): Literature Review

Matías García(⊠) and Sandra Cano

School of Computer Engineering, Pontificia Universidad Católica de Valparaíso, Valparaíso, Chile

Abstract. User eXperience (UX) shapes the way how users interact with products, systems, and services therefore it is necessary to be able to accurately evaluate how this interaction behaves. We propose a review of some of the most relevant articles and how their discoveries impact this line of research including a list of metrics that jointly translate the raw variables captured by an eye tracker to results related to aspects of the UX model of a product. It is also discussed how state of the art Passive Eye tracking Technologies (PET) offer a cheap and simple way of implementing these experiments within the budget limit constraints, reviewing their pros and cons for future investigations that wish to apply this technology for evaluating UX and even carry out massive Eye Tracking studies with experiments done remotely with the help of cloud technology.

Keywords: Eye tracking · User eXperience (UX) · Human-computer interaction

1 Introduction

The growth of technology has made it possible for users to access information immediately. Thus, accessing or having information about a product, system or service involves technology. However, product designers need to understand how users visually perceive the attributes of a product [1]. Therefore, User eXperience (UX) is related to how people feel when interacting with a product or service, as well as their perception when using it [2].

People perceive information from the environment through different stimuli captured by sensory systems. The visual stimulus is one of the sensory channels that most research has conducted to understand how people process the information received from many contexts, such as: multimedia learning [3], advertising, psychology-neuroscience [4], human-computer interaction [5], diagnostic interpretation, neuro-marketing, among others.

When the user is interested in a product, the user fixes his visual attention with greater intensity [6]. Studies have shown that pupil dilation increases when there is a greater intensity of the visual stimulus [7]. Eye tracking devices are used as tools to capture the visual behavior of a person when interacting with a system/product or service.

© The Author(s), under exclusive license to Springer Nature Switzerland AG 2022
G. Meiselwitz (Ed.): HCII 2022, LNCS 13315, pp. 134–145, 2022.
https://doi.org/10.1007/978-3-031-05061-9_10

Eye tracking is a technique used to follow the movement of a person's eyes, to know where they are looking and for how long [8]. Therefore, eye fixation location is related with attention and eye fixation duration is related with processing difficulty and amount of attention. Eye fixation duration can change according to the type of information (texts, audio, or graphics) and type of tasks (problem solving or reading). Therefore, human gaze can contain complex information about a person's interests, hobbies, and intentions.

Several studies have used the Eye tracking technique for different purposes. A study made by Toreini, Langner and Maedche [9] used eye tracking for visual attention feedback (VAF), where they examined types of VAF using a Tobii 4-C eye tracker. Meanwhile Socas et al. [10] used different techniques on the evaluation of User eXperience such as direct observation, questionnaires, and eye tracking, where with the device recorded the movement of eyes and analyzed the user's attention. Another study was conducted by Katharina et al. [11] which used eye tracking with patients with advanced amyotrophic, which was used to evaluate the usability of eye tracking computer devices. Therefore, the Eye tracking technique is used to study visual behavior. A study made by Hwang and Lee [12] to evaluate the purchase intention in mobile environments where a device Tobii X2–30. However, a study made by Bott et al. [13], which is used in a web camera and algorithms of recognition of eye movements: Therefore, the study investigates a novel methodology for administering a visual paired comparison (VPC) decisional task using a web camera. Different Eye Tracking techniques are used to detect the eyes movement, which studies have found have used smart devices or web cameras.

Eye Tracking devices combine a camera and infrared light (IR) sources to estimate the position using the IR light to position the eyes in relation to the camera. However, recently the gaze interaction using cameras has been used to estimate the eye gaze [14–16]. Therefore, these methods use algorithms of computer vision to detect the eyes, which is known as a video-based gaze tracking system. A feature-based method extracts features such as the iris/pupil center, eye corners and iris/pupil contours. The method is proposed by Xiao et al. [16] which consists of three parts for the iris centers, anchor point and head pose calculations. The head pose is estimated based on the six facial landmarks of eye corners, nose tip, mouth corners and chin using the OpenCV iterative algorithm [17]. The algorithm uses intensity differences between pixels to estimate the positions of 68 facial landmarks. Therefore, one of the primary stages in eye tracking based on video is face recognition. Viola and Jones' face detector uses Haar features corresponding to the eyes for face detection [18]. In addition, the distance between the eyes is often utilized for face normalization, for the localization of other facial landmarks as well as in filtering out structural noise. The method video-based requires eye detection and tracking, which includes three aspects such as: detect the existence of eyes, interpret eye positions in the images and finally for video images, the detected eyes are tracked from frame to frame. The eye position is commonly measured using the pupil or iris center.

This study reviews the literature on the implemented eye tracking algorithms, libraries using computer vision systems, which can be trained to track movements of the eye. This video-based method uses a webcam, thus being low-cost because it does not require an eye tracking device.

2 Background

The eyes are a vital part of human physiology, studying characteristics of the eye such as speed, regularity of blinks, lengths of fixations, among others. These characteristics are used to evaluate the behavior of a person when responding to any kind of visual stimulus [19]. Eye tracking technology consists of recording and detecting eye movements. There are technologies that require specialized commercial hardware and software that are expensive. However, there is a growing need for cheap and ubiquitous methods to obtain information about human gaze. Therefore, video-based methods are using web cameras currently to detect the eyes and gaze tracking. Some challenges are that web cameras are subject to background light levels and the possibilities to zoom are limited. Therefore, the quality of the image depends on light conditions and the distance between the camera and the participant. The algorithms used in web cameras are different from classical eye tracking algorithms, which use infrared cameras and an additional infrared light source to track the eyes. In literature the termination associated with eye tracking is gaze estimation, which is applied to measure eye positions and movements.

A study conducted by Zeng et al. [20], present a set of stages of the gaze estimation module, which consists of (1) face detection, (2) iris position and pupil center detection, (3) estimating the ratio of each pupil center, (4) one-point calibration, and (5) the five gaze directions estimation. In steps 1 and 2, they use the 68-point face detection method implemented by the open-source Python Dlib library [21]. The authors used the GaVe interface to explore usability, where they found that this interface is effective and easy to use for most participants.

Several types of eye movements are studied in eye gaze research to collect information about user behavior. Some characteristics of eye movements are fixations, saccades, scanpath, gaze duration and pupil size [22]. Fixations related variables as fixation duration, mean fixation duration, fixation spatial density, number of areas fixated, fixation sequences and fixation rate. Saccades are involuntary eye movements that occur between fixations, some variables related are saccade number, amplitude, and fixation-saccade ratio. Scanpath includes a map of series of short fixations and saccades alternating before the eyes reach a target location on the screen, some variables in Scanpath are direction, duration, length, and area covered. Gaze duration refers to the sum of all fixations made in an area of interest before the eyes leave that area and the proportion of time spent in each area. Pupil size is used to evaluate the cognitive workload.

3 Research Method

This study presents a review of the literature, which was applied based on a framework and literature review in software engineering proposed by [23, 24], which sets out stages to be followed such as: (1) identify the research question; (2) Identify relevant studies; (3) study selection and (4) charting the data.

3.1 Identify the Research Questions

The first step was to define the research questions approached in this literature review (Table 1).

Table 1. Research questions

ID	Question
RQ1	What research topics are being addressed with Eye tracking?
RQ2	What is PET (Passive Eye-tracking Technologies)?
RQ3	What metrics are used in PET to evaluate UX?

3.2 Identify Relevant Studies

The keywords used were "Eye tracking", "eye movement", "user experience", "thinking aloud" and "product design evaluation". The search was done on the Web of Science (WoS) and Scopus databases, and its results were filtered by the inclusion criteria shown in Table 2.

Table 2. Inclusion criteria

Article	Criteria
Subject	The article must be related with the Eye Tracking technique, its algorithm, or its applications
Publisher	It must be published in a scientific journal or be in a conference
Experimentation and results	It must prove or disprove its hypothesis with the results obtained from an applied experiment. These results must be detailed and concrete
Language	The article must be written in English
Publication year	It must be published from 2010 to 2021

The results obtained (see Fig. 1) from this search performed on both Scopus (blue colour) and WoS (green colour) databases showcase an evident peak of articles in 2020. However, there has been a decrease in the number of published articles even though the availability of Eye Tracking technology is greater than before, such as the high-quality Eye Tracking devices offered by Tobii. This might be due to the lack of interest in buying these expensive eye trackers, also considering how hard it is to apply an Eye Tracking experiment during the COVID-19 pandemic that made it difficult for people to physically meet and comply with the health regulations established.

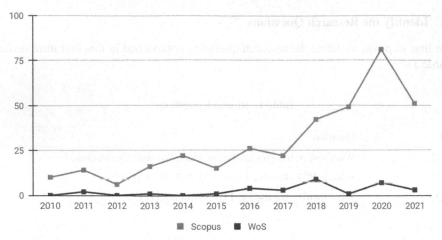

Fig. 1. Search results by year. The blue line represents the number of articles recovered from the Scopus database while the green line represents the articles recovered from Web of Science database. (Color figure online)

Nevertheless, 2021 also saw an unexpected boom in the use of virtual conferencing platforms, allowing for people to interact from their homes. In parallel, this represents an opportunity for PET, since these Eye Tracking technologies rely on using video-oculography (VOG), basically a video-based method, to capture eye movements. In the case of PET, VOG is applied making use of a webcam integrated to most of the commercially available notebooks. In the worst-case scenario, it might be necessary to buy a webcam, but it remains cheaper than acquiring an expensive eye tracker.

3.3   Study Selection

The articles gathered were then filtered and we extracted their most relevant discoveries to sort them in the table below, offering a brief run through on the current state of Eye Tracking research shown in Table 3.

Table 3.   Eye tracking line of research

Article	Progress
Djamasbi et al. [25]	Eye Tracking reveals differences in user behavior in two generational groups of participants when interacting with a web interface, affecting their perception of its design, its contents and how they accomplish tasks. These differences support the importance of aesthetics in usability research
Guo et al. [26]	Products with higher UX have a shorter time to first fixation, having less time for people to realize they are present during a visual stimulus. Products with higher UX also evoke larger pupil size and longer fixation time

(continued)

Table 3. (*continued*)

Article	Progress
Qu et al. [27]	The metrics obtained by Eye Tracking (time to fixation, fixation count and blink count) strongly describe aspects of website usability in a website such as elements that require a higher cognitive effort
Joseph and Murugesh [28]	From this article we extract the metrics present in state-of-the-art PET that describe the human-computer interaction
Xu and Zhang [29]	They propose a combination of UX evaluation methods (Eye Tracking and questionnaires), effectively implementing both attitudinal and behavioral evaluation to quantitatively describe and quantitatively measure aspects of interface design
Kuo et al. [30]	The visual analysis of a product's components impacts how the user perceives its attributes, in this case, attributes used in the study were selected based on two models: Kano's model and Norman's model
Zammarchi et al. [31]	Eye Tracking metrics can be used to highlight elements or tasks in a web interface that require re-design to improve its usability
Joseph et al. [32]	Eye Tracking metrics (fixation duration, maximum fixation duration, standard deviation of pupil dilation) allow estimating the cognitive load that a user requires to interact with a mobile app interface. In the case of people aged between 50 and 60 + years old, it revealed higher cognitive load involved in copy-paste tasks compared to younger participants
Guo et al. [33]	The level of complexity in an interface affects users' eye behavior and its satisfaction generated when completing a task on it. Understanding visual complexity allows for designers to create better guidelines for developing new products with an appropriate complexity layout and finally, improve user satisfaction

3.4 Answering the Research Questions

RQ1: What Research Topics are Being Addressed with Eye Tracking? The research topics addressed by this literature review on Eye tracking broadly vary, but the common factor between these articles is ultimately their focus to describe aspects of the Human-Computer Interaction through the observation of eye behavior, applying an UX model to evaluate how participants react and feel to visual stimuli, thus the main research topic involved is Computer Science. Eye tracking also reveals valuable metrics that allow us to describe and evaluate product design, increasing the interest of industrial, graphic and web designers to use Eye tracking experiments in for example: package design and interface design. As such, this technique also heavily dives into Engineering. However, primarily our goal is to synthesize the literature that involves or can be related to aspects of the UX model [38] in products or services to help describe not only their usability, but also its findability and how desirable these products might be to the public.

RQ2: What are PET (Passive Eye-Tracking Technologies)? Passive Eye-tracking Technologies (PET) are software tools that use a webcam as an eye tracker to monitor the movement of the eye, becoming popular by being completely remote and non-intrusive. Therefore, this technology consists of software tools that read eye movements without physically mounting devices on the participant and instead use VOG, a camera records the user's face and through a series of pattern recognition techniques, these PET recognize the different types of eye movements and the projection of the gaze to the screen identifying the different points or often described in experiments as Areas of Interest (AOI) that the user is observing on a computer screen during a visual stimuli.

Passive eye monitoring is the science of teaching a computer to automatically determine, with digital video footage, the spatial location, 3D gaze direction and pupil activity of a person's eye in real-time [34]. The steps in passive video-based eye tracking include user calibration, capturing video frames of the face and eye regions of the user, eye detection and mapping with gaze coordinates on screen [22].

PET is a low cost video-based method because it does not require specialized commercial hardware. However, the main problem with using a webcam is that they work only in visible light and they typically have a wide-angle lens with limited possibilities to zoom. Therefore, algorithms used are programmed differently than classical eye tracking algorithms, which utilize infrared cameras and an additional infrared light course that produces a glint to track the eyes.

Some examples of PET are WebGazer [39], PyGaze [40], OpenGazer [41] and GazeRecorder [42], each with its own cons and pros due to their design and eye tracking algorithm.

RQ3: What Metrics are used in PET to Evaluate UX? The value of Eye Tracking metrics lies on the quantitative data captured with this technique based on the observation and recording of eye movements during the real time interaction between user and product. This data has the advantage of being able to be subject of statistical analysis, however the data output of an Eye Tracking experiment is harder to interpret compared to the data output of attitudinal UX evaluation methods such as Card Sorting and Think Aloud. The Eye Tracking data is basically a dataset of spatial coordinates (the distance in relation to an "X" and "Y" axis set across the computer screen as in a cartesian 2D map) associated with their current timestamp. With this information, Eye Tracking technologies formulate the Eye Tracking metrics. These metrics help to understand the user behavior when they are watching any kind of visual stimulus.

This study proposes and synthesizes some of the Eye Tracking metrics that are commonly included in state-of-the-art PET (Passive Eye-tracking Technologies) (see Table 4).

Table 4. Eye tracking metrics in PET

Metrics	Indicator
Fixation duration	A longer fixation duration describes issues related to extracting information, or it indicates that the object is more appealing [35]
Fixations per Area Of Interest (AOI)	More fixations on a specified area signify that it is more perceptible, or more significant to the viewer than other areas [36]
Time to first fixation	Products with higher user experience can evoke a shorter time to first fixation [37]
Dwell time	Time spent in the same position and area. It indicates measuring the time the user remains at a search result after a click [28]

3.5   Charting Data

In this section, we organized and charted all the articles obtained by the search query on the Scopus database. We chose to use Scopus (354 articles in total) over WoS (31 articles) for our analysis due to Scopus having a significantly bigger amount of articles, plus most of the articles were present in both databases (Fig. 2).

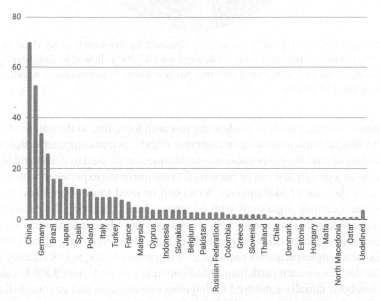

Fig. 2. Eye tracking articles published by each country in Scopus. It highlights how China, Germany and Brazil take the lead on the amount of Eye tracking articles published compared to the rest of the world.

China takes the lead with the most articles published (73 articles) related to Eye tracking research attributed to 160 Chinese authors, including the most relevant papers, while the average in other countries don't surpass 20 authors. This fact might be partially caused by the high prices of commercially available eye trackers that require researchers to allocate a significant amount of their investigation budget to acquire these devices. This limits the amount of research that can be carried out in countries where Eye tracking doesn't have much interest, but also raises the opportunity for PET to be used instead for Eye tracking experiments thanks to its much lower price and with acceptable accuracy, it helps to democratize the Eye tracking technique in the world (Fig. 3).

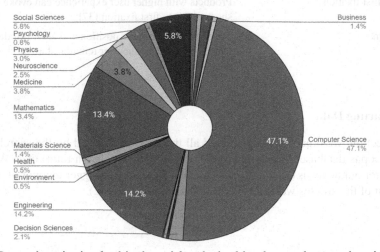

Fig. 3. Research topics involved in the articles obtained by the search query done in Scopus database. The top three topics are Computer Science with 47.1%, followed by Engineering 14.2% and Mathematics with 13.4%. It's worth pointing out how Social Sciences takes a significant 5.8% of the total number of articles.

Computer Science clearly is the leading research topic due to the nature of the Eye tracking technique implementation on computer interfaces and computer systems. Engineering comes second mainly because this technique can be used to detect and highlight the elements in a design that can be improved to optimize the experience with a product or service. In the case of Mathematics, VOG can be used to test regression models in their Machine Learning algorithms with its output data, comparing their accuracy and effectiveness. Social Sciences with 5.8% showcases how Eye tracking can be used to explain and describe the relation between individuals in a demographic group, but carrying out Eye tracking experiments requires a lot of resources due to individually placing eye tracker devices on each participant. Medicine takes an important 3.8% because Eye tracking has been initially employed to diagnose eye diseases and cognitive disorders, mainly using electro-oculography (EOG) with specialized equipment.

4   Conclusions

This research follows the Eye tracking line of research and how it has found how specific metrics of eye behavior can be used to evaluate aspects of UX in products or services. In the case of designers, this capability is an important tool to improve their designs, whether it's a website, an application or a new package for a product. This line of research has had its peak of published articles in 2020, going downwards in 2021 evidenced by the results taken from two academic databases, Scopus and Web of Science. One of the relevant factors comes from the fact that while Eye tracking technology is widely available in the market, its prices are high and this discourages future researchers to spend a significant amount of their budget into acquiring the devices needed to implement an Eye tracking experiment. In this research, we propose PETs that require no more expensive equipment than a webcam (that mostly comes integrated into notebooks), getting acceptable results compared to the accuracy of other commercially available eye trackers.

An important weakness of Eye tracking comes from the complexity of its data to be analyzed, however we summarize a quick guide for these metrics to be translated to aspects of the UX model, such as usability, findability and desirability, including the cognitive load required to carry out complex tasks through an interface. These metrics help understand and describe how the interaction between an user and a product works and how the user reacts when exposed to its design in visual stimuli. The complexity of interpreting Eye tracking data can also be complemented with other UX evaluation methods in the same experiment, while extracting a behavioral data that is less subject to be consciously or unconsciously altered by the participant, this being a strong point in favor of Eye tracking.

References

1. Shi, A., Huo, F., Hou, G.: Effects of design aesthetics on the perceived value of a product. Front Psychol. **12**, 670800 (2021). https://doi.org/10.3389/fpsyg.2021.670800
2. ISO (2008). ISO 9241-210:2008, Ergonomics of human system interaction - Part 210: Human centred design for interactive systems. Geneve: ISO
3. Takacs, Z.K., Bus, A.G.: How pictures in picture storybooks support young children's story comprehension: an eye-tracking experiment. J. Exp. Child Psychol. **174**, 1–12 (2018). https://doi.org/10.1016/j.jecp.2018.04.013, ISSN 0022-0965
4. Gibbons, A.: Multimodality, Cognition, and Experimental Literature, 1st edn. Routledge (2011). https://doi.org/10.4324/9780203803219
5. Lukander, K.: A short review and primer on eye tracking in human computer interaction applications (2016)
6. Roda, C., Thomas, J.: Attention aware systems: Theories, applications, and research agenda. Comput. Hum. Behav. **22**, 557–587 (2006)
7. Eriksen, C.W., Yeh, Y.-Y.: Allocation of attention in the visual field. J. Exp. Psychol. Hum. Percept. Perform. **11**, 583–597 (1985)
8. Just, M.A., Carpenter, P.A.: A theory of reading: from eye fixations to comprehension. Psychological Rev. **87**, 329–354 (1980)
9. Toreini, P., Langner, M., Maedche, A.: Using eye-tracking for visual attention feedback. In: Davis, F.D., Riedl, R., vom Brocke, J., Léger, P.-M., Randolph, A., Fischer, T. (eds.) Information Systems and Neuroscience. LNISO, vol. 32, pp. 261–270. Springer, Cham (2020). Doi: https://doi.org/10.1007/978-3-030-28144-1_29

10. Socas, V., González, C., Caratelli, S.: Emotional Navigation in nonlinear narratives. In: Proceedings of the XV International Conference on Human Computer Interaction - Interacción '14 (2014). https://doi.org/10.1145/2662253.2662271

11. Linse, K., Rüger, W., Joos, M., Schmitz-Peiffer, H., Storch, A., Hermann, A.: Usability of eye-tracking computer systems and impact on psychological wellbeing in patients with advanced amyotrophic lateral sclerosis. Amyotrophic Lateral Sclerosis Frontotemporal Degeneration 19(3-4), 212–219 (2018). https://doi.org/10.1080/21678421.2017.1392576

12. Hwang, Y.M., Lee, K.C.: Using eye tracking to explore consumers' visual behavior according to their shopping motivation in mobile environments. Cyberpsychol. Behav. Soc. Networking 20(7), 442–447 (2017). Doi:https://doi.org/10.1089/cyber.2016.0235

13. Bott Nicholas, T., Alex, L., Dorene, R., Elizabeth, B., Paul, Zola Stuart

14. Web Camera Based Eye Tracking to Assess Visual Memory on a Visual Paired Comparison Task. Frontiers in Neuroscience, vol 11 (2017). https://doi.org/10.3389/fnins.2017.00370

15. Ansari, M.F., Kasprowski, P., Obetkal, M.: Gaze tracking using an unmodified web camera and convolutional neural network. Appl. Sci. 11, 9068 (2021). https://doi.org/10.3390/app11199068

16. Brächter, T., Gerhardt, D.: Camera image based method of real time gaze detection and interaction. Int. J. Sci. Res. Publ. (IJSRP) 10(11) (2020)

17. Xiao, F., Zheng, D., Huang, K., Qiu, Y., Shen, H.: A single camera gaze tracking system under natural light. J. Eye Move. Res. 11(4) (2018). https://doi.org/10.16910/jemr.11.4.5. https://doi.org/10.16910/jemr.11.4.5

18. Kaehler, A., Bradski, G.: Learning OpenCV 3: Computer Vision in C++ with the OpenCV Library. O'Reilly Media, Inc., CA (2016)

19. Viola, P., Jones, M.: Robust real-time face detection. In: Proceedings of International Conference Computer Vision, vol II, p. 747 (2001)

20. Gholami, Y., Taghvaei, S.H., Norouzian-Maleki, S., Sepehr, R.M.: Identifying the stimulus of visual perception based on eye-tracking in urban parks: case study of Mellat Park in Tehran. J. For. Res. 26(2), 91–100 (2021). https://doi.org/10.1080/13416979.2021.1876286

21. Zeng, Z., Liu, S., Cheng, H., Liu, H., Li, Y., Feng, Y.: Feliz Wilhelm Siebert. GaVe: A Webcam- Based Gaze Vending Interface Using One-Point Calibration (2022). https://arxiv.org/abs/2201.05533

22. Lame, A.: Eye tracking library easily implementable to your projects, February 2019. https://github.com/antoinelame/GazeTracking

23. Kar, A., Corcoran, P.: A review and analysis of eye-gaze estimation systems, algorithms and performance evaluation methods in consumer platforms. IEEE Access 5, 16495–16519 (2017). https://doi.org/10.1109/ACCESS.2017.2735633

24. Arksey, H., O'Malley, L.: Scoping studies: towards a methodological framework. Int. J. Soc. Res. Methodol. 8(1), 19–32 (2005). https://doi.org/10.1080/1364557032000119616

25. Kitchenham, B.A.: Procedures for undertaking systematic reviews. Joint Technical report, Computer Science Department, Keele University (TR/SE- 0401) and National ICT Australia Ltd. (0400011T.1) (2004)

26. Djamasbi, S., Siegel, M., Skorinko, J., Tullis, T.: Online viewing and aesthetic preferences of generation Y and the baby boom generation: testing user web site experience through eye tracking. Int. J. Electron. Commer. 15(4), 121–158 (2011). https://doi.org/10.2753/jec1086-4415150404

27. Guo, F., Ding, Y., Liu, W., Liu, C., Zhang, X.: Can eye-tracking data be measured to assess product design? visual attention mechanism should be considered. Int. J. Ind. Ergon. 53, 229–235 (2016). https://doi.org/10.1016/j.ergon.2015.12.001

28. Qu, Q.X., Guo, F., Duffy, V.G.: Effective use of human physiological metrics to evaluate website usability. Aslib J. Inf. Manag. 69(4), 370–388 (2017). https://doi.org/10.1108/ajim-09-2016-0155

29. Joseph, A.W., Murugesh, R.: Potential eye tracking metrics and indicators to measure cognitive load in human-computer interaction research. J. Sci. Res. **64**(01), 168–175 (2020). https://doi.org/10.37398/jsr.2020.640137

30. Xu, J., Zhang, Z.: Research on user experience based on competition websites. J. Phys: Conf. Ser. **1875**(1), 012014 (2021). https://doi.org/10.1088/1742-6596/1875/1/012014

31. Kuo, J.Y., Chen, C.H., Koyama, S., Chang, D.: Investigating the relationship between users' eye movements and perceived product attributes in design concept evaluation. Appl. Ergon. **94**, 103393 (2021). https://doi.org/10.1016/j.apergo.2021.103393

32. Zammarchi, G., Frigau, L., Mola, F.: Markov chain to analyze web usability of a university website using eye tracking data. Stat. Anal. Data Mining ASA Data Sci. J. **14**(4), 331–341 (2021). https://doi.org/10.1002/sam.11512

33. Joseph, A.W., Jeevitha Shree, D.V., Saluja, K.P.S., Mukhopadhyay, A., Murugesh, R., Biswas, P.: Eye tracking to understand impact of aging on mobile phone applications. In: Chakrabarti, A., Poovaiah, R., Bokil, P., Kant, V. (eds.) ICoRD 2021. SIST, vol. 221, pp. 315–326. Springer, Singapore (2021). Doi: https://doi.org/10.1007/978-981-16-0041-8_27

34. Guo, F., Chen, J., Li, M., Lyu, W., Zhang, J.: Effects of visual complexity on user search behavior and satisfaction: an eye-tracking study of mobile news apps. Universal Access in the Information Society. Published (2021). https://doi.org/10.1007/s10209-021-00815-1

35. Hammoud, R.I.: Passive Eye Monitoring: Algorithms, applications and experiments. Springer. https://doi.org/10.1007/978-3-540-75412-1

36. Just, M.A., Carpenter, P.A.: The role of eye-fixation research in cognitive psychology. Behav. Res. Methods Instrum. **8**, 139–143 (1976)

37. Poole, A., Ball, L.J., Phillips, P.: In search of salience: a response time and eye movement analysis of bookmark recognition. In: Fincher, S., Markopolous, P., Moore, D., Ruddle, R. (eds.) People and Computers XVIII-Design for Life: Proceedings of HCI 2004. Springer-Verlag Ltd., London (2004)

38. Byrne, M.D., Anderson, J.R., Douglas, S., Matessa, M.: Eye tracking the visual search of click-down menus. In: Proceedings of CHI 99. pp. 402–409. ACM Press, NY (1999)

39. Morville, P.: User Experience Design. Semantic Studios (2004). http://semanticstudios.com/user_experience_design/

40. Papoutsaki, A., Sangkloy, P., Laskey, J., Daskalova, N., Huang, J., Hays, J.: WebGazer: scalable webcam eye tracking using user interactions. In: Proceedings of the 25th International Joint Conference on Artificial Intelligence (IJCAI-16), pp. 3839–3845 (2016). https://www.ijcai.org/Abstract/16/540

41. Dalmaijer, E.S., Mathôt, S., Van der Stigchel, S.: PyGaze: an open-source, cross-platform toolbox for minimal-effort programming of eyetracking experiments. Behav. Res. Methods **46**(4), 913–921 (2013). https://doi.org/10.3758/s13428-013-0422-2

42. Zieliński, P.: Opengazer: open-source gaze tracker for ordinary webcams. OpenGazer (2009). http://www.inference.org.uk/opengazer/

43. GazeRecorder. (21 2021 septiembre). GazeCloudAPI | Real-Time online Eye-Tracking API. GazeCloud. https://gazerecorder.com/gazecloudapi/. Accessed 12 Nov 2021

Psychological Characteristics Estimation from On-Road Driving Data

Ryusei Kimura[1], Takahiro Tanaka[2], Yuki Yoshihara[2], Kazuhiro Fujikake[3],
Hitoshi Kanamori[2], and Shogo Okada[1(✉)]

[1] Japan Advanced Institute of Science Technology, Nomi, Japan
{s2110059,okada-s}@jaist.ac.jp
[2] Institutes of Innovation for Future Society, Nagoya University, Nagoya, Japan
{tanaka,y-yuki,hitoshi_kanamori}@coi.nagoya-u.ac.jp
[3] School of Psychology, Chukyo University, Nagoya, Japan
fujikake@lets.chukyo-u.ac.jp

Abstract. Automobiles are essential to society, while they cause some problems. In particular, traffic accidents by older drivers are a serious problem. Driving assistance systems are promising solutions to this problem. However, it is not easy to develop acceptable and comfortable driving assistance systems for all drivers. To realize such systems, we need to consider the driver characteristics or personality of the drivers.

In this paper, we predict the psychological characteristics of older drivers from on-road driving data and propose a classification model. We posit that road types are important information for estimation and that important driving behaviors appear not only in whole driving but also in partial driving. Under these hypotheses, our feature extraction method segments time-series data using road types and further segments data into various duration sequences. Experimental results show that some items can be predicted with high accuracy and validate the efficacy of the segmentation. We use a dataset that includes time-series driving data, the Driving Style Questionnaire scores, and the Workload Sensitivity Questionnaire scores of 24 older drivers. Also, this study gives a new perspective to the prediction of individual characteristics.

Keywords: Driver characteristics · Driving style · Driving workload · Driving assistance systems · Social signal processing · Multimodal

1 Introduction

Automobiles are beneficial and indispensable to society, but they cause some problems. In particular, traffic accidents by older drivers are a serious problem. According to the report by the Centers for Disease Control and Prevention, approximately 8,000 older adults were killed in traffic crashes, and more than 250,000 were treated in emergency departments for crash injuries in 2019 [1]. Driving assistance systems are one of the solutions to this problem, but it is not easy to develop acceptable and comfortable driving assistance systems for

G. Meiselwitz (Ed.): HCII 2022, LNCS 13315, pp. 146–159, 2022.
https://doi.org/10.1007/978-3-031-05061-9_11

all drivers. Exiting systems are usually designed based on average driver characteristics [2] even though there are various types of drivers. Therefore, some people feel uncomfortable with feedback on driving and ignore systems. To realize acceptable and individualized driving assistance systems, we need to consider driver's characteristics. Thus, driver's characteristics recognition is an important task. Representative characteristics of drivers are driving style, and a lot of previous research has focused on driving style recognition from driving data. However, recognition of psychological aspects of driving style has not been focused on. Although sensitivity to driving workload is also one of the important characteristics of drivers, similar to the psychological driving style, psychological sensitivity to driving workload has not been focused on, and no study estimates it from driving data. Several studies showed the relationship between driving performance and personality [3,4], and stress [5,6].

Based on these results, we posit that psychological driving style and psychological sensitivity to driving workload can be estimated from driving data. Our goal is to estimate the psychological driving style and the psychological sensitivity to the driving workload of older drivers from on-road driving data. As metrics of them, we use Driving Style Questionnaire (DSQ) [7], and Workload Sensitivity Questionnaire [8], which are measured based on a self-report questionnaire. We use the dataset in [9], which includes time-series driving data and scores of DSQ and WSQ of older drivers.

This paper proposes a model for estimating psychological driving style and sensitivity to driving workload. We incorporate two hypotheses to our model that road types are important information for estimation and that important driving behaviors appear not only in whole driving but also in partial driving. Our model can consider different driving scenes and capture important driving behavior for estimation by segmenting time-series driving data. In addition to the estimation, we analyze road type-specific differences of effective sensors and the duration of important driving behaviors. These analyses give clues to further study on estimation of drivers' psychological characteristics.

In the area of social signal processing, psychological characteristics or personality traits are predicted from a great variety of data sources, for example, multi-party meetings [10], social media [11], speech [12], and game [13]. However, predicting psychological characteristics from driving data has not been shed light. This study gives a new data source and a new method to predict psychological characteristics to social signal processing.

The rest of this paper is organized as follows. In Sect. 2, we describe existing related studies. Then, Sect. 3 details the dataset used in our experiments. We explain the feature extraction method in Sect. 4. The experimental setting is presented in Sect. 5, and the experimental results are presented in Sect. 6. In Sect. 7, we discuss the results of the experiments. Finally, Sect. 8 concludes this paper.

2   Related Works

This study focuses on estimating drivers' psychological characteristics from driving data. No previous study has addressed this estimation, but several studies analyzed the relationship between driving and personality. We hypothesize that psychological characteristics can be estimated from driving data based on the previous study. This research is also related to driving assistance systems.

2.1   Relationship Between Driving and Personality

Some studies revealed that driving performance is related to personality. Adrian et al. [3] investigated the relationship between driving performance and personality traits among older drivers. It was reported that personality (extraversion) is negatively related to driving performance. In [14], aggressive driving, crashes, and moving violations are predicted by the driver's personality. Guo et al. [15] found that driver's personality traits affect accident involvement and risky driving behavior. Also, an association between driving stress and personality was indicated in [5].

2.2   Driver Assistance Systems

To date, various types of driver assistance systems have been proposed, and they are grouped into two broad categories: those that promote safe driving and those that improve fuel efficiency. There are various ways in which the driver assistance systems provide feedback. Stoichkov et al. [16] proposed a visual feedback system that lowers fuel consumption and the risk of traffic accidents. Fazeen et al. [17] proposed audio feedback systems using a smartphone for safety awareness. These feedback systems are easy to implement and cause less discomfort than other approaches but may be ignored by a driver and cannot always force them to acknowledge the feedback. Xu et al. [18] developed a system that stiffens the accelerator pedal according to the discrepancy between the actual vehicle speed and the desired vehicle speed. Although such systems feature enforcement against the drivers, some drivers may feel uncomfortable and disable them. Syed et al. [19] analyzed drivers' acceptance of the feedback systems. The automatic estimation of the driver's personality traits enables such assistance systems to support the individual driver adaptively.

3   Dataset

In this study, we use the dataset provided by the Institute of Innovation for Future Society of Nagoya University [9]. The dataset is collected from driving tests, and 24 older drivers participated. The participants drove a car equipped with various sensors including a GPS sensor in the driving tests. They drove two times each, but we used 38 driving data (23 drivers) because the seven driving data have large missing parts. Hence, there are drivers whose two driving data are used and one driving data is used. The driving tests were conducted on public roads around Nagoya city. In the tests, all participants departed from Nagoya University first, then drove on the arterial road, and then circumnavigated the residential area, finally, returning to Nagoya University. The total mileage and the total driving time are different for each participant. The driving duration ranged from 2245 s to 4762 s, with an average of 2885 s. The Mileage ranged from 10079 m to 14810 m, with an average of 12109 m. Figure 1 shows a car used in the tests. In addition to the driving tests, the drivers answered questions about driving style and workload.

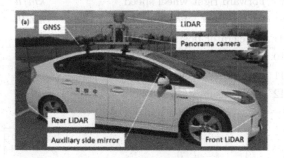

Fig. 1. The car used in driving test.

3.1   In-vehicle Sensor Data

The Participants drive public roads with cars which equip with 38 in-vehicle sensors. The driving data is obtained through Controller Area Network (CAN). We use 12 in-vehicle sensors which are detailed in Table 1 plus GPS sensor. The GPS sensor is only used for preprocessing of driving data. The cars did not equip with a jerk sensor. Hence, jerk values are calculated by the first-order difference of estimated acceleration. The example of the time-series data of steering angle is shown in Fig. 2.

3.2   Driving Style Questionnaire and Workload Sensitivity Questionnaire

Driving Style Questionnaire (DSQ) and Workload Sensitivity Questionnaire (WSQ), which are based on a self-report questionnaire are introduced by [7] [8] for characterizing drivers from a psychological aspect. In [8], the relevance between car following behavior and DSQ are validated. Table 2 and Table 3 detail the items of DSQ and WSQ. DSQ has eight items on a scale from 1 to 4, and WSQ has 10 items on a scale from 1 to 5. We classify drivers with high and low scores of DSQ and WSQ items.

Table 1. In-vehicle sensors.

	Sensor	Unit
1	Steering angle	deg
2	Electronic power steering (EPS) torque	Nm
3	Forward acceleration	m/s^2
4	Lateral acceleration	m/s^2
5	Yaw rate	deg/sec
6	Speed	km/h
7	Forward right wheel speed	km/h
8	Forward left wheel speed	km/h
9	Accelerator position	$\%$
10	Brake pressure	MPa
11	Estimated acceleration	m/s^2
12	Fuel consumption	ml

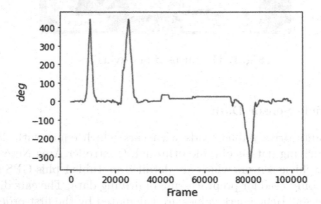

Fig. 2. The example of the time-series data of steering angle.

Table 2. Driving Style Questionnaire (DSQ).

	Item
1	Confidence in driving skill
2	Hesitation for driving
3	Impatience in driving
4	Methodical driving
5	Preparatory maneuvers at traffic signals
6	Importance of automobile for self-expression
7	Moodiness in driving
8	Anxiety about traffic accidents

Table 3. Workload Sensitivity Questionnaire (WSQ).

	Item
1	Understanding of traffic conditions
2	Understanding the road conditions
3	Interference with concentration
4	Decline in physical activity
5	Disturbance on the pace of driving
6	Physical pain
7	Path understanding and search
8	In-vehicle environment
9	Control operation
10	Driving posture

4 Feature Extraction

To classify drivers, we extract features from time-series sensor data listed in
Table 1 plus a jerk. Our feature extraction method is based on two hypotheses.
The first hypothesis is that driving behaviors that are related to drivers' psychological characteristics and workload are different among road types. Since
driving behavior is strongly related to road type, several studies on driving style
recognition from driving data used road type information [20,21]. Besides, mental workload in driving depends on road context [22] and visibility [23]. Based
on this hypothesis, we segment time-series driving data into two road types,
namely, arterial roads and intersections. Then, features are extracted from each
road type. The driving data is segmented by the car's position obtained through
the GPS sensor. We use four intersections where all drivers' data were recorded
accurately. We treat these four intersections as the same road types, but features
are extracted separately because the visibility and ease of driving are not equal.
Figure 3 shows the driving route in the tests.

The second hypothesis is that differences in drivers' psychological characteristics appear not only in whole driving but also in partial driving. We further
segment time-series driving data into many sequences with various durations
because we have no a priori knowledge about when or where the important
driving behaviors for predicting psychological characteristics appear. The time-
series data of arterial roads are segmented so that the duration of each segment
is equal and the average duration for each segment is [All, 60, 30, 15, 10, 5, 3].
"All" means no division, i.e., whole arterial road (with an average of 355 s).
For intersections, since time-series data are too short, they were segmented into
the first and second halves of each intersection. This segmentation capture both
long-term and short-term driving behavior. Statistics (mean, median, variance,
maximum, kurtosis, and skewness) are calculated from each sequence and used
as input features of machine learning models. An overview of the model is shown
in Fig. 4.

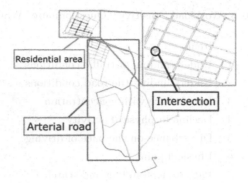

Fig. 3. Driving test route (red line). (Color figure online)

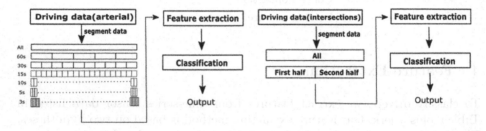

Fig. 4. Overview of the model. The left side shows an overview of the arterial road and the right side shows an overview of the intersections.

5   Experimental Setting

We classify drivers with high and low scores of DSQ and WSQ and evaluate the accuracy of the classification models. We use logistic regression with L2 regularization, linear support vector machine, random forest as classification models. Deep neural network models such as a 1-D CNN and LSTM are suitable for time-series data. However, we do not use them in the present study because the amount of data is too small to train prediction models. The regularization parameter values of the logistic regression and linear support vector machine are selected from [0.001, 0.01, 0.1, 1, 10, 100]. The maximum depth of the tree of the random forest is selected from $[3, 5, 7, 9, 11]$. These hyperparameters are tuned in the training set. As an evaluation criterion of classification models, we report an F1-score because a class imbalance occurs in some items. We use leave-one-person-out cross-validation to evaluate classification models. To avoid overfitting, we apply feature selection for each fold, and features that correlate with scores of each item with $|r| > 0.1$ are used. Scales of DSQ and WSQ are different; we split these scores based on the median value to create binary classification labels and then, conduct binary classification.

6   Results

To evaluate our models and the efficacy of two types of segmentation by road types and sequences of various duration, we compare three models, namely, (i) a model with both road type and various duration segmentation, (ii) a model only with road type segmentation, and (iii) a model without any segmentation. Model (i) and (ii) predict drivers' psychological characteristics separately on the arterial roads and at the intersections.

6.1   Comparison Between Various Duration Segmentation Models

First, we focus on the classification accuracies of the two models with various duration segmentation and compare them. Tables 4 and Table 5 respectively show the classification results of DSQ and WSQ. Columns 2 to 4 in Tables 4 and 5 show the accuracies of the models with two types of segmentation (i) and columns 5 to 7 show the accuracies of the models without various duration segmentation (ii). LR, SVM, and RF denote logistic regression, support vector machine, and random forest. The bold values indicate the highest accuracy among all models and road types for each item.

Concerning DSQ, the model (i) achieved the best accuracies in six items while the model (ii) achieved the best accuracies in four items. For all DSQ items, the best F1 scores were above 0.7. In particular, the best F1 scores of confidence in driving skill, impatience in driving, and anxiety about traffic accidents were 0.831, 0.825, and 0.848, respectively. These scores were comparably high and exceeded 0.8. Concerning WSQ, the model (i) achieved the best accuracies in five items, and the model (ii) achieved the best accuracies also in five items. In comparison with the result of DSQ, the accuracies for WSQ were low and only two items, control operation and driving posture had the best F1 scores above 0.7.

For both DSQ and WSQ, all best F1 scores were higher than 50%, random-assignment baseline. According to this result, our models worked well for estimating the driver's psychological aspect, but, for some items, the accuracy was not high, particularly WSQ items. The various duration segmentation improved the F1 scores of importance of automobile for self-expression (DSQ) and decline in physical activity (WSQ) by 0.143 and 0.104, respectively. However, in other items, segmentation resulted in slight improvement or degradation.

Table 4. Classification accuracy (F1-score) of the model with and without various duration segmentation for DSQ.

DSQ	With segmentation			Without segmentation		
	LR	SVM	RF	LR	SVM	RF
Arterial road						
Confidence in driving skill	0.754	0.754	**0.831**	**0.831**	0.56	0.774
Hesitation for driving	0.618	0.593	0.733	0.754	0.625	0.618
Impatience in driving	0.794	0.774	0.812	**0.825**	0.746	0.794
Methodical driving	0.702	0.712	**0.774**	0.714	0.615	0.724
Preparatory maneuvers at traffic signals	0.512	0.565	0.5	0.714	0.667	0.636
Importance of automobile for self-expression	0.566	0.593	0.69	0.538	0.553	0.56
Moodiness in driving	**0.759**	0.746	0.733	0.75	0.565	0.692
Anxiety about traffic accidents	**0.848**	0.831	**0.848**	0.813	0.75	0.787
Intersections						
Confidence in driving skill	0.733	0.618	0.812	0.754	0.727	0.812
Hesitation for driving	**0.755**	**0.755**	0.654	0.627	0.642	0.667
Impatience in driving	0.759	0.737	0.774	0.737	0.75	0.774
Methodical driving	0.69	0.468	0.691	0.643	0.56	0.69
Preparatory maneuvers at traffic signals	0.727	0.739	0.652	**0.766**	**0.766**	0.682
Importance of automobile for self-expression	**0.708**	**0.708**	0.655	0.553	0.565	0.542
Moodiness in driving	0.692	0.706	0.702	0.733	0.565	0.702
Anxiety about traffic accidents	0.774	0.787	0.831	0.831	0.82	**0.848**

6.2 Comparison with Road Type Segmentation Models

Second, we compare the results of the models with and without road type segmentation. Tables 6 and 7 respectively show the classification results of DSQ and WSQ of the model without any segmentation (iii). The bold values indicate the F1 scores that are higher than the best F1 scores of the models with road type segmentation (i) and (ii). For the model (iii), only anxiety about traffic accidents (DSQ) was predicted more accurately with the F1 score of 0.871, and the accuracies of other items were not more than the accuracies of the models (i) and (ii). Therefore, road type segmentation worked well to estimate drivers' psychological characteristics.

Table 5. Classification accuracy (F1-score) of the model with and without various duration segmentation for WSQ.

WSQ	With segmentation			Without segmentation		
	LR	SVM	RF	LR	SVM	RF
Arterial road						
Understanding of traffic conditions	0.625	0.625	**0.667**	0.51	0.324	0.654
Understanding road conditions	0.5	0.583	0.64	0.458	0.5	0.455
Interference with concentration	0.429	0.419	0.458	**0.679**	0.381	0.35
Decline in physical activity	0.429	0.368	0.439	0.417	0.429	0.381
Disturbance on driver's pace	0.526	0.571	**0.609**	0.538	0.537	0.455
Physical pain	0.333	0.409	0.458	0.531	0.537	0.35
Path understanding and search	0.372	0.341	0.103	0.41	0.462	0.25
In-vehicle environment	0.372	0.612	0.56	0.652	0.696	0.625
Control operation	0.605	0.571	0.462	0.41	0.389	0.474
Driving posture	0.69	0.577	0.746	**0.774**	0.531	0.69
Intersections						
Understanding of traffic conditions	0.565	0.545	0.52	0.565	0.5	0.52
Understanding road conditions	**0.667**	0.619	0.612	0.591	0.578	**0.667**
Interference with concentration	0.522	0.45	0.489	0.578	0.622	0.489
Decline in physical activity	**0.683**	0.6	0.667	0.571	0.537	0.579
Disturbance on driver's pace	0.524	0.585	0.545	0.571	0.512	0.571
Physical pain	0.5	0.476	0.5	0.533	**0.591**	0.488
Path understanding and search	0.564	0.632	0.5	0.649	**0.684**	0.667
In-vehicle environment	**0.698**	0.714	0.609	0.652	0.683	0.545
Control operation	0.619	0.488	0.619	0.558	0.537	**0.718**
Driving posture	0.511	0.533	0.593	0.667	0.612	0.577

Table 6. Classification accuracy of the model without road type segmentation for WSQ.

WSQ	LR	SVM	RF
Confidence in driving skill	0.831	0.618	0.754
Hesitation for driving	0.549	0.609	0.679
Impatience in driving	0.812	0.615	0.794
Methodical driving	0.667	0.68	0.655
Preparatory maneuvers at traffic signals	0.651	0.667	0.553
Importance of automobile for self-expression	0.636	0.651	0.667
Moodiness in driving	0.724	0.625	0.727
Anxiety about traffic accidents	**0.871**	**0.852**	**0.862**

Table 7. Classification accuracy of the model without road type segmentation for DSQ.

WSQ	LR	SVM	RF
Understanding of traffic conditions	0.566	0.591	0.553
Understanding road conditions	0.622	0.667	0.609
Interference with concentration	0.6	0.6	0.444
Decline in physical activity	0.455	0.564	0.308
Disturbance on driver's pace	0.364	0.476	0.476
Physical pain	0.542	0.45	0.263
Path understanding and search	0.381	0.55	0.486
In-vehicle environment	0.681	0.634	0.512
Control operation	0.378	0.524	0.579
Driving posture	0.774	0.653	0.69

7 Discussion

We investigated the contributions of features or sensors to classification, and then describe the effectiveness of two types of segmentation.

7.1 Contribution of Each Sensor

We analyze which sensors were effective for classification and reveal that the importance of features depended on road types. We focus on the results of the model with two types of segmentation and compare important sensors to classify confidence in driving skill between different road types. This item was predicted accurately in both road types. We regard the mean decrease in the impurity of random forest for each feature as the feature importance to the prediction. Then, they summed up to calculate the importance of each sensor. Table 8 shows the five most important sensors for the arterial roads and the intersections. The arterial roads and the intersections have common effective sensors, EPS torque and yaw rate. Sensors for acceleration are important for arterial roads, while sensors for speed are important for intersections. We confirm that important sensors depended on road types also in other items. These differences occurred due to the difference in driving behavior among road types. Thus, it is assumed that the classification models with road type segmentation can capture this difference and improve accuracy.

Table 8. The five most important sensors for classification.

	Arterial roads	Intersection
1	EPS torque	EPS torque
2	Yaw rate	Yaw rate
3	Forward acceleration	Forward left wheel speed
4	Lateral acceleration	Forward right wheel speed
5	Estimated acceleration	Speed

7.2 Contribution of the Segmentation

To verify the efficacy of various duration segmentation, We analyze which duration of segments worked well for classification. As in the previous subsection, we focus on the results of confidence in driving skill of the model with two types of segmentation. Feature importance is calculated in the same way as in the previous subsection and summed up the importance for each segment duration. Table 9 shows relative proportions of feature importance for each segment duration. We compare the feature importance with normalization by the number of segments or without normalization because the number of segments is different depending on the duration of segments.

Without normalization, the proportions of features of 3 s and 5 s were larger than that of other segment duration. With normalization, proportions of short duration segments became small, while proportions of long duration segments became large. This tendency was also seen in other items. This result indicates that short-duration driving behaviors have a lot of contributions for estimating drivers' psychological characteristics, but important features are only a part of them. This result demonstrates that important driving behaviors appear not in whole driving but in partial driving. Additionally, many features of short-duration segments are too localized and are not robust due to sensor noise.

Table 9. Relative proportion of importance of each segment duration.

Duration of segment	Unnormalized	Normalized
All	0.3%	11.0%
60 s	2.3%	12.9%
30 s	5.2%	14.7%
15 s	9.7%	15.2%
10 s	16.0%	17.1%
5 s	25.5%	14.5%
3 s	41.0%	14.6%

8 Conclusion

In this paper, we addressed a challenging task, estimating the psychological characteristics of drivers from on-road driving data. We presented a model for estimation and it could estimate with high accuracy. In particular, confidence in driving skill, impatience in driving, and anxiety about traffic accidents were accurately classified with F1 scores of 0.831, 0.825, and 0.848, respectively. In addition to this estimation, we revealed that important sensors depended on road types and that important driving behaviors had various duration. This study gives a baseline of estimation of psychological characteristics from driving data and benefit analysis.

References

1. Centers for Disease Control and Prevention. Older adult drivers (2021). https://www.cdc.gov/transportationsafety/older_adult_drivers/index.html. Accessed 11 Jan 2022
2. Martinez, C.M., Heucke, M., Wang, F.-Y., Gao, B., Cao, D.: Driving style recognition for intelligent vehicle control and advanced driver assistance: a survey. IEEE Trans. Intell. Transp. Syst. **19**(3), 666–676 (2017)
3. Adrian, J., Postal, V., Moessinger, M., Rascle, N., Charles, A.: Personality traits and executive functions related to on-road driving performance among older drivers. Accident Anal. Prevention **43**(5), 1652–1659 (2011)
4. Classen, S., Nichols, A.L., McPeek, R., Breiner, J.F.: Personality as a predictor of driving performance: an exploratory study. Transp. Res. Part F: Traffic Psychol. Behav. **14**(5), 381–389 (2011)
5. Matthews, G., Dorn, L., Glendon, A.I.: Personality correlates of driver stress. Personality Individ. Differ. **12**(6), 535–549 (1991)
6. Dorn, L., Matthews, G.: Two further studies of personality correlates of driver stress. Personality Individ. Differ. **13**(8), 949–951 (1992)
7. Ishibashi, M., Okuwa, M., Doi, S., Akamatsu, M.: Indices for characterizing driving style and their relevance to car following behavior. In: SICE Annual Conference 2007, pp. 1132–1137. IEEE (2007)
8. Ishibashi, M., Okuwa, M., Doi, S., Akamatsu, M.: Indices for workload sensitivity of driver and their relevance to route choice preferences. In: The Second International Symposium on Complex Medical Engineering, pp. 71–74, May 2008
9. Yoshihara, Y., Takeuchi, E., Ninomiya, Y.: Accurate analysis of expert and elderly driving at blind corners for proactive advanced driving assistance systems. In: Transportation Research Board 95th Annual Meeting, no. 16–1992 (2016)
10. Mana, N., et al.: Multimodal corpus of multi-party meetings for automatic social behavior analysis and personality traits detection. In: Proceedings of the 2007 Workshop on Tagging, Mining and Retrieval of Human Related Activity Information, pp. 9–14 (2007)
11. Philip, J., Shah, D., Nayak, S., Patel, S., Devashrayee, Y.: Machine learning for personality analysis based on big five model. In: Balas, V.E., Sharma, N., Chakrabarti, A. (eds.) Data Management, Analytics and Innovation. AISC, vol. 839, pp. 345–355. Springer, Singapore (2019). https://doi.org/10.1007/978-981-13-1274-8_27

12. Polzehl, T., Möller, S., Metze, F.: Automatically assessing personality from speech. In: 2010 IEEE Fourth International Conference on Semantic Computing, pp. 134–140. IEEE (2010)
13. Yaakub, C.Y., Sulaiman, N., Kim, C.W.: A study on personality identification using game based theory. In: 2010 2nd International Conference on Computer Technology and Development, pp. 732–734 (2010)
14. Dahlen, E.R., Edwards, B.D., Tubré, T., Zyphur, M.J., Warren, C.R.: Taking a look behind the wheel: an investigation into the personality predictors of aggressive driving. Accid. Anal. Prev. **45**, 1–9 (2012)
15. Guo, M., Wei, W., Liao, G., Chu, F.: The impact of personality on driving safety among Chinese high-speed railway drivers. Accid. Anal. Prev. **92**, 9–14 (2016)
16. Stoichkov, R.: Android smartphone application for driving style recognition. Department of Electrical Engineering and Information Technology Institute for Media Technology (2013)
17. Fazeen, M., Gozick, B., Dantu, R., Bhukhiya, M., González, M.C.: Safe driving using mobile phones. IEEE Trans. Intell. Transp. Syst. **13**(3), 1462–1468 (2012)
18. Li, X., Jie, H., Jiang, H., Meng, W.: Establishing style-oriented driver models by imitating human driving behaviors. IEEE Trans. Intell. Transp. Syst. **16**(5), 2522–2530 (2015)
19. Syed, F., Nallapa, S., Dobryden, A., Grand, C., McGee, R., Filev, D.: Design and analysis of an adaptive real-time advisory system for improving real world fuel economy in a hybrid electric vehicle. Technical report, SAE Technical Paper (2010)
20. Murphey, Y.L., Milton, R., Kiliaris, L.: Driver's style classification using jerk analysis. In: 2009 IEEE Workshop on Computational Intelligence in Vehicles and Vehicular Systems, pp. 23–28. IEEE (2009)
21. Aguilar, J., Aguilar, K., Chávez, D., Cordero, J., Puerto, E.: Different intelligent approaches for modeling the style of car driving. In: Proceedings of the 14th International Conference on Informatics in Control, Automation and Robotics - Volume 2: ICINCO, pp. 284–291. INSTICC, SciTePress (2017)
22. Bongiorno, N., Bosurgi, G., Pellegrino, O., Sollazzo, G.: How is the driver's workload influenced by the road environment? Procedia Eng. **187**, 5–13 (2017)
23. Baldwin, C.L., Freeman, F.G., Coyne, J.T.: Mental workload as a function of road type and visibility: comparison of neurophysiological, behavioral, and subjective indices. In: Proceedings of the Human Factors and Ergonomics Society Annual Meeting, vol. 48, pp. 2309–2313. SAGE Publications Sage CA, Los Angeles, CA (2004)

FLAS: A Platform for Studying Attacks on Federated Learning

Yuanchao Loh[1], Zichen Chen[2], Yansong Zhao[1], and Han Yu[1(✉)]

[1] School of Computer Science and Engineering, Nanyang Technological University,
Singapore, Singapore
zichen_chen@ucsb.edu
[2] The University of California, Santa Barbara, USA
{yloh028,yansong001,han.yu}@ntu.edu.sg

Abstract. Smartphones have become a part of everyday life, and users are contributing to Machine Learning with a simple touch (ML). Federated Learning (FL) is a new collaborative learning technique that preserves privacy and addresses the problem of traditional ML. Despite this, it has a large attack surface area and is vulnerable to privacy attacks. Studying the impact of such attacks on the resulting FL models is an important research topic. Currently, there is a lack of an experimental platform to conduct such studies. We attempt to bridge this gap in this paper by proposing the Federated Learning Attack Simulation (FLAS) platform. It is a web-based application designed with an easy-to-use workflow for non-experts and the ability to accelerate testing and analysis for Federated Learning (FL) professionals. Preliminary evaluations have demonstrated the effectiveness of FLAS in supporting the study of common privacy attacks on FL.

Keywords: Simulation platform · Federated Learning · Privacy attacks

1 Introduction

Machine Learning (ML) models trained by traditional methods rely on participants relaying sensitive training data to a central server for model learning and analysis, placing users at great risk of privacy infringement. Federated Learning (FL) was introduced by Google in [2] and has since become popular with applications in healthcare [3], transportation, and financial services [4], sixth-generation (6G) networking, and text prediction [5]. It is proposed that machine learning models train in devices and send updates to a central data centre only for aggregation. Therefore, users' private information is kept secure on their devices. In the rapidly evolving digital age, we accumulate high levels of confidential data, making it imperative to adopt FL globally, as it guarantees secure machine learning.

Despite the privacy-preserving properties of FL, these networks and ML models are vulnerable to adversarial privacy attacks which can negatively affect

G. Meiselwitz (Ed.): HCII 2022, LNCS 13315, pp. 160–169, 2022.
https://doi.org/10.1007/978-3-031-05061-9_12

model performance (e.g., poisoning attacks) or expose private information (e.g., membership inference attacks). Resolving such vulnerabilities of FL can have significant impact on real-world applications [6]. Since [2], FL researchers have made significant advances in counter measures to adversarial attacks. For example, Secure Multiparty Computation and Homorphic Encryption improve the technology's privacy-preserving capabilities and result in significant communication and computational overhead [7].

Existing research on this topic requires significant effort from FL specialists in order to manually develop FL systems for model analysis. Furthermore, as it requires precise technical knowledge, existing research procedures only allow researchers to participate in the area of focus. There is a need for a new research strategy that makes it easy for non-specialists to train FL models and conduct customized research studies in attacks on FL in order to continuous improve its robustness [7]. To achieve this objective, an easy-to-use research tool for rapidly experimenting and analyzing FL attacks is required.

In this paper, we bridge this gap for proposing the Federated Learning Attack Simulation (FLAS) platform. It is a web-based simulator that allows users to easily configure experiments about attacks on FL and examine the consequences of these attacks. FLAS includes a graphical interface that allows selection of attack vectors and parameters without manually developing the attacks. Following attack simulations, the results are stored and visualized on a dashboard with key information including the attacker ratio, proportion of information leaked, model precision and other factors.

FLAS has the potential streamline the FL attack studies by using automated simulations and a dashboard that speeds up testing and analysis. It can also be used to help educate researchers from other fields of study by exposing them to the concept of privacy attacks on federated learning, thereby encouraging them to contribute to FL without significant prior technical knowledge.

2 Related Work

2.1 Importance of FL Research

The growth of the Internet of Things (IoT) and social networking applications in recent years has resulted in an enormous increase in the amount of data generated at the network edge. IoT systems include, but are not limited to, agriculture, education, healthcare, self-driving cars, and smart homes [8].

Sensors in modern IoT networks gather, respond, and adjust to incoming data in real time. Insufficient computation and communication resources at network edges (i.e., memory shortage and long training time) pose challenges for ML, as do data security concerns [8]. Conventional machine learning algorithms require participants' private training data to be sent to a central server for training the model and analysis, placing users at risk of privacy leakage.

FL has grown in popularity in response to these challenges of traditional ML. It enables device-generated data to be stored and processed locally, allowing edge computing devices to collectively train the model while only sending intermediate updates to a central server [2,6]. FL has been applied in sensitive information infrastructure and services such as healthcare and finance [6]. New research works on sixth-generation (6G) networks [5] and SARS-COV-2 predictions on future oxygen requirements for infected patients are being developed across 20 participating institutes [11] are also emerging. FL is a promising approach to protect users' sensitive data [8].

2.2   FL Vulnerabilities and Existing Strategies

Despite its benefits and broad applications in various fields, there have been growing concerns about how transmitting model updates may disclose sensitive data to third parties or a malicious central server [10]. FL is prone to various attacks such as data poisoning on model results and model poisoning via manipulated gradient updates. Class representatives, membership, properties, training inputs, and labels are all subject to inference attacks (all uses gradient updates one way or another). A curious model aggregator can conduct membership inference attacks on other local learners, such as parameter, input, and attribute inferences [7,10].

While there are strategies that improve FL privacy-preserving capabilities, such as Secure Multiparty Computation (SMC), Homomorphic Encryption, and Differential Privacy [6,7,10], these strategies degrade model performance [7,10]. As a result, [7,10] argue that there is an urgent need for diverse research communities (i.e., an interdisciplinary effort) to take advantage of existing benchmarking tools and implementations and to make FL safe from adversarial attacks. FL, in particular, requires a tool that can perform rapid testing and analysis in order to speed up the research process.

2.3   Limitations of Current FL Attack Research Techniques

Many research works on the implementation of adversarial attacks on FL systems, such as data poisoning attacks [12], model poisoning with backdoor capabilities [13,14], and membership inference attacks [15] have been reported. However, the underlying operating systems and Python libraries used in each of the papers mentioned above differ, necessitating rigorous efforts even from FL experts to leverage the models to perform attack analysis. [12] defines poisoning attacks using PyTorch version 1.2.0, whereas [13] defines poisoning attacks using PyTorch version 1.7.0 and [14] defines poisoning attacks using TensorFlow 2.3. TensorFlow 2.0, on the other hand, is used in the membership inference attacks defined in [15].

Re-implementing these models, depending on the researchers' resources and capabilities, necessitates tedious debugging rather than attack modification and analysis exploration. Furthermore, this approach discourages learner researchers from quickly comprehending, studying, and partaking in the experiments. FLAS provides an easy-to-use alternative tool for configuring and conducting such studies, especially for novices who are starting to explore this topic.

3   FLAS System Design

The goal of FLAS is to provide users with an easy-to-use platform that can simulate and analyze different attacks on an FL learning process. It is designed to allow researchers to change the parameters in a federated learning algorithm through the user interface, and configure and execute pre-defined attacks. The results reflecting the effects of the selected attacks on the FL model trained are store in a dashboard to support research analysis.

3.1   FLAS Overview

Figure 1 depicts the FLAS site map, which provides an overview of the Federated Learning Attack simulation system. Before reaching the FLAS home page, the user must first log into the system on the landing page. Users can learn more about the design of FL and its functionality in terms of attacking simulation and analysis by visiting the overview page.

Next, users can simulate FL attacks by customizing the federated attack parameters and starting the attack on the attack configuration page. FLAS saves the attack results to a database and redirects users to the dashboard page, where they can begin analyzing the attack details through a high-level overview or an in-depth view.

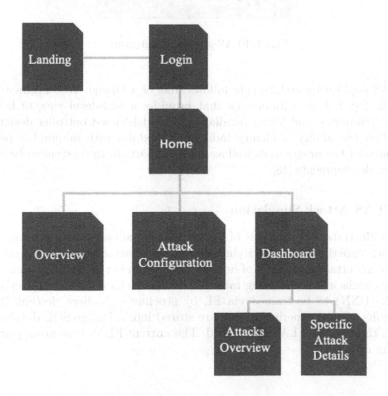

Fig. 1. FLAS site map

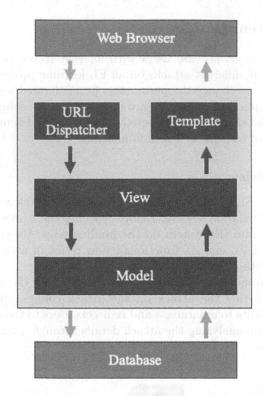

Fig. 2. FLAS system architecture

FLAS application architecture follows that of a Django Web Framework as shown in Fig. 2. It is a framework that provides a high-level view of how the Models, Templates, and Views (similar to a Model-View-Controller design pattern) allow the ability to change individual modules with minimal impact on other parts of the program, as well as allowing parts of the system to be reused in future developments [18].

3.2   FLAS Attack Simulation

Figure 3 illustrates the workflow of using FLAS. It allows a user to configure an FL attack experiment by specifying the datasets, attack types (e.g., inference or poisoning attacks), number of honest clients, number of attackers, number of training epochs and the learning rate. Currently, the base model is a deep neural network (DNN) to be trained via FL by stochastic gradient descent (SGD). The results of each experimentation are stored into a PostgreSQL database for analysis through the FLAS Dashboard. The current FLAS system supports the following attacks:

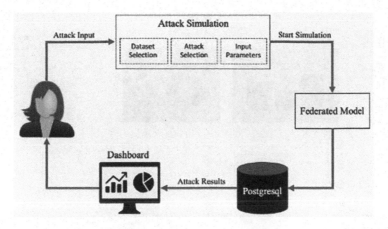

Fig. 3. FLAS system workflow

Data Poisoning Attack. In [12], users are allowed to set the percentage of attackers who inject noise into the learning process via label flipping. The attack is customizable according to different user-defined scenarios covering the choice of datasets (e.g., Fashion-MNIST or CIFAR-10), training rounds, and the learning rate.

Model Poisoning Attack. In [21], a stealth attacker is involved who sends poisoned model updates to the FL server. The attack is customizable according to the number of agents, training rounds and learning rate.

4 FLAS Dashboard Design

The dashboard is divided into two interfaces, giving users a high-level view (Fig. 4) and a comprehensive view of the simulated attack statistics (Fig. 5).

The provided main page focuses on displaying a quick summary of simulated attacks using various datasets and attack types. The page displays a reverse-chronological list of attacks with the most critical details such as the attack title, simulated date/time, and final model accuracy. The user can choose one of the attacks and be taken to the attack statistics page.

Users can view the dataset, attack type, learning rate, number of clients, number of workers, number of poisoned workers, epoch accuracy, classification report and confusion matrix on the in-depth statistics interface. The page includes a pin help users to pin or unpin attack shortcuts in the right hand column for easier switching and faster review of simulated attacks.

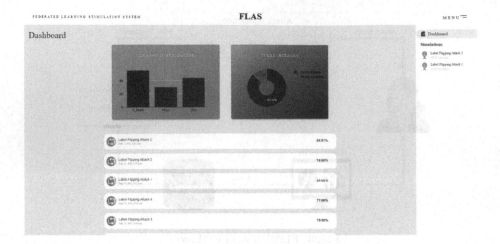

Fig. 4. FLAS main dashboard

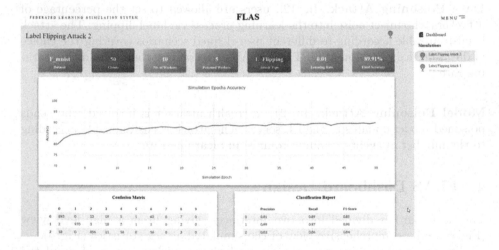

Fig. 5. FLAS results visualization

5 Results and Discussion

Table 1 is the results obtained from three separate label flipping attack experiments performed by FLAS over fifty epochs, with five, ten and twenty attackers, respectively. Note that attackers were set to perform label flipping with the same source (0: T-shirt/Top) and target labels (6: Shirt).

The results show that label flipping attacks have significant effects on the target class performance. An inverse correlation between the number of attackers with the source label recall values can be observed. As the number of attackers doubled, the decrease in source class recall values also doubled: 6% drop with

Table 1. Experiment results with different # of attackers.

# of Attackers	Class	Precision	Recall	F1-score
5	0	0.81	0.89	0.85
5	6	0.76	0.66	0.70
10	0	0.90	0.78	0.84
10	6	0.73	0.71	0.72
20	0	0.95	0.54	0.69
20	6	0.63	0.76	0.69

five attackers, 12% drop with ten attackers, and 24% drop with twenty attackers. The F1-Scores (i.e. weighted average of Precision and Recall) also showed a 16% drop with twenty attackers. Under all attack scenarios, the target class (Class 6) experienced significant reductions in precision. With the FLAS tool, users can repeat these experiments with different parameter configurations to study various attack scenarios with ease.

6   Conclusions and Future Work

The privacy-preserving capabilities of FL can impact the robustness of the resulting models. This work aims to aid experiments on FL attacks by developing a tool to simplify the discovery process and encourage interdisciplinary participation in this field of study. Federated Learning Attack Simulation (FLAS) is a unique web-based application to simulate federated attacks and supports dashboard visualization analysis of attack results. This system is designed to be user-friendly for inexperienced users, as well as maintainable and portable across multiple servers. Through a carefully designed user interface, FLAS allows for the customization of attack parameters and simulation of attacks with no or little no technical knowledge. Furthermore, it has the tools necessary for faster analysis, which caters to expert researchers of the study.

FLAS is still in its early stages of development, with users unable to upload datasets or models. Future research will look into how to improve the system so that it can incorporate user self-upload capabilities, allow for more complex federated models, and simulate other adversarial attacks to accommodate more sophisticated FL attack research.

Acknowledgments. This research is supported by the National Research Foundation, Singapore under its AI Singapore Programme (AISG Award No: AISG2-RP-2020-019); the Joint NTU-WeBank Research Centre on Fintech (Award No: NWJ-2020-008); the Nanyang Assistant Professorship (NAP); and the RIE 2020 Advanced Manufacturing and Engineering (AME) Programmatic Fund (No. A20G8b0102), Singapore. Any opinions, findings and conclusions or recommendations expressed in this material are those of the author(s) and do not reflect the views of National Research Foundation, Singapore.

References

1. Aldhaban, F.: Exploring the adoption of Smartphone technology: literature review. In: 2012 Proceedings of PICMET 2012: Technology Management for Emerging Technologies, pp. 2758–2770. IEEE (2012)
2. Konečný, J., McMahan, H.B., Ramage, D., Richtárik, P.: Federated optimization: distributed machine learning for on-device intelligence. arXiv preprint arXiv:1610.02527 (2016)
3. Qian, F., Zhang, A.: The value of federated learning during and post-COVID-19. Int. J. Qual. Health Care 33(1), mzab010 (2021)
4. Lim, W.Y.B., et al.: Federated learning in mobile edge networks: a comprehensive survey. IEEE Commun. Surv. Tutor. 22(3), 2031–2063 (2020)
5. Yang, Z., Chen, M., Wong, K.-K., Poor, H.V., Cui, S.: Federated learning for 6G: applications, challenges, and opportunities. arXiv preprint arXiv:2101.01338 (2021)
6. Yang, Q., Liu, Y., Chen, T., Tong, Y.: Federated machine learning: concept and applications. ACM Trans. Intell. Syst. Technol. 10(2), 12 (2019). https://doi.org/10.1145/3298981
7. Lyu, L., et al.: Privacy and robustness in federated learning: attacks and defenses. arXiv:2012.06337 (2021)
8. Yang, Q., Liu, Y., Cheng, Y., Kang, Y., Chen, T., Yu, H.: Federated learning. Synth. Lect. Artif. Intell. Mach. Learn. 13(3), 1–207 (2019). https://doi.org/10.2200/S00960ED2V01Y201910AIM043
9. Wang, S., et al.: Adaptive federated learning in resource constrained edge computing systems. IEEE J. Sel. Areas Commun. 37(6), 1205–1221 (2019). https://doi.org/10.1109/JSAC.2019.2904348
10. Li, T., Sahu, A.K., Talwalkar, A., Smith, V.: Federated learning: challenges, methods, and future directions. IEEE Signal Process. Mag. 37(3), 50–60 (2020). https://doi.org/10.1109/MSP.2020.2975749
11. Flores, M., et al.: Federated learning used for predicting outcomes in SARS-COV-2 patients. Research Square (2021)
12. Tolpegin, V., Truex, S., Gursoy, M.E., Liu, L.: Data poisoning attacks against federated learning systems. In: Chen, L., Li, N., Liang, K., Schneider, S. (eds.) ESORICS 2020, Part I. LNCS, vol. 12308, pp. 480–501. Springer, Cham (2020). https://doi.org/10.1007/978-3-030-58951-6_24
13. Bagdasaryan, E., Veit, A., Hua, Y., Estrin, D., Shmatikov, V.: How to backdoor federated learning. In: International Conference on Artificial Intelligence and Statistics, pp. 2938–2948. PMLR (2020)
14. Bagdasaryan, E., Shmatikov, V.: Blind backdoors in deep learning models. arXiv preprint arXiv:2005.03823 (2020)
15. Hitaj, B., Ateniese, G., Pérez-Cruz, F.: Deep models under the GAN: information leakage from collaborative deep learning. CoRR arXiv preprint arXiv:1702.07464 (2017)
16. Samek, W., Müller, K.-R.: Towards explainable artificial intelligence. In: Samek, W., Montavon, G., Vedaldi, A., Hansen, L.K., Müller, K.-R. (eds.) Explainable AI: Interpreting, Explaining and Visualizing Deep Learning. LNCS (LNAI), vol. 11700, pp. 5–22. Springer, Cham (2019). https://doi.org/10.1007/978-3-030-28954-6_1
17. Xu, J., Glicksberg, B.S., Su, C., Walker, P., Bian, J., Wang, F.: Federated learning for healthcare informatics. J. Healthc. Inform. Res. 5(1), 1–19 (2021)
18. Leff, A., Rayfield, J.T.: Web-application development using the model/view/controller design pattern. In: Proceedings Fifth IEEE International Enterprise Distributed Object Computing Conference, pp. 118–127. IEEE (2001)

19. Semantic-Org, A.: Semantic-Org/Semantic-UI: semantic is a UI component framework based around useful principles from natural language. GitHub (2014). http://github.com/semantic-org/semantic-ui/. Accessed 30 May 2021
20. AAAI: AAAI-22 demonstrations program. In: AAAI 2022 Conference. https://aaai.org/Conferences/AAAI-22/aaai22demoscall/. Accessed 14 Oct 2021
21. Bhagoji, A.N., Chakraborty, S., Mittal, P., Calo, S.: Analyzing federated learning through an adversarial lens. In: International Conference on Machine Learning, pp. 634–643. PMLR (2019)

A Study on the Influencing Factors of User Interaction Mode Selection in the Short Video Industry: A Case Study of TikTok

Haoxuan Peng, Xuanwu Zhang, and Cong Cao(✉) ⓘ

Zhejiang University of Technology, Hangzhou, China
congcao@zjut.edu.cn

Abstract. In recent years, with the popularity and development of the Internet and mobile terminals, social short videos, as a new form of social media, have become more and more popular among consumers. By using emerging technologies such as big data and artificial intelligence, the short video platform can analyse user needs more accurately and push services and content in a targeted manner. At the same time, users can send barrage and share information with other users while watching short videos. With accurate user portraits and information push, as well as various interaction modes, TikTok has reached one billion monthly active users. In this paper, taking the TikTok short video platform as an example and examining it from the perspective of its users, a research model is built based on the ternary reciprocal determinism model of social cognitive theory. A questionnaire survey was used to gather 176 valid samples. Partial least squares structural equation modelling (PLS-SEM) was used to analyse the model and the research hypothesis. The results show that individual interaction is influenced by the individual, environment, social presence, public opinion, personal traits, trust properness and other factors. At present, there is little literature on the driving factors of short video application users' participation behaviour. Therefore, this paper aims to fill the gaps in related fields to a certain extent.

Keywords: Short video industry · Driving factors · Interactive behaviour

1 Introduction

Users can capture impressive moments through short video applications and create short-format videos with a duration of several seconds to several minutes. Users can also use simple video editing tools to easily create entertainment videos, edit the created videos, add music and words and modify them into their favourite styles, such as humorous short films, dancing, singing, cooking, physical exercise or other daily activities. Users can also forward these videos to other social media platforms to get "likes" and fans. According to CNN, TikTok, a short video application originating in China, has become popular all over the world [1]. TikTok is a social media application that enables users to make and share short videos. It allows users to quickly and easily create and upload 15-s videos and share them with friends, family members or users all over the world. In the

© The Author(s), under exclusive license to Springer Nature Switzerland AG 2022
G. Meiselwitz (Ed.): HCII 2022, LNCS 13315, pp. 170–184, 2022.
https://doi.org/10.1007/978-3-031-05061-9_13

United States, TikTok has been downloaded almost 80 million times, with 800 million downloads worldwide [1]. In November 2018 [1], it beat other popular applications such as Snapchat, Facebook, Instagram and YouTube and was ranked first among photo and video applications in the Apple App Store. As of April 2019, TikTok had gained a 200% market share in the two-year growth period: the app had 1.17 million ratings in the Apple App Store, 9.67 million ratings on Google Play (April 2019) and more than 9.67 million downloads. With the rapid popularisation and acceptance of social video sharing applications, it may form an ecology at the intersection of Snapchat and Instagram and create a unique user experience [2]. Due to the pandemic situation, many people have reduced stayed home more and sought new forms of entertainment [3]. This has led to this trend becoming increasingly obvious.

Since the rise of TikTok short video, it has gradually formed a relatively fixed development model, which can be divided into the following types: marketing with the personal account as the main body, carrying out marketing activities with the official account of the publishing organisation as the main body, and creating a good atmosphere with TikTok activities to win users' attention and realise traffic. To put it simply, the core of these marketing models is the platform's users. However, when watching short videos, users often exhibit certain interactive behaviours. According to the research, there are four kinds of interactive behaviours that users can display: posting text comments, exhibiting consumer behaviours such as giving gifts and buying promotional products, sending live barrage [4] and liking the content [5].

The driving factors behind these interactive behaviours are the focus of this paper. At present, there is little literature on the driving factors of short video users' participation behaviour. Most of the research focuses on the users or the short video applications themselves. Therefore, this study puts forward the following questions: What factors influence users' choice of interaction behaviour when watching short mobile videos, and how do these factors affect their willingness to interact? In order to answer the above questions, this paper explores the ternary reciprocal determinism model based on social cognitive theory.

In this study, taking the ternary reciprocal determinism model in social cognitive theory as a theoretical reference, the driving factor model of users watching short videos is constructed, and the samples are counted by a questionnaire collection method. The 176 sample data items collected by partial least squares structural equation modelling (PLS-SEM) are analysed, and the research model and hypotheses proposed in this paper are tested. The results of this study show that short video users will show a strong tendency to follow the crowd when watching videos, and their initial attitudes will be significantly influenced by the personality of the video creator and the social presence of the user in the live broadcast environment. At the same time, the short video users' personal factors, such as their personality traits and their trust in the short video platform, will also have a certain impact on related interaction behaviours. These results will help readers gain a comprehensive understanding of the driving factors of related interactive behaviours when users watch short videos, thus providing some theoretical reference for related research in the future.

The research ideas and contents of this paper are as follows: first, the research and results related to short videos and interactive behaviours are organised and summarised. Second, based on the interactive behaviour model of the ternary reciprocal determinism model of social cognition theory and related literature, the framework of this study is put forward, and the corresponding research hypotheses are constructed. Then, the questionnaire is designed according to the scale of existing literature. Relevant data were collected through online questionnaires, and the proposed research models and hypotheses were evaluated and tested using PLS-SEM. Finally, the corresponding research conclusions and theoretical and practical significance, as well as the limitations of the research and future research directions, are put forward.

2 Literature Review

The short video platform (taking TikTok as an example) is a new online video sharing model that can satisfy self-expression, a sense of accomplishment, social interaction and escapism [6]. Previous studies have shown that individual differences are related to various online activities [7], particularly demographic characteristics and personality traits [1]. These traits are emphasised as the main factors that affect an individual's choices in short video interaction. In modern personality psychology, the theory of personality traits defines traits as the neural characteristics of individuals that have the ability to dominate individual behaviours and make individuals give consistent responses in the changing environment. Personality traits have been generally accepted as the driving factors that influence people's behavioural response to the environment. At present, however, there are relatively few articles on the driving factors of participation behaviour from the perspective of personality traits, and the related empirical studies are generally based on external factors and from the perspective of individual behaviour to deduce individual personality traits. There is still a relative lack of literature on how to reflect the individual's behaviour choice by studying personality traits from within the individual.

In Finn and Barnes-Holmes's articles on interactive psychology, the authors state that interactive behaviour essentially deals with the interaction between individuals and stimulus objects and is dependent upon their previous interactions with stimulus objects. An important aspect of this construction is to regard the incentive and response as a whole. Stimulus does not cause a response but participates in the stimulus–response function. This function is bidirectional and constantly changes over time [8]. When the content launched by the short video platform meets the user's psychological expectation of participation, it also constitutes a stimulus–response behaviour, and the user's interactive behaviour is readily expressed. In the process of interaction, social media acts as the main body of trust. Only when users trust can they choose to interact. Cheng et al. started their comprehensive research in 2017 and expanded the trust mode within social media by introducing three interaction modes: interpersonal, group and public. They found three main conclusions about the antecedents of trusting social media. In summary, they found that saving time and information quality are very important for mass media (TikTok). Common topics and convenience are of great significance to group communication, and ability plays the most prominent role in interpersonal communication [9].

In general social interaction research, a common experience obstacle is identifying the social interaction effect of endogenous behaviour, that is, the causal effect of social cognitive behaviour of commenting on something on the individual cognitive behaviour of commenting on something [10]. Under the influence of social stereotypes, individuals are thinking about the social attributes of their own behaviours. When judging whether to continue this behaviour, it is usually habitually interrupted by traditional social views and inevitably falls into the "cage dilemma" created by society.

This study takes short video social media, specifically TikTok, as an example. There are many short video applications available. Although their overall functions are similar, there are still different characteristics among the various applications. For example, the difference between TikTok and other short video applications lies in creativity, which can stimulate users' innovative mood. Short-format video applications also provide users with additional technologies and functions (for example, special filters, smooth video editing and pitch conversion) to allow them to create personalised and entertaining videos [1]. The epidemic in 2021 confirmed the three pillars of TikTok's success: "hyper-personalised algorithm, virtualisation and employment of influencers" [11].

3 Theoretical Background and Hypotheses

In this paper, the research theory of the influencing factors of interaction behaviour mainly relies on the ternary reciprocal determinism model of social cognitive theory. In our research, a questionnaire survey and depth analysis approach are adopted.

The ternary reciprocal determinism model, proposed by American psychologists Bandura and Walters [12], is based on behaviourism, humanism and cognitive psychology and has several distinctive features. "Ternary" refers to the internal factors, external environmental factors and human behaviour factors. They are independent of each other, connected with each other and determined by each other, and there is a dynamic deterministic relationship between them [13]. The research into short video user behaviour presented in this paper is based on personality traits. Therefore, it is necessary to choose a substantive research method. Based on this, this paper studies the factors that influence the interaction behaviour of users in the short video industry. We believe that there are two factors that affect user interaction: personal factors and environmental factors. We subdivide personal factors into trust propensity and personal traits, and divide environmental factors into public opinion and social presence, as shown in Fig. 1. Social expectations mean that users of TikTok should interact with the specific content in Tik-Tok according to their own judgement. Structural assurance refers to whether TikTok users can interact with each other with full cognitive ability.

An individual's attitude towards behaviour is determined by his or her belief in the implementation of certain behaviours and an evaluation of the results [14, 15]. Individual users' choice of interaction behaviour is based on their own judgement of content, and choosing an appropriate interaction behaviour according to individual cognition is a self-commitment to individual participation in interaction. Individual differences play a key role in rational decision making. Whether in a single-factor model or multi-factor model, the similarities and differences of individuals will always change the narrow decision-making interval [16]. Is the centre of individual behaviour choice, and can only individuals cause the difference of interactive choice? To address these questions, this paper puts forward the following hypothesis:

H1: *Individual attitude has a significant positive influence on the behaviour of TikTok users in choosing an interaction mode.*

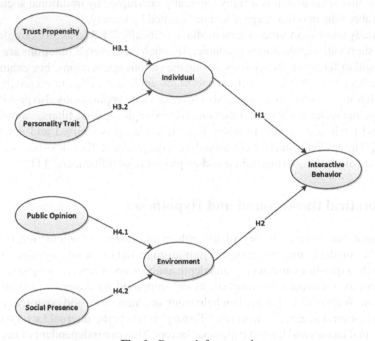

Fig. 1. Research framework.

Many of our decisions are made in the social environment. In the social environment, we are willing to accept others' behaviour infection, and we do not reject imitating others' decisions and behaviours. Our decisions are rarely made in complete isolation [17]. Many theoretical studies generally believe that in an emergency, "people show a tendency of mass behaviour, that is, do what others do" [18]. When we interact with another person, we form attitudes and interaction patterns based on their feedback in environmental communication [19]. Individuals are gradually infected by their environment and use feedback from the environment to conceive whether their choice of interaction behaviour is gregarious and correct. Therefore, this paper puts forward the following hypothesis:

H2: *The positive environment has a significant positive influence on the behaviour of TikTok users in choosing an interaction mode.*

Users' trust tendency refers to the willingness to trust apps like TikTok, which is reflected in the trust degree of users (trusters) to software information managers (trustees), even before evaluating and analysing the trustworthiness and ability of trustees [20]. Trust tendency helps TikTok users to authorise TikTok in a wide range of situations, spontaneously based on their trust in the app while using it. On the one hand, the choice of user interaction mode can enhance users' trust tendency; on the other hand, it can also be counterproductive for the server to provide users with content services that are more dependent on users. Therefore, this paper puts forward the following hypothesis:

H3.1: *Trust tendency has a significant positive influence on the behaviour of TikTok users in choosing an interaction mode.*

Previous studies have shown that the personality traits of individuals can influence their use of behaviours in some applications and social media [21]. In the process of short video platform operation and maintenance, information managers usually associate personality traits with goals [22]. By analysing the preference behaviours of people with different personality traits in choosing interactive behaviour patterns, we can push short videos for individuals more pertinently. Therefore, this paper puts forward the following hypothesis:

H3.2: *Personality traits have a significant positive influence on the behaviour of TikTok users in choosing an interaction mode.*

According to the research of social psychology theory, most public opinions will oppress the individual's behaviour choice. According to this theory, individuals' willingness to express their opinions depends on how they view or judge the opinions of others (that is, the opinions of the majority) according to their own opinions [23]. Afraid of being isolated, when individuals think that their views are different from those of most people, they are unlikely to express their views [24]. Then, when the individual's opinion accords with that of most public opinions, more respect for the individual's inner behaviour choices will naturally appear. Therefore, this paper puts forward the following hypothesis:

H4.1: *Public opinion has a significant positive influence on the behaviour of TikTok users in choosing an interaction mode.*

Social presence was first proposed by Short et al. and aimed at explaining the degree to which a person is regarded as a "real person" and the perceived degree of contact with others in the process of communication through media [25]. Interpersonal interaction with the help of media tools can help both parties understand each other's characteristics and perceive each other's existence in the process of information exchange. Therefore, increasing social telepresence will help individuals obtain information more easily in the process of communication. The short video platform gives users more interactive possibilities, including live broadcast, video, comments, gifts and other modes. Social telepresence will not only affect participants' perception of atmosphere, but also affect their interaction in an immersive virtual environment [26]. Strengthening the telepresence attribute can help users establish a close connection with the platform, thus psychologically affecting the user's willingness to interact. Therefore, this paper puts forward the following hypothesis:

H4.2: *Social telepresence has a significant positive influence on the behaviour of TikTok users in choosing an interaction mode.*

4 Research Methodology

In this study, data are collected through online questionnaires, and models are built to prove the rationality of relevant assumptions. The scale design of the questionnaire is based on the mature scale in the relevant literature.

After fully considering the characteristics of short video users and the characteristics of interactive behaviour, some improvements and adjustments have been made. In the initial questionnaire, in addition to the basic information on the respondents, other items were measured using the Likert five-point scale. Simply put, the scores of these questions are given in a range from 1 to 5, with 1 indicating strong opposition and 5 indicating strong approval. After completing the preliminary questionnaire design, in order to ensure that the validity of the questionnaire met the experimental requirements, we first conducted a small-scale pre-test in related fields. A total of 20 questionnaires were collected in the initial stage to check whether the semantic and grammatical expressions of the options in the questionnaire were easy to understand and whether the reliability and validity met the requirements. At the same time, according to the respondents' feedback, some expressions use in the questionnaire were modified, and finally a formal questionnaire was formed. Formal questionnaires were mainly distributed through an online survey platform, and 206 questionnaires were collected. From these, 176 valid questionnaires were obtained by eliminating the questionnaires with too short an answer time or too concentrated selection options. Because PLS-SEM has fewer restrictions on experimental data, this study decided to use this method to evaluate the research model and verify the relevant assumptions. For example, PLS has good explanatory power, even when the sample size is small. In addition, PLS does not require the data to be normally distributed and can effectively deal with reflective–formative measurement models.

According to the principle of 10 times, the minimum sample size of this study should be 60, so the dataset of 176 valid samples met these requirements. The demographic characteristics of the survey participants are shown in Table 1. Among all the participants, gender distribution was relatively balanced, with 52% of participants being male and 48% female. Therefore, gender differences did not excessively affect the accuracy of questionnaire information collection. According to the age structure of the participants, the age range of the participants was mainly 18–40 years old, accounting for about 90% of the total number. In addition, most of the participants had received higher education, and about 66% of the participants had a bachelor's degree or above.

Table 1. Demographic profile of respondents, N = 176.

Measure	Category	N	Percent
Gender	Male	91	52%
	Female	85	48%
Age	18–25	51	29%
	26–30	64	36%
	30–35	22	13%

(*continued*)

Table 1. (*continued*)

Measure	Category	N	Percent
	36–40	21	12%
	Over 40	18	10%
Education	College	36	20%
	Undergraduate	59	34%
	Postgraduate	81	46%

5 Results

5.1 Measurement Model

In this study, internal consistency reliability was used to test the reliability of each construction measurement method. As shown in Table 2, the composite reliability (CR) and Cronbach's alpha coefficient (CA) of each construct are both greater than 0.80, which indicates that the measurement of each construct in this study has good internal consistency reliability [27, 28].

Table 2. Descriptive statistics for the constructs.

	CA	CR	AVE
Interactive behavior (Inter)	0.932	0.957	0.880
Individual (Indiv)	0.929	0.955	0.876
Environment (Envir)	0.908	0.943	0.845
Trust propensity (Trust)	0.976	0.984	0.954
Personality trait (Trait)	0.974	0.983	0.951
Public opinion (Publi)	0.904	0.954	0.912
Social presence (Socia)	0.892	0.949	0.903

The structural validity of this study is mainly evaluated by discrimination validity, aggregation validity and content validity. Because the scales of this study are adapted from the existing literature, they have good content validity. It can be seen from Table 3 that the standardised external load of each structure is greater than 0.900, and the average variance extraction (AVE) of each structure in Table 2 is greater than 0.5 [29, 30]. Therefore, it can be seen that all the structures have good convergence. As shown in Table 4, the square root of the AVE of any structure in the model is greater than the correlation value with other structures. In addition, as shown in Table 3, the standardised external load of the structure to which each index belongs is greater than its cross-load. This shows that the measurement of different structures in this study has sufficient judgement validity [27, 28].

Table 3. Factor loadings and cross loadings.

	Inter	Indiv	Envir	Trust	Trait	Publi	Socia
Inter.1	**0.928**	0.274	0.633	−0.051	0.411	0.016	0.788
Inter.2	**0.955**	0.335	0.701	0.028	0.469	0.104	0.843
Inter.3	**0.931**	0.274	0.660	−0.041	0.446	0.114	0.796
Indiv.1	0.281	**0.942**	0.017	0.430	0.674	−0.179	0.286
Indiv.2	0.313	**0.934**	0.065	0.430	0.630	−0.128	0.328
Indiv.3	0.290	**0.932**	0.006	0.398	0.641	−0.127	0.316
Envir.1	0.681	0.039	**0.924**	0.235	0.231	0.318	0.733
Envir.2	0.646	0.024	**0.904**	0.244	0.242	0.301	0.708
Envir.3	0.629	0.023	**0.930**	0.313	0.151	0.386	0.705
Trust.1	0.006	0.439	0.303	**0.975**	0.369	0.299	0.225
Trust.2	−0.012	0.454	0.272	**0.977**	0.361	0.299	0.199
Trust.3	−0.057	0.419	0.265	**0.978**	0.321	0.317	0.171
Trait.1	0.464	0.693	0.227	0.387	**0.977**	−0.256	0.495
Trait.2	0.465	0.685	0.225	0.348	**0.977**	−0.253	0.500
Trait.3	0.452	0.648	0.211	0.315	**0.971**	−0.222	0.493
Publi.1	0.093	−0.155	0.344	0.293	−0.245	**0.954**	0.203
Publi.2	0.069	−0.142	0.352	0.302	−0.233	**0.956**	0.203
Socia.1	0.852	0.295	0.762	0.168	0.464	0.171	**0.953**
Socia.2	0.786	0.335	0.715	0.221	0.504	0.235	**0.947**

Note: Bold number indicate outer loading on the assigned constructs

Table 4. Correlations among constructs and the square root of the AVE.

	Inter	Indiv	Envir	Trust	Trait	Publi	Socia
Inter	**0.938**						
Indiv	0.315	**0.936**					
Envir	0.710	0.031	**0.919**				
Trust	−0.021	0.448	0.287	**0.977**			
Trait	0.472	0.693	0.227	0.359	**0.975**		
Publi	0.085	−0.155	0.364	0.312	−0.250	**0.955**	
Socia	0.863	0.331	0.778	0.204	0.509	0.213	**0.950**

Note: Bold number represent the square roots of the AVEs

5.2 Structural Model

In order to test the statistical significance of the T value of the path coefficient in the research model, bootstrapping in Smart PLS 3.3.7 was adopted in this study. The number of original samples in this study was 176. According to the general recommendation of PLS research, this paper sets the guiding subsample at 5,000. The path coefficient and statistical significance test results are shown in Fig. 2. The six hypotheses put forward in this paper all passed the statistical significance test. The empirical results show that the personal trait, trust properness, public opinion and social presence of short video users have a significant impact on the choice of interactive behaviour. Finally, this study found that different factors have different degrees of influence on users' choice of interaction behaviour. Among them, social presence has the greatest influence among the environmental factors, followed by personal traits. Trust propensity and public opinion are also among the personal factors of users.

When evaluating a structural model, the determination coefficient (R^2) is usually used to evaluate the predictive ability of the model [30]. R^2 is usually between 0 and 1. The higher the numerical value, the stronger the predictive ability. In this study, the R^2 of short video user interaction reached 0.589, the R^2 of environmental factors reached 0.674 and the R^2 of personal factors reached 0.526, all of which met the relevant requirements. This shows that the model proposed in this study has good explanatory power.

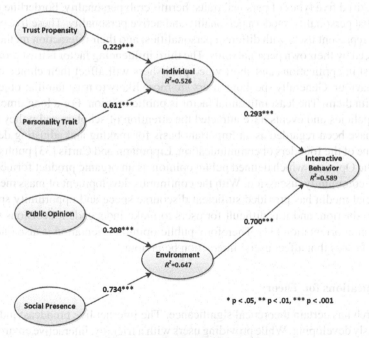

Fig. 2. PLS-SEM analysis results.

6 Discussion and Future Research

6.1 Summary of Results

This study attempts to explore the driving factors that influence short video users' interaction behaviour. First, based on the existing research results, we summed up two macro factors that affect the interaction behaviour of users, namely, individual and environment. Based on this, we constructed a research model and put forward assumptions. We used PLS-SEM to evaluate the research model and hypotheses. Based on the statistical analysis of 176 samples, the six hypotheses put forward in this paper are all valid, and both environmental factors and personal factors have significant positive effects on users' interaction behaviour. These important factors include social presence, public opinion, personal trait and trust propriety, among which personal trait and trust propriety are personal factors, while public opinion and social presence are environmental factors.

According to the experimental data, the influence of environmental factors is slightly greater than that of personal factors. Based on empirical data and user interviews, we find that social presence has the most significant positive impact on interactive behaviour and that technology can greatly promote social telepresence. TikTok has a series of powerful algorithms that can recommend related videos according to users' preferences, immerse users and have a strong sense of social presence in video content, thus affecting interactive behaviour. The second is personal traits. When we included it in the questionnaire, we summarised five types of personal traits: hermit crab personality, borderline personality, pacifist personality, modern personality and active personality. These five personal trait types represent users with different personalities, and their interaction methods will be influenced by their own personal traits. The third influencing factor is trust propensity. Users' trust in applications and short video publishers will affect their choice of interaction behaviour. Generally speaking, users are more likely to trust familiar objects and interact with them. The least influential factor is public opinion. For a long time, public views on policies and events have attracted the attention of scholars and policy makers [31] and have been regarded as an important basis for making and adjusting decisions [32]. As one of the founders of communication, Lippmann and Curtis [33] published the book Public Opinion, which defined public opinion as an organic product formed in the process of community discussion. With the continuous development of mass media, the rise of social media has provided sufficient discourse space and opportunity space for opinion production, and it is difficult for users to make independent decisions without public opinion orientation [34]. Therefore, public opinion orientation is also one of the important factors that affect users' interaction behaviour.

6.2 Implications for Theory

Our research has certain theoretical significance. The internet live broadcast industry is continuously developing. While providing users with a friendly, interactive environment, the whole industry is also trying to eliminate various drawbacks, such as trust and watching addiction [1]. Therefore, major short video platforms are trying to solve related problems by relying on live broadcast technology and recommending algorithms. For example, when a user watches a short video for a certain period of time, the platform will intelligently display relevant tips, suggesting that the user rest and relax. When users are preparing to buy products in the live broadcast room, the platform will also prompt users to pay attention to discrimination. The results of this study are aimed at further exploring the interactive behaviour mechanism of short video users and promoting the development of related theories. In previous research, researchers investigated the development status of short video platforms and analysed the relationship between the technology of short video platforms and the interaction behaviour of users. However, there is little detailed discussion on the drivers of short video user interaction behaviour. This study attempts to synthesise various influencing factors and present the influencing factors of short video users' willingness to interact with each other with a complete system structure. The research results aim to provide a more comprehensive perspective for understanding user interaction behaviour. This helps researchers to further understand the internal mechanisms of related content.

6.3 Implications for Practice

The influencing factors of mobile short video users' usage behaviour obtained in the research process can reveal the current focus and usage trend of China Mobile short video users to a certain extent. It can also provide more accurate information services and guidance for the optimisation of the marketing model and product structure of the mobile short video industry. In an era when content is king, a short video blogger with a large number of fans must be able to attract the attention of the public. Therefore, in the early stage of related short video marketing activities, targeted design should be carried out according to the six factors mentioned in this article to obtain greater benefits. For example, at present, many enterprises choose to broadcast live on TikTok and sell their products or services. However, not all live broadcast rooms are profitable, so it is necessary to target design according to users' behavioural characteristics.

6.4 Limitations and Future Research

Although this study provides a useful discussion on the influencing factors of mobile short video users' interaction behaviour, it inevitably has its own limitations; therefore, it is necessary to improve it in future research. First, the data sample size of this study was 176. Although this meets the minimum sample size requirement of PLS-SEM, increasing the sample size would effectively increase the accuracy of the model and evaluation. Second, in the process of this empirical study, the moderating effects of cultural factors and product categories were not considered. At present, some frontier research points out that the difference in product types may change the user's choice of interaction mode. In addition, users in different regions have different cultures and customs, and they may adopt different behaviours and interaction strategies. Therefore, in future research, we should consider the moderating effects of product types and cultural factors on mobile short video users' interaction behaviours, so as to further expand the research results.

7 Conclusion

In recent years, the growth rate of users in the short video industry has become evident, and a large number of individuals, organisations and enterprises have entered this emerging industry. To some extent, due to the fast pace of modern life, the trend of consuming shorter and shorter videos has emerged [35]. TikTok short video is the darling of the internet era and can satisfy users in their often-fragmented leisure time. It has also made outstanding contributions to the spread of culture and entertainment. Therefore, it is a way to push the short video industry and even the traditional online video industry towards a new era exploring new ways to attract users and promote user interaction. At present, the research on short video platforms such as TikTok mainly starts with the internal technology of the short video industry; research on the two-way interactive communication between individual users and short video platforms is still scarce. Based on empirical research methods, this paper uses PLS-SEM to evaluate the research models and assumptions. The results show that short video users' interactive behaviour choices are mainly influenced by two factors: individual factors and environmental factors. Among individual factors, we mainly study the bias control of trust properness and personality trait on the choice of interaction behaviour. Among the environmental factors, we find that public opinion and social presence also have positive effects on the choice of individual interaction behaviour. The research results of this paper can help short video platform managers build a more credible exchange environment, shorten the interaction distance with users and promote two-way development between users and platforms. It plays an active role in promoting the new development of the short video field. At the same time, it also lays a foundation for follow-up scholars to study the role of short video platforms in the digital economy environment.

Acknowledgments. The work described in this paper was supported by grants from the Zhejiang University of Technology Humanities and Social Sciences Pre-Research Fund Project (GZ21731320013), and the Zhejiang University of Technology Subject Reform Project (GZ21511320030).

References

1. Zhang, X., Wu, Y., Liu, S.: Exploring short-form video application addiction: socio-technical and attachment perspectives. Telematics Inform. **42**, 101243 (2019)
2. Wang, Y.: Humor and camera view on mobile short-form video apps influence user experience and technology-adoption intent, an example of TikTok (DouYin). Comput. Hum. Behav. **110**, 106373 (2020)
3. Kim, J., Merrill, K., Jr., Collins, C., Yang, H.: Social TV viewing during the COVID-19 lockdown: the mediating role of social presence. Technol. Soc. **67**, 101733 (2021)
4. Hao, X., Xu, S., Zhang, X.: Barrage participation and feedback in travel reality shows: the effects of media on destination image among Generation Y. J. Destin. Mark. Manag. **12**, 27–36 (2019)
5. Anthony, K.: iDisorder: understanding our obsession with technology and overcoming its hold on us. Br. J. Guid. Couns. **41**(5), 609–611 (2013)

6. Omar, B., Wang, D.: Watch, share or create: the influence of personality traits and user motivation on TikTok mobile video usage. Int. J. Interact. Mob. Technol. (iJIM) **14**(4), 121 (2020)
7. Teo, T.S.H.: Demographic and motivation variables associated with internet usage activities. Internet Res. **11**(2), 125–137 (2001)
8. Finn, M., Barnes-Holmes, D.: In support of reacquainting functional contextualism and interbehaviorism. J. Contextual Behav. Sci. **19**, 1–5 (2021)
9. Cheng, X., Fu, S., de Vreede, G.-J.: Understanding trust influencing factors in social media communication: a qualitative study. Int. J. Inf. Manag. **37**(2), 25–35 (2017)
10. Manski, C.F.: Identification of endogenous social effects: the reflection problem. Rev. Econ. Stud. **3**, 531–542 (1993)
11. Feldkamp, J.: The rise of TikTok: the evolution of a social media platform during COVID-19. In: Hovestadt, C., Recker, J., Richter, J., Werder, K. (eds.) Digital Responses to Covid-19. SIS, pp. 73–85. Springer, Cham (2021). https://doi.org/10.1007/978-3-030-66611-8_6
12. Bandura, A., Walters, R.H.: Social Learning Theory. Prentice Hall, Englewood Cliffs (1977)
13. Bandura, A., Bandura, A.: Social Foundations of thought and Action: A Socio-Cognitive Theory (1986)
14. Sutanonpaiboon, J., Abuhamdieh, A.: Factors influencing trust in online consumer-to-consumer (C2C) transactions. J. Internet Commer. **7**(2), 203–219 (2008)
15. Walczuch, R., Lundgren, H.: Psychological antecedents of institution-based consumer trust in e-retailing. Inf. Manag. **42**(1), 159–177 (2004)
16. Berthet, V., Autissier, D., de Gardelle, V.: Individual differences in decision-making: a test of a one-factor model of rationality. Pers. Individ. Differ. **189**, 111485 (2022)
17. Baddeley, M.: Herding, social influence and economic decision-making: socio-psychological and neuroscientific analyses. Philos. Trans. R. Soc. B Biol. Sci. **365**(1538), 281–290 (2010)
18. Helbing, D., Farkas, I., Vicsek, T.: Simulating dynamical features of escape panic. Nature **407**(6803), 487–490 (2000)
19. Hackel, L.M., Doll, B.B., Amodio, D.M.: Instrumental learning of traits versus rewards: dissociable neural correlates and effects on choice. Nat. Neurosci. **18**(9), 1233–1235 (2015)
20. Schoorman, F.D., Mayer, R.C., Davis, J.H.: An integrative model of organizational trust: past, present, and future. Acad. Manag. Rev. **32**(2), 344–354 (2007)
21. Peltonen, E., Sharmila, P., Opoku Asare, K., Visuri, A., Lagerspetz, E., Ferreira, D.: When phones get personal: predicting big five personality traits from application usage. Pervasive Mob. Comput. **69**, 101269 (2020)
22. Reisz, Z., Boudreaux, M.J., Ozer, D.J.: Personality traits and the prediction of personal goals. Pers. Individ. Differ. **55**(6), 699–704 (2013)
23. Noelle-Neumann, E.: The spiral of silence a theory of public opinion. J. Commun. **24**(2), 43–51 (1974)
24. Lee, M.J., Chun, J.W.: Reading others' comments and public opinion poll results on social media: social judgment and spiral of empowerment. Comput. Hum. Behav. **65**, 479–487 (2016)
25. Short, J., Williams, E., Christie, B.: The Social Psychology of Telecommunications. Wiley, Toronto, London, New York (1976)
26. Dubosc, C., Gorisse, G., Christmann, O., Fleury, S., Poinsot, K., Richir, S.: Impact of avatar facial anthropomorphism on body ownership, attractiveness and social presence in collaborative tasks in immersive virtual environments. Comput. Graph. **101**, 82–92 (2021)
27. Fornell, C., Larcker, D.F.: Evaluating structural equation models with unobservable variables and measurement error. J. Mark. Res. **18**(1), 39–50 (1981)
28. Hair, J.F., Sarstedt, M., Ringle, C.M., Mena, J.A.: An assessment of the use of partial least squares structural equation modeling in marketing research. J. Acad. Mark. Sci. **40**(3), 414–433 (2012)

29. Barclay, D., Higgins, C., Thompson, R.: The partial least squares (PLS) approach to casual modeling: personal computer adoption and use as an illustration. Technol. Stud. Special Issues Res. Methodol. **2**(2), 285–309 (1995)
30. Chin, W.W., Newsted, P.R.: Structural equation modeling analysis with small samples using partial least squares. In: Hoyle, R. (ed.) Statistical Strategies for Small Sample Research, pp. 307–341. Sage Publication, Beverly Hills (1999)
31. Krugman, H.E.: The impact of television advertising: learning without involvement. Public Opin. Q. **29**(3), 349–356 (1965)
32. Sofalvi, A.J., Airhihenbuwa, C.O.: An analysis of the relationship between news coverage of health topics and public opinion of the most important health problems in the United States. Am. J. Health Educ. **23**(5), 296–300 (1992)
33. Lippmann, W., Curtis, M.: Public Opinion. Routledge, New York (2017)
34. Snow, B.D.A.: Framing processes and social movements: an overview and assessment. Ann. Rev. Sociol. **26**, 611–639 (2000)
35. Wright, C.: Are beauty bloggers more influential than traditional industry experts? J. Promot. Commun. **5**(3) (2017)

User Preferences for Organizing Social Media Feeds

Kristine M. Rogers[✉] [iD]

University of Maryland, College Park, MD 20742, USA
krogers@umd.edu

Abstract. How do users prefer to have social media posts organized?
This research looks at the impact of the type of activity on social media
users' preferences for organizing posts, focusing on four use cases: change
detection (following a topic over time), experiential (following an event
as it happens), browsing, and searching. The hypothesis studied is that
users' views on the most appropriate ordering would vary based on task.
Survey responses from 188 social media users were analyzed to identify display preferences for social media posts. All four use cases were
found to be common, although responses regarding browsing (the most
common category) indicated that the respondents engaged in two recognizably distinct types of browsing activity: time-oriented browsing
for updates (often focused on friends, family, and news) and general
browsing (e.g., for entertaining or humorous content). Respondents who
engaged in change detection, experiential, and time-oriented browsing
activities expressed strong preferences for chronological sorting. Those
who engaged in general browsing and searching preferred relevant documents first. The paper discusses the implications of these findings for
task-based design of information retrieval systems for social media feeds.

Keywords: Social media · Sort orders · Updates · Change detection ·
Experiential · Browsing · Search · Information retrieval

1 Introduction

*"Listened to Agatha Christie audio book today but all the chapters played in
random order so it turned into more of a mystery than intended."*[1] In January 2012, cross-country skier Felicity Aston posted this tweet while chronicling
a groundbreaking solo cross-country ski trip across Antarctica. This comment
seems humorous because we expect stories to be presented in the "right" order,
as laid out by the author. When presented in an order different than what the
reader expects, this can cause confusion or angst. And yet this is the experience
that many social media users have come to expect when visiting sites such as

---

[1] Tweet available at https://twitter.com/felicity_aston/status/156917318293794816.

We would like to acknowledge Dr. Douglas Oard (UMD) for support with the research
and comments on this manuscript.

G. Meiselwitz (Ed.): HCII 2022, LNCS 13315, pp. 185–204, 2022.
https://doi.org/10.1007/978-3-031-05061-9_14

Twitter and Facebook in search of updates. Instead of following individual narratives, users see posts in an order that an unknown algorithm provides. While the algorithm may be "right" on a granular level (people do respond to the posts that are ranked as highly relevant), the overall interaction can feel disjointed. Social media users often consume content in the order presented, but some users post frustrated comments on social media sites, including asking for the return of additional sort options, such as chronological sort.

We approach this case of users seeking updates on people and topics as being like the act of following a story over time. To the user, a social media feed represents a complex mix of multiple queries; a typical feed includes a combination of people, organizations, groups, and topics, each of which a user follows for a different reason. A user might follow US football teams and players to get information on the current game, friends and family members to get an ongoing sense of what is happening in their lives, or a photography hashtag to get inspiration for an art project. Which aspect they wish to see during any given social media session will depend on their intent at that moment. For instance, if a football game is taking place, the user might wish to focus primarily on commentary from fellow fans that they follow during the event.

We studied the impact of the type of activity in which a user is engaged on their preferences for organizing posts, focusing on four use cases: change detection (following a topic over a long period of time), experiential (following an event as it happens), browsing, and searching. We were particularly interested in determining whether time-bounded social media behaviors are insufficiently addressed by social media sites. The hypothesis studied is that users' views on the most appropriate ordering would vary based on task, with users preferring chronological sort ordering for update-related activities such as change detection and experiential use. To understand when one organization might be preferred over another, we surveyed 188 social media users to understand their sort order preferences for specific use cases. The results indicate that users had distinct organization preferences for different use cases.

All four types of activities were found to be fairly common among the set of social media users who responded to the survey, with browsing (both time-oriented and general) being the most popular category; 93% of respondents indicated that they have performed some browsing tasks. Searching was the next most common, being performed by 63% of surveyed users. Change detection followed close behind, with 61% of the respondents following a specific topic over time on social media. Experiential use was the least common, though nearly half of respondents indicated that they have followed a specific event on social media.

The survey results identified some of the characteristics of social media use for each of the analyzed use cases; these insights can help to guide the design of new system capabilities, and to anticipate some of the challenges that may arise when introducing new capabilities. Further analysis revealed that respondents engaged in two recognizably distinct types of browsing: time-oriented browsing for updates (often on friends, family, and news) and general browsing (e.g., for entertaining or humorous content). This is noteworthy because feeds offered by many social media sites are better suited for general browsing, which was less common among respondents. We also noted that time-oriented browsing shares some characteristics with change detection.

Respondents preferred chronological ordering for change detection, experiential, and time-oriented browsing. Many of the respondents who perform searches prefer relevance-ranked results, though a large percentage also indicated interest in newer posts. For general browsing, respondents wanted relevant or popular posts. For most use cases, respondents appeared to support the idea of clustering; only users in time-oriented browsing preferred no grouping. There was a preference for recall-oriented feeds for the change detection, experiential, and time-oriented browsing use cases; users indicated a fear of missing out on something interesting—they wanted to see all posts related to the topic or new theme.

We asked about both presentation order and about clustering preferences. We included questions about clustering to adjust for users' optimism about their ability to keep up with masses of streaming information. In Present Shock, Rushkoff describes the "quest for digital omniscience," in which users believe they can catch up on what they missed; they actually won't (or more precisely, can't) because of overwhelming volumes [19]. Clustering by account, event, or other alternatives are ways to help users get a sense for what is happening, even if not every individual post is present.

2 Related Work

According to a 2021 Pew Research Center poll, roughly 70% of the adult population of the United States uses social media [2]. The U.S. Census Bureau estimates that, as of July 2018, there were approximately 332 million people in the United States, 77.7% of whom are adults [23]. This means that as many as 181 million adults are current social media users. Understanding what users want to accomplish in social media can help to identify and meet their needs.

Many social media sites' initial designs used chronological sort orders for displaying users' posts. Over the course of time, many of the popular sites transitioned from chronological sort orders to relevance rankings. Figure 1 provides the dates that four of the major social media sites—Facebook, Twitter, Instagram, and Snapchat—made this shift. The result has included negative reactions from parts of the user base, with sites like Instagram, Facebook, and Twitter continuing to field complaints about sort orders. In late 2018, Twitter responded to user comments by broadly reintroducing a capability to resort posts chronologically— though the default remains relevance ranking.

Researchers have looked at the fragmentation of messages in modern media for many years. In *Amusing Ourselves to Death*, Postman observed that the advent of the telegraph ushered in an era of discontinuous, fragmented messages—a "decontextualized information environment" lacking social or intellectual context [18]. Sadler extended this concept into the social media world, examining the implications of fragmented stories across sites such as Twitter [20].

Fig. 1. Timing of social media moves to relevance ranking

People use social media to learn, share, or be entertained. Others visit to participate in a "mass shared experience," such as live tweeting a sporting event. In these cases, a device becomes a "second screen," in which the user simultaneously watches a television show (e.g., British TV show "The X-Factor") and participates in a worldwide online conversation [14]. Some people visit social media because they are bored or alone. One study demonstrated that pictures on social media sites can help reduce feelings of loneliness [17].

To understand some of the user behaviors in social media, along with associated sort implications, one study looked at learning to rank in social media to predict which posts were most memorable, so these items could be preserved for future reminiscence [15]. Kim et al. performed a survey to understand students and their use of social media; they found that students' social media activities ranged from academic research to entertainment [12].

Many studies in the field of information retrieval have compared approaches for relevance ranking, focusing on augmenting relevance with such features as recency, document diversity, sentiment, and popularity. However, we were unable to find studies that address the underlying assumption that relevance ranking is preferable over non-relevance sort approaches, such as chronological orderings of posts. There has also been work on reviewing historical posts to aid in storytelling—for instance, gathering content about a user from their friends' pages to produce a more robust story of their life [21]. However, there seems to be less of a focus on how to meet users' needs as they process pieces of the many stories they encounter in social media. Other research has looked at searching of social media data in crisis situations to verify what is happening and support crisis decision making, using natural language processing techniques to discover and track events and subevents [10].

Research has characterized the cognitive impact of the large volumes of information provided to users as they review social media. Users' divided attention can reduce the likelihood that a user will find, much less act upon, an individual post by a friend [9]. Within only a few seconds, users of sites such as Twitter make decisions about whether a post is of interest. Some of the indicators provided on these sites—for instance, likes and retweets—may distract the user rather than aiding in focusing their attention on the content [6]. Kim and Sin highlighted the importance of understanding the task that the user is trying to perform so as to offer "purpose-based personalization" [11].

The behavior of users when performing an update task may be repetitive. In the case of change detection, they may run a daily search for the same topic;

when following an event, the user may run a search for a term or a hashtag over and over while the event is happening. These repeated queries are like a session search that runs over an extended period of time—days, months, or years. Session search considers queries in an evolving way; users update and refine their searches based on what they find [5]. Similarly, users who are getting updates on social media may adjust their searches for hashtags, people, and other things based on specifics at a given moment.

One might think of the user's process for following updates on social media as a reading comprehension task. Reading is a "constructive act" in which the reader creates mental representations of the material [8]. Given the active, participatory nature of the medium, social media users can be thought of as "reading-to-write." In other words, users often are poised to react to something they view or read (liking, commenting, and more). Flower et al. note, "the reader as writer is expected to manipulate information and transform it to his or her own purposes" [7]. Put into this context, users' expressions of a desire for chronological ordering may not be surprising. "To achieve a proper understanding of the situation described by a text, the reader needs to know when the described events took place both relative to each other and relative to the time at which they were narrated" [24]. Social media feeds do not appear to be optimized for comprehension. The burden currently is on the user to determine how a new post connects to an overarching story about a person, group, or topic. Social media adds a further cognitive burden, as users often weigh credibility when putting the post into a broader context [3].

3   Research Methods

To understand sort order preferences in social media, we surveyed adult social media users in the United States. People engage in many behaviors on social media; we scoped our study to focus on update tasks—specifically change detection and experiential uses—to help us learn the extent to which users are satisfied with how sites enable them to follow a story over time. We also asked questions about more general social media use such as browsing and searching, to allow us to see how similar or different the actions and expectations of users are across those scenarios. This helps us understand users' social media sort preferences and distinguish whether posts airing frustrations are isolated events (e.g., immediately after an interface change), or if they represent large-scale or long-term concerns. This is not intended to be a comprehensive study of all user behaviors in social media; our general focus is on tasks that relate to following a story over either a short or a long period of time.

Surveys are a widely used research method that can be useful for gaining a qualitative understanding of users' attitudes, intent, and information about their experiences across a population [16]. An online survey is a low-cost method to get input from a broad set of users. This approach can also provide insight into which social media systems users associate with specific tasks. Additionally, a survey allows us to gain some insight into why the users prefer a certain approach, and what users perceive as lacking in other social media sites. This is an initial step toward understanding the scope of story-driven update tasks across social media, which can be further investigated in complementary ways, such as with user studies.

3.1 Survey Design

Our survey on sort orders in social media included five sections. First, users viewed a consent page containing an overview of the purpose of the study and information about Institutional Review Board approval. They were then asked to consent to participate. Second, we asked three screening questions to validate that participants were adult social media users located in the United States. Third, we asked demographic questions focusing on location, gender, ethnicity, education, and employment. Fourth, we asked for background on which social media sites the respondents used, how frequently, and which of the four tasks they performed. The fifth section consisted of sets of questions for each use case. Users were only required to complete one set of use case questions. After answering for one, they could optionally answer questions for additional use cases. Upon completion of the survey, interested users could submit their email address for an optional raffle for one of five $25 gift cards. While most of our questions were consistent across the use case sections, we included additional questions specific to the update tasks. For example, in the change detection section we asked respondents about their level of expertise on the topic, and the amount of time they have followed it. Questions about sort satisfaction and sort vs. clustering preference were in all use case sections. The survey questions included a mix of closed- and open-ended questions.

We distributed the survey online, using a convenience sampling approach. We reached out to potential participants via email and social media sites— specifically, on Twitter, Facebook, and Reddit. This included advertising to users during events with significant online followings, such as the Super Bowl. In addition to social media posts about our survey, we sent targeted posts to specific Twitter and Facebook users who had made comments about social media sort orders. Our email campaign focused on US universities, in particular on programs with an emphasis on communications, journalism, and library and information science.

3.2 Research Questions

We had four research questions (RQ) for this study. We believed that the use cases were all prevalent across our respondents, and that there would be a distinct pattern of sort and clustering preferences for each use case.

RQ1: How Prevalent are Each of the Use Cases? We hypothesized that these use cases are common among social media users. We anticipated that browsing would be the most common, followed by searching. Of the update tasks, we believed that change detection would be more common than experiential use of social media.

RQ2: How do Users Prefer to have Results Sorted for a Specific Task? Our hypothesis was that the update tasks would prefer chronological ordering. We thought that users engaging in browsing were there for entertainment value and would have no sort preference. We hypothesized that users engaged in searching would prefer relevance ranking.

RQ3: Would Users Accept Clustering as an Approach for Organizing a Feed? We thought that users would be more interested in clustering for some tasks than others. For change detection and experiential use, we believed users would like to have new posts on the topic or event clustered by theme or development. We did not expect that users focused on browsing or searching would be as interested in clustering posts by theme, user, etc.

RQ4: How Many Posts do Users Feel they need to see for Each Task? Our hypothesis was that users engaged in update tasks would only want to see a few posts per event or development. We anticipated that users engaged in browsing would want to see more, and searchers would only want a few posts.

4 Survey Results

We received 193 valid responses from the survey participants. Five of these respondents did not perform any of the four tasks being studied; we provide the results for the remaining 188 responses in this paper. Despite the potential bias that can be a factor in surveys, our findings provide a starting point for future studies to build upon. Given that there are approximately 181 million social media users in the US, at a confidence level of 95%, our results would have a margin of error (confidence interval) of 7% if our sampling was random.[2]

4.1 Overview of Responses

All four of our use cases were prevalent among the respondents, as detailed in the *Frequency* column in Table 1; this is the percentage of the 188 respondents who said they perform the task. Browsing was the most common task, with approximately 96% of users indicating that they browse social media without a specific task or goal. RQ1 involved understanding how prevalent each use case is within social media. Even accounting for the margin of error for the question of how frequent each use case is performed, it appears that many US users perform the tasks studied in the survey. Table 1 displays the percentage of the 188 respondents who indicated that they perform each of the four studied tasks.

Note that not all respondents completed all four of the more detailed use case sections. Participants were required to provide detailed responses for at least one use case section. They were sent to a section containing a task that they indicated

---

[2] Due to our convenience sampling approach there may be sample bias in which we oversampled for people from certain fields, or individuals who had preexisting strong opinions about sort orders. This caveat applies to all confidence level statements throughout this article.

Table 1. Total respondents who perform each task.

Use case	Frequency
Change detection	66%
Experiential	49%
Browsing	96%
Searching	65%

they perform, and we preferentially assigned users to Change Detection when possible because of our interest in that use case. After completing one section, they were asked whether they wished to answer another section for another task they perform (if they performed multiple tasks).

As we had hoped, we have the tightest confidence intervals for change detection because that use case was disproportionately sampled. Use case sections answered by fewer participants have broader confidence intervals. For the change detection section we received a total of 116 complete responses (Conf. Interval ±9.1%). There were 55 responses for the experiential section (Conf. Interval ±13.2%). 93 users completed the Browsing section (Conf. Interval ±10.2%). The searching section received the smallest number of responses, with 31 completing that section (Conf. Interval ±17.6%).

4.2   Demographics of Respondents

The survey asked questions on the respondents' age, gender, race and ethnicity, location, highest level of education, current employment, level of computer use, and social media site use.

Age: Most respondents were under age 45. The most common group was 25–34 (32%), followed by 28% for users aged 35–44. Another 19% were in the 18–24 age group. 11% of the respondents were in the 45–54 age group, 5% in the 55–64 range, 4% were aged 65–75, and 1% between 75–84.

Gender: We received more responses from women than men. 67% of the respondents were female, while 29% were male. The remaining 4% identified as "other," or chose not to respond.

Race and Ethnicity: Most of the respondents indicated that they were white or Caucasian (77%). Asian respondents (12%) were the next most frequent, followed by black or African American (5%) and Hispanic or Latinx (3%). The remaining 4% selected other or prefer not to answer.

Location: We asked respondents to self-report their location by providing their U.S. zip code. More than half of the respondents were located east of the Mississippi river. The general locations of respondents are displayed geographically in Fig. 2.

Education: Survey respondents were relatively well-educated. On their highest level of education completed, 39% said they possess Master's Degrees, and 31%

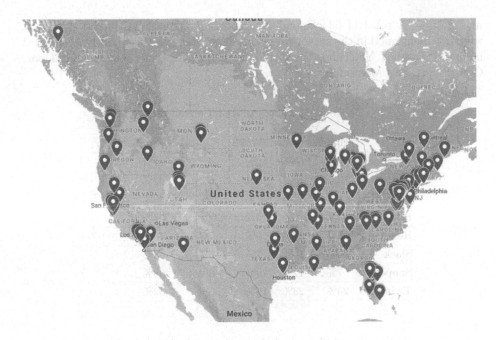

Fig. 2. Locations of survey respondents.

indicated that they have Bachelor's Degrees. 9% of respondents have Doctorates, 7% have attended some college but have no degree, 5% possess a professional degree such as a J.D. or M.D., 5% have a high school degree or equivalent, 3% have an Associate's Degree, 1% attended trade or vocational training, and 1% have less than a high school diploma.

Employment: 54% of respondents have full-time employment, and 16% are students. Another 14% have part-time jobs. 6% are self-employed, 4% are unemployed, 2% are retired, and 2% identify as homemakers. The remaining 2% selected "other." The industries employing the largest numbers of respondents were education and libraries, technology, government, and healthcare.

Computer Use: Many respondents indicated that they are online for significant time outside of work or school. 37% are online 2–4 hours per day. 28% spend 1–2 hours per day online, and 20% are online 4–8 hours per day. 9% spend 8–12 hours per day online, 6% spend only 10 min to 1 h online, and 1% spend more than 12 h online.

Looking at the way respondents access social media sites, 44% access these systems via their phones. Approximately 31% of respondents use laptop computers, 14% use desktop computers, and 12% visit social media sites from tablets.

Social Media Site Use: The most commonly used social media sites among respondents were Facebook, Instagram, and Twitter. Social media sites that many of the respondents said they use rarely or never were Snapchat, Slack, and Reddit. Table 2 lists the frequency that respondents make use of these social media sites. A total of 63 respondents reported using social media sites that were not mentioned in the survey questions. The sites most frequently mentioned included Tumblr, YouTube, Discord, WhatsApp, message boards, NextDoor, GoodReads, and Mastodon.

Table 2. Frequency of social media use.

Site	Hourly	Daily	Weekly	Monthly	Rarely/never
Facebook	15%	57%	12%	4%	11%
Instagram	13%	35%	15%	6%	32%
LinkedIn	2%	11%	21%	27%	40%
Pinterest	1%	6%	15%	21%	57%
Reddit	12%	17%	9%	7%	55%
Slack	10%	10%	7%	5%	68%
Snapchat	7%	12%	9%	3%	68%
Twitter	24%	23%	12%	9%	32%

One question that could be asked about our analysis is whether user opinions regarding organization might be systematically influenced by the organization of the social media sites that they currently use. To test for this, we performed chi-squared tests looking at the relationship between social media sites and respondents' level of satisfaction regarding current organization of posts, preferred sort order for the use case, clustering preference, and views on the number of posts that should be displayed. There were no statistically significant relationships between social media site use and any of these characteristics for any of our use cases. Thus, we have no reason to believe that users' social media ordering preference is derived from their use of a specific existing site.

4.3 Change Detection

When asked whether they use social media to follow a specific topic or theme over a long period of time, approximately 66% of respondents indicated that they perform change detection tasks. Common topics that change detection-focused users follow over time include news and politics (19%), television shows (12%), movies (10%), the music industry (9%), sports (8%), and art (8%). Other frequently mentioned topics included science and technology, research on their area of expertise, and travel. One user explained that they look for "TV shows and what people are talking about them. I search for them using hashtags."

When asked why users perform change detection tasks, the primary motivators were curiosity (26%), entertainment (24%), and a desire not to miss out on something interesting (22%). The vast majority were driven by personal reasons; few indicated that they follow topics over time because they are required by work or school (7%). That said, in text responses several users noted that, while they are not specifically required to stay up to date on their topic, they do so for professional learning.

The most common social media sites that respondents use for change detection tasks are Twitter (29%), Facebook (21%), Instagram (16%), and Reddit (15%). They typically get updates late in the day, with 28% of respondents focusing on change detection tasks in the evening, and 18% at night. Early morning (16%) and around lunchtime (15%) were less common.

Our survey included several questions specific to change detection, designed to gauge user expertise and time spent focusing on their topic. Most respondents indicated at least an intermediate level of knowledge, with 22% at the expert level, 40% self-assessing that they have an advanced understanding, and 35% at the intermediate level. No respondents rated themselves as having only a basic understanding of their topic. A substantial majority have been following their topic over a long period of time, and 42% have been keeping up with their topic for more than 5 years. 27% of respondents have followed their topic for 2–5 years, and an additional 16% for 1–2 years. 42% of respondents spend 10 min to 1 h getting updates each day, 25% less than 10 min a day, and 23% 1–2 hours a day. Most respondents had checked on their topic recently before taking the survey; 30% within the past hour, another 35% within the past day, and 26% within the past week.

4.4   Experiential

Experiential use cases were the least common, though still prevalent within the respondent groups. 49.5% of respondents indicated that they have used social media to follow or interact with a live event.

The kinds of topics users follow live are conferences and talks (20%), sporting events (19%), political events (17%), television shows (10%), and live streaming video (10%). Respondents also mentioned award shows, emergencies such as natural disasters, and professional events. Even though the primary experience took place during the event itself, some users also looked for context surrounding the event. As one respondent said about watching college football bowl games, "... I wanted see what others were thinking leading up to the game and after the football game ended."

The main actions that users take when interacting with live events are to browse posts of friends and people they follow (27%) and browse other users' posts (27%). 21% of respondents also interact with their friends and people they follow, and 20% interact with other users' posts.

Twitter was the most common social media site that respondents use for engaging with a live event, with 44% indicating usage of that site. The next most prevalent sites for experiential use were Facebook (30%) and Instagram (13%). 29% of respondents for this section indicated that they follow events that take place in the evening (31%), at night (20%), or in the afternoon (17%). 40% of respondents said they spend 10 min to 1 h focused on a live event; 24% spend 1–2 h, 22% spend less than 10 min following the event, and another 11% spend 2–4 h on the event.

4.5   Browsing

The most common category, 96.3% of responses indicated some type of browsing activity. Respondents' answers to an open-ended question about what they expected to see when browsing led us to the decision to split the category into two parts. Even after the split in browsing results, time-oriented browsing remained a very common task across respondents; 69 of the use case section responses provided details about the goals of this update task. Only 24 of the participants appeared to focus on general browsing.

Time-Oriented Browsing. The most frequent word that appeared within the category was "update." Even though the original question framed browsing as an activity without a specific goal, many users' responses indicated that there actually was a goal: to get updates on friends, family, and news. Further research is needed to determine why users responded to this as a distinct activity from change detection, since there are similarities between the two. These users might not know from session to session which items would be updated, but they had a sense that it would fall into a finite set of categories. These users were also interested in topics such as art, health, movies, television shows, weather, financial information, music, and the outdoors. One respondent said, "It's everything. I follow specific bloggers, friends, family, local news, national news, international news, hobbies, and interests."

The main reasons why respondents said they perform time-oriented browsing were out of curiosity (31%), for entertainment (26%), out of a desire not to miss something interesting (15%), because their friends are talking about something (14%), and because their family is talking about something (7%). A common response supplied in the "other" section was that these users are motivated by boredom.

The main social media sites that people performing time-oriented browsing use are Facebook (30%), Instagram (18%), Twitter (16%), Reddit (12%), and Pinterest (9%). People perform this task in the evening (33%), at night (22%), in the early morning (17%), around lunchtime (12%), or in the afternoon (9%).

General Browsing. After splitting the browsing category, this became the least common, with only 24 of the participants providing detailed responses about general browsing. These users mentioned a variety of non-time-oriented topics of interest. For example, common topics included fashion, photography, memes, articles, and interesting ideas. News was still a popular topic within this set of respondents. One user stated, "I browse fashion, home decor, recipes, memes, and more on Pinterest on a daily basis." Another referred to searches for odd or overdone humorous content with their comment, "Unironically, dank memes."

Respondents engaged in general browsing undertake this activity out of curiosity (35%), for entertainment (30%), out of a desire not to miss something interesting (14%), because their friends are talking about something (10%), and because their family is talking about something (6%).

The social media sites that respondents indicated were most used for general browsing were Facebook (24%), Twitter (21%), Reddit (19%), Instagram (16%), and Pinterest (9%). This browsing activity tends to happen later in the day, with 27% of respondents doing general browsing at night, 26% in the evening, or 15% around lunchtime.

4.6   Searching

Searching was a common use case, with 65.4% of respondents indicating that they have run searches on social media sites. Respondents run searches in social media for a wide variety of reasons. The most common explanations included learning more about a recent event (18%), looking up organizations and places (17%), and looking for specific users' profiles (15%). Respondents also use social media search capabilities to research events, news, trending topics, and to find articles. These searches are often driven by other content being posted to social media. For instance, as one user noted, they "...search for trending hashtags. Or look at someone's profile." Another looks "...for more information about someone who has commented."

Social media searchers run queries out of curiosity (30%), for entertainment (19%), because their friends are talking about something (17%), because they don't want to miss out on something (16%), because their work requires it (9%), or because their family is talking about it (7%).

The most common social media sites on which respondents ran searches were Facebook (26%), Twitter (21%), Pinterest (15%), Reddit (13%), and Instagram (10%). Respondents typically run searches in the evening (33%), at night (23%), in the afternoon (14%), or in the early morning (12%).

5   Organization Preferences

Respondents expressed the largest amount of dissatisfaction with the way posts are organized for browsing, something we did not expect given that we initially thought of browsing as an entertainment-focused task. That said, when asked specifically what orders they prefer for organizing social media posts, users preferred chronological order for update and browsing tasks. Clustering preferences were more varied.

5.1   Current Sort Orders

While not a formal research question, we wanted a sense of the prevalence of user frustrations with social media sort orders, to put social media posts asking for different sorting into broader perspective. For all use cases except browsing, the most prevalent response to this question was "neutral." Figure 3 details the levels of satisfaction for each sort order. Users expressed the highest amount of dissatisfaction with sort orders for time-oriented browsing. 52% of respondents indicated that they were unsatisfied or very unsatisfied, while only 17% said

they were satisfied or very satisfied. In contrast, only 33% of users who perform
general browsing were unsatisfied with current sort orders, equal to the number
who expressed that they were satisfied or very satisfied.

Respondents who perform change detection tasks were statistically significantly more unsatisfied with current social media sort orders (t-value: 2.92, DF:
44.31, p-value: 0.01).[3] 37% of respondents indicated either being unsatisfied or
very unsatisfied, compared to 19% selecting satisfied or very satisfied. Users
engaged in experiential tasks had more neutral to positive feelings about current
social media sort orders when following live events, with 55% of respondents
indicating that they are neutral. 26% described themselves as satisfied or very
satisfied, and only 20% said they are unsatisfied or very unsatisfied.

Interestingly, for the two update tasks there was a statistically significant
difference between responses of different age groups. Respondents aged 35–54
were statistically significantly more likely to be negative than those aged 18–34
about current sort orders for performing change detection tasks (t-value: 3.51,
DF 102.16, p-value <0.01). For experiential tasks, respondents from 35–54 were
more likely to be unsatisfied than respondents from 18–34, though when a Bon-
ferroni correction for multiple tests was applied, this result was not statistically
significant (t-value: 2.39, DF: 44.05, p-value: 0.02).

Users engaged in searching had more positive than negative responses to
questions about the status quo. 29% of these respondents were satisfied with
their experience, and 25% were unsatisfied or very unsatisfied.

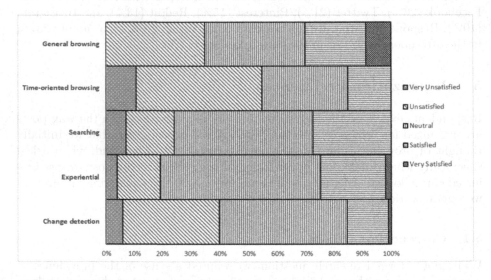

Fig. 3. Satisfaction with current organization of posts.

<hr>

[3] For statistical significance tests in this section, two-tailed one sample t-tests were
applied.

5.2   Preferred Sort Orders

To answer RQ2, we asked respondents about their sort order preferences in two ways: first, an open-ended question, followed by a closed-ended set of selections. After coding text responses, we found that users provided consistent responses to both questions. For change detection, experiential, and time-oriented browsing, users prefer chronological orders by a wide margin. Even accounting for margin of error, users engaged in these activities were significantly more likely to prefer chronological sort. Users who are running searches tend to prefer relevance ranked results, though these responses were not statistically significantly more frequent than those who wanted search results to be chronological. Those respondents who performed general browsing wanted results to be either relevance ranked or ranked by popularity. Figure 4 displays the responses to the closed-ended questions about sort order preferences.

In their text responses, a number of users expressed their preference in terms of what they did not want. For example, 11 users responding to change detection noted that they do not want sorting to be based upon proprietary algorithms, or approaches using popularity as the driving factor. A common trend was expression of a desire not to miss out. "The algorithm-based feeds are harder to follow over long periods of time because you are guaranteed to miss something," according to one user. Another user was more blunt about preferences: "Literally in order of when they were posted. No freaking algorithms." A user describing organization preferences for browsing said, "...I just want a straight up timeline style feed. I want to see things as they happen, not two days later."

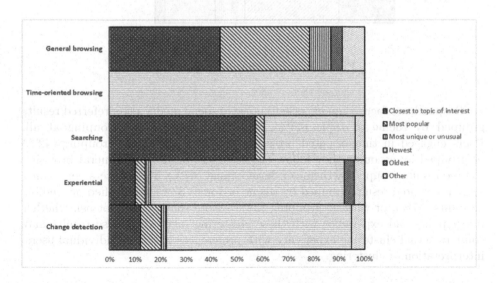

Fig. 4. Sort order preferences.

5.3   Clustering

To address RQ3, we asked users whether they would like to have results grouped in some way. Figure 5 shows the most popular options, which for most use cases was a grouping by theme or development. None of the differences for this section were statistically significant, but they provide a general sense for possible preferences that could be explored further. For change detection tasks, users preferred to have results grouped by themes or developments relating to the overall topic (43%), with the next most popular being grouped by followed accounts (23%) and no grouping (20%).

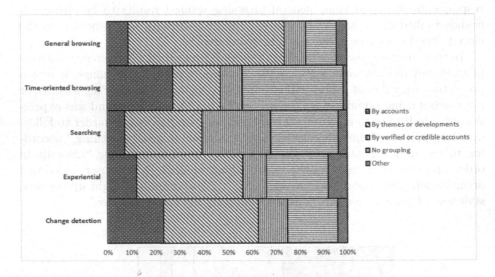

Fig. 5. Preferred clustering approach.

Respondents focused on experiential use of social media also preferred results grouped by theme or topic (45%), with 26% wanting no grouping at all. Users engaged in time-oriented browsing tended to prefer no grouping (42%) or grouped by accounts they follow (28%). Users performing general browsing wanted results grouped by theme or development (67%). 37% of users performing searches wanted results grouped by theme or topic (37%), by verified or credible accounts (27%), or with no grouping (27%). Note that we did not ask whether participants had experience with these clustering methods. These results may conflate actual clustering experience with preferences based on individual users' interpretation of clustering.

5.4   Numbers of Posts

We asked respondents to answer two questions to help us understand RQ4: would they prefer to see all posts vs. only the best posts for certain categories, and what number of posts would they like to see for new developments. Differences in responses for these questions were not statistically significant, but provide general insights into preferences that can be tested further with more specific experiments. Users engaged in change detection prefer to see all posts by people they follow (49%), with "only the best posts related to the topic I follow" as a distant second (19%). As for the number of posts, they prefer 5–10 posts per theme (31%) or all posts on the theme (31%). For experiential use cases, respondents want to see all posts related to the topic (43%) or all posts by people they follow (21%). Time-oriented browsing users were the most focused on seeing everything—they would like to see all posts by the people they follow (70%), and all posts on new themes that come up (43%), or 2–4 posts per theme (a distant second, at 24%). Users engaged in general browsing were split between wanting only the best posts by people they follow (21%) and only the best posts for each theme or development (21%). These users wanted to see 5–10 posts per theme (42%). Search-focused respondents wanted to see either the best posts related to the topic (28%) or all posts about themes or developments (28%). 34% want to see all posts on the theme.

5.5 Ending a Session

We asked how respondents decide they are done with their session for each use case. The results for this section provide general insights, but differences are not statistically significant. Change detection-focused respondents were split between "when I have found the relevant themes and developments" (25%), "when I have read all of the relevant posts" (25%), and "when I have answered my questions" (23%). For experiential use cases, respondents were generally done when the event was over (26%) or when they have read all the relevant posts (23%). Time-oriented browsers were done when they ran out of time (33%) or when they have read all the relevant posts (32%). General browsers were done when they ran out of time (42%) or met some other criteria (e.g., got bored, got an idea, or found something interesting to follow up on) (17%). Users searching were done when they answered their questions (48%), read all relevant posts (24%), or ran out of time (21%).

6 Implications for System Design

In many of the use cases studied here, the user intends to have a deep interactions with social media content. Responses about update tasks indicated that users engaging in these activities at least claim to favor recall over precision. Generally, users wanted to see "all" of something (all posts on the topic, all by people they follow, all on the theme, etc.). They may favor an interface that provides some amount of clustering with at least the ability to get to all relevant posts in the appropriate order (likely chronological).

Thinking of each user as a bundle of interests, a single feed can contain a large amount of unrelated information. For users with a specific intent—for instance, following posts about a live event—it can be difficult to find relevant commentary when the user sees all posts (relevant to the event or not) flowing into the same stream, organized by relevance to the user's overall interests, popularity, and other factors. Popular search engines have made a transition from a single stream of ranked results to blocks of related content, each organized by relevance to user needs (aggregated search). We suggest taking a cue from aggregated search, in case such presentation approaches can be integrated into social media feeds as well. Aggregated search involves grouping together like sets of "information nuggets." Relevant content within these blocks (referred to as "verticals") are sorted, and then the blocks themselves are also sorted [13]. Disentangling a user's social media feed is a more complex undertaking than providing responses to a query. That said, based on our survey results, we believe there may be tacit indicators to help identify and adjust the organization of the feed for a specific activity, in a way that connects an individual post to an overarching story. Sites like Twitter provide topic and event-centric views in addition to user-curated lists. Twitter's Explore tab includes content customized based on user interests, as well as topics—to include feeds related to live events [22].

The large numbers of "neutral" responses about current ordering of social media feeds may indicate some flexibility in accepting new arrangements, especially considering the mismatch between the current views and the preferred sort for update tasks (chronological). Users may not feel tied to a specific organization approach. Bron et al. looked into design of aggregated search for multi-session search tasks; this revealed that some users tend to prefer tabbed interfaces over blended interfaces [4]. Arguello et al. found that more complex tasks resulted in higher search and content interaction. Unlike Bron et al., Arguello et al. found a preference for blended aggregated interfaces, which could be attributed to users' technical sophistication levels [1]. This kind of comparison could serve as an initial starting point for update-focused interface design, with distinct views for a specific topics, themes, or events, while making it easy for the user to return to a general feed for serendipitous discovery.

7   Limitations

We recognize that there are limitations inherent in using surveys. First, a survey represents only a snapshot in time. Behaviors change over time, and approaches shift as capabilities of systems change or users adopt new technologies. Only a limited number of people are willing to take surveys, and—based on our convenience sampling approach—our respondents could over-represent users who have strong feelings about sort orders within social media (sample bias). Given the length and complexity of our survey, we may have further limited the number of respondents; we made a tradeoff decision to opt for a greater level of detail in the questions we asked of the social media user base.

We attempted to overcome these limitations in several ways. We shared the survey through a variety of channels to gather responses outside our local demographic. We also asked questions multiple ways to compare responses—for instance, respondents were asked to provide a textual description of their sort preferences, then later were able to select from a predefined set of categories. We also ran a small pilot of the survey involving read-aloud and online versions of the survey, which we used to clarify wording and adjust the overall survey flow.

8 Conclusion and Future Work

We ran a survey to understand user behavior and user preferences relating to four distinct use cases within social media. We focused on update tasks, which were divided into change detection and experiential use cases. The survey results revealed that users have distinct behaviors and data organization preferences for each of the use cases. Additionally, analysis of results indicated that browsing preferences could be split into two subcategories—general browsing and time-oriented browsing. The results from this study could be used to inform future interface design for social media systems where users perform these activities, to improve users' experience with these systems.

Based on the findings from this study, we plan to run a hands-on user study focusing on the change detection needs of a user who follows the same topic over a long period of time. The follow-on study tests whether the preferred organizational approaches identified in this study can lead to successful task completion. Users will complete change detection tasks in a prototype interface, and system designs will be evaluated based on how well they aid the user in completing the task. This will give us insights into optimizing arrangements of clustered results. Additionally, we will test the notion of the recall-oriented nature of update tasks as it relates to a fear of missing out. The survey results reveal that users generally prefer to see "all" data. To what extent do they strive to see every post, driven by a quest for digital omniscience? We will test how well their need is satisfied if they are unable to view all posts about key developments on their topic of interest due to data volume and time constraints.

References

1. Arguello, J., Wu, W.C., Kelly, D., Edwards, A.: Task complexity, vertical display and user interaction in aggregated search. In: SIGIR 2012, p. 435. ACM Press (2012). https://doi.org/10.1145/2348283.2348343
2. Auxier, B., Anderson, M.: Social media use in 2021 (April 2021). https://www.pewresearch.org/internet/2021/04/07/social-media-use-in-2021/. Accessed 2 June 2022
3. Barzillai, M., Thomson, J., Schroeder, S., van den Broek, P.: Learning to Read in a Digital World. John Benjamins Publishing Company, Netherlands (2018)
4. Bron, M., van Gorp, J., Nack, F., Baltussen, L.B., de Rijke, M.: Aggregated search interface preferences in multi-session search tasks. In: SIGIR 2013, pp. 123–132. ACM (2013). https://doi.org/10.1145/2484028.2484050
5. Carterette, B., Clough, P., Hall, M., Kanoulas, E., Sanderson, M.: Evaluating retrieval over sessions: the TREC session track 2011–2014. In: SIGIR 2016, pp. 685–688. ACM (2016). https://doi.org/10.1145/2911451.2914675

6. Counts, S., Fisher, K.: Taking it all in? Visual attention in microblog consumption. In: AAAI Conference on Weblogs and Social Media (2011)
7. Flower, L., Stein, V., Ackerman, J., Kantz, M.J., McCormick, K., Peck, W.C.: Reading-to-write: exploring a cognitive and social process. Oxford University Press, Oxford (1990)
8. Haas, C., Flower, L.: Rhetorical reading strategies and the construction of meaning. Coll. Compos. Commun. **39**(2), 167–183 (1988). https://doi.org/10.2307/358026
9. Hodas, N.O., Lerman, K.: Attention and visibility in an information rich world. In: 2013 IEEE International Conference on Multimedia and Expo Workshops (ICMEW), pp. 1–6 (July 2013). https://doi.org/10.1109/ICMEW.2013.6618396
10. Imran, M., Castillo, C., Diaz, F., Vieweg, S.: Processing social media messages in mass emergency: a survey. ACM Comput. Surv. **47**(4), 67:1-67:38 (2015). https://doi.org/10.1145/2771588
11. Kim, K.S., Sin, S.C.J.: Perceived usefulness of social media features/elements: effects of coping style, purpose and system. Proc. ASIS&T **54**(1), 722–723 (2017). https://doi.org/10.1002/pra2.2017.14505401131
12. Kim, K.S., Sin, S.C.J., He, Y.: Information seeking through social media: impact of user characteristics on social media use. Proc. ASIS&T **50**(1), 1–4 (2013). https://doi.org/10.1002/meet.14505001155
13. Kopliku, A., Pinel-Sauvagnat, K., Boughanem, M.: Aggregated search: a new information retrieval paradigm. ACM Comput. Surv. **46**(3), 1–31 (2014). https://doi.org/10.1145/2523817
14. Lochrie, M., Coulton, P.: Mobile phones as second screen for TV, enabling inter-audience interaction. In: International Conference on Advances in Computer Entertainment Technology - ACE, p. 1. ACM Press (2011). https://doi.org/10.1145/2071423.2071513
15. Naini, K.D., Kawase, R., Kanhabua, N., Niederée, C., Altingovde, I.S.: Those were the days: learning to rank social media posts for reminiscence. Inf. Retr. J. **22**, 159–187 (2018). https://doi.org/10.1007/s10791-018-9339-9
16. Olson, J.S., Kellogg, W. (eds.): Ways of Knowing in HCI. Springer, Heidelberg (2014). https://doi.org/10.1007/978-1-4939-0378-8
17. Pittman, M., Reich, B.: Social media and loneliness: why an Instagram picture may be worth more than a thousand Twitter words. Comput. Hum. Behav. **62**, 155–167 (2016). https://doi.org/10.1016/j.chb.2016.03.084
18. Postman, N.: Amusing Ourselves to Death: Public Discourse in the Age of Show Business. Penguin, London (1985)
19. Rushkoff, D.: Present Shock: When Everything Happens Now. Current (2013)
20. Sadler, N.: Fragmented Narrative: Telling and Interpreting Stories in the Twitter Age. Routledge, London (2021)
21. Saini, M.K., Al-Zamzami, F., Saddik, A.E.: Towards storytelling by extracting social information from OSN photo's metadata, WISMM 2014, pp. 15–20. ACM (2014). https://doi.org/10.1145/2661714.2661721
22. Twitter Help: How to use the Explore tab. https://help.twitter.com/en/resources/twitter-guide/topics/how-to-get-started-with-twitter/how-to-use-the-explore-tab-twitter-help. Accessed 2 June 2022
23. US Census Bureau (July 2021). https://www.census.gov/quickfacts/fact/table/US/PST045219. Accessed 2 June 2022
24. Zwaan, R.A., Radvansky, G.A.: Situation models in language comprehension and memory. Psychol. Bull. **123**(2), 162 (1998)

A Property Checklist to Evaluate the User Experience for People with Autism Spectrum Disorder

Katherine Valencia[1,2](✉) [iD], Federico Botella[2] [iD], and Cristian Rusu[1] [iD]

[1] Pontificia Universidad Católica de Valparaíso, Av. Brasil 2241, 2340000 Valparaíso, Chile
katherine.valencia.c@mail.pucv.cl
[2] Universidad Miguel Hernández de Elche, Avenida de la Universitat s/n, 03202 Elche, Spain

Abstract. Autism Spectrum Disorder (ASD) refers to a neurodevelopmental disorder which is characterized by repetitive patterns, and difficulties with social interaction and communication. People with ASD have, an affinity with technology; many studies designed and/or developed specific systems for this kind of users. Some studies have evaluated the usability and UX of such systems through general evaluation methods. It is important to consider adaptations to the UX evaluation methods when evaluating systems designed for people with ASD, because these users have a diversity of characteristics, affinities and needs that traditional methods do not consider. Given the need for specific evaluation methods, in this paper we propose a property checklist for interactive systems designed for people with ASD, based on our previous proposal of 9 UX factors. Two versions of the property checklist are proposed, a full version and a compact version, which include 9 categories (directly related to the 9 UX factors), and 50 and 24 items, respectively. These property checklists are intended to assess compliance with specific recommendations and may help evaluating UX in systems designed for people with ASD.

Keywords: Autism Spectrum Disorder · User eXperience · Property checklists · UX factors for ASD

1 Introduction

Autism Spectrum Disorder (ASD) is a neurodevelopmental disorder that people can suffer from, which is characterized by restricted repetitive patterns of behavior, interests, activities, and alterations in social communication and social interaction [1]. After a systematic review of the literature [2] carried out in previous works, we have been able to show an increase in the interest of research in developing and designing systems for people with ASD, mainly because technology provides the users with a reliable and safe environment. Because people with ASD may have a diversity of characteristics, needs, and affinities, we have reviewed in the literature [3] how studies consider these characteristics when developing and designing systems for these users. By analyzing the studies found [2, 3], we have realized that concepts such as usability and/or User

© The Author(s), under exclusive license to Springer Nature Switzerland AG 2022
G. Meiselwitz (Ed.): HCII 2022, LNCS 13315, pp. 205–216, 2022.
https://doi.org/10.1007/978-3-031-05061-9_15

eXperience (UX) are concepts considered and applied by the authors, and some of them have evaluated these concepts through multiple evaluation methods. However, these studies have evaluated their systems through assessment methods without considering the characteristics, affinities, and needs of people with ASD.

Using particularized evaluation methods, that consider the characteristics, needs and affinities of people with ASD, is necessary when evaluating systems designed for said users. Given the need for evaluation methods and a particular evaluation methodology for people with ASD, we have described a preliminary proposal for a methodology to evaluate UX in systems for people with ASD [4]. The methodology recommends a set of user testing and inspection methods that can be applied. Our proposed methodology recommends that the first inspection method to be carried out, when evaluating systems designed for people with ASD, is the property checklist method.

In this paper we propose two particularized property checklists to preliminarily evaluate systems for people with ASD. These checklists are presented as a full version and a compact version, which include 9 categories, and 50 and 24 items, respectively. We have based these property checklists on our proposal of 9 UX factors for people with ASD [5]. These factors are: engaging, predictable, structured, interactive, generalizable, customizable, sense-aware, attention retaining and frustration free.

This rest of this paper is organized as follows. Section 2 presents a theoretical background. Section 3 presents and describes a proposed property checklist based on the nine UX factors for people with ASD. Finally, Sect. 4 presents conclusions and future work.

2 Theorical Background

2.1 Autism Spectrum Disorder

Autism Spectrum Disorder is a condition characterized by repetitive patterns, difficulties with social interaction and communication, as defined in the fifth edition of the Diagnostic and Statistical Manual of Mental Disorders (DSM-5) [1]. DSM-5 remarks that this condition is characterized by two domains of deficits, such as: (1) social communication and social interaction, and (2) restricted repetitive patterns of behavior, interests, and activities. Additionally, depending on the degree of support that people with ASD require, they are categorized under three categories of severity: level 1 "Requiring support", level 2 "Requiring substantial support" and level 3 "Requiring very substantial support".

2.2 User Experience

The ISO 9241-11 standard defines usability as "extent to which a system, product or service can be used by specified users to achieve specified goals with effectiveness, efficiency and satisfaction in a specified context of use" [6]. User eXperience (UX) extends the concept of usability beyond effectiveness, efficiency and satisfaction. The ISO 9241-210 standard defines UX as "user's perceptions and responses that result from the use and/or anticipated use of a system, product or service" [7]. That is, UX is the degree of "satisfaction" that users obtain when interacting with the system or product.

2.3 Property Checklist

A property checklist is a list of criteria or elements organized in a systematic and logical way [8], allowing users to verify the presence or absence of listed elements to guarantee that all are considered or completed. Property checklists are related to a series of high-level design properties, such as consistency or feedback, and low-level design properties, such as the color or size of the characters on the screen [9].

A property checklist is an inspection method, which means that it is a review carried out by a set of evaluators, who are generally experts, based on their own judgment [10]. There is no final user involvement. The fulfillment of the items detailed in the property checklist are verified by the evaluators. On some occasions the elements of the property checklist can give clues to design solutions [9].

3 Property Checklist Proposals

3.1 Developing the Property Checklist

In a previous systematic literature review [2] we have found that the UX evaluation methods used in multiple studies related to systems designed for people with ASD, were not adequate for these users and needed a particularization. People with ASD present characteristics which must, undoubtedly, be considered when designing solutions for them and evaluating said solutions. It is important to have methods and/or methodologies that consider the characteristics, affinities and needs of people with ASD. For the same reason, in previous works we have proposed a methodology to evaluate the user experience in systems designed for people with ASD [4]. The methodology considers 3 sequential stages, these are: "planning stage", "execution stage" and "results analysis stage". The first step in the "execution stage" is the execution of a property checklist, which considers the execution of the property checklist method, which has the role of providing a quick first view of the shortcomings that the evaluated system may have.

To define our specific property checklist for use in systems for people with ASD, we have relied on studies that have developed checklists focused on evaluating the usability of different systems [11–13]. These studies detail that, to design the questionnaire for a specific property checklist, a usability criteria or factors must be used as the basis to define the questions defined in the checklist.

These criteria must be specific to the context in which the checklist is to be used, so to design a specific questionnaire for systems for people with ASD, we have based it in our proposal of 9 particularized UX factors for people with level 1 ASD [5] (according to the severity levels of the DMS-5 [1]). These 9 particularized UX factors have been created based on Morville's UX factors [14], characteristics, affinities and needs of people with ASD, and guide design recommendations found in the literature. The 9 particularized UX factors are: engaging, predictable, structured, interactive, generalizable, customizable, sense-aware, attention retaining and frustration free.

Considering these 9 UX factors we have created a proposal for a property checklist to evaluate the UX for people with ASD. For each item we are specifying the original references. In some cases, references are missing, as some items are based on our own judgement. This property checklist proposal is presented below.

1. **Engaging.** This category focuses on evaluating how the systems provide elements that commit the user to interact with them. The category is defined as: "The system commits the user to interact with it". Elements like feedback, rewards, and motivational elements are at the core of this category [3, 15–23]. To evaluate compliance with these elements, we propose 6 items:

   1.1. The system delivers constant feedback to the user in a clear and concise manner. This must be not only through text but also through visual elements or audio [19].
   1.2. The system does not deliver demotivating messages to users in the event of a mistake made [18].
   1.3. The system provides rewards to the user for positive actions or good performance [3].
   1.4. The system has a history of the actions performed by the user.
   1.5. The system allows users to see the performance of users over a period of time.
   1.6. The system allows users to view past activities, so they can perform them again.

2. **Predictable.** The predictable category focuses on the structured thinking of people with ASD. Systems need to provide a predictable environment, which is why this category is defined as: "The system has a predictable environment, generating an environment of trust among users". Concepts such as the repetition of actions, control over the system and consistency in the system, have been considered for the fulfillment of this category [3, 15, 18, 20–25]. A total of 4 items details the predictable category:

   2.1. The system doesn't have sudden and unexpected actions.
   2.2. The content of the system is predictable and consistent [3].
   2.3. The system allows the user to pause, restart and/or cancel actions [22].
   2.4. The system allows repeating actions, tasks or activities [15].

3. **Structured.** Category defined as: "The system is structured". Having a clear, logical, and simple structure is important when designing solutions for people with ASDs [3, 15, 18, 19, 21, 22, 24–26]. Considering this, 3 items have been established that address these aspects:

   3.1. The system's navigational setup, aesthetics, and content are structured and consistent [26].
   3.2. Navigation in the system is simple and logical [24].
   3.3. The expressions and language used in the system are consistent and intuitive.

4. **Interactive.** The interactive category focuses on the importance of considering the characteristics, affinities and needs that people with ASD present when interacting with the system. This category is defined as: "The system generates interactions taking into account the characteristics, affinities and needs of the users, as well as their difficulties in social interactions". Memory load, tasks that grow in complexity

and instructions with a clear and simple objective, are some of the aspects considered [3, 15–22, 24, 26, 27]. We propose 6 items for this category:

4.1. The system includes a variety of tasks that are simple, concise, and grow in complexity based on the progress of the user [22].
4.2. Each task has a clear and explicit objective [18].
4.3. Instructions are provided to users from the first interaction.
4.4. Instructions on the system are clear, simple, brief and context-appropriate [20].
4.5. The user can access the instructions at any time.
4.6. The system provides elements to minimize the memory load of the user (for example: grouping and delivering the necessary and concise information to the user) [19].

5. **Generalizable.** Category defined as: "The system is familiar enough and similar to real life to facilitate generalizing skills". Having content, visual and audio elements in the system that are easy to interpret is important when working with people with ASD [3, 19, 20, 26]. Four items detail the generalizable category:

5.1. Activities, tasks or information in the system are based on previously learned activities, tasks or information [20].
5.2. The interaction with the system is familiar and similar to the real life of the users.
5.3. Visual aesthetics, audio and touch inputs are similar to real life.
5.4. The system has language, phrases, and concepts familiar to users [19].

6. **Customizable.** Considering that people with ASD have a diversity of characteristics, affinities and needs, this category focuses on aspects that help personalize the system [3, 15–23, 26, 27]. This category is defined as "The system can be customized considering the needs, abilities and preferences of people with ASD". A total of 5 items have been considered:

6.1. The system allows the user to customize frequent actions [19].
6.2. The system considers personalization in the event of possible fine motor problems in users [22].
6.3. The system allows users and tutors to customize aspects quickly, easily and effortlessly. (eg. disable sounds, configure the level and intensity of the sounds, modify the color palette, font type, size, layout and activity times) [26].
6.4. The system adapts to the level of expertise, needs and/or preferences of the user [19].
6.5. The system has a predefined basic configuration that considers the characteristics, affinities and needs of users with ASD [22].

7. **Sense-aware.** ASD people may have hyper- or hypo-reactivity to sensory input, so it is necessary for systems to consider these characteristics. The sense-aware category is defined as: "The system considers the senses of users with ASD". Having a clear,

understandable and relevant design, graphics and audio are some important aspects to consider [3, 15–29]. Below are 10 items to consider:

7.1. The system provides information to the user through multimedia, texts, among others [22].
7.2. The system provides light and dark color modes, and does not have bright colors, and contrasts that can be uncomfortable [26].
7.3. The visual and sound elements are clear, meaningful, functional, non-disruptive and legible.
7.4. Visual elements (such as icons) are clear and aesthetically pleasing and are provided to communicate ideas [18].
7.5. Each clickable element has a clear functionality and is easily selectable and recognizable [22].
7.6. The actions and states of the system are clear and simple.
7.7. The system interface is clear, simple and minimalist [22].
7.8. There is a prudent number of functionalities, images, texts, animations, among others, in each view of the system [22].
7.9. Texts have clear messages, readable font size and type, and are free of rhetorical figures.
7.10. The system uses a clear, familiar, precise and appropriate language.

8. **Attention Retaining.** The attention retaining category is defined as: "The system retains users' attention and manages time appropriately". Time management of transitions and activities, generating stimuli that help retain attention, avoiding elements that cause distraction, are relevant in systems designed for people with ASD [15, 18–22, 25, 27]. In order to evaluate this category, 5 items have been created:

8.1. The waiting time between transitions, tasks or activities is minimal [27].
8.2. The system responds to user actions in real time and without delays [19].
8.3. Transitions are simple, with no distracting sounds or animations.
8.4. The system provides dynamic stimuli such as animations and/or controlled music to attract users' attention [18].
8.5. The system views do not provide distracting elements [25].

9. **Frustration Free.** The frustration free category is defined as: "The system tries to avoid the frustration of its users during their interaction". Preventing, recognizing and recovering from errors are key elements in order not to frustrate users. In addition, it is important to keep in mind that the information communicated in this error handling must be clear, precise, with a close and simple language [3, 19, 20, 22, 25, 26]. Below is a total of 7 items to consider:

9.1. The system allows you to confirm, cancel or repair unwanted actions [22].
9.2. Impossible actions to be carried out in the system are notified in a timely, clear and simple manner [19].
9.3. Errors are presented in a controlled way, avoiding "strong" colors and loud noises.

9.4. Users are able to quickly find information and solutions to any problem, which are clearly presented and easy to execute [22].

9.5. The system displays error messages in plain language, accurately indicates the problem and constructively suggests how to avoid such errors [22].

9.6. The documentation and help provided by the system is provided in a visual, textual, concrete, not extensive and structured way.

9.7. The system asks the user clearly, precisely, simply and in real time to confirm actions [19].

3.2 An Abbreviated Property Checklist

Considering that the time and resources can be limited when carrying out the UX evaluation, we propose two property checklists:

- **Full version**: proposal that includes the 9 categories and 50 items detailed above. It is recommended to use this version if you have enough time and resources to be able to evaluate the compliance of each of the items and submit any comments.
- **Compact Version**: proposal that contemplates the 9 categories and 24 items. These 24 items have been selected from the full version list, and we believe that they represent the essential elements to comply in each of the categories. We recommend using this version if time and resources are limited.

The compact version of the property checklist is presented below.

1. **Engaging.** The system commits the user to interact with it. Category has 3 items in its compact version. These items have been chosen because they focus on feedback, rewards and progress. These elements are necessary according to the definition of the engaging UX factor [5].

    1.1. The system delivers constant feedback to the user in a clear and concise manner. This must be not only through text but also through visual elements or audio [19].

    1.2. The system provides rewards to the user for positive actions or good performance [3].

    1.3. The system has a history of the actions performed by the user.

2. **Predictable.** The system has a predictable environment, generating an environment of trust among users. Category that has 2 items in its compact version. These items have been chosen because they focus on concepts such as repetition and predictability. These elements are necessary according to the definition of the predictable UX factor [5].

    2.1. The content of the system is predictable and consistent [3].

    2.2. The system allows repeating actions, tasks or activities [15].

3. **Structured.** The system is structured. Category that has 2 items in its compact version. These items have been chosen because they focus on aspects such as structure, consistency and simplicity. These elements are necessary according to the definition of the structured UX factor [5].

   3.1. The system's navigational setup, aesthetics, and content are structured and consistent [26].
   3.2. Navigation in the system is simple and logical [24].

4. **Interactive.** The system generates interactions taking into account the characteristics, affinities and needs of the users, as well as their difficulties in social interactions. Category that has 4 items in its compact version. These items have been chosen because they focus on the characteristics that tasks, instructions and content must have, and minimizing memory load. These elements are necessary according to the definition of the interactive UX factor [5].

   4.1. The system includes a variety of tasks that are simple, concise, and grow in complexity based on the progress of the user [22].
   4.2. Each task has a clear and explicit objective [18].
   4.3. Instructions on the system are clear, simple, brief and context-appropriate [20].
   4.4. The system provides elements to minimize the memory load of the user (for example: grouping and delivering the necessary and concise information to the user) [19].

5. **Generalizable.** The system is familiar enough and similar to real life to facilitate generalizing skills. Category that has 2 items in its compact version. These items have been chosen because they focus on how tasks, activities, content and interaction with the system should be familiar and previously learned by the users. These elements are necessary according to the definition of the generalizable UX factor [5].

   5.1. Activities, tasks or information in the system are based on previously learned activities, tasks or information [20].
   5.2. The interaction with the system is familiar and similar to the real life of the users.

6. **Customizable.** The system can be customized considering the needs, abilities and preferences of people with ASD. Category that has 2 items in its compact version. These items have been chosen because they focus on the customization of system elements in a flexible and fast way. These elements are necessary according to the definition of the customizable UX factor [5].

   6.1. The system allows users and tutors to customize aspects quickly, easily and effortlessly. (eg. Disable sounds, configure the level and intensity of the sounds, modify the color palette, font type, size, layout and activity times) [26].
   6.2. The system has a predefined basic configuration that considers the characteristics, affinities and needs of users with ASD [22].

7. **Sense-aware.** The system considers the senses of users with ASD. Category that has 4 items in its compact version. These items have been chosen because they focus on the importance of having visual, audio and language elements that are clear, legible, understandable, minimalist and functional. These elements are necessary according to the definition of the sense-aware UX factor [5].

   7.1. The visual and sound elements are clear, meaningful, functional, non-disruptive and legible.
   7.2. The actions and states of the system are clear and simple.
   7.3. The system interface is clear, simple and minimalist [22].
   7.4. The system uses a clear, familiar, precise and appropriate language.

8. **Attention Retaining.** The system retains users' attention and manages time appropriately. Category that has 3 items in its compact version. These items have been chosen because they focus on elements to retain attention, elements that are not distracting and the response time of the system. These elements are necessary according to the definition of the attention retaining UX factor [5].

   8.1. The system responds to user actions in real time and without delays [19].
   8.2. The system provides dynamic stimuli such as animations and/or controlled music to attract users' attention [18].
   8.3. The system views do not provide distracting elements [25].

9. **Frustration Free.** The system tries to avoid the frustration of its users during their interaction. Category that has 2 items in its compact version. These items have been chosen because they focus on error handling (preventing, recognizing and recovering from errors) and communicating errors, solutions and documentation in a clear, simple and close manner. These elements are necessary according to the definition of the frustration free UX factor [5].

   9.1. The system displays error messages in plain language, accurately indicates the problem and constructively suggests how to avoid such errors [22].
   9.2. The documentation and help provided by the system is provided in a visual, textual, concrete, not extensive and structured way.

3.3 Using the Property Checklist

When using any of the proposed property checklists, we recommend:

- Having evaluators who are familiar with the context and the system to be evaluated. Considering the complexity and objective of the evaluation to be carried out. If required, and if you have enough time and resources, carry out one or two pilot tests
- Having the support of at least 3 or 5 UX experts.
- Having the support of at least 3 or 5 experts in the ASD domain, such as: experts in the areas of psychology, speech therapy and special education.

- For each item, the evaluator must indicate compliance with it through a 5-point scale. Please note that some items may not be assessed until the end of the assessment.
- The evaluators will be able to grant general observations for each one of the 9 categories.

The evaluators should identify the compliance with each item through a Likert scale of 5 options: "Totally not compliant" (1), "Not compliant" (2), "Neutral" (3), "Compliant" (4) and "Totally compliant" (5). We expect the evaluators to take notes regarding the problems associated to each item and indicate where they were found in the system.

Once compliance with each of the items has been identified, both in the full version or in the compact version, a qualitative and quantitative analysis can be carried out. Obtaining as a result a percentage of satisfaction of the system designed for people with ASD, charts that visually represent said results and observations of the evaluators.

4 Conclusions and Future Work

Several studies have evaluated the usability and/or UX of systems designed for people with ASD through non-specific evaluation methods. People with ASDs have a diversity of characteristics, needs, and affinities that are not considered in standard assessment methods. As a first step to develop particularized evaluation methods for systems designed for people with ASD, we have created a property checklist proposal. Studies [8–10] mention that for the creation of a property verification list it is necessary to have usability criteria or factors that provide a theoretical basis for the items or elements to be considered.

We have created two proposals for property checklists, a full version and a compact version, which have the objective of evaluating the UX in systems designed for people with ASD. These proposals are categorized into 9 categories, which are directly related to our proposal of 9 UX factors (engaging, predictable, structured, interactive, generalizable, customizable, sense-aware, attention retaining and frustration free) for systems designed for people with ASD [5]. The full version property checklist has 50 items, which is recommended to use in case of having enough time and resources to be executed. The compact version property checklist has 24 essential items from each category which is recommended to use if there is not enough time and resources to be executed.

Additionally, we have provided some recommendations that should be considered when using the property checklist. An example of these recommendations is having the support of professionals in the domain, such as experts in the areas of psychology, speech therapy and special education, and experts in usability and/or UX.

We hope that our property checklist will be useful for designers and developers of systems for people with ASD. This checklist can be used to design systems based on the different items presented, as well as be used to evaluate existing systems, and thus ensure a minimum level of quality and satisfaction for users with specific needs such as people with ASD. When applied as a first step in our methodology to evaluate UX in systems designed for people with ASD, this property checklist will provide evaluators a quick first view of the shortcomings that the evaluated system may have.

In future work, we want to validate and refine the property checklist proposal and use it as an inspection method in our proposed methodology to evaluate the UX of people with ASD.

Acknowledgments. Katherine Valencia is a beneficiary of ANID-PFCHA/Doctorado Nacional/ 2019-21191170.

References

1. American Psychiatric Association: Diagnostic and Statistical Manual of Mental Disorders. 5th edn. American Psychiatric Publishing, Arlington (2013)
2. Valencia, K., Rusu, C., Quiñones, D., Jamet, E.: The impact of technology on people with autism spectrum disorder: a systematic literature review. Sensors **19**, 4485 (2019)
3. Valencia, K., et al.: Technology-based social skills learning for people with autism spectrum disorder. In: Meiselwitz, G. (ed.) HCII 2020. LNCS, vol. 12195, pp. 598–615. Springer, Cham (2020). https://doi.org/10.1007/978-3-030-49576-3_44
4. Valencia, K., Rusu, C., Botella, F.: A preliminary methodology to evaluate the user experience for people with autism spectrum disorder. In: Meiselwitz, G. (ed.) HCII 2021. LNCS, vol. 12774, pp. 538–547. Springer, Cham (2021). https://doi.org/10.1007/978-3-030-77626-8_37
5. Valencia, K., Rusu, C., Botella, F.: User experience factors for people with autism spectrum disorder. Appl. Sci. **11**(21), 10469 (2021)
6. International Organization for Standardization, Ergonomics of Human System Interaction— Part 11: Definitions and Concepts. https://www.iso.org/obp/ui/#iso:std:iso:9241:-11:ed-2: v1:en. Accessed 09 Dec 2021
7. International Organization for Standardization, Ergonomics of Human System Interaction— Part 210: Human-Centered Design for Interactive Systems. https://www.iso.org/obp/ui/#iso: std:iso:9241:-210:ed-2:v1:en. Accessed 09 Dec 2021
8. Hales, B.M., Pronovost, P.J.: The checklist—a tool for error management and performance improvement. J. Crit. Care **21**(3), 231–235 (2006)
9. Jordan, P.W.: Designing Pleasurable Products. An Introduction to the New Human Factors. 1st edn. Taylor & Francis, London (2000)
10. Fernández, A., Insfran, E., Abrahão, S.: Usability evaluation methods for the web: a systematic mapping study. Inf. Softw. Technol. **53**(8), 789–817 (2011)
11. Almeida, R., Andrade, R., Darin, T., Paiva, J.: CHASE: checklist to assess user experience in Iot environments. In: International Conference on Software Engineering, South Korea, pp. 41–44. IEEE (2020)
12. Kalawsky, R.: VRUSE - a computerised diagnostic tool: for usability evaluation of virtual/synthetic environment systems. Appl. Ergon. **30**(1), 11–25 (1999)
13. Johnson, G., Clegg, C., Ravden, S.: Towards a practical method of user interface evaluation. Appl. Ergon. **20**(4), 255–260 (1989)
14. Semantic Studios. http://semanticstudios.com/user_experience_design/. Accessed 12 Oct 2019
15. Bartoli, L., Garzotto, F., Gelsomini, M., Oliveto, L., Valoriani, M.: Designing and evaluating touchless playful interaction for ASD children. In Proceedings of the 2014 Conference on Interaction Design and Children, New York (2014)
16. Tsikinas, S., Xinogalos, S.: Design guidelines for serious games targeted to people with autism. In: Uskov, V.L., Howlett, R.J., Jain, L.C. (eds.) Smart Education and e-Learning 2019. SIST, vol. 144, pp. 489–499. Springer, Singapore (2019). https://doi.org/10.1007/978-981-13-8260-4_43

17. Tsikinas, S., Xinogalos, S.: Towards a serious games design framework for people with intellectual disability or autism spectrum disorder. Educ. Inf. Technol. **25**, 3405–3423 (2020)
18. Carlier, S., Paelt, S.V., Ongenae, F., Backere, F.D., Turck, F.D.: Empowering children with ASD and their parents: design of a serious game for anxiety and stress reduction. Sensors **20**(4), 966 (2020)
19. Khowaja, K., Salim, S.: Heuristics to evaluate interactive systems for children with Autism Spectrum Disorder (ASD). PLoS ONE **10**(7), e0136977 (2015)
20. Higgins, K., Boone, R.: Creating individualized computer-assisted instruction for students with autism using multimedia authoring software. Focus Autism Other Dev. Disabil. **11**(2), 69–78 (1996)
21. UKEssays. https://web.achive.org/web/20160913195408/http://www.ukessays.co.uk/essays/design/autism.php. Accessed 01 July 2021
22. Aguiar, Y., Galy, E., Godde, A., Tremaud, M., Tardif, C.: AutismGuide: a usability guidelines to design software solutions for users with autism spectrum disorder. Behav. Inf. Technol. 1–19 (2020). https://www.tandfonline.com/doi/full/10.1080/0144929X.2020.1856927
23. Barakat, H., Bakr, A., El-Sayad, Z.: Nature as a healer for autistic children. Alex. Eng. J. **58**(1), 353–366 (2019)
24. Chung, S., Ghinea, G.: Towards developing digital interventions supporting empathic ability for children with autism spectrum disorder. Univ. Access Inf. Soc. **21**, 275–294 (2020). https://doi.org/10.1007/s10209-020-00761-4
25. Winograd, T.: Understanding natural language. Cogn. Psychol. **3**(1), 1–191 (1972)
26. Raymaker, D., Kapp, S., McDonald, K., Weiner, M., Ashkenazy, E., Nicolaidis, C.: Development of the AASPIRE web accessibility guidelines for autistic web users. Autism Adulthood **1**(2), 146–157 (2019)
27. Hailpern, J., Harris, A., Botz, R.L., Birman, B., Karahalios, K.: Designing visualizations to facilitate multisyllabic speech with children with autism and speech delays. In: Proceedings of the Designing Interactive Systems Conference, DIS 2012, Newcastle Upon Tyne, UK (2012)
28. Mcallister, K., Maguire, B.: Design considerations for the autism spectrum disorder-friendly key stage 1 classroom. Support Learn. **27**(3), 103–112 (2012)
29. Quezada, A., Juárez-Ramírez, R., Jiménez, S., Ramírez-Noriega, A., Inzunza, S., Munoz, R.: Assessing the target' size and drag distance in mobile applications for users with autism. In: Rocha, Á., Adeli, H., Reis, L.P., Costanzo, S. (eds.) WorldCIST'18 2018. AISC, vol. 746, pp. 1219–1228. Springer, Cham (2018). https://doi.org/10.1007/978-3-319-77712-2_117

Operation Strategies for the PUGC Social Media Video Platform Based on the Value-Chain Perspective—A Chinese Case 'Bilibili'

Mu Zhang and Han Han[✉]

Shenzhen University, Shenzhen, China
{zhangmu,han.han}@szu.edu.cn

Abstract. With the rapid development of digital technology and Internet industry, the threshold of video creation and communication has been lowered continuously, and the monopoly of traditional media on the right to speak has been weakened, and the decentralization of video content manufacturing has been realized. More and more network users began to participate in the production of content products. In this process, a number of video platforms with UGC and PUGC as content production modes were formed. YouTube meets the needs of different users with professional and diversified video content, and has a number of loyal users. Twitter has grown into a composite video platform through mergers and acquisitions. China's short video software TikTok has achieved great success in overseas markets, and the download volume ranks first in Europe. Social network platform giants such as Instagram, Facebook and Snapchat also laid out the video field according to the characteristics of their original products. With the increasing competition, many changes in the supply and demand of video content products have prompted the reconstruction of the value chain of video platform. Therefore, video platform urges to stand out in today's highly competitive media market in constantly upgrading and optimizing its own value chain and adopting corresponding strategies to promote the production of content providers and the experience needs of content consumers. Under such context, this paper carries out a discussion based on the research of China's representative PUGC video platform Bilibili, so as to generate some possible operation strategies in perspective of value chain for the future development of PUGC video platform.

Keywords: Value chain in HCI · PUGC platform · Operation management

1 The Development Status and Trend of PUGC Video Platform

1.1 Video Content Ecology is Getting Mature

Making and Sharing Video Contents Become More Professional and Popular. There are two basic modes for video-based content sources, PGC (Professional Generated Content) and UGC (User Generated Content). PGC producers have professional

G. Meiselwitz (Ed.): HCII 2022, LNCS 13315, pp. 217–226, 2022.
https://doi.org/10.1007/978-3-031-05061-9_16

knowledge and qualifications, mainly including vertical experts, traditional media practitioners, self-media teams and professional entertainment and film teams. Their professional level ensures the video quality and enriches the video content in various vertical fields. The number of such professional teams is small and the production cost is high. The other mode UGC means that users contribute and create content. For creators, it generally has the functions of entertainment and social interaction, and most of the self-made content is full of all kinds of imagination, creativity and freshness. This kind of video is usually uploaded spontaneously by users, and the quality varies. With the increasing demand of consumers for video quality, content payment is recognized by the society, and the technical threshold is lowered by new tools, more people no longer share content works just because of interest and social interaction, but gain attention and traffic through well-made content, forming their own influence and gaining economic benefits at the same time. Under this trend, the video platform of UGC mode is rapidly adjusted and upgraded, from the free-growing stage of UGC to PUGC, forming a sticky and attractive PUGC, which is a compound production mode that contains both user attributes and professional attributes. The video produced in this mode meets the public's demand for content, which is flexible and diverse; and at the same time, it can also comply with the content promotion needs of different activities and themes of the platform, which greatly enriched the platform content and enhanced the user stickiness.

The Development of Video Platform Greatly Promotes the Production and Dissemination of Content. At present, PUGC video platform has established a relatively mature encouragement mechanism and algorithm push, which can screen out high-quality video content through data analysis, so that creators' videos can be effectively delivered to target consumers. After a large amount of capital poured into the short video industry, the platform rewarded with traffic and subsidies to further stimulate the creators to reproduce the content. Take YouTube as an example, it attaches great importance to the creator's income, e.g., the creators can get higher share in its Partners Program. These strategies form a virtuous circle of PUGC video platform content production.

The Video Content Structures Virtual Subdivisions. As the video platform has entered a mature stage, the classification of content has become more diversified and refined. For creators, vertical subdivision not only makes it easier to form their own characteristics, but also enhances users' stickiness. Capital and short video platforms are more willing to support creators who produce content seriously and focus on profundity rather than those random uploaders, which also encourages more vertical content creators to be more willing to create high-quality video content. Support for vertical authors is the focus of YouTube creator system. And content creator tags also encourage users to form a circle. Just like the collaborative evolution in the history of human development, the video content market has begun to show its large diversity and comprehensiveness.

1.2   A Brand-New Consumer Group—Generation Z

Generation Z, the most important group of Internet content consumption, plays a very important role in the rapid development of PUGC video platform.

The term Generation Z is originally defined by American scholars and practitioners, following the Generation X and Y, it specifically refers to the people who were born between 1997 to 2012. The people in this generation have been closely related to information generation since their birth, and are greatly influenced by the Internet, instant messaging, smart phones and social media. Therefore, they are also called "network generation", "Internet generation" or "digital media generation", as they are the first generation who grow up with complete access to the internet and social media. According to the "2019 Generation Z Report" released by ZebraIQ, at present, there are about 2.4 billion people in the global Generation Z, accounting for 32% of the global population [1].

The rapidly rising Generation Z has become the most concerned group in the world. This is not only because of their huge number and strong spending power, but also because they have distinctive consumption ideas. However, the content consumption in the field of short video and medium video is also significantly different from other generations.

First of all, Generation Z, as an affluent generation, has less experience of material shortage compared to the previous generations especially in China. Therefore, for this generation, more needs are turned to spiritual and entertainment requirements. They are more concerned about "experience", and are keen to try to reach the best-matched value and service. Therefore, in terms of video content consumption, Generation Z is more copyright-conscious than other generations. They have a higher recognition of content consumption and are willing to pay for high-quality content.

In addition, the obvious difference between Generation Z and other generations in video consumption is that when they receive new video content or browse video content they are interested in, they will have a strong demand for interaction, communication, expression and sharing, which shows a strong sense of participation in information. The survey data of Tencent and other institutions also show that the users of the Generation Z have high degree of attachment and activeness in playing specific short videos. Among all kinds of likes, reposts and comments, 36% of the interactions come from the users of the Generation Z. They not only actively participate in short video content comments, but also have a strong sense of rules, citizenship and individuality [2].

Moreover, in the process of growing up with the Internet, the indiscriminate bombing of massive information to some extent make Generation Z acquire natural false recognition ability. Therefore, they tend to be favour in the more democratic short video platform, and their demands for authenticity are undoubtedly stronger. Whether it is the construction of particular figure's character, the display of different types of content, or the video marketing of various brands, compared with posturing, the content authenticity becomes an essential factor that builds up a sense of substitution and resonates with Generation Z users.

2   Definition and Characteristics of Video Content Platform Value Chain

"Value Chain Theory" was originally put forward by Michael Porter in 1985 [3], which emphasizes that in the upper, middle and lower reaches of the industrial chain, enterprises need to create value in all aspects such as manufacture, production, distribution and sales, so as to achieve the business goal by creating a series of value-bounded activities, for the co-creation process of products, in which way to meet with the needs of customers and markets and leverage the maximization of interests.

While for the video industry, its value chain is a set of relationships based on the business logic relationships among the design, production, operation and dissemination of video content products, and it is a complex of the internal and external industrial chains of video content enterprises. Video content products circulate in every link of the industrial chain from production to dissemination, and each link can add value in the transaction, and promote the constant circulation of value-creation process. Therefore, if the video platform wants to gain a competitive advantage in the market, it needs to constantly innovate the operation mode and optimize all links in the value chain.

For PUGC video platform, in the process of creating value, users and video platform are the main parts of the value chain. On the one hand, video creators become the core of the value chain, providing content products for the video platform, which has a decisive influence on the survival and sustainable development of the platform. On the other hand, video consumers are increasingly involved in the creation and dissemination of products, which makes them take roles as both consumers and producers and are closely related to the upstream, middle and lower reaches of the value chain. Therefore, in terms of the operation of PUGC video platform, the user's value proposition should be taken as the beginning of the linear value logic of PUGC video platform business model, which means that only by completing the value creation and value transmission activities can enterprises construct and acquire value to sustain its business. Hence, in this linear value logic, realizing the customer's value is the starting point of its business model, value creation and value transmission are the intermediate links, and value acquisition is thereby its ultimate goal [4].

3   The Generating Logic of Bilibili's Value Chain

In recent years, China's video platforms have developed rapidly under the background of capital and technology empowerment, and industry giants such as TikTok, KUAISHOU, WeChat Video, etc., whose monthly active number exceeds 200 million have been exhibited. In the fierce competition, Bilibili Video Platform (Bilibili) has become a dark horse among all the video platforms in China with its unique positioning and development path, and it has achieved double growth in terms of user quantity and word of mouth in a short time.

3.1   Development Status of Bilibili

In addition, with the rapid integration of Generation Z, which grew up in the era of rapid economic development in China, into the mainstream society, they, as the main force of current consumption, release the new demographic dividend and inject new vitality into the market. With the irresistible development momentum of "video", a market full of potential has been emerging. According to iiMediaResearch [5], by 2025, the number of video users will be close to 1.2 billion, and the video market will exceed 1.8 trillion Yuan.

Bilibili seized this opportunity, firmly grasped the younger generation, and became the preferred platform for them to consume and create videos. Bilibili was founded on June 26th, 2009, and was positioned as a niche video website for creating and sharing ACG content (Anime, Comic, Game) in the early stage. However, through the accumulation of rich content resources and the change of operation strategy, Bilibili has grown into a cultural paradise covering more than 7,000 interest circles. The number of active users on Bilibili has increased more than tenfold in the past five years. In the first quarter of 2020, the monthly average number of active users has reached 172 million, and the users' stickiness is constantly increasing, and the average daily usage time of users reaches 87 min, most of them are under the age of 25 [6]. According to China's Internet statistics report, it means that one out of every two young people in China is a Bilibili user. Bilibili can become a highly concentrated cultural community and video platform for young generations in China, and the construction of its value chain has certain reference value for the development of PUGC video platform.

3.2 Bilibili's Value Chain

The rapid development of Bilibili is mainly due to the development of the company's value chain through value co-creation. The concept of value co-creation was first put forward by management masters Prahalad and Ramaswamy who believe that the future competition of enterprises will depend on a new way of value creation—creating value together through the heterogeneous interaction between consumers and other stakeholders and enterprises [7]. Bilibili is a typical value co-creation platform, as where consumers, video creators and platforms participate in the company's value manufacturing together, and create new values in the process of mutual value exchange.

Bilibili shows its strong efforts in building a value system for the growth of high-quality video content by series of approaches: First, encourage and promote users to produce high-quality videos, and build their own profit matrix through diversified product line design; Afterwards, by setting the data analysis mechanism, the content products will be accurately distributed, and users will be guided to screen out creators and works that can get their approval; Then, according to the feedback from users, the creators will be encouraged to produce higher quality products; In addition, Bilibili organized integrated marketing activities to enhance brand influence, attract more high-quality creators to the platform and strengthen consumer stickiness. This series of approaches forms a positive cycle system, which presents the logic of Bilibili's continuous value creation. In this system, Bilibili's value chain can be divided into five stages: product production, product line design, product distribution, product consumption and branding & marketing.

In product production, based on platform advantages, Bilibili gathered a large number of creators to produce video products. The creator is the core unit of content production and the most loyal user in Bilibili, who provide 90% PUGC content for Bilibili. Bilibili has been committed to helping video creators to produce, distribute and make profits of content. In 2018, Bilibili launched the Excellent Uploader Incentive Plan to give greater support to video content creators. If the number of fans of the creators reaches 1,000 or the cumulative number of videos plays is more than 100,000, they can apply to join the plan. After joining, the newly delivered original manuscripts can get incentive income when the number of plays reaches 1,000. The plan aims to support the originality of the content, and protect the new creators to ensure their copyright and exposure. This has attracted more and more creators to join, and also promoted the emergence of more high-quality content. In addition to the support for video creators, Bilibili has designed rich interactive functions in the interfaces, so that the users who watch videos on Bilibili are not only content consumers, but also content creators by interactive actions. Through various interactive forms such as like, collection, coin-reward, comments (and bullet chats), etc., users can show their opinions, preferences and attitudes to all the public audience, and get emotional satisfaction in watching experience and social interaction. Even some language symbols created by user groups in the interactive process form self-sufficient cultural products other than video content. In particular, bullet chats (the instant comments running on the screen while playing), being as an interactive form, wins great popularity in Bilibili, which has become a unique digital culture of the platform, and the bullet chats to large extent even determines the value of video along with the content itself. In 2020, the highest bullet-chat interaction on Bilibili video has exceeded 2.5 million pieces, and such kind of interaction has become a special way of communication between video consumers and creators on Bilibili, which forms a specific scenario of Bilibili's video-watching, giving new value to the platform.

In Terms of Product Line Design, apart from the main PUGC video service, the commercial structure of Bilibili is relatively rich, and it provides varieties of cultural products and services for young people. In recent years, Bilibili has made some progress in the field of live broadcast, e-commerce and paid classroom, besides taking game agents as a way to make profits. For example, the member purchase section takes the sales of ACG-content-derived products as the core profit of its e-commerce service; the pay-for-knowledge section invites KOL as lecturers to produce and sell online courses, etc. Moreover, Bilibili has also made many attempts in advertising business—through copyrighted content, large-scale offline activities and self-produced short film and reality shows, it is now able to provide varieties of choices for advertisers to join the business. On a larger scale, Bilibili also try bring innovation in terms of business scenarios, which empower advertisers' integrated-marketing capabilities by providing series of standard-ization and modelled service, e.g. the upgrade of enterprise account on the platform and the establishment of B-DATA brand database for the enterprise users, which embeds commercial service into each single video consumption scenario. At the AD TALK Marketing Conference 2020, Bilibili shared its marketing methodology of content creators, and also announced the upgraded integrated-marketing strategy called "Z+ Plan" that clearly targets to the people in Generation Z, which initiates more marketing scenarios and establishes more links between brands and the young people, meanwhile, encour-ages the enterprise users build its own brand account by building the B-DATA platform, which in turns, improves the efficiency in launching new profit centers empowering different industries such as transportation, e-commerce and FMCG around its business span.

At the Level of Product Distribution Logic, we can observe from the homepage of Bilibili that at present, interest-based recommended videos can be generally divided into these categories: watched videos, highly praised videos, rising-star programs, watched-people-praised videos, advertisements and interactive videos. These tags push various products to users based on the tags of content or users. From the perspective of video platform, the usual way to maximize the benefits of the platform is to get more traffic for the platform. However, Bilibili's practice is different. In terms of traffic distribution strategy, it does not take whether the content can get more clicks as the key consideration, but adopts relatively fair traffic distribution standards, relies on the distribution mechanism of fan relationship and interest push, and makes customized recommendations according to the data of users' concerns and other topics. Bilibili's recommendation system is based on the data of users' liking for content to judge whether a product is worthy of recommendation—the main data source of the recommendation system is users' positive feedback, including "one click with three links" (like, coin-reward, collection), positive bullet-chat comment and regular comment content. The logic behind this data selection for distribution is that the click volume of a video is not necessarily related to whether consumers like the video or not as it might be the title or other reasons that causes the high click performance, but the likes and rewards must have a higher correlation with the degree of positive preference. Under this more accurate data strategy, 70% of the traffic in Bilibili is distributed to small and medium-sized creators, even unknown creators. This traffic distribution strategy, which is not based on short-term interests, helps Bilibili to expand the number of creators and ensure the quality of product resources in a more sustainable way.

In the Stage of Product Consumption, Bilibili mainly adopts the content consumption mode based on community economy. Bilibili is like a nation where most users do not visit the video platform like tourists, but become residents of a certain city and participate in the construction of the city. Bilibili has a classification of official users, which requires the accomplishment of 100 questions about community rules and ACG-related knowledge before sending bullet chats, writing regular comments, scoring, etc., while other types of users only have the limited rights of viewing without interaction. The high entry barrier prevents outsiders who have little knowledge of ACG from participating, maintains the integrity of the interactive ceremony chain, improves the sense of belonging of users and ensures the community atmosphere. This is an important feature makes Bilibili distinctive from other video platforms. Therefore, community economy is the basis of its development. In essence, community economy is based on the media platform, which brings together a group of people with similar interests, similar cognition and cultural identity, and generates commercial value in the process of emotional communication, information dissemination and value sharing. Bilibili has built a content-based interactive platform. By publishing content, creators attract people with similar interests to share and communicate together. When the accumulation and attention of these categories reach a certain level, they will be "independent" and gradually form a number of high-viscosity vertical network communities. Users can find their own "circle" here, and cooperate with other users in the "circle" to produce, exchange information and communicate emotionally, so as to get spiritual benefits. Such strategy greatly enhances the platform's appeal to the users, and keep the users churn rate in Bilibili far lower than other video platforms.

In the Aspect of Brand Communication, due to the growth of the Generation Z and the intergenerational media content discourse power, Bilibili began to break the "traditional circle" and focus on reaching a wider audience. Through a series of marketing activities, Bilibili exported a brand-new brand image to the public, which enhanced the commercial value of the platform—it was its first time in launching a series of homemade variety shows through festivals: On the New Year's Eve in 2019, the New Year's Eve party planned for the preferences of the Generation Z gained a lot of traffic even outside the community; On the eve of the National Youth Day in 2020, Bilibili, together with many traditional media in China's most influential schools, launched a speech video dedicated to the new generation of young people (named as Houlang meaning 'the wave behind' in Chinese), which helps Bilibili attract great attention from the majority and heated public discussion. During the epidemic period, Bilibili also make abundant effort in developing various activities and programs such as online classrooms, online music festivals and online exhibitions to attract young people's participation. Since 2018, Bilibili has held the "Top 100 UP-loaders (video content creators) Awards Ceremony" every year, which not only reward the outstanding creators of the year, but also further expand the influence of the platform into the circle of public media and entertainment. At the same time, around the latest slogan "Videos You are Interested in, All on Bilibili", Bilibili has shot and put in a large number of advertisements, setting up a new image of a comprehensive video community with diverse themes and a wider reach of people.

4 Reflections for the Operation Strategy of PUGC Video Platform

4.1 Optimize the Content Creation Ecology

High-quality content is the core resource of PUGC video platform. Therefore, promoting the ecological and healthy development of platform content is the key to maintain the vitality of PUGC video platform. Internet users, especially Generation Z consumers, have increasingly obvious demand for specialization and diversification of cultural products. Therefore, the video platform should focus on maintaining the diversity of platform content products, improving the content quality of each vertical segment, supporting and guiding newly-built partitions and communities, and promoting the growth of minority interest groups.

The video platform should further provide creators with more sufficient creative support. On the one hand, through the design of product logic and the formulation of rules, it would be more able to strengthen the protection of copyright of creators' works and support the growth of high-quality creators. On the other hand, it is essential to speed up the development of products that can improve the commercialization efficiency, enrich the business matching path, and promote more creators to realize the virtuous circle of commercial realization of content.

4.2 Enhance the Application Means of Science and Technology

As the centralized content distribution and information cocoon causes a lot interfere for the users, the algorithm recommendation of PUGC video platform needs to shift its emphasis from gaining traffic to promoting positive experience. Since PUGC video platform produces a large number of content products every day, how to distribute these products to users in need, so that users can get rich experience and the content products they need becomes crucial for the sustainable development of PUGC platform. PUGC platform needs to establish a scientific data process and traffic distribution mechanism with high accuracy, constantly improve data mining and analysis, and integrate the relationship between content providers and consumers by means of artificial intelligence and other technological approaches, so as to ensure the objectivity and timeliness of video content and the maximum utilization and development of resources. In addition, the platform should conduct in-depth observation and research on the target users, focus on the user experience, and implement refined operation according to different communities and users' natural attributes, interests, usage scenarios and consumption preferences, so as to provide users with personalized video content.

4.3 Strengthen the Maintenance of Community Atmosphere

Based on the case of Bilibili, we can analyze that PUGC video website is not only a content platform, but also has social attributes. Promoting social situations is conducive to enhancing the platform's user attachment and forming brand competitiveness. Therefore, the platform should promote the communication and interaction between different interest groups. Encourage cross-group content production, develop the linkage between works and fans among different groups, and form content products and consumption points shared by various interest groups by holding comprehensive activities and jointly developing products, so as to promote the integration of various group cultures and subcultures. Precise settings should be made for group identification, and the user rating system (new users and existing users, members and non-members, etc.) should be improved, so as to give different use privileges to internal members, promote diversified experiences, and build a community with high inclusiveness and participation.

It is important to be pointed out that although there is still a huge gap in revenue and user scale between Bilibili and super PUGC video sites like YouTube, from the growth rate in recent years, Bilibili has great growth potential in the future especially under the context formed by the Generation Z in China. In the battlefield of video platforms, Bilibili has established its own dominant position by means of high-level content generalization, positive community atmosphere and decentralized creator diversion. More essentially, its good content ecology and community atmosphere need long-term accumulation and maintenance for the formulation of environment. Therefore, its special development strategy discussed in this paper provides certain reference value for the PUGC video platforms both in China and in the globe. The discussion raised in this article also expects to initiate future analysis with more quantitative data for comparative studies among different platforms under diverse socio-cultural context.

Acknowledgment. The author Mu Zhang and corresponding author Han Han would like to express sincere appreciation to all the participants who contribute advise and data to this study.

References

1. Generation Z report in 2019 [EB/OL]. https://www.sohu.com/A/328359912_728306. Accessed 30 Dec 2021

2. Gao, F.: Analysis of Short Video Consumption Characteristics of Generation Z, News Lovers (2020)
3. Porter, M., Advantage, C.: Creating and Sustaining Superior Performance. Collier MacMillan, London (1985)
4. Zhu, M., Li, C.: Research on elements, framework and evolution of business model value logic: review and prospect. Sci. Technol. Prog. Countermeasures **2020**(12), 149–160 (2020)
5. iiMediaResearch.   https://report.iimedia.cn/m-report.jsp?reportId=39194&acPlatCode=IIM Report&acFrom=recomBar_1061&iimediaId=77180. Accessed 30 Dec 2021
6. Bilibili Inc 2020 Fourth Quarter and Fiscal Year Financial Report. https://ir.bilibili.com/static-files/09f30d5d-5de5-4338-b767-921ce1a07a47. Accessed 20 Jan 2022
7. Prahalad, C.K., Ramaswamy, V.: Co-creation experiences: the next practice in value creation. J. Interact. Mark. **18**(3), 5–14 (2004)

Behind the Scenes: Exploring Context and Audience Engagement Behaviors in YouTube Vlogs

Hantian Zhang(✉)

Sheffield Hallam University, Sheffield S1 1WB, UK
Hantian.Zhang@shu.ac.uk

Abstract. Famous video bloggers (vloggers) on YouTube can develop large audiences, which can be related to the gaining of audience engagement (AE), manifested by the viewers' participation and consumption on YouTube. Studies have unveiled vloggers' behaviors for engaging audiences, or audience engagement behaviors (AEBs), in their videos, including interacting with viewers via comments, disclosing self-information, giving rewards, and offering other information. Meanwhile, video blogs (vlogs) are produced under "vlogging context" - situational elements involved in vlog production. Studies have shown the effect of context on the content of online media. However, while it can be argued that context can affect vlog content produced, the contextual factors that may shape vloggers' AEBs within the content have not been explicitly explored. This research aims to propose contextual factors that can condition vloggers' AEBs on YouTube. A qualitative case study on three popular vloggers was implemented. A thematic analysis was performed on sampled vloggers' videos to identify contextual factors that can condition the three vloggers' AEBs. The results propose that personal, environmental, and medium context are three main contextual factors that condition the three vloggers' AEBs. This research argues that how vloggers' AEBs are presented to the audience depends on their vlogging context. It expands the understanding of YouTube vloggers or similar streaming media creators' practices for AE by considering the role of context.

Keywords: Context · Vlogs · YouTube · Audience engagement

1 Introduction

Video blogs (vlogs) are a type of online streamed media uploaded to the Internet, in which video bloggers (vloggers) document their daily activities or cover specific topics [1]. The founding of YouTube in 2005 encouraged vlog consumption [2] since it allows users to easily distribute online videos [3]. Vloggers who are operating on YouTube are also called YouTubers [4]. Today's vloggers on YouTube are producing videos covering various topics, including beauty, comedy, and gaming. Recent years has also seen a widespread consumption of vlogs [5]. Some vloggers have developed large audiences with millions of subscribers on YouTube [6]. Vloggers' success in terms of viewership

G. Meiselwitz (Ed.): HCII 2022, LNCS 13315, pp. 227–244, 2022.
https://doi.org/10.1007/978-3-031-05061-9_17

can be related to the gaining of *audience engagement (AE)*, manifested by viewers' participation (e.g., liking and commenting on videos) and the consumption of video content [7–9], creating continuous connections between vloggers and viewers. Previous research found that vloggers implement audience engagement behaviors (AEBs) to establish AE including responding to comments [10], disclosing personal stories [11, 12], promoting information that viewers may find useful [9] and providing rewards [13]. Those AEBs are existing in the vlog "content" – audio-visual information that is directly available to the audience in vlogs as in media [14].

This paper shifts attention from the content of vlogs that contains AEBs to the vlogging "context", which has rarely been discussed in vlogging. In general, context relates to situational elements that are critical to interpreting an object [15]. In information science, context was defined as "the quintessence of a set (or group) of past, present and future situations" [16, p. 3], for which an "information horizon" consisting of a range of resources is determined for an individual to seek related information [16, p. 8]. Context also refers to the information that indicates the situation of entities in Human-Computer Interaction (HCI) [17]. In (online) media, context relates to situational elements involved in content production, consumption, and distribution [18–20]. So, while vlog content refers to factors in vlogs that are directly received by the audience, this paper sees vlogging context as elements that reflect the circumstances of different entities involved in the vlog production process. The contextual factors can relate to the situations of vloggers, film locations, audiences, and the online platforms. However, while it can be argued that context can affect vlog content based on previous studies (e.g., [21, 22]), the contextual factors that may affect vloggers' AEBs within the content have rarely been explored.

This paper aims to explore the question: "What are the contextual factors that can condition vloggers' audience engagement behaviors in their videos?" The research implemented a qualitative case study of three popular vloggers on YouTube. In the study, contextual factors behind the vloggers' AEBs in their videos were observed. As mentioned, existing research mainly focused on the delivery of AEBs in vlog content. This paper, however, emphasizes the crucial role of context behind those AEBs. So, this research took an initial step to widen the understanding of YouTube vloggers or similar streaming media creators' practices for AE by considering the role of context in addition to content. Also, since the concept of context is rarely discussed in vlogging, this paper contributes to establishing an initial exploration of context in vlogging. Therefore, this paper contributes to the literature that explores context, content, and AE in streaming media. For practical implications, this study can help vloggers and potentially creators on other steaming media to build audience bases using strategies based on their production context.

2   Related Work

2.1   Vlogging Context

Limited studies have discussed context in vlogs specifically. Snelson [23] sees context as filming locations for vlogs. However, research in other media implies a broader concept of context in vlogs.

For example, in HCI, context is defined as any information that can indicate the situation of entities, including locations, people and object that is related to the user-computer interaction [17, 24]. Matuk et al. [15] defined context in computer-supported collaborative learning (CSCL) as situational elements involved in the CSCL process, which has focal, immediate, and peripheral layers. Each layer relates to elements like study tools, participants' status, and institutional environment. Context also relates to situational factors in the consumption, production, and distribution of (online) media. For example, in television studies, context is considered to be the environment where the audience consume television programs [25] or the presence of co-viewers (i.e., whether viewers are watching television on their own or with others) [19]. Regarding production, Lena's [18] research in music production refers "social context" to the music market environment, namely whether the market is dominated by independent or major labels. Furthermore, for distribution, Jaakonmäki et al. [20] see context or "contextual features" as relating to when and where the content is posted on the social media platforms.

Using YouTube to watch or distribute vlogs can be seen as a form of HCI between users and YouTube. Vlogs are also audio-visual media that are consumed and produced. Therefore, while vlog content refers to elements the audience can directly see or hear from a vlog including the vlogger, their (non)verbal behaviors and objects shown, based on the above definitions, the context in vlogs can be considered generally as factors that indicate the situations of the entities involved in the consumption, production, and distribution of vlogs on related online platforms.

This paper focuses on the effect of context on AEBs within the vlog content produced. Therefore, the author views vlogging context as elements that indicate the situations of entities that are involved in the vlog production process.

2.2 Vloggers' Audience Engagement Behaviors (AEBs)

Vloggers were found to encourage AE in their videos through different behaviors, especially AEBs. Four AEBs that can be identified in previous research are *interaction, self-disclosure, information offering*, and *rewards* [9–13].

Interaction relates to vloggers' behaviors that can trigger viewers' actions beyond watching videos, or vloggers' responses to viewers' actions. For instance, vloggers on YouTube were found to encourage viewers to comment on videos [9], ask viewers to suggest new videos ideas [10], and respond to viewers' comments [26].

Self-disclosure refers to vloggers' disclosure of personal information in their videos. For example, vloggers may show their daily life activities in videos [27, 28], or even talk about their life struggles [11]. Self-disclosure is associated with the feeling of authenticity [11, 12]. This makes viewers feel a deep connection with vloggers [28], driving their ongoing YouTube activities [29].

Rewards are offered by vloggers to reward audiences' actions. Research has shown that vloggers ask viewers to like the videos [9] which could allow the audience to get more similar content. Some vloggers also announce giveaways of certain products to engage their viewers [13].

Finally, information offering refers to vloggers' providing information viewers may need or find useful. It is shown by previous research that vloggers promote information such as their social media sites [9, 10] in videos. Users' participation and consumption on social media are found to be affected by the need for information [30, 31]. Therefore, vloggers' information offering may engage viewers by satisfying such a motivation.

2.3 Vloggers' AEBs and Vlogging Context

Previous research suggests the effect of context on user engagement strategies on social media sites. Typically, Jaakonmäki et al. [20] suggested marketing professionals can choose online influencers and launch marketing campaigns by considering the context (i.e., days and hours) on social media. This suggests the impact of context on marketers' decisions on the engagement strategies, in the form of social media content. However, there is a lack of discussion regarding the contextual factors in vlogs on YouTube that may affect vloggers' AEBs.

On the other hand, studies show that context can affect (online) media production. Research suggests that vloggers' features such as their personalities and production skills may condition their non-verbal behaviors (e.g., gaze, facial expressions) in the videos, and further affect viewership of the vlogs [32]. Those features can be seen as vlogging context that affects vlog content. Context has also been seen as affecting the content uploaded to YouTube. Yarosh et al.'s [22] research showed that due to YouTube's moderation policies and age restrictions, youth-authored content on YouTube is less likely to be inappropriate compared with those on Vine. Similarly, Rieder et al. [33] indicated that YouTube's algorithmic structure can affect creators' production strategies including making longer videos. These studies reflect the effect of the context of YouTube on the content user produced. Research in written blogs also unveiled the effect of context on content. For instance, researchers found several motivations of bloggers, including documenting personal lives, expressing emotions, and presenting opinions [21, 34]. These motivations can be seen as connecting to the context of bloggers that affect blog creation. Furthermore, context also affects traditional media production. Typically, Lena's [18] research on the market context and content of rap music showed that song lyrics in the market dominated by independent labels are different from the ones in the market dominated by major labels. Since vlogging context relates to situational factors involved in vlog production, it can also be argued that vlogging context can affect the vlog content produced. Meanwhile, as AEBs are delivered by the content of the vlogs, it can be argued that the vlogging context can also condition vloggers' AEBs that are conveyed by the content.

However, there is still little research that has specifically examined the vlogging context that may influence vloggers' AEBs on YouTube. Therefore, this paper aims to establish a starting point to fill this gap by exploring the question: "What are the contextual factors that can affect vloggers' audience engagement behaviors in their videos?".

3   Research Method

The research focuses on exploring contextual factors behind the four vloggers' AEBs derived from existing research mentioned, which are interaction, self-disclosure, rewards, and information offering. It is worth noting that AE is also related to the audience-centered Uses and Gratifications Theory (UGT), in which audiences engage with media due to their needs including social interaction, information seeking, entertainment, and personal identity [35, 36]. Audiences can also be producers on social media like vloggers, whose production behaviors are driven by their own needs and motivations [36]. It can be argued that audience motivation and gratification is important for AE on YouTube [8]. However, instead of focusing on the categories from UGT, the above AEBs the author chose to investigate are specific behaviors vloggers are implementing that may help them to gain AE on YouTube, although some of them may still relate to UGT.

A qualitative case study on three popular vloggers was implemented. A case study allows researchers to investigate subjects in detail in real-life situations [37], without separating them from their environments [38]. Since vlogging context is behind the production of vlog content, a way to identify those contextual factors is to have an in-depth observation of vloggers' natural practices of AEBs within YouTube. Therefore, a case study fits the research requirement.

Three popular vloggers on YouTube were selected: Zoe Sugg (beauty vlogger), Daniel Middleton (game vlogger), and Lilly Singh (comedy vlogger) [39]. Each vlogger had reached over 10 million subscribers in 2017 on YouTube. As of 2020, only 700 out of 37 million YouTube channels recorded have 10 million subscribers [6]. The subscribers reflect those vloggers' high AE, since subscribing to a YouTube channel "demonstrates that a user desires a continued relationship with that YouTube personality" [28, p. 89]. The three vloggers are focusing on different topics namely beauty, gaming, and comedy, presenting the nature of vloggers' topic diversity on YouTube. This makes the shared patterns discovered from the vloggers more important than focusing on only one vlogger. All vloggers started uploading videos from 2012 or earlier. This reflects these vloggers' long production histories and ensures the richness of information obtained from them [40, 41].

3.1 Sample Collection

The study targeted the vloggers' YouTube channels: *Zoella* by Zoe Sugg [42], *DanTDM* by Daniel Middleton [43], and *Lilly Singh* by Lilly Singh [44]. Videos uploaded by the vloggers before August 2017 on their channels were collected. The collection tool was Link Klipper, a Google Chrome extension. In total, 3,495 videos were extracted. Among them, 346 are from Zoella, 2,535 are from DanTDM, and 614 are from Lilly Singh. Each vlogger features multiple video types. Hence, to achieve a sufficient vision of the featured videos on the vloggers' channels, the videos have been categorized based on their format and topics [39]. After the categorization, one video was selected from the beginning or close to the beginning of each year in each video type, up to the year 2017. In total, 200 videos were collected as the final samples. There are 76 videos from Sugg, 50 videos from Middleton, and 74 from Singh [39].

One thing that the author would like to address is that the collection of the main video data happened back in 2017. This creates a potential limitation regarding the timeliness of the results. On the other hand, each vlogger had already reached an extremely high AE reflected by their subscriber amount back in 2017, which is also maintained. Since AEBs are implemented by vloggers to encourage AE, it can be argued that the AEBs of the vloggers up until that period are extremely important for their overall success in terms of audience base. This also makes the contextual factors behind those AEBs equally important. Therefore, the results from the data are still relevant by unveiling critical contextual factors behind the AEBs of those vloggers during the period when their high AE has already been built. This also opens future research opportunities to compare the results with later data. Furthermore, research rarely directly addresses the relationships between contextual factors and vloggers' AEBs. Hence, this research will still contribute to the field.

3.2 Examination

The video samples were analyzed through thematic analysis (TA). According to Braun and Clarke [45], TA is used to find patterns in a range of texts that is crucial and relevant to the research question. Although Braun and Clarke mainly introduced TA in psychology such as analyzing interviews and focus group transcripts, they mentioned that the method is widely used beyond psychology and can be used in audio-visual works. In this case, using TA is suitable to identify important patterns of both AEBs and related contextual factors.

The examination took two steps. The first step is identifying vloggers' AEBs in their videos. The research is focusing on pre-defined behaviors in vlogs based on existing literature (interaction, self-disclosure, information offering, rewards). A deductive thematic analysis was adapted on sampled videos to identify these behaviors in the vlog data. All video samples were watched in full, and the four AEBs were applied to relevant content as codes. If the AEBs were delivered verbally (e.g., asking for comments as interaction), transcripts were extracted and coded. Related non-verbal factors (e.g., showing daily activities as self-disclosure, and providing some information in the video description) were transformed into the textual description and coded [39].

Step two was following a more inductive TA approach to identify the merge of themes that reflect the influence of contextual factors on the AEBs. The vlogs were re-evaluated with coded AEBs. Since vlogging context can relate to any element in the vlog production, the evaluation was done by identifying how each AEB can be affected by elements within the three broad layers of context adapted from Matuk et al. [15]: focal, immediate, and peripheral context. The focal layer involves elements that are essential to the vlog content that may influence the AEBs. These may include the activities, people, objects that are directly presented in the video, and YouTube as a tool for distributing vlogs. The immediate layer involves elements that are outside the focal context but may still be important for the AEBs. These may include the experience of vloggers, and their relationships with other people when producing the video. Finally, peripheral context involves broader environments when vlog productions take place. These three layers of context were originally introduced in CSCL [15]. However, the author considers these three layers of context can be adapted as a general guideline to evaluate the vlogging context. It is because producing and distributing vlogs on YouTube for viewers to consume is also a form of computer-mediated communication with similar entities as CSCL such as participants (e.g., audience and vloggers), tools (e.g., YouTube), and the environment.

Based on this guideline, contextual factors that may influence the AEBs were primarily identified from the video content analyzed, including the AEBs themselves and other content delivered around the behaviors. Resources besides video content were also used in the evaluation for more evidence, such as other videos on the vloggers' channels, their social media sites, and the YouTube environment. For instance, if the vloggers indicated a specific environment have affected their video production, such as trends, further exploration on the internet was also conducted to provide additional evidence of this trend. After the evaluation, codes were applied as descriptions indicating how each AEB were affected by the elements within the vlogging context. Further comparisons between the descriptions were drawn to identify patterns. Similar descriptions of the contextual factors' effect on the AEBs were grouped and given a new code. For example, if vloggers' relationships with other people were seen as affecting the AEBs, the contextual factor was coded as *social relationships* within the *vlogger context* [39].

3.3 Reliability

The author conducted the solo coding process without the second coder, which may cause potential problems regarding reliabilities of the results. To minimize the issue, strategies have been implemented. One is the repeated review and analysis of the data [46]. The whole analysis process was executed at least twice with a time gap of at least two weeks in between each analysis [47]. This allows the author to evaluate similarities and differences in the coding results to further justify outcomes. The second strategy is using triangulation by referring to multiple resources for data interpretation [48, 49], meaning the evidence from other related resources, including vloggers' social media, YouTube comment sections, and other videos were also used to justify the results.

4   Results

Three main contextual factors that can affect the three vloggers' AEBs were identified through the coding process: *personal, environmental,* and *medium context*. Within these three main factors, multiple contextual factors were identified.

4.1   Personal Context

Personal context involves *vlogger* and *audience context*.

Vlogger Context. Vlogger context refers to vloggers' situations during the vlog production. Factors in the vlogger context that can affect AEBs mainly refers to the situations of vloggers' *social relationships, personal experiences*, and *social characteristics*.

Social relationships relate to the situation in which vloggers has specific relationships with other people, such as being family members and friends with others. The results show that all vloggers have involved other people in the video due to their relationships, which affect their AEBs, especially self-disclosure and information offering.

For example, in Sugg's vlogs, she involves other people to answer questions set by her. In a video, she asks her boyfriend and brother about her past (v1)[1], including her first childhood holiday and her first job. However, in another similar video with her boyfriend and another friend, the questions are related to her status such as her favorite food and zodiac sign (v2). The involvement of these questions that will unveil Sugg's personal information can be seen as a form of self-disclosure as an AEB. The choice of questions may be conditioned by the people she involved. Therefore, in this case, social relationships as a contextual factor conditioned Sugg's self-disclosure.

Singh also involves other people in her video production. For instance, she invites her family members and asks them about her childhood stories (v3). Asking the questions that will unveil Singh's childhood stories can be seen as self-disclosure, conditioned by her relationship with her family. Singh also involves other vlogger friends in videos but resulted in different types of content, such as comedy sketches (v4). In those videos, she promotes her friends' channels. This can be seen as information offering conditioned by Singh's relationships with those vloggers. Promoting other vloggers' channels may not only drive viewers to subscribe to those vloggers but also lead viewers to continuously engage with her content to get more related information regarding what other vloggers they might be interested in subscribing to.

In the sampled data, there is one video of Middleton meeting his friend. The whole video shows him picking up his friend and taking part in different activities together. The video is a self-disclosure of his life (v5). However, the specific events disclosed in the video would not exist or be different if Middleton had not built a friendship with the person involved. Therefore, social relationships also condition Middleton's AEBs in his video.

Personal experiences refer to situations in which vloggers are experiencing or have experienced something when making vlogs. It was found to affect AEBs, especially interaction and self-disclosure. For example, all vloggers show their life activities in their videos. Sugg started a life vlog series on her channel Zoella and now constantly updates it on her other channel *Zoe Sugg*. Singh also records her experience on her channel Lilly Singh, with some latest similar videos uploaded on her second channel *Lilly Singh Vlogs*. Middleton also records footage of him attending gaming events in his early videos. These videos directly disclose the vloggers' life to their audiences as self-disclosure. However, without the context in which the vloggers are experiencing the activities, the vlogs would not be made in the first place. Their experiences may also affect their decisions of what to show to viewers in the videos.

Furthermore, all vloggers have answered questions sent by viewers about their experiences as interaction. For example, Sugg answers questions about the special dream she used to have (v6). Singh was asked about her travel experience in New Zealand (v7). Middleton answers a question regarding his video-making process (v8). Vloggers can choose the questions to answer when producing the video. If the vloggers did not have related experiences, certain questions might not be picked or answered in the video to interact with their viewers.

---

[1] Videos are referenced using their numbers in the Appendix. For the full list of videos referenced in the paper, please see Table 1 in the Appendix.

Finally, vloggers' social characteristics relate to their social features such as interests and hobbies. The results showed that the social characteristics lead to the three vloggers' self-disclosure. For example, in one video, Sugg discloses multiple things about herself, including her interests such as favorite food (v9). Middleton in a gameplay video mentions that the game mode he is playing is one of his favorites (v10). Singh also discloses her interests in videos, such as asking her family about her favorite drink (v3). It can be argued that the action of sharing these interests as self-disclosure is driven by their context of having these interests in the first place.

Audience Context. The results propose that the situations of the three vloggers' audiences, or audience context, also affects their AEBs. The results showed that the audience context that can affect AEBs are related to the situations of *audience experiences* and *interests.*

Audience experiences refer to the situation in which viewers experienced things that vloggers may be aware of when making the videos. All vloggers' productions have been affected by audience experiences, which also affect their AEBs. For example, in a video about skincare, Sugg says that the reason she has made this video is that people have similar skin problems to hers (v11). This indicates that the video production was driven by viewers' experience. This has been further evidenced by messages people sent to Sugg on Twitter regarding their skin problems that may influence Sugg's decision to make the video. This leads to her disclosing her skin issues as self-disclosure and information offering regarding the skincare techniques.

Certain comedy videos produced by Singh are about relations between parents and children. At the end of the video, she usually asks viewers to comment under the videos to say if they can relate to the content (v12). This may indicate that she has noticed that her audience has similar experiences when she makes such a video. The evidence was shown in her interview with CBC in 2014, in which she indicates that when she talks about her experiences with her parents, her fans always mention their similar experiences [50]. Encouraging comments in those videos can be seen as a form of interaction. However, it can be argued that such an AEB was conditioned by Singh's awareness of audience experiences.

Middleton usually asks for gaming advice from his audience, which is the interaction for encouraging AE. For example, in a gameplay video, he has encountered some difficulties when doing an in-game task. He then asks his viewers what he has done wrong (v13). It indicates that Middleton knows his viewers are playing the same game. This has been further evidenced by the viewers' advice provided in the comment section in his previous gameplay video of the same game (v14). This reflects the context of audience experience affecting Middleton's interaction with viewers in the video.

Audience interests refer to the situation in which viewers are interested in specific video content. This contextual factor was identified to mainly affect the interaction and rewards. For instance, at the start of a video in which Sugg showcases items she kept in her bag (v15), she says this video is highly requested. Singh, at the beginning of a video where she reviews the Grammy Awards, also mentioned the video was requested by the viewers (v16). It can be argued that the requests from the audience are conditioned by their viewers' interests in certain video content. This consequently drives vloggers' responses to the requests as the interaction for AE. Middleton in a gameplay video also indicates he made this video because the viewers liked the last one (v17). This indicates viewers' interests has led to Middleton making more similar videos as rewards to viewers' likes.

The effect of audience interest can also be identified from the audience's actions towards vloggers. An example is when vloggers answer viewers' questions. Without audience interest, the questions might not be asked, and the vloggers might not be able to pick those questions to answer and hence interact with their viewers.

4.2 Environmental Context

Environmental context relates to the situation of vloggers and audiences' surroundings that can affect the production of videos, which involves the situations of *social* and *physical environments*. Both environments may alter vloggers' productions and hence their AEBs.

Social Environment. Social environment refers to the situation in which some social activities happen around vloggers and audiences. The analysis found that the situational factors within this context that can condition vloggers' AEBs are mainly *YouTube trends* and *social events*.

YouTube trends refer to periods when creators are making a similar type of video on the site due to its popularity. All vloggers have made videos following certain trends, resulting in different AEBs. For instance, Sugg made a video series called "My Brother Does My Make-up" (v18). The whole video series was initially made as a response to viewers' requests. On the other hand, the "who does whose makeup" has been a popular challenge since 2010 [51]. This trend may have driven the audience to request the video from Sugg, resulting in her response to the request as a form of interaction. Similarly, one video shows Singh doing her makeup, but with her vlogger friend as the voice-over (v19). The video was made in the context of another trending happened on YouTube [52]. The video results in Singh promoting the friend vlogger's channel, which can be seen as an information offering. Middleton made a video to show his reaction to fan-made remix videos of him (v20). The production of this fan-made video may be driven by the popularity of making vlogger-related remix videos. This trend may then drive Middleton to react to the fan creations as a form of interaction.

Social events typically refer to situations in which there are public or popular events, such as festivals and (inter)national days. For example, Sugg uploaded an Easter DIY video (v21), containing different AEBs including providing guides for Easter DIYs as information offering, asking viewers to like the videos for more similar videos as rewards, and encouraging viewers to tweet her their reproduction of her DIYs as interaction. The existence of these AEBs can be seen as being affected by Easter as a social event that drives Sugg's video production. Singh finished a New Year's video by using her audiences' clips (v22) as a form of collaboration or interaction. The key contextual factor that drives such a collaboration is the New Year as a social event. One of Middleton's videos is about him attending a gaming event (v23). This video is his self-disclosure of his experience to his viewers, and the primary driver of video production is the gaming event as a social event.

Physical Environment. Physical environment represents situations of artificial or natural environments. This specifically refers to the *locations* around the vloggers during the vlog production.

The results show that locations lead all vloggers to make videos that disclose their life activities. For instance, a video from Sugg presents her activities in different locations, such as her friend's home (v24). Similarly, Middleton shows his trip to Australia, including showing his hotel view (v25). Singh's video in which she travels to New Jersey shows several locations, such as showing a sports stadium (v26). Sugg and Middleton also filmed tours of their offices to disclose their production environment (v27, v28). In these cases, it is the features of the locations that drive vloggers to show particular footage to their audience as self-disclosure.

4.3 Medium Context

Finally, based on the coding process, the results show that YouTube, as the medium for vlogging, also has its context that can affect vloggers' AEBs. For instance, YouTube offers functions that allow vloggers to engage viewers with interaction, such as using comment sections. YouTube also has features that directly affect the ways of producing and consuming videos. For example, Singh and Middleton did live streams on YouTube. The live stream function gives vloggers the chance to interact with viewers in real-time. It can be speculated that YouTube's algorithmic structure also plays an important role in vlog production. For example, as mentioned, vloggers have created videos following YouTube trends, leading to different AEBs embedded in the video content. The making of these videos can be affected by the trends as social environmental context. However, vloggers may also know that trending videos can be recommended to the viewers by the algorithm. Overall, it can be argued that the medium context of YouTube regarding its functions is critical for vloggers to engage their audience.

5 Discussion and Implications

What are the contextual factors that can condition vloggers' AEBs in their videos? Based on the results from the case study on three popular vloggers, this paper proposes that personal, environmental, and medium context are three critical contextual factors that affect vloggers' AEBs in their videos.

Personal context involves vlogger and audience context. Vlogger context refers to the situations of vloggers in vlog production. These involve the situations of vloggers' social relationships, personal experiences and social characteristics as three vlogger contextual factors. The results propose that the situation in which vloggers have different social relationships with other people can drive vloggers' content production and condition their AEBs due to the specific content produced. The situation of vloggers' personal experiences when making the video also determines what content they will show to the viewers, resulting in AEBs that tights to those content. Finally, the situation of vloggers' social characteristics, such as their interests and hobbies also condition their content produced, and hence the AEBs within those content. Audience context refers to situations of audience-related factors that vloggers are aware of in vlog production, namely the situations of audience experiences and interests as two audience contextual factors. The results propose that vloggers make specific content that accommodates the status of their audience experience and interests, within which related AEBs are embedded.

Similar to previous research, the above results indicate the effect of personal context on vlog content. For example, research by Biel and Gatica-Perez [32] indicates that vloggers' non-verbal behaviors such as their facial expressions, eye contact, and the distance to the cameras when making the videos can be conditioned by vloggers' features including personalities and production skills. Researchers in blogs also found that different motivations of bloggers such as the willingness to document their lives, discuss opinions or form communities, lead to different types of written blog content [21, 34]. These are similar to how vlogger context affects vlog content in the results. Furthermore, Pries et al. [53] in their research on youth usage of YouTube, found that different purposes such as being entertained or learning lead young audiences to consume different content on YouTube. This could potentially lead YouTubers to make videos to accommodate those needs, which can be seen as similar to audience context affecting the vlog content in the results. However, instead of just showing the effect of personal context on vlog content, the above results further propose the specific personal contextual factors that can condition vloggers' AEBs within the content.

Environmental context relates to the situations of the surroundings in video production, which compromises social and physical environment. The contextual factors in the social environment are the situations of YouTube trends and social events. The results propose that the context in which different YouTube trends and social events are happening drives vloggers to produce specific content to follow those events and trends, leading to the implementation of AEBs in those videos. The physical environment mainly refers to the situation of locations for vlog production. The results propose that the status of the locations, such as their features, drive vloggers to make video content about the location, leading to AEBs in those content.

The results are similar to previous research that showed the effect of environmental factors on media production. For instance, Lena's [18] research in rap music showed that the music content is influenced by the market context in which whether major or independent labels dominate the market. This can be seen as a form of the social environment around the media production, similar to the YouTube trend or social events that affect the production of the vlogs. However, the results in this paper further propose the factors in the social environment that can affect vloggers' AEBs within the content.

Locations have also been specifically emphasized as an important contextual factor in vlogging. Snelson [23] studied students' vlogging behaviors and categorized vlogging context as different filming locations. The researcher found that vlogs made in classrooms are involving vloggers and other people showing school life, while vlogs filmed at home involves vloggers talking about school experience such as making complaints. Similarly, the results on the three vloggers show that the vloggers content is conditioned by the filming locations. However, the results further propose the effect of locations on vloggers AEBs within the content.

Finally, medium context refers to the features of YouTube as an online medium for delivering vlogs to the viewers including its functions such as living streaming and technical structures such as algorithms. According to Calder et al. [54], users' experience of online media "is thought to be more active, participatory and interactive" (p. 323). Similarly, YouTube as an online medium provides vloggers and viewers with functions to engage with each other, just like other platforms such as Twitch and Mixer (now known as Facebook Gaming) [55]. Previous research has also shown the effect of YouTube context on its content. For instance, Yarosh et al.'s [22] research on YouTube and Vine indicate that the platform policies of YouTube, such as its age restrictions and the content moderation method led to less inappropriate content on YouTube than Vine. Reider et al. [33] indicate YouTube's algorithmic structure affects not only creators' strategies in video production but also their overall behaviors on the site. These include making longer videos, networking with other channels, and changing publishing timetables. While results in this paper are comparable to the existing research where the context of YouTube as a medium altered the creators' behaviors and content, this paper proposes the medium context's effect on vloggers' AEBs.

When it comes to implications, previous research mainly focused on vloggers' AEBs in the video content. This paper shifts the attention to the vlogging context, which can help to build a widened understanding regarding the building of AE between vloggers and audiences on YouTube. The results argue that although vloggers can employ different strategies to engage their viewers, how or whether these factors are presented to the audience depends on various contextual factors involved during the vlog production. This reflects the context-dependent nature of AE in vlogging. It highlights the importance of context on AEBs and the vloggers' building of audiences, and how an audience can be influenced by the vloggers. This paper opens new directions on researching vloggers and similar streaming service producers' behaviors for building their audiences. It contributes to the existing literature in content, context, and AE in streaming media.

Researchers have also suggested marketing professionals take context into account when implementing user engagement strategies on social media (e.g., [20]). Therefore, for practical implications, this study may help vloggers and potentially creators on other steaming media to consider their production context when building audience bases using different strategies. The research can contribute to creating a production guideline for creators to implement AEBs based on their production context for AE. In addition to guidelines, the research can also contribute to the design of video templates for the YouTube platform for creators to foster AEBs based on their context.

The paper also has certain limitations. First, the main video data were gathered in 2017. Analyzing these data initially addresses the research aim to explore important contextual factors for the three vloggers' AEBs, based on their practices during that period in which their audience base has already been built. As YouTube is a fast-changing platform, future research could focus on more recent vlogging practices and compare them with the current research outcome, to see whether there are similar or new discoveries to be made. Second, future research could examine the vloggers from other fields. Third, the results might not indicate the levels of efficiency of those contextual factors. For example, whether the personal context is more effective than the environmental context on AEBs. Future research could consider comparing these factors. Finally, contextual factors that may affect the creator's AEBs on other streaming or video platforms such as Twitch or Tiktok, can also be considered in the future study.

Appendix

Table 1. List of the videos referenced in the paper

Video no	Video title	URL
v1	Boyfriend VS Brother	Zoella	https://www.youtube.com/watch?v=aLP l2G-epfw
v2	Best Friend VS Boyfriend	Zoella	https://www.youtube.com/watch?v=aal yr7v0t14
v3	My Family Answers Questions About Me	https://www.youtube.com/watch?v= WXYdIarPN2s
v4	The 5 Stages to Becoming a Fangirl (ft. Grace Helbig)	https://www.youtube.com/watch?v=mvV z4p7LCbQ
v5	WE SAVED A RABBIT.	https://www.youtube.com/watch?v=BQQ mUso6eZE
v6	The Questions I've Never Answered	Zoella	https://www.youtube.com/watch?v=7Y9 hldc2ZFs
v7	#AskSupermanLIVE (01/11/16)	https://www.youtube.com/watch?v=2Av pV7o6vEM
v8	YOUTUBER CONFESSIONS	TDM Vlogs #29	https://www.youtube.com/watch?v=UnS ztuERMS4
v9	50 Facts About Me	Zoella	https://www.youtube.com/watch?v=9NG Qm9i33Mc
v10	Minecraft	YOU WANT EGGS WITH THAT BACON?!	https://www.youtube.com/watch?v=UNN jFQZ25HQ
v11	My Makeup Routine For Problem Skin Days	Zoella	https://www.youtube.com/watch?v=1Vg LebKIqDU
v12	The Difference Between You and Your Parents	https://www.youtube.com/watch?v=6Bx baVjnMSM
v13	"OIL EXTRACTION"	Diamond Dimensions Modded Survival #65	Minecraft	https://www.youtube.com/watch?v=UhJ zbbtU9aM
v14	"I BUILT A ROCKET!"	Diamond Dimensions Modded Survival #64	Minecraft	https://www.youtube.com/watch?v=Vaw UPTt_DZI
v15	What's In My Bag?	Zoella	https://www.youtube.com/watch?v=x59f-EPEaFY
v16	Jay Z Almost Poops	Grammys 2015 Review	https://www.youtube.com/watch?v=n3l GVDc7qVo
v17	SCARIEST OLD MAN IN MINECRAFT!!!	https://www.youtube.com/watch?v=2PI kgL_Tsd8

(continued)

Table 1. (*continued*)

Video no	Video title	URL
v18	My Brother Does My Make-up	https://www.youtube.com/watch?v=-6-axi3jprE
v19	Boy-FRIEND Does My Makeup Voiceover (ft. Ryan Higa)	https://www.youtube.com/watch?v=Wo-Ux6fmQyk
v20	DANTDM SINGS?!?!	https://www.youtube.com/watch?v=lASIY-vZOQ4
v21	6 Quick & Easy Easter Treats | Zoella	https://www.youtube.com/watch?v=pxKJPlk2GX4
v22	2016…That Is A Wrap! (ft. #TeamSuper)	https://www.youtube.com/watch?v=xrZlIVqdMrQ
v23	EUROGAMER 2013 EVENT MONTAGE! – TheDiamondMinecart	https://www.youtube.com/watch?v=hLkraPc6faI
v24	VLOG: My week with Louise (feat. FleurdeForce & Baby Glitter)	https://www.youtube.com/watch?v=TRTBg_BfbAI
v25	AUSTRALIAN JET BOAT RIDE!!!	https://www.youtube.com/watch?v=pqkrZqKGT00
v26	Jersey Vloggity!	https://www.youtube.com/watch?v=nc3m2zGKMXE
v27	OFFICE TOUR!! | TheDiamondMinecart	https://www.youtube.com/watch?v=qiyPLS_phN4
v28	My Office Tour 2016 | Zoella	https://www.youtube.com/watch?v=gTUi5iUqQ_I

References

1. Zhang, H.: Evoking presence in vlogging: a case study of U. K. beauty blogger Zoe Sugg. First Monday **23**(1) (2018). https://doi.org/10.5210/fm.v23i1.8107
2. Kaminsky, M.S.: Naked Lens-Video Blogging and Video Journaling to Reclaim the YOU in YouTube: How to Use Online Video to Increase Self Expression, Enhance Creativity, and Join the Video Regeneration. Organik Media Incorporated, New York (2010)
3. Weaver, A.J., Zelenkauskaite, A., Samson, L.: The (non) violent world of YouTube: content trends in web video. J. Commun. **62**(6), 1065–1083 (2012). https://doi.org/10.1111/j.1460-2466.2012.01675.x
4. Corrêa, S.C.H., Soares, J.L., Christino, J.M.M., de Sevilha Gosling, M., Gonçalves, C.A.: The influence of YouTubers on followers' use intention. J. Res. Interact. Mark. **14**(2) (2020). https://doi.org/10.1108/JRIM-09-2019-0154
5. DataReportal, Hootsuite, We Are Social.: Share of internet users worldwide watching vlogs weekly as of 3rd quarter 2021, by age and gender. Statista (2022). https://www-statista-com.hallam.idm.oclc.org/statistics/1254829/age-gender-reach-worldwide-watching-vlogs/. Accessed 08 Feb 2022
6. Funk, M.: How many YouTube channels are there? (2020). https://www.tubics.com/blog/number-of-youtube-channels. Accessed 16 Jan 2021

7. Burgess, J., Green, J.: YouTube: Online Video and Participatory Culture, 2nd edn. Politya, Cambridge (2018)
8. Khan, M.L.: Social media engagement: what motivates user participation and consumption on YouTube? Comput. Hum. Behav. **66**, 236–247 (2017). https://doi.org/10.1016/j.chb.2016.09.024
9. McRoberts, S., Bonsignore, E., Peyton, T., Yarosh, S.: Do it for the viewers! Audience engagement behaviors of young YouTubers. In: Read, J.C., Stenton, P. (eds.) Proceedings of the 15th International Conference on Interaction Design and Children, pp. 334–343. ACM, New York (2016). https://doi.org/10.1145/2930674.2930676
10. Tarnovskaya, V.: Reinventing personal branding building a personal brand through content on YouTube. J. Int. Bus. Res. Mark. **3**(1), 29–35 (2017). https://doi.org/10.18775/jibrm.1849-8558.2015.31.3005
11. Jerslev, A.: Media times‖ in the time of the microcelebrity: celebrification and the YouTuber Zoella. Int. J. Commun. **10**, 5233 (2016)
12. Marôpo, L., Jorge, A., Tomaz, R.: "I felt like I was really talking to you!": intimacy and trust among teen vloggers and followers in Portugal and Brazil. J. Child. Media **14**(1), 22–37 (2020). https://doi.org/10.1080/17482798.2019.1699589
13. Rybaczewska, M., Jebet Chesire, B., Sparks, L.: YouTube vloggers as brand influencers on consumer purchase behaviour. J. Intercult. Manag. **12**(3), 117–140 (2020). https://doi.org/10.2478/joim-2020-0047
14. Odden, L.: What is Content? Learn from 40+ definitions (2013). http://www.toprankblog.com/2013/03/what-is-content/. Accessed 04 Feb 2022
15. Matuk, C., DesPortes, K., Hoadley, C.: Conceptualizing context in CSCL: cognitive and sociocultural perspectives. In: Cress, U., Rosé, C., Wise, A.F., Oshima, J. (eds.) International Handbook of Computer-Supported Collaborative Learning. CCLS, vol. 19, pp. 85–101. Springer, Cham (2021). https://doi.org/10.1007/978-3-030-65291-3_5
16. Sonnenwald, D.H.: Evolving perspectives of human behavior: contexts, situation, social networks and information horizons. In: Wilson, T., Allen, D. (eds.) Exploring the Contexts of Information Behaviour, pp. 176–190. Taylor Graham, London (1999)
17. Dey, A.K.: Understanding and using context. Pers. Ubiquit. Comput. **5**(1), 4–7 (2001). https://doi.org/10.1007/s007790170019
18. Lena, J.C.: Social context and musical content of rap music, 1979–1995. Soc. Forces **85**(1), 479–495 (2006). https://doi.org/10.1353/sof.2006.0131
19. Bickham, D.S., Rich, M.: Is television viewing associated with social isolation?: roles of exposure time, viewing context, and violent content. Arch. Pediatr. Adolesc. Med. **160**(4), 387–392 (2006). https://doi.org/10.1001/archpedi.160.4.387
20. Jaakonmäki, R., Müller, O., Vom Brocke, J.: The impact of content, context, and creator on user engagement in social media marketing. In: Bui, T.X., Sprague, R. (eds.) Proceedings of the 50th Hawaii International Conference on System Sciences, pp. 1152–1160 (2017). https://doi.org/10.24251/HICSS.2017.136
21. Nardi, B.A., Schiano, D.J., Gumbrecht, M., Swartz, L.: Why we blog. Commun. ACM **47**(12), 41–46 (2004). https://doi.org/10.1145/1035134.1035163
22. Yarosh, S., Bonsignore, E., McRoberts, S., Peyton, T.: YouthTube: youth video authorship on YouTube and Vine. In: Proceedings of the 19th ACM Conference on Computer-Supported Cooperative Work & Social Computing, pp. 1423–1437. ACM, New York (2016). https://doi.org/10.1145/2818048.2819961
23. Snelson, C.: Vlogging about school on YouTube: an exploratory study. New Media Soc. **17**(3), 321–339 (2015). https://doi.org/10.1177/1461444813504271
24. Farahbakhsh, F., Shahidinejad, A., Ghobaei-Arani, M.: Context-aware computation offloading for mobile edge computing. J. Ambient. Intell. Humaniz. Comput. 1–13 (2021). https://doi.org/10.1007/s12652-021-03030-1

25. Rubin, A.M., Rubin, R.B.: Age, context and television use. J. Broadcast. Electron. Media **25**(1), 1–13 (1981). https://doi.org/10.1080/08838158109386424
26. Tur-Viñes, V., Castelló-Martínez, A.: Commenting on top Spanish YouTubers: "no comment." Soc. Sci. **8**(10), 266 (2019). https://doi.org/10.3390/socsci8100266
27. Berryman, R., Kavka, M.: 'I guess a lot of people see me as a big sister or a friend': the role of intimacy in the celebrification of beauty vloggers. J. Gend. Stud. **26**(3), 307–320 (2017). https://doi.org/10.1080/09589236.2017.1288611
28. Ferchaud, A., Grzeslo, J., Orme, S., LaGroue, J.: Parasocial attributes and YouTube personalities: exploring content trends across the most subscribed YouTube channels. Comput. Hum. Behav. **80**, 88–96 (2018). https://doi.org/10.1016/j.chb.2017.10.041
29. de Bérail, P., Guillon, M., Bungener, C.: The relations between YouTube addiction, social anxiety and parasocial relationships with YouTubers: a moderated-mediation model based on a cognitive-behavioral framework. Comput. Hum. Behav. **99**, 190–204 (2019). https://doi.org/10.1016/j.chb.2019.05.007
30. Buf, D.M., Ștefăniță, O.: Uses and gratifications of YouTube: a comparative analysis of users and content creators. Rom. J. Commun. Public Relations **22**(2), 75–89 (2020). https://doi.org/10.21018/rjcpr.2020.2.301
31. Kamboj, S.: Applying uses and gratifications theory to understand customer participation in social media brand communities: perspective of media technology. Asia Pac. J. Mark. Logist. **32**(1), 205–231 (2019). https://doi.org/10.1108/APJML-11-2017-0289
32. Biel, J.I., Gatica-Perez, D.: Vlogsense: conversational behavior and social attention in YouTube. ACM Trans. Multimedia Comput. Commun. Appl. (TOMM) **7**(1), 1–21 (2011). https://doi.org/10.1145/2037676.2037690
33. Rieder, B., Coromina, Ò., Matamoros-Fernández, A.: Mapping YouTube: a quantitative exploration of a platformed media system. First Monday **25**(8) (2020). https://doi.org/10.5210/fm.v25i8.10667
34. Jolly, J.L., Matthews, M.S.: Why we blog: homeschooling mothers of gifted children. Roeper Rev. **39**(2), 112–120 (2017). https://doi.org/10.1080/02783193.2017.1289579
35. Blumler, J.G., Katz, E.: The Uses of Mass Communications: Current Perspectives on Gratifications Research. Sage, Newbury Park (1973)
36. Zimmer, F., Scheibe, K., Stock, W.G.: A model for information behaviour research on social live streaming services (SLSSs). In: Meiselwitz, G. (ed.) SCSM 2018. LNCS, vol. 10914, pp. 429–448. Springer, Cham (2018). https://doi.org/10.1007/978-3-319-91485-5_33
37. Zainal, Z.: Case study as a research method. Jurnal Kemanusiaan **5**(1), 1 (2007)
38. Yin, R.K.: Case Study Research and Applications: Design and Methods, 4th edn. Sage, Thousand Oaks (2009)
39. Zhang, H.: Data sets, analysis and descriptions for the case study. Zenodo (2022). https://doi.org/10.5281/zenodo.5975892
40. Patton, M.Q.: Purposeful sampling. In: Patton, M.Q. Qualitative Evaluation and Research Methods, pp. 169–186. Sage, Beverly Hills (1990)
41. Perry, C.: Processes of a case study methodology for postgraduate research in marketing. Eur. J. Mark. **32**(9/10), 785–802 (1998). https://doi.org/10.1108/03090569810232237
42. Zoella [YouTube Channel]. https://www.youtube.com/user/zoella280390
43. DanTDM [YouTube Channel]. https://www.youtube.com/channel/UCS5Oz6CHmeoF7vSad0qqXfw
44. Lilly Singh [YouTube Channel]. https://www.youtube.com/channel/UCfm4y4rHF5HGrSrqbvOwOg
45. Braun, V., Clarke, V.: Using thematic analysis in psychology. Qual. Res. Psychol. **3**(2), 77–101 (2006)
46. Pyett, P.M.: Validation of qualitative research in the "real world". Qual. Health Res. **13**(8), 1170–1179 (2003). https://doi.org/10.1177/1049732303255686

47. Anney, V.N.: Ensuring the quality of the findings of qualitative research: looking at trustworthiness criteria. J. Emerg. Trends Educ. Res. Policy Stud. (JETERAPS) 5(2), 272–281 (2014)
48. Krefting, L.: Rigor in qualitative research: the assessment of trustworthiness. Am. J. Occup. Ther. 45(3), 214–222 (1991). https://doi.org/10.5014/ajot.45.3.214
49. Patton, M.Q.: Enhancing the quality and credibility of qualitative analysis. Health Serv. Res. 34(5 Pt 2), 1189 (1999)
50. Q on CBC: Lilly Singh: Superwoman of the internet (2014). https://www.youtube.com/watch?v=T9STyy1KNgY. Accessed 12 Dec 2017
51. RandomMan: My boyfriend does my makeup (2020). https://knowyourmeme.com/memes/my-boyfriend-does-my-makeup. Accessed 11 June 2021
52. Sasso, S.: The best video challenge since boyfriends putting on makeup (2016). https://www.refinery29.com/2016/11/131177/jenna-marbles-boyfriend-voiceover-challenge. Accessed 13 Sep 2018
53. Pires, F., Masanet, M.J., Scolari, C.A.: What are teens doing with YouTube? Practices, uses and metaphors of the most popular audio-visual platform. Inf. Commun. Soc. 24(9), 1175–1191 (2019). https://doi.org/10.1080/1369118X.2019.1672766
54. Calder, B.J., Malthouse, E.C., Schaedel, U.: An experimental study of the relationship between online engagement and advertising effectiveness. J. Interact. Mark. 23(4), 321–331 (2009). https://doi.org/10.1016/j.intmar.2009.07.002
55. Zimmer, F., Scheibe, K., Zhang, H.: Gamification elements on social live streaming service mobile applications. In: Meiselwitz, G. (ed.) HCII 2020. LNCS, vol. 12194, pp. 184–197. Springer, Cham (2020). https://doi.org/10.1007/978-3-030-49570-1_13

Text Analysis and AI in Social Media

Development of Bilingual Sentiment and Emotion Text Classification Models from COVID-19 Vaccination Tweets in the Philippines

Nicole Allison Co[✉][iD], Maria Regina Justina Estuar, Hans Calvin Tan,
Austin Sebastien Tan, Roland Abao, and Jelly Aureus

Ateneo de Manila University, Metro Manila, Philippines
allison.co@obf.ateneo.edu
https://www.ateneo.edu/

Abstract. Social media can be used to understand how the public is responding to the ongoing nationwide COVID-19 vaccination campaign, allowing policymakers to respond effectively through informed decisions. However, conducting social media analysis in the Philippine-context presents a challenge because natural informal conversations make use of a combination of English and local language. This study addresses this challenge by including part-of-speech tags, frequency of code switching and language dominance features to represent bilingualism in training machine learning models with COVID-19 vaccination-related Tweets for sentiment and emotion analysis. Results showed that the English-Tagalog Logistic Regression sentiment classification model performed better than Textblob, VADER and Polyglot with an accuracy of 70.36%. Similarly, the English-Tagalog SVM emotion classification model performed better than Text2emotion, NRC Affect Intensity Lexicon and EmoTFIDF with an average mean-squared error of 0.049. The added bilingual features only improved these performance metrics by a small margin. Nevertheless, SHAP analysis still revealed that sentiment and emotion classes exhibit varying levels of these bilingual features, which shows the potential in exploring similar linguistic features to distinguish between classes better during text classification for future studies. Finally, Tweets from September 2021 to January 2022 shows negative, mainly anger and sadness, perceptions towards COVID-19 vaccinations.

Keywords: Covid-19 · Social computing · Social media · NLP · Sentiment analysis · Emotion analysis

1 Introduction

Studying public perceptions behind public health issues provides a way for policymakers to listen and respond more effectively through informed decisions to solve the country's biggest health crises. The government has to enlist the participation of the Filipino citizens in decision-making to develop strong trust and collaboration between the two parties in facing nationwide health challenges. Governments are then faced with the problem of how to engage the public at a big, nationwide scale [19, 30].

An example is the ongoing nationwide COVID-19 vaccination campaign. The Department of Health (DOH), along with local government units (LGUs), continue to organize mass vaccination programs since November 2020, with the goal of achieving herd immunity. As of the end of January 2022, only around 53% of the Filipino population has been vaccinated against COVID-19 [22]. For measles and polio, the herd immunity threshold defined by the World Health Organization (WHO) is 95% and 80% of the population, respectively. Since there is no evidence that immune response from natural infection lasts and natural immunity can cause unnecessary cases and deaths, the WHO supports further increasing COVID-19 vaccination rates to fight the virus [1]. Given this, the success of COVID-19 vaccination programs is a joint effort of the government and the people. Therefore, the challenge is to find a way of knowing the level of cooperation of the public to the government's policies and initiatives relating to COVID-19 vaccinations in the Philippines. This way, the government can adjust their response based on what works and what does not, as experienced and expressed by the Filipinos themselves. Social media is a good avenue to understand how the public are reacting to this event, to be able to respond better to their needs.

However, conducting social media analysis, specifically Twitter analysis in the Philippine, public health context presents a challenge because of the preferred usage by Filipinos of English and local language in both informal and formal conversations. In this study, the focus is on narratives from those who are known to use a combination of English and Tagalog (the main language spoken by Filipinos), called "Taglish," in day-to-day communication. Thus, existing classification models often trained with English, non-health data, does not produce optimal results when provided with bilingual, health-related inputs [2].

For these reasons, this study added part-of-speech tags, frequency of code switching and language dominance to commonly used text features in training machine learning models for sentiment and emotion classification on health-related English-Tagalog bilingual Tweets. The study used COVID-19 vaccination-related keywords to extract Tweets for training, given that it is an ongoing problem with an abundance of Tweets relating to this topic at the time of the study. The developed models were analyzed using SHapley Additive exPlanations (SHAP) to provide a deeper explanation on how predictions were derived, through the identification of the top influential features per sentiment and emotion category. Lastly, the models were applied to entire COVID-19 vaccine-related Tweets data set collected from September 2021 to January 2022 to understand the perception of Filipinos on COVID-19 vaccinations over time.

2   Review of Related Literature

2.1   COVID-19 Vaccine Perception

Some research has been done to understand how people perceive the current COVID-19 vaccinations. The first theme in COVID-19 vaccine perception is the risk perception of the disease. If one perceives COVID-19 to pose a higher risk to one's health, then the motivation to take preventive measures, like vaccination increases [5,30]. Risk perception for COVID-19 is high in most circumstances, given that it is a new virus, that individuals, health authorities and governments have major difficulties controlling and managing the spread of [10].

Moreover, accessibility of the COVID-19 vaccine also has a role in how they are perceived. If COVID-19 vaccines are readily available to the public, then they would be more willing to take it [16]. However, not all nations believe that their country's authorities are capable of procuring and distributing for population-wide vaccinations [5]. Moreover, for a lower-middle income country like the Philippines, free COVID-19 vaccinations from the government encourages vaccine uptake [14,30].

Lastly, there is hesitancy surrounding COVID-19 vaccine safety and efficacy. Since COVID-19 vaccines are still newly developed, there is lack of information and evidence with regards to its safety [16]. Fear that the COVID-19 vaccine may alter genetics and lead to issues is an example of a vaccine safety concern [30].

2.2   Relationship of Perception and Social Media

Social media is a good avenue to continue to study public perception on COVID-19 vaccinations. This is because social media not only allows users to express their perceptions online, but they tend to influence the perceptions of other users as well through the spread of unverified information and the appeal of personal narratives.

First, online disinformation campaigns make users believe that vaccinations are not safe, which lowers the mean vaccination coverage. [36]. Distinction between informed and uninformed, and pro-vaccine and anti-vaccine users is blurred on social media [8,28]. The danger of misinformation on social media lies in how fast it spreads, given how easy it is to share a post with just a click. Moreover, these unverified posts often come from trusted family and friends belonging to one's social network, giving the information a false sense of trustworthiness [23]. The potential spread of misinformation is exacerbated by "information silos" or "echo chambers." Social media platforms are known to tailor the content one sees to the user's usage patterns. This introduces biases because it discourages users from viewing posts with differing viewpoints that challenge their existing beliefs [8].

Second, personal narratives abundant on social media provide an emotional appeal towards vaccine acceptance or hesitancy, in contrast to the rational appeal of facts and statistics from the previous themes [20,24]. Personal narratives are persuasive because they provide empathy and social and emotional support through their conversational and relatable nature [39]. Highly emotional replies are a characteristic of anti-vaccination posts. The strong emotions and toxicity present in anti-vaccine supporters' replies can be comparable to effective propaganda strategies [24].

2.3   Models for Studying Perception Through Social Media

Given the relationship of perception and social media, the following studies present methodologies for gathering, processing and analyzing social media data.

Twitter as a Social Media Data Source. Twitter API is used to collect data from Twitter, a popular social media data source [2,13,27]. The language feature can be used to extract country-specific data [2]. Given the casual nature of a social media site like Twitter for majority of Filipino users, including these Tagalog or "Taglish" Tweets in the study would provide a more accurate representation of the population.

Pre-processing for Tweets. After Tweet extraction, data pre-processing is done to eliminate noise and inconsistencies with raw data [15]. Pre-processing usually starts with transforming all texts into lowercase for uniformity [34]. Then, Natural Language Toolkit (NLTK), a Python library, provides functions for tokenizing, filtering, stemming, lemmatizing and part of speech tagging. Tokenizing splits a Tweet into smaller units, such as words or sentences, which makes it easier to find patterns in the text [32]. Filtering can eliminate irrelevant data in a collection of texts. Emojis are useful in understanding sentiment, but are not recognizable by machine learning models. Thus, common emojis can be replaced with their plain text meaning [34]. Unnecessary data like numerical data, URLs, special characters and punctuation can be filtered out [2,15]. Stop words that provide no insight, such as "the", "to" and "of" should also be removed [29]. Stemming and lemmatizing combines words with the same meaning. Stemming is a process of cutting off the ending of words, usually derivational affixes [29]. On the other hand, lemmatizing takes the base dictionary form of a word [34]. Part of speech tagging can also be applied with lemmatizing to increase accuracy.

Feature Extraction on Textual Data. Once noise has been eliminated from the Tweets, the data can already be converted into numerical features so that it can be recognized by models. One way is the bag-of-words approach which counts the frequency that a word appears in a text. Grammar and word sequence are both disregarded [11]. However, counting frequency per Tweet may not be as accurate when some Tweets are shorter or longer than the others. Term frequency - inverse document frequency (TF-IDF) puts weight on each word by counting how many times it appears in the current text, over the total number of times it appears in the whole data set. The TF-IDF value, as a feature, represents the importance of a word in a Tweet [11]. Finally, N-gram models can be used to predict the most probable Nth word in the sequence, given a string of N-1 words, allowing related words to be grouped together. A bigram looks at the previous word and a trigram looks at the two previous words to make predictions [12].

Performing Sentiment Analysis. With the features generated from the Tweets, supervised learning can be used to train models to classify Tweets based on their polarity (positive, negative or neutral). SVM and Naive Bayes are two common supervised learning algorithms for sentiment analysis in the context of vaccine-related Tweets [26,27,35]. For an SVM classifier, the algorithm creates boundaries in an n-dimensional space, to classify data into classes [17]. On the other hand, a Naive Bayes model assumes that all variables are independent from one another. Probabilities are predicted based on the past attributes of each class [31]. Furthermore, ensembling-based approaches can improve performance by combining predictions from multiple models. Good results have been achieved from using a hybridization of Naive Bayes, SVM, Linear Regression and SGD classifiers to perform sentiment analysis on bilingual data [37,38].

As opposed to developing a new model, some studies used existing tools such as Valence Aware Dictionary and sEntiment Reasoner (VADER) and Polyglot for sentiment analysis. VADER is a lexicon-based and rule-based sentiment analyzer. A lexicon-based approach labels a list of lexical features, like words, with their polarity. VADER is known to work well with social media data, as it recognizes slang, emoji, capitalization and writing style when expressing sentiment intensity [6]. Polyglot also offers polarity lexicons, but for multiple languages. This library can identify polarity at both a word and sentence level [7].

Performing Emotion Analysis. There is often the need to understand more nuanced information about people's perceptions online, using emotions, despite the popularity of sentiment analysis. For instance, a negative post may pertain to sadness, anger or fear, all of which entail different interventions [4]. Supervised learning algorithms and ensembling have also been explored for emotion analysis prediction models [37,38].

There are several emotion classification models developed by psychologists. The first approach to representing emotions is the categorical approach, which places emotions into distinct classes. These include Paul Ekman's model (happiness, fear, sadness, surprise, disgust and anger), Robert Plutchik's model (surprise verus anticipation, joy versus sadness, anger versus fear and trust versus disgust) and Orthony, Clore, and Collins (OCC)'s model (envy, relief, appreciation, self-reproach, shame, reproach, pity, admiration, disappointment, grief, gratification, fears-confirmed, gloating, hope, like and dislike). The second is the dimensional approach, which considers emotion states to be bound to each other. Russell's circumplex model (arousal and valence dimensions), Plutchik's wheel of emotions (combinations of Plutchik's model's primary emotions), Russell and Mehrabian's model (pleasantness, arousal and potency dimensions) and Parrot's Emotion Taxonomy (primary, secondary and tertiary emotions with love, joy, surprise, anger, sadness, and fear as the primary emotions) belong to the second approach. The last approach is the appraisal based approach which addresses emotion changes through the significant components such as cognition, expressions, physiology, motivation, motor, reactions, and feelings [3, 18, 25].

SHAP Values for Explainable Text Classification Models. The trained models can be analyzed further to gain insights into what makes up each sentiment and emotion class. This form of analysis is referred to as "explainable AI." Explainable AI is a growing field of study, such that the development of machine learning models are not just concerned with extraction and prediction, but also explanation. This enables transparent models that allows for a "human in the loop" perspective to using and improving trained models [9]. SHAP is a model-agnostic, game theory approach for interpreting the findings of any machine learning model. SHAP values can be used to identify the which tokens are responsible to classify into each given category [33]. Features' SHAP values are computed by taking subsets of features, training the model with and without this subset of features, comparing the predictions to the current value to evaluate the features' effect, and then representing that effect in a numerical value [21]. The higher the computed SHAP value for a token is, the higher is the impact of that token in calculating the probability of classifying an input into a certain class [33].

3   Methodology

3.1   Data Extraction on Twitter

This study collected COVID-19 vaccine-related Tweets from September 2021 to Janaury 2022, limited to the Philippine context, through the Twitter API. Keywords that were used are vaccination, vaccinations, vaccine, vaccines, immunization, vaccinate, vaccinated, and bakuna (Tagalog word for vaccine), #covidvaccineph, #covid19vaccineph, and #covaxph. This study used the language (lang: "tl") to determine whether the Tweet is of Philippine-context, given that most Tweets do not store the geographical location from which they were Tweeted, but Twitter automatically classifies all Tweets into their respective languages. Tweets from established news were removed, given that the goal of this study is to analyze sentiments and emotions in human perception, not news reports.

3.2   Manual Annotation of the Data Set

Two taggers were recruited to annotate the same set of 10,000 Tweet samples from September 2021 to October 2021. A coding manual and orientation session were used to give instructions. Taggers were asked to go through each Tweet and mark it as positive, negative, neutral or invalid for the sentiment. On the other hand, emotion labels, namely, love, joy, sadness, fear, anger, and/or surprise, were arranged using a checkbox format, where taggers can check one or more emotions that are present in each Tweet.

For the sentiment, positive Tweets pertain to Tweets that express willingness to get vaccinated, persuade others to get vaccinated or show satisfaction over vaccination programs. Negative Tweets pertain to Tweets that express disappointment or disagreement with COVID-19 vaccines, vaccination programs or other's opinions on them. Neutral Tweets pertain to Tweets that simply state facts, ask questions, or imply the user is still unsure about his/her stance on vaccinations. Invalid Tweets are those that do not make any sense, like if they are gibberish, written in a completely different language, or contains just links.

Parrot's emotion taxonomy is chosen for this study's emotion classification given that it allows for distinct classes of emotions and covers a wide variety of emotion states. Parrot defined six emotion categories as shown in Table 1. Neutral and invalid from the sentiment portion were no longer tagged for emotion.

Table 1. Emotion categories based on Parrot's emotion taxonomy

Emotion	Description
Love	Affection, lust, longing, caring, tenderness, or compassion
Joy	Happiness, satisfaction, enthusiasm, excitement, hope, optimism, or relief
Sadness	Suffering, hurt, hopelessness, grief, disappointment, shame, regret, isolation, neglect, humiliation, insecurity, or sympathy
Fear	Alarm, panic, anxiety, nervousness, apprehension, worry, distress, or dread
Anger	Annoyance, frustration, rage, bitterness, hate, digust, or envy
Surprise	Amazement, or astonishment

After the tagging, inter-rater reliability score was computed. Only Tweets that have the same sentiment annotation for both taggers were included in the final data set for sentiment analysis. On the other hand, tags for each emotion category, where True was given a value of 1 and False was given a value of 0, were averaged to obtain a final data set containing emotion probability weights for emotion analysis.

3.3 Data Pre-processing

This study used the NLTK to pre-process the remaining raw Tweets. Line breaks, mentions, URLs, punctuation and special characters were filtered out. All Tweets were converted to lowercase for uniformity. Tokenization was performed on the normalized Tweets. Then, lemmatization was used to reduce the words to their core meaning. Stop words and numbers were then filtered out. The study used English stop words from NLTK and tagalog stop words downloaded from stopwords-iso's collection.

3.4   Feature Extraction

This study added bilingual features to commonly used text features like bag-of-words, TF-IDF and n-grams. Performance metrics of models with and without the bilingual features were compared to determine whether they made significant impact.

Text Features. Bag-of-words with n-gram modelling, bag-of-words without n-gram modelling, TF-IDF with n-gram modelling and TF-IDF without n-gram modelling, were used to represent words in vectors to training supervised learning models to see which one would produce the best results. The sklearn package was used for bag-of-words (CountVectorizer) and TF-IDF (TfidfVectorizer) and the gensim package was used for n-gram modelling.

Bilingualism Features. Given that text features distinguish between unique words and not the language it is in, this study added additional features that describe the bilingualism in a text to provide the model with more context in training. POS tagging using NLTK was used, such that the frequency of nouns, verbs, adverbs and adjectives in a Tweet were added as features. Next is the frequency of code-switching. This is defined as the number of times the user switches from using English to Tagalog or Tagalog to English in a Tweet. The second is language dominance. This is represented by the number of words that are in English and the number of words that are in Tagalog within a Tweet.

3.5   Developing a Bilingual Sentiment and Emotion Text Classification Models

Existing sentiment and emotion analysis packages were first applied to the data set to define a baseline performance for this study to improve on. Two variations of the Philippine COVID-19 vaccinations Tweets data set were used for this, namely, the original bilingual data set and a translated version. Google Translate was used for Tagalog to English translation. The study chose two English-optimized sentiment analyzers, TextBlob and VADER, and one multilingual analyzer, Polyglot, for sentiment analysis and text2emotion, NRC Affect Intensity Lexicon and EmoTFIDF for emotion analysis. The predictions of these models were compared to the actual tags, from the manual annotation, to calculate for the accuracy and mean squared error for sentiment and emotion analysis, respectively.

For this study's development of sentiment and emotion text classification models optimized for bilingual health-related data, StratifiedKFold cross validation, with cv = 5 was used to split the data set into training and test sets. For each iteration, the training set was used as input into supervised learning models (Naive Bayes, SVM, Logistic Regression, SGD, and Random Forest for sentiment analysis and SVM, Random Forest and Gradient Boosting Regressor for emotion analysis) and an ensemble on 3 of these supervised learning models (Logistic Regression, SVM and Random Forest for sentiment analysis and SVM, Random Forest and Gradient Boosting Regressor for emotion analysis). The model were then evaluated with the test set to get the accuracy, precision, recall and F1 score for sentiment analysis and mean squared error for emotion analysis. The weights of the best model for sentiment and emotion classification were then exported using the pickle Python package for the trained model to be reusable for new data.

3.6   SHapley Additive ExPlanations (SHAP) Analysis of Models

SHAP analysis was applied to the best models for sentiment and emotion classification to calculate the contribution of each feature to the model's prediction. This allowed the study to identify top features that influence the probability of the models to predict each sentiment and emotion category given a new Tweet input. Shapley values and plots were derived using a sample of Tweets in the data set, given the resource intensive nature of running SHAP analysis.

3.7   Applying Sentiment and Emotion Text Classification Models in Understanding COVID-19 Vaccine Perception

Lastly, the sentiment and emotion text classification models were applied to the whole COVID-19 vaccine-related Tweets data set collected from September 2021 to January 2022 to understand the Filipino perception on COVID-19 vaccination in the country, as seen on Twitter. The overall percentage of positive, negative, and neutral sentiments and average score of love, joy, sadness, fear, anger and surprise emotions revealed by the predictions were presented, along with a time-series view of the said results.

4   Results

4.1   Inter-rater Reliability Score

Table 2 shows the inter-rater reliability scores for the sentiment and emotion labels annotated by the two taggers who participated in the study.

Table 2. Inter-rater Reliability Scores

	Inter-rater reliability score
Sentiment labels	57.99%
Emotion labels	51.42%

The other Tweets that were assessed differently for sentiment by the two taggers were removed from the data set. Moreover, 20.64% of the Tweets were tagged as "Invalid" and thus, were also removed from the data set. This left 3735 valid Tweets for sentiment classification with a distribution of 635 Tweets for positive, 1298 Tweets for negative, and 1802 Tweets for neutral.

For emotion analysis, the emotion tag (True = 1 or False = 0) for each emotion category were averaged instead to derive weights that represent the probability of a Tweet to exhibit that emotion category. Since neutral and invalid Tweets, which consist 38.66% of the data set, do not have emotions, they were removed from the data set for emotion classification, leaving 6134 valid Tweets.

Table 3. Accuracy of existing sentiment text classification models

Package	Original data set	Translated data set
TextBlob	48.10%	47.30%
VADER	49.50%	48.80%
Polyglot	50.00%	51.10%

4.2 Performance of Existing Sentiment Text Classification Models

As seen in Table 3, the highest accuracy achieved is 51.10% from Polyglot with the translated data set. This value serves as the baseline performance measure that this study aims to improve on for sentiment analysis.

4.3 Performance of Existing Emotion Text Classification Models

Table 4. Mean squared error of existing emotion text classification models

Package	Joy	Sadness	Fear	Anger	Surprise
Original data set					
Text2Emotion	0.1435	0.1547	0.1415	0.1451	0.1089
NRC Affect Intensity Lexicon	0.1434	0.1448	0.0925	0.1362	0.0269
EmoTFIDF	0.1027	0.1284	0.0625	0.1326	0.0112
Translated data set					
Text2Emotion	0.1256	0.1686	0.1851	0.1493	0.1271
NRC Affect Intensity Lexicon	0.2342	0.1736	0.1764	0.1433	0.0497
EmoTFIDF	0.1029	0.1187	0.0667	0.1278	0.0153

Both the existing packages and the annotated data set of this study used a range of 0–1 for the emotion scores, making them comparable when evaluating the mean squared error. However, it is important to note that these packages and the annotated data set used to evaluate them all used different emotion classification theories. Similar emotions, like "Happy" and "Joy", have been mapped to one another, while other emotions that have no equivalent in the tagged data set's classifications, like "Trust", were ignored. This limitation may have affected the resulting metrics of the models. Moreover, none of the existing analyzers had "love" as an emotion category, while this study chose an emotion taxonomy that has "love" as one of the categories. Hence, it was not possible to get a baseline value for the "love" category. The lowest mean squared errors per emotion category are 0.1027 for joy, 0.1187 for sadness, 0.0625 for fear, 0.1278 for anger and 0.0112 for surprise. These values serve as the baseline performance measure that this study aims to improve on for emotion analysis (Table 4).

4.4 Performance of Developed Sentiment Text Classification Models

The performance of training six different models for sentiment classification are shown in Fig. 1. The figure shows a comparison of the accuracy obtained using different text feature extraction methods, with or without the added features of this study, namely, part-of-speech (POS) tags, frequency of code switching and language dominance features. As seen in the figure, the addition of these new features slightly improved the accuracy for 3/4 of the SVM and Logistic Regression models. This shows that bilingualism may be a significant factor in sentiment classification, but more fine tuning is needed in future studies to determine the best representation for these features and the best models to understand these features.

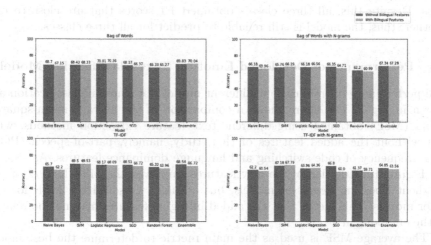

Fig. 1. Accuracy of developed sentiment text classification models

Accuracy is used as the main metric to determine the best model for sentiment classification, given that all three sentiment categories have equal important in practice. The models may be evaluated from a general perspective, rather than on a per class basis. The Logistic Regression model trained with bag of words, POS tags, frequency of code switching and language dominance features got the highest accuracy with 70.36%. This accuracy shows a significant improvement from the baseline accuracy of 51.10% obtained from applying the existing Polygot package on the same data set. This model can be examined further with its precision, recall and F1-score as shown in Table 5.

Table 5. Precision, recall and F1 score of best sentiment model

	Positive	Negative	Neutral
Precision	48.03%	65.65%	81.63%
Recall	54.88%	68.75%	75.68%
F1-Score	64.20%	72.54%	70.66%

It can be observed that the model consistently performs better, in terms of precision and recall, for predicting the "neutral" class followed by the "negative" class then the "positive" class. This may be attributed to the fact that the original annotated data set also contained more neutral Tweets than negative Tweets and more negative Tweets than positive Tweets. The model performance may be improved in future iterations by supplying the model with more data with "negative" and "positive" classifications to reduce bias for the "neutral" class. Despite this, all three classes obtained F1 scores that are close to one another, thus, the model is still reliable to predict for all three classes.

4.5 Performance of Developed Emotion Text Classification Models

The performance of training four different models for emotion classification are shown in Fig. 2. The figure shows a comparison of the average mean-squared errors (MSE) obtained using different text feature extraction methods, with and without the added features of this study, namely, part-of-speech (POS) tags, frequency of code switching and language dominance features. As seen in the Figure, the addition of the new features improved the MSE for 1/2 of the Gradient Boosting Regressor models, but actually worsened the MSE for the other models. This means that the added bilingual features may not be related to the emotions in a text.

The average MSE is used as the main metric to determine the best model for emotion classification because it measures how far the predicted probabilities are from the actual values. Since the difference in predicted and actual values are squared, bigger errors, which implies the model failed to predict the right level of emotion, are highlighted. The SVM model trained with bag of words,

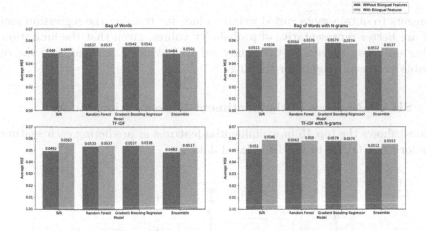

Fig. 2. Average MSE of developed emotion analyzers

POS tags, frequency of code switching and language dominance features got the lowest average MSE of 0.0499. This model produced mean squared errors per emotion category found in Table 6.

Table 6. Mean squared error of the best emotion model

MSE of	Best model	Existing package	Difference
Love	0.0539	–	–
Joy	0.0577	0.1027	−0.0450
Sadness	0.0766	0.1187	−0.0421
Fear	0.0383	0.0625	−0.0242
Anger	0.0681	0.1278	−0.0597
Surprise	0.0047	0.0112	−0.0065

As seen in Table 6, the developed emotion model had lower mean squared errors compared to the baseline values taken from running existing Python packages, which indicates better performance. This is with the exception of love, which does not have a baseline value, given the absence of this category in existing packages. With the given mean squared errors of the developed emotion model above, it can be calculated that the margins of error are 0.23, 0.24, 0.28, 0.20, 0.26 and 0.07 for love, joy, sadness, fear, anger and surprise probability values, respectively. While these margins of error may seem large for a 0–1 scale, it is important to note that the goal of the model, emotion classification, is concerned more with detecting the level of presence of an emotion in a text, rather than the exact probability value of an emotion. For instance, obtaining an emotion score of 0.70 and 0.90 for the love category both signify that love is a prevailing emotion in the text, despite the difference in exact values. However, these may also be improved upon in future iterations by increasing the number of taggers to obtain more varied weight values for training the regression model that are better representative of probability values, given that the limitation of having only two taggers in this study led to having 0, 0.5 and 1 as the only possible averaged weights for training.

4.6 SHAP Analysis of the Sentiment Classification Model

Figure 3 shows the top 15 most influential features in predicting each sentiment class for the COVID-19 vaccine-related data set.

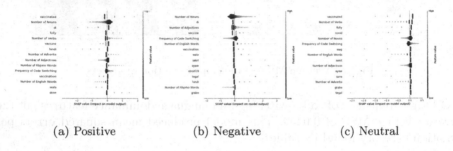

(a) Positive   (b) Negative   (c) Neutral

Fig. 3. Top influential features per sentiment classification

For text features, words that imply getting vaccinated, such as "vaccinated" and "fully" increases the chances of predicting positive while "di/hindi" (not) decreases the chances of predicting positive. Conversely, "di/hindi" (not) and related negative words like "wala" (none), "sakit" (pain), "ayaw" (dislike) and "tagal" (long duration), point to the negative sentiment.

It can also be observed that low number of nouns and high number of verbs are attributed to the positive sentiment class. This implies that positive Tweets mostly describe actions. On the other hand, high number of nouns and adjectives are attributed to the negative sentiment class. This means that negative Tweets are mostly about entities. Furthermore, the trained model recognizes that positive Tweets have more Tagalog than English words, while negative Tweets have more English than Tagalog words. High frequency of code switching is also observed to be a characteristic of positive Tweets, while negative Tweets do not seem to exhibit high levels of English-Tagalog bilingualism. Lastly, neutral Tweets are shown to be Tweets that contain low values of the top features that are in the positive and negative sentiment classes.

4.7 SHAP Analysis of the Emotion Classification Model

Figure 4 shows the top 15 most influential features in predicting each emotion class for the COVID-19 vaccine-related data set.

For text features, "vaccinated" is the top word for the love class, showing that encouraging vaccination is one way people show care during COVID-19

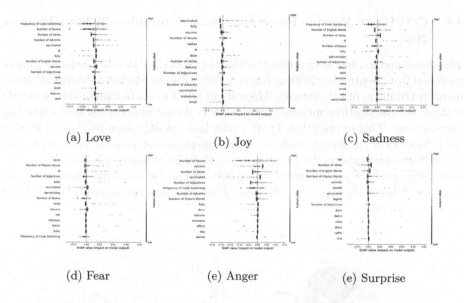

(a) Love   (b) Joy   (c) Sadness

(d) Fear   (e) Anger   (e) Surprise

Fig. 4. Top influential features per emotion classification

times. "Vaccinated", "fully", "vaccine", "wakas" (finally), "dose", and "bakuna" (vaccine) contribute positive SHAP values to this joy class, showing that joy is attributed to the act of getting vaccinated. For sadness, top words were "di" (not), "pakiramdam" (feeling), "effect", and "sakit" (pain), linking the sadness class to discomfort potentially caused by vaccinations. "Covid" remained the top word for fear, showing high perceived risk for the virus. In the anger graph, it can be observed that vaccine and time related (i.e., vaccine, vaccinated, mamaya (later)) words actually have negative SHAP values, while "bansa" is the only word in the top 15 features that has a positive SHAP value. It can be inferred that anger is mostly directly to national authorities, and not COVID-19 vaccines and processes themselves. Lastly, "bat" (short for "bakit" meaning why) is the top word in surprise, showing that surprise often leads to inquiry.

In terms of POS tags, love contains more nouns, verbs and adverbs, which signify directing an action towards an entity. Joy is characterized with high number of verbs, because it is attributed to the act of getting vaccinated as stated earlier. Sadness has more nouns and less verbs, implying that sadness is linked to entities. Fear, on the other hand, is high in number of adjectives. Like sadness, anger is also characterized by high number of nouns and adjectives, implying that anger is also linked to users' perception of entities. Lastly, surprise shows low verb count. Code-switching is observed as a feature of love, sadness and anger related Tweets, while code-switching does not seem to impact other emotion categories. Love and sadness are also dominated by more English than Tagalog words, while anger, fear and surprise have more Tagalog than English words. Similar to code-switching, language dominance does not have a significant impact on the joy class as well.

4.8   COVID-19 Vaccine Perception from September 2021 to January 2022

The developed models were applied to 288,944 COVID-19 vaccine-related Tweets gathered from September 2021 to January 2022. Figure 5a shows the overall sentiment distribution of this data set. Majority of the Tweets belong to the "neutral" sentiment classification, meaning, most Tweets do not express a stance regarding vaccinations. This implies that Twitter may have mostly been used as an avenue for information gathering, such as asking questions or sharing facts regarding COVID-19 vaccinations, by Filipino users, than expressing opinions. Moreover, comparing positive and negative sentiments, it can be seen that there are more negative Tweets than positive Tweets. This means that generally, COVID-19 vaccines and related processes are not well-received by the citizens.

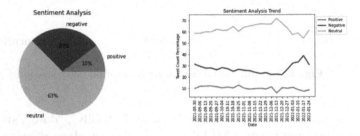

(a) Overall Sentiment Distribution (b) Change in Sentiment Over Time

Fig. 5. Sentiment distribution from September 2021 to January 2022

This can be examined further with the sentiment analysis trend graph in Fig. 5b. Positive trend line stayed generally consistent throughout the timeline, with a slight dip in early January. On the other hand, the negative trend line seemed to be decreasing at small increments from September to mid-December, then had a sudden spike up from mid-December to mid-January, but eventually decreased again towards the end of January. These patterns may be attributed to the improving COVID-19 case numbers until mid-December, then the sudden surge of the Omicron variant afterwards.

The average emotion scores shown in Fig. 6a give a deeper analysis of the sentiment insights above. It can be seen that the negative sentiments mostly stem from feelings of anger and sadness, rather than fear, as these two emotions are the top prevailing emotions among the collected Tweets.

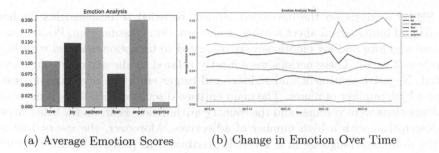

(a) Average Emotion Scores   (b) Change in Emotion Over Time

Fig. 6. Emotion distribution from September 2021 to January 2022

From the emotion analysis trend graph in Fig. 6b, the most noticeable changes are the sudden rise in "anger" and drop in "joy" on the week of December 13 to 19, 2021. Upon investigating the Tweets during this period, it was found that this was because of increased negative discussions surrounding a potential mandatory vaccine roll-out, triggered by this topic being raised as a question for Miss Philippines in the Miss Universe event held in that week. An insight from this is that when a COVID-19 vaccination-related protocol is brought up on an worldwide level, Filipinos are not confident that the country's response is comparable to international standards. Following this, the increasing trend in anger and decreasing trend in joy from mid-December to mid-January, followed by an improvement by end of January, in the graph may be due to the surge of the Omicron variant, as mentioned in the sentiment analysis.

5 Conclusion

In conclusion, this study showed the possibility of using new features that represent bilingualism in creating sentiment and emotion text classifiers for conducting social media analysis in the Philippine context, due to the regular use of "Taglish" (a combination of English and Tagalog) in casual communication channels. These new features are part-of-speech (POS) tags, frequency of code switching and language dominance, which were used in conjunction with the common text feature extraction methods in previous studies. COVID-19 vaccine-related Tweets were extracted for training the new text classifiers. Initial study showed that these new features may possibly improve sentiment classification accuracy, but was found to be less relevant in emotion classification. Nevertheless, these initial results opened the doors for syntax, choice of language and other linguistic features to be explored as text classification features.

In terms of performance, the developed English-Tagalog Logistic Regression sentiment and SVM emotion text classification models performed better than existing models with an accuracy of 70.36% and mean squared errors of 0.0539 (love), 0.0577 (joy), 0.0766 (sadness), 0.0383 (fear), 0.0681 (anger), and 0.0047 (surprise), respectively.

SHAP analysis on the developed models revealed several insights on how words and bilingualism affect sentiment and emotion classification. Positive sentiment and love and joy emotions were observed to be action-oriented with high number of verbs. These actions were mostly related to the act of getting vaccinated. Negative sentiment and sadness and anger emotions are entity-focused, with a high number of nouns. The main entities discussed with these sentiments and emotions were vaccines and the country authorities. Fear, on the other hand, is descriptive, with a high number of adjectives. Moreover, the use of English as the dominant language in a Tweet is attributed to positive sentiment, while Tagalog was dominant in negative Tweets. In emotions, love and sadness were English-dominant, while anger, fear and surprise were Tagalog-dominant. Code-switching was also observed as a characteristic of the positive sentiment class. In emotions, love, sadness and anger classes were characterized by code-switching.

The text classification models were applied to COVID-19 vaccine-related Tweets gathered from September 2020 to January 2022. Predictions from these models revealed that Filipinos generally had negative, focused on sadness and anger, perceptions on COVID-19 vaccinations. While a slow but steady improvement towards more positive perceptions can be seen over time, this was reversed during the Omicron surge, which saw a surge in negative and angry Tweets starting mid-December as well. Despite this, the end of January shows promising progress towards better COVID-19 vaccine-related perceptions in the Philippines. The use of the sentiment and emotion models in studies like these provides opportunity for decision makers to understand public sentiment in relation to policies and programs being implemented on the ground.

References

1. Coronavirus disease (covid-19): Herd immunity, lockdowns and covid-19. https://www.who.int/news-room/questions-and-answers/item/herd-immunity-lockdowns-and-covid-19
2. Abrigo, A.B.C., Estuar, M.R.J.E.: A comparative analysis of n-gram deep neural network approach to classifying human perception on Dengvaxia. In: 2019 IEEE 2nd International Conference on Information and Computer Technologies (ICICT), pp. 46–51. IEEE (2019)
3. Bandhakavi, A., Wiratunga, N., Massie, S.: Emotion-aware polarity lexicons for twitter sentiment analysis. Expert. Syst. **38**(7), e12332 (2021)
4. Bianchi, F., Nozza, D., Hovy, D.: FEEL-IT: emotion and sentiment classification for the Italian language. In: Proceedings of the Eleventh Workshop on Computational Approaches to Subjectivity, Sentiment and Social Media Analysis, pp. 76–83 (2021)
5. Biasio, L.R., Bonaccorsi, G., Lorini, C., Pecorelli, S.: Assessing COVID-19 vaccine literacy: a preliminary online survey. Hum. Vaccines Immunother. **17**(5), 1304–1312 (2021)
6. Bonta, V., Janardhan, N.K.N.: A comprehensive study on lexicon based approaches for sentiment analysis. Asian J. Comput. Sci. Technol. **8**(S2), 1–6 (2019)
7. Chen, Y., Skiena, S.: Building sentiment lexicons for all major languages. In: Proceedings of the 52nd Annual Meeting of the Association for Computational Linguistics (Short Papers), pp. 383–389 (2014)

8. Chou, W.Y.S., Oh, A., Klein, W.M.: Addressing health-related misinformation on social media. JAMA **320**(23), 2417–2418 (2018)
9. Cirqueira, D., et al.: Explainable sentiment analysis application for social media crisis management in retail (2020)
10. Cori, L., Bianchi, F., Cadum, E., Anthonj, C.: Risk perception and COVID-19 (2020)
11. Cotfas, L.A., Delcea, C., Roxin, I., Ioanăş, C., Gherai, D.S., Tajariol, F.: The longest month: analyzing COVID-19 vaccination opinions dynamics from tweets in the month following the first vaccine announcement. IEEE Access **9**, 33203–33223 (2021)
12. Devopedia: N-gram model, March 2021. https://devopedia.org/n-gram-model
13. Garay, J., Yap, R., Sabellano, M.: An analysis on the insights of the anti-vaccine movement from social media posts using k-means clustering algorithm and VADER sentiment analyzer. IOP Conf. Ser. Mater. Sci. Eng. **482**, 012043 (2019)
14. Harapan, H., et al.: Willingness-to-pay for a COVID-19 vaccine and its associated determinants in Indonesia. Hum. Vaccines Immunother. **16**(12), 3074–3080 (2020)
15. Hemalatha, I., Varma, G.S., Govardhan, A.: Preprocessing the informal text for efficient sentiment analysis. Int. J. Emerg. Trends Technol. Comput. Sci. (IJETTCS) **1**(2), 58–61 (2012)
16. Karlsson, L.C., et al.: Fearing the disease or the vaccine: the case of COVID-19. Pers. Individ. Differ. **172**, 110590 (2021)
17. Khakharia, A., Shah, V., Gupta, P.: Sentiment analysis of COVID-19 vaccine tweets using machine learning. Available at SSRN 3869531 (2021)
18. Kusal, S., Patil, S., Kotecha, K., Aluvalu, R., Varadarajan, V.: AI based emotion detection for textual big data: techniques and contribution. Big Data Cogn. Comput. **5**(3), 43 (2021)
19. Lasco, G., Yu, V.G.: Communicating COVID-19 vaccines: lessons from the dengue vaccine controversy in the Philippines. BMJ Glob. Health **6**(3), e005422 (2021)
20. Limaye, R.J., et al.: Social media strategies to affect vaccine acceptance: a systematic literature review. Expert Rev. Vaccines **20**, 959–973 (2021)
21. Lundberg, S.M., Lee, S.I.: A unified approach to interpreting model predictions. In: Proceedings of the 31st International Conference on Neural Information Processing Systems, pp. 4768–4777 (2017)
22. Mathieu, E., et al.: A global database of COVID-19 vaccinations. Nat. Hum. Behav. **5**, 947–953 (2021)
23. Merchant, R.M., South, E.C., Lurie, N.: Public health messaging in an era of social media. JAMA **325**(3), 223–224 (2021)
24. Miyazaki, K., Uchiba, T., Tanaka, K., Sasahara, K.: The strategy behind anti-vaxxers' reply behavior on social media. arXiv preprint arXiv:2105.10319 (2021)
25. Murthy, A.R., Kumar, K.A.: A review of different approaches for detecting emotion from text. IOP Conf. Ser. Mater. Sci. Eng. **1110**, 012009 (2021)
26. Mustofa, R., Prasetiyo, B.: Sentiment analysis using lexicon-based method with Naive Bayes classifier algorithm on# newnormal hashtag in Twitter. J. Phys. Conf. Ser. **1918**, 042155 (2021)
27. Piedrahita-Valdés, H., et al.: Vaccine hesitancy on social media: sentiment analysis from June 2011 to April 2019. Vaccines **9**(1), 28 (2021)
28. Puri, N., Coomes, E.A., Haghbayan, H., Gunaratne, K.: Social media and vaccine hesitancy: new updates for the era of COVID-19 and globalized infectious diseases. Hum. Vaccines Immunother. **16**(11), 2586–2593 (2020)

29. Raghupathi, V., Ren, J., Raghupathi, W.: Studying public perception about vaccination: a sentiment analysis of tweets. Int. J. Environ. Res. Public Health **17**(10), 3464 (2020)

30. Robledo, D.A., Lapada, A.A., Miguel, F.F., Alam, Z.F.: COVID-19 vaccine confidence and hesitancy among schools' stakeholders: a Philippine survey. J. Cardiovasc. Dis. Res. **12**(3), 29–35 (2021)

31. Rochmawati, N., Wibawa, S.: Opinion analysis on Rohingya using Twitter data. IOP Conf. Ser. Mater. Sci. Eng. **336**, 012013 (2018)

32. Sanketh, R.: Text preprocessing with NLTK. Towards Data Science, May 2021

33. Satu, M.S., et al.: TClustVID: a novel machine learning classification model to investigate topics and sentiment in COVID-19 tweets. Knowl.-Based Syst. **226**, 107126 (2021)

34. Shamrat, F.J.M., et al.: Sentiment analysis on twitter tweets about COVID-19 vaccines using NLP and supervised KNN classification algorithm. Indonesian J. Electr. Eng. Comput. Sci. **23**(1), 463–470 (2021)

35. Villavicencio, C., Macrohon, J.J., Inbaraj, X.A., Jeng, J.H., Hsieh, J.G.: Twitter sentiment analysis towards COVID-19 vaccines in the Philippines using Naïve Bayes. Information **12**(5), 204 (2021)

36. Wilson, S.L., Wiysonge, C.: Social media and vaccine hesitancy. BMJ Glob. Health **5**(10), e004206 (2020)

37. Yadav, K., Lamba, A., Gupta, D., Gupta, A., Karmakar, P., Saini, S.: BI-LSTM and ensemble based bilingual sentiment analysis for a code-mixed Hindi-English social media text. In: 2020 IEEE 17th India Council International Conference (INDICON), pp. 1–6. IEEE (2020)

38. Yadav, K., Lamba, A., Gupta, D., Gupta, A., Karmakar, P., Saini, S.: Bilingual sentiment analysis for a code-mixed Punjabi English social media text. In: 2020 5th International Conference on Computing, Communication and Security (ICCCS), pp. 1–5. IEEE (2020)

39. Zhao, Y., Zhang, J.: Consumer health information seeking in social media: a literature review. Health Inf. Libr. J. **34**(4), 268–283 (2017)

Fine Grained Categorization of Drug Usage Tweets

Priyanka Dey[✉] and ChengXiang Zhai

University of Illinois at Urbana-Champaign, Champaign, IL, USA
pdey3@illinois.edu

Abstract. Drug misuse and overdose has plagued the United States over the past decades and has severely impacted several communities and families. Often, it is difficult for drug users to get the assistance they need and thus many usage cases remain undetected until it is too late. With the booming age of social media, many users often prefer to discuss their emotions through virtual environments where they can also meet others dealing with similar problems. The widespread use of social media sites creates interesting new opportunities to apply NLP techniques to analyze content and potentially help those drug users (e.g., early detection and intervention). To tap into such opportunities, we study categorization of tweets about drug usage into fine-grained categories. To facilitate the study of the proposed new problem, we create a new dataset and use this data to study the effectiveness of multiple representative categorization methods. We further analyze errors made by these methods and explore new features to improve them. We find that a new feature based on tweet tone is quite useful in improving classification scores. We further explore possible downstream applications based on this classification system and provide a set of preliminary findings.

Keywords: Categorization · Social media analytics · Drug usage · Public health

1 Introduction

The epidemic of drug misuse and overdose has left several communities and families devastated across the United States. Deaths from drug overdoses have risen sharply in recent years as shown in Fig. 1. According to the Centers for Disease Control and Prevention (CDC), there have been almost 1 million deaths induced from drug overdose since 2000 [1]. Moreover, research from the CDC shows that drug usage is becoming increasingly detrimental; the 2020 rate of drug overdose deaths has accelerated and increased 31% since 2019 [1].

Although this issue is widespread, it is difficult to find ways to help such individuals. One of the biggest barriers in reducing this epidemic is that many users prefer to hide their drug addictions from others. Users may fear criminalization or perhaps fear judgment. Thus, it becomes increasingly difficult to identify

G. Meiselwitz (Ed.): HCII 2022, LNCS 13315, pp. 267–280, 2022.
https://doi.org/10.1007/978-3-031-05061-9_19

patterns in drug usage and oftentimes many such drug usage cases remain unde-tected until it is too late. However, recent state-of-the-art research shows that social media can help in identifying drug discussion. With the booming age of technology and social media, members of a community often prefer to discuss their emotions through a virtual environment (e.g. social media) where they can also meet others dealing with similar problems. [2–4] have found that users often discuss the abusive use of prescription drugs on social media sites such as Twitter.

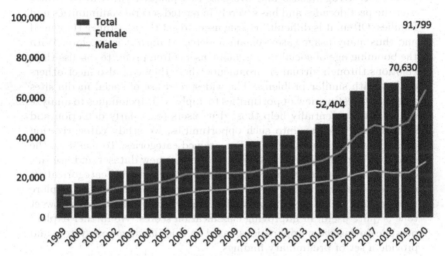

Figure 1. National Drug-Involved Overdose Deaths*
Number Among All Ages, by Gender, 1999-2020

*Includes deaths with underlying causes of unintentional drug poisoning (X40–X44), suicide drug poisoning (X60–X64), homicide drug poisoning (X85), or drug poisoning of undetermined intent (Y10–Y14), as coded in the International Classification of Diseases, 10th Revision. Source: Centers for Disease Control and Prevention, National Center for Health Statistics. Multiple Cause of Death 1999-2020 on CDC WONDER Online Database, released 12/2021.

Fig. 1. Total deaths by overdoses from 1999–2020 in the US

Much of past research work has focused on identifying drug abuse on social media. One research work [2] showed that using n-grams and specified drug-related keywords (using a drug-slang lexicon, synonym expansion, and word clus-ters) allows for the creation of an effective system to classify tweets as "abuse" or "non-abuse" tweets. Another work [3] presented a supervised machine learning technique to identify prescription-drug abuse tweets [5] presented a monitor-ing system of tweets using machine learning to identify drug abuse in real-time. Another work [6] focused on opioid abuse and designed a binary classifier for rel-evant tweets. Although such tweets facilitate the study of drug-abuse patterns, this coarse categorization does not allow us to fully understand and capture the myriad effects drug usage induces on a community. Thus, a more fine-grained categorization of tweets discussing drug usage is necessary. This further classification can then help us better study a community (e.g. crimes influenced by various drugs, reactions and opinions of users about drug usage).

To address this limitation, we present a novel task of identifying drug usage on social media and study how to categorize such tweets into fine-grained categories to enable many downstream applications such as real-time monitoring of drug users in a community, companion agents for drug users to help them on a daily basis, and drug-induced crime hotspot models. To facilitate this study, we generate a novel dataset consisting of drug usage tweets and study the effectiveness of multiple representative categorization methods using this dataset. We further analyze errors made by these methods and explore new features to improve them. We find that a new feature based on tweet tone is effective in improving classification scores. We then explore possible downstream applications based on this classification system and provide a set of preliminary findings.

2   Research Aim

The main contribution of this paper is to introduce and formulate a new tweet categorization problem so as to obtain a more detailed understanding of the tweets discussing drug usage. As previous work has studied how to identify tweets discussing drug abuse, we define our problem to further classify the tweets into refined categories. To study this, we introduce the following research questions:

1. How should we create a fine-grained categorization of drug-usage tweets?
2. How can we construct a new dataset to facilitate quantitative evaluation of classification techniques?
3. Which machine learning algorithms work best for this new task?
4. How can we leverage this categorization system to support further applications?

In the following sections, we first look at RQ1 and RQ2 where we detail how we have designed the various labels for drug usage and created our dataset. We then continue with RQ3 to identify the best algorithms to assist in the classification task. We follow with RQ4 and provide examples of possible use cases and some preliminary findings. We conclude with potential future research directions.

3   Designing Fine-Grained Labels for Drug-Usage Tweets

For RQ1, we first propose a new drug usage taxonomy which helps to better capture drug discussion in a multitude of aspects. After a manual analysis of drug-related tweets, we identified 3 broad categories to describe this discussion. Firstly, many tweets discuss the effects of drug usage in a community, i.e. drug-induced crimes or statistics related to drug usage in a community. We categorize such tweets as News/Research as they provide a recollection of events and statistics related to drug usage in a community. Tweets in this category are particularly useful as they can give us insight into which areas of a state have common drug-induced crimes e.g. theft or battery. Authorities can then use these annotations to determine future crime hotspots and help reduce drug-related casualties.

From our manual analysis, we also observed that many tweeters discuss their personal and active use of drugs. Thus, we use Usage as our second category. Detection of tweets discussing drug usage provides several applications. For example, if a user makes a post about using a drug, health officials can monitor the user and prevent them from a possible overdose. Furthermore, psychologists can learn more about a patient's behavior based on the user's tweets posted while under the influence. In order to further understand usage, we divide this category based on a temporal aspect i.e. Past Usage, Current Usage, and Intent to Use. This further subcategorization of usage tweets enables more effective monitoring. For example, health and addiction officials can build a real-time monitoring system to determine if a tweet discusses current usage. Officials can then extract tweet information such as the user's region and determine the next best steps in order to prevent a drug overdose and/or casualty. Past usage tweets can provide cues as to whether a person plans on using again; Intent to Use tweets can help officials take measures to provide early, preventative care.

We also identified a third category "Reactions/Opinions" based on tweets that express opinions and reactions regarding drug usage. Identification of such tweets can be useful, e.g. psychologists can monitor annotated tweets classified as reactions/opinions to see how the members of a community feel about some event or general drug use (e.g. members conversing about the legalization of marijuana in a state). These tweets can then provide insight into members' possible future actions (e.g. more drug usage). Tools designed to help these individuals (e.g. companion chatbots) can then be built based on these data.

To summarize, we present a total of 6 fine-grained categories:

1. **News/Research**: Discuss any events related to drug use (e.g. arrests, crimes)
2. **Usage**: Discuss user using drugs. These tweets are further sub-classified into 3 categories:
   (a) **Past Usage**: Refer to events where the user has used in the past
   (b) **Current Usage**: Refer to user using (at the time the tweet was written)
   (c) **Intent to Use**: Refer to those who are planning on using in the near future
3. **Reactions/Opinions**: Refer to a response (reaction or opinion) to some drug-induced act

In order to generate a classification system for these categories, we design two separate classification models: a topical model to understand tweet context (News/Research, Usage, Reactions/Opinions) and a temporal model to further understand drug usage (Past Usage, Current Usage, Intent to Use).

4   Dataset Construction

This research is a first attempt to study fine-grained drug usage on social media; thus, for this task, we generate our own dataset. We collect drug usage tweets using the Twitter Search API. We gather tweets from November 2016 to December 2016; we filter based on drug keywords which include names of drugs e.g. heroin, marijuana, weed, cocaine, fentanyl, ketamine, methamphetamine and words related to addiction and abuse such as opioidcrisis, addiction, overdose, compulsions.

We identified approximately 5% of the collected tweets between the 2 months to be relevant for our study. Table 1 shows the 10 most frequently found drug-related words in the tweets. After collection, all tweets were manually annotated into one of the categories listed in the previous section. Table 2 shows the number of tweets per category and corresponding example tweets. In total, we have generated a dataset of 921 annotated tweets.

Table 1. Most common words in drug-related tweets

Word	Frequency	Relative frequency
Weed	519	38.73%
Marijuana	238	17.76%
Addiction	161	12.01%
Cannabis	80	5.97%
Cocaine	78	5.82%
Opioid	76	5.67%
Heroin	68	5.07%
Booze	53	3.96%
Overdose	43	3.21%
Nicotine	11	0.82%

5 Classification Methods

As an initial study of the new task and for RQ3, we set our goal as to establish some baseline results by evaluating a few popular representative multi-class classifiers that have shown success in several classification tasks.

1. Multinomial Naive Bayes (NB): modification of the Naive Bayes model and is popularly used for text document classification

2. Logistic Regression (LogR): uses the logistic function to determine whether or not data falls into a particular category

3. Support Vector Machine (SVM): goal is to calculate the maximal margin hyperplane(s) separating the categories of the data

4. Random Forests (RF): uses an average of decision tree predictions to determine the best predictive accuracy

5. Multi-Layered Perceptron (MLP): feed-forward artificial neural network where training data is split into multiple layers of information.

Table 2. Tweet categories: examples and counts

Tweet Category	Example	Count
News	"Ex-Ravens cheerleader charged with rape, supplying booze to minor set for hearing" " Remember, rape isnt always male to female."	301
Past usage	"I told my mom I do methamphetamine and she responded with" "So thats why your face is so f**ked up"	104
Current usage	"smokin weed in the backstage photo booth with Duchovny."	130
Intent to use	"Ill find some weed tonight."	195
Reactions	"in debate we had to choose from 4 topics and i chose marijuana. Im debating on the CON side because marijuana is SO stupid."	191

6. Extremely Randomized Trees (Extra): conceptually similar to Random Forests but splitting the tree is based on a random attribute rather than highest information gain.

7. Bidirectional Encoder Representations from Transformers (BERT): transformer-based machine learning model popularly used in NLP tasks and can be fine-tuned for various downstream tasks. Research has shown this model can work quite well for various classification tasks including text using BERT contextualized embeddings [7].

We generated features using word2vec [8] by transforming tweets into a set of vectors and then computing a mean vector for each tweet for the first 6 models. We used the default BERT embeddings for the BERT model. We then used the Python sklearn package [9] to train each classifier to best fit the training data. 80% of the annotated dataset was used as training data, and the remaining 20% was used for testing. 5-fold cross-validation was used to calculate the accuracy, precision, and recall scores.

5.1   Comparison of Classifiers for Topical Model

The results for the topical model are summarized in Table 3. The News and Usage categories were the easiest to categorize as can be seen by their relatively high values of accuracy, precision, and recall scores. The SVM seemed to work best for the News Category whereas the LogR and NB models worked best for the Usage category. The Reactions category, however, proved to be difficult to classify with most classifiers having extremely low recall scores. The best models appear to be BERT, the RF, and the Extra. Overall, the BERT model performed the best across all categories and provided the highest scores for the Reactions category.

Table 3. Class accuracy, precision, and recall scores for topical model

Classifier	News			Usage			Reactions		
	Acc	Precision	Recall	Acc	Precision	Recall	Acc	Precision	Recall
Naive Bayes	0.65	0.94	0.80	**0.77**	**0.80**	**0.91**	0.39	0.50	0.16
Random forest	0.68	0.70	0.86	0.64	0.64	0.94	0.48	1.00	0.32
SVM	**0.81**	**0.88**	**0.86**	0.65	0.64	0.94	0.41	0.64	0.17
LogR	0.71	0.70	0.92	**0.71**	**0.76**	**0.86**	0.30	0.40	0.19
MLP	0.77	0.77	0.92	0.69	0.68	0.95	0	0	0
Extra	0.68	0.66	0.94	0.68	0.70	0.86	0.45	0.72	0.34
BERT	0.79	0.79	**0.88**	**0.90**	**0.90**	0.81	**0.54**	**0.54**	**0.61**

5.2 Comparison of Classifiers for Temporal Model

Results for the temporal model are summarized in Table 4. This classification proved to be a much more difficult task. Intent to Use and Current Usage were easiest to identify as seen by their relatively higher accuracy, precision, and recall scores. The RF and Extra models seem to work best for these categories. The LogR model also worked well for classifying Intent to Use tweets. On the other hand, the Past Usage was extremely difficult to identify, and most classifiers did not work well, suggesting that there is much room for future research on this categorization task. Overall, the LogR model provided high scores for all categories.

Table 4. Class accuracy, precision, and recall scores for temporal model

Classifier	Past			Current			Possible Intent to Use		
	Acc	Precision	Recall	Acc	Precision	Recall	Acc	Precision	Recall
Naive Bayes	0.29	0.10	0.17	0.51	0.40	0.71	0.53	0.79	0.52
Random forest	0.25	0.10	0.12	**0.52**	**0.41**	**0.70**	0.54	0.78	0.52
SVM	0.40	0.24	0.42	0.44	0.32	0.57	0.56	0.84	0.54
LogR	0.42	0.19	0.37	0.48	0.40	0.59	**0.69**	**0.74**	**0.80**
MLP	0.31	0.10	0.20	0.48	0.24	0.75	0.51	0.84	0.48
Extra	0.33	0.13	0.33	**0.53**	**0.40**	**0.77**	0.53	0.84	0.50
BERT	0.21	0.15	0.18	0.42	0.47	0.44	0.41	0.29	0.34

5.3 Error Analysis and Novel Tweet Tone Feature

After determining the best ML algorithms for the topical and temporal models, we explored examples of misclassified tweets to obtain insights for possible solutions to improve classification performance.

For each of the categories, we have identified some common tweets misclassified by many of the ML algorithms. We present some examples in Table 5. Based on a manual lexical analysis of these tweets, we found that each of the categories seemed to have a unique emotion or tone exhibited by their tweets.

For example, News/Research tweets tend to discuss drug-induced crimes or difficult stories about recovering addicts, both of which are despairing topics. On the other hand, emotions such as joy can be seen in Usage (e.g. a tweeter was able to buy his favorite weed); a pensive tone can be seen in Reactions/Opinions tweets (e.g. lack of support for legalizing weed). Based on these findings, we explored the use of a novel tweet categorization feature - tweeter tone (detected by a user's choice of words in a tweet).

Table 5. Examples of misclassified tweets

Category	Tweet
News	"kilo of cocaine only worth 1500 in Columbia $77k in Britain tho"
Past usage	"i smell like pussy money n weed n she said Ooo i like yo cologne"
Current usage	"I think somehow we all took a syph of Toms weed and are just tripping the **** out"
Intent to use	"lmfao this sophomore turned to me and asked for legal advice about smoking and i just gave him so bs answers"
Reactions	"if kids these days understood the effects drugs can have on your brain, maybe the world would be a better place"

For each category, each training tweet's tone was determined using the Watson Tone Analyzer API [10] which categorizes a tweet's tone as anger, fear, joy, tentative, analytical, sad, confident, or none. This additional tone feature was then added to each training and testing sample's feature vector. The ML algorithms were then retrained using the updated feature vector to measure the new classification scores. We specifically chose to analyze the BERT and LogR models since they provided the best scores for the topical and temporal models. Tables 6 and 7 show a comparison of the precision and recall scores with the addition of the tone feature.

From these tables, we can see that the tone feature can indeed lead to performance improvement in both precision and recall. The News/Research category has an extremely high recall as well as a great improvement in precision. The Reactions/Opinions category has a higher precision score as well. Most usage categories are still difficult to identify, but the tone feature did not seem to lower the precision and recall scores significantly. Thus based on this experiment, it seems the tone of a tweet message can improve scores for this classification task.

Table 6. Comparison of precision and recall scores in topical model (BERT) w/tone feature

Category	Original precision	New precision	Original recall	New Recall
News	0.79	0.88	0.85	0.91
Usage	0.90	0.90	0.81	0.83
Reactions	0.54	0.74	0.61	0.74

Table 7. Comparison of precision and recall scores in temporal model (LogR) w/tone feature

Category	Original precision	New precision	Original recall	New recall
Past usage	0.19	0.33	0.37	0.43
Current usage	0.40	0.42	0.41	0.52
Possible intent	0.74	0.87	0.80	0.78

6 Exploration of Possible Applications Using Fine-Grained Drug Usage Tweet Classification

The classification models and fine-grained drug usage taxonomy that we have introduced can further lead to the creation of downstream applications. Although a full exploration of possible applications is beyond the scope of this paper, we propose 3 different future applications and perform some preliminary work for each:

1. **Lexical Analysis of Usage Tweets**: Illicit drug usage is oftentimes difficult to identify. By viewing Usage tweets containing illicit drug keywords, we may be able to find related drug cases.
2. **Sentiment Analysis of Reactions/Opinions Tweets**: Such an analysis can be beneficial in predicting community changes: legalizing marijuana for recreational use.
3. **Geographical Analysis of Usage Tweets**: This analysis can be useful in determining whether certain areas of the United States have similar drug usage patterns.

In order to perform initial experiments for each of the proposed applications, we first use our topical model (BERT) and temporal model (LogR) to classify additional tweets. We classify an additional 14,746 tweets gathered from the Twitter API between the months January to March of 2021. We were able to classify a total of 4,850 News/Research tweets, 8,506 Usage tweets (3,534 Past Usage, 3,180 Current Usage, and 1,92 Intent to Use), and 1,390 Reactions/Opinions tweets. We then extract locations from these tweets based on users' locations using the Twitter API. We were able to obtain 13,189 geo-tagged tweets which we use for the proposed applications' analyses.

6.1   Lexical Analysis of Usage Tweets

One possible application is analyzing the classified usage tweets to determine the usage of illicit drugs. Illicit drug usage can oftentimes be difficult to identify as users may fear imprisonment or other types of criminalization. However, as shown through Sarker et al. and Shutler et al., many users do indeed discuss illegal drug usage on social media. Thus, further analysis of Usage tweets can help identify patterns. Moreover, the fine-grained analysis of Usage tweets can give us an additional temporal feature.

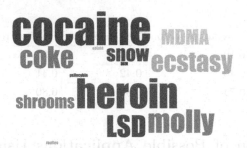

Fig. 2. Word cloud for most common illicit drugs in drug usage tweets

To identify illicit drug usage in the United States via the Usage tweets, we first filter tweets that contain names and slang for illegal drugs (e.g. "coke", "roofies", "cocaine"). In Fig. 2, we present a word cloud of the most commonly discussed illegal drugs. We identify "heroin" and "cocaine" to be the most commonly misused drugs. We then generate a choropleth map to further show which states

Fig. 3. Choropleth map of United States: counts of illicit drug usage tweets

Sentiments for Drug Usage Across the United States

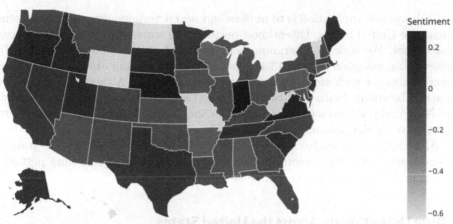

Fig. 4. Choropleth map of United States: sentiments of drug usage extracted from reactions/opinions tweets

have the most illicit drug usage tweets. We present our findings in Fig. 3. Based on this map, we see that Idaho, Indiana, and Maine have the highest illicit drug usage tweets in the United States.

6.2 Sentiment Analysis of Reactions/Opinions Tweets

Another possible application is to analyze the classified Reactions/Opinions tweets to determine the public's opinion on drug usage. Such an application can be particularly useful when deciding effective drug control policies. For example, if marijuana usage is favorable in a particular region, severe legal restrictions against weed might instigate an increase in its black market.

To determine sentiments, we leverage the VADER (Valence Aware Dictionary for Sentiment Reasoning) python package [11] which provides scores on a document for various sentiments. We apply this method to the 13,189 geo-tagged tweets we obtained from further classification and extract scores for the positive and negative sentiments. To calculate sentiment for a US state, we compute the mean of sentiments for all tweets retrieved from a state. We present our findings in a choropleth map in Fig. 4; positive scores show an overall positive reaction towards drug usage in that state and negative scores show an overall negative reaction. Some states also had mixed opinions towards drug usage and thus mean scores around 0. We identify Indiana, Virginia, and Nebraska to have the highest positive sentiments and Wisconsin, Arkansas, and Alabama to have the highest negative sentiments.

6.3   Geographical Analysis of Usage Tweets

A third possible application is to understand and determine drug usage patterns across the United States. Oftentimes, neighboring states may have similar drug usage rates. We seek to determine location-based patterns on drug usage. To achieve this, we again use the 13,189 geo-tagged tweets and observe drug usage tweet counts for each state. We present results in Fig. 5. Although many states have similar counts (with most tweets centered around California, Texas, Florida, and New York), we can see that many of the Southeastern states and Mid-western states have similar amounts of Usage tweets.

Although not currently explored in this preliminary analysis, the fine-grained categorization of Usage tweets further allows us to monitor real-time patterns for Current Usage and Intent to Use tweets.

Drug Usage Counts Across the United States

Fig. 5. Choropleth map of United States: drug usage counts across the United States

7   Conclusion

In this paper, we introduced a novel and useful classification task to better understand drug discussion on Twitter by categorizing tweets into fine-grained categories: News/Research, Usage (Past Usage, Current Usage, and Intent to Use) and Reactions/Opinions. We have created a new annotated dataset which we use to evaluate several popular representative categorization methods (NB, LogR, SVM, RF, MLP, Extra, BERT) . We further analyzed common errors and identified tweet tone to improve classification results. We then explored some downstream applications and presented some initial work based on our proposed classification system.

One limitation of our work is that the dataset we have constructed is small, which is mainly due to the limited resources available to us. Although we are able to make some interesting findings in this paper, an important future work is to further construct larger datasets. Our baseline results also show that some categories, notably Past Usage, are difficult to classify, and in general, there is much room for additional research to improve the accuracy of this task.

Finally, the trained classifiers using our annotated data set can already be potentially useful for building novel applications as we have shown in Sect. 6. This work can further be extended to generate a real-time system for monitoring drug usage tweets. Such a system can then be used to extract common emotions among users. These tones can then be used to build a companion chatbot that can provide users with a customized experience. These common emotions can also be used to build a system that allows psychologists to monitor behavior patterns in drug users and in turn help them develop effective and personalized drug treatments.

Acknowledgments. This material is based upon work supported by the National Science Foundation under Grant No. 1801652 and by the National Institutes of Health under Grant 1 R56 AI114501-01A1.

References

1. CDC: Now Is The Time To Stop Drug Overdose Deaths Article. https://www.cdc.gov/drugoverdose/featured-topics/overdose-prevention-campaigns.htm. Accessed 1 Feb 2022
2. Sarker, A., O'Connor, K.: Social media mining for toxicovigilance: automatic monitoring of prescription medication abuse from twitter. Drug Safe. **39**(3), 231–240 (2015). https://doi.org/10.1007/s40264-015-0379-4
3. Shutler, L., Nelson, L.: Drug use in the twittersphere: a qualitative contextual analysis of tweets about prescription drugs. J. Addict. Dis. **34**(4), 303–310 (2015)
4. Chary, M., Genes, N.: Leveraging social networks for toxicovigilance. J. Med. Toxicol. **9**(2), 184–191 (2013)
5. Phan, N., Chun, S.: Enabling real-time drug abuse detection in tweets. In: 2017 IEEE International Conference on Data Engineering (ICDE), pp. 1510–1514, https://doi.org/10.1109/ICDE.2017.221
6. Flores, L., Young, S.: Regional variation in discussion of opioids on social media. J. Addict. Dis. **39**(3), 315–321 (2021). https://doi.org/10.1080/10550887.2021.1874804
7. Sun, C., Qiu, X., Xu, Y., Huang, X.: How to fine-tune Bert for text classification? In: Sun, M., Huang, X., Ji, H., Liu, Z., Liu, Y. (eds.) CCL 2019. LNCS (LNAI), vol. 11856, pp. 194–206. Springer, Cham (2019). https://doi.org/10.1007/978-3-030-32381-3_16
8. Mikolov, T., Chen, K.: Efficient estimation of word representations in vector space. In: Bengio, Y., LeCun, Y. (eds.) Proceedings of 1st International Conference on Learning Representations (ICLR) 2013, Workshop Track. IEEE, Arizona, USA (2013)
9. Pedregosa, F., Varoquaux, G.: Scikit-learn: machine learning in python. J. Mach. Learn. Res. **12**, 2825–2830 (2011)

10. Agrawal, A., Sonawane, S.: Tone analyzer. Int. J. Eng. Sci. Comput. **7**(10), 15060–15064 (2017)
11. Hutto, C., Gilbert, E.: VADER: a parsimonious rule-based model for sentiment analysis of social media text. In: Proceedings of the International AAAI Conference on Web and Social Media, ICWSM, vol. 8, pp. 216–225, Michigan, USA (2014)

Text Mining for Patterns and Associations on Functions, Relationships and Prioritization in Services Reflected in National Health Insurance Programs

Maria Regina Justina Estuar[1]([⊠]) [iD], Maria Cristina Bautista[2] [iD],
Christian Pulmano[1] [iD], Paulyn Jean Acacio-Claro[2] [iD], Quirino Sugon[1],
Dennis Andrew Villamor[1], and Madeleine Valera[2] [iD]

[1] School of Science and Engineering, Ateneo de Manila University,
Quezon City, Philippines
restuar@ateneo.edu
[2] Ateneo Graduate School of Business, Ateneo de Manila University,
Makati City, Philippines
mcbautista@ateneo.edu

Abstract. The new Philippine Universal Health Care (UHC) Act was adopted in February 2019 and was slated for implementation in January 2020. However, the ongoing pandemic has thrown the national health insurance program under close scrutiny as the single biggest source of funding for hospitalization or personal health care. An inductive study using text analysis is utilized to discover implementation patterns over time of the national health insurance program's (NHIP) policies extracted as functions and relationships in publicly issued executive circulars. Standard text mining methods were used to extract publicly available PhilHealth Circulars from 2000–2020. Principal component analysis was used to determine clusters per time period, where time period covered years served each administration from 2000–2020. Result of PCA was used to determine functions. Within each cluster, frequent pattern algorithm (FP growth) was used to determine words that are most frequently associated with each other. This methodology was used to determine policy focus on the following services: membership, benefits, contributions, collection, and accreditation. Results indicate that the first four periods prioritized on accreditation while the remaining three periods focused on membership and benefits. The results also suggest insights on the use of text mining on policy-settings, on change trajectories, if any, and implications for the implementation of the new UHC Act.

Keywords: Text mining · Universal health coverage transition · Cluster analysis · Health insurance program · Policy science

University Research Council, Ateneo de Manila University.

1   Introduction

The COVID pandemic that has gripped the world, beginning 2020, has thrown health system reforms in disarray and raised questions on the extent to which the Philippine Universal Health Care (UHC) Act enacted in February 2019, is a game-changer that can contribute to a much nimbler, twenty-first century health system. One of the two main implementing partners to the law is the Philippine Health Insurance Corporation (known as PhilHealth), the biggest single purchaser of healthcare in the country. PhilHealth was established in 1995, carved out from one of the earliest social security agencies in the region and known as Medicare, to one of the latecomers in terms of population coverage and financial protection. A study noted that at its inception in 1995, health insurance coverage of the population was under 50% and rose to about 90% in 2017, the year when discussions on the UHC Act were deliberated in the Philippine Congress [9]. This contrasts with some countries in the region like Taiwan and Thailand, which also had their health financing reforms 1995 and 1991, respectively but achieved universal coverage in 1997 and 2002, respectively [7,14]

A key difference for the Philippines is the voluntary nature of its NHIP. From its origins in the social security system, participation is through regular premium payments to make one eligible for benefits. The Philippine labor force is dominated by the informal sector (more than 50%) and this made premium collection a challenge [2]. The capacity of the system to provide for benefits is dependent on its ability to raise revenues from premium payments. The portion of health costs covered by PhilHealth was 55% in a 2017 survey [8]. This has led one health sector review to observe that what PhilHealth managed to do was "pay a little bit of everything for all Filipinos" [13]. Even collecting from the government the payment of premiums of civil servants was often delayed. The new UHC veers away from this 'membership' view to one of universal health coverage, making participation an entitlement of citizenship. This can be a big ask for an organization that has not leveraged its being a single payer to drive health sector performance, as per various health sector assessments [1,3,9].

Twenty-five years' implementation of the NHIP, and a further 25 years before as Medicare, raises a question on whether patterns can be gleaned in terms of developed capacities and relationships. This paper looks into the patterns of policy setting in the country's major health financing institution, PhilHealth. It asks the question of what has preoccupied PhilHealth in terms of its policy setting functions. How was it building systems and relationships in performing its critical role of financing health care and improving health system performance? Through the application of text mining, word association and principal component analysis, we seek to identify values, associations in functional and relational areas, reflected in the following service areas: membership, benefits, contribution, collection and accreditation. The study seeks to look into areas that can improve functioning and enhance decision-making capacities for UHC implementation.

2   Salient Features of the Philippine Health Care System and the New UHC Act

Designed to promote and protect the right of all citizens with access to care at no financial hardship to poor, vulnerable and geographically isolated groups in the country, Republic Act 11223 or the UHC Act has its equity objective expressed intently. It holds to the premise and promise of universal access to an essential basic package of care through a gate keeping primary care, without needing any insurance membership card. The Implementing Rules and Regulations in support of the Act was approved in November 2019. A staggered implementation program that spans across 10 years is envisioned. Priority activities include the registration of every citizen to a primary care provider, capitation of outpatient care and primary practitioners' (general practitioners) licensing - to be undertaken within the first year of implementation. Managerial and financial integration are planned within three and six years, respectively. Integration activities are intended to streamline activities away from the current fragmented system to the increased role of the Department of Health (DOH), initially project managing the transition to service delivery networks at provincial levels, to funding health workers at the primary levels.

The UHC Act views the challenge of reforms in terms of allocating responsibilities between individual and population-based health to two state institutions- the Philippine Health Insurance Corporation (PhilHealth) and the Department of Health (DOH), respectively. PhilHealth retains paying for individual members' care, with out of pocket expenses specified to zero for poorer segments, and seeks to improve on the 56% support value (defined as shares of members' hospital bills covered by PhilHealth) prior to the Act's implementation. The new UHC Act has clauses on strengthening the corporation in its governing board composition, the installation of a provident fund for its employees and permitting funds surplus or reserves to be used for investments in hospital facilities and other investment options. These are similar functions to what Medicare, prior to 1995, as part of the social security agency, was performing. The source of funds for PhilHealth under the new UHC Act will move from members' premiums to a large portion coming from the taxpayers in the form of government subsidies for special population groups - the elderly and the targeted poor. Earmarked taxes from tobacco, gambling and sugar-content are also allocated for PhilHealth's UHC implementation. This moves the corporation, a quasi-government body, to some form of public financial management, with consequences to the macroeconomy.

For its population-based role, the Department of Health will establish bureaus for health promotion, health technology assessment, drug reference pricing, appoint representatives to the local health boards, spearhead the operations and transition to UHC through project management, health workforce development activities and service delivery integration. This is expected to facilitate some of PhilHealth's functions in benefit prioritization and reference rate settings for reimbursements.

Slightly more than half (53%) of health care providers are from the private sector [3]. Private health care is viewed to deliver better quality care at higher costs. Employees use private health insurance to top up PhilHealth coverage. The extent of the relationship with the private sector can be considered an enduring one for PhilHealth, an institution that mediates the pooling of funds and the payment to service providers. Private health insurers, which top-up PhilHealth's first peso coverage, are impacted by whatever PhilHealth does to pay for health care for members. The new UHC Act has no specific directives on this beyond mention of incentives and contracting [12] and it will be PhilHealth's primary role to negotiate this relationship with providers, for their reimbursements to the costs of care provided.

Nearly half of providers are public facilities, except for a few military hospitals and DOH-managed facilities in the regions and the specialty centres (heart, lung, kidney, children's). Health facilities at sub-regional levels are owned and managed by the local government units (LGUs) . PhilHealth's programs for the poor before and presently under the new UHC Act, are coordinated with the LGUs who are responsible for pro-poor programming and allocating internal revenue and social health insurance funds to their constituencies. LGU relations form another critical function for PhilHealth, in the light of its capitation payment scheme, which pays on a per head basis, regardless of use, for the beneficiaries of the government's targeted anti-poverty program (known as 4Ps) managed by the social welfare department.

3   Universal Healthcare Coverage Concepts

The Philippine Universal Health Care (UHC) Act can be viewed as part of the rubric of reforms espoused as universal health coverage by the global community, led by the World Health Organization. Universal health coverage can be narrowly defined in terms of providing all people with the financial protection to meet their need for quality health services. Its three main dimensions, often illustrated as a cube, encompass three areas: width, to refer to population coverage; length, to refer to health services available for all, and depth or height, to refer to extent of financial protection afforded [10]. In the Philippines, early references to greater financial protection for all were in relation to the national health insurance program, hence the word 'coverage', where it meant that more and more of the population will be protected from the catastrophic financial consequences due to ill-health, through cover provided by wider risk pools from health insurance.

The Philippine UHC Act involves a process beyond social health insurance involving a pooling of funds and enrollment to participate [4]. The new Act holds a human rights based approach to the promotion and protection of the right to health for all citizens to have access to care, with no financial hardship for prioritized groups, vulnerable and poor populations in geographically isolated areas. In Sect. 3 of the Act, the general objective is to realize universal health care in the Philippines through the provision of equitable access of all Filipinos to quality, affordable and cost-effective health care goods and services. The country's

UHC's three-fold objectives of equity, quality and cost-effectiveness are within the ambit of universal health coverage. Hence the reference to universal health coverage, for the Philippines, is through its universal health care law, to therefore mean as one and the same. To guarantee financial protection amidst the promise of comprehensive health care, the new Act adopts a systemic approach and sets out to clearly delineate roles and responsibilities of agencies and stakeholders to improve the performance of the health system.

Transitioning to universal health coverage, for a corporation like PhilHealth, will entail some understanding of its systems and value proposition, contained in the policies it has in place which impact the Philippine health system. A glimpse at its products, in terms of 'circulars', allows the determination of some change trajectories and the identification of areas that need strengthening for the new UHC Act's implementation.

4 Methodology

4.1 Data Preparation

A total of 669 PhilHealth circulars from 2000–2020 were extracted online using a web scraper program written in Python. The tool utilized the urllib and BeautifulSoup modules to fetch all the circulars titles in the PhilHealth website. Table 1 shows the distribution of circulars per year. PhilHealth circular titles were stored in a data frame.

Table 1. PhilHealth circulars by administration

Group	Administration	Year covered	Number of circulars
1	Duque	2000–2005	190
2	Fajardo	2006–2007	57
3	Aquino	2008–2010	110
4	Banzon	2011–2012	85
5	Padilla	2013–2015	116
6	Aristoza	2016–2018	76
7	Morales/Giran	2019–2020	35

4.2 Data Analysis

Figure 1 presents the text mining pipeline for the study. Standard preprocessing techniques were used including removal of stopwords, tokenization to remove unnecessary texts. Term Frequency-Inverse Document Frequency (TF-IDF) scores were computed for each word. The analysis is limited to the results of the principal component analysis (PCA) to arrive at initial themes that depict

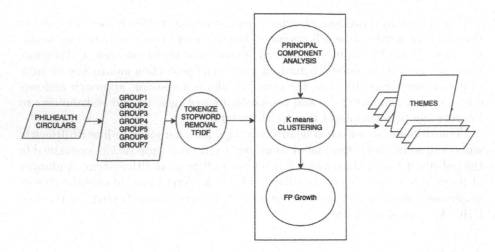

Fig. 1. Text mining pipeline

concentration of activities and policy foci within each term. Subsequent research is still in progress in verifying and analyzing the themes presented.

PCA was used on the TF-IDF sparse matrix to reduce the dimensions then K-means algorithm was used to obtain clusters with maximum number of clusters limited to k = 4. Total variance accounted for or total variance explained defines the percentage of activities performed for that time period. FP Growth algorithm was used to determine the frequent item sets in the data then the association rules were computed based on the frequent item sets. Confidence scores are used to determine strength of association between paired terms.

5   Results

Figures 2, 3, 4, 5, 6, 7 and 8 show the results from the PCA analysis. Results of PCA for the first five years accounted for 64% of the variance covering activities involving: **provision of positive drug list, amendment of circulars**, and healthcare and **renewal of accreditation**. Association tree with corresponding confidence scores are shown in Fig. 2. The cluster represents prioritization on accreditation services.

Confidence scores for the first cluster in Group 1 show strong association between the term positive and list at 100% which indicates that these two terms are used together 100% of the time, followed by a 73% frequency score for the paired term, drug and list.

For Group 2, a total of 98% of the activities of PhilHealth was focused on three areas: **implementation of circulars, amendment of circulars** and development of **programs for paying** members. Figure 3 shows the association tree with corresponding confidence scores indicating perfect association between circular and implementation, circular and amendment, and program

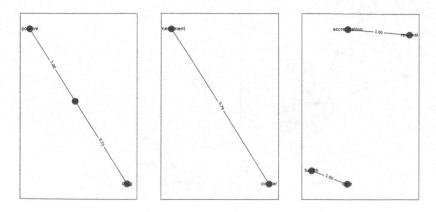

Fig. 2. Group 1: 2000–2005 (total variance explained = .64)

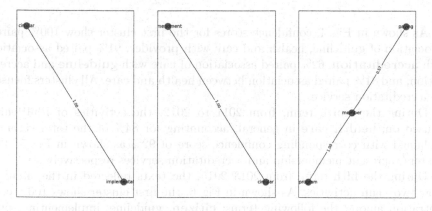

Fig. 3. Group 2: 2006–2007 (total variance explained = .98)

and members with average association between paying and members. The clusters represent membership, collection and accreditation respectively.

PCA output for Group 3 produced two clusters, with the first cluster containing terms that center on **provider**, with perfect association with the following terms: guideline, care, health and strong association with **accreditation**. The second cluster shows moderate association between **hospital** and accreditation.

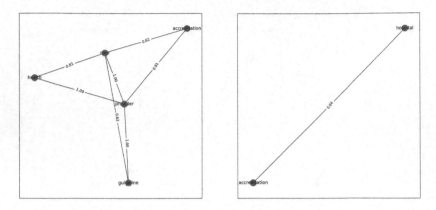

Fig. 4. Group 3: 2008–2010 (total variance explained = .75)

As shown in Fig. 4, confidence scores for the first cluster show 100% paired association of guideline, health and care with provider, 91% paired association with **accreditation**, 62% paired association of care with **guideline** and accreditation, and 81% paired association between health and care. All clusters focused on accreditation service.

During the fourth term, from 2011 to 2012, the activities of PhilHealth focused on: **health care** in general, accounting for 84% of the total variance explained with corresponding confidence score of 92% as shown in Fig. 5. The clusters represent membership and accreditation services respectively.

During the fifth term, from 2013–2015, the texts reflected in the circulars show two main activities. As shown in Fig. 6, the first cluster shows 100% concentration among the following terms: **citizen**, guideline, implementing, pursuant, **mandatory**, coverage, **senior** with 50% frequency on paired association with the terms: republic act, pursuant and **revision**. The second cluster focuses

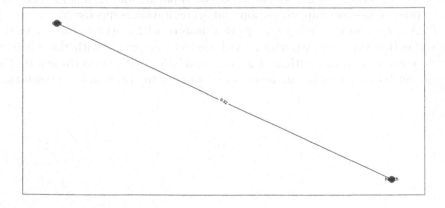

Fig. 5. Group 4: 2011–2012 (total variance explained =.84)

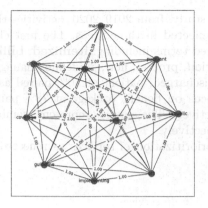
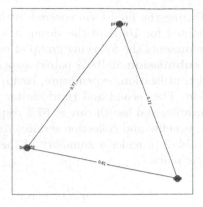

Fig. 6. Group 5: 2013–2015 (total variance explained = .70)

on activities related to the benefit package with confidence scores of 77%, 72% and 61% on paired words respectively. The two clusters account for 70% of the variance explained. The clusters represent membership and benefit services respectively.

During the 2016–2018 term, as shown in Fig. 7, 89% of the concentration of activities can be seen in two main areas. The first cluster represents activities related to: corporation, policy, management, reference, diagnosis, care, ensuring, statement, and **quality** with 100% paired associations among all pairs and 80% with the inclusion of the term: adult. Cluster 2 represents activities related to the following terms: **special and privilege** at 100% paired association and 67% association among the following terms: event, affected, fortuitous, privilege, lifting, and area. The clusters represent membership and benefit services respectively.

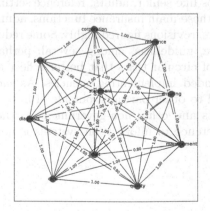
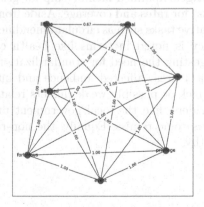

Fig. 7. Group 6: 2016–2018 (total variance explained = .89)

During the final term covered in the study, from 2019–2020, activities that accounted for 100% of the variance is reflected in three areas. The first cluster represents the following group of paired terms including: **itemized, billing** and **submission** at 100% paired association, process, flow, guideline, managed, report, utilization, expenditure, covid, discharged, patient at 50% paired association. The second and third cluster focus on benefit package at 78% paired association and health care at 67% respectively. The clusters represent membership, benefits and collection services respectively.

Table 2 provides a summary of the prioritization of services as reflected in the circulars.

Table 2. Prioritization on services

Period	Membership	Benefits	Contribution	Collection	Accreditation
2000–2005	–	–	–	–	+
2006–2007	+	–	–	+	+
2008–2010	–	–	–	–	+
2011–2012	–	+	–	–	+
2013–2015	+	+	–	–	–
2016–2018	+	+	–	–	–
2019–2020	+	+	–	+	–

6   Discussions

It is observable that in terms of setting policies, basic social health insurance activities were prominent, such as accreditation, payment mechanism, benefit package, expanded membership for groups like senior, adults, reference settings (lists) for rates and coverage. Aside from these main insurance functions, administrative tasks such as circular amendments, revisions figured highly. Some redundancy is noted in terms like health care, guidelines, among several, perhaps suggesting the need for standardization of circular titles that better reflect the subject area, affected audience and intended implementing agencies. As such, the lack or possibly weak clusters related to other partners, outside of hospital providers, like the local government units and other non-government partners shows some remiss. Equally, the non-reference to information systems is noteworthy.

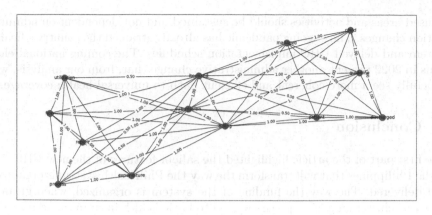

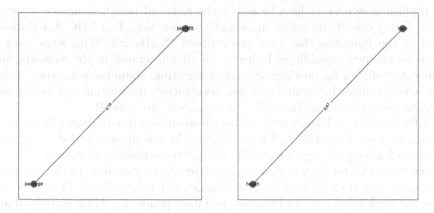

Fig. 8. Group 7: 2019–2020 (total variance explained = 1.00)

The policy process across the years show different prioritization in relation to services provided by the national health insurance program. The first four periods were consistent in providing accreditation services but varying prioritization on other services including membership, collection and benefits. The last three periods provided more focus on membership, benefits and contributions.

Different administrations prioritized different UHC-related activities. It remains to be seen if these policies can be viewed as building blocks for the full implementation of the law or may fragment the health system further. Linking these to administrative periods provides some insights into the priority focus of the particular administration, if not insights into governance styles. The groups with the low number of circulars and relatively lower variance (with the exception of group 2) appear to be headed by those coming or rising from inside the corporation. Perhaps they were able to do more diverse activities, executing earlier policies, than just setting policies. To achieve universal health coverage, focused areas and activities should be sustained and not dependent on administration changes. The ongoing pandemic has already strained the country's health system and delayed UHC implementation schedules. The coming national elections in 2022 present another administrative change that, from our analysis, will hopefully see a more coherent and efficient path to universal health coverage.

7   Conclusion

The first part of the article highlighted the salient features of the new UHC Act in the Philippines that will transform the way the PhilHealth system is organized and delivered. The way the funding of the system is organized, when gleaned from our cluster analysis, appears welded to basic health insurance and may not move much from its present first-peso coverage, still with high out of pocket costs, reimbursement and provider payment systems. A number of credible health sector performance reviews have looked at the national health insurance program which forms one of the major drivers of the new law. The UHC Act removes some of the functions that have preoccupied PhilHealth. With some capacities released, new capabilities hitherto not demonstrated in the circulars, like in incentive design for provider network integration, commissioning/contracting new benefits under integrated systems, negotiation, monitoring and evaluation, information systems and knowledge management, are needed.

Fifty years of social and national health insurance programming form a generation or two. The trajectories for change do not appear varied, suggestive perhaps of some path dependence [5]. The patterns gleaned from cluster analysis are complex and examining for path dependence can alert to the processes needed to transition to and manage change and innovation [5,11]. Achieving universal health coverage for Filipinos, and post-pandemic, will require bold and innovative ideas to improve the financing of health care, with performance and systems improvement at its core.

A 'look-back' at the past in terms of the released circulars provide future UHC implementation with knowledge or a memory bank. Text mining and cluster analysis provided some associative learning, which even just through words, highlighted some categorization and some observable differences in categories [6]. The linkage to administration provides one such category. Linkage to financial performance of the corporation is subject of future research. Member utilization, provider behavior and influence of financing on over-all sector performance cannot be gleaned from current information systems. Hence, on the cusp of UHC implementation and with the recent experience of the pandemic, information technology and systems must be in place to improve monitoring and evaluation to support decision making and policy-setting.

Acknowledgement. The paper is part of a research project funded by the University Research Council of Ateneo de Manila University in 2020. The authors would like to thank the Ateneo Social Computing Science Laboratory for the support provided in healthcare research.

References

1. World Bank: Philippine Health Sector Review. World Bank (2011)
2. Chakraborty, S.: Philippines' government sponsored health coverage program for poor households (2013)
3. Dayrit, M.M., Lagrada, L.P., Picazo, O.F., Pons, M.C., Villaverde, M.C.: The Philippines health system review (2018)
4. Doetinchem, O., Carrin, G., Evans, D., Organization, W.H., et al.: Thinking of introducing social health insurance? Ten questions. World Health Organization, Tech. rep. (2009)
5. Fouda, A., Paolucci, F.: Path dependence and universal health coverage: the case of Egypt. Front. Public Health **5**, 325 (2017)
6. Hilgard, E.R., Bower, G.H.: Theories of learning (1966)
7. Hsiao, W.C.: Taiwan's path to universal health coverage-an essay by William C Hsiao. BMJ 367, l5979 (2019)
8. Lam, H.Y.: How protected are philhealth members and beneficiaries? The 2017 philhealth support value survey. Tech. rep., National Institutes of Health, University of the Philippines, Manila (2019)
9. Obermann, K., Jowett, M., Kwon, S.: The role of national health insurance for achieving UHC in the Philippines: a mixed methods analysis. Glob. Health Action **11**(1), 1483638 (2018)
10. World Health Organization: Tracking universal health coverage: first global monitoring report. World Health Organization (2015)
11. Perello-Marin, M.R., Marin-Garcia, J.A., Marcos-Cuevas, J.: Towards a path dependence approach to study management innovation. Management Decision (2013)
12. Schuhmann, R., Bautista, M.: Contracting for health care services under the new Philippine universal health care act. Vru/WCL **54**(1), 98–115 (2021)
13. Solon, O.J.C., Herrin, A.N.: The challenge of reaching the poor with a continuum of care: a 25-year assessment of health sector performance. Risk, Shocks, Building Resilience, p. 85 (2016)
14. Tangcharoensathien, V., Suphanchaimat, R., Thammatacharee, N., Patcharanarumol, W.: Thailand's universal health coverage scheme. Econ. Polit. Wkly. **47**, 53–57 (2012)

Analyzing Change on Emotion Scores of Tweets Before and After Machine Translation

Karin Fukuda[1] and Qun Jin[2(✉)]

[1] Graduate School of Human Sciences, Waseda University, Tokorozawa, Japan
karinfukuda@fuji.waseda.jp
[2] Faculty of Human Sciences, Waseda University, Tokorozawa, Japan
jin@waseda.jp

Abstract. Many of the texts posted on Twitter are broken sentences, and the translated sentences may not be accurate. An inaccurate translation may spoil the meaning of the original text and induce miscommunication between the poster and the reader who uses the machine translation. Since many sentences tweeted on Twitter contain emotional expressions, this study uses sentiment analysis to calculate and compare the sentiment scores of the original and translated sentences to investigate the change in sentiment before and after machine translation. As a result of using dictionaries to classify tweets before and after translation, it was found that the classification of positive sentences tended to be more likely the same before and after translation. In addition, the results of the sentiment analysis of "joy", "like", "relief" and "excitement" by machine learning showed that the sentiment of "joy" was particularly increased when translated from Japanese into English.

Keywords: Sentiment analysis · Emotion score · SNS · Twitter · BERT · Machine translation

1 Introduction

With the development of the information society, it is now possible for anyone to transmit and share information. In particular, the number of active users of social networking services (SNS), which allow people to easily transmit information, continues to increase. It is expected that there will be an increase in interaction among users all over the world, regardless of where they live. One of the leading social networking services is Twitter, which has 210 million daily active accounts as of 2021 [1]. Since Twitter incorporates Google's translation function, it is possible for users to receive tweets sent in a different language and read in their native language with the help of machine translation. However, many of the texts posted on Twitter are broken sentences, and the translated sentences may not be accurate. An inaccurate translation may spoil the meaning of the original text and induce miscommunication between the poster and the reader who uses the translation. Since many sentences tweeted on Twitter contain emotional expressions, in this study, we use sentiment analysis approach to calculate and compare the sentiment

G. Meiselwitz (Ed.): HCII 2022, LNCS 13315, pp. 294–308, 2022.
https://doi.org/10.1007/978-3-031-05061-9_21

scores of the original and translated sentences to investigate the change in sentiment before and after machine translation.

Saif M. Mohammad [2] et al. compared the accuracy of sentiment analysis for Arabic and English texts respectively and investigated why the sentiment values have changed. The result showed that regardless of the translation method, many texts lost some of their positive or negative emotions when translated and were classified as neutral. From this result, it can be expected that when machine translating between Japanese and English, emotional words may be not correctly translated, making the translated text more likely to be classified as neutral. To analyze and investigate such a change in detail, after confirming the tendency of the sentences before and after the machine translation to change to positive, negative, or neutral by using sentiment dictionaries, we conduct a multi-class sentiment analysis, in which the sentiment categories are not limited to positive, negative, and neutral, but also include more detailed sentiment categories. In this study, we use the 10 emotions model proposed by Nakamura [3] (i.e., sadness, shame, anger, disgust, fear, surprise, like, excitement, relief, and joy) as the classification categories.

The remainder of this paper is structured as follows. Section 2 overviews related work on classification of emotions and multilingual sentiment analysis. Section 3 outlines the research approach in this study and describes the emotional word dictionaries we use. Section 4 presents experiment results including detailed procedures and discussions on the results. Section 5 summarizes this study and highlights future work.

2   Related Work

2.1   Categories of Emotions

Currently, many sentiment categories have been proposed in sentiment analysis, and the two most common categories to sentiment analysis in various languages are positive and negative. These are the simplest and most opposite axes, and many studies have been conducted on them as the basis of emotion analysis. However, the actual emotions that people have are more complicated and complex. Therefore, more diverse emotion classifications have been proposed to date.

Firstly, there is the Prutik's wheel of emotions [4], which systematically shows the emotions that we have. It is a three-dimensional model in which eight primitive emotions (i.e., admiration, ecstasy, terror, amazement, grief, loathing, rage, and vigilance). They are thought to have evolutionary origins, exist as primary colors, and more complex emotions, which are a mixture of these colors, and arranged in a cone around the primary colors.

As for the sentiment categories in Japanese, Akira Nakamura's Dictionary of Emotional Expressions [3] has ten categories: joy, anger, sorrow, fear, shame, like, dislike, excitement, anxiety, and surprise. The Emotional Expression Dictionary contains words which belong to these 10 emotions. Most of the multi-label sentiment analysis in Japanese is centered on these 10 emotions. Yamamoto et al. [5] used the 10 emotion categories in an experiment to determine the role of emoticons in sentences posted on Twitter. They obtained useful results by classifying the 10 categories into two groups: "joy, like, relief, and excitement" as positive, "sadness, fear, anger, disgust, surprise, and shame" as negative, and emotions that were not classified in these two groups as neutral.

In this study, in addition to positive, neutral, and negative categories, "joy, like, relief, and excitement" were further investigated, referring to Yamamoto [5] and others out of the ten emotion axes of Nakamura [3].

2.2 Multilingual Sentiment Analysis

The main purpose of multilingual sentiment analysis is to prove the usefulness of data diversion to languages with less textual data. In a study by Saif M. Mohammad et al. on how translation alters sentiment for Arabic and English texts [2], they compared the accuracy of each sentiment analysis result and investigated why and how the sentiment values changed. In addition, the results of machine translation and manual translation were performed in sentiment analysis, and it was shown that even though machine translation did not work accurately between English and Arabic, the machine translation resulted in fewer score changes. In contrast, in a study by Boxing Chen et al. [6] by Boxing Chen et al., sentiment analysis between English and Chinese was conducted. In this study, they conducted dictionary-based sentiment analysis on Chinese and English translated texts to study the consistency of sentiment analysis, and incorporated it into the translation model to improve the translation accuracy.

Another study by Mika Saeki et al. compared sentiment in Japanese and English [7]. Firstly, this study focused on adverbs in the book "Modern Adverbial Usage Dictionary" by Tobita and Asada [8]. In this book, adverbs are assigned an "image value", which is an emotion score between -3 and $+3$. They analyzed the Japanese adverbs with different image values and their corresponding English translations from the bilingual database of Japanese and English. As a result, they found that adverbs for emotional expressions in Japanese sentences tend to be expressed by other part of speech in English. In addition, it was shown that by using the imagery of English words, the equivalence of images after translation from Japanese to English is maintained., In this sense, maintaining the results of sentiment analysis before and after translation between Japanese and English will lead to improved translation accuracy.

2.3 The Position of This Study

In this study, we investigate and quantify the changes in emotions before and after machine translation by using a series of processes: classification of emotions into "positive, neutral, and negative" by a dictionary, evaluation by human evaluators, and analysis of emotions into multi-classes of "joy, like, relief and excited" by machine learning. As for the positive/negative categorization, Mohammad [2] et al. found that the translated text was more likely to be classified as neutral than the original text. In this study, we use a sentiment dictionary to verify the results between English and Japanese.

3 Analysis of Change on Sentiment Scores Before and After Machine Translation on Tweets

The basic framework of sentiment analysis and experiment design is shown in Fig. 1. We collected about 10,000 tweets each in Japanese and English, and for these tweets, we used sentiment dictionaries to classify the tweets into three categories: negative, neutral, and positive. We recruited three human evaluators to reclassify the tweets in each language as negative, neutral, or positive. For those that were classified as positive, we further classified the texts based on "joy, like, relief and excited" and used the results as fine-tuning data for BERT [9] to perform sentiment analysis on the sentences that were classified as positive before translation.

3.1 Dictionaries for Sentiment Analysis

In this study, we conducted sentiment analysis using the following dictionaries to investigate whether differences in sentiment scores between dictionaries can detect changes in nuance before and after machine translation, and to confirm the tendency of previous studies to become neutral after translation.

We used SentiWords [10] as a dictionary of English sentiment words, which is pre-polarized by assigning different sentiment scores to the same word for each part of speech. It contains a rich and balanced vocabulary of about 155,000 words: 23,000 negative, 11,400 neutral, and 19,000 positive words, which are assigned an emotion score between −1 and 1. It should also be considered that about 60,000 of these words are compound words.

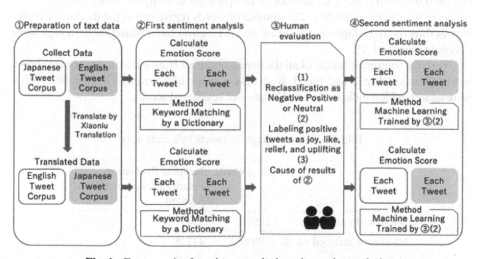

Fig. 1. Framework of sentiment analysis and experiment design

We used the Japanese Evaluation Polarity Dictionary [11] as a dictionary of Japanese emotion words. In this experiment, we limited to only use the "Japanese Evaluation Polarity Dictionary (Noun Edition) ver. 1.0 (December 2008 version)". The noun version contains about 5000 negative words, 3000 positive words, and 5000 neutral words.

3.2 Tweet Acquisition and Processing

In this experiment, we prepared more than 10,000 tweets of politicians in both English and Japanese. In the sentiment analysis using the dictionary, the sentiment score of the text is calculated by comparing the words in the text with the words registered in the dictionary and adding up the scores of the matching words. Since the parts of speech are also registered in SentiWords and in the noun section of the Japanese Evaluation Polarity Dictionary, we also matched parts of speech. We considered them as a group of words when the words were returned to their original form in a specific sequence because the number of compound words in SentiWords is more than 60,000. Since the parts of speech are not accurately classified for compound words, we did not perform part-of-speech matching only for compound words.

In addition, we referred to the quantification method of emotion scores proposed by Tago and Jin [12] for the calculation of sentiment scores.

4 Experiment Results and Discussions

4.1 Scrutinizing Dictionaries

For each dictionary, the total number of morphological analyses, word matching, and coverage are shown in Table 1. The number of words registered in the dictionary corresponds directly to the high coverage rate, and in English, SentiWords shows a high word coverage rate of 41.1%.

Next, the sentiment scores of all the tweets obtained from each dictionary are shown as histograms in Fig. 2 and Fig. 3. Both have a shape like a normal distribution with a peak near 0, which indicates a good balance of registered words.

Table 1. Word coverage for tweets by a dictionary

	SentiWords	Japanese Evaluation Polarity Dictionary (Noun Edition)
Number of morphemes	169120	234718
Number of matched words	109861	41725
Coverage rate	41.1%	17.8%

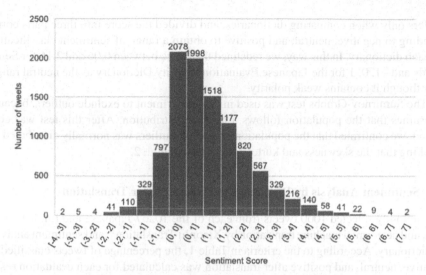

Fig. 2. Distribution of sentiment scores by sentiWords

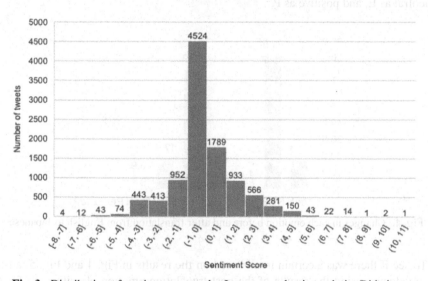

Fig. 3. Distribution of sentiment scores by Japanese evaluation polarity Didctionary

4.2 Redefining the Neutral Score

SentiWords has 155,000 words, while the Japanese Evaluation Polarity Dictionary has 13,000 words, which means that the probability of word matching is about 12 times higher when using SentiWords. In addition, SentiWords assigns a continuous sentiment score between –1 and 1, while the Japanese Evaluation Polarity Dictionary takes discrete values of –1, 0, and 1. From these facts, it is expected that a simple score comparison will produce biased results depending on the registered words. Therefore, we excluded outliers only when comparing dictionaries, and divided the score into three parts corresponding to negative, neutral, and positive to obtain a range of sentiment classification for each dictionary. In this way, we redefined the range between 0.06 and 2.16 for SentiWords and −1, 0, 1 for the Japanese Evaluation Polarity Dictionary as the neutral range, even though it contains weak polarity.

The Sumirnov-Grubbs test was used in this experiment to exclude outliers, because it assumes that the population follows a normal distribution. After this test was conducted, we confirmed that the population excluding outliers was normally distributed by checking that the skewness and kurtosis were less than ± 2.

4.3 Sentiment Analysis by Dictionary Before and After Translation

We randomly selected 1000 tweets from each of the 10267 Japanese and English tweet data, and performed machine translation by Xiaoniu Translation[1], and sentiment analysis by dictionary. According to the criteria in Table 1, the percentage of tweets classified as negative, neutral, and positive after translation was calculated for each evaluation result before translation of 1000 tweets and shown in Fig. 4 and Fig. 5. Negative is denoted as N, neutral as E, and positive as P.

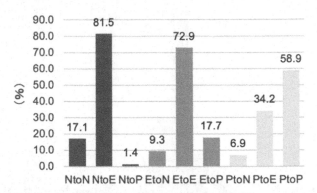

Fig. 4. Changes in each emotion before and after translation from English to Japanese

To see if there was a certain trend based on the results in Fig. 4 and Fig. 5, a t-test was conducted using the results of the post-translation emotion classification for the two most common destinations for each emotion category before translation. A one-tailed test was conducted on 1/10th of the pre-translation number of each category to see if there was a tendency to be categorized as positive or negative after translation. In addition, a two-tailed test was conducted to see if there was a significant difference from the median of the redefined neutral scores to confirm whether there was sufficient neutral categorization. Examples of the null hypothesis (H0) and alternative hypotheses (H1) and the results of the tests are shown in Table 2 and Table 3. The rejection area was set at 5%.

---

[1] https://niutrans.com.

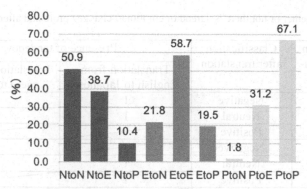

Fig. 5. Changes in each emotion after translation from Japanese to English

< Examples of null hypothesis and alternative hypothesis >

H_{NtoE0}: The post-translation sentiment score of the tweets classified as NtoE is not different from 0.

H_{NtoE1}: The post-translation sentiment score of the tweets classified as NtoE is different from 0.

H_{NtoN0}: The post-translation sentiment score of the tweets classified as NtoE is not less than −1.

H_{NtoN1}: The post-translation sentiment score of the tweets classified as NtoE is less than −1.

Table 2. The null hypothesis and its consequences for English-to-Japanese translation

	H_{NtoE0}	H_{NtoN0}	H_{EtoE0}	H_{EtoP0}	H_{PtoP0}	H_{PtoE0}
Number of Samples	21	21	56	56	23	23
Variance	20	20	55	55	22	22
t-value	−1.8530	−3.7986	1.5925	−3.1092	3.1384	4.8571
p-value	0.0787	0.0006	0.1170	0.0015	0.0024	0.0001
Result		Reject		Reject	Reject	Reject

Table 3. The null hypothesis and its consequences for Japanese-to-English translation

	H_{NtoN0}	H_{NtoE0}	H_{EtoE0}	H_{EtoN0}	H_{PtoP0}	H_{PtoE0}
Number of Samples	11	11	72	72	17	17
Variance	10	10	71	71	16	16
t-value	−0.7079	−5.0271	−1.1128	7.9992	4.6783	7.7534
p-value	0.2476	0.0005	0.2696	0.000	0.0001	0.000
Result		Reject		Reject	Reject	Reject

Table 4. Whether or not there is a trend of emotional change before and after translation

Classification before translation	Classification after translation	Presence of tendency	
		Translation from English to Japanese	Translation from Japanese to English
Negative	Negative	1	0
	Neutral	1	0
	Positive		
Neutral	Negative		1
	Neutral	1	1
	Positive	1	
Positive	Negative		
	Neutral	0	0
	Positive	1	1

As for the tweets for which the null hypothesis was not rejected, although the fact that the null hypothesis was not rejected is not an indicator that the null hypothesis is correct, we assume that the null hypothesis is correct because there is no inconsistency in the results of Fig. 3 and Fig. 4. Table 4. summarizes the presence or absence of a trend of emotional change before and after translation.

It was found that in both the Japanese-to-English and English-to-Japanese translations, what was classified as positive in the source text tended to remain positive after translation, and what was classified as neutral tended to remain the same before and after translation.

4.4 Human Evaluation

To determine the accuracy of the results of word matching using a dictionary including parts of speech, to create training data for sentiment analysis after machine translation using BERT, and to determine the causes of any particular differences in sentiment analysis using a dictionary before and after machine translation, we obtained evaluations of 359 tweets in Japanese and 316 tweets in English from three evaluators who are fluent in Japanese and English.

Questions. Three questions were set for the evaluator. In Question 1, we asked the target tweets to be reclassified into positive, neutral, and negative to evaluate the accuracy of sentiment analysis using a dictionary. Question 2 aims to be used as training data for sentiment analysis by machine learning after that, and the tweets that the evaluator judged to be positive in Question 1 are classified into four types: "joy, like, relief or excitement ". In Question 3, we asked the tweet pairs whose scores were particularly different in the sentiment score by the dictionary about the presence or absence of nu-ance changes and the cause of the changes in a questionnaire format with options. As the questionnaire options, four options were set with reference to the questionnaire evaluation in [2].

Results of Human Evaluation. We compared the results of three evaluators for Question 1 and used the most selected classification as the emotion classification of the tweet. For tweets that did not receive any of the three answers, we used our own data and added to the judgment. There were 8 such data in Japanese and 15 in English. By comparing these with the results of the dictionary, the concordance rate between the results of sentiment analysis by SentiWords and the Japanese Evaluation Polarity Dictionary and the evaluation by human evaluators was calculated as the classification accuracy. Table 5 shows the accuracy of each sentiment classification and the accuracy of the entire dictionary.

Table 5. Accuracy of emotion classification by a dictionary

	SentiWords			Japanese evaluation polar dictionary (Noun Edition)		
	Positive	Neutral	Negative	Positive	Neutral	Negative
Accuracy for each emotion classification	63.0%	18.5%	54.5%	49.4%	57.7%	51.7%
Overall classification accuracy	42.9%			55.3%		

The criteria for the classification score used when comparing dictionaries were used, but when only 0 was judged to be neutral for SentiWords, only the neutral accuracy increased by about 25%, and the overall score was 52.7%. This means that many weak tweets with relatively low scores of 0.66 or more and less than 2.16 were classified as non-neutral in human evaluation. On the other hand, if only 0 is set to neutral in the Japanese Evaluation Polarity Dictionary, the accuracy of classification to neutral increases by about 5%, but the accuracy of classification to the other two decreases by about 10%, totaling 46.4%. This indicates that the reduction in the total number of neutrals in the dictionary reduced the concordance rate with the results of evaluation by humans, and the classification is more accurate when the neutral range is widened for Japanese.

As a result of Question 1, 85 tweets in Japanese and 94 tweets in English were classified as positive. In Question 2, we asked them to classify the tweets as "joy, like, relief or excitement". In addition, we asked five more evaluators to perform the same evaluation of Question1 and Question2 for 606 tweets in English and 467 tweets in Japanese to increase training data. Table 6 shows the number of tweets assigned to each emotion.

Table 6. Breakdown of people's evaluation of positive tweets

	Joy	Like	Relief	Excitement
Tweet in English (606)	326	188	79	349
Tweet in Japanese (467)	218	176	108	231

In both Japanese and English, the number of tweets assigned to "relief" was the lowest, indicating concern about learning bias. In addition, in these four categories, the classification results of two or more people coincided in 62.3% of all classifications. Next, we show the evaluation results of Question 3. First, for the translation pairs that showed a particular difference in score (21 pairs in Japanese and 27 pairs in English), the results of the three evaluators were compared to see whether there was nuance change or not, and the one with the higher score was adopted. As a result, 85.7% of the translation pairs from Japanese to English and 70.4% of the translation pairs from English to Japanese were considered to have nuance changes. If the classification balance of the registered lexicon is maintained, the results of sentiment classification by the dictionary can detect the change in nuance between Japanese and English before and after translation. Table 7 shows the results of the evaluation of the translation pairs that were judged to have nuance changes.

Table 7. Reasons for changes in nuance

	Mistranslation (A)	Missing translation (B)	Exact translation but cultural differences (C)
Tweet pair of translation from Japanese to English (21)	16	1	15
Tweet pair of translation from English to Japanese (27)	17	5	9

In Table 7, (A) and (C) were often mentioned at the same time, in 12 cases for Japanese-to-English translation pairs and in 8 cases for English-to-Japanese translation pairs. Since (A) and (C) are contradictory options that may or may not be mistranslations, it is likely that the evaluators were divided in their judgment on whether a translation with a difference in perception was a mistranslation. The number of responses to (B) was higher than the number of responses to (C). In addition, the number of responses to (B) was particularly low for the Japanese-to-English translation pairs. From this, it can be expected that the possibility of a translated sentence being classified as neutral due to no translation of an emotional word is low when translating from Japanese to English.

Finally, Table 8 shows the two examples of corresponding tweet pairs and the evaluations from the evaluators who selected "Others" in Question 3.

The first pair of tweets is a media headline tweet, and the first half of the sentence describes the action taken in response to the second half. In the translated sentence, the first half of the sentence is the cause of the second half, resulting in a possible misunderstanding by the reader. As the evaluator said, when translating a limited number of words, such as headlines, the sentence break is not recognized during translation, and it is thought that there is a change in nuance.

The second pair of tweets, "Sorry Fake News," is literally translated as an apology for being fake news, and the connection to the reason for the tweet, "I'm recording it," is lost. In addition, the "long" in "long before" is not reflected in the translation.

Table 8. Reasons for changes in nuance - free answers

Before translation	岸田氏自衛隊法の改正検討　アフガン退避初動に遅れ（フジテレビ系（FNN））
After translation	Kishida's discussion on amending the SDF law delayed the initial withdrawal from Afghanistan (Fuji TV Group (FNN))
Evaluation	Sentence breaks are not recognized.
Before translation	Sorry Fake News, it's all on tape. I banned China long before people spoke up. Thank you @OANN
After translation	フェイクニュースで申し訳ありませんが、録画しておきました。人々が話す前に、私は中国を禁止した。ありがとう@OANN
Evaluation	There are some parts untranslated and cultural differences in how they are received.

4.5   Results of the Sentiment Analysis by BERT

We performed translation on sentences that were classified positively in Question 2 and used the results of Question 2 of the human evaluation to perform multi-label classification using BERT. BERT is the Google's neural language model published by Jacob Devlin et al. in 2018 [9]. After pre-training, BERT can be used for various natural language processing (NLP) tasks by learning task-specific labeled data from relatively small amounts of data through fine-tuning. BERT can handle a variety of natural language processing tasks. In this experiment, we used cl-tohoku/bert-base-japan-whole-world – masking [13], which was created by Tohoku University using Wikipedia articles, as a pre-training model for Japanese, and bert-base-uncased [9], which was trained using BookCorpus [14] and contains unpublished novels, as a pre-training model for English. For these pre-training models, we performed multi-label classification by fine-tuning using 80% of the data obtained in Question 2 as training data and 10% each as validation and test data. The model was evaluated again on the pre-translation data, and the accuracy, fit rate, recall rate, and F-measure were calculated by comparing the results with those of the human evaluations. Table 9 and Table 10 show the results of the calculations and the accuracy for all the emotions: joy, happiness, relief, and excitement.

We could not calculate Precision for the English model for "like" and "relief" and the Japanese model for "relief" because there were no predicted classifications. Figures 6 and 7 shows the results of the comparison between the human evaluation of pre-translation and BERT classification of the post-translation for Japanese-to-English and English-to-Japanese translations.

Table 9. English model: accuracy, precision, recall, and F-measure for each emotion

	Joy	Like	Relief	Excitement	Overall
Accuracy	0.765	0.706	0.902	0.627	0.353
Precision	0.725			0.581	
Recall	0.967	0	0	0.750	
F-measure	0.829			0.655	

Table 10. Japanese model: accuracy, precision, recall, and F- measure for each emotion

	Joy	Like	Relief	Excitement	Overall
Accuracy	0.737	0.711	0.895	0.737	0.368
Precision	0.615	1.00		0.765	
Recall	9.615	0.083	0	0.684	
F-measure	0.615	0.154		0.722	

The results show that the classification of positive emotions in the dictionary tended to remain the same before and after translation, but the emotions of "joy" and "excitement" were lost in many cases. However, for the emotion of "joy" when translated from Japanese to English, the number of classifications increased significantly after translation compared to before translation.

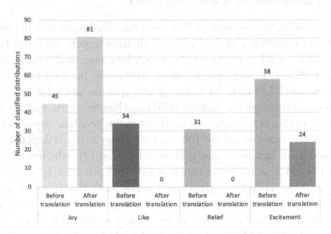

Fig. 6. Comparison of sentiment classification using BERT for Japanese-to-English

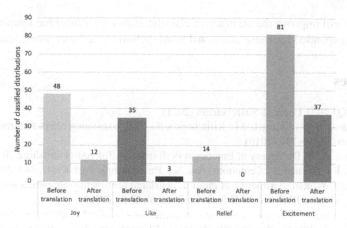

Fig. 7. Comparison of sentiment classification using BERT for English-to-Japanese

5 Conclusion

In this study, we first compared sentiment classification between Japanese and English using a sentiment word dictionary. The results of the sentiment classification by the dictionary show that tweets that are classified as neutral before machine translation tend to be classified as neutral after translation. Moreover, regardless of the language before translation, tweets classified as positive before translation tend to be classified as positive after translation, and there is no general tendency for tweets to be classified as neutral after translation. The sentiment dictionary is expected to achieve higher accuracy in sentiment classification by a complete set of the number of words in the dictionary, the score scale, and the parts of speech to be handled in advance. As shown in this study, the difference in sentiment scores between English and Japanese has a high probability of detecting changes in nuance, and as shown by Boxing Chen et al. [6], maintaining the sentiment scores before and after translation can help increase translation accuracy.

Furthermore, we investigated whether there was a change in nuance before or after machine translation between Japanese and English, based on the results of sentiment analysis by the dictionary and human evaluations. As a result, 85.7% and 70.4% of the translated tweet pairs with different scores showed nuance changes. This indicates that the machine can detect nuance changes even when comparing sentiment scores from different dictionaries. In this study, we also attempted to quantify the change in sentiment before and after the translation of tweets that were classified as positive before the translation by using sentiment analysis based on machine learning with a more detailed sentiment axis: "joy, like, relief and excitement." The results showed similar trends in both languages, but "joy" tended to increase especially in the translation from Japanese to English. However, the overall accuracy was low. In particular, the data for "like" and "relief" were severely lacking and could not be verified, so it is essential to create a larger corpus.

As for our future work, we will well prepare for the difference in recorded vocabulary, the registration of parts of speech and the scale of emotion scores, explore more subdivided emotional axis, and limit evaluation conditions, such as the "mistranslation of words". We will improve the accuracy of classification with a machine learning method by using an expanded Japanese emotional word dictionary and increasing learning data.

References

1. Twitter. Q3 2021 Letter to Shareholders (2021)
2. Mohammad, S.M., Salameh, M., Kiritchenko, S.: How translation alters sentiment. J. Artif. Intell. Res. **55**, 95-130 (2016)
3. Nakamura, A.: A Dictionary of Emotional Performance. Tokyodo Publishing, Tokyo (1993)
4. Plutchik, R.: The nature of emotions. Am. Sci. Res. Triangle Park **89**(4), 344–350 (2001)
5. Yamamoto, Y., et al.: Extracting of multi-dimensional sentiment from tweets using twittism. In: Proceeding of IPSJ Kansai-Branch Convention, G-01 (2014)
6. Chen, B., Zhu, X.: Bilingual sentiment consistency for statistical machine translation. In: Proceedings of the 14th Conference of the European Chapter of the Association for Computational Linguistics (2014)
7. Saeki, M., et al.: Japanese-english target analysis of japanese adverbs focusing on emotional expression. In: Proceedings of the Association for Natural Language Processing Annual Convention, A1–9 (2005)
8. Tobita, H., Asada, Y.: Modern Adverbial Usage Dictionary. Tokyodo Publishing, Tokyo (1994)
9. Devlin, J., Chang, M.-W., Lee, K., Toutanova, K.: BERT: Pre-training of Deep Bidirectional Transformers for Language Understanding. arXiv (2018)
10. Gatti, L., Guerini, M., Turchi, M.: SentiWords: Deriving a high precision and high coverage lexicon for sentiment analysis. IEEE Trans. Affect. Comput. **7**(4), 409–421 (2016)
11. Kobayashi, N., Inui, K., Matsumoto, Y., Tateishi, K., Fukushima, T.: Collecting evaluative expressions for opinion extraction. In: Su, K.-Y., Tsujii, J., Lee, J.-H., Kwong, O.Y. (eds.) IJCNLP 2004. LNCS (LNAI), vol. 3248, pp. 596–605. Springer, Heidelberg (2005). https://doi.org/10.1007/978-3-540-30211-7_63
12. Tago, K., Jin, Q.: Influence analysis of emotional behavious and user relationships based on twitter. Tsinghua Sci. Technol. **23**(1), 104–113 (2018)
13. cl–tohoku/bert–base–Japanese–whole–world–masking, Hugging Face. https://huggingface.co/cl-tohoku/bert-base-japanese-whole-word-masking
14. Zhu, Y., et al.: Aligning Books and Movies: Towards Story-like Visual Explanations by Watching Movies and Reading Books. Arxiv (2015)

Automatic Meme Generation
with an Autoregressive Transformer

Denis Gordeev[1,3,4](✉) and Vsevolod Potapov[2]

[1] Russian Presidential Academy of National Economy and Public Administration,
Moscow, Russia
gordeev-di@ranepa.ru
[2] Centre of New Technologies for Humanities, Moscow, Russia
[3] Russian Foreign Trade Academy, Moscow, Russia
[4] Moscow State Linguistic University, Moscow, Russia

Abstract. In this paper we study the way to automate meme generation from textual prompts using autoencoders based on the Transformer
architecture. For this we collected a dataset of about 5000 meme images.
Then we run an OCR (optical character recognition) library Tesseract
on top of these images to get English texts from them. We filtered poorly
recognised texts using FastText language identification library to get only
images with English texts and scores above 0.9. Then we trained image
generation models using our dataset.

Keywords: Optical character recognition · OCR · Internet meme ·
Image classification · DALLE · Autoencoder · Transformer

1 Introduction

Multimodal problems have recently attracted increased attention from
researchers. Partially this interest may be attributed to this year's works on
applying Transformers to the visual domain by OpenAI [12,13]. However, many
prior works also focused on multimodal problems, including such tasks as image
captioning [3] and visual question answering [1]. Some researchers think that
machine learning models should be available to take multimodality into consideration for better natural language understanding and progress in other similar
fields [8]. Multimodal tasks pose increasing interest due to their closeness to real
world problems and situations.

Multimodal problems often pose difficulties for modern machine learning
methods. They either focus on only one modality (e.g. text) and ignore all other
modalities [5] or get lackluster scores for problems where one modality is not
enough for robust predictions [8,19].

Internet memes are a popular means of multimodal communication via the
Internet. The word meme has first been coined by Richard Dawkins [4]. It was
used to describe the cultural counterpart of human genes [9]. However, the word
"meme" was adopted into common language and got a new connotation as in

G. Meiselwitz (Ed.): HCII 2022, LNCS 13315, pp. 309–317, 2022.
https://doi.org/10.1007/978-3-031-05061-9_22

the phrase "Internet meme". An Internet meme is usually defined as a unit of information that can replicate itself via Internet in a shape of some media (image, video, text, audio) [2]. Moreover, Internet memes can contain more than one media type at the same time, image + text being the most characteristic for the common perception of them.

Memes pose an interesting multimodal fusion problem: Consider a sentence like "When you say H20 instead of water". This sentence can be accompanied by an infinite number of possible images. Moreover, the same images may illustrate opposing point of views. Memes often express their pragmatics in an implicit way and considering only text or image is often not enough for understanding meme pragmatics and meaning. See Fig. 1 for an example.

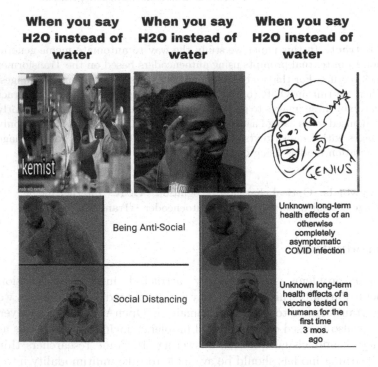

Fig. 1. Examples of adversal varying modalities in a meme.

In this paper we study the way to automate meme generation from textual prompts (that may be accompanied with images) using a variational autoencoder and an autoregressive autoencoder based on the Transformer [20] architecture. For this we collected a dataset of about 5000 meme images. Then we run an OCR (optical character recognition) library Tesseract [17] on top of these images to get English texts from them. We filtered poorly recognised texts using FastText language identification library to get only images with English texts and scores above 0.9^1. These two steps filtered out the majority of images and we were left with 2082 images. All monocodal images have been removed as they do not contain texts. Then we trained image generation models using our dataset.

2 Related Work

Fig. 2. RuDolph meme generations (the prompt is Harry Potter meme: Severus Snape)

There are some works concerning meme classification. There has been conducted a shared task on Troll Meme Classification in Tamil [18]. However, the multimodality posed serious difficulties for the contestants and the best performing solution reached only 55 points in F1-score. Another contest has been arranged by Facebook. Klela: We find that state-of-the-art methods perform poorly compared tohumans, illustrating the difficulty of the task and highlighting the challenge that this important problem poses to the community.

Some researchers have also [14] constructed a system for meme generation. They treat automatic image meme generation as a translation process. For a given input sentence, an image meme is generated by combining a meme template image and a text caption where the meme template image is selected from a set of popular candidates using a selection module, and the meme caption is generated by an encoder-decoder module.

---

1 https://fasttext.cc/docs/en/language-identification.html.

Fig. 3. Dalle meme generations (the prompt is Harry Potter meme: Severus Snape)

There has also been a considerable effort in creating Internet memes datasets. Suryawanshi et al. have leveraged memes related to the 2016 U.S. presidential election for creating a multimodal meme dataset for offensive content detection [19]. They have also trained baseline models that combined text and visual information. For textual data they have used a Stacked LSTM (long-short term memory) model [6], while for images they have fine-tuned VGG16 [16].

There has also been recently some progress in the field of image generation from text prompts. DALLE is one of such works [13]. First, it trains a variational autoencoder on images. Then, an autoregressive transformer model was trained. The autoregressive model used concatenated image and text embeddings as its input and output. So its task was to create a compressed representation of the text and image latent space. Unfortunately, the Transformer model was not released to the public and there is only a single community adoption of its results[2]. Sber's DALLE model is the most advanced implementation nowadays but even it struggles with text generations. However, it should be noted that it successfully manages to capture typical meme structures and even manages to generate characters in zero-shot manner (see Fig. 3). An important piece of work was done by Phil Wang[3] who has set up the code for standalone DALLE training.

There have also appeared works that use multi-domain Transformer models for various image and text-related tasks: including image generation and visual question answering. The most notable example is the model called RuDOLPH

---

[2] https://github.com/sberbank-ai/ru-dalle.
[3] https://github.com/lucidrains.

[15]. However, even such models struggle with text generations and require a lot off manual filtering (see Fig. 2).

This work tries to study DALLE implementation on a small domain of meme generation.

3 Dataset

We have collected a dataset of 5026 English meme images. Most of them (4900) are polycodal and contain both visual and textual information, the 1026 left images are monocodal. They have been manually collected and are of various topics that depict not only traditional Internet discussion themes (like Harry Potter), but are also concerned with modern trends (zoom, stonks memes, Trump and Biden).

Table 1. Database statistics

Class name	Polycodal memes	Monocodal memes	Total	Total after filtering	Ratio of before/after filtering (%)
Animals	734	0	734	214	29
Gym fitness health	954	62	1016	410	40
Harry Potter	691	39	730	207	28
Madagascar	99	0	99	32	32
Naruto	131	0	131	36	27
Rick and Morty	63	0	63	42	67
Russia	42	0	42	8	19
Shrek	497	16	513	216	42
Spongebob	1049	1	1050	619	59
Stonks	432	0	432	210	49
The Incredibles	104	0	104	55	53
Trump	28	0	28	5	18
Zoom	76	0	76	28	37
Total	4900	118	5018	2082	41

"Deep annotation" method has been partially used for the dataset labelling and collection process. Deep annotation method has been proposed by Rodmonga Potapova [11] during the study of multimodal polycode communication in social networks and its effects on the transformation of psychophysiological and cognitive characteristics of the personality of Internet users [10]. Deep annotation is an annotation algorithm that follows a scheme of questions and answers which consider sample modality, its paraverbal and verbal characteristics.

Then we run an OCR (optical character recognition) library Tesseract [17] on top of these images to get English texts from them. We filtered poorly recognised texts using FastText language identification library [7] to get only images with English texts and scores above 0.9^4. These two steps filtered out the majority

---

[4] https://fasttext.cc/docs/en/language-identification.html.

of images and we were left with 2082 images. All monocodal images have been removed as they do not contain texts. As can be seen from Table 1 cartoon Internet memes were most OCR-friendly.

4 Experiments

All our code is based on the open-source implementation of DALLE[5]. At first we trained a variational autoencoder (VAE) to produce a compact vector representing our image. Then using concatenated embeddings from texts and VAE we have trained an autoregressive Transformer.

4.1 Variational Autoencoder

We trained a discrete variational encoder using the all our images as they do not depend on correct text reconstructions. We have also used the same 256×256 images as the input as was done in the original DALLE paper, unlike 128×128 inputs used in its open-source implementation by default. We have trained the model for 40 epochs. All other hyperparameters are not changed from the default ones in the open-source implementation (Fig. 4).

We did not change aspect ratio of an image as we deemed it to be important for the correct representation of characters and texts. So we used centered crop to make all images have the same aspect ratio. As can be seen in Fig. 5, it has partially cropped texts in some cases.

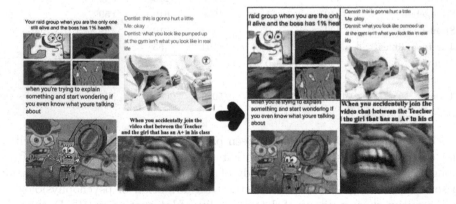

Fig. 4. Examples of VAE reconstructions

---

[5] https://github.com/lucidrains/DALLE-pytorch.

4.2 Autoregressive Transformer

As the second stage we have trained an autoregressive Transformer. We have trained it for 40 epochs. The model depth has been set to 8, the learning rate was set to 3e−4 for the first 30 epochs, then it was lowered to 3e−5. All texts longer than 80 tokens were truncated. For texts and images reconstructions cross entropy loss function has been used. We have used 8 heads with 64. Image generation results can be seen in Figure.

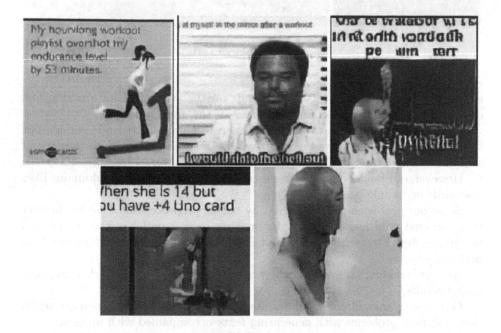

Fig. 5. Examples of generated images

As the input and target representations we have used the results of the VAE model trained at the previous step. To all Tesseract outputs we have prepended the class of the image from Table 1 and the word "meme".

As the number of data samples was not enough we have also tried using images of texts on various backgrounds for model training. Random shapes were added to these images to represent noise (see Image Fig. 6).

5 Results

We have managed to build a model that can generate meme images for a limited set of domains. Mainly we are able to generate memes with recurring characters in similar positions. Text generation results are unstable but in many cases they

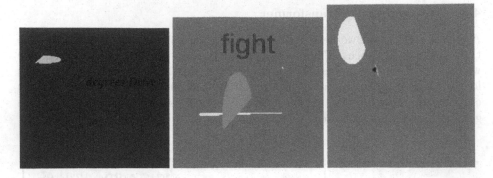

Fig. 6. Examples of texts

are correct. We suppose it mainly depends on the amount of information lost during image transformation for the images of the given type.

Our model mainly fails to properly generate human faces and animals but those who are memes "themselves". It can be mainly attributed to the lack of samples in the dataset and the diversity of faces and animal kinds.

However, our model manages to generate memes for some fixed domains likes postcards or "stonks" memes.

Some our problems may be solved by better preprocessing. Due to the fact that centered crop has been used for making images uniform, the modal had to remove first letters from texts and lines which might have deteriorated its performance.

Also pre-training on synthetically generated memes can be used to improve model results.

Our models are still noise and despite partial training on just images of texts, they still have problems with generating texts accompanied with images.

Still we get results that show that DALLE can be trained in a low resource domain.

6   Conclusion

In this paper we study the way to automate meme generation from textual prompts using variational autoencoders based on the Transformer architecture. We have used DALLE architecture for our model and trained on joint Tesseract OCR predictions and Internet meme images. We get better results for cartoon characters and drawings than for people. In future work we are going to investigate meme synthetic data and better preprocessing schemes to provide more robust results.

Acknowledgments. This research was supported by the Russian Science Foundation (RSF) according to the research project № 22-28-01050.

References

1. Antol, S., et al.: VQA: visual question answering. In: Proceedings of the IEEE International Conference on Computer Vision, pp. 2425–2433 (2015)
2. Castaño Díaz, C.M.: Defining and characterizing the concept of Internet meme. Ces Psicología **6**(2), 82–104 (2013)
3. Chen, X., et al.: Microsoft coco captions: data collection and evaluation server. arXiv preprint arXiv:1504.00325 (2015)
4. Dawkins, R., Davis, N.: The selfish gene. Macat Library (2017)
5. Devlin, J., Gupta, S., Girshick, R., Mitchell, M., Zitnick, C.L.: Exploring nearest neighbor approaches for image captioning. arXiv preprint arXiv:1505.04467 (2015)
6. Gers, F.A., Schmidhuber, J., Cummins, F.: Learning to forget: continual prediction with LSTM. Neural Comput. **12**(10), 2451–2471 (2000)
7. Joulin, A., Grave, E., Bojanowski, P., Douze, M., Jégou, H., Mikolov, T.: Fasttext.zip: compressing text classification models. arXiv preprint arXiv:1612.03651 (2016)
8. Kiela, D., et al.: The hateful memes challenge: detecting hate speech in multimodal memes. arXiv preprint arXiv:2005.04790 (2020)
9. Milner, R.M.: The world made meme: discourse and identity in participatory media. Ph.D. thesis, University of Kansas (2012)
10. Potapova, R., Dzhunkovskiy, A.: Deep polycode text annotation: language instruction technique for the tech field. Vestnik Moskovskogo gosudarstvennogo lingvisticheskogo universiteta. Obrazovanie i pedagogicheskie nauki (2(835)) (2020). (in Russian)
11. Potapova, R., Potapov, V., Dzhunkovskiy, A.: The new method of formatting and annotation of a polycodal multimodal corpus (the case of Internet social networks). Vestnik Moskovskogo gosudarstvennogo lingvisticheskogo universiteta. Gumanitarnye nauki (7(823)) (2019). (in Russian)
12. Radford, A., et al.: Learning transferable visual models from natural language supervision. arXiv preprint arXiv:2103.00020 (2021)
13. Ramesh, A., et al.: Zero-shot text-to-image generation. arXiv preprint arXiv:2102.12092 (2021)
14. Sadasivam, A., Gunasekar, K., Davulcu, H., Yang, Y.: memebot: towards automatic image meme generation. arXiv preprint arXiv:2004.14571 (2020)
15. Shonenkov, A., Konstantinov, M.: Rudolph: one hyper-modal transformer can be creative as dall-e and smart as clip (2022)
16. Simonyan, K., Zisserman, A.: Very deep convolutional networks for large-scale image recognition. arXiv preprint arXiv:1409.1556 (2014)
17. Smith, R.: An overview of the tesseract OCR engine. In: Ninth International Conference on Document Analysis and Recognition (ICDAR 2007), vol. 2, pp. 629–633. IEEE (2007)
18. Suryawanshi, S., Chakravarthi, B.R.: Findings of the shared task on troll meme classification in Tamil. In: Proceedings of the First Workshop on Speech and Language Technologies for Dravidian Languages, pp. 126–132 (2021)
19. Suryawanshi, S., Chakravarthi, B.R., Arcan, M., Buitelaar, P.: Multimodal meme dataset (multioff) for identifying offensive content in image and text. In: Proceedings of the Second Workshop on Trolling, Aggression and Cyberbullying, pp. 32–41 (2020)
20. Vaswani, A., et al.: Attention is all you need. arXiv preprint arXiv:1706.03762 (2017)

Multimodal Emotion Analysis Based on Visual, Acoustic and Linguistic Features

Leon Koren$^{(\boxtimes)}$ (iD), Tomislav Stipancic (iD), Andrija Ricko (iD), and Luka Orsag (iD)

Faculty of Mechanical Engineering and Naval Architecture, University of Zagreb, Zagreb, Croatia

`leon.koren@fsb.hr`

Abstract. In this paper, a computational reasoning framework that can interpret social signals of the person in interaction by focusing on the person's emotional state is presented. Two distinct sources of social signals are used for this study: facial and voice emotion modalities. As a part of the first modality, a Convolutional Neural Network (CNN) is used to extract and process the facial features based on live stream video. The voice emotion analysis containing two sub-modalities is driven by CNN and Long Short-Term Memory (LSTM) networks. The networks are analyzing the acoustic and linguistic features of the voice to determine the possible emotional cues of the person in interaction. Relying on the multimodal information fusion, the system then fuses data into a single hypothesis. Results of such reasoning are used to autonomously generate the robot responses which are shown in a form of non-verbal facial animations projected on the 'face' surface of the affective robot head PLEA. Built-in functionalities of the robot can provide a degree of situational embodiment, self-explainability and context-driven interaction.

Keywords: Nonverbal behavior · Multimodal interaction · Artificial intelligence · Cognitive robotics · Social signal processing

1 Introduction

Human-robot interaction is an interdisciplinary field of science aimed at studying the interaction and interaction strategies between humans and robots. One way to achieve and maintain their connection is through nonverbal communication using a two-way communication channel. An effective nonverbal communication often implies that the robot can interpret: (i) communication signals, including the emotional state of the person in communication, and (ii) information-enriched environment in which the interaction takes place [1]. In this case, the robot must constantly analyze and respond in a timely manner to environmental changes to maintain the mutual understanding. The realization of these requirements is primarily based on classical theses on human behavior and interpretation of human actions and emotions, but also on the latest knowledge about

© The Author(s), under exclusive license to Springer Nature Switzerland AG 2022
G. Meiselwitz (Ed.): HCII 2022, LNCS 13315, pp. 318–331, 2022.
https://doi.org/10.1007/978-3-031-05061-9_23

the psychology of the human mind [2]. Previous research related these issues can be classified depending on the methods used to reason about emotions and feelings. These methods can be a single modal (audio, video, text) or multi-modal. The multi modal approach combines two or more modalities to achieve results that are more appropriate to the current context [3].

The approaches based on a video modality analyze facial expressions, body movements and/or head movements. Emotions acquired from facial expressions can be determined using spatial positions of characteristic points defined by the Facial Action Coding System – FACS [4]. For example, using the FACS classification, Tarnowski et al. [5] developed a system that relies on k-NN classifiers and MLP neural networks. In contrast to the analysis of facial expressions, the connection between emotions and bodily gestures [6] is much less explored area. These methods are therefore mostly used as a supplement to the methods based on facial expressions. The latest methods use a 3D convolutional neural network to acquire data from images, as described in Poria et al. [7].

Audio modality relies on the prosodic and linguistic elements of speech. Subsequent research shows that although methods of extracting prosodic elements give reliable results, they also depend on personality of the individual. Guided by these conclusions, researchers define new methods of sound analysis to achieve objectivity [8]. To facilitate the extraction of sound features, a development framework called OpenSMILE was designed [9]. Using this framework, two different approaches have been established: (1) speaker-dependent and (2) speaker-independent approach. The speaker-dependent approach gives much better results, but it cannot be applied to many interlocutors at the same time. In these cases, a speaker-independent could be a viable choice providing a slightly lower accuracy in recognition [10]. Deep neural networks are proving to be also a desirable choice in extracting audio elements, but they are not appropriate in analyzing the temporal audio characteristics. To solve this problem, the Long Short-Term Memory (LSTM) networks [11] can be used especially when manual sample sets are analyzed. Automatic extraction of these samples, as in the video modality, is possible with Convolutional Neural Networks (CNN) [12].

The textual modality, which falls under the field of artificial intelligence known as Natural Language Processing (NLP), uses: (1) Bag of Words (BoW) methods [13], (2) large emotion lexicons [14] and (3) statistical approaches, in which the labelled data should be available [15]. The analysis of emotions in text can also be categorized into two additional groups: (1) the knowledge and (2) the statistical-based analysis. The most recent approaches are hybrid systems composed of deep learning techniques, logical thinking, and linguistics. Such systems allow computers to better understand information on Internet primarily designed for humans.

Contemporary approaches favor the multimodal approach over the single modal analysis. The results driven by many modalities are proven to be more robust and resistant to environmental changes [16]. The information fusion techniques are methods used to merge information acquired by more than one modality into a single hypothesis. The two main procedures for information fusion are: (1) fusion at the level of characteristics (early fusion) [17] and (2) fusion at the decision level (late fusion) [18]. The hybrid fusion method [19], which combines both procedures, also emerges from these two processes. Until recently, most multi-modal systems were based on a fusion of video and audio modalities, but recent research shows that adding textual modality can be useful.

Recent methods use neural networks in fusion processes, primarily deep learning methods like CNN, RNN or LTSM network architectures [20]. The deep multimodal fusion is mostly used in the early fusion. The main disadvantage of this model is inability to perceive connections within modalities [21]. This problem has been addressed by the MFN (Memory Fusion Network) algorithm that uses a special attention mechanism to process connections within modalities [22]. On the other hand, the disadvantage of this algorithm is its reliance on the attention mechanism which can be deceiving, as shown by Pruthi et al. [23]. A more reliable algorithm is MAN (Multi-attention network) which uses multiple attention algorithms and an upgraded LSTM algorithm [24]. The Deep-HOSeq algorithm (deep network with higher order Common and Unique Sequence information) provides even more reliable results [25]. This algorithm relies on the LSTM algorithm to analyze interactions between modalities and advanced network architectures in analyzing each modality. This model also uses the additional layer to perform the final fusion.

These studies show promising results in all areas, but it is important to note that most of the data sets used in testing are based on simulated situations. In real situations, people are much more spontaneous showing a more than one feeling at a time, which impairs the accuracy of the results. The constant analysis of the new data is therefore mandatory to get more reliable results. The next step is also using the interaction techniques to guide the interaction and to strengthen the current reasoning hypothesis.

2   System Design

The system is comprised of acoustic, linguistic, and visual modalities. Those modalities are combined through an algorithm for information fusion based on weighted factors. The visual modality uses the face detection algorithm ResNet 300 [26] to extract information from a live camera feed. Additionally, it uses an algorithm based on work of Savchenko [27] to perform an emotion recognition. The acoustic modality is designed to extract features from audio recordings using spectrogram and *Mel-scale Frequency Cepstral Coefficients*. A classification is performed by a pre-trained convolutional neural network. The linguistic modality is based on a hand-crafted feature extractor and the bag-of-words approach. A recognition is achieved with the Long-Short Term Memory algorithm. Additionally, external electronics is used to isolate speech from background noise and to detect with whom the algorithm is interacting (Fig. 1).

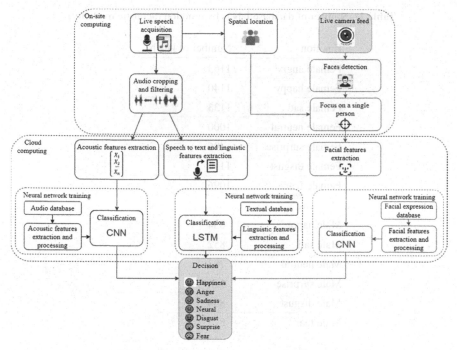

Fig. 1. Graphical representation of the system

2.1 Visual Modality

The algorithm for visual modality comprises of two neural networks, a residual neural network for face detection and a convolutional neural network for facial expression recognition. Former one extracts area around faces that is then cropped and resized for a compatible size. The image is then processed in a pre-trained model based on Savchenko [27] using a pretrained MobileNet algorithm that is fine-tuned considering the age, gender, and expression detection. Although the algorithm possesses the face detection capability and can be used as a standalone system, in this way the system can filter non-speaking interactors before performing the facial expression detection. Following this approach, the algorithm only needs to compute an expression for one face within the stream, which accelerates the overall process.

2.2 Acoustic Modality

The acoustic modality is based on an algorithm described in Koren et al. [28]. The main difference between the original algorithm and the one used within this work is in the number of emotion labels. While the original can distinguish just four of them, the one used in this work can detect seven, which is the same as for the visual modality. To achieve this, the output layer is expanded to fourteen outputs (7 emotions for both genders) and the network is retrained with an additional dataset from the open-source databases: (1) SAVEE [29], (2) CREMA-D [30], (3) RAVDNESS [31], (4) TESS [32] and (5) Emo-DB [33]. The number of labels for each emotion is shown in Table 1.

Table 1. Number of database labels by emotion (acoustic modality)

Emotion	Number of labels
Female angry	1163
Female happy	1140
Female sad	1133
Female neutral	1000
Female surprise	496
Female disgust	1131
Female fear	1096
Male angry	887
Male happy	854
Male sad	852
Male neutral	782
Male surprise	156
Male disgust	838
Male fear	827

The model validation is done with the help of the k-Fold cross validation using 10 folds. Withing the next step the best performing fold is used during the recognition process. Additionally, the validation accuracy is used inside the cross-validation procedure to grade the best performing model. To visualize the performance of the algorithm, a confusion matrix is created.

2.3 Linguistic Modality

Following the procedure applied to the acoustic modality, the process of development the new linguistic modality is also inspired with the work of Koren et al. [28].

Within the new linguistic modality, the number of labels is also increased to seven. Another difference between the acoustic and linguistic modalities is in the output layer. As it is known, the gender of a writer cannot be always detected from written sentences, so the gender is often omitted. The original algorithm is using only the Daily Dialog dataset. On contrary, the new algorithm is updated by the ISEAR dataset bringing the more precise results in recognition. The number of labels for each emotion is shown in Table 2.

Table 2. Number of database labels by emotion (linguistic modality)

Emotion	Number of labels
Angry	1550
Happy	1550
Sad	1550
Neutral	1550
Surprise	1550
Disgust	1448
Fear	1269

2.4 Information Fusion

The new algorithm is making decisions based on all three modalities together through the information fusion procedure. The algorithm is using the weighted factors in combination with the base factors in equal ration for all three modalities. All generated decisions acquired by modalities are multiplied by a given weight factor and at the end all three values are summed together. This kind of an algorithm is proven to be an effective in real-time applications where computational speed is priority. The algorithm for information fusion is shown in Table 3.

Table 3. Information fusion algorithm

1: $d(t) \leftarrow LinguisticModality(f(t))$ 2: $d(a) \leftarrow AcousticModality(f(a))$ 3: $d(v) \leftarrow VisualModality(f(v))$	*Decision making in both modalities*
3: $Procedure\ InformationFusion(T, A, V)$ 4: $for\ i\ in\ 1\ to\ N\ do:$ 5: $F(T, A, V)_i = W_t * T_i + W_a * A_i + W_v * V_i$ 6: $return\ F(T, A, V)$	*Procedure for information fusion where N is the number of labels used in sentiment recognition*
7: $Procedure\ Clasification(X)$ 8: $return\ (\text{label}(\max(X)))$	*Classification based on a result from a fusion*
9: $F \leftarrow InformationFusion(d(t), d(a), d(v))$ 10: $C \leftarrow Clasification(F)$	*Fusion Classification*

Additionally, weights can be altered programmatically through the usage of additional algorithms, which can evaluate the quality of input data depending on environmental conditions. For example, the dark room for visual modality will lessen the grade and corresponding weight [34], while the silent room with one speaker can improve the grade and weight for linguistic and acoustic modalities. Example of algorithm described in above text is shown in Table 4.

Table 4. Weight changing algorithm (room brightness)

1: $MaxValue \leftarrow Power(2,\ Bits(Img)) - 1$	*Get maximum possible value for pixel*
2: $Procedure\ GetBrightness(Img)$	
3: $for\ i\ in\ 1\ to\ Height(Img)\ do:$	*Procedure for getting*
4: $for\ k\ in\ 1\ to\ Width(Img)\ do:$	*brightness from acquired*
5: $BrightnessSum+=Value(Img(i,k))$	*image trough the average*
6: $Dimension = Height(Img) *$	*pixel value*
$Width(Img)$	
7: $return\ BrightnessSum/Dimension$	*Get brightness*
8: $Brightness \leftarrow GetBrightness(Img)$	*Procedure for normaliz-*
9: $Procedure\ NormalizeBrightness(B, MV)$	*ing the brightness*
10: $return\ abs(B/MV * 2 - 1)$	*Get normalized bright-*
11: $NormalizedBright \leftarrow$	*ness*
$NormalizedBrightness(Brightness, MaxValue)$	
12: $Procedure\ GetWight(NB, MW)$	*Procedure for calculat-*
13: $return\ MW * abs(NW - 1)$	*ing the weight of visual mo-*
	dality depending on aver-
14: $NewWeigth \leftarrow$	*age brightness of image*
$GetWeight(NormalizedBright, MaxWeight)$	
15: $W_v \leftarrow NewWeight$	*Put the new weight to al-gorithm for the information fusion*

2.5 Early Information Fusion

In addition to the information fusion method through the weighted factor late fusion, an algorithm also employs the method of early fusion based on the work of Koren et al. [35]. The system takes information from a camera and a sound localization device to create hypothesis about interacting person and to provide a response. This kind of early fusion can enable the algorithm to express a human-like interaction behavior which in turn can create better connections between the person and the system.

This approach reduces the need for computing power on centralized station because in any given moment there is only one interacting person. System is implemented on site as a plug and play solution with architecture shown on Fig. 2.

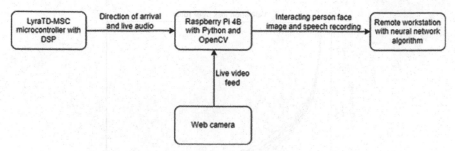

Fig. 2. System architecture for an early fusion and person localization

3 Results

Training for both, acoustic and linguistic modalities is done by using the k-fold algorithm (e.g., 10-fold algorithm). After the training, the best fold is determined and the results are analyzed with the F1-score method, the validation accuracy and the confusion matrix. The visual modality is used as a pretrained model.

3.1 Acoustic Modality

After the k-fold training of modality the validation accuracy and the loss graphs are generated. The validation loss graph is shown on Fig. 3. The results showed that all folds are similar, but the one colored in cyan achieves lowest score of 0.99. This also can be seen on the accuracy graph (Fig. 4) where fold colored in cyan achieves 64.3% score. Additionally, this fold achieves 60.7% on F1-score.

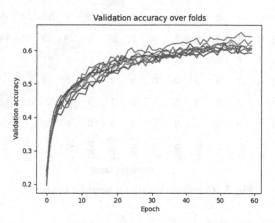

Fig. 3. Validation accuracy by folds for the acoustic modality (Color figure online)

Fold with the best score, both in accuracy and loss, is chosen for the final algorithm and based on these results the confusion matrix is generated. Because the acoustic modality is trained for both genders, the matrix is doubled in its size as seen on Fig. 5.

Validation loss over folds
Validation loss
Epoch
2.4
2.2
2.0
1.8
1.6
1.4
1.2
1.0
0
10
20
30
40
50
60

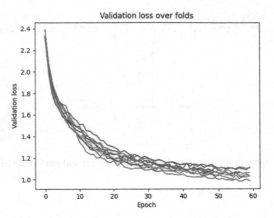

Fig. 4. Validation loss by folds for the acoustic modality (Color figure online)

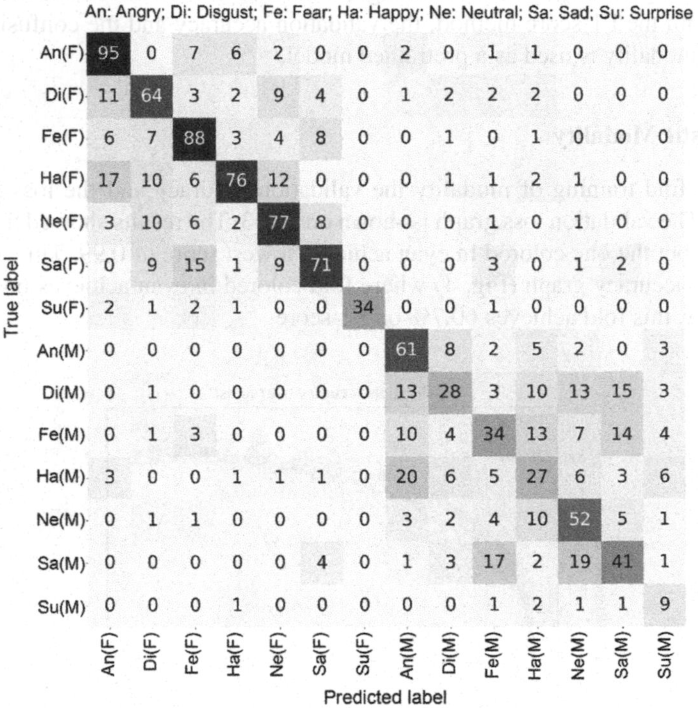

An: Angry; Di: Disgust; Fe: Fear; Ha: Happy; Ne: Neutral; Sa: Sad; Su: Surprise
True label
Predicted label
An(F) Di(F) Fe(F) Ha(F) Ne(F) Sa(F) Su(F) An(M) Di(M) Fe(M) Ha(M) Ne(M) Sa(M) Su(M)
An(F) 95 0 7 6 2 0 0 2 0 1 1 0 0 0
Di(F) 11 64 3 2 9 4 0 1 2 2 2 0 0 0
Fe(F) 6 7 88 3 4 8 0 0 1 0 1 0 0 0
Ha(F) 17 10 6 76 12 0 0 0 1 1 2 1 0 0
Ne(F) 3 10 6 2 77 8 0 0 0 1 0 1 0 0
Sa(F) 0 9 15 1 9 71 0 0 0 3 0 1 2 0
Su(F) 2 1 2 1 0 0 34 0 0 1 0 0 0 0
An(M) 0 0 0 0 0 0 0 61 8 2 5 2 0 3
Di(M) 0 1 0 0 0 0 0 13 28 3 10 13 15 3
Fe(M) 0 1 3 0 0 0 0 10 4 34 13 7 14 4
Ha(M) 3 0 0 1 1 1 0 20 6 5 27 6 3 6
Ne(M) 0 1 1 0 0 0 0 3 2 4 10 52 5 1
Sa(M) 0 0 0 0 0 4 0 1 3 17 2 19 41 1
Su(M) 0 0 0 1 0 0 0 0 0 1 2 1 1 9

Fig. 5. Confusion matrix for acoustic modality

3.2 Linguistic Modality

The K-fold training is performed during a development of the linguistic modality. By following the similar procedure as before while developing other modalities, the graphs for the validation accuracy (Fig. 6) and the loss (Fig. 7) are generated. Accuracy is higher compared to the acoustic modality (67.1%), while the loss is similar (1.02). The best fold is also colored as before in cyan on both graphs, and the F1-score is on 66%.

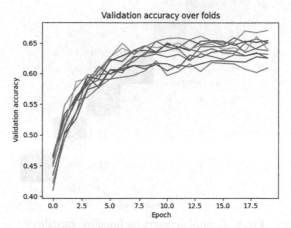

Fig. 6. Validation accuracy by folds for the linguistic modality (Color figure online)

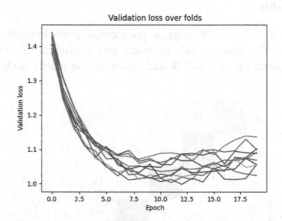

Fig. 7. Validation loss by folds for the linguistic modality (Color figure online)

Same as for the acoustic modality, the best fold is chosen as a final algorithm and the confusion matrix is generated, as shown in Fig. 8.

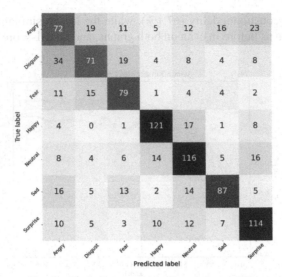

Fig. 8. Confusion matrix for linguistic modality

3.3 Visual Modality

The visual modality is created using a pre-trained convolutional neural network, MobileNet v7. For that reason, only accuracy and confusion matrix are given. The accuracy on the validation set is 64.7% and the confusion matrix is shown at Fig. 9.

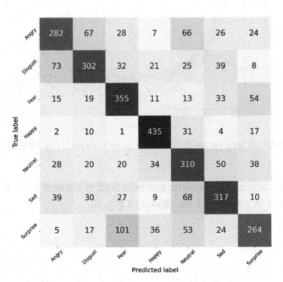

Fig. 9. Confusion matrix for visual modality

4 Discussion and Conclusion

The man goal of this work was to develop a framework for the social robot PLEA using three weak classifiers. The final system behavior is achieved through the process of information fusion. The additional goal was to present diverse types of information fusion techniques with the help of the person localization model and the weighted factor late fusion.

A framework that is developed will be used for evaluation of more complex modalities and fusion algorithms in the future, as well to collect and pre-annotate the data collected during the interaction of the robot PLEA with people. Furthermore, within the next step this algorithm will be evaluated through the real-world tests with students where the interaction experiences of the students will be gathered through questionaries. The final conclusions will be achieved after the thorough statistical analysis.

Analyzing the results, similar problems as described in [28] can be recognized. The main difference is in the fact that the previous system works with four emotions (three + neutral) which is a smaller number compared to the number of emotions used in this work (six + neutral).

In this work, the neutral emotion happens less frequent by the system because of higher granularity in available emotions. The neutral emotion represents those cases in which the system does not detects the emotional state of the person. The "not detected" case is therefore specific for transient cases when the person is changing the face expression from one emotion to the other.

Within the future work, individual modalities will be additionally improved by using more optimized data. Additional improvements are also expected in the information fusion procedure, in which some context-aware approach based on probability calculus will be applied to ensure adaptability of the system under the conditions of uncertainties [36]. A development of the methodology will also move forward toward the research on new interaction strategies where the system is going to be able to strengthen or weaken the current reasoning hypothesis by asking questions or by provoking the person in interaction to provide more information about the current environmental conditions in which this person is seen as a part of the environment.

Acknowledgements. This work has been supported in part by the Croatian Science Foundation under the project "Affective Multimodal Interaction based on Constructed Robot Cognition—AMICORC (UIP-2020-02-7184)."

References

1. Stipancic, T., Koren, L., Korade, D., Rosenberg, D.: PLEA: a social robot with teaching and interacting capabilities. J. Pac. Rim Psychol. **15** (2021). https://doi.org/10.1177/183449092 11037019
2. Barrett, L.F.: How Emotions are Made: The Secret Life of the Brain (2017)
3. Stipancic, T., Rosenberg, D., Nishida, T., Jerbic, B.: Context driven model for simulating human perception – a design perspective. In: Design Computing and Cognition DCC 2016 (2016)

4. Wathan, J., Burrows, A.M., Waller, B.M., McComb, K.: EquiFACS: the equine facial action coding system. PLoS ONE (2015). https://doi.org/10.1371/journal.pone.0131738
5. Tarnowski, P., Kolodziej, M., Majkowski, A., Rak, R.J.: Emotion recognition using facial expressions. In: International Conference on Computational Science (2017). https://doi.org/10.1016/j.procs.2017.05.025
6. Melzer, A., Shafir, T., Tsachor, R.P.: How do we recognize emotion from movement? Specific motor components contribute to the recognition of each emotion. Front. Psychol. **10** (2019). https://doi.org/10.3389/fpsyg.2019.01389
7. Poria, S., Chaturvedi, I., Cambria, E., Hussain, A.: Convolutional MKL based multimodal emotion recognition and sentiment analysis. In: 16th IEEE International Conference on Data Mining (ICDM), Barcelona (2016). https://doi.org/10.1109/ICDM.2016.0055
8. Koolagudi, S.G., Murthy, Y.V.S., Bhaskar, S.P.: Choice of a classifier, based on properties of a dataset: case study-speech emotion recognition. Int. J. Speech Technol. **21**(1), 167–183 (2018). https://doi.org/10.1007/s10772-018-9495-8
9. Eyben, F., Schuller, B., openSMILE:): the Munich open-source large-scale multimedia feature extractor. ACM SIGMultimedia Rec. **6**, 4–13 (2015). https://doi.org/10.1145/2729095.2729097
10. Swain, M., Routray, A., Kabisatpathy, P.: Databases, features and classifiers for speech emotion recognition: a review. Int. J. Speech Technol. **21**(1), 93–120 (2018). https://doi.org/10.1007/s10772-018-9491-z
11. Zhao, J., Mao, X., Chen, L.: Speech emotion recognition using deep 1D & 2D CNN LSTM networks. Biomed. Sig. Process. Control **47**, 312–323 (2019). https://doi.org/10.1016/j.bspc.2018.08.035
12. Anand, N.: Convolutional and recurrent nets for detecting emotion from audio data. Convoluted Feelings (2015)
13. Savigny, J., Purwarianti, A.: Emotion classification on Youtube comments using word embedding. In: 2017 International Conference on Advanced Informatics, Concepts, Theory, and Applications (2017). https://doi.org/10.1109/ICAICTA.2017.8090986
14. Bandhakavi, A., Wiratunga, N., Padmanabhan, D., Massie, S.: Lexicon based feature extraction for emotion text classification. Pattern Recogn. Lett. **93**, 133–142 (2017). https://doi.org/10.1016/j.patrec.2016.12.009
15. Oneto, L., Bisio, F., Cambria, E., Anguita, D.: Statistical learning theory and ELM for big social data analysis. IEEE Comput. Intell. Mag. **11**(3), 45–55 (2016). https://doi.org/10.1109/MCI.2016.2572540
16. Poria, S., Cambria, E., Bajpai, R., Hussain, A.: A review of affective computing: from unimodal analysis to multimodal fusion. Inf. Fusion **37**, 98–125 (2017). https://doi.org/10.1016/j.inffus.2017.02.003
17. Poria, S., Cambria, E., Hussain, A., Huang, G.B.: Towards an intelligent framework for multimodal affective data analysis. Neural Netw. **63**, 104–116 (2015). https://doi.org/10.1016/j.neunet.2014.10.005
18. Li, J., Qiu, T., Wen, C., Xie, K., Wen, F.Q.: Robust face recognition using the deep C2D-CNN model based on decision-level fusion. Sensors **18**(7) (2018). https://doi.org/10.3390/s18072080
19. Amer, M.R., Shields, T., Siddiquie, B., Tamrakar, A., Divakaran, A., Chai, S.: Deep multimodal fusion: a hybrid approach. Int. J. Comput. Vision **126**(2–4), 440–456 (2017). https://doi.org/10.1007/s11263-017-0997-7
20. Gao, J., Li, P., Chen, Z., Zhang, J.: A survey on deep learning for multimodal data fusion. Neural Comput. **32**(5), 829–864 (2020). https://doi.org/10.1162/neco_a_01273
21. Zhu, H., Wang, Z., Shi, Y., Hua, Y., Xu, G., Deng, L.: Multimodal fusion method based on self-attention mechanism. Wirel. Commun. Mob. Comput. (2020). https://doi.org/10.1155/2020/8843186

22. Zadeh, A., Liang, P.P., Mazumder, N., Poria, S., Cambria, E., Morency, L.P.: Memory fusion network for multi-view sequential learning. In: Thirty-Second AAAI Conference on Artificial Intelligence, 5634–5641 (2018). arXiv:1802.00927

23. Pruthi, D., Gupta, M., Dhingra, B., Neubig, G., Lipton, Z.C.: Learning to deceive with attention-based explanations. ACL (2020). https://doi.org/10.18653/v1/2020.acl-main.432

24. Xu, Q., Zhu, L., Dai, T., Yan, C.: Aspect-based sentiment classification with multi-attention network. Neurocomputing **388**, 135–143 (2020). https://doi.org/10.1016/j.neucom.2020.01.024

25. Verma, S., et al.: Deep-HOSeq: deep higher order sequence fusion for multimodal sentiment analysis. In: 2020 IEEE International Conference on Data Mining (2020). https://doi.org/10.1109/ICDM50108.2020.00065

26. Karandeep, S.G., Aleksandr, R.: Face Detection OpenCV (2021). https://github.com/grover kds/face_detection_opencv. Accessed 31 Jan 2022

27. Savchenko, A.V.: Facial expression and attributes recognition based on multi-task learning of lightweight neural networks. In: IEEE 19th International Symposium on Intelligent Systems and Informatics, Subotica (2021). https://doi.org/10.1109/SISY52375.2021.9582508

28. Koren, L., Stipancic, T.: Multimodal emotion analysis based on acoustic and linguistic features of the voice. In: Meiselwitz, G. (ed.) HCII 2021. LNCS, vol. 12774, pp. 301–311. Springer, Cham (2021). https://doi.org/10.1007/978-3-030-77626-8_20

29. Haq, S., Jackson, P.: Speaker-dependent audio-visual emotion recognition. In: AVSP (2009)

30. Cao, H.W., Cooper, D.G., Keutmann, M.K., Gur, R.C., Nenkova, A., Verma, R.: CREMA-D: crowd-sourced emotional multimodal actors dataset. IEEE Trans. Affect. Comput. 377–390 (2014). https://doi.org/10.1109/TAFFC.2014.2336244

31. Livingstone, S.R., Russo, F.A.: The Ryerson audio-visual database of emotional speech and song (RAVDESS): a dynamic, multimodal set of facial and vocal expressions in North American English. PLoS ONE (2018). https://doi.org/10.1371/journal.pone.0196391

32. Pichora-Fuller, M.K., Dupuis, K.: Toronto Emotional Speech Set (TESS). Toronto Emotional Speech Set (TESS), Toronto (2020). https://doi.org/10.5683/SP2/E8H2MF

33. Burkhardt, F., Paeschke, A., Rolfes, M., Sendlmeier, W., Weiss, B.: A database of German emotional speech. In: 9th European Conference on Speech Communication and Technology, Lisabon, pp. 1517–1520 (2005)

34. Stipancic, T., Jerbic, B.: Self-adaptive vision system. In: CamarinhaMatos, L.M., Pereira, P., Ribeiro, L. (eds.) DoCEIS 2010. IAICT, vol. 314, pp. 195–202. Springer, Heidelberg (2010). https://doi.org/10.1007/978-3-642-11628-5_21

35. Koren, L., Stipancic, T., Ricko, A., Orsag, L.: Person localization model based on a fusion of acoustic and visual inputs. Electronics (2022). https://doi.org/10.3390/electronics11030440

36. Stipančić, T., Jerbić, B., Ćurković, P.: Bayesian approach to robot group control. In: International Conference in Electrical Engineering and Intelligent Systems, London (2012). https://doi.org/10.1007/978-1-4614-2317-1_9

Linguistic and Contextual Analysis of SNS Posts for Approval Desire

Erina Murata[1] , Kiichi Tago[2] , and Qun Jin[3]($\boxtimes$)

[1] School of Human Sciences, Waseda University, Tokorozawa, Japan
erinamurata@ruri.waseda.jp
[2] Department of Information and Network Science, Chiba Institute of Technology, Narashino, Japan
tago@net.it-chiba.ac.jp
[3] Faculty of Human Sciences, Waseda University, Tokorozawa, Japan
jin@waseda.jp

Abstract. In recent years, SNS has become a service that everyone uses. In this study, we analyze Twitter, one of the most popular SNSs, which allows users to post their daily events and feelings within 140 characters and is used by people all over the world. In this study, we investigate the relationship between SNS posts and latent approval needs. The linguistic features of tweets and their contextual features are analyzed using information such as the frequency of posts and the number of characters in tweets, and the degree of desire for approval is defined and quantified based on the results of the analysis of tweets. The experiment results show that the agreement between the naïve Bayes classifier and human ratings was about 60%. It is found that users with a high percentage of posts for approval desire tend to post less frequently and with a higher average number of characters. This indicates that it may be because these users post for approval desire when it is important or when they really want to say something.

Keywords: Approval desire · Internet psychology · SNS · Twitter · Naïve Bayes · Text mining

1 Introduction

In recent years, many people have been using social networking services (SNSs), and there are many services available as SNS platforms, such as Twitter, Instagram, and Facebook. In this study, we focus on tweets on Twitter, which is a system for posting daily events and feelings in 140 characters or less. On the other hand, people who view tweets can access them just as easily as those who post them. If we look at the posted tweets, we can see that there are many different types of tweets, such as conversations with friends or tweets that are posted for the people around to recognize them.

In this study, the relationship between SNS posts and latent desire for approval will be explored. The linguistic features of SNS posts (tweets, posts in Twitter) and their contextual features will be analyzed in terms of post frequency and the number

G. Meiselwitz (Ed.): HCII 2022, LNCS 13315, pp. 332–344, 2022.
https://doi.org/10.1007/978-3-031-05061-9_24

of characters in a tweet, and the degree of desire for approval will be investigated and quantified. Then, we analyze and discuss what characteristics a user with high or low approval needs may have.

First, tweets are collected and a small portion of them are used by a human evaluator to make a judgment. The resulting labeled data set is used to build a machine learning model. Next, the same evaluator further evaluates a small subset obtained from the machine learning results. This improves the accuracy of the machine learning model and allows us to assess and determine the desire for approval of a post. Furthermore, statistical tests will be conducted to verify the effectiveness of the proposed language and context analysis and approval desire quantification method. Through this research, we try to establish a computational method for determining whether a post is seeking approval from peers in SNS.

This paper is organized as follows. Section 2 overviews related work and discusses the position of this study. Section 3 describes the methods and details experiment design of this study. Section 4 shows the experiment results and discusses on the results. Finally, in Sect. 5, the conclusion of this study and future works are given.

2 Related Work

2.1 SNS and the Desire for Approval

The desire for approval is one's desire to be recognized by others and to recognize oneself as valuable, also known as the desire for respect and self-esteem. There are more and more studies on the influence of social media on human self-esteem and approval desire in recent years.

A study by Kato [1] shows that some people use functions such as "Like" more than necessary to create a good impression of themselves to others, which suggests that there may be a relationship between the need for approval and the use of SNS. People who "like" other people's posts are more likely to like their own posts and expect others to "like" their posts. Another study by Kanoh [2] found that those with a high desire for approval are more likely to use functions, such as "Retweet" and "Like", which are specially implemented for approval desire.

A more recent study by Valkenburg et al. [3] concluded that for adolescents whose self-esteem is particularly dependent on peer approval, the experience of positive values in social media can influence their self-esteem relatively easily. In addition, a study by Nemoto et al. [4] states that a major reason why people use SNS is because the desire to be approved by others and the desire to connect with others are closely related. They state that there is a strong affinity between SNS and the desire for approval. They also cite emoji, decorative characters, hashtags, and URLs as factors that may pose challenges when using SNS to gather information.

2.2 Analysis of SNS and Users

J. Kim et al. [5] showed that SNS is composed of three elements: individuals, emergency agencies, and organizations, in terms of SNS users. They also stated that the core of an

SNS consists of many individuals. Social networking sites have been useful for analysis in various fields. In recent years, there has been an increase in research on social media and COVID-19. Some studies have focused on the role of public key persons in the social networks of COVID-19. S. Yum [6] found that governments need to understand the characteristics of public key persons to properly inform COVID-19. They stated that topic-based and person-based networks were found to play different roles in social networks. W. Ahmed et al. [7] conducted a study analyzing 5G conspiracy theories in the context of COVID-19 on Twitter and stated that a combination of rapid and targeted interventions is important to reduce the influence of fake news. They also stated that many social media platforms offer users the function to report inappropriate content, which should be utilized.

In another study on the use of SNS in the field of education, M. Saqr et al. [8] found that social network analysis can improve understanding of the process of collaboration, predict the under-achievers by means of learning analytics, and clarify the division of roles between learners and teachers. By using SNS analysis indicators, they were able to classify students according to their academic ability with high accuracy. This shows that the use of SNS is effective in the field of education.

On how to accurately assess the reliability of information provided on the Internet, N. Jing et al. [9] used an approach using Tree Augmented Naïve (TAN) Bayes Classifier and the PageRank algorithm to evaluate the information reliability of user profiles in online professional social networks. After comparing the two classification approaches, they stated that the integrated approach of TAN Bayes Classifier and PageRank algorithm performed better than the TAN Bayes Classifier.

Samsir et al. [10] implemented text mining and document-based sentiment on Twitter data using a naïve Bayesian machine learning approach, and showed that expectation, sadness, and anger are very dominant in sentiment analysis. K. Tago et al. [11] used sentiment scores determined by naïve Bayes, by which defining groups with high scores as positive groups and groups with low scores as negative groups. They further conducted statistical tests to see if there were differences in scores among users.

2.3   Position of This Research

However, most of the research on approval desire focused on the psychological side and used questionnaire survey and statistical analysis method. Few works on analysis of relevance between the approval desire and the content itself and its context, such as the intention and frequency of a post in social media, have been done, using a computational approach based on data analysis.

In this study, we use machine learning methods to conduct linguistic and contextual analysis of tweets that make people feel the need for approval. In addition, by introducing a degree index, Degree of Approval Desire (DAD), regarding the presence or absence of the desire for approval, we further use text mining and statistical analysis methods to examine what linguistic and contextual characteristics exist between users with high and low degrees of approval desire.

3 Methods and Experiment Design

3.1 Overview of the Framework

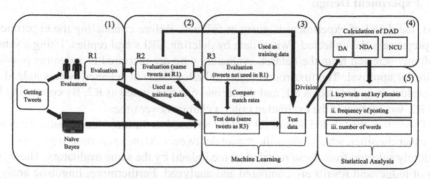

Fig. 1. Overview of the framework

The framework of our proposed analysis method and experiment design is shown in Fig. 1. The descriptions and procedures are detailed as follows.

1. Posts are collected from social media public accounts. A small set of posts is randomly selected and judged by human evaluators as "Desire for approval (DA) ", "No desire for approval (NDA)", or "No clear/unknown (NCU)".
2. In consideration of the fact that the judgement discrepancy may occur over time, we design the experiment that a subset of the same data is evaluated again by the same evaluators after a certain period.
3. The results of these two evaluations are used before and after as training data to build a machine learning model. To assess the results judged by the machine learning, we randomly select a small set of posts judged by the model, let the evaluators judge, and compare these two sets of results.
4. Meanwhile, we analyze the linguistic and contextual features of posts labeled as "Desire for approval (DA) "and "No desire for approval (NDA)" by a text mining tool. We further define the degree of approval desire (DAD) for a specific social media user, which is expressed by Eq. (1) in terms of the number of DA and NDA posts in a certain period.

$$DAD = DA/(DA + NDA) \tag{1}$$

5. Based on the degree of approval desire, we determine two groups of users with high or low desire for approval. Thereafter, we test the following three hypotheses to see if there is a difference by a statistics method.

   i. There is a difference in the number of keywords and key phrases that feature the desire for approval between the high and low groups.
   ii. There is a difference in the frequency of posting between the high and low groups.

iii. There is a difference in the number of post characters between the high and low groups.

3.2 Experiment Design

Next, the detail of experiment is given as follows. Before conducting the experiment, we preprocess the collected Twitter data by deleting URLs and replies. Using a subset randomly selected from the dataset, evaluators who are recruited to judge the possible desire for approval. The first-round result (labelled data) of selected tweets judged by the evaluators is denoted as R1, and the second one is denoted as R2. By comparing R1 and R2, we confirm if the evaluation criteria changes over time.

Thereafter, the labeled data of R1 and R2 are used as training data to build the Naïve Bayesian classifier, which judges the remainder tweets. At the same time, a part of tweets randomly selected from these remainders are judged by the same evaluators. These two sets of judgement results are compared and analyzed. Furthermore, linguistic analysis and network analysis based on text data analytics, such as co-occurrence network, are applied to extract and analyze the linguistic and contextual features.

Finally, the degree of approval desire for each of the Twitter users is calculated, which is used to determine the two groups of high and low desire. In this experiment, referring to the method used in [11], the top five of all users is regarded as the high group, while the bottom five as the low group. For these two groups, three hypotheses described above are tested by statistical methods.

4   Experiment Results and Discussions

4.1   Data Acquisition and Survey Results

To collect data from Twitter, we set the following conditions and collected tweets from 20 users.

- Language: Japanese
- Posted more than 1,000 tweets in the period from the account creation to the date of data collection
- Period: Nov. 25, 2015 – Oct. 3, 2021
- Number of tweets collected: Approximately 400 tweets per person

According to these conditions, we first collected 7116 tweets. Duplicates were then removed, resulting in 7087 tweets. We randomly selected 600 tweets from the 7087 tweets as training data for the model to judge the desire for approval.

Then, to determine the posters' desire for approval from the tweets, we conducted an evaluation to classify them into three categories: desire for approval (DA), no desire for approval (NDA), and not clear or unknown (NCU). In the evaluation, the same four evaluators evaluated the selected tweets twice over a period. If the evaluations resulted in a tie by majority result, no judgment could be made about the presence or absence of the desire for approval, so it was removed from the results. The number of evaluations

Table 1. Judgement results by the number and percentage per evaluator (R1)

Number (%)	Evaluator A	Evaluator B	Evaluator C	Evaluator D	Majority Result
DA	175 (29%)	141 (24%)	181 (30%)	222 (37%)	160 (27%)
NDA	308 (51%)	326 (54%)	150 (25%)	117 (20%)	176 (29%)
NCU	117 (20%)	133 (22%)	269 (45%)	261 (44%)	118 (20%)
Total	600 (100%)	600 (100%)	600 (100%)	600 (100%)	454 (76%)

Table 2. Judgement results by the number and percentage per evaluator (R2)

Number (%)	Evaluator A	Evaluator B	Evaluator C	Evaluator D	Majority result
DA	174 (29%)	108 (18%)	125 (21%)	199 (33%)	135 (23%)
NDA	335 (56%)	370 (62%)	158 (26%)	146 (24%)	202 (34%)
NCU	91 (15%)	122 (20%)	317 (53%)	255 (43%)	87 (15%)
Total	600 (100%)	600 (100%)	600 (100%)	600 (100%)	424 (71%)

and their percentages for each evaluator are shown in Table 1 for the first evaluation and Table 2 for the second evaluation.

The difference in the number of evaluations between the first and second times is shown in Table 3.

For all evaluators, the number of responses for NDA increased for the second evaluation. On the other hand, the number of DA responses decreased significantly for all evaluators. The smallest discrepancy in responses was observed for Evaluator A, and the largest for Evaluator C.

4.2 Evaluation of Classification Results by Naïve Bayesian Classifier and Five-Cross Validation

Using the results in Table 1 and Table 2 as training data, a machine learning model was constructed. In this study, we used a naïve Bayesian classifier. The test data was 600 new randomly selected tweets. The training data was constructed by each of the following six methods to train the naïve Bayesian classifier.

1. R1 majority result

Table 3. Difference in the number of responses and their percentages for each evaluator between the first and second evaluations (R2-R1)

Number (%)	Evaluator A	Evaluator B	Evaluator C	Evaluator D	Majority Result
DA	− 1 (0%)	− 33 (−6%)	− 56 (−9%)	− 23 (−4%)	− 25 (−4%)
NDA	+ 27 (+5%)	+ 44 (+7%)	+ 8 (+1%)	+ 29 (+5%)	+ 26 (+4%)
NCU	− 26 (−4%)	− 11 (−2%)	− 48 (−8%)	− 6 (−1%)	− 31 (−5%)
Total Change	54 (9%)	88 (15%)	112 (19%)	58 (10%)	82 (14%)

2. R2 majority result
3. Both R1 and R2 majority results
4. R1 results of all four evaluators
5. R2 results of all four evaluators
6. Both R1 and R2 results of all four evaluators

The evaluator evaluated the 600 tweets used as test data and compared the results of the naïve Bayes classifier with the evaluator's results. Here the data obtained by the evaluator's results is denoted as R3. However, we excluded tweets whose evaluations were tied by majority result (e.g., two for DA and two for NDA). The number of tweets excluded in this study was 117 (20%).

When we constructed a naïve Bayesian classifier for each of the above 1 to 6, the highest rate of agreement between the classification result and R3 result was "R2 majority result", which was 63%.

Table 4 shows a summary of the classification results of the naïve Bayesian classifier constructed using R2 majority result as the training data and the evaluation results of R3.

Table 4. Classification results by classifier using R2 majority result as training data and R3 results.

		Human Evaluation (R3)			
		DA	NDA	NCU	Total
Naïve Bayes	DA	53 (11%)	57 (12%)	31 (6%)	141 (29%)
	NDA	25 (5%)	241 (50%)	39 (8%)	305 (63%)
	NCU	6 (1%)	22 (5%)	9 (2%)	37 (8%)
	Total	84 (17%)	320 (66%)	79 (16%)	483 (100%)

The results were closer to the human evaluation when the second time was used as the training data for the naïve Bayesian classifier than the first time. The reason for this may be that evaluators have become accustomed to the evaluation and their standards have been established.

Next, to check the accuracy of the naïve Bayesian classifier, we constructed a naïve Bayesian classifier using the majority results of both R2 and R3 as training data and conducted a five-cross validation. However, 293 tweets (24%) that could not be judged by majority result were excluded. As a result, the mean of the agreement rate was 0.60.

Finally, using the majority result of R2 and R3 as the train data, 5887 tweets that were not used as training data in the previous experiments were classified using a naïve Bayesian classifier. The training data included 176 tweets from R2 and 117 tweets from R3 that could not be determined by the majority result. For those tweets that could not be determined by the majority result (e.g., two labeled DA and two labeled NDA), we set them both labels. The results of the classification are shown in Table 5.

Table 5. Classification results of tweets

	DA	NDA	NCU	Total
Number (%)	846 (14%)	4257 (72%)	784 (13%)	5887 (100%)

4.3 Validation by Statistical Analysis Methods

To analyze the linguistic and contextual features of posts labeled "Desired Approval (DA)" and "Not Desired Approval (NDA)," we use the degree of approval desire (DAD) defined in Eq. (1). The closer the value is to 0, the less likely they are to post tweets that are considered approval-seeking, and the closer it is to 1, the more likely they are to tweet about approval-seeking. The results of DAD for the top five and bottom five users are shown in Table 6.

Table 6. DAD of the top five and bottom five

Top Five	DAD	Bottom Five	DAD
1	0.387	16	0.077
2	0.381	17	0.075
3	0.352	18	0.074
4	0.292	19	0.050
5	0.291	20	0.042

To test whether there is a difference in the number of keywords and key phrases that characterize the desire for approval between groups with high and low DAD, we conducted co-occurrence networks for each of the groups with the top five and bottom

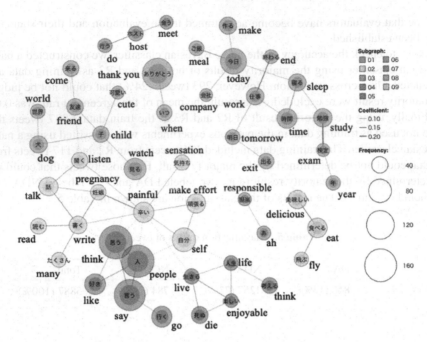

Fig. 2. Co-occurrence network for the top five users

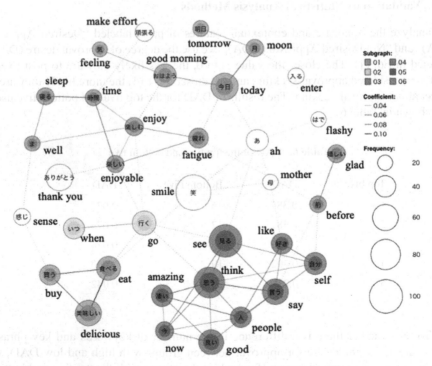

Fig. 3. Co-occurrence network for the bottom five users

five DAD. The co-occurrence networks for the top five and bottom five users are shown in Fig. 2 and Fig. 3, respectively.

From the figures, we see that the group with higher DAD often used words that would attract the sympathy of others, such as "painful" and "make effort," as well as words related to home and family, such as "pregnancy," "life," and "die." On the other hand, the group with lower DAD posted more greetings, such as "good morning" and "smile " added to the end of a post.

To verify if there is a difference in posting frequency between the groups with high and low DAD. The posting frequency indicates the number of posts per day and is defined by Eq. (2).

$$\text{Posting frequency} = \text{Number of days between the first and last post} \qquad (2)$$

The average frequency of posting for the top five was 3.86, while the average for the bottom five was 26.06. This result implies that users with a high percentage of tweets indicating a desire for approval post less frequently. To further validate the results, a two-tailed t-test without correspondence was conducted. This is an analysis to verify if there is a significant difference between the numbers obtained in the two groups, and if the p-value is lower than 0.05, it can be considered a significant difference. As a result of verifying the posting frequency for each of the five top and bottom users, the p-value was 0.12, and it could not be determined that there was a significant difference. The results of the regression analysis are shown in Fig. 4. The frequency tended to be higher for users who did not post many tweets indicating their desire for approval. The coefficient of determination in this analysis was 0.21.

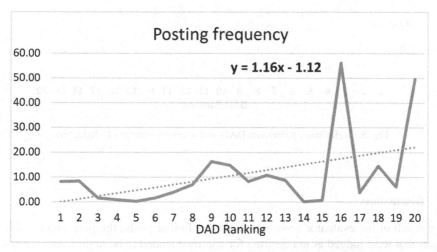

Fig. 4. Relationship between DAD and posting frequency

From this result, we can conclude that the percentage of tweets that show the desire for approval will be low because people who post frequently tend to post less important tweets such as greeting "good morning". On the other hand, for those who post less frequently, they post when it's important or when they really want to say something. Therefore, the percentage of these tweets that show the desire for approval will increase.

Finally, to test if there is a difference in the number of characters in the posts between the groups with high and low DAD. The mean of character number was 49.6 for the top five group with the higher DAD and 30.8 for the bottom group with the lower DAD. The median was 42.8 for the top five and 25.1 for the bottom five. From the above results, it was clear that the users who tweeted less for the approval desire had fewer characters, while the users who tweeted more for the approval desire had more characters. In addition, a two-tailed t-test with no correspondence showed that the mean number of characters was 0.10 and the median was 0.17, either of which was not significantly different. Since t-tests are less likely to show significant differences when there is less data, it is possible that more detailed clarification can be achieved by increasing the number of users to be analyzed.

The results of the regression analysis using DAD and the average number of characters are shown in Fig. 5. From the results, it is likely that the posts from users with a higher percentage of tweets indicating the desire approval tend to have more characters. The coefficient of determination for this analysis was 0.11.

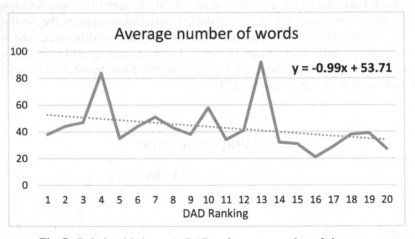

Fig. 5. Relationship between DAD and average number of characters

4.4 Discussions

As a result of the evaluator's assessment of the Twitter posts, the percentage of NCU tweets that were judged as not desiring for approval tended to be higher. And the result was closer to the human evaluation (R3) when the second result (R2) was used as the training data for the naïve Bayesian classifier than the first result (R1). The reason for this may be that the evaluators became accustomed to the evaluation and settled on a standard. It was also found that users whose tweets indicating the desire for approval posted less frequently. Furthermore, such users tweeted more characters.

In this study, we introduced and defined DAD, the degree of approval desire, to show how much a user tends to post tweets that implies the desire for approval. However, we have not been able to quantify the strength of the desire for approval in a single tweet. While some tweets may make everyone feel the desire for approval, different people may have different evaluation standards. In the future, we will further develop this study to quantify the degree of desire for approval from the perspective of tweet content.

Evaluating the desire for approval involves not only short-term criteria, such as the timing of the evaluation and temporary feelings, but also long-term values, such as personality and outlook on life. In addition, even when looking at the tweets of a single user, sometimes there are fewer tweets indicating the desire for approval and some other times there are more tweets indicating the approval desire. Therefore, it is difficult to accurately grasp whether a tweeter is seeking the degree of approval, because the evaluator's criteria may change over time. Judging users based on their temporary nature is probably one of the most misleading factors in understanding the characteristics of Twitter users.

5 Conclusion

In this study, we analyzed the linguistic and contextual features of tweets indicating the desire for approval, focusing on the frequency of posts and the number of characters in tweets. Evaluators rated whether the tweets showed the desire for approval, and a model was constructed using train data for a naïve Bayesian classifier to discriminate the presence or absence of the desire for approval. The experiment results show that the agreement between the created naïve Bayes classifier and human evaluations was about 60%. Furthermore, we explored the relationship between users with a high desire for approval in SNS and the characteristics of their tweets.

In addition, from the experiment results, it is found that users with a high percentage of posts for approval desire tend to post less frequently and with a higher average number of characters. This implies that such users post for approval desire when it is important or when they really want to say something on SNS.

As for future work, we plan to improve the experiment with the framework proposed in this study using a larger social media dataset. In addition, we will try to grasp and clarify the characteristics of tweets indicating the desire for approval by conducting a dependency analysis.

References

1. Kato, C.: SNS use of young people as seen from praise seeking need and rejection avoidance need. Memoirs Hokuriku Gakuin Univ. 7, 315–323 (2014). (in Japanese)
2. Kanoh, H.: The relationship between approval desire and social media use. Res. Rep. Inform. Educ. 1, 18–23 (2019). (in Japanese)
3. Valkenburg, P.M., Pouwels, J.L., Beyens, I., van Driel, I.I., Keijsers, L.: Adolescents' social media experiences and their self-esteem: a person-specific susceptibility perspective. Technol. Mind. Behav. 2(2) (2021)
4. Nemoto, T., Fujimoto, T.: The mechanism of approval seeking posting guided by present SNS analysis. Inf. Eng. Expr. Int. Inst. Appl. Inform. 5(2), 36–45 (2019)

5. Kim, J., Hastak, M.: Social network analysis: characteristics of online social networks after a disaster. Int. J. Inf. Manage. **38**(1), 86–96 (2018)
6. Yum, S.: Social network analysis for coronavirus (COVID-19) in the United States. Soc. Sci. Q. **101**(4), 1642–1647 (2020)
7. Ahmed, W., Vidal-Alaball, J., Downing, J., Seguí, F.L.: COVID-19 and the 5G conspiracy theory: social network analysis of twitter data. J. Med. Internet Res. **22**(5), e19458 (2020)
8. Saqr, M., Fors, U., Nouri, J.: Using social network analysis to understand online problem-based learning and predict performance. PLoS ONE **13**(9), e0203590 (2018)
9. Jing, N., Wu, Z., Lyu, S., Sugumaran, V.: Information credibility evaluation in online professional social network using tree augmented naïve Bayes classifier. Electron. Commer. Res. **21**(2), 645–669 (2019). https://doi.org/10.1007/s10660-019-09387-y
10. Samsir, et al.: Naives Bayes algorithm for twitter sentiment analysis. J. Phys. Conf. Ser. **1933**(1), 012019 (2021)
11. Tago, K., Takagi, K., Kasuya, S., Jin, Q.: Analyzing influence of emotional tweets on user relationships using Naive Bayes and dependency parsing. World Wide Web **22**(3), 1263–1278 (2018). https://doi.org/10.1007/s11280-018-0587-9

Social Media Engagement Anxiety: Triggers in News Agenda

Kamilla Nigmatullina(✉) 🆔 and Nikolay Rodossky 🆔

Saint Petersburg State University, 7/9 Universitetskaya Emb, 199034 Saint Petersburg, Russian Federation
k.nigmatulina@spbu.ru

Abstract. In this study, we focus on decrease in public activity of news media audiences on social networks and on possible media strategies of "comfortable involvement" of socially anxious people into commenting. We draw upon cognitive research that describes a vicious circle of user experience. The environment for the anxiety scenarios in relation to user comments creates favorable conditions for the growth of digital escapism. We hypothesize that: 1) Users who are looking for getting rid of anxiety on social networks are drawn into an even greater spiral of anxiety, interacting with other people in the context of news stories that provoke the whipping up of negative emotions; 2) The dynamics of the cascade of messages can depend on the characteristics of the emotions embedded in the message and its context.

We took 10 Russian regional media and their content on the VKontakte social network published from November 2020 to November 2021 and processed all commented posts in each media. Frequency analysis of messages demonstrated no specific pattern for anxiety. Number of comments and likes on comments correlated with specific regional issues and local news agenda more than with coronavirus agenda.

Our results showed that engaged users comment more than 2 times in a period but drop out of the discussion in 2–3 moths if agenda is not escalated by news outlet itself. Emotional triggers in news stories containing reasons for anxiety depends on a combination of factors from the site's functionality to the type of news media and local information policy.

Keywords: Social media · Anxiety · User engagement · Social isolation · News media

1 Introduction

1.1 Negative Emotions in Social Media

The era of the coronavirus pandemic has shown more clearly than ever the leading role of emotions in the use of social media. Emotional contagion has more than ever contributed to the spread of information and disinformation, and has also played a critical role in shaping the scenarios of the future circulated in social networks [1]. The rise of the

G. Meiselwitz (Ed.): HCII 2022, LNCS 13315, pp. 345–357, 2022.
https://doi.org/10.1007/978-3-031-05061-9_25

importance of emotions in public communication in the new media era was predicted by Marshall McLuhan when he spoke of retribalization. McLuhan called the era of electric media "the Age of Anxiety… that compels commitment and participation" [2]. This connection between anxiety and participation, or engagement, has been the subject of theoretical and empirical works by media researchers since then.

A prominent media philosopher Jaron Lanier has pointed out that social media, while provoking new and new emotions, kills our ability to empathize with others. From his point of view, social networks deliberately provoke negative emotions in the first place, since a user who is in an embittered or anxious state is more likely to buy the advertised product or service and spend more time in front of the screen, because the negative emotion is retained in the person longer, than positive [3]. Nicholas Carr, one of the most influential media researchers, wrote that with the constant updating of feeds and statuses on social networks, the identity of a user, for example, Facebook, is constantly "at risk", resulting in a permanent fear of being "invisible" online significantly increases the intensity of our engagement in the media [4]. John Keane emphasizes that in our era, which he calls the era of communication abundance, social exclusion and political anxiety are felt more strongly than in all previous times [5].

Social networks are equipped with special emotional affordances that help users express their feelings [6]. These affordances help not only users, but also the ecosystem of the platforms themselves: content that is saturated with human emotions is shared by users much faster, information is spread more widely, and coverage increases [7]. As was observed, "it makes sense for digital companies to try to promote emotion expression because emotions keep people engaged on the platforms and engagement means more opportunities to present ads and gather data… For instance, Twitter messages about cancer that included joy, sadness, and hope are liked more than others, and tweets that contain joy and anger are retweeted more than others" [8]. Moreover, it is negative emotions that better contribute to the transfer of information. One study looked at Centers for Disease Control and Prevention Facebook account, which had more posts about health than about the Ebola outbreak that originated in Africa at the time. On a sample of almost 1,000 posts, the researcher concluded that the Ebola posts received much more attention and generated more engagement. From the researcher's point of view, the reason for this lies in the feeling of anxiety generated by the reports of Ebola [9].

The need for social isolation in the era of the coronavirus pandemic has forced people to actively use social media. Although some researchers believe that social media will provide people with the necessary illusion of co-presence in the community, empathy and the exchange of positive supportive emotions [10] and that they will help to get rid of the stigmatization of the disease [11] they are not always that helpful. In the first year of the pandemic, it was fear and anxiety that were the most common emotions on social media [12]. A February 2020 study in China, where the pandemic originated, found that 42.6% of people suffer from anxiety symptoms [13].

The notion that positive or negative emotions are virally transmitted through social media has changed the strategies of many companies and public figures. In particular, business structures have begun to take into account the positive emotional engagement of the consumer [14] or the employee [15] in social networks as one of the most important performance indicators. Equally much attention to emotional involvement in social media is used by state structures, especially if we are talking about authoritarian or quasi-authoritarian political systems. Actively expressed in social networks, negative sentiments can help the authorities track down the pain points of the regime, address real social problems, and indignant citizens can thus "let off steam" as an alternative to street protests. Researchers of post-Soviet states believe that, as a result, the negative emotional engagement of citizens in social networks is rather beneficial for authoritarian rulers and, moreover, the latter are aware of this [16].

1.2 Anxiety in Social Media

Back in the mid-2000s, John Suler demonstrated what he called the online disinhibition effect. The point was that during communication on the Web, users reveal their emotions to their interlocutors much more freely, since due to a complex of mental mechanisms, such as a sense of one's own anonymity, impunity, etc., the usual communication barriers are removed. And while this sometimes helps to express innermost feelings and provide the support needed, there is often a toxic disinhibition, expressed in insults, bullying, sexual harassment [17]. For many people, especially teenagers, the feeling that potential sexual predators are always online causes a long-term feeling of anxiety, which can remain with them after leaving the online [18]. At the same time, it is worth noting that recent studies state that there is still no exact data on whether contacts on social networks contribute to real cases of harassment or, on the contrary, make potential victims more able to make the problem public [19].

Some media philosophers believe that anxiety is at the root of the Web phenomenon itself. Thus, Annette Wong is sure that the development of information technology and cyberspace is changing not only reality, but also our perception of reality. The conceptual frameworks - "frames" - that we used earlier are no longer suitable for grasping the order of things as they are. In many ways, this refers to the oppositions that we still use in thinking: male - female, man - machine, image - reality. Today, these frames are becoming both more blurred, but also more clear, as a result of the reaction, which causes a gap between different conceptual grids and increases the level of anxiety in society [20].

The connection between anxiety and social media has never been 100% transparent, and in the era of the pandemic, it could become even more confusing. Cross-sectional studies cannot definitively show whether social media use causes anxiety or whether people prone to anxiety disorders "are more likely to spend more time on social media; have addictive and problematic social media use behavior; have negative interactions with social media; and invest on social media" [21]. In the context of the pandemic, there is increased time spent using social media and increased levels of anxiety, but again there are difficulties in determining cause and effect. Is coronavirus anxiety causing one to spend more time on social media? Or is information and misinformation backed up by emotional contagion raising offline levels of anxiety about covid? [22] A study conducted among adult residents of Italy during the pandemic showed that at first social networks became an outlet for self-isolated citizens, emotionally positively influenced their condition, but then, as the lockdown continued, "the facilitated and increased access to social media during the COVID-19 pandemic risked to further increase anxiety, generating a vicious cycle that in some cases may require clinical attention". As a result, the author of the study calls the relationship between anxiety and the use of social networks "a bidirectional (i.e., reciprocal) relationship" and considers the use of social media "an adaptive coping behavior" which in this case is "creating new social interactions" [23]. Other researchers studying the psychological state of adolescent users of social networks during the pandemic crisis have come to the conclusion that using social media as a substitute for physical social relations makes users feel less happy [24]. Some scientists point to a direct link between anxiety and the fact that the basis of social networks (such as Facebook) is self-presentation and, thus, the constant review of the user's self-esteem can shake his or her mental stability [25]. At the same time, although the connection between social networks and anxiety is confirmed by many studies, a number of authors who agree with the relationship between time spent on social media and increased anxiety suggest using social network engagement *mutatis mutandis* as a clinical therapy for anxiety disorders [26].

1.3 Features of Anxious Social Media Engagement

Some researchers argue that it is not the time spent on social media, but the specific patterns of participation in online communications that determine the level of user anxiety [27]. For example, activities such as "repeated checking for messages, personal investment, and addictive and problematic use" can produce symptoms of an anxiety disorder in a person who spends time on social media [21], while problematic use is usually understood as an exaggerated concern about one's profile in social media, excessive motivation to use them, in which this use can negatively affect the work, study, personal life of the user [28]. Particularly associated with anxiety is the so-called passive use of social media, which includes editing and updating one's profile, lurking, detached scrolling, with possible expressing emotions in form of "likes". It is believed that socially lonely people are more prone to passive use of social networks, so the pandemic, forcing many to become isolated, has contributed to a sharp increase in the number of passive users. The reverse of passive use is active use, which includes posting, commenting, sharing, participation in public discussion [29]. Although it is commonly believed that there are slightly more female among people prone to online anxiety, studies show that passive use of social media "is related to greater anxiety symptoms for both genders" [25]. At the same time, such actions as sharing photos, engaging in communities, and communicating with real life friends are considered to be safer for users' mental health [30].

Other causes of anxiety on social media include negative self-perceptions, according to which "I'm nobody offline, but online I am someone" or "the online world is safer than the real one". Also, in a number of European countries, a relationship was found between anxious and addictive behavior on the Web and the use of recreational psychoactive substances, problematic alcohol use and general psychological stress among the population [31]. An interesting observation was made regarding the fact that users who gravitate towards homophily (attraction to those who are similar to themselves) and tend to express their religious beliefs online show a greater propensity for anxiety behavior [32].

The key factor in the emergence of anxiety among users is online aggression which is expressed in a number of interactive anti-social behaviors such as trolling, flaming, abuse, bullying, and hate speech and became possible thanks to the illusion of invisibility in social media [33]. An important feature of aggressive behavior on the Web, which must be taken into account in the analysis, is its memetic nature, the reason for which is the uncritical, often viral repetition of what was seen earlier on social networks and perceived as normative online behavior [34]. The manifestation of bullying and other types of aggression on the network sometimes became the cause of virtual (in the form of deactivation of user accounts) and even, in extreme cases, real suicide [35]. Another psychological phenomenon of social networks is Fear of Missing Out (FoMO) syndrome. FoMO has been called the main type of problematic and addictive social media use, and it consists of a constant desire to check your social media account and compulsively follow the updates of the feed and friends' accounts. It is believed that FoMO reduces the ability to self-control, distorts a person's self-esteem and contributes to the formation of anxious-aggressive behavior [36]. Among the psychological effects of FoMO are also "a lack of sleep, reduced life competency, emotional tension, negative effects on physical well-being, anxiety and a lack of emotional control" [37]. In addition, depression and online burnout are also directly related to social media anxiety and have also been on the rise in the past two years [38].

2 Methodology

We relied on the conceptual provisions of existing studies of message distribution patterns in various social networks, given that VKontakte has not yet become the object of such a study. Scholars focus on the patterns of distribution of specific meanings in messages using the example of cascades in Twitter discussions [41], on predicting the development of discussions [42], on the characteristics of the subjects of dissemination of messages, on the ideological potential of messages and their authors [43], on collective behavior and attitudes when sending messages [44], information distortion models [45], and other aspects. Most often, researchers study the process of increasing messages on a particular topic, emotional content, or polarization potential. It becomes important to develop models of infection, message distribution, which would make it possible to predict an epidemic of comments and the involvement of specific types of users in it - the central nodes of the network. At the same time, users who act as "silent witnesses" of such cascades and epidemics of messages remain outside the scope of such studies. Our hypothesis was that the spiral of anxiety provokes users to drop out of commenting due to the feeling of the impossibility of collectively overcoming anxiety, due to increased anxiety due to repetitive meanings and scenarios in user comments on sensitive topics. Posts about covid-19 become such a litmus test for most social media users, since, as we have studied earlier, comments on this topic are characterized by typical scenarios for unfolding discussions about the future development of events.

Regional media in Russia today are more engaging for their audience in social networks. Russian researchers mainly turn to comments on the VKontakte and Instagram platforms as the most used, however, in recent years there has been an overall decline in user engagement. The author introduces the concept of "tired audience", whose fatigue is caused, among other things, by the abundance of repetitive messages, media formats and the emotional background of these messages. In some regions, activity is shifting from social media platforms to the space of messengers, especially in Telegram, which recently introduced the function of chats and commenting on posts, as well as using emoji in reactions. The study of Twitter in Russia in this context is completely irrelevant, since the platform is steadily losing an active audience in Russia, against the backdrop of the rapid growth of Instagram and TikTok. At the same time, VKontakte has matured with its audience, and today people aged 35–50 are the most active on this platform. This age group is characterized by a culture of negative and plaintive commenting, expressing dissatisfaction and saying "in the name of justice". The growing activity of the state and state-owned media in tracking complaints and dissatisfaction in VKontakte additionally provokes the activity of subscribers of the appropriate age, and the use of bots often turns these discussions into polarized and destructive ones.

According to an internal study by the media measurement platform Popsters, for VKontakte, text formats show good engagement, second only to photos [46]. Another interesting conclusion of the analytical company BrandAnalytics is that starting from 2018, the number of aggressive posts by Russian-speaking users on social media is decreasing [47], at the same time, VKontakte messages do not fall into the top 10 aggressive sites (the only social network on this list is Twitter). In November 2021, the same company recorded an increase in the virality of regional media content on social media [48]. Thus, according to the data of media measurement companies, it can be argued that Russian-speaking users are interested in text and photo formats in regional media, which for the most part do not cause negative emotions.

Given this context, we selected 10 popular regional media accounts in Russian regions and selected all comments for the year in each media using the tool PepperNinja [49]. Next, we checked how often comments from the same users occur during this period. We also additionally interviewed editors and journalists of these media about the trends that they themselves observe in the comments to their posts.

The following resources were included in the sample, in total 161,067 messages and 28,633 active users for the period 11/01/2020–11/01/2021 (conditional boundaries of the second and fourth waves of covid-19 in Russia) (Table 1):

Table 1. Sample, news accounts in VK.

News group	Link	N (subscribers in thsnds)	N (comments)	N (active users, at least 1 comment)
City51 I Novosti Murman-ska i ob-lasti (Mur-mansk)	https://vk.com/city51ru	95+	2790	1503
Region 29 (Arkhangelsk)	https://vk.com/region29news	108+	1302	623

(continued)

Table 1. (*continued*)

News group	Link	N (subscribers in thsnds)	N (comments)	N (active users, at least 1 comment)
161.ru (Rostov-on-Don)	https://vk.com/news161	94+	2677	602
nn.ru (Nizhny Novgorod)	https://vk.com/newsnnru	38+	1122	406
72.ru (Tyumen)	https://vk.com/news72ru	230+	34080	4888
Jantarnyj DLB (Kalinigrad)	https://vk.com/amberbolt	259+	6109	1861
Irkutsk onlajn (Irkutsk)	https://vk.com/irkutsk_vonline	31+	5371	1799
Golos Dagestana (Dagestan)	https://vk.com/golos_dagestan	108+	100000	13885
SIA-press (Surgut)	https://vk.com/siapress	43+	2242	588
Vestnik Petrozavodska (Karelija)	https://vk.com/vestnikpetrozavodsk	62+	5374	2478

3   Results

The frequency analysis of the comments showed that the main vocabulary consists of neutral words without any particular anxiety connotations: *person, work, vaccine, city, administration, nation, child*. We counted the number of users who commented on posts more than once in the sample, and it turned out that 50% of the authors on average of the entire sample belong to this group. Next, we checked the distribution of comments during the selected time period from users who left comments more than once. It turned out that, on average, user activity fades after 3 months. We also checked the number of likes of other users to comments and found that the most engaging user messages belong to the following types of messages: expression of aggression in the form of a wish for harm to a specific individual or an abstract subject, a detailed description of a possible solution to the problem, a detailed description of the situation in which the commenter found himself. According to LiveDune instrument photo content attracts more likes and comments than video and texts, and more comments are engaged by posts longer than 2000 characters.

The context in which commenting on a news item unfolds most often appears in the newsgroup not for the first time. They involve exactly recurring themes specific for a particular regional news group. We did not see specific topics that equally involve people in different regions, most often these are universal topics: quality of life, utilities, incidents, and the activity of local authorities. As for the topic of the pandemic, it is distributed unevenly throughout the year, but more or less equally for all groups, depending on the general federal agenda, which in the November 2020-November 2021 segment was mainly associated with vaccination and QR codes for the vaccinated citizens.

Using the example of the "Irkutsk Online" group, we see that after leaving the first comment, people quickly lose their motivation to write more. Less than 1% of users reach 8+ comments. Only two subscribers from the entire sample made more than 100 comments in 13 months. The TOP-10 accounts for 699 comments, or 13% of all entries. The keywords of the message core from the top 10 posts with the maximum number of comments also do not contain a specific vocabulary: person, nation, year, child, place, son, knee (Table 2).

Table 2. Actively commenting users of "Irkutsk online" group.

Number of comments in the database per year	5371
Number of unique users	1799
Share of users with 1 comment	58,70%
Share of users with 2 comments	16,70%
Share of users with 3 comments	8%
Share of users with 4 comments	3,70%
Share of users with 5 comments	2,90%
Share of users with 6 comments	1,60%
Share of users with 7 comments	1,40%
Percentage of users with 8 or more comments	0,80%

Using the example of the "Region 29" group, we can say that, nevertheless, the covid agenda (or rather, the topic of vaccination) still attracted more comments than the classic local news - crimes, housing and communal services, salaries, holidays (Table 3).

Table 3. Thematic analysis of "Region 29" news page top comments.

Month	Nov. 21	Oct. 21	Sep. 21	Aug. 21	Jul. 21	Jun. 21	May 21
Top post 1	142 covid	504 covid	288 covid	129 holidays	291 covid	91 holidays	321 Victory Day
Top post 2	108 covid	246 covid	79 crimes	128 holidays	164 hospital renovation	72 employment	148 covid
Top post 3	95 covid	240 covid	73 covid	82 covid	158 homicide	68 weather	118 weather

(*continued*)

Table 3. (*continued*)

Month	Nov. 21	Oct. 21	Sep. 21	Aug. 21	Jul. 21	Jun. 21	May 21
Top post 4		236 covid					72 covid
Top post 5		225 covid					

Month	Apr. 21	Mar. 21	Feb. 21	Jan. 21	Dec. 20	Nov. 20
Top post 1	76 garbage	259 sports	326 heating	282 garbage	416 sports	239 sports
Top post 2	62 interview MP	226 sports	215 sports	126 sports	193 sports	226 sports
Top post 3	56 animals	167 sports	199 sports	98 garbage	177 sports	177 covid
Top post 4		108 health	181 sports		153 housing	127 holidays
Top post 5		102 transport	123 sports		112 holidays	64 holidays

Moreover, if the keywords of the comments do not contain explicit references to anxiety, then the words in the texts of the posts themselves are interesting for further qualitative analysis. For example, in the top keywords of the most commented posts in the Region 29 group, there are such as: *wait, start, mandatory, die.*

4 Discussion

The life circumstances in which users of Russian-speaking social networks found themselves during the pandemic have undoubtedly changed. The general uncertainty and anxiety of the situation makes people emotionally react to the deterioration of the situation in various areas of social, political and economic life. At the same time, for each region, specific topics can be identified that traditionally involve citizens in commenting, as they relate to protracted unresolved problems, local cultural codes or features of local politics. Subscribers have no motives for regular commenting in local news groups, their activity is associated with specific reasons. In our study, we, as well as analysts from media metering companies, observed an overall decrease in subscriber activity on traditional social media platforms as VK.

At this stage of the study, we cannot claim that a special pattern of anxiety has appeared in commenting on the VKontakte platform. Perhaps this is due to the age of users of this resource, and to the limitations of the platform itself. In general, researchers of the accounts of regional news media VKontakte note the low activity of the audience in relation to news and somewhat greater involvement in visual content. We also saw that photo and video posts, and especially live broadcasts from city events (sports, political, cultural) make people communicate with each other about the event more than text formats.

News triggers that provoke emotional reactions are not associated with specific vocabulary, which can be described as an "anxiety dictionary". We can assume that visual images in this case may have a more serious role than vocabulary combinations, but this requires further research.

Our sample consisted mainly of private independent groups, which in general today serve as the main platform for emotional exhaust in cities. According to a respondent from Tyumen (editor-in-chief of a private publication), *"people browse social networks more than official media. For example, the "ChS Tyumen" group on VKontakte collects hundreds of comments under a "complaining" post. While in the regional media - 50–100 maximum"*. This quote refers to a news group that is not run by the professional news media and its editors, but by amateur activists, as opposed to the official media, most often owned by the local administration. *"People criticize the government everywhere, but more actively in independent groups,"* agrees a respondent from Surgut (administrator of the independent public VK page). Thus, in our sample of independent news accounts, we generally see more intense commenting than the regional average.

5 Conclusion

Social networks initially provided a functional space for communication in local communities about the news agenda. During the pandemic, the need to express digital emotions has become relevant for most users, including subscribers to news accounts in Russian regions. In addition to the coronavirus-related agenda, people also met content typical of their regions. In those periods when the federal agenda related to vaccination dominated, users were more actively involved in its discussion. But in "calm" periods, local communities reacted equally violently to the triggers that provoked them even before the pandemic. It can be assumed that part of the anxiety caused by the general global situation could be offset by the severity of local news, such as ecology, homeless animals, urban planning, or criminal cases. That is why the activity of users in the comments does not exceed several months - exciting situations end or fade for a while.

Our study is limited, first, by the nature of the sample - the characteristics of the regions and independent news media. In the future, the study will include such parameters as the ideological affiliation of the editorial board, visual formats of messages, as well as the presence of bot communication in the comments.

Acknowledgements. This research is supported in full by Russian Science Foundation, project 21-18-00454.

References

1. Nigmatullina, K., Rodossky, N.: Pandemic discussions in VKontakte: hopes and fears. In: Meiselwitz, G. (ed.) HCII 2021. LNCS, vol. 12775, pp. 407–423. Springer, Cham (2021). https://doi.org/10.1007/978-3-030-77685-5_30
2. McLuhan, M.: Understanding Media: The Extensions of Man. McGraw Hill Education, New York (1964)
3. Lanier, J.: Ten Arguments for Deleting Your Social Media Accounts Right Now. Henry Holt and Co., New York (2018)
4. Carr, N.: The Shallows: What the Internet is Doing to Our Brains. W. W. Norton and Co., New York (2011)
5. Keane, J.: Democracy and Media Decadence. Cambridge University Press, Cambridge (2013)
6. Moreno, M.; D'Angelo, J. Applying an affordances framework to social media intervention approaches. J. Med. Internet Res. **21**(3) (2019)
7. Stieglitz, S., Dang, X.L.: Emotions and information diffusion in social media—sentiment of microblogs and sharing behavior. J. Manag. Inf. Syst. **29**, 217–248 (2013)
8. Steinert, S.: Corona and value change. The role of social media and emotional contagion. Ethics Inf. Technol. **23**(1), 59–68 (2020). https://doi.org/10.1007/s10676-020-09545-z
9. Strekalova, Y.: Health risk information engagement and amplification on social media: news about an emerging pandemic on Facebook. Health Educ. Behav. **44**(2), 332–339 (2017)
10. Waxman, S.: Can social media and technology come to the rescue in a pandemic? WrapPRO (2020). https://www.thewrap.com/jaron-lanier-technology-social-media-pandemic/
11. Chandrashekhar, V.: The burden of stigma: from leprosy to COVID-19, how stigma makes it harder to fight epidemics. Science **369**(6510), 1419–1423 (2020)
12. Wheaton, M., Prikhidko, A., Messner, G.: Is fear of COVID-19 contagious? The effects of emotion contagion and social media use on anxiety in response to the coronavirus pandemic. Front. Psychol. **11**, 3594 (2021)
13. Wiederhold, B.: Using social media to our advantage: alleviating anxiety during a pandemic. Cyberpsychol. Behav. Soc. Netw. **23**(4), 197–198 (2020)
14. Di Gangi, P.M., Wasko, M.M.: Social media engagement theory: exploring the influence of user engagement on social media usage. J. Org. End User Comput. **28**(2), 53–73 (2016)
15. Parry, E., Solidoro, A.: Social media as a mechanism for engagement? In: Bondarouk, T., Olivas-Luján, M.R. (eds.) Social Media in Human Resources Management, pp. 121–141. Emerald Group Publishing, Bingley (2013)
16. Toepfl, F., Litvinenko, A.: Transferring control from the backend to the frontend: a comparison of the discourse architectures of comment sections on news websites across the post-Soviet world. New Media Soc. **20**(8), 2844–2861 (2018)
17. Suler, J.: The online disinhibition effect. Cyberpsychol. Behav. **7**(3), 321–326 (2004)
18. Boyd, D.: It's Complicated: The Social Lives of Networked Teens. Yale University Press, New Haven (2014)
19. Gundersen, K., Zaleski, K.: Posting the story of your sexual assault online: a phenomenological study of the aftermath. Fem. Media Stud. **21**(5), 840–852 (2021)
20. Wong, A.: Cyberself: identity, language and stylisation on the internet. In: Gibbs, D., Krause, K.-L. (eds.) Cyberlines 2.0: Languages and Cultures of the Internet, pp. 259–260. James Nicholas Publishers, Melbourne (2006)
21. Keles, B., McCrae, N., Grealish, A.: A systematic review: the influence of social media on depression, anxiety and psychological distress in adolescents. Int. J. Adolesc. Youth **25**(1), 79–93 (2020)
22. Drouin, M., et al.: How parents and their children used social media and technology at the beginning of the COVID-19 pandemic and associations with anxiety. Cyberpsychol. Behav. Soc. Netw. **23**(11), 727–736 (2020)

23. Boursier, V., Gioia, F., Musetti, A., Schimmenti, A.: Facing loneliness and anxiety during the COVID-19 isolation: the role of excessive social media use in a sample of Italian adults. Front. Psychiatry, 1380 (2020)
24. Cauberghe, V., et al.: How adolescents use social media to cope with feelings of loneliness and anxiety during COVID-19 lockdown. Cyberpsychol. Behav. Soc. Netw. 24(4), 250–257 (2021)
25. Thorisdottir, I., et al.: Active and passive social media use and symptoms of anxiety and depressed mood among Icelandic adolescents. Cyberpsychol. Behav. Soc. Netw. 22(8), 535–542 (2019)
26. Vannucci, A., Flannery, K., Ohannessian, C.: Social media use and anxiety in emerging adults. J. Affect. Disord. 207, 163–166 (2017)
27. Shensa, A., et al.: Social media use and depression and anxiety symptoms: a cluster analysis. Am. J. Health Behav. 42(2), 116–128 (2018)
28. Schou Andreassen, C., Pallesen, S.: Social network site addiction: an overview. Curr. Pharm. Des. 20, 4053–4061 (2014)
29. O'Day, E.B., Heimberg, R.G.: Social media use, social anxiety, and loneliness: a systematic review. Comput. Hum. Behav. Rep. 3, 100070 (2021)
30. Malik, A., Dhir, A., Nieminen, M.: Uses and gratifications of digital photo sharing on Facebook. Telematics Inform. 33(1), 129–138 (2016)
31. Henzel, V.; Håkansson, A.: Hooked on virtual social life. Problematic social media use and associations with mental distress and addictive disorders. PLOS ONE 16(4), e0248406 (2021)
32. Shaw, A.M., Timpano, K.R., Tran, T.B., Joormann, J.: Correlates of Facebook usage patterns: the relationship between passive Facebook use, social anxiety symptoms, and brooding. Comput. Hum. Behav. 48, 575–580 (2015)
33. Aroyehun, S.T., Gelbukh, A.: Aggression detection in social media: using deep neural networks, data augmentation, and pseudo labeling. In: Proceedings of the First Workshop on Trolling, Aggression and Cyberbullying, pp. 90–97. Association for Computational Linguistics, Santa Fe (2018)
34. Sparby, E.M.: Digital social media and aggression: memetic rhetoric in 4chan's collective identity. Comput. Compos. 45, 85–97 (2017)
35. Kumar, R., Ojha, A.K., Malmasi, S., Zampieri, M.: Benchmarking aggression identification in social media. In: Proceedings of the First Workshop on Trolling, Aggression and Cyberbullying, pp. 1–11. Association for Computational Linguistics, Santa Fe (2018)
36. Eraslan, L., Kukuoglu, A.: Social relations in virtual world and social media aggression. World J. Educ. Technol. Curr. Issues 11(2), 1–11 (2019)
37. Alutaybi, A., Al-Thani, D., McAlaney, J., Ali, R.: Combating fear of missing out (FoMO) on social media: the fomo-r method. Int. J. Environ. Res. Publ. Health 17(17), 6128 (2020)
38. Bettmann, J.E., Anstadt, G., Casselman, B., Ganesh, K.: Young adult depression and anxiety linked to social media use: assessment and treatment. Clin. Soc. Work J. 49(3), 368–379 (2021)
39. Hoffmann, C.P., Lutz, C.: Spiral of silence 2.0: political self-censorship among young Facebook users. In: Proceedings of the 8th International Conference on Social Media and Society, pp. 1–12. Association for Computing Machinery, New York (2017)
40. Syahputra, I.: Expressions of hatred and the formation of spiral of anxiety on social media in Indonesia. SEARCH J. Media Commun. Res. 11(1), 95–112 (2019)
41. Hui, C., et al.: Information cascades in social media in response to a crisis: a preliminary model and a case study. In: Proceedings of the 21st International Conference on World Wide Web, pp. 653–656 (2012)
42. Cao, Q., et al.: DeepHawkes: bridging the gap between prediction and understanding of information cascades. In: Proceedings of the 2017 ACM on Conference on Information and Knowledge Management, pp. 1149–1158 (2017)

43. Monti, C., et al.: Learning ideological embeddings from information cascades. In: Proceedings of the 30th ACM International Conference on Information and Knowledge Management, pp. 1325–1334 (2021)

44. Rosas, F., Hsiao, J.H., Chen, K.C.: A technological perspective on information cascades via social learning. IEEE Access **5**, 22605–22633 (2017)

45. Ermakova, L., Nurbakova, D., Ovchinnikova, I.: Covid or not covid? Topic shift in information cascades on Twitter. In: 3rd International Workshop on Rumours and Deception in Social Media (RDSM) Collocated with COLING 2020, pp. 32–37 (2020)

46. https://popsters.ru/blog/post/aktivnost-auditorii-socialnyh-setey-v-2021-godu. Accessed 2 Nov 2022

47. https://br-analytics.ru/blog/social-aggressiveness-2021/. Accessed 2 Nov 2022

48. https://br-analytics.ru/blog/top-100-media-november-2021/. Accessed 2 Nov 2022

49. https://pepper.ninja/. Accessed 2 Nov 2022

14 Days Later: Temporal Topical Shifts in Covid-19 Related Tweets After Pandemic Declaration

Hamzah Osop[1]([✉])[iD], Basem Suleiman[2][iD], and Abdallah Lakhdari[2]

[1] Agency of Science Technology and Research (A*STAR), IHPC, Singapore, Singapore
hamzah_bin_osop@ihpc.a-star.edu.sg
[2] School of Computer Science, University of Sydney, Sydney, Australia
{basem.suleiman,abdallah.lakhdari}@sydney.edu.au

Abstract. On 11 March 2020, the World Health Organisation (WHO) declared the Covid-19 situation as a pandemic. Twitter became a popular platform for information sharing as evident in the increase in the number of tweets generated in March 2020. Studying topics discussed right after the pandemic declaration is crucial in understanding the impacts Covid-19 may have on people. As time passes, it is also beneficial to analyse the changes in topics and perceptions. We analysed over 1.4 million tweets to uncover key main topics discussed over the period from 25 March 2020 to 22 April 2020. Using a combination of BERT, LDA and K-Means clustering, three main topics of *Preventive Measures*, *News* and *Financial Support* were identified. Over a period of five different time points, we analysed the temporal change in main topics being discussed, reflecting the changes in perceptions of the people as Covid-19 virus progressed.

Keywords: Covid-19 · Twitter · Topic modelling · Topical shifts · Information sharing · Information seeking

1 Introduction

The World Health Organisation (WHO) declared the Covid-19 situation as pandemic on 11 March 2020, after the number of cases and affected countries continued to grow due to the virus. The global Covid-19 pandemic caused an exceedingly high number of deaths and created a sense of panic, fear, and even discrimination across communities. Worldwide, countries quickly introduced strict restrictions including social distancing, limited movement, closing international borders and travel to prevent the spread of the virus. As a result, people had to isolate themselves and work from home, with many spending most of their time online, using social networks and teleconferencing tools to facilitate communication and to socialise. The accessibility to social media tools has facilitated information sharing and seeking behaviours among users [1]. Among the many social networking platforms, Twitter has become one of the commonly used tools to engage in communication and sharing of information online, primarily due to its simplicity and ease of use [1].

G. Meiselwitz (Ed.): HCII 2022, LNCS 13315, pp. 358–369, 2022.
https://doi.org/10.1007/978-3-031-05061-9_26

Posts or messages published on Twitter, otherwise known as tweets, reflect users' timely opinions. The act of sharing information amongst Twitter users, termed *information diffusion* [2], in fact, grew rapidly during the pandemic. The number of COVID-19 related tweets increased significantly, beginning from March 12, 2020, right after the pandemic declaration [3]. From a modest 320 million tweets per day in November 2018 to a high of 500 million tweets by March 2020 [3]. This roughly corresponded to about 350,000 tweets per minute. Twitter has indeed become a channel where people get to voice out their anxieties and worries about the pandemic. For example, individuals and social media influencers have participated not only in spreading Covid-19 related topics but also their own opinions and views about it.

Consequently, users and even organisations become heavily reliant on this fast-growing information shared on Twitter. They follow key important topics related to the pandemic, and continuously spread the information. With information being diffused on Twitter over time, society's perception may also change during that time. Thus, it becomes important to identify and analyse different topics that may arise from the pandemic situation. Being able to identify people's perception and its changes over time can help decision-makers, such as healthcare and government agencies, launch relevant campaigns, like supporting mental health, sharing trusted information, addressing core topics that concern the masses and reducing risks of misinformation about the pandemic which could negatively impact more people.

Hence, the goal of this paper is to study Covid-19 related tweets shared by users and identify key topics and study its changes over time. Particularly, we employed machine learning models to classify the text information in tweets at different time points during the pandemic and analyse subsequent temporal shifts in topics.

2 Related Works

Prior works have shown the potential of Twitter being used as an effective platform to disseminate and share information [4–6]. The peak in the number of tweets generated during the Covid-19 pandemic is testimony to the popularity of Twitter as the platform to share information. Furthermore, Twitter's *#hashtag* feature makes it easy for users to follow each other and be kept updated with the latest news and developments. Organisations such as government agencies, health, news, and media outlets have also utilised Twitter's *#hashtag* feature to share updates to their followers in relation to sharing and disseminating important topics and news.

As a source of user-generated data, tweets have also been purposefully leveraged upon for surveillance [7], analysis of sentiments [8–10] and decision support [11]. Given the real-time nature of a tweet, certain events can potentially be detected live [12] and information surveillance be conducted at current [13]. Analysing data contained in tweets can provide meaningful information such as identifying common themes or main topics being discussed. Topic modelling, which refers to the task of identifying underlying topics discussed within a collection of documents or text [14], can potentially explain the impacts that Covid-19 may have on users' physical being and mental health conditions [15].

In addition, analysing the temporal components associated with tweets can further unfold critical details pertaining to the spread of Covid-19 related information. In the study by Gunaratne et al. [16], the incorporation of temporal effects helped explain the change in perception of people from being anti-vaccination to being pro-vaccination [16]. Therefore, implementing topic modelling with temporal effect not only identify changes in topics being discussed during the pandemic, but also explain why changes took place. For example, from these unstructured and scattered tweets, society's perception on evolving key important topics such as the effectiveness of wearing masks or the development of vaccines could be identified.

3 Methodology

The aim of the study is to identify the topics being tweeted at different time points during the pandemic using machine learning and analyse the shifts in topical trends.

```
<class 'pandas.core.frame.DataFrame'>
RangeIndex: 61989 entries, 0 to 61988
Data columns (total 32 columns):
 #   Column                     Non-Null Count   Dtype
---  ------                     --------------   -----
 0   created_at                 61989 non-null   datetime64[ns, UTC]
 1   id                         61989 non-null   int64
 2   id_str                     61989 non-null   int64
 3   full_text                  61989 non-null   object
 4   truncated                  61989 non-null   bool
 5   display_text_range         61989 non-null   object
 6   entities                   61989 non-null   object
 7   metadata                   61989 non-null   object
 8   source                     61989 non-null   object
 9   in_reply_to_status_id      3110 non-null    float64
 10  in_reply_to_status_id_str  3110 non-null    float64
 11  in_reply_to_user_id        3660 non-null    float64
 12  in_reply_to_user_id_str    3660 non-null    float64
 13  in_reply_to_screen_name    3660 non-null    object
 14  user                       61989 non-null   object
 15  geo                        11 non-null      object
 16  coordinates                11 non-null      object
 17  place                      406 non-null     object
 18  contributors               0 non-null       float64
 19  retweeted_status           48633 non-null   object
 20  is_quote_status            61989 non-null   bool
 21  quoted_status_id           2643 non-null    float64
 22  quoted_status_id_str       2643 non-null    float64
 23  retweet_count              61989 non-null   int64
 24  favorite_count             61989 non-null   int64
 25  favorited                  61989 non-null   bool
 26  retweeted                  61989 non-null   bool
 27  lang                       61989 non-null   object
 28  possibly_sensitive         17146 non-null   float64
 29  extended_entities          4886 non-null    object
 30  quoted_status              623 non-null     object
 31  withheld_in_countries      13 non-null      object
dtypes: bool(4), datetime64[ns, UTC](1), float64(8), int64(4), object(15)
memory usage: 13.5+ MB
```

Fig. 1. Tweet attributes collated via tweepy

3.1 Data Collection

The Covid-19 related tweets were retrieved using a Twitter Streaming API called *tweepy*. We identified the top five Covid-19 related hashtags that were trending during the retrieval period, from Twitter main website (https://twitter.com), namely *coronavirus* (C1), *coronavirusoutbreak* (C2), *covid-19* (C3), *covid19* (C4) and *coronaviruses* (C5). Using these hashtags as search terms, we retrieved individual collection of tweets related to the search term used which were stored in a single JSON format. In total, we collated over 1.4 million tweets during a period of 25 March 2020 till 22 April 2020. Information contained together with the full tweeted text includes the tweet creation date, user details information, geolocation as well as the retweeted status. A full detail of the tweet attributes is illustrated in the Fig. 1 above.

3.2 Modelling the Topics from Tweet Corpus

We designed and adopted our Topic Modelling Implementation Process flow as illustrated in Fig. 2. The flow identified key implementation processes that includes (1) Data Collection, (2) Data Preprocessing, (3) Text Conversion, (4) Dimensionality Reduction, (5) Clustering, and (6) Analysis through Keyword Extraction.

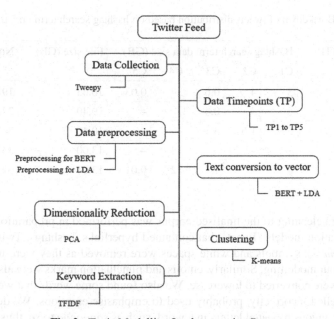

Fig. 2. Topic Modelling Implementation Process

Data Preprocessing and Cleaning. Tweets retrieved were in a JSON format which was not entirely a convenient format to analyse with. The raw JSON data was first converted into a CSV format, where individual tweet was represented in a single row with each of its attributes corresponding to columns within the row. This resulted in a fairly large dataset with exactly 1,439,058 individual tweets. However, we noticed that in certain days during the retrieval period, there were either no tweets matching the search terms used or no tweets retrieved on those dates.

To facilitate the investigation on different topics being discussed during the pandemic, the tweets were first grouped based on the tweet's *'created_at'* date, corresponding to an estimate of one-week periods. However, to accommodate to tweets that were not available on certain days, we resorted to having a similar distribution of tweets within the groups, which resulted in tweets being categorised into five different time points, TP1 to TP5. The dates corresponding to the time points are as follows:

- TP1: 25 March to 28 March 2020
- TP2: 2 April to 3 April 2020
- TP3: 14 April to 16 April 2020
- TP4: 17 April 2020
- TP5: 20 April to 22 April 2020

Table 1 illustrates the breakdown of tweet corpus collated, based on the hashtag used as search terms.

Table 1. Breakdown of tweets distribution based on hashtag search term on 5 timepoints

Time points (TP)	Hashtag search term data size (GB)					File size (GB)	No. of rows
	C1	C2	C3	C4	C5		
TP1	3	4	0.6	3	0.03	10.63	198,886
TP2	6	3	0.4	–	–	9.40	274,682
TP3	5.8	1.6	3	–	–	10.40	307,173
TP4	–	–	–	13	–	13.00	337,460
TP5	–	–	1	12	0.01	13.01	320,857

Next, data cleaning to the finalised corpus was performed in preparation for use in text classification models. Tweets that contained hyperlinks, hashtags, Twitter handles (*<@username>*), symbols and white spaces were removed as they were not going to be useful in our modelling. Similarly, emojis and punctuation marks were also removed, and all text were converted to lowercase. We also found some words that were repeated or words spelled incorrectly, probably used to emphasise emotions. We deleted more than two continuous repeated letters in a word, such as *sooooooo* to *so*, thus facilitating subsequent analysis and processing.

Stop words were next removed from the tweet corpus. Additionally, we expanded the stop words to include words that were closely related to the Covid-19 virus and the search terms we used as they seemed to appear frequently in the tweets. By removing them from the tweet corpus, we aimed at eliminating some text biases and capture a better representation of the topics discussed in the tweets. To account for the informal nature of tweet text and potential spelling errors, we improved the accuracy of words used by using a dictionary package in Python. Finally, the corpus was lemmatised using the package from Scikit-learn. An example of the result from preprocessing exercise is illustrated in Fig. 3, highlighting the differences for input to BERT and LDA models.

Topic Modelling. An unsupervised learning method to cluster the tweets at different time points and to model the topics based on the clusters generated was adopted. To

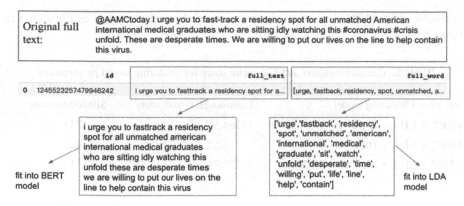

Fig. 3. Data preprocessing results for fitting into BERT and LDA model

model the topics from the tweet corpus, two text classification models were identified, (1) the Bidirectional Encoder Representations from Transformer (BERT) and (2) the Latent Dirichlet Allocation (LDA). BERT algorithm computes the text embeddings of words by taking into consideration the context of all other words in a sentence [17]. Hence, the output of BERT is a vector representation of each word sequence in a sentence. LDA is a probabilistic text modelling technique in machine learning that assumes each topic is a mixture of the underlying set of words, and each document is a mixture over a set of topic probabilities [18]. LDA outputs the weights of each word found in the corpus that makes up a topic. We used both BERT [19] and LDA [20] model to convert the tweets into vectors and combined the advantages from each model to achieve better results [19, 21].

To reduce the dimensionality of the corpus while maintaining the context of the tweets, Principal Component Analysis (PCA) algorithm was proposed [22]. With the outputs from BERT and LDA, three different clustering models were proposed to cluster all the tweets based on similar topics within each time points. The three models were (1) K-Means [23, 24], (2) Density-Based Spatial Clustering of Applications with Noise (DBSCAN) [25, 26], and (3) Minibatch K-means [27].

Clustering. We combined the outputs from BERT and LDA models using a weight hyperparameter and applied it to K-Means algorithm, instead of DBSCAN and Minibatch K-Means, to cluster the tweets. The selection for K-Means was based on the evaluation of the best performing models in clustering tweets, using the Calinski-Harabasz and Silhouette score index. The Silhouette analysis is useful in obtaining the optimal clustering effect which refers to the difference between "average distance within the cluster and minimum distance between the clusters" [28]. A value closer to one thus indicating a better split between clusters, or neighbouring clusters are far away from each other [29]. The Calinski-Harabasz index, also known as the Variance Ratio Criterion, measures degree of dispersion between clusters [28]. Similarly, the higher the score, the better separated are the clusters, indicating better model performance.

We applied both scoring indexes on the three clustering models. Both the Calinski-Harabasz and Silhouette scored highest using K-Means clustering, with a score of

14264.59 and 0.17425 respectively, when compared to other models as illustrated in Table 2.

Table 2. Calinski-harabasz and Silhouette score for clustering model performance

Vector + Clustering model	Calinski-Harabasz score	Silhouette score
BERT + LDA + K-Means	14264.59	0.1743
BERT + LDA + DBSCAN	6.52	−0.2422
BERT + LDA + Minibatch K-Means	14256.74	0.1740

Based on the clustering results, top 100 words from each cluster were analysed to determine the topics using the Term Frequency-Inverse Document Frequency (TF-IDF) algorithm [30].

All analyses were performed using Python, leveraging on the Scikit-learn library, and data were uploaded to Google Collab which could handle our large dataset.

4 Results and Discussion

4.1 Number of Clusters

The optimal number of clusters were determined by both the Elbow method for K-Means clustering and perplexity score for LDA. Based on the Elbow method and perplexity score results, the optimal number of clusters for each time point was given as three (Fig. 4).

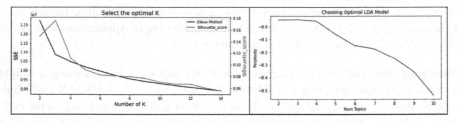

Fig. 4. Elbow method results (L) and perplexity score (R)

4.2 Determining Topics in Clusters

We used the Term Frequency-Inverse Document Frequency (TF-IDF) algorithm to filter out the top 100 words in each cluster. We manually analysed the top 100 words from each cluster for all the time points and developed a list of unigrams (Table 3) to assist in defining the topics discussed in clusters within the time points. Using voting by majority method, we determined the final topic for each cluster as listed in Table 4.

We uncovered three main topics of *Preventive Measures, News* and *Financial Support* from the clusters across different time points. For example, the topics discussed in Cluster 1, looked to defer slightly. At TP1 and TP2, the particular topic being discussed was *Global News,* while at TP3 to TP5, the topic was identified as just *News.* Perhaps, this could be explained with tweets that were posted in late March till early April, were associated with news sources originating from all over the world as the Covid-19 virus spread worldwide. As the situation worsened, tweets tend to revolve more around local news. Nonetheless, we concluded that the overall topic discussed still fall under the main topic of *News.* Similarly for Cluster 2 and 3, where the overall topic covers about *Preventive Measures* and *Financial Support* respectively. To evaluate the potential shift in topics, we calculated the distribution of topics discussed within each time points based on the proportion of tweets covering each topic.

Table 3. Unigrams defining the topic discussed

Topic	Unigrams
Preventive measures	Stay, home, mask, face, protect, ventilator, symptoms, quarantine, drug, isolation
News	Case, death, news, day, report, China, country, update, confirm, spread, world, outbreak, rate, American, India, Spain, Italy, record, recover, data
Financial support	Help, need, work, business, want, support, government, pay, free, share, money, fund, food, service, online

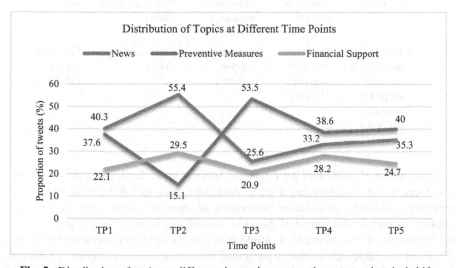

Fig. 5. Distribution of topics at different time points suggesting temporal topical shifts

Approximately two weeks after the pandemic declaration, the main topic discussed at TP1 was *Financial Support* (40.3%), followed closely by *Preventive Measures* (37.6%)

and *News* (22.1%). By mid-April 2020 onwards, at TP4, the topical shifts stabilised with *Preventive Measures* (38.6%) being the most discussed topic followed by *Financial Support* (33.3%) and *News* (28.2%). At the final time-point TP5, the topic distribution was *Preventive Measures* (40%), *Financial Support* (35.3%) and *News* (24.7%). Figure 5 clearly illustrates the temporal shifts in topics discussed over the five time points.

The proportion of *News* topic was relatively consistent, with a low of 21% to a high of 29.5%, across the different time points. This perhaps indicated a regular transmission of news during that period with no major news happening worthy of discussion, especially after the breaking news of the pandemic declaration happened two weeks prior. The highest proportion for *News* topic was in TP2 (29.5%), where we found several main keywords related to 'American', 'york', 'lie' and 'trump'. This could have indicated an increase in concern by residents in the United States, perhaps those in New York with circumstances surrounding the Covid-19 situation.

Table 4. Distribution of topics at different time points

Cluster	Time points (TP)				
	TP1	TP2	TP3	TP4	TP5
1	Global news (22.08%)	Global news (29.5%)	News (20.9%)	News (28%)	News (with health-related content) (24.68%)
2	Preventive measures (37.67%)	Preventive measures & treatment (15.1%)	Epidemic in America & preventive measures (53.5%)	Preventive measures & American news (38%)	Preventive measures (40%)
3	Trump influence & financial support in America (40.26%)	Impact & financial support (55.4%)	Government support (25.6%)	Financial support (34%)	Financial support (35.32%)

It was also noticeable that the *Financial Support* topic displayed a trend similar to *News* topic but on a slightly larger scale. With the known fact that Covid-19 virus was spreading worldwide, government institutions all over the world were taking serious measures to prevent the spread of the virus as well as ensuring their population were well taken care of. For example, a simple search on Google on government policies at TP3, listed several search results. In particular, mission updates on Singapore's Ministry of Foreign Affairs website by Consulate of the Republic of Singapore in New York, shared updated government guidelines enforced in the United States by the then Trump administration [31]. The guidelines which were released on 16 March 2020, coinciding with TP3, included advice to work or study from home, avoid gatherings of more than

10 people, and to avoid random travel, shopping, or social events. After about 10 days, at TP4, the reflection and impact of the lockdown policy had become a hot topic for discussion. The lockdown policy brought both positive and negative impacts, where virus spread and infection was brought to the minimal as possible, but at the expense of economy shutting down, small companies going bankrupt and people losing their jobs. Hence, the reflection in the proportion of tweets on the *Financial Support* topic, such as discussing government response and financial support in TP2 which had the largest proportion.

Interestingly, the distribution on *Preventive Measures* topic seemed to be the opposite to the *Financial Support* topic, especially at TP2 and TP3. The social and economic impacts were more apparent at the beginning of the lockdown as the number of infected had not grown exponentially. However, as the virus spread rapidly and infected more people, the potential health impacts and dangers of Covid-19 virus became known to more people. People were now more concerned about how to prevent themselves from being infected, thus the reflection on the proportion of tweets discussing *Preventive Measures* topping at 53.5% in TP3 and stabilising to 40% at TP5.

5   Conclusion and Future Work

The uncovering of three main topics in our analysis and noticeable significant shifts in the levels of discussions over time, revealed genuine real-time concerns of the people especially early in the pandemic phase. In fact, in a similar study conducted by Zhang et al. [32], that were based on tweets generated on 2 March 2020 till 18 March 2020, the authors also reported uncovering similar topics such as *Preventive Measures* and *Government Help and Support*, thus validating the outcome in our topic modelling.

Our findings suggest that public concerns evolve as new government policies and regulations are quickly introduced to the masses. This raised many concerns such as the state of the medical systems having to cope with the rapid increase in the number of patients. Understanding the topics or themes in tweets can be used to target intervention programs such as supporting those who suddenly found themselves unemployed due to safety measures being enforced. Our research thus has significance for government and social science research. We get to understand society's perception changes throughout the pandemic. Perhaps government policies and medical institutions are able to monitor the sentiments and trends of public opinions and help to guide the public.

Additionally, this study has the potential of extracting health-related information that can be useful in designing government policies as well as for healthcare decision makers. There were several tweets within the clusters that showed evidence of health-related topics. In Cluster 1 at TP5 and Cluster 2 at TP2, key words like *symptoms*, *treatment*, *drug,* and *vaccine* emerged in *News* and *Preventive Measures* topics.

However, this study is not without its limitations. A classification method using supervised learning could provide a better topic modelling result. However, an attempt to adopt this approached failed as we could not defined an effective way to label the tweets for a training dataset due to the large differences in labelling results when performed manually among the research team members. While we performed statistical analysis to determine the optimal number of clusters, but due to our extremely large text corpus, we could probably develop better topic definition with more clusters. This was evident in some clusters containing multiple discussion topics such as in Cluster 1 at TP5, Cluster 2 at TP2 and Cluster 3 at TP1.

For our future works, using a modified LDA model such as Gibbs Sampling Dirichlet Multinomial Mixture (GSDMM) [33] for example can address the challenges posed by short texts or documents such as a tweet. The high dimensionality and sparsity in short text potentially reduce the performance of LDA and K-Means model. Hence exploring more complex model, implementing feature extraction methods and employing better data cleaning methods may iteratively improve the final clustering and topic modelling results.

Acknowledgement. We would like to thank Yuncong Li, Jing Zhang, Lan Luo, Yingjiayue Lu and Chengjiu Liu for assisting with the data analysis and visualization conducted in this study.

References

1. Heverin, T., Zach, L.: Twitter for city police department information sharing. Proc. Am. Soc. Inf. Sci. Technol. **47**(1), 1–7 (2010)
2. Jung, A.-K., Mirbabaie, M., Ross, B., Stieglitz, S., Neuberger, C., Kapidzic, S.: Information diffusion between Twitter and online media (2018)
3. Project, G.: Visualising Twitter's evolution 2012–2020 and how tweeting is changing. In: The Covid-19 Era (2020). https://blog.gdeltproject.org/visualizing-twitters-evolution-2012-2020-and-how-tweeting-is-changing-in-the-covid-19-era/. Accessed 2021
4. Thelwall, M., Thelwall, S.: Retweeting for COVID-19: consensus building, information sharing, dissent, and lockdown life. arXiv preprint arXiv:2004.02793. **10** (2020)
5. Sinnenberg, L., Buttenheim, A.M., Padrez, K., Mancheno, C., Ungar, L., Merchant, R.M.: Twitter as a tool for health research: a systematic review. Am. J. Publ. Health **107**(1), e1–e8 (2017)
6. Missier, P., et al.: Tracking dengue epidemics using twitter content classification and topic modelling. In: Casteleyn, S., Dolog, P., Pautasso, C. (eds.) ICWE 2016. LNCS, vol. 9881, pp. 80–92. Springer, Cham (2016). https://doi.org/10.1007/978-3-319-46963-8_7
7. Jordan, S.E., Hovet, S.E., Fung, I.C.-H., Liang, H., Fu, K.-W., Tse, Z.T.H.: Using Twitter for public health surveillance from monitoring and prediction to public response. Data **4**(1), 6 (2019)
8. Joyce, B., Deng, J.: Sentiment analysis of tweets for the 2016 US presidential election. In: 2017 IEEE MIT Undergraduate Research Technology Conference (URTC) (2017)
9. Rane, A., Kumar, A.: Sentiment classification system of Twitter data for US airline service analysis. In: 2018 IEEE 42nd Annual Computer Software and Applications Conference (COMPSAC) (2018)
10. Lu, Y., Zheng, Q.: Twitter public sentiment dynamics on cruise tourism during the COVID-19 pandemic. Curr. Issue Tour. **24**(7), 892–898 (2021)
11. Phan, H.T., Nguyen, N.T., Tran, V.C., Hwang, D.: An approach for a decision-making support system based on measuring the user satisfaction level on Twitter. Inf. Sci. **561**, 243–273 (2021)
12. Hasan, M., Orgun, M.A., Schwitter, R.: Real-time event detection from the Twitter data stream using the TwitterNews+ Framework. Inf. Process. Manag. **56**(3), 1146–1165 (2019)
13. Colditz, J.B., et al.: Toward real-time infoveillance of Twitter health messages. Am. J. Publ. Health **108**(8), 1009–1014 (2018)

14. Jónsson, E., Stolee, J.: An evaluation of topic modelling techniques for Twitter. University of Toronto (2015)
15. Garcia-Gasulla, et al.: Global data science project for covid-19 summary report (2020)
16. Gunaratne, K., Coomes, E.A., Haghbayan, H.: Temporal trends in anti-vaccine discourse on Twitter. Vaccine **37**(35), 4867–4871 (2019)
17. Roitero, K., Bozzato, C., Della Mea, V., Mizzaro, S., Serra, G.: Twitter goes to the doctor: detecting medical tweets using machine learning and BERT. In: SIIRH@ ECIR (2020)
18. Tong, Z., Zhang, H.: A text mining research based on LDA topic modelling. In: International Conference on Computer Science, Engineering and Information Technology (2016)
19. Devlin, J., Chang, M.-W., Lee, K., Toutanova, K.: BERT: pre-training of deep bidirectional transformers for language understanding. arXiv. abs/1810.04805 (2019)
20. Negara, E.S., Triadi, D., Andryani, R.: Topic modelling Twitter data with latent Dirichlet allocation method. In: 2019 International Conference on Electrical Engineering and Computer Science (ICECOS) (2019)
21. Osas, U.: Topic extraction from tweets using LDA (2019). https://medium.com/@osas.usen/topic-extraction-from-tweets-using-lda-a997e4eb0985. Accessed 5 Dec 2020
22. Anjaria, M., Guddeti, R.M.R.: Influence factor based opinion mining of Twitter data using supervised learning. In: 2014 Sixth International Conference on Communication Systems and Networks (COMSNETS) (2014)
23. Sechelea, A., Huu, T.D., Zimos, E., Deligiannis, N.: Twitter data clustering and visualization. In: 2016 23rd International Conference on Telecommunications (ICT), p. 1–5 (2016)
24. Oktarina, C., Notodiputro, K.A., Indahwati, I.: Comparison of K-means clustering method and K-medoids on Twitter data. Indonesian J. Stat. Appl. **4**(1), 189–202 (2020)
25. Indah, R.N.G., et al.: DBSCAN algorithm: Twitter text clustering of trend topic pilkadapekanbaru. J. Phys. Conf. Ser. **1363**(1), 012001 (2019)
26. Capdevila, J., Pericacho, G., Torres, J., Cerquides, J.: Scaling DBSCAN-like algorithms for event detection systems in Twitter. In: Carretero, J., GarciaBlas, J., Ko, R.K.L., Mueller, P., Nakano, K. (eds.) ICA3PP 2016. LNCS, vol. 10048, pp. 356–373. Springer, Cham (2016). https://doi.org/10.1007/978-3-319-49583-5_27
27. Alanezi, M.A., Hewahi, N.M.: Tweets sentiment analysis during COVID-19 pandemic. In: 2020 International Conference on Data Analytics for Business and Industry: Way Towards a Sustainable Economy (ICDABI) (2020)
28. Wang, X., Xu, Y.: An improved index for clustering validation based on Silhouette index and Calinski-Harabasz index. IOP Conf. Ser. Mater. Sci. Eng. **569**(5), 052024 (2019)
29. Mamat, A.R., Mohamed, F.S., Mohamed, M.A., Rawi, N.M., Awang, M.I.: Silhouette index for determining optimal k-means clustering on images in different color models. Int. J. Eng. Technol. **7**(2), 105–109 (2018)
30. Rathi, M., Malik, A., Varshney, D., Sharma, R., Mendiratta, S.: sentiment analysis of tweets using machine learning approach. In: 2018 Eleventh International Conference on Contemporary Computing (IC3) (2018)
31. Affairs, M.o.F.: Update on Covid-19 (16 March 2020) (2021). https://www.mfa.gov.sg/Overseas-Mission/New-York-Consul/Mission-Updates/2020/03/16032020-UPDATE-ON-COVID-19. Accessed 15 Dec 2021
32. Zheng, H., Goh, D.H.-L., Lee, C.S., Lee, E.W.J., Theng, Y.L.: Uncovering temporal differences in COVID-19 tweets. Proc. Assoc. Inf. Sci. Technol. **57**(1), e233 (2020)
33. Yin, J., Wang, J.: A Dirichlet multinomial mixture model-based approach for short text clustering. In: Proceedings of the 20th ACM SIGKDD International Conference on Knowledge Discovery and Data Mining, pp. 233–242. Association for Computing Machinery, New York (2014)

Development of a Text Classification Model to Detect Disinformation About COVID-19 in Social Media: Understanding the Features and Narratives of Disinformation in the Philippines

Hans Calvin Tan(✉)(iD), Maria Regina Justina Estuar(iD), Nicole Allison Co,
Austin Sebastien Tan, Roland Abao, and Jelly Aureus

Ateneo de Manila University, Metro Manila, Philippines
hans.tan@obf.ateneo.edu
https://www.ateneo.edu/

Abstract. As of January 02, 2022, the Philippines is combating another surge in COVID-19 cases. With vaccinations still ongoing, the country remains vigilant and the government continues to promote compliance to minimum health standards as preventive measures to minimize the spread. Disinformation remains a challenge especially if compliance to minimum health standards and adoption of health interventions are necessary to curb the spread of COVID-19. Incorrect and unverified information about the virus increased as well which continues to run rampant in social media and with minimal models to detect disinformation in a Philippine context. The study aimed to understand the features of disinformation of COVID-19 in a Philippine context with the goal of creating a text classification model to detect disinformation of COVID-19 in social media to promote vaccine usage in the country. The usage of social network analysis was performed to understand the narratives present regarding COVID-19 disinformation. Words related to vaccines, government corruption, and government mismanagement were prevalent under the disinformation categories of "False" and "Mostly False" while words related to health information such as cases or vaccine counts were prevalent under the "Mostly True" and "True" category. Linear SVM text classification model performed the best through accuracy, precision, and recall in detecting disinformation by using TF-IDF as a feature compared to using both TF-IDF and n-grams. Disinformation narratives revolved around the idea of COVID-19 cases/vaccines, government mismanagement, and regulations. Results showed that disinformation caused distrust of the government's management over the pandemic. Moreover, the spread of disinformation was contained to the user itself and spread to at least one other user.

Supported by the Ateneo Social Computing Science Laboratory, Ateneo de Manila University.

G. Meiselwitz (Ed.): HCII 2022, LNCS 13315, pp. 370–388, 2022.
https://doi.org/10.1007/978-3-031-05061-9_27

Keywords: COVID-19 · Philippine disinformation · TF-IDF · N-Grams · Text classification model · Social network analysis

1   Introduction

As of this writing, the Philippines is combating another surge in COVID-19 cases. In February, total cases have surpassed two million, where the current case numbers is increasing at an average of 10%. As of February, 04 2022, vaccine statistics indicated that over 59.8 million people had been fully vaccinated which resulted in over 54.6% of the population vaccinated [14]. Transmission reduction is affected by compliance of the public to minimum health standards including social distancing and wearing of masks. Increased vaccination coverage through vaccine acceptance will also reduce transmission of COVID-19 cases. Both interventions rely on public's trust in the government's policy and implementation plans.

Filipinos rely on social media such as Facebook for reading the news as it has been reported that 60% to 90% of the redirects coming from Facebook lead to news websites. Filipinos are known to use social media at a great scale, where around 97% of the population currently has signed up under the social media platform, Facebook [13]. Hence, this also means that information about preventive measures and practices have become more widespread in social media as Filipinos relied on social media to be informed about the latest news [20].

This reliance on social media for information also provided opportunity for other users to spread unverified facts about different topics including information on the COVID-19 virus. This act of providing misleading information that has the function to mislead is known as disinformation [16]. This may either come from individual posts, and both fake or real news about the main topic, such as COVID-19, but also related topics such as vaccines [9]. Once the spread of unverified or incorrect news is proliferated, this disinformation may have negative effects with regards to the infection rates, and vaccination rates in the Philippines [8,22,28].

Recently, social media has enabled a feature where users can see a link leading to credible health info sources regarding COVID-19 when articles about the pandemic are in their social media feed. However, this feature does not tag the post on whether or not it contains disinformation about the pandemic [1]. Though there have been several studies developed to create models to combat disinformation [7,26,32], only minimal research has been done to understand this in a Philippine context using a text classification modeling approach.

Given this, the study aims to understand COVID-19 disinformation in the Philippines by identifying the features of true or false COVID-19 information present in social media. Moreover, the creation of a text classification model to detect COVID-19 disinformation from social media in the context of the Philippines is also performed to serve as a future tool in detecting disinformation. Lastly, the usage of social network analysis is also performed determine patterns in narratives including its transformation.

2   Review of Related Literature

2.1   Understanding Disinformation

Disinformation is defined as misleading information that has the function of misleading. It is argued that disinformation has three main features in itself. Firstly, disinformation is information, where information is understood as a representation of the world as being a certain way. This is otherwise known as semantic content. Information then can be referred to as representational content that is true, or representational content as false [16].

Secondly, disinformation is misleading information likely to cause misbelief. It is important to understand that disinformation does not have to succeed in misleading someone on any given occasion. Thus, disinformation still contains the propensity to mislead a person, and may place people at risk of suffering from harm if it were not to be stopped [16].

Lastly, disinformation is non-accidentally misleading information. This is what separates disinformation from honest mistakes, or satires made by people. Thus, disinformation focuses on the intention of people creating misleading information rather than just checking for the errors in the information [16].

2.2   Disinformation and Social Media

With the definition of disinformation, it can be understood that disinformation can be present so long as people create and share this type of information. This process of creation and sharing is possible online, which leads to the assumption that the creation of online disinformation, where it is false or misleading information that is intentionally spread to people, is also present. Online disinformation coincides with the definition of fake news, where fake news is the creation of news articles that are verified to be false, and intentionally created to mislead readers [5]. This includes unverified facts as well given that fake news falls under the same topic. Unverified facts may be included under disinformation as it both falls under the definition of disinformation, and its relation to fake news [10]. The spreading of fake news is one of the ways a person can spread disinformation online, especially in social media [12].

The reliance of using social media as a news source has increased in the past few years. Survey has shown that 74% of Twitter users use the platform as a way to read the news, and 68% of Facebook users also use the platform to read the news [30]. It has been reported that at least 97% of the population has a Facebook account in the Philippines. The reliance of using social media as a news source has grown largely in the Philippines. Around 60 to 90% of the redirects found in the website lead into news sites [13]. Moreover, there has been a growing reliance on the use of social media to gain information about the pandemic as well. It has been reported that people create groups to inform each other about the pandemic, and has been used to help the population in response to the lack of action from the government [23]. This has been a citing concern for the people as it can disinformation can easily be spread as information cannot be immediately verified [20,23].

With the pandemic still on the horizon, there has been a huge reliance on obtaining information to prevent being affected by COVID-19. People rely on multiple sources of information as truth to determine the events of the COVID-19 pandemic.

Statistics show that the people in the United States during the COVID-19 pandemic mostly rely on the government as a source for understanding the disease. However, only half the people who rely on the government as a source for understanding the disease actually trust the government's information [4]. Moreover, people rely on traditional news companies next and their level of trust depends on the news outlet and political view of the person. Hence, it can be inferred that the government and traditional news outlets can be used as sources of truth regarding the pandemic [4].

Another study delved into the research of understanding fact-checking, and which sources provide the right information about the pandemic. An estimate of 900% increase occurred with fact-checking instances during the pandemic, where most people rely on fact-checking tools such as the International Fact-Checking Network, and Google Fact-Checking Tools. These were used to debunk disinformation present in mostly social media posts [11].

Another approach was through the usage of the PolitiFact's APIs where the LIAR dataset has been made. This contained several real and fake news found in the website through categorizing them as true, mostly true, half-true, mostly false, and false. It can be understood that mostly true and mostly false are seen as perceived truth and perceived false so as to differentiate the two from true and false [33]. This would help in the understanding the features of the different categories by knowing the implications on how these tweets would be categorized.

2.3   Feature Set of COVID-19 Disinformation

The different categories of disinformation would imply different sets of features. Studies have been made to detect disinformation about COVID-19. Two studies approached this by obtaining sources from fact-checking websites such as factcheck.org, snopes.com, and such by feeding in keywords about COVID-19 [15,25]. The latter approached this by collecting correct information through obtaining tweets from official and verified Twitter handles of official government accounts, medical institutes, and news channels. Information collected is then used to create a binary classification for distinguishing real and fake information disinformation on COVID-19 related posts [27].

Another method probes deeper into parts of speech. This feature selection method removes stop words found in the data set followed by keeping verbs, nouns, and adjectives to limit the vector size [15]. Further analysis on the dataset on a token level was performed to identify the 10 most frequent tokens used in either fake, real, or ca combination of both. Results showed that there are significant similarities between the important words in fake news and real news.

Both studies approached the feature extraction by involving the use of term frequency-inverse document frequency (TF-IDF) [15,27]. In addition, the use of n-grams was also included as a feature selection method in text classification. This allowed the understanding of themes behind the sequences of words which may create a pattern noticeable within the different disinformation categories. Studies have shown the usage of word n-grams to be useful in detecting disinformation where remarkable accuracy was obtained from the TF-IDF based machine learning models. [6,29].

2.4 Text Classification Models to Detect COVID-19 Disinformation

A supervised learning approach was used to model the detection of disinformation, where multiple supervised learning models were tested to determine which model generated the highest accuracy to detect disinformation [15,27]. If a class were to have underrepresented samples, then the usage of Synthetic Minority Oversampling Technique (SMOTE) was done onto the training set of the model to over represent the underrepresented samples. SMOTE is the process of creating new samples based on the training set to improve the training of underrepresented classes. This aided in the creation of the model by avoiding over-fitting from the other classes [21]. It has been cited that the Linear Regression produces the best results [15]. The latter cited that Support Vector Machines produced the highest accuracy, while Linear Regression was the second highest [27]. Thus, the usage of Support Vector Machines or Linear regression was recommended if similar studies were to approach a similar approach [27]. It should be noted that both studies had different approaches with feature selection and extraction, hence results vary in the context of their methods. This study will perform similar approach in comparing text classification models.

2.5 Approaches in Social Network Analysis of Disinformation

Another approach in understanding disinformation, specifically, how it spreads is performed using social network analysis. A social network analysis assumes that a person's social life is created from relations and the patterns formed by these relations. It involves both, a social network, and network analysis. Social networks are a set of nodes or members that are connected by other types of relations [17,19,24].

The development of a graph for social network analysis can be done in multiple methods. One approach has proposed the usage of NetworkX and iGraph for creating the graph of the network. These both are available in the python library which can be easily be implemented on social media data [3,31]. A performance comparison was done to determine which library can create a graph the most efficient way, and results show that iGraph leads ahead against the other two libraries available [3]. Another method has proposed the usage of Gephi, which is a program in Java made to create network graphs as well. This was done by mapping out the person sharing the tweet as a node, and the people tagged in the tweet as edge nodes. Color codes for each cluster of data were created to

understand the interaction occurring in a specific cluster faster [18]. Based on
the discussed methods, the study would go for the usage of Gephi as a graphing
tool because this was used on Twitter data as discussed in this section.

2.6   Summary

Disinformation is misleading information that has the function of misleading. It
is likely to cause misbelief, and it does not have to succeed in misleading someone.
It is also non-accidental. This separates disinformation from honest mistakes, or
satires. Hence, this coincides with the definition of fake news, which has been
prevalent in social media. This has been a concern for countries who rely on
social media for news. Fact-checking tools are present online to counteract this
problem. Moreover, popular datasets such as the LIAR dataset has been made to
open the avenue of creating machine learning tools in combatting disinformation.
Several feature sets have been used in analyzing COVID-19 disinformation. The
usage of TF-IDF and n-grams were used for text classification models. Models
such as Linear Regression, or Support Vector Machines were used in these text
classification models where Linear Regression yielded higher accuracy compared
to the other model. SMOTE can also be used as a way to further boost the
underrepresented classes during training to avoid over-fitting. The usage of social
network analysis, alongside the creation of a text classification model, also helps
in the understanding of COVID-19 disinformation in social media. Tools such as
NetworkX, iGraph, or Gephi are useful in creating these graphs but Gephi has
been proved capable of generating graphs from Twitter data.

3   Methodology

3.1   Social Media Data Extraction

The study focused on mining or extracting data from Twitter for this research.
Twitter API was used to access the data. Keywords such as #covid19ph,
#covid19, covid19, coronavirus, bakuna, and resbakuna were used to search
for the tweets that contain any of these keywords. The language filter was set to
Filipino for the keywords #covid19ph, #covid19, covid19, coronavirus, bakuna,
and resbakuna. Moreover, the language filter for the keyword #covid19ph would
also be used for English. Usernames were mapped to the Account IDs of each
tweet so as to identify the users during Social Network Analysis.

3.2   Representational Labeling

Extracted information was parsed into sentences and prepared in a matrix for
labeling. A coding manual was created to serve as guide in tagging or labeling
of sentences. Two volunteers tasked to label the tweets based on the guide. Only
tweets that were labeled similar were retained for the final data set.

The information was labeled as true, mostly true, half-true, mostly false,
false, and invalid. This approach is similar to the approach made on the LIAR

dataset to determine if the information was either mostly true, or mostly false. Mostly true and mostly false would mean either perceived truth or perceived false. This can be understood as assumptions that present subjectively true or false information. True or false would mean factual data that was either correct or wrong. Half-true would then be the split between the true and false information. This verdict column was used as a basis to detect the difference between real information, and disinformation.

3.3   Data Cleaning

The cleaning process includes adding new columns into the dataset named source and target. Source contained where the tweet came from, and target would be the mentions of the user. However in the event of a retweet, the source would be the user found after the string "RT" in the tweet, while target would be the user who shared the tweet.

The tweets are cleaned by removing the string "RT", and mentioned users once the source and target column has been removed to just contain the content of the tweet itself. The author ID, and time created were also removed as this would not be necessary for the feature extraction process. Rows containing multiple target users because of multiple tweet mentions were duplicated as this would be treated as the source targeting each user one by one. The dataset is also broken down further by removing the stopwords found in the text of the dataset by using the NLTK passage with contains the stopwords in the English language and a local package for Filipino stop words.

Collection words or search words were also removed as this may skew the results from feature selection later on given that these tweets contained words that matched with the keyword for querying. Links were removed as well so as to maintain the tweet in context of the English and Filipino language. Lastly, the tweets were converted to lower case to ensure words that are capitalized would be the same as the lowercase counterparts. A CSV file is exported once the data cleaning has finished. The CSV file is duplicated feature selection and extraction, and social network analysis.

3.4   Feature Selection and Extraction

The dataset is duplicated such that one dataset is used for N-gram tokenization while the other dataset remains the same. Both datasets were transformed to TF-IDF vectors as input to classification model. Word cloud generation and Performance measures were used to determine best features suitable for detecting the different disinformation categories using the methods made by multiple studies [2, 15, 27, 29].

N-grams is used to understand the frequent cluster of words present in each tweet. This can come in the form of bigrams or trigrams, where it is frequency of 2 or 3 words respectively. Using the dataset, the frequency of two and three clusters of words are used. Gensim package in Python was used for the creation of the n-grams.

The parameter used for the model was min_count as 1, meaning this ignored all the bigrams or trigrams made with a count below this parameter. This allowed the analysis of the difference between the combination of words found in disinformation verdict category, rather than using each word as itself.

After n-grams are processed, the process of doing Text Frequency - Inverse Document Frequency (TF-IDF) begins. The algorithm is available through the SKLearn package by importing the TfidfVectorizer package. Under this package, a model is available for training the TF-IDF process, before it extracted this information from the entire dataset. The dataset is used to train the model and transformed right after to create a matrix from the TF-IDF process. A bar chart is created to visualize the top words used in each disinformation category by using the TF-IDF value of each approach. The top 25 words are visualized for both approaches (where one contained n-grams, while the other contained no n-grams). This is used to determine the common theme in each disinformation category.

Using the get_feature_name_out() method of TfidfVectorizer, the study obtained the feature words extracted from each disinformation verdict category which can be used for further analysis on what contributes to each category. This was done on the model that produced the best accuracy and recall through comparing the results based on the two pre-processed datasets. Word clouds were generated composed of the top 25 words found in each category. The top 25 words were obtained by using the weights/values based on the TF-IDF values obtained. The creation of the word cloud was made possible through the usage of the WordCloud package in python.

3.5   Creating the Text Classification Model

The machine learning models used for text classification include: logistic regression, random forest classifier, and linear support vector machine. The data set is split into a 60% training set and 40% validation set. SMOTE was applied to the training set to boost the underrepresented classes inside the dataset. Each underrepresented class was brought up to match the class with the highest samples inside the dataset. SMOTE was not applied for the validation set. The SKLearn library was used for the model given that pre-made models and functions are available in this library. Moreover, the imbalanced learning package was used for the SMOTE process. A confusion matrix was generated to further aid the understanding of the recall and accuracy of each class, and the model overall.

3.6   Social Network Analysis of COVID-19 Disinformation

The same data set was used for the Social Network Analysis using the NetworkX package in python. The data used the source column as the nodes, and the target column as the edges for the network graph. Each node included the row index of each tweet so as to later analyze the clusters of networks, and how it spreads from one node to another. Moreover, the node includes the type of disinformation as classified by the model. This allowed the analysis of the spread of disinformation if a user happened to spread disinformation. The degree of each node was computed as well through the usage of NetworkX, excluding self-loops.

Louvain Community Detection package was used to determine the grouping of the nodes to determine the clusters of nodes in the network graph. The creation of clusters is used to understand each narrative present in the social network, and analyze its spread throughout the network. A CSV file is generated as an output and used in Gephi, a graph mapping tool.

After the data has been imported in Gephi, the modularity is computed to further detect sub-communities in the network graph. The layout algorithm, ForceAtlas 2, is used to layout the network graph and consequently, the social network graph. Labels are included into the network graph to identify the nodes that have the most connections, and how the narratives transpire from one node to another. The disinformation categories "True/Mostly True", "Half True", "Mostly False", and "False" are used as the categories to determine the color the node should receive. The nodes would also have colors depending on the cluster of the nodes based on their interaction. The output graph did not include the labels so as to protect the identities of the users.

4   Results

4.1   COVID-19 Philippine Disinformation Dataset

A total of 9137 tweets composed of multiple tweets ranging from news outlets, user tweets, and such were collected through the Twitter API. The dates of the tweet were from the week of September 16, 2021 until September 21, 2021. A total of 5219 retweets were found and labeled as retweets to be used in determining the connections between users for social network analysis. The dataset is stored in a matrix with the following feature including: author_id, username, created_at, geo, id, language, like_count, quote_count, reply_count, retweet_count, source, and tweet. Moreover, the tweet may serve as weights onto the model given that multiple retweets may mean a higher significance for those topics.

This dataset is processed to only include the username, created_at, and the actual tweet as these were the only required information for the tagging. Table 1 shows the results of the inter-rater reliability. A total of 4779 remained, equivalent to 52.30% of the dataset remained after going through the inter-rater reliability process.

4.2   Features of the Different COVID-19 Disinformation Categories

The creation of two different sets of charts based on the two different feature approaches, namely: TF-IDF and TF-IDF with n-grams, were made to understand the common words occurring in the tweets depending on the disinformation category. Figures 1 and 2 both show multiple bar charts where Fig. 1 shows the frequency from each disinformation category while Fig. 2 shows the frequency with n-grams from each disinformation category.

Table 1. Count of each disinformation category

Verdict	Count
True	1976
Mostly true	198
Half true	37
Mostly false	96
False	42
Invalid	2430

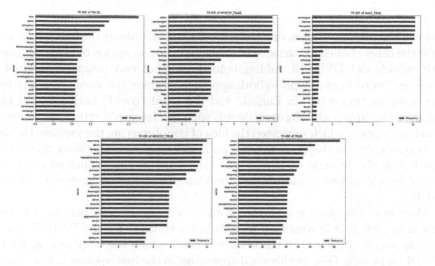

Fig. 1. TF-IDF: top 25 words

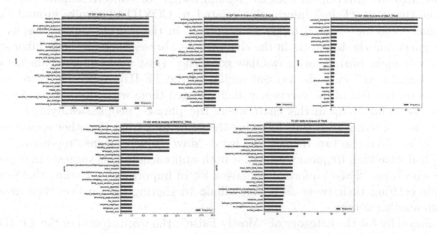

Fig. 2. TF-IDF with N-Gram: top 25 words

Fig. 3. Word cloud (TF-IDF)

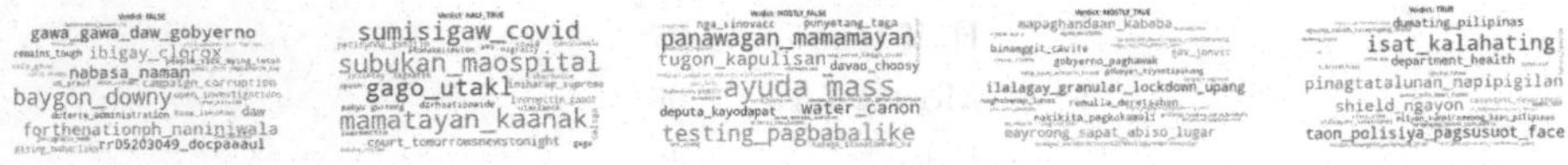

Fig. 4. Word cloud (TF-IDF with N-Gram)

First, by looking at the "False" category, there were two different themes between the two approaches where one focused mostly about the idea of vaccines, while the other challenged the idea of solely just the corruption of the government with regards to COVID-19. Adding onto the idea of vaccines, terms related to "vaccines" word were present in both approaches where the words "gawa" which may mean as "made up" in English, and the word "covid" both incurred high counts in this approach. Moreover, words such as "baygon", "true", or "science" would also appear which suggested the idea of counteracting the vaccines through alternatives such as "baygon" which is an insect repellent. The word "mamatay" (death) was also brought up in relation to the word vaccines which reinforces the idea of people further counteracting the idea of vaccines as protection (Figs. 3 and 4).

Moreover, the presence of the word "corruption" was present for both approaches where it is combined with "administration" for the TF-IDF with n-gram approach compared to the word "vaccine" which was only present for the TF-IDF approach. Despite this word appearing in the first approach, it was more coupled with the words found in the n-gram approach where words or phrases related to the government such as "remain_tough" or "duterte_administration" suggesting that "False" information related to COVID-19 revolve around this theme. "Vaccine" related topics were present in the n-gram version despite it not garnering the top spots in the chart where phrases such as "hoax_lukohan" (hoax & make fun), "gawa_gawa_daw_gobyerno" (just made up government) or "baygon_downy" were also present similar to the TF-IDF approach where people were skeptical of the vaccines, and challenged them with alternatives. Death was also brought up in this chart where the phrase "people_sick_dying_totoo" (people sick dying real) thus following the same idea from the other approach.

Aside from the two themes, the word "daw" which means "reportedly" in English have high frequency values for both approaches. This suggests that people who tweet "False" information gave a lot of importance including this word while crafting their tweet as it may allude to sharing information they heard from another source.

Secondly for the category of "Mostly False", the words found in the TF-IDF approach and the words/phrases found in the n-gram approach were similar to one another where the words from the TF-IDF approach were the building pieces towards the phrases in the n-gram approach. Given this understanding, the common words between the two approaches were related to the politics within the country. Words brought up using the TF-IDF approach did not make sense such as "water", "canon", or "tugon" (instruct) while it showed more relevant meanings in the TF-IDF with n-grams approach. For example, "water_canon" and using the tweets as context, it involved the usage of a "water canon" to counteract people demanding for vaccines, cash aid, or mass testing. This then would give significance to why words or phrases such as "tugon_kapulisan" (instruct_police), "kapulisan", or "ayuda_mass" were present in both approaches. Hence, the common theme that can be seen from both approaches would be related to how the government acted with regards to the COVID-19 pandemic in the Philippines. Another theme present is related to vaccines, but more specifically the vaccines procured from China and the city of Davao, where words such as "Davao_Choosy", or "Sinovac" were present in both charts which suggested the idea on how people did not have a positive sentiment with regards to either procuring or not procuring vaccines from China.

Thirdly for the category of "Half True", the prevailing words found at the top lists of both approaches tend to revolve around the idea of "contracting COVID-19", and "alternative medication". Words and phrases such as "covid", or "subukan_maospital" (try being hospitalized) supported the idea of "contracting COVID-19" as a theme which suggested people possibly informing each other the possibility of being hospitalized because of COVID-19 if they were to follow false information. This was further emphasized when the word "ivermectin" was brought up especially with multiple debates regarding the medication being ineffective for COVID-19. "Death" was brought up once again in this category where the phrase "mamatayan_kaanak" (relative's death) where this may coincide with the idea of "contracting COVID-19" where they may have possibly used relatives as a way to convince them not to push through with those false information.

Fourthly for the category of "Mostly True", the common theme found in this category was more unified in both approaches where the top words mostly discussed ideas related to having the quarantine restrictions implemented to prevent the further spread of COVID-19. Words and phrases such as "mayroong_granular lockdown" (implemented granular lockdown), "sapat" (enough), "abiso" (advice), or "mapaghandaan_kababaan" (preparation & going down) garnered the higher echelon of the charts which suggested the idea that people mostly discussed about the quarantine lockdown restrictions that have occurred in the country to prevent the further spread of COVID-19. Aside from this, the discussions related to having enough rules or preparation were also discussed as these topics mostly coincide with the idea of issuing a lockdown in a country. Thus, this reinforced the idea of people having a perceived truth of the lockdowns being enough to bring down the number of cases in the country.

Lastly for the category of "True", the prevalent idea common between the two charts and in the category itself would be the discussion of either the cases of COVID-19, number of vaccine doses, or other enforcement rules implemented into the country. This may be seen in words such as "cases", "kaso" (cases), "doses", and "Philippines" where it all discussed the factual events happening in the country. Hence, this further emphasized the point of the "True" category having factual ideas as it is hard to refute the number of cases or the number of vaccine doses the country had during the time when the tweets were obtained. Moreover, phrases such as "taon_polisya_pagsuot_face" (year and "ngayon_shield", which meant "police wearing face shield today", signify the implementation of wearing a face shield within the country. This was further highlighted during these months especially since the government emphasized the usage of these items given the number of cases ravaging the country at that time period.

4.3   Text Classification Model Performance

The two datasets with the two different feature approaches were used to train the different machine learning models. Table 2 shows the performance of each model based on the two approaches made for the study. Results show that the feature of TF-IDF overall performed better compared to the feature of TF-IDF with n-grams. This can be observed on all the models, and across all the metrics such as accuracy, precision, recall, and F1 score. Moreover, the model that performed the best is the model that used Logistic Regression while Linear SVM performed the next best among all the models.

Table 2. Text classification model performance

Model	Feature	Accuracy	Weighted average precision	Weighted average recall	Weighted F1 score
Logistic regression	TF-IDF	94.72%	94%	95%	94%
	TF-IDF & N-Grams	93.06%	93%	93%	92%
Random forest	TF-IDF	94.09%	94%	94%	94%
	TF-IDF & N-Grams	93.16%	93%	93%	92%
Linear SVM	TF-IDF	94.30%	94%	94%	94%
	TF-IDF & N-Grams	93.27%	93%	93%	93%

The goal of this study was to create a model to detect disinformation. Results presented in Table 3, show precision and recall scores for "Mostly False" and "False" category of both Logistic Regression and Linear SVM. Linear SVM

Table 3. Text classification model classification report

Classification	Model	Feature	Precision	Recall	F1 score
False	Logistic regression	TF-IDF	80%	57%	60%
	Random forest		100%	57%	73%
	Linear SVM		100%	57%	73%
	Logistic regression	TF-IDF & N-Grams	92%	52%	67%
	Random forest		92%	52%	67%
	Linear SVM		92%	57%	71%
Mostly false	Logistic Regression	TF-IDF	88%	56%	69%
	Random Forest		93%	61%	74%
	Linear SVM		93%	61%	74%
	Logistic regression	TF-IDF & N-Grams	88%	56%	69%
	Random forest		79%	56%	66%
	Linear SVM		82%	56%	67%
Half-true	Logistic Regression	TF-IDF	100%	80%	89%
	Random forest		100%	80%	89%
	Linear SVM		100%	80%	89%
	Logistic regression	TF-IDF & N-Grams	100%	80%	89%
	Random forest		100%	80%	89%
	Linear SVM		100%	80%	89%
Mostly true	Logistic Regression	TF-IDF	94%	73%	82%
	Random forest		100%	65%	79%
	Linear SVM		100%	65%	79%
	Logistic regression	TF-IDF & N-Grams	89%	62%	73%
	Random Forest		98%	60%	74%
	Linear SVM		90%	63%	74%
True	Logistic regression	TF-IDF	95%	100%	98%
	Random forest		94%	100%	97%
	Linear SVM		94%	100%	97%
	Logistic regression	TF-IDF & N-Grams	93%	99%	96%
	Random forest		93%	100%	96%
	Linear SVM		94%	100%	97%

yielded the highest precision and accuracy for these two disinformation categories despite having a lower accuracy compared to Logistic Regression. This meant that under the circumstance of solely detecting for disinformation, then Linear SVM performs the best out of all the models assuming it uses the TF-IDF feature only. Figure 1 shows the feature set of words used in the model.

The low training size may have been the leading factor to why results performed similar across each model, especially the "Half-True" category. The performance of each model may be further improved by adding more data into the training set, especially under the "False", "Mostly False", and "Half True" categories. With the further addition into these categories, additional variances may further improve the text classification model in detecting disinformation of COVID-19 in the Philippines.

4.4   Social Network Analysis of Philippine COVID-19 Disinformation

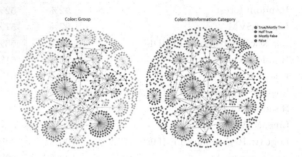

Fig. 5. Social network analysis map (Color figure online)

Figure 5 shows multiple clusters present in the graphs, more evident in the left graph. These clusters mostly revolved around news outlets or prominent health experts within the country, where people retweeted information shared by these nodes. For instance, clusters under the color pink, black, orange, green, cyan, flesh, and blue represent news outlets or health expert clusters. These are representation of how these news outlets or health experts started multiple tweet, and was retweeted by multiple users. Hence, this is why one node can be found in the middle with multiple edges found in these clusters. For instance, information regarding COVID-19 cases updates, or vaccine doses updates were retweeted multiple times hence this can be seen in the graph where there were instances of one node having multiple edges. This may be represented as well with the green color as showing "True/Mostly True" information given that these updates from the news outlets overall do not hold any disinformation.

These narratives were present throughout these clusters, where some narratives coincide with the other clusters as clusters were grouped based on node interaction rather than tweet information. The cyan, pink, black, blue, and green clusters had instances of sharing the same narratives because they report the news occurring in the country. Hence, multiple narratives exists within the different clusters with respect to the different events that happened in the Philippines.

This however did not stop the creation of disinformation where users created "Half True", "Mostly False", or "False" information after retweeting from these news outlets or prominent health experts. For instance, there were several instances where users warned the usage of vaccines as useless because people would still get infected from COVID-19 despite having vaccines. These tweets were created after the news of receiving multiple vaccine doses of different brands.

The narrative of "ineffective vaccines" was immediately marked as "False" or "Mostly False" which explains the multiple red or orange nodes found in the graph near the major clusters. This was coupled with the topic of the approval of vaccines for ages 12–17 thus people were then concerned even further about this topic.

Another narrative which caused the creation of disinformation tweets from the other nodes would be the issues of corruption and mismanagement of COVID-19 funds from the government. Information about the investigation of this topic from news outlets were marked as true given that it was an on-going case. This then spawned multiple government propaganda citing that no mismanagement has occurred within the government despite multiple sources indicating so. This did not stop here as some nodes retweeted the information and created new tweets about these but this time as "False" or "Half True" where these nodes tried to refute the claim based on the disinformation presented.

A similar narrative which also caused the creation of disinformation was caused by the issue of government protocols regarding COVID-19, specifically lock downs and rules. Information about this topic was circulating within the graph as news outlets reported the updates regarding the lockdown situation in the Philippines, and the rules that go alongside it. However, this created multiple tweets within the same topic but as disinformation. There were several instances of users who created new rules to confuse other users and go against the government. Moreover, there were other instances of people going against these rules for security and health concerns. However, this was combined with disinformation as some users cited false information as a reason to go against these rules. This was evident on the proposal of having on-site classes as users expressed the concern of contracting COVID-19 given that the vaccines were ineffective in protecting the people.

These nodes (containing disinformation) tend to mention multiple users in the tweet, hence this explained the reason why these nodes may have multiple edges despite the other nodes not marked with any of the disinformation labels. Moreover, the spread of disinformation was mostly contained and did not spread further than one degree. There were several instances where the nodes that contained disinformation had no edges. This showed how disinformation remained mostly to the user itself, or at least one other user. If the node contained disinformation but was surrounded by multiple edges, this can be analyzed as retweets as the tweet was shared by multiple people. There was one instance of this in the graph where they brought up the topic of using water canons to deter people requesting for COVID-19 aid funds. People supported this "Mostly False" tweet by retweeting it but based on the lower left corner of the verdict graph, the cluster itself was contained and it did not spread further than the retweet.

It is important to note that the disinformation captured in these different narratives fit in the definition of disinformation, and the features discussed in related literature. The different narratives contained information despite being disinformation in itself. This can be seen in the falsely cited rules circulating, and discussed earlier. Moreover, it caused misbelief to readers where it did not have to succeed in misleading the person. This was seen where disinformation targeted multiple users despite not spreading to those users. Disinformation only went up to at least one user in the event the disinformation succeeded in spreading. For instance, this was seen in the discussion of using waters to deter people requesting for COVID-19 aid funds. Lastly, these tweets were non-accidental compared to satires or honest mistakes people make. Hence, disinformation found in the dataset matched the definition and features of disinformation.

Overall, the narratives regarding COVID-19 revolved around the idea of COVID-19 cases, vaccine doses/drives, government mismanagement, and COVID-19 regulations. The cause of disinformation was due to the distrust of the people towards the government's management regarding the pandemic. This was seen in the tweets containing disinformation where it mostly revolved around judging and counteracting the government's actions during the pandemic as seen through the different narratives discussed, and whether or not their actions were effective enough to address the issues brought by the pandemic. Moreover, disinformation found in these different narratives fit in the definition and feature of the word itself.

5   Conclusion

The feature of using TF-IDF yielded an overall higher accuracy, precision, and recall over using TF-IDF with n-grams for all models. Logistic Regression yielded a higher accuracy, precision, and recall over the other two models. However, Linear SVM yielded the highest accuracy when it comes to detecting disinformation. The usage of Linear SVM trained on the TF-IDF feature is therefore recommended for the text classification model of detecting disinformation based on the results in this study. N-gram method is best used to understand the feature words of each category.

The features of disinformation under TF-IDF used in the text classification model vary depending on the disinformation category the feature was associated with. False information is associated with words such as "gawa", "daw", "baygon", or "clorox" thus suggesting a common theme of vaccinations to be associated with the category because it infers to people calling COVID-19 or vaccines as a hoax, and using alternative medication instead of vaccines. Moreover, corruption was a common theme as well based on the words "corruption" and "administration" suggesting the mismanagement of the government during the pandemic. Mostly False is associated with words such as "kapulisan", "pananawaga", "canon", or "ayuda" which can be inferred as COVID-19 policies to be associated with this verdict. This is possible because of the policy of "ayuda", and the "kapulisan" enforcing the policies. Moreover, Half True is associated with words such as "maospital", "subukan", "kaanak", or "mamatayan" which infers a common theme of contracting COVID-19, or death from the disease. Mostly True is associated with "mayroong", "ilalagay", "abiso", "sapat", or "mapaghandaan" which were words common in the theme of lockdowns and government policies. Lastly, True is associated with words found with factual data such as "kaso", "cases", "health", "department", "Philippines", or "doses" where these words form up tweets related to information about the COVID-19 and vaccine situation in the Philippines.

Social Network Analysis was used to then determine the narratives of COVID-19, and how it transformed from one narrative to another. The common narratives are: idea of COVID-19 cases, vaccine doses/drives, government mismanagement, and COVID-19 regulations. The spread of disinformation remained mostly to the user itself, or at least one other user. This was caused by the people's distrust towards the government's management regarding the pandemic. Hence, it was a leading cause to why disinformation was created from these narratives.

References

1. Coronavirus information hub for media. https://www.facebook.com/formedia/solutions/coronavirus-resources
2. Ahmed, W., Vidal-Alaball, J., Downing, J., Seguí, F.L., et al.: Covid-19 and the 5g conspiracy theory: social network analysis of Twitter data. J. Med. Internet Res. **22**(5), e19458 (2020)
3. Akhtar, N.: Social network analysis tools. In: 2014 Fourth International Conference on Communication Systems and Network Technologies, pp. 388–392. IEEE (2014)
4. Ali, S.H., Foreman, J., Tozan, Y., Capasso, A., Jones, A.M., DiClemente, R.J.: Trends and predictors of Covid-19 information sources and their relationship with knowledge and beliefs related to the pandemic: nationwide cross-sectional study. JMIR Publ. Health Surveill. **6**(4), e21071 (2020)
5. Allcott, H., Gentzkow, M.: Social media and fake news in the 2016 election. J. Econ. Perspect. **31**(2), 211–36 (2017)
6. Amjad, M., Sidorov, G., Zhila, A., Gómez-Adorno, H., Voronkov, I., Gelbukh, A.: "bend the truth": Benchmark dataset for fake news detection in Urdu language and its evaluation. J. Intell. Fuzzy Syst. **39**(2), 2457–2469 (2020)
7. Bang, Y., Ishii, E., Cahyawijaya, S., Ji, Z., Fung, P.: Model generalization on Covid-19 fake news detection. arXiv preprint arXiv:2101.03841 (2021)
8. Barua, Z., Barua, S., Aktar, S., Kabir, N., Li, M.: Effects of misinformation on Covid-19 individual responses and recommendations for resilience of disastrous consequences of misinformation. Progr. Disaster Sci. **8**, 100119 (2020)
9. Bernadas, J.M.A.C.: Reimagining the "public" in public health: exploring the challenges of and opportunities for public relations research in public health in the Philippines. Publ. Relat. Rev. **47**(3), 102043 (2021)
10. Bondielli, A., Marcelloni, F.: A survey on fake news and rumour detection techniques. Inf. Sci. **497**, 38–55 (2019)
11. Brennen, J.S., Simon, F., Howard, P.N., Nielsen, R.K.: Types, sources, and claims of Covid-19 misinformation. Reuters Inst. **7**(3), 1 (2020)
12. Colliander, J.: "this is fake news": Investigating the role of conformity to other users' views when commenting on and spreading disinformation in social media. Comput. Hum. Behav. **97**, 202–215 (2019)
13. David, C.C., San Pascual, M.R.S., Torres, M.E.S.: Reliance on Facebook for news and its influence on political engagement. PloS ONE **14**(3), e0212263 (2019)
14. DOHPH: Coronavirus pandemic (Covid-19). Our World in Data (2020). https://doh.gov.ph/covid19tracker/

15. Elhadad, M.K., Li, K.F., Gebali, F.: Detecting misleading information on Covid-19. IEEE Access **8**, 165201–165215 (2020)
16. Fallis, D.: What is disinformation? Libr. Trends **63**(3), 401–426 (2015)
17. Freeman, L.: The development of social network analysis. Study Sociol. Sci. **1**(687), 159–167 (2004)
18. Gruzd, A., Mai, P.: Going viral: how a single tweet spawned a Covid-19 conspiracy theory on Twitter. Big Data Soc. **7**(2), 2053951720938405 (2020)
19. Knoke, D., Yang, S.: Social Network Analysis. Sage Publications, Upper Saddle River (2019)
20. Lacsa, J.E.M.: Covid-19 infodemic: the role of social media and other digital platforms. J. Publ. Health (2021)
21. Liu, Y., Boukouvalas, Z., Japkowicz, N.: A semi-supervised framework for misinformation detection. In: Soares, C., Torgo, L. (eds.) DS 2021. LNCS (LNAI), vol. 12986, pp. 57–66. Springer, Cham (2021). https://doi.org/10.1007/978-3-030-88942-5_5
22. Loomba, S., de Figueiredo, A., Piatek, S., de Graaf, K., Larson, H.J.: Measuring the impact of exposure to Covid-19 vaccine misinformation on vaccine intent in the UK and US. MedRXiv (2020)
23. Maravilla, M.I.: Covid-19 survivors Philippines: towards the promotion of public health during the Covid-19 pandemic. J. Publ. Health (2021)
24. Marin, A., Wellman, B.: Social network analysis: an introduction. SAGE Handb. Soc. Netw. Anal. **11**, 25 (2011)
25. Chen, B., et al.: Transformer-based language model fine-tuning methods for COVID-19 fake news detection. In: Chakraborty, T., Shu, K., Bernard, H.R., Liu, H., Akhtar, M.S. (eds.) CONSTRAINT 2021. CCIS, vol. 1402, pp. 83–92. Springer, Cham (2021). https://doi.org/10.1007/978-3-030-73696-5_9
26. Paka, W.S., Bansal, R., Kaushik, A., Sengupta, S., Chakraborty, T.: Cross-sean: a cross-stitch semi-supervised neural attention model for Covid-19 fake news detection. Appl. Soft Comput. **107**, 107393 (2021)
27. Patwa, P., et al.: Fighting an infodemic: Covid-19 fake news dataset. arXiv preprint arXiv:2011.03327 (2020)
28. Pierri, F., et al.: The impact of online misinformation on us Covid-19 vaccinations. arXiv preprint arXiv:2104.10635 (2021)
29. Pizarro, J.: Using n-grams to detect fake news spreaders on Twitter. In: CLEF (Working Notes) (2020)
30. Tandoc, E.C., Jr., Lim, D., Ling, R.: Diffusion of disinformation: how social media users respond to fake news and why. Journalism **21**(3), 381–398 (2020)
31. Tsvetovat, M., Kouznetsov, A.: Social Network Analysis for Startups: Finding Connections on the Social Web. O'Reilly Media, Inc., Newton (2011)
32. Vijjali, R., Potluri, P., Kumar, S., Teki, S.: Two stage transformer model for Covid-19 fake news detection and fact checking. arXiv preprint arXiv:2011.13253 (2020)
33. Wang, W.Y.: "liar, liar pants on fire": A new benchmark dataset for fake news detection. arXiv preprint arXiv:1705.00648 (2017)

A Comparison of Web Services for Sentiment Analysis in Digital Mental Health Interventions

Toh Hsiang Benny Tan[1](✉) (iD), Sufang Lim[2] (iD), Yang Qiu[1,2] (iD),
and Chunyan Miao[1,2] (iD)

[1] School of Computer Science and Engineering, Nanyang Technological University, Singapore,
Singapore
{bennytanth,qiuyang,ascymiao}@ntu.edu.sg
[2] Joint NTU-UBC Research Centre of Excellence in Active Living for the Elderly, Nanyang
Technological University, Singapore, Singapore
sufang@ntu.edu.sg

Abstract. The use of web services allows for an easy and cost-effective way to implementation natural language processing capabilities such as sentiment analysis in digital interventions such as those used in mental healthcare. To the best of our knowledge, the majority of studies to date focus on the use of sentiment analysis for the analysis of user reviews and social platforms. This study thus aims to explore the use of 18 currently available web services in the analysis of user submitted content from a digital mental health intervention. The web services are compared on the basis of their accuracy, precision, recall, f-measures and mean square error. Given the sensitive nature of user content from digital mental health interventions, we also explored how the various web services handled the data submitted to them for analysis. The results of the study provide other researchers with a better idea of the performance and suitability of the various web services for use in digital mental health interventions.

Keywords: Sentiment analysis · Mental health intervention · Web service

1 Introduction

In today's high pressure work environment, mental health issues are a growing concern, affecting multiple aspects of a person, such as their physical well-being [1], and imposing an increasingly heavy burden on their host countries [2]. As traditional treatment methods are manpower intensive, involving mental health care practitioners such as psychiatrists, psychologists, therapists or counsellors, significant portions of people suffering from mental health issues do not receive treatment. Of those that do receive treatment, extended periods of "clinical whitespace" between visits [3], result in delays in identifying critical changes in a patient's mental state, and preventing timely intervention [4].

It is fortunate then, that through the use of technology, mental health services can now by easily accessed, bypassing barriers to traditional service such as availability, geographic location, cost, and confidentiality [5, 6]. While early use of technology

G. Meiselwitz (Ed.): HCII 2022, LNCS 13315, pp. 389–407, 2022.
https://doi.org/10.1007/978-3-031-05061-9_28

mainly involved the collection of data for analysis by clinicians [7], technology has evolved to the point that the digitalization of mental health interventions such as mood and behavior journals are now easily accessible [8].

With advent of technologies such as machine learning and artificial intelligence, the use of technologies such as natural languagepProcessing (NLP) techniques in digital mental health interventions are now being explored. One such NLP technique being explored is that of sentiment analysis. While studies have shown the potential for sentiment analysis to be used for the personalization of content [9] or the identification of critical mental state changes such as suicide ideation [10, 11], a recent review on the use of machine learning and NLP in mental health [12] found that NLP techniques in mental health are currently most frequently used to analyze existing data from medical records or social media, rather than directly in digital mental health interventions.

One reason for this is lack of presence in the digital mental health intervention space is that sentiment analysis has traditionally been viewed as a computer science domain. However, with the rise of web services, there is now the possibility for sentiment analysis to be integrated into digital mental health interventions without the technical know-how or cost required to develop and implement a sentiment analysis web service inhouse [9].

While there are many web services available which can carry out the task of sentiment analysis, there are very few studies comparing the performances of these web services in analyzing content extracted from digital mental health interventions. In addition, a rapidly changing web service landscape means that information regarding existing web services can quickly become outdated within a few years. As such, this work aims to fill achieve two purposes.

The first purpose is to provide up to date information on currently available sentiment analysis web services, while the next purpose is to compare the performance of these web services using data collected from the clinical trial of a digital mental health intervention for parents of children with chronic life-threatening illness [13]. The remainder of this paper is structured as follows: Sect. 2 explores the main concepts and techniques currently used in sentiment analysis. Section 3 describes the web services which were used in this study focusing on their sentiment analysis service. Section 4 describes how the study was carried out, covering data collection, extraction, preprocessing, implementation, and analysis of the data, as well as how the web services were evaluated. Section 5 presents the findings and discussion, while Sect. 6 and 7 details the limitations and conclusion respectively.

2 Background

Sentiment analysis is a natural language processing technique, which aims to understand an author's opinion towards different subjects, and can target subjects at different levels ranging from broad subjects such as an experience, to specific subjects such as an object [14]. Sentiment analysis is considered a classification problem [15], and classification normally takes place based on the polarity of the input. Such classification can either take the form of a binary sentiment classification where sentiment is classified into two categories, positive and negative, or a multi-class sentiment classification where sentiment is classified into more than two categories, such as positive, neutral and negative [16].

Just as the subject of a user's opinion can vary in scope, sentiment analysis can take place at either the document, sentence, or entity/aspect level. At the document level, the entire document is considered as a single information unit and analyzed. One common approach to document level sentiment analysis is decomposing the document into smaller units such as sentences and entities, classifying the sentiment of these smaller units, then combining the results to obtain a single classification. As sentences can be considered short documents, the approaches to document and sentence level sentiment analysis can be considered the same [14]. For sentence level sentiment analysis, the sentence is considered as a single information unit. It is then decomposed into individual words, the sentiment orientation of the individual words classified, then the results combined. At the aspect/entity level, sentiment analysis involves not only identifying the sentiment orientation of each word, but also the target towards which the sentiment is directed. This target can be an entity such as a cup, or an aspect or feature of the entity such as its handle.

2.1 Sentiment Analysis Approaches

Based on a survey of sentiment analysis algorithms and applications by Yousef, Medhat and Mohamed [15], sentiment analysis techniques can be broadly categorized into either machine learning approaches or lexicon-based approaches as can be seen from Fig. 1.

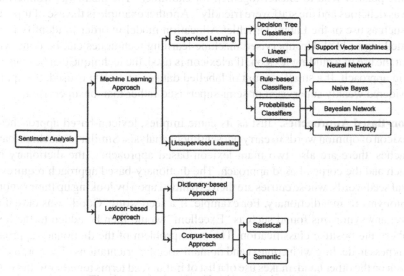

Fig. 1. Sentiment classification techniques [15]

Machine Learning Approaches. Machine learning approaches treat sentiment analysis as a classification problem and tries to solve this problem by apply machine learning techniques to syntactic and linguistic features. Machine learning approaches can be grouped into two main categories, supervised and unsupervised. Supervised approaches are normally used when there is a large amount of labeled training data available.

Supervised Machine learning approaches can generally be broken down in to 5 distinct steps [17]. The first step in the process is data pre-processing. In this stage text is tokenized into n-grams in order to create a bag-of-words. The next step involves feature generation in order to capture patterns in the data. N-gram presence and frequency are two of the most widely used features, while other common features include number of negations, length of message, as well as number of superlative and comparative adjectives. As N-grams are often the main features in a classifier, the larger the dataset, the larger the dimensionality of the dataspace. Due to this, the third step is feature selection, in which only features with high predictive power are retained and the rest discarded. Proper selection of features serves to both improve the efficiency of the algorithm while preventing overfitting. The fourth step is to train the selected algorithm. In sentiment analysis, some commonly used techniques include Support Vector Machine, Naïve Bayes and Neural Network. Once the model has been trained, the fifth and final step is the evaluation of the model. Several metrics commonly used for evaluating the effectiveness of a classification model include its accuracy, precision, recall and f-measure.

Given that the amount of labeled training data can vary significantly between domains, in the event that labelled training data is lacking, unsupervised machine learning is used. Unsupervised machine learning requires at minimum, unlabeled input samples. Once this is available, various unsupervised learning techniques can be applied. One example of such techniques includes clustering, whereby the main idea is that words of the same polarity tend to cluster together, for example, "The meal was wonderful, the food was delicious and the staff were friendly". Another example is the use of topic modeling such as use of the Latent Dirichlet Allocation model in order to identify relevant or related topics [18]. Unsupervised machine learning techniques can be done with or without the use of a guiding lexicon. If a lexicon is used, the technique can be considered a hybrid approach. If a small amount of labelled data or semi data is used, the approach is considered weakly supervised or semi-supervised rather than unsupervised.

Lexicon-Based Approaches. Just as its name implies, lexicon-based approaches rely on a lexicon of opinion words to carry out sentiment analysis. Similar to machine learning approaches, there are also two main lexicon-based approaches, the dictionary-based approach and the corpus-based approach. The dictionary-based approach requires a set of initial seed words, whose entries are then expanded upon by looking up their synonyms and antonyms from a dictionary. For example, if a seed word "Good" was classified as positive, any synonyms found such as "Excellent" would then be added to the lexicon and inherit the positive classification. The main problem of the dictionary approach is its weakness in dealing with context and domain specific orientations. The corpus-based approach on the other hand, makes use of a list of initial seed terms together with syntactic patterns to identify other opinion words and grow its lexicon. For example, starting with a list of adjectives as seed terms, one might use linguistic constraints such as connectives to identify the orientation of conjoined adjectives. For example, if a seed adjective "Good" was classified as being positive, when analyzing a corpus, any adjective following an "And" connective would also be classified as positive, such as "Good and Cheap". On the other hand, any adjective following a "But" connective would be classified as having the opposite negative orientation, for example, "Good but expensive". The benefit of the corpus approach is its ability to better handle context and domain specific orientations.

Methods of identifying patterns include statistics based methods such as Latent Semantic Analysis (LSA) [19], or semantics based approaches such as assigning a sentiment value to a set of words, then calculating the similarity between words and assigning a sentiment value based on semantic similarity [20].

Lexicon-based approaches are carried out in three steps. The first step is preprocessing where various Natural Language Processing tasks such as part of speech tagging, stemming, stop-word removal, negation handling and tokenization and are carried out on the input. Next each token is evaluated against the lexicon to assign it a polarity, normally between −1 and 1. Lastly, the polarity of all words with a non-zero polarity value are added together and divided by the number of words with non-zero polarity, to arrive at the average polarity of the input.

Unlike machine-based approaches whose classification quality depends on the quality of the training data and feature selection; the classification quality of a lexicon-based approach is dependent on the quality of its lexicon. Because of this, advance techniques, such as machine learning are now being used to create the lexicon. Such techniques can be considered a hybrid approach, and an example of this is the use of supervised machine learning to generate the lexicon. To do this, words are tokenized as per machine learning requirements. Following this, each token is compared against the labelled sentiment of its containing sentence. Tokens then take on the polarity of the polarity of the sentences which frequently feature them. For example, if the word "expensive" often appears in sentences labelled negative in the training dataset, the word "expensive" will be assigned with a negative polarity.

2.2 Related Works

Sentiment analysis has long had a history of being applied to not only product reviews, but also in the analysis of news articles [21] and in the prediction of stock performances [22]. In recent years, the use of technological innovations such as mobile devices and machine learning has been seen as a possible means of scaling up healthcare services, and one such possibility is the use of natural language processing techniques to scale up mental health analysis.

While studies have been done in the use of sentiment analysis for mental wellbeing, such as via the monitoring of intra-company communications [3], or the analysis of diary entries [23], the majority of such studies deal with the analysis of the target content in an offline format using offline implementations of Valence Aware Dictionary for Sentiment Reasoning (VADER) or Linguistic Inquiry and Word Count (LIWC) to analyze collated data in a non-real time manner.

Studies which make use of sentiment analysis in a real-time manner to customize content have started to appear [9, 24], for these studies that make use of web services to carry out such sentiment analysis, beyond acknowledging the cost effectiveness and ease of implementation the use of such services afford them, few explain how the web service implemented was selected among the services available.

One reason for this is that while studies have been done on the suitability of the various available web services for the purposes of analyzing content from social networking sites [25], tourism sites [26], and Massive Open Online Courses (MOOCs)

[27], there are currently no studies done on the performances of the various available web services in terms of analyzing responses of the type which might be obtained from digital interventions.

In addition, in this age of rapid technological growth, the technological landscape changes just as rapidly. For example, in 2015, Serrano-Guerrero et al. carried out a review of 15 sentiment analysis web services for the purpose of analyzing movie review, data from twitter as well as amazon product reviews [28]. Today, just 6 years later, 7 of the 15 web services no longer operate or provide dedicated sentiment analysis services, and new web service providers have sprung up to take their place. This coupled with the fact that machine learning algorithms continually change depending on the data used to train them, highlights the need to periodically review the sentiment analysis web service landscape in order to stay updated and make relevant informed decisions.

3 Web Services

For the purpose of this study, numerous web services for automatic sentiment analysis were considered. In order to obtain comparable results, the web services in this study were selected based on the following criteria.

The first criteria is that the sentiment analysis web service had to be available in the form of a stand-alone service. The reason for this is that as this study is looking at the feasibility of making use of such services in the context of digital mental health interventions, hence sentiment analysis which was tightly coupled with email or social media monitoring service would not be suitable.

The second criteria is the ability for the service to accept input in the form of text of varying lengths. For example, some sentiment analysis web services found were meant to evaluate content such as blogs, and only accepted input in the form of a universal resource locator (URL), while other sentiment analysis services specialized in analysis Twitter content, thus limiting the length of the text input to 280 characters. Such services were deemed unsuitable as responses from digital mental health interventions were most likely to be in the form of text, while open ended responses to such intervention activities could easily exceed the 280-character limit imposed by such services.

Lastly, for the purpose of this study, free-of-charge access to the web service should be available in the form of either a free usage tier, or a free trial. Services which did not provide either a free usage tier or a free trial were not included in this study. The lone exception to this is the LIWC API. LIWC or Linguistic Inquiry and Word Count, is a computerized text analysis software, used in numerous studies for the analysis of textual data. The web service version is known as LIWC API and is only available commercially, while the academic version is only available as an offline software and is known as LIWC2015. While there were no trials available, an analyst at Receptiviti, the company working on the LIWC API, confirmed that the only minor difference between the LIWC API and LIWC2015 was the way in which words were counted. In the LIWC API for example, "10,000" would be counted as one word, however, in LIWC2015, it would be considered as two words, "ten thousand". As the LIWC API's pricing model was based on a word count rather than character count or API calls, such a change, while inconsistent with LIWC2015, could be considered beneficial to the end user from a utilization and cost perspective. Given that LIWC is often considered the gold standard in existing research regarding the psychometric properties of textual data, and that there was only minor difference between the web service and LIWC2015, the research team felt that the inclusion of LIWC was important, and that LIWC2015 was a reasonable substitute to evaluate the sentiment analysis performance of the LIWC API.

After excluding web services which did not fulfill the above criteria, 17 web services and LIWC2015 was selected for inclusion in this study. The web services included are Amazon Comprehend, Cloudmersive, Dandelion, Google Cloud Natural Language, IBM Watson Natural Language Understanding, Komprehend, Lexalytics Semantria, MeaningCloud, Microsoft Azure Text Analytics, Rosette Text Analytics, Salesforce Einstein, Sentigem, Sentiment140, Text-processing, Text2Data, Textgain, and Twinword.

4 Methodology

In order to compare the web services, the following 3-stage process shown in Fig. 2 was used. Each stage consists of a number of tasks and outputs, which would be used in the later stages.

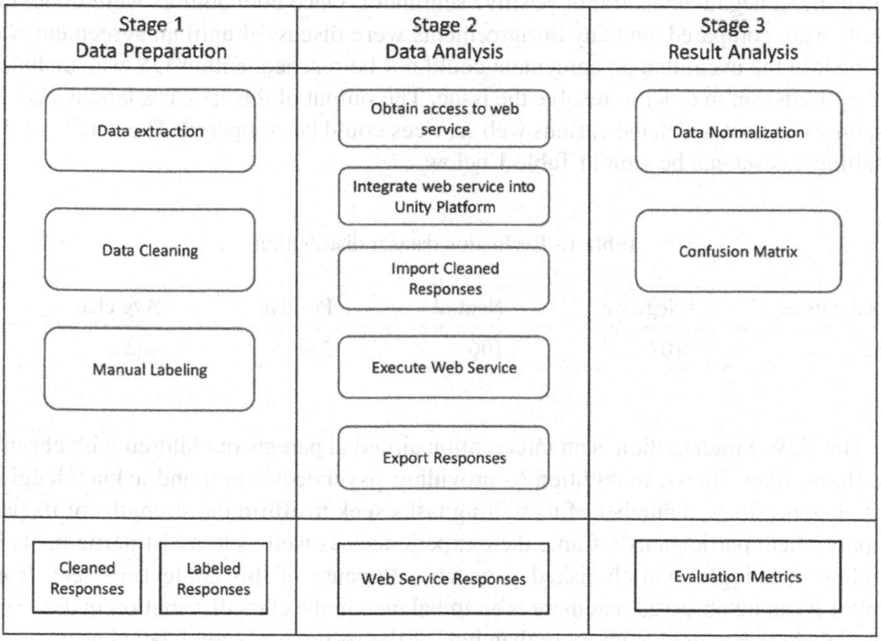

Stage 1 Data Preparation	Stage 2 Data Analysis	Stage 3 Result Analysis
Data extraction	Obtain access to web service	Data Normalization
Data Cleaning	Integrate web service into Unity Platform	Confusion Matrix
	Import Cleaned Responses	
Manual Labeling	Execute Web Service	
	Export Responses	
Cleaned Responses Labeled Responses	Web Service Responses	Evaluation Metrics

Fig. 2. 3-Stage sentiment analysis comparison process

4.1 Data Preparation

The data from this study, comes from the Narrative E-Writing Intervention for Parents of Children With Chronic Life-Threatening Illnesses pilot trial. The first task in stage

1 is the extraction of the relevant data to be used the master dataset. Data included consisted of participant responses to their assigned e-writing tasks, as well as the response identification number of each response for reference.

Once the data had been successfully extracted, the next task was the cleaning of data. To keep within the service limitations of the various web services, only responses with lengths between 10 characters to 8000 characters were included. Next, due to the sensitive nature of the data, the remaining data was reviewed for proper de-identification of the participants. Once this was done, the data was once again reviewed to remove content such as emoticons, special characters such as "#" or "@", unnecessary space, duplicate words and web-links. For the purposes of this study, stop words and punctuation were not removed, as such processing should be handled by the sentiment analysis service. The output of this task is a clean set of responses for use with the web services in stage 2.

Once a clean set of responses was available, the final task in stage 1 was the manual labeling of sentiment for all the responses. In NLP research, human-annotated data is often used as the "gold standard", in order to provide ground truth about the data. Whenever a tool differs from this annotation, it can be considered wrong. In order to do this, authors TTHB and LS carried out the labeling separately, labeling each response with either a negative, neutral or positive sentiment. Once both authors were done, the results were compared, and any disagreements were discussed until an agreement was reached. In the event that an agreement could not be reached, author QY was included in the discussion in order to resolve the issue. The output of this task is a labelled set of responses against which the various web services could be compared. The details of the resulting dataset can be seen in Table 1 below.

Table 1. Evaluation dataset distribution

Total entries	Negative	Neutral	Positive	Avg char
452	107	106	239	462

The NeW-I intervention is an intervention aimed at parents of children with chronic life-threatening illness. In addition to providing psychoeducation and acknowledging participants efforts, a number of its writing tasks seek to affirm the strengths of its participants, help participants reframe their experiences as well as to assist participants in building meaningful and cherished memories. Because of this context, as seen from Table 1 it can be observed that there is an imbalance in the class distribution in the manually labeled responses, with more than half of the responses being labeled as positive, while only about a quarter were labeled as negative.

4.2 Analysis of Data

This stage of the process involves obtaining access to the various web services, integrating it into the Unity platform and running the web service on the cleaned data. With the exception of Sentiment140 and LIWC2015 which was an offline software, all other services required registering and obtaining an API Key in order to make use of their web service. While Sentiment140 did not require registration, registration was encouraged. In the event of suspected abuse, users of the Sentiment140 web service without registration will be directly blocked from accessing the service due to lack of contact information.

For integration into the Unity platform, for Amazon Comprehend, Cloudmersive, Google Cloud Natural Language, IBM Watson Natural Language Understanding, Komprehend, Lexalytics Semantria, Microsoft Azure Text Analytics, Rosette Text Analytics and Text2Data, the provided.Net SDKs were used, while for Dandelion, MeaningCloud, Salesforce Einstein, Sentiment140, Sentigem, Textgain, Text-processing and Twinword, their REST APIs were used directly.

Once the web services had been successfully integrated, the cleaned responses were imported in their CSV form, and the web service was called on each participant response record. While a number of services had multiple models and Lexalytics Semantria had Singlish as a language, in order to ensure consistency between the services, all web services were called at the document level, using their default general model, at their default language and settings. In the event of service limitations such as Text2Data's daily API call limit of 50 API calls per day, the data analysis was carried out over multiple days as required to complete the analysis.

The responses from the web service were written to an SQLite database for easier data comparison and review. Once all responses were verified as having been successfully labeled, the database table was exported to a CSV file. These web service sentiment analysis responses are the final output of stage 2.

4.3 Comparison of Web Services

Once all the services have completed their analysis, in order to facilitate comparison, the results have to be normalized. The selected format for normalization is the ternary format of negative, neutral and positive. For Cloudmersive, Dandelion, IBM Watson Natural Language Understanding, Komprehend, Lexalytics Semantria, Rosette Text Analytics, Salesforce Einstein, Sentigem, Sentiment140, Text2Data, Text-Processing and Twinword which already provided the polarity in a ternary form, the results were used directly.

For Amazon comprehend, polarities of positive, negative and neutral were kept unchanged. In the event that the response received a polarity of "Mixed", the polarity with the highest confidence was used instead.

For Google Cloud Natural Language, negative scores were taken as negative sentiment, positive scores as positive sentiment. Scores of 0 could either indicated mixed sentiments or neutral sentiment, however, as no means was provide to disambiguate between the two, a neutral score was assigned. In addition, as per documentation, responses with low score and low magnitude could also indicate a neutral score. For our study, we took this as a score of between -0.2 and 0.2, and a magnitude of less then 1.

For Meaningcloud, which provided two levels of magnitude for each polarity, P+ and P were taken as positive, N+ and N were taken as negative, and both Neu and None were both taken as neutral.

For Microsoft Azure Text Analytics, polarity was given in terms of negative, neutral, positive, mixed, where a mixed polarity represents the presence of both positive and negative sentiment in a document. A confidence score for the negative, neutral and positive polarity is also given, and in the event a "Mixed" polarity is assigned to a response, the confidence score is used to disambiguate with the response being assigned the polarity with the highest confidence.

For Textgain, polarity is given as a score ranging from -1 to 1. As per the documentation, negative scores are assigned a negative polarity label, and positive scores are assigned a positive polarity label. As no information is given about the neutral label, only scores of 0 are assigned the neutral polarity label.

For LIWC2015, the emotional tone summary dimension returns score on a scale of 0 to 100. In the official documentation, it is noted that a number around 50 can indicate a lack of emotional tone. As such, for the purpose of this study, Scores greater than 60 are taken as having a positive polarity, scores less then 40 are taken as having a negative polarity, and scores between 40 and 60 are taken as having a neutral polarity.

Once all web service responses have been normalized, the next task is the generation of a confusion matrix for each web service. A confusion matrix is a table often used to describe the performance of an algorithm or classification model, comparing the models results against the expected results. In our study, the results of the sentiment analysis web services are compared against the results of the manually labeled sentiment. From the confusion matrix, we can calculate the number of true positives (TP), true negatives (TN) and true neutrals (T0) where the web service's classifications matches the actual classification, as well as the number of false positives (FP), false negatives (FN), and false neutrals (F0) where the web service's classifications do not match the actual classification.

Once the confusion matrix has been generated, the information can then be used to calculate the actual evaluation metrics by which the web services will be compared. With reference to recent works [25, 28], the accuracy, precision, recall, f-measures and mean square error (mse) have been selected as the metrics by which to evaluate and compare the web services.

Accuracy represents how often a classifier is correct, and the formula used for the calculation of accuracy is as follows:

$$Accuracy = \frac{TP + TN + T0}{Total\ Responses} \quad (1)$$

where total responses is the total number of responses used in this study, and TP, TN and T0 represent the number of true positive, true negative and true neutral classifications as derived from the confusion matrix of each web service.

Precision indicates how often a classifier is correct whenever it predicts a particular polarity. The calculation of precision is done individually for each polarity (positive, negative or neutral). The formula for the calculation of positive precision is as follows:

$$Positive\ Precision = \frac{TP}{TP + FP} \quad (2)$$

where TP represents the number of true positives, and FP represents the number of false positives. The negative and neutral precisions are calculated in a similar manner.

Recall indicates how often a classifier is correct, whenever it is presented with a response with a particular polarity. Similar to precision, the calculation for recall is done individually for each polarity (positive, negative or neutral). The formula for the calculation of positive recall is as follows:

$$Positive\ Recall = \frac{TP}{Total\ Actual\ Positive} \tag{3}$$

where TP represents the number of true positives, and Total Actual Positive represents the total number of actual positives in the test dataset. The negative and neutral recall values are calculated in a similar manner.

The f-measure indicates the harmonic mean of precision and recall. Accuracy is preferred when true positives and true negatives are important, while the f-measure is preferrable when false positives and false negatives are important. Another instance where f-measure may be preferrable is when the class distribution is imbalanced, such as in the dataset used in this study. Calculation for the f-measures are also done individually for each polarity (positive, negative or neutral), and the formula for the calculation of the positive f-measure is as follows:

$$Positive\ F\text{-}Measure = \frac{2 * Positive\ Precision * Positive\ Recall}{Positive\ Precision + Positive\ Recall} \tag{4}$$

where the values for positive precision and positive recall are as calculated from the formulas above. The negative and neutral f-measure are calculated in a similar manner.

The mean square error represents the mean of the squared difference between the actual classification and the predicted classification. For the calculation of the mean square error, both the overall mean square error for the service as well as the mean square error for the individual polarities are calculated. The formula used for the overall mean square error is as follows:

$$mse = n^{-1} \sum_{i=1}^{n} (x_i - y_i)^2 \tag{5}$$

where n represents the total number of entries, x represents the response actual classification, and y represents the web service response classification. The positive, negative and neutral mean square error are calculated in a similar manner. The final output of stage 3 are the above evaluation metrics. The result of these evaluation metrics will be presented and discussed in the next section.

5 Findings

5.1 Results

Overall, when dealing with open ended responses between 10 and 8000 characters in lengths, all services successfully analyzed all responses. Table 2 provides a summary of the classification results for the various web services, while Table 3 provides the MSE of the various web services. Observations regarding the performance of the various web services are as follows.

In terms of accuracy and f-measure, AWS Comprehend and Google Cloud Natural Language performed well, with both achieving the highest accuracy of 71%, and both placing in the top 4 for positive, neutral and negative f-measure performance rankings. In terms of precision, both AWS and Google once again performed well with both placing in the top 3 in terms of positive and negative precision performance, and both being the only services to score over 75% for both positive and negative precision. In terms of recall, most services performed very well in terms of positive recall, with only 8 services scoring below 75%. While Cloudmersive and Textgain had the highest Positive Recall scores of 94%, this was due to an inclination to classify input as positive. This resulted in them having the highest number of false positives causing them to performing poorly and ranking in the bottom 3 in the areas of positive precision, negative recall, and negative mean square error.

Table 2. Summary of classification results.

	Acc	P+	P0	P−	R+	R0	R−	F1+	F10	F1−
AWS	**0.71**	0.76	0.51	**0.77**	0.88	0.47	0.55	0.81	**0.49**	0.64
CM	0.62	0.62	0.56	0.63	**0.94**	0.17	0.35	0.75	0.26	0.45
Dandelion	0.56	0.74	0.67	0.38	0.59	0.24	0.80	0.66	0.36	0.51
Google	**0.71**	0.78	0.43	0.76	0.90	0.37	0.60	**0.83**	0.40	**0.67**
IBM Watson	0.66	0.74	0.67	0.54	0.84	0.13	0.80	0.79	0.22	0.64
Komprehend	0.51	0.68	0.25	0.41	0.57	0.12	0.76	0.62	0.16	0.53
Lexalytics	0.59	0.73	0.34	0.59	0.71	0.39	0.52	0.72	0.36	0.55
MC	0.59	0.65	0.33	0.61	0.84	0.23	0.40	0.74	0.27	0.48
MS Azure	0.67	0.71	**0.72**	0.57	0.85	0.20	0.74	0.77	0.31	0.64
Rosette	0.60	0.75	0.46	0.45	0.67	0.17	**0.85**	0.71	0.25	0.59
Salesforce	0.48	0.71	0.49	0.34	0.41	0.37	0.76	0.52	0.42	0.47
Sentigem	0.64	0.67	0.53	0.58	0.89	0.19	0.53	0.77	0.28	0.55
S140	0.55	**0.85**	0.38	0.49	0.45	0.60	0.73	0.59	0.46	0.59
TP	0.35	0.70	0.24	0.46	0.24	**0.69**	0.29	0.36	0.36	0.36
Text2Data	0.58	0.75	0.36	0.49	0.64	0.33	0.72	0.69	0.34	0.58
Textgain	0.62	0.61	0.67	0.61	0.94	0.13	0.37	0.74	0.22	0.46
Twinword	0.58	0.66	0.28	0.60	0.85	0.21	0.39	0.74	0.24	0.47
LIWC2015	0.59	0.66	0.28	0.46	0.86	0.07	0.49	0.75	0.11	0.47

On the other hand, when dealing with negative recall, only 5 services managed to obtain a score of 75% and above. Rosette Text Analytics, IBM Watson, Dandelion, Salesforce Einstein and Komprehend with negative recall scores of 85%, 80%, 80%, 76% and 76% respectively. It should be noted that these services also had the highest number of false negatives, resulting in low negative precision score.

As a whole, all web services did not perform well when presented with neutral responses. The service with the highest neutral recall rate was Text-processing with a neutral recall rate of 69%. 11 out of 18 of the web services had a neutral recall rate of less than 25%, meaning that less than a quarter of neutral responses were labelled correctly, and the average neutral recall rate of all services was only 28%.

In terms of mean squared error, Google displayed the best performance in terms of overall and positive mean squared error, Sentiment140 displayed the best performance in terms of negative mean squared error, and Text-processing the best in terms of neutral mean squared error.

Table 3. Summary of MSE.

	MSE	MSE+	MSE0	MSE−
AWS	0.12	0.05	0.13	0.25
CM	0.21	0.05	0.21	0.59
Dandelion	0.29	0.40	0.19	0.14
Google	**0.10**	**0.04**	0.16	0.16
IBM Watson	0.17	0.15	0.22	0.16
Komprehend	0.27	0.34	0.22	0.17
Lexalytics	0.16	0.13	0.15	0.24
MC	0.19	0.07	0.19	0.43
MS Azure	0.18	0.15	0.20	0.23
Rosette	0.22	0.28	0.21	0.12
Salesforce	0.34	0.52	0.16	0.14
Sentigem	0.19	0.08	0.20	0.40
S140	0.20	0.29	0.10	**0.10**
TP	0.21	0.25	**0.08**	0.23
Text2Data	0.20	0.22	0.17	0.16
Textgain	0.22	0.05	0.22	0.60
Twinword	0.18	0.07	0.20	0.42
LIWC2015	0.22	0.11	0.23	0.45

5.2 Discussion

Performance. Taking into account the performances of the various web service relative to each other across all evaluation metrics without focusing on a single metric, the top performing sentiment analysis web services were Google, AWS and Azure in that order. The top two companies Google and AWS performed particularly well, with Google placing in the top 4 position in 12 out of 14 metrics, and AWS placing in the top 4 position in 11 out of 14 metrics. While 3rd place Azure did not rank particularly highly on any particular metric, neither did it score particularly lowly on any metrics. One common factor among these three top performing sentiment analysis web services was that they potentially had access to large amounts of first-hand data via the front facing products and services provided their parent companies. This highlights the importance of data access, data quality and data quantity when it comes to machine learning and artificial intelligence techniques.

While analyzing the results presented by Textgain and Cloudmersive, it was observed that they were extremely similar. A deeper look at the original web service responses of both services revealed that both services return the same polarity scores, but with Textgain returning the scores in 2 decimal figures and Cloudmersive returning the score in 15 decimal places. Minor differences in the way in which normalization and polarity labeling was carried out resulted in the minor differences in results. For example, in Cloudmersive, a score of −0.04 was given a neutral label by the web service, however, for Textgain, because no range was given for a neutral score in the official documentation, such a score was normalized to negative as per the normalization process documented in Sect. 4.3. Overall the differences in final performance due to the normalization process was minimal. As Textgain is a spin-off of University of Antwerp, utilizing the University's open-source NLP Library, Pattern, it can be deduced that Cloudmersive makes use of a similar implementation in their sentiment analysis web service.

The performance of LIWC2015 in this study also provides some food for thought. In past studies, LIWC is often taken as the gold standard and even today, is often used as the sole means of evaluation and analysis in many studies [29–32]. In this study however, LIWC performed extremely poorly overall, ranking in the bottom 5 for all metrics except positive recall, Positive f-measure, and positive mean square error. This is an interesting observation as LIWC2015 relies heavily on a lexicon-based approach, and may indicate that the LIWC2015 emotional tone summary variable based on the default LIWC lexicon library may not be suitable for use in the analysis of data such as journal entries and mental health intervention responses, demonstrating the heavy reliance of lexicon-based approaches on their dictionary. This also supports earlier studies which show that machine learning based approaches tend to perform better then lexicon-based approaches [33].

Mental Health Interventions. When considering which web service to select for use in a mental health intervention, a web services accuracy and f-measures may not always be the main deciding factor. For example, when dealing with mental health intervention responses, very often, the ability to correctly classify sentiment when presented with a negative response is of great importance. This is especially so when dealing with interventions targeting stress, anxiety or depression (SAD), due to the possibly serious consequences of mishandling such responses. As such, web services with high negative recall such as Rosette Text Analytics, IBM Watson, and Dandelion should be considered.

In addition, the negative mean square error of the services is another important metric, indicating how far the web services predictions for negative responses deviate from the actual classification. For this metric, the top services were Sentiment140, Rosette Text Analysis and Dandelion, with IBM and Google coming in 4th.

Based on negative recall and negative mean square error, if correct classification of negative responses is the most important factor, with no regard for false negatives, Rosette Text Analytics, IBM and Dandelion would be good web services to consider.

However, if in addition to correctly classifying negative responses, the number of false negative needs to be minimized, IBM would be the best choice, due to having the highest precision among the services considered.

If correct classification of positive content is important to an intervention, such as when trying to identify experiences considered positive by the intervention participants, positive recall and positive mean square error are important. Of the web services evaluated, Google, AWS, Cloudmersive, Textgain and Meaningcloud displayed good positive recall of over 80% as well as good positive MSE scores of less than 0.10. Of these Google and AWS had the highest positive precisions rates of 0.78 and 0.76 respectively.

In the event that the correct classification of both positive and negative sentiments were equally important, the IBM Watson NLU web service should be considered. While it did not place first in any particular metric, its was the only web service to obtain score of over 80 for both positive and negative recall. It also ranked 7th and 4th in terms of positive and negative mean square error. It should however be noted that in terms of neutral recall, it ranked 2nd from the last, with a neutral recall score of only 0.13.

Ethical AI. While web services such those evaluated in this study provide an easy way for researchers to make use of artificial intelligence and machine learning technologies such as sentiment analysis in their projects, care should be taken in considering the ethical aspect of implementing such technology, especially when dealing with sensitive information such as those found in mental health interventions. Ethical AI refers to the development, implementation and use of AI technologies which maintain human dignity and do not cause harm to people. This is often done by taking into account fundamental human values including but not limited to transparency, fairness, privacy and individual rights.

Of the 17 web services in this study, only 6 services, Amazon, Google, IBM, Azure, Salesforce and Textgain provided documentation of having considered ethics, fairness and responsibility in the design, implementation and provision of their web services. One example of how these documented consideration have been operationalized is how data sent to them for analysis is handled. By nature of being a web service, in order for the text to be analyzed, it needs to be sent to the web service provider. Transparency in terms of how such data is handled is important, especially if the web service is to be used to processes sensitive information such as those found in mental health interventions. Of the 6 services listed above, Amazon, IBM and Salesforce log submitted content for use in improving delivered services by default. However, this is clearly and explicitly stated, while also providing clear documentation on how users could opt-out of such data collection in order to ensure privacy and protect user rights. Google, Azure and Textgain on the other hand do not collect or use user submitted content for purposes such as the improvement of services. Microsoft stores collected data in an encrypted form for 48 h for analysis by restricted personnel in the event of a catastrophic failure which users can opt out from, while Google and Textgain process submitted web requests in memory and do not store the content submitted at all. All 6 services also provided clear documentation that users retained ownership of content submitted to their servers was provided in their terms of service and privacy policy.

Of the remaining services, which do not provide documentation of having considered ethics in the design, implementation and provision of their web services, 4 explicitly state that they collect submitted data for service improvements yet provide no documented way to opt out, 1 explicitly state that they do not collect any submitted data, and 6 do not provide any clear indication of how submitted data is handled.

Given the sensitive nature of the data and the technology use, when considering which web service to make use of for digital mental health interventions, consideration of the ethical issues involved in the design, development and implementation of such AI technologies should be factored in, and web services should not be evaluated solely based on their performance.

6 Limitations

When discussing this study, there are some important limitations which should be considered in future work. The first limitation pertains to the manual labeling of sentiments in the creation of a "gold standard". It should be noted that while three authors were involved in the labeling process, it is well known that human agreement on sentiment can vary greatly [34, 35], and that none of the authors were trained mental health clinicians. Given that at least four non-expert reviewers are required in order to achieve the equivalent accuracy of one expert reviewer [36], the accuracy of the labeling might still be improved. This could be improved in future studies either through the involvement of expert reviewers, or by crowd sourcing the manual labeling process in order to increase the number of reviewers.

The next limitation of the study is the language of the dataset. The dataset used in this study comes from the clinical trial of a digital mental health intervention carried out in the country of Singapore. While only English-speaking participants were recruited in the pilot trial, it should be noted that in Singapore, the language most often used is Singlish, an English based creole language. While generally based on British English, it is possible to find differences in terms of the grammar and vocabulary used [37]. Being a creole language, Singlish was not available as a input language option, and the closest alternative English was used in this study. In the preprocessing phase, while non-English words were replaced, the grammatical structure of the data was retained in order to obtain a better understanding of the performance of the various web services in the local context. The only web service with Singlish as an input language option was Lexalytics, however, this option was not available in their trial.

The last limitation is the source and size of the dataset. Just as when training a classifier, a larger data and more varied dataset allows for a better understanding of a classifier's performance under difference circumstances. In practice however, challenges such as privacy, prevalence, as well as legal and organizational barriers make the collection of medical data difficult [38]. This is even more so when dealing with mental health, where due to the impact of stigma [6], such barriers are even more prominent and sensitive. If possible, future works should include datasets from other digital mental health interventions in order to provide a more holistic overview of the various web services.

7 Conclusion

In this paper, we explore 18 sentiment analysis web services, comparing their performance in the analysis of digital mental health intervention data. Based on our results, we can conclude that while no one web service seems to be perfect, web services by larger technology companies such as Google, Amazon, Microsoft and IBM, in general perform better then then those from smaller technology companies. However, if the goal is to find the best performing web service for a specific metric smaller sentiment analysis web services should be considered as well.

Given the ease with which these web services can be integrated into various platforms and applications, the ethical use and implementation of such technology should also be considered. These results may help other researchers, especially those who do not possess the expertise to develop their own algorithms and implement their own services, understand the feasibility of applying existing sentiment analysis web services to their own digital mental health intervention applications and studies. It also provides them with up-to-date information on the various available web services and their performances to help in their evaluation and selection process.

As a closing remark, something to consider is that while this study serves as a good starting point, analysis in this study was done only at the document level for ease of comparison. Many of these web services offer multiple levels of analysis such as entity and aspect level sentiment analysis and these can provide a much greater level of detail and information with which to understand the users.

Acknowledgements. This research is supported in part by the Joint NTU-UBC Research Centre of Excellence in Active Living for the Elderly (LILY), Nanyang Technological University, Singapore.

References

1. Hollis, C., et al.: Technological innovations in mental healthcare: harnessing the digital revolution. Br. J. Psychiatry **206**(4), 263–265 (2015)
2. Cuijpers, P., Riper, H., Andersson, G.: Internet-based treatment of depression. Curr. Opin. Psychol. **4**, 131–135 (2015)
3. Coppersmith, G., Hilland, C., Frieder, O., Leary, R.: Scalable mental health analysis in the clinical whitespace via natural language processing, pp. 393–396. IEEE (2017)
4. Gkotsis, G., et al.: Characterisation of mental health conditions in social media using informed deep learning. Sci. Rep. **7**(1), 1–11 (2017)
5. Miller, E., Polson, D.: Apps, avatars, and robots: the future of mental healthcare. Issues Ment. Health Nurs. **40**(3), 208–214 (2019)
6. Renn, B.N., Hoeft, T.J., Lee, H.S., Bauer, A.M., Areán, P.A.: Preference for in-person psychotherapy versus digital psychotherapy options for depression: survey of adults in the US. NPJ Digit. Med. **2**(1), 1–7 (2019)
7. Reid, S.C., et al.: A mobile phone application for the assessment and management of youth mental health problems in primary care: a randomised controlled trial. BMC Fam. Pract. **12**(1), 1–14 (2011)
8. Qu, C., Sas, C., Roquet, C.D., Doherty, G.: Functionality of top-rated mobile apps for depression: systematic search and evaluation, JMIR Ment. Health **7**(1), e15321 (2020)

9. D'alfonso, S., et al.: Artificial intelligence-assisted online social therapy for youth mental health. Front. Psychol. **8**, 796 (2017)
10. Birjali, M., Beni-Hssane, A., Erritali, M.: Machine learning and semantic sentiment analysis based algorithms for suicide sentiment prediction in social networks. Procedia Comput. Sci. **113**, 65–72 (2017)
11. Madhu, S.: An approach to analyze suicidal tendency in blogs and tweets using sentiment analysis. Int. J. Sci. Res. Sci. Eng. **6**(4), 34–36 (2018)
12. Le Glaz, A., et al.: Machine learning and natural language processing in mental health: systematic review. J. Med. Internet Res. **23**(5), e15708 (2021)
13. Ho, A.H.Y., et al.: A novel narrative e-writing intervention for parents of children with chronic life-threatening illnesses: protocol for a pilot, open-label randomized controlled trial. JMIR Res. Protoc. **9**(7), e17561 (2020)
14. Liu, B.: Sentiment analysis and opinion mining. Synth. Lect. Hum. Lang. Technol. **5**(1), 1–167 (2012)
15. Medhat, W., Hassan, A., Korashy, H.: Sentiment analysis algorithms and applications: a survey. Ain Shams Eng. J. **5**(4), 1093–1113 (2014)
16. Tang, H., Tan, S., Cheng, X.: A survey on sentiment detection of reviews. Expert Syst. Appl. **36**(7), 10760–10773 (2009)
17. Kolchyna, O., Souza, T.T., Treleaven, P., Aste, T.: Twitter sentiment analysis: lexicon method, machine learning method and their combination. arXiv preprint arXiv:1507.00955 (2015)
18. Xianghua, F., Guo, L., Yanyan, G., Zhiqiang, W.: Multi-aspect sentiment analysis for Chinese online social reviews based on topic modeling and HowNet lexicon. Knowl.-Based Syst. **37**, 186–195 (2013)
19. Hu, Y., Li, W.: Document sentiment classification by exploring description model of topical terms. Comput. Speech Lang. **25**(2), 386–403 (2011)
20. Maks, I., Vossen, P.: A lexicon model for deep sentiment analysis and opinion mining applications. Decis. Support Syst. **53**(4), 680–688 (2012)
21. Xu, T., Peng, Q., Cheng, Y.: Identifying the semantic orientation of terms using S-HAL for sentiment analysis. Knowl.-Based Syst. **35**, 279–289 (2012)
22. Hagenau, M., Liebmann, M., Neumann, D.: automated news reading: stock price prediction based on financial news using context-capturing features. Decis. Support Syst. **55**(3), 685–697 (2013)
23. Tov, W., Ng, K.L., Lin, H., Qiu, L.: Detecting well-being via computerized content analysis of brief diary entries. Psychol. Assess, **25**(4), 1069 (2013)
24. Catania, F., Di Nardo, N., Garzotto, F., Occhiuto, D.: Emoty: an emotionally sensitive conversational agent for people with neurodevelopmental disorders (2019)
25. Basmmi, A.B.M.N., Abd Halim, S., Saadon, N.A.: Comparison of web services for sentiment analysis in social networking sites, p. 012063. IOP Publishing (2020)
26. Gao, S., Hao, J., Fu, Y.: The application and comparison of web services for sentiment analysis in tourism, pp. 1–6. IEEE (2015)
27. Pinto, H.L., Rocio, V.: Combining sentiment analysis scores to improve accuracy of polarity classification in MOOC posts. In: Moura Oliveira, P., Novais, P., Reis, L.P. (eds.) EPIA 2019. LNCS (LNAI), vol. 11804, pp. 35–46. Springer, Cham (2019). https://doi.org/10.1007/978-3-030-30241-2_4
28. Serrano-Guerrero, J., Olivas, J.A., Romero, F.P., Herrera-Viedma, E.: Sentiment analysis: a review and comparative analysis of web services. Inf. Sci. **311**, 18–38 (2015)
29. Park, M., Cha, C., Cha, M.: Depressive moods of users portrayed in Twitter (2012)
30. Choi, D., Kim, P.: Sentiment analysis for tracking breaking events: a case study on twitter. In: Selamat, A., Nguyen, N.T., Haron, H. (eds.) ACIIDS 2013. LNCS (LNAI), vol. 7803, pp. 285–294. Springer, Heidelberg (2013). https://doi.org/10.1007/978-3-642-36543-0_30

31. Sell, J., Farreras, I.G.: LIWC-ing at a century of introductory college textbooks: have the sentiments changed? Procedia Comput. Sci. **118**, 108–112 (2017)
32. Syah, T., Apriyanto, S., Nurhayaty, A.: Student's prevailing, confidence, and drives: LIWC analysis on self-description text, pp. 295–299. Atlantis Press (2020
33. Annett, M., Kondrak, G.: A comparison of sentiment analysis techniques: polarizing movie blogs. In: Bergler, S. (ed.) AI 2008. LNCS (LNAI), vol. 5032, pp. 25–35. Springer, Heidelberg (2008). https://doi.org/10.1007/978-3-540-68825-9_3
34. Wilson, T., Wiebe, J., Hoffmann, P.: Recognizing contextual polarity in phrase-level sentiment analysis, pp. 347–354 (2005)
35. Bermingham, A., Smeaton, A.F.: A study of inter-annotator agreement for opinion retrieval, pp. 784–785 (2009)
36. Snow, R., O'connor, B., Jurafsky, D., Ng, A.Y.: Cheap and fast–but is it good? Evaluating non-expert annotations for natural language tasks, pp. 254–263 (2008)
37. Harada, S.: The roles of singapore standard english and singlish. Inf. Res. **40**, 70–82 (2009)
38. Althnian, A., et al.: Impact of dataset size on classification performance: an empirical evaluation in the medical domain. Appl. Sci. **11**(2), 796 (2021)

An Extendable Sentiment Monitoring Model for SNS Considering Environmental Factors

Yenjou Wang[1](✉) ⓘ, Neil Yen[2] ⓘ, and Qun Jin[3] ⓘ

[1] Graduate School of Human Sciences, Waseda University, Tokorozawa, Japan
`yjwjennifer2021@ruri.waseda.jp`
[2] School of Computer Science and Engineering, Aizu University, Aizu-Wakamatsu, Japan
`neilyyen@u-aizu.ac.jp`
[3] Faculty of Human Sciences, Waseda University, Tokorozawa, Japan
`jin@waseda.jp`

Abstract. Rapid growth of social network service (SNS) has drawn significant attention from the publics. Existing research indicated that emotional state and behavioral tendency of SNS users can be identified and predicted through sentiment analysis. However, it is found that not only the posts can express user's emotions but the overall environmental conditions faced by the user may lead to the generation of different emotions based on the cognitive theory of emotion and observation. Therefore, it may lead to bias between the sentiment analysis results and the actual situation if only analyzing the post. This study targets to propose an extendable sentiment monitoring model which considers the actual environment of users in SNS. Through this model, the result of sentiment analysis is closer to reality. By analyzing the content of users' continuous posts, the sentiment analysis can take into account the pre- and post-textual relationships. The classification result of external affecting sentiment factors by K-means is used as criteria for weighting method to adjust the results of sentiment analysis based on BERT. Finally, the time series analysis is used to predict sentiment tendency monitor sentiment changes. The experiment results show that the training and validation accuracy are 89.24% and 84.00%, respectively. By our weighting method to revise the BERT results, the F1 score is improved from 0.839 to 0.850.

Keywords: Sentiment analysis · Emotion model · K-means · BERT · Machine learning · Human behavior analysis

1 Introduction

With the popularity of the Internet around the world, the Social Networking Service (SNS) is rapidly expanding. Within only three years, the number of tweets posted every day has risen from 500 million in 2019 to 800 million in 2022 [1]. Nowadays, SNS is a platform not only for people to share life but also to represent life. In addition, its influence is enough to initiate various social phenomena. As a result, the topic of SNS analysis has drawn great attention from the public. Moreover, many forums, comments and posts have been extracted to analyze and explain people's opinions or tendencies towards specific items. Among them, emotional expression becomes the core of human

© The Author(s), under exclusive license to Springer Nature Switzerland AG 2022
G. Meiselwitz (Ed.): HCII 2022, LNCS 13315, pp. 408–421, 2022.
https://doi.org/10.1007/978-3-031-05061-9_29

behavior. Therefore, using machines to understand emotions has always been a long-term target [2], which is known as sentiment analysis or opinion mining.

In the past decade, researchers have devoted to conducting emotional recognition and modeling to the contents in SNS [3–5] so as to further understand users' emotional status. According to Solomon's Cognitive Theory of Emotion, emotions are judgments [6], which are generated from human cognition of overall situational stimuli. In the actual situation of SNS, users can be easily affected by the comments of other users. Therefore, we believe that it is necessary to add environmental factors in SNS sentiment analysis. However, the existing studies focus more on judging the emotion of the post itself. A single emotion judgment point may cause a bias in the judgment result of emotion or a gap in reality. Beyond that, the models cover finite emotional classifications which cannot be expanded.

Considering above issues, we proposed a multi-dimensional emotion model in our previous study [7]. Specifically, it adopts the emotion theory called Plutchik's wheel of emotion [8] as the research basis. Defining emotions through sets allows complex emotional changes to be calculated by computers. Beyond that, the emotion definition is mapped to the polar coordinate system tried to solve the issue that the emotion model cannot be extended in the existing research. In this study, to further improve the previous model, we propose an extendable sentiment monitoring model, which includes a weight calculation method for environmental factors. The overall interactive environment in SNS including actual factors that may affect the users' emotions are considered. We extract data from GoEmotions dataset, which is provided by Google [9]. Each data is tagged with the related situation. In addition, the K-means is used to classify these data and generalize the similar situation in an emotion to calculate the basis weight. Lastly, these weights are used to adjust the BERT analysis results to make the result of determining emotion closer to the users' actual emotional experience. The main contributions of this study are given as follows.

- A systematic method for calculating the weight of SNS environmental factors based on K-means is proposed, which can help to understand the relationship between SNS users' feelings and external factors.
- A weighting method is proposed to revise the BERT results and reduce the deviation of sentiment analysis results from reality.

The remainder of this paper is organized as follows. The related work on NLP and semantic analysis in SNS is overviewed in Sect. 2. In Sect. 3, the process and method of research architecture in details are described. In Sect. 4, we present and evaluate the experiment results of our proposed new model and discuss the results. Finally, this study is summarized, and future work directions are highlighted in Sect. 5.

2 Related Work

Emotion is a major source of research on human behavior. Currently, many applications detect human emotions from speech, expressions, texts, and other forms of message sources. While there is a rich tracking record for context in the emotion research field, it is also one of the most difficult sources of human emotion characteristics to obtain compared with other sources. Sentiment analysis utilizes natural language processing (NLP) techniques to analyze a person's opinions; therefore, it is also known as opinion mining [10]. At present, the research of NLP sentiment analysis generally develops in the following directions. A social orientation analysis by Himelboim et al. [11]. They collected tweet conversations on 10 controversial political topics and plotted their network connections and understood the political orientation of users through sentiment analysis. This study belongs to the field of application of sentiment analysis. Demszky et al. [9] and Wang et al. [12] are dedicated to creating a large dataset with emotional labels and verifying it with machine learning methods, providing a reliable data source for modeling on sentiment analysis. Lee et al. [13] proposed a new sentiment analysis method, in which a stacked autoencoder and LSTM recurrent neural network are used to extract emotional feature signals. The results of their method show that more useful signals can be obtained than the previous analysis methods, and the results of sentiment classification are more accurate. The practical application of sentiment analysis has always been the goal of all researchers. However, different sentiment analysis methodologies directly affect the analysis results. In order to serve as a basis for practical applications, we focus on the area of context analysis, adding contextual factors to the variables of sentiment analysis.

In the field of emotion models, Rout et al. [14] used both unsupervised and supervised machine learning methods. Besides automatically identifying the sentiment of tweets obtained from the Twitter public domain, different machine learning methods have also been applied to examine the effectiveness of various feature combinations. This hybrid approach can effectively improve the performance of machine learning for sentiment analysis of unstructured data sources such as tweets. Seyeditabari et al. [15] have used a bidirectional GRU network in Recurrent Neural Networks (RNNs) to analyze the tweets. The aim is to capture information about the contextual relationships of texts and to classify texts into seven emotions based on the emotion theory proposed by Shaver et al. In addition to the sentiment analysis of the main language, Kanclerz et al. [16] used the Polish language corpus BiLSTM for transfer learning through LASER to predict the sentiment of other texts of high-resource language, e.g., English, Portuguese, thus strengthening the effectiveness of sentiment analysis across different languages.

According to the previous discussion, the current research has gone beyond just using a traditional machine learning approach to classify sentiment into indicators of different polarities, but also has moved towards combining theoretical foundations or complex machine learning algorithms through cross-comparison to make the sentiment analysis results more accurate. Despite this, there are still many questions about the extensibility of the model, and the problem of how well the model fits the real situation needs to be solved and improved. Therefore, this study not only improve the definition of emotion proposed in the previous work, but also maintain the extensibility of the model. The actual use environment of SNS users is also considered. All of factors affecting SNS users are used as parameters to revise the results of sentiment analysis. analysis. We propose an extendable sentiment monitoring model and thus reduce the difference between results of sentiment analysis and reality through a systematic method.

3 An Extendable Sentiment Monitoring Model

This section describes and discusses the methods applied in our model. We aim to focus on the relationship between SNS user sentiment and external environmental factors. NLP basically provides two ways to analyze data, one is to manually define rules to label text into various aspects (e.g., positive, or negative) as a basis for distinguishing sentence polarity, and the other is to employ machine learning techniques automatically classify sentences. In this study, we combine the above two methods to propose our model. There are five main parts in our approach: data preparation, weighting by scalable external factors, emotion model training, emotion calibration and emotion monitoring. An overview of the proposed framework is illustrated in Fig. 1.

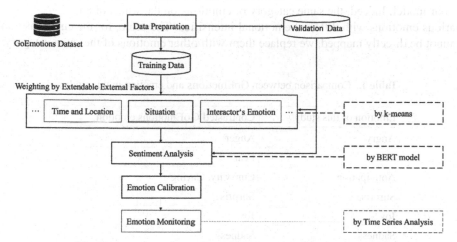

Fig. 1. Overview of an extendable sentiment monitoring model

In this study, we use the dataset which called GoEmotions as our data source. Section 3.1 explains how those tweets from GoEmtions are filtered and collected. Section 3.2 describes how external factors affecting SNS user sentiment are classified and how the systematic weighting method works. Section 3.3, a machine learning method has been utilized to analyze emotion and provide the eigenvalues of personal emotion. Section 3.4, the results of sentiment analysis by BERT are calibrate based on environmental factor weights. Section 3.5, time series analysis is used to predict sentiment tendency to judge whether the emotional change is within the expectations. Since this study focuses on analyzing the relationship between external factors of SNS and emotion, the bold line pattern in Fig. 1 represents the completed part of this paper.

3.1 Data Preparation

According to the current research, sentiment is usually mostly classified into positive, negative, or few amount of emotion variety. So it is difficult to further analyze the reasons for mood shifts. In order to enable the results of sentiment analysis to be interpreted in more detail, a stable and emotionally distinct source of information is needed to support model training. Furthermore, research shows that a good data source also affects the accuracy of the model for the verification data [17]. To solve this problem, we used a dataset that called GoEmotions dataset, which is created by Google, as our data source for the subsequent model analysis. GoEmotions contains 58k Reddit comments that are extracted from popular English-language subreddits and labeled with 27 sentiment categories. It is the largest fully annotated English fine-grained sentiment dataset to date, which completely includes positive, negative and neutral emotions. At the same time, in order to maintain the extensibility of the research model, we followed the definition of emotion based on Plutchik's wheel of emotion in the previous work that includes 8 kinds of emotions (i.e., joy, surprise, anger, sadness, fear, trust, and anticipation). As explained in Table 1, there are several emotions in the GoEmotions dataset that can be mapped to our model. Indeed, the same category of emotions on the wheel of emotion contains various emotions with different emotional intensities. Therefore, for the emotions that cannot be directly mapped, we replace them with other emotions of the same category.

Table 1. Comparison between GoEmotions and emotion in this study

Emotion of this study	Contrasted GoEmotions emotion
Angry	Anger
Fear	Fear
Anticipation	Curiosity, Caring
Surprise	Surprise
Joy	Joy
Sadness	Sadness
Trust	Approval
Disgust	Disgust
Angry	Anger

Finally, based on emotion definition, 8,000 posts from the GoEmotions dataset (1,000 per emotion category) are randomly chosen as our dataset, and the noise data, such as "[NAME]", emoji, at the posts are removed. In order to understand the relationship between the posts and environmental factors. In the data preprocessing stage, we use a human-annotated method to label each post with the corresponding situation for subsequent analysis.

3.2 Weighting by Extendable External Factors

In NLP analysis, we often face the problem of multiple semantics, where the same word can have different semantic meanings for various reasons. As explained in the Introduction, it is clear that human emotion can be influenced by the surrounding environment. Therefore, K-means is used to classify the eigenvalues of the eight basic emotions. At the same time, the relationship between the eight emotions and the external factors is identified. Then, a systematic method for calculating the weighting of external factors is described in detail in this section. The weighting of external factors is used as the basis for weighting calibration of the BERT results, leading to a more realistic sentiment discrimination result. The details of each process are given in the following subsections below.

Classification of External Factors. In fact, users are in different situations when they submit their posts. Before the illustration of the relationship between the eight emotion categories and external factors, it is necessary to summarize which situations are contained in each emotion, and which situations are similar to the emotional impact of SNS users. Therefore, in this stage, K-means is adopted to classify and generalize each sentiment-similar situation. In this study, situation is mainly defined as the state of SNS users' contributions, e.g., communicating with people, blessing others or encountering trouble, etc. In this study, the situation is defined as the type of set, so it can be extended and added other external factors that affect SNS users, such as time, location, and emotion of users who communicate the user in SNS (we call it interactor's emotion in this study). The similar situations of emotion are divided into 4 main contexts (called clusters in classification). We analyze each sentiment separately. The posts containing the situation tag in each emotion are classified by K-means, and the similar situation is determined through the analysis results of K-means. Since human emotions are very complex, emotion also has deviation even under the same situation. Because of this, the same situation may appear in different clusters of emotion in the analysis results. For a clear explanation, a simple case study is provided as shown in Table 2.

Table 2. An example K-means classification result (before being sorted by proportion)

Cluster	Situation	Proportion of each situation		
		Talking	*Bless*	*Joking*
Cluster 0	Talking, Talking, Talking, Bless, Joking, Joking	1/2	1/6	1/3
Cluster 1	Talking, Bless, Bless	1/3	2/3	–
Cluster 2	Talking, Talking, Joking	2/3	–	1/3
Cluster 3	Talking, Bless, Joking	1/3	1/3	1/3

In order to solve the above problems, we calculate the proportion of each situation located in each cluster to understand which cluster the situation prefers. Despite this, it still occurs that the situation has the same proportion in different clusters, e.g. the situation - Joking in Table 3. In this case, a random process is used to solve this problem.

Table 3. An example of K-means classification result after sorting based on proportion

Cluster	Situation
Cluster 0	Joking
Cluster 1	Bless
Cluster 2	Talking
Cluster 3	–

Finally, term frequency-inverse document frequency (tf-idf) which is a statistical method to evaluate the importance of a word to a file set or a file in a corpus, is used to extract the important keywords of each cluster as the judgment basis for calibration of the BERT result.

Weight Calculation Method for External Factors. In the previous section, we discussed the use of K-means to classify the sentiment. In this section, we discuss how to calculate the weight of each cluster as the calculation benchmark for calibration of the BERT result.

Even under the same situation, there may be different emotions due to the user's own conditions. On the other hand, if a situation is owned by only one emotion, then the situation has a high chance to represent a certain emotion. Therefore, the K-means result of each emotion is compared with each other. If a cluster contains a set of situations that intersect with other sentiments, it means that the cluster is not so important to this sentiment. The weight of this cluster is reduced accordingly. We propose a systematic approach to define and describe the process of weight calculation as follows.

X is used to represent the number of emotions in this study. The cluster of each emotion has a relative weight, W representing the overall weight set, which contains the weight set of each emotion w. It can be represented as Eq. (1).

$$W = \{w_1, w_2, w_3, \ldots w_X\} \tag{1}$$

There are 4 clusters in each emotion, and the weight of each cluster (wC) of any single emotion (w) is represented as Eq. (2).

$$w_i = \{wC_{i_1}, wC_{i_2}, wC_{i_3}, wC_{i_4}\} \tag{2}$$

In the relationship between the emotion and situation, each emotion (AC_i) has corresponding situations, which is derived from the input data. For each emotion (AC_i), we categorize all situations under this emotion into four categories. It can be represented as Eq. (3). In addition, as described in the previous section, we use K-means to divide each emotion situation into four clusters (aC_{i_j}) as represented in Eq. (4).

$$SC = \{AC_1, AC_2, AC_3, \ldots AC_X\} \tag{3}$$

$$AC_i = \{aC_{i_1}, aC_{i_2}, aC_{i_3}, aC_{i_4}\} \ \forall i > 0 \ \& \ i \leq X \tag{4}$$

Therefore, the weight of calculation method on target cluster ($wC_{emotion_{cluster}}$) can be represented as Eq. (5).

$$wC_{emotion_{cluster}} = \log_2(\sigma) - \log_2\left(\sum_{\substack{1 \leq i \leq X \\ 1 \leq j \leq 4}} aC_{emotion_{cluster}} \cap aC_{i_j}\right) \qquad (5)$$

where σ is a set as the basic weight of aC_{i_j}, which $\sigma \in \mathbb{N}$ and $\sigma > \log(X)$. In this study, we use $\sigma = 27$, which represents the maximum number of repetitions of a situation. Figure 2 describes the actual flow of weight calculation. First, the target cluster is compared with all clusters of other emotions by a double loop. If there is an intersection, it means that the cluster contains the same situation, and the repeat value (count) is increased. Finally, the initial weight is subtracted $\log_2(count)$ to get the final weight.

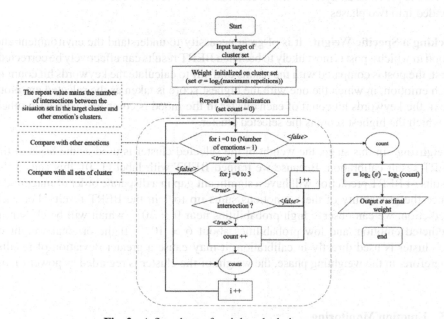

Fig. 2. A flowchart of weight calculation process

Through this process, the clusters in each emotion are given corresponding weights. This weight represents the influence of the situation in the cluster on emotions.

3.3 Sentiment Analysis

According to Plutchik's wheel of emotion theory, there are various possibilities for the combination of human emotions. The composition of emotions would be more complex than previous emotion models if we propose an emotion model based on the wheel of emotion and keep the expansion. Furthermore, Furthermore, an SNS post, as unstructured data, often has semantic issues in different contexts. In this phase, Bi-directional

Encoding Representation for a Transformer (BERT) model is used for semantic analysis. BERT can not only solve semantic problems by considering the context but also can help the model to achieve higher accuracy more easily because it is trained on a large number of article, which makes it easier for the model to have higher accuracy. In this study, we adopt the bert-base-uncased model which inherits from the Pretrained Model to learn the possible implied sentiment categories in each post.

3.4 Emotion Calibration

In order to make the result of emotional analysis closer to the reality, in Sect. 3.2, we explained how to use K means to classify situation and calculate the weights and keywords for each cluster. Based on the previous section, this section describes t how to use the previous analysis results to calibrate BERT results. The calibration process is divided into two phases.

Picking a Specific Weight. It is of great necessity to understand the environment and mood in which a post is most likely to be. Then BERT results can effectively be corrected. First, the post is compared with the results of tf-idf to calculate the keywords hit count in each emotion, in which the one with the highest score is taken as the selected emotion. Next, the keywords hit count of each cluster in the mood is compared with each other, in which the highest score is the selected cluster.

Weighting. In this stage, the weight of the selected cluster is used as the basis of the BERT result calibration. Because we trained BERT with label [1, 0], this causes the results of BERT prediction will have a significant gap in a disparate order of magnitude. Since the probability of the 8 emotions sums up to 1 in the BERT results. For each prediction, we can observe high probabilities near 9×10^{-1} which will be chosen as predicted emotion, and low probabilities about 6×10^{-6}. If the original weight of the cluster is used directly in calibration, it may cause a greater deviation of results. Therefore, in the weighting phase, the weight of the cluster is regarded as power of 10.

3.5 Emotion Monitoring

In order to better understand of the relationship between external factors and sentiment. The time series analysis is used to predict the tendency of emotion. Since the human emotion remains for a period, a series of tweets are analyzed at the beginning and a series of emotion prediction results are generated. Furthermore, this result is compared with the results of sentiment analysis based on BERT to determine whether the sentiment change is within the predicted range. Moreover, each unpredicted change in sentiment is also recorded to understand the reasons for unpredicted changes in the sentiment.

4   Experiment Result and Discussion

This section presents the experiment setup, results, and discussions on the results.

4.1 Experiment Setup

This is a highly integrated development model based on Python's Jupyter Notebook, which at the time of writing provides a runtime environment consisting of Python 3.7, NumPy 1.19.5, scikit-learn 1.0.2, transformers 4.16.2, and a randomly assigned rental GPU (Tesla P100) for training.

The dataset was extracted from the GoEmotions project of Google Research team, and the excel power query was used for fuzzy analysis to remove over-similarity and under-featured sentences for pre-processing. Our model uses the current version of the bert-base-uncased and uses two AdamW optimizers, one for BERT and one for linear layer. A linear scheduler with a warmup strategy and perform 1000 steps warmup. Finally, we use seed = 13, epochs = 10 to start our Bert fine-tune flow.

4.2 Results and Discussion

Firstly, we discuss the evaluated results from our proposed experimental setup with three metrics, i.e., accuracy, loss and F1 score.

BERT Implementation Result. As the mainstream of current popular general-purpose NLP, there are strong semantic analysis capabilities for BERT. However, the architecture of BERT is huge that needs a lot of costs for training process. Fortunately, the open-source nature of BERT allows many researchers to focus on offering training of BERT for various tasks and reducing the training costs of subsequent researchers. This study randomly selected 80% posts of dataset for training and 20% posts of dataset for validation. According to the results, by adjusting the parameters, the training accuracy of final model has reached 89.24%, and the validation accuracy has reached 84.00%, while the training loss is 1.46% and validation loss is 1.43% respectively. The learning curves are shown in Figs. 3 and 4. In Fig. 3, the blue curve labeled 'train_acc' is the accuracy of training curve, and the orange curve labeled 'val_acc' is the accuracy of validation curve. And in Fig. 4, the blue curve labeled 'train_loss' is the loss of training curve, and the orange curve labeled 'val_loss' is the loss of validation curve. Table 4 summarizes the results of BERT.

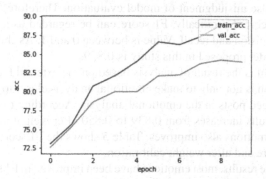

Fig. 3. Training curve graph of accuracy (Color figure online)

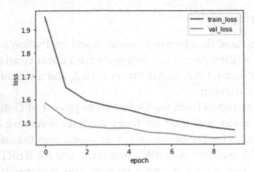

Fig. 4. Training curve graph of loss (Color figure online)

Table 4. The results of BERT.

	Accuracy	*Loss*
Training dataset	89.24	1.46
Validation dataset	84.00	1.43

It can be seen from the results since the data structure of our data source has been optimized and verified by BERT. Therefore, even we remove emoji that may help with emotion recognition, and a new label is added to the data structure. The training accuracy of the research model reaches 73.09% in the training of the first epoch, and the model tends to be stable when the epoch reaches the 6–8 layer. It is a good foundation for subsequent weight adjustment.

F1 Score and Weight Adjustment Results. In order to better understand the quality of the model proposed in this study, we usually evaluate the quality of a model through recall, precision and F1 score. Recall is the proportion of positive samples that can be predicted among all positive samples. Precision represents how many positive samples should be correctly predicted among all positive samples. However, if either of them is outliers, it may cause misjudgment of model evaluation. Therefore, this study focuses on observing F1 score. Specifically, F1 score can be regarded as a kind of harmonic mean of model precision and recall. Value is between 0 and 1. As shown by results, the F1 score of the model proposed in this study is 0.839.

Since the weight is the result of analysis based on the external factors. The purpose of weight correction is not only to make emotional analysis closer to reality, but also to correct the misjudged posts in the emotional analysis. According to the results, the F1 score of BERT results increases from 0.839 to 0.850 after weight adjustment, and the F1 score of most emotions also improves. Table 5 shows the F1 score of our model and each emotion before and after weight calibration.

As shown by the results, most emotions have been improved in F1 scores after weight correction. However, for more complex emotions, such as anticipation and trust, since they may be combined from several emotions, their situation mostly overlaps with other emotions. The weighted correction method also did not substantially improve their F1

Table 5. F1 score results table of the overall model and each emotion.

	F1 score	F1 score (after weight calibration)
Anger	0.758	0.766
Fear	0.880	0.893
Anticipation	0.895	0.871
Surprise	0.873	0.886
Joy	0.900	0.917
Sadness	0.849	0.865
Trust	0.765	0.786
Disgust	0.780	0.798
Overall of our model	**0.839**	**0.850**

score or even decreased, such as anticipation. Notably, the negative emotions are easily discriminated in the previous emotion analysis, e.g., anger and disability did not reach the expected score. Although we repeatedly extracted data from these two emotions in a random way to create a new dataset, the results still have no significant change. We think that the lack of many negatives swear words in the data may have influenced this result. This is because people usually express strong emotions in the context of swear words.

5 Conclusion

Emotion status can be influenced by the external environment when interactions take place on SNS platform. Emotion is often an important factor in human decision-making process. Therefore, the improvement of sentiment analysis model is important for the overall social decision making and related tendency. The overall framework of this study is to build a scalable sentiment monitoring model for SNS posts based on K-means and BERT to analyze external environmental factors that affect user sentiment and revise the results of sentiment analysis. To improve the issue that the results of sentiment analysis do not match the true state of SNS contributors. In this paper, K-means is used to classify posts with situation tags to understand relationships between sentiment and situation. These relationships are used as benchmarks for correcting BERT results to achieve the main purpose of this study to consider external factors in the sentiment analysis process. Finally, a time series analysis is used to predict emotional tendencies. The time series analysis results and BERT results of the same post is compared to understand whether the emotional changes of a SNS user is under the expected range.

Despite a wide range of emotional theoretical models that have been defined, we learned that the audience in SNS is usually the whole world. This means that the user's communication is continuously carried out 24 h a day, and the conversion of media framing is also very fast. Users may have different emotions when they receive new topics. Moreover, differences in personal experience or background may also produce the cognition difference on the same word.

In future work, we will improve the analytical ability for the real-time posts in our model. The personal feature (e.g., health status, country) will be also added to our model as the basic parameters to improve the usability of the model for different users.

References

1. Worldometers.info website: Real Time Statistics Project. Worldometers.info., Dover, Delaware, USA. http://www.internetlivestats.com/. Accessed 19 Jan 2022
2. Picard, R.W.: Affective Computing. MIT Press, Cambridge (1997)
3. Kishinani, K.: Emotion detection from text using PyTorch and federated learning. Open-source project on GitHub. https://github.com/karankishinani/Emotion-detection-from-text-using-PyTorch-and-Federated-Learning. Accessed 19 Jan 2022
4. Devlin, J., Chang, M.W., Lee, K., Toutanova, K.: BERT: pre-training of deep bidirectional transformers for language understanding. arXiv preprint arXiv:1810.04805 (2018)
5. Yang, G., Baek, S., Lee, J.W., Lee, B.: Analyzing emotion words to predict severity of software bugs: a case study of opensource projects. In: Symposium on Applied Computing (SAC 2017), New York, USA, pp. 1280–1287 (2017)
6. Solomon, R.C.: On emotions as judgments. Am. Philos. Q. 183–191 (1988)
7. Wang, Y.J., Yen, N., Hung, J.C.: Design of a multi-dimensional model for UGC-based emotion analysis. In: 2019 International Conference on Computational Science and Computational Intelligence Proceedings, pp. 1383–1387. IEEE, USA (2019)
8. Plutchik, R.: The nature of emotions: human emotions have deep evolutionary roots, a fact that may explain their complexity and provide tools for clinical practice. Am. Sci. 89(4), 344–350 (2001)
9. Demszky, D., Movshovitz-Attias, D., Ko, J., Cowen A., Nemade, G., Ravi, S.: GoEmotions: a dataset of fine-grained emotions. In: The 58th Annual Meeting of the Association for Computational Linguistics Proceedings, pp. 4040–4054. Association for Computational Linguistics (2020)
10. Alsaeedi, A., Khan, M.Z.: A study on sentiment analysis techniques of Twitter data. Int. J. Adv. Comput. Sci. Appl. 10(2), 361–374 (2019)
11. Himelboim, I., Sweetser, K.D., Tinkham, S.F., Cameron, K., Danelo, M., West, K.: Valence-based homophily on Twitter: network analysis of emotions and political talk in the 2012 presidential election. New Media Soc. 18(7), 1382–1400 (2014)
12. Wang, W., Chen, L., Thirunarayan, K., Sheth, A.P.: Harnessing Twitter "Big Data" for automatic emotion identification. In: International Conference on Privacy, Security, Risk and Trust and 2012 International Conference on Social Computing Proceedings, Amsterdam, pp. 587–592 (2012)
13. Lee, Y.J., Huang, J.J., Wang, H.Y., Zhong, N.: Study of emotion recognition based on fusion multi-modal bio-signal with SAE and LSTM recurrent neural network. J. Commun. 38(12), 109–120 (2017)
14. Rout, J.K., Choo, K.-K., Dash, A.K., Bakshi, S., Jena, S.K., Williams, K.L.: A model for sentiment and emotion analysis of unstructured social media text. Electron. Commer. Res. 18(1), 181–199 (2017). https://doi.org/10.1007/s10660-017-9257-8
15. Seyeditabari, A., Tabari, N., Gholizadeh, S., Zadrozny, W.: Emotion detection in text: focusing on latent representation. arXiv preprint arXiv:1907.09369 (2019)

16. Kanclerz, K., Miłkowski, P., Kocoń, J.: Cross-lingual deep neural transfer learning in sentiment analysis. Procedia Comput. Sci. **176**, 128–137 (2020)
17. Arpit, D., et al.: A closer look at memorization in deep networks. In: Proceeding of 34th International Conference on Machine Learning (ICML), vol. 70, pp. 233–242 (2017)

Empirical Evaluation of Machine Learning Ensembles for Rumor Detection

Andrés Zapata[1], Eliana Providel[1,2(✉)], and Marcelo Mendoza[2]

[1] Escuela de Ingeniería Informática, Universidad de Valparaíso, Valparaíso, Chile
andres.zapata@alumnos.uv.cl, eliana.providel@uv.cl
[2] Departamento de Informática, Universidad Técnica Federico Santa María,
Santiago, Chile
marcelo.mendoza@usm.cl

Abstract. Rumor detection is a recent and quite active topic of multi-disciplinary research due to its evident impact on society, which can even result in physical harm to people. Despite its broad definition and scope, most existing research focuses on Twitter, as it makes the problem a bit more tractable from the point of view of information gathering, processing, and further retrieval of social network features. This paper presents an empirical study of novel machine learning ensembles for the rumor classification task on Twitter. As it has been observed that certain neural models perform better for specific veracity labels, we present a study on how the combination of such classifiers in different kinds of ensemble results in a new classifier that has better performance. Using benchmark data, we evaluate three groups of models (two groups of deep neural networks and a control group of classical machine learning methods). In addition, we study the performance of three ensemble strategies: Bagging, Stacking, and Simple Soft Voting. After varying several parameters of the models, such as the number of hidden units and dropout, among others experimental factors, our study shows that the LSTM, Stacked LSTM (S-LSTM), Recurrent Convolutional Neural Networks (RCNN), and Bidirectional Gated Recurrent Unit (Bi-GRU) yields the best results reaching an accuracy of 0.93 and an average precision of 0.97.

Keywords: Rumor detection · Fake news · Ensembles · Machine learning · Deep learning

1 Introduction

Social networks have become a widely used medium of dissemination of news, allowing information to arrive instantly to many users in a short time. Although information spreads daily through social media, its veracity is not commonly known or verified. This situation can be exploited to spread misinformation, eventually harming users [24,26]. Given the amount of published information and its rapid propagation through social media, it has become difficult for specialists or fact-checkers to review or determine the veracity of online information [19].

G. Meiselwitz (Ed.): HCII 2022, LNCS 13315, pp. 422–436, 2022.
https://doi.org/10.1007/978-3-031-05061-9_30

Hence, it is essential to have automatic mechanisms to detect rumors or fake news.

How to automatically detect rumors or fake news on social media has been a widely studied problem, where several models of machine learning and deep learning have been used for this classification task [8,14,18,22]. For the classification it is possible to use binary labels: *true rumor* and *false rumor* [6], or multi-class labels such as *rumor, non-rumor, true rumor*, and *false rumor* [27], among others, taking into account different features such as text, temporal, and propagation-based features, among others. Generally speaking, these approaches are based on automatic text classification [16], testing different data sources that can be helpful for the task.

Despite the variety of models and approaches, it has been only recently observed that some architectures perform better for certain classes than for the rest [18]. Although, in general, this fact decreases the average precision of the classifier, we aim to empirically assess whether the combination—technically, the ensembles—of several such architectures can yield a combined classifier with better general performance. For the experiments, we used the widely-used Twitter16 dataset [14], most commonly used in experiments about fake news classification, using the text of each message as well as the propagation structure on Twitter. We implement different ensemble techniques such as bagging [25], model stacking [21] and Simple Soft Voting [25]. For each strategy, we use deep learning models (LSTM, Stacked LSTM, RCNN, Bidirectional GRU, Bidirectional Stacked GRU) and classic machine learning models (Naive Bayes, Random Forest, SVM). Best results are obtained when stacking several models of neural networks, obtaining an accuracy of 0.93 and an average precision of 0.97. In addition to these results, a key technical contribution of this work is the use of four veracity categories, in contrast to existing results that use at most three categories.

The rest of this paper is structured as follows: first, we review related work grouped by the machine learning methods used, as well as by the usage of the Twitter16 dataset (Sect. 2); then, we describe the architecture of the classifiers empirically evaluated (Sect. 3). Next, Sect. 4 describes the experimental setup and evaluation. Section 5 presents the results obtained, and finally, we present the conclusions and discussion in Sect. 6.

2   Related Work

How to detect disinformation or how to detect rumors in social media is a problem that has been extensively researched from different angles and aspects. Indeed, as the literature is quite extensive in the area, we only present studies directly related to this work, or that serve as a seminal point of reference. We group the articles into two groups: the first group describes the machine learning or the ensemble method used, independent of the dataset used. On the other hand, the second group describes works that use the Twitter16 [14] dataset, which is the same used in this work. All the related works presented in this section are based on Twitter content.

Learning Methods. Vijeev *et al.* [22] work with different learning machines such as Naive Bayes, Random Forests and Support Vector Machines. In their experiments they consider features of the content of the post (sentiment, POS tags, word-count, reply type, among others), and user-based features (age, if the account is verified, listed count, average number of posts, and geolocation enabled). The best results are obtained using Random Forest and a specific combination of features (namely, user-follower-count, user-friends-count, favorite-count, retweet-count, age, average-number-posts, and breaking-count). Hunt *et al.* [8] addresses the problem studying if the machine learning algorithms can predict the veracity of a tweet and if it is possible to use models trained with historical events to predict a new event. Specifically they work with six crisis events. They present an approach to track misinformation associated with the crisis event. Using different machine learning methods, the best results are obtained using SVM.

Ahmad *et al.* [2] use four datasets and the Linguistic Inquiry and Word Count (LIWC) lexical resource to explore textual properties. To detect rumors, they use ensemble techniques such as Bagging Classifiers, Boosting Classifiers and Voting Classifiers. They show that ensembles obtain better results than individual machines. Following this line, Kim *et al.* [10] work with a weighted soft voting strategy, assembling Random Forest, XGBoost, and Multilayer perceptron. Using content, propagation and user-based features, they expose the importance of properly choosing the learning method, because it can negatively affect the results of the classification.

Agarwal and Dixit [1] use SVM, Convolutional Neural Networks (CNN), Long Short-Term Memory (LSTM), K-Nearest Neighbors (KNN), and Naive Bayes for an ensemble that learn representations of news, authors and titles jointly. The best results are obtained using LSTM. Similarly, Kumar *et al.* [12] study the behavior of different neural networks and ensemble models for detecting rumors. They conclude that CNN with Bidirectional LSTM-based ensembles combined with attention mechanisms achieve the best results. Kaur *et al.* [9] propose a method to merge machine learning classifiers based on false prediction rate. In the model, they use different methods for extraction of features, which are then used in an ensemble multi-level Voting Classifier.

Kotetti *et al.* [11] use ensembles of neural networks with temporal series. As input they use Twitter conversations and its reactions, converting them to time-series vectors. The neural networks explored are Recurrent Neural Network (RNN), Gated Recurrent Unit (GRU) and LSTM, among others. Al-Rakhami and Al-Amri [4] analyze the credibility associated with data from COVID-19, using two categories: *credible* and *non-credible*. For the model they use tweet-level and user-level features, exploring six learning methods for the ensemble. The best results are obtained using Stacking ensemble with SVM and Random Forest. Other work by Akhter *et al.* [3] focuses in the Urdu language. They use five learning ensembles, trained on character-level, word-level, and statistical-based features. They obtain the best results combining SVM, Decision Trees, and Naive Bayes, using Selection and Voting ensembles.

Related Work Based on the Twitter16 Dataset. Tian *et al.* [20], tackle the task of early detection of rumor, using the content of the tweet and the comments associated. They propose a transfer learning model considering the stance in the post, using Stance-Bert and Stance-CNN-LSTM transfer learning models. Dong *et al.* [7], addressed rumor detection using information of the user and the sentiment in the post at word-level and tweet-level, with hierarchical attention networks. Vu and Jung [23] use the structure of propagation of a post, which is represented using a graph convolutional neural network. Another work by Malhotra and Vishwakarma [15] explores text-based features using LSTM and RoBERTa vector representations, and a Graph Convolutional Network using user-based features to obtain propagation structure. Following the same line, Lin *et al.* [13] propose an encoder-decoder-detector model based on graph convolutional neural networks. The encoder codifies text and propagation features, which are used by the decoder to learn overall structural information. Finally, the detector takes the decoder output to classify events as fake or not fake.

3   Classifier Architecture

The general classifier architecture for rumor detection that we use has three principal stages, presented in Fig. 1. These stages are *Preprocessing*, *Vectorization*, and *Classification*. In the following, we detail each of these stages.

1. **Preprocessing**: this stage prepares the text for vectorization, removing or replacing symbols or characters that do not contribute to the representation of the text. Preprocessing techniques, their description, and a example of text before and after preprocessing are presented in the Table 1.
   Considering the different techniques of text preprocessing, we form two groups for the experiments. The first, called Preprocessing-Group1, with techniques P1 and P2; and the Preprocessing-Group2 with techniques P3, P4, P5, P6, and P7. We made these groups with the objective of reducing the number of experiments, having Preprocessing-Group1 remove non-meaningful tokens from tweets, and Preprocessing-Group2 reduce the number of words for a more efficient vectorization.
2. **Vectorization**: in this stage we represent every tweet and their propagation tree into an embedding that will be used for classification. The embedding used is word2vec [17] of length 300, following the same configuration as in previous work [18]. As the neural models require a fixed-size input, but the tweet propagation information is of varied depth, we settle on creating sequences of average length (in this case, 36) by trimming or padding the input sequences.
3. **Classification**: in this stage the input is the embedding of every propagation tree and the output is a specific class with their label of veracity. Here, we first train classical learning methods, as well as five neural networks to use them in the ensembles, as we show in Fig. 1. We use standard techniques for training each model. More specifically, we create three ensembles using the following techniques:

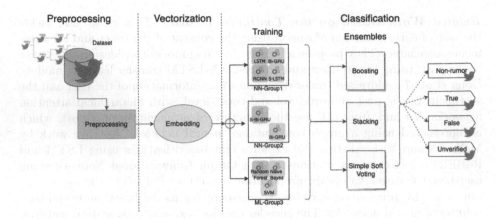

Fig. 1. General classifier architecture with three stages: preprocessing, vectorization, and classification. The initial input is the tweet timeline and their propagation structure, while the final output is a veracity label.

Table 1. Preprocessing techniques and examples of their application.

ID	Description	Ex. before preprocessing	Ex. after preprocessing
P1	Replace the URL with the string URL	@Twitter This is a test!. Or is it?? 😊 https://t.co/ #Test	@Twitter This is a test!. Or is it?? 😊 URL #Test
P2	Replace mentions with the string REF	@Twitter This is a test!. Or is it?? 😊 https://t.co/ #Test	REF This is a test!. Or is it?? 😊 https://t.co/ #Test
P3	Remove the # symbols in the hashtags	@Twitter This is a test!. Or is it?? 😊 https://t.co/ #Test	@Twitter This is a test!. Or is it?? 😊 https://t.co/ Test
P4	Remove emojis and special characters, such as: exclamation marks, double quotes, punctuation, emojis, symbols & pictographs, among others.	@Twitter This is a test!. Or is it?? 😊 https://t.co/ #Test	@Twitter This is a test Or is it httpstco #Test
P5	Convert the text to lowercase	@Twitter This is a test!. Or is it?? 😊 https://t.co/ #Test	@twitter this is a test!. or is it?? 😊 https://t.co/ #test
P6	Remove stopwords	@Twitter This is a test!. Or is it?? 😊 https://t.co/ #Test	@Twitter This test ! . Or ?? 😊 https://t.co/ # Test
P7	Apply Stemming and Lemmatization	Programmers program with programing languages	programm program with program languag

- ○ *Stacking*, which consists in combining the predictions of a base group of machine learning models using another machine learning method, also referred to as a meta-model [21].
- ○ *Bagging*, this technique consists of training each model with a random sample of the dataset, known as a bootstrap sample, and then combining the predictions, which in the implementation done in this study is taking the average of the predictions [5].
- ○ *Simple Soft Voting* [25], is based on training the machine learning models with the whole training dataset and averaging the predictions to get the final result.

Learning Techniques Groups. The machine learning models used to make the ensembles were divided into three groups:

o The first one, *NN-Group1*, is composed of the following neural networks: LSTM, Stacked LSTM (S-LSTM), Recurrent Neural Networks (RCNN) and Bidirectional GRU (Bi-GRU).
o The second group, *NN-Group2*, is composed of the following neural networks: Bidirectional GRU and Stacked Bidirectional GRU (S-Bi-GRU).
o The third group, *ML-Group3*, was composed of the following classic machine learning models: Naive Bayes (NB), Random Forest (RF) and Support Vector Machine (SVM).

The models of the first and second group were chosen due to previous work [18], because each model obtained the best results for a particular class. Hence, as is the main motivation of this work, we first aim to have a first group that combines the best models for each class. On the other hand, the second group aims to evaluate how the quantity of neural networks in the ensemble affect the predictive performance; given that we have five neural networks, and to not repeat all the models, two were chosen to make the ensemble—one that wasn't used in the first group and another that was. Finally, the third group was chosen to have a comparison between neural networks and classical machine learning algorithms.

4   Experimental Evaluation

This section presents both the description of the Twitter16 dataset, as well as the experimental design used to evaluate the ensemble models previously described.

4.1   Data Description

We use data from the Twitter16 dataset, first presented in the investigation of Ma *et al.* [14]. This dataset contains 818 tweets along with their respective conversational threads, which are technically known as *propagation trees*. A propagation tree has an initial post (root of the tree) along with the replies and retweets related to the root. Propagation trees capture sequential and temporal information regarding the reactions from other users to the root message, thus enabling a richer analysis provided for instance by recurrent or graph-based neural networks. Figure 2 shows a depiction of a propagation tree. To summarize, from the dataset we analyze 753 propagation trees, which correspond to a total of 21,741 post and 297,301 retweets. However, retweets were not considered as input data, as they provide no additional information.

Regarding classification labels, the dataset has four categories of veracity: *non-rumor*, *true rumor*, *false rumor* and *unverified rumor*. The first, with 205 root posts, corresponds to a post whose veracity is known at the moment of publication. The second, with 188 root posts, label posts where the veracity is corroborated some period of time after its publication. The third, with 180 root posts, corresponds to posts where the falsehood of the post is established some period of time after its publication. Finally, the fourth label, with 180 root posts, is assigned to posts where the veracity is unknown or cannot be confirmed.

4.2   Experimental Design

For the experimental evaluation, we compare the performance of various techniques and parameters shown below. A more detailed summary of the experiments is presented in the Table 2.

- Using different techniques of preprocessing, as shown before in Table 1, with Preprocessing-Group1 and Preprocessing-Group2.
- Each preprocessing group was tested separately, to have a better understanding of how the techniques affect the results of the ensembles.
- Three ensembles were used: Stacking, Bagging and Simple Soft Voting.
- For these ensembles, we use the model groups described before (Sect. 3), that is NN-Group1, NN-Group2, and ML-Group3, where each model of each group is trained individually. It is important to highlight that the meta-model for Stacking is a LSTM with 256 units.
- For the neural networks, we modify the parameters by (i) using different units in the hidden layer, and (ii) using different dropout rates. In all cases, we use the sigmoid activation function along with categorical cross-entropy as loss function.
- All experiments on neural networks were trained with 200 epochs and a batch size of 128.

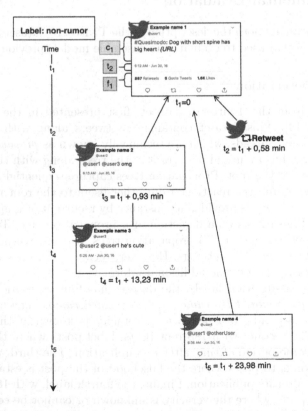

Fig. 2. Example of propagation tree for a non-rumor label.

In summary, we tested 4 parameters for each experiment. We had to test 2 groups of neural networks, 3 ensemble techniques, 2 groups of preprocessing techniques, and 4 values for units of dimensionality in the neural networks. This results in a total of 48 experiments. Then we did another round of experiments, but with the units of dimensionality fixed on the best values given by previous work [18], which added another 12 experiments. Finally, we tested one group of classic machine learning algorithms, varying just the 3 ensemble techniques and the 2 groups of preprocessing techniques, adding 6 more experiments, totaling 66 experiments.

Once we did these experiments, we decided to test different values for dropout on the neural networks but, to reduce the number of experiments, we made these tests only in the best ensemble configurations. We tested the values: 0.2, 0.3, 0.4, 0.6, 0.7, and 0.8, resulting in 6 experiments corresponding each one, to one value of dropout. However, as these results show little variation overall, we consider only the 66 aforementioned experiments and do not present any details of the dropout rate variation in the results table (Table 6).

Table 2. Experimental configurations, regarding models, parameters, and preprocessing groups applied.

Ensemble	Preprocessing	Machine learning or neural network	Parameters
Stacking or Bagging or Simple Soft Voting	Preprocessing-Group1: P1 and P2	NN-Group1: LSTM Stacked LSTM RCNN Bidirectional GRU	Units: 64, 128, 256, 512 Activation function Sigmoid Lost func.: Categ. Cross Entropy Dropout: 0.5
		NN-Group2: Bidirectional GRU Stacked Bidirec. GRU	Units: 64, 128, 256, 512 Activation function Sigmoid Lost func.: Categ. Cross Entropy Dropout: 0.5
		ML-Group3: Naive Bayes Random Forest SVM	–
	Preprocessing-Group2: P3, P4, P5,P6, and P7	NN-Group1: LSTM Stacked LSTM RCNN Bidirectional GRU	Units: 64, 128, 256, 512 Activation function Sigmoid Lost func.: Categ. Cross Entropy Dropout: 0.5
		NN-Group2: Bidirectional GRU Stacked Bidirectional GRU	Units: 64, 128, 256, 512 Activation function Sigmoid Lost func.: Categ. Cross Entropy Dropout: 0.5
		ML-Group3: Naive Bayes Random Forest SVM	–

5   Results

This section presents the results of four different experiments, based on the combination of models, ensemble techniques, and parameters, as it is shown in Table 2. The first experiment consists in evaluating the results obtained for each ensemble, as it is shown in Fig. 3. For this configuration and for every preprocessing group, we evaluate the three ensembles. More specifically, Fig. 3(a) presents the results when using Preprocessing-Group1, whereas Fig. 3(b) shows the results for Preprocessing-Group2. Here, we can see that the best results, accuracy of 0.938, are obtained when using the NN-Group1 ensemble, Stacking technique, Preprocessing-Group1, and quantity of hidden units 256 for LSTM, Stacked LSTM, and Bi-GRU, and 512 for RCNN (Fig. 3(a)). The second-best result was produced by the Soft Voting ensemble which showed an accuracy of 0.920.

If we study the results by class, we can see that the best results are obtained with the Stacking technique and with NN-Group1, except in the recall metric for false rumor where a slightly higher result is obtained with Simple Soft Voting technique. In the Table 3, 4, and 5, we present a summary of the best results, that is, using the Stacking technique in terms of Precision, F1-score, and Recall. In general, Stacking present the best results, showing an accuracy of 0.93 and an average precision of 0.97, being the non-rumor label the one with lowest precision, with a value of 0.885.

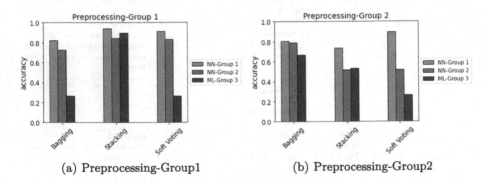

(a) Preprocessing-Group1   (b) Preprocessing-Group2

Fig. 3. First experiment results: ensemble techniques and preprocessing groups for each group of models.

Table 3. Precision by class considering the best results of an ensemble.

Machine learning or neural network	Ensembles	Preprocessing group	Precision (true)	Precision (false)	Precision (unveri-fied)	Precision (non-rumor)	Avg. precision
NN-Group1	Stacking	Group1	1.000	1.000	1.000	0.885	0.971
NN-Group1 (512)	Soft voting	Group1	1.000	0.815	1.000	0.882	0.970

Table 4. F1-score by class, considering the best results of an ensemble.

Machine learning or neural network	Ensemble	Preprocessing group	F1-score (true)	F1-score(false)	F1-score (unverified)	F1-score (non-rumor)	Avg. F1-score
NN-Group1	Stacking	Group 1	1.000	0.870	0.918	0.939	0,938
NN-Group1 (512)	Soft voting	Group1	0.980	0.863	0.903	0.938	0,921

Table 5. Recall by class, considering the best results of an ensemble.

Machine learning or neural network	Ensembles	Preprocessing group	Recall (true)	Recall (false)	Recall (unverified)	Recall (non-rumor)	Avg. recall
NN-Group1	Stacking	Group1	1.000	0.800	0.933	1.000	0,933
NN-Group1(512)	Soft voting	Group1	0.960	0.917	0.824	1.000	0,925

The second experiment consists of adjusting the parameters of the neural network, that is, number of hidden units and dropout, while keeping the default parameters for the classical machine learning algorithms ensemble. Also, we use three ensemble techniques. Figure 4(a) presents the results when modifying the dropout parameters, while Fig. 4(b) shows the variation of hidden units, from 64 to 512. Both figures show the performance of the NN-Group1 with units of dimentionality 256 for LSTM, Stacked LSTM, and Bi-GRU, and 512 for RCNN. As we can see in the two graphics, the results have small differences in performance.

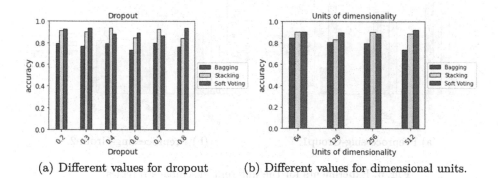

(a) Different values for dropout   (b) Different values for dimensional units.

Fig. 4. Second experiment results: variation of neural model hidden units and dropout.

The third experiment consists of exploring how the preprocessing technique affects the results in each ensemble. In short, as illustrated in Fig. 5(a), Stacking provides the best results in all configurations, but Bagging—when using Preprocessing-Group2—also shows good results.

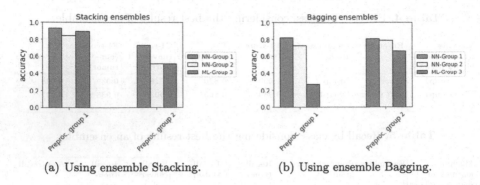

(a) Using ensemble Stacking.   (b) Using ensemble Bagging.

Fig. 5. Third experiment results: effect of the preprocessing strategies.

Finally, the fourth experiment focuses on machine learning and neural networks groups (NN-Group1, NN-Group2, and ML-Group3). Here we evaluate the behavior of the ensembles associated with the learning machines or neural networks that are used. We present the results separated by preprocessing group, Fig. 6(a) and 6(b). As we can see, NN-Group1 and NN-Group2 obtain, generally speaking, better results than ML-Group3 for the three ensembles. And NN-Group1 obtained a slightly better result than NN-Group2. In addition, these results can also corroborate the result presented in the previous experiment, where the best performance is obtained when using Preprocessing-Group1.

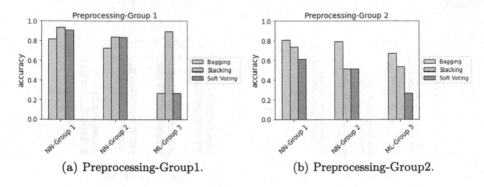

(a) Preprocessing-Group1.   (b) Preprocessing-Group2.

Fig. 6. Experiments for the different ensemble strategies.

As we present above (Sect. 4), we compute 66 experiments that we summarize in this section. In Table 6 we present all the results.

Table 6. Experimental results by metrics (Acc.: Accuracy, Prec.: Precision, Rec.: Recall, F1: F1-Score). Model group abbreviations: NN-Group1 (NN1), NN-Group2 (NN2), ML-Group3 (ML). Pre-processing group abbreviations: Preprocessing-Group1 (PPG1), Preprocessing-Group2 (PPG2). Veracity labels abbreviations: True (T), False (F), Non-Rumor (NR), Unverified (U). Ensemble method abbreviations: Bagging (B), Stacking (S), Simple Soft Voting (V).

Ex.	Group	En.	Prep. Group	Acc.	Prec. (T)	Prec. (F)	Prec. (U)	Prec. (NR)	Rec. (T)	Rec. (F)	Rec. (U)	Rec. (NR)	F1 (T)	F1 (F)	F1 (U)	F1 (NR)
1	NN1	B	PPG1	0.818	0.887	0.821	0.716	0.857	0.851	0.771	0.866	0.795	0.869	0.795	0.784	0.825
2	NN1	B	PPG2	0.804	0.840	0.765	0.750	0.848	0.851	0.729	0.785	0.848	0.846	0.747	0.767	0.848
3	NN1	S	PPG1	0.938	1.000	1.000	1.000	0.885	1.000	0.800	0.933	1.000	1.000	0.870	0.918	0.939
4	NN1	S	PPG2	0.735	0.793	0.680	0.926	0.562	0.852	0.654	0.694	0.750	0.821	0.667	0.794	0.643
5	NN1	V	PPG1	0.912	0.929	0.912	0.920	0.885	0.963	0.939	0.885	0.852	0.945	0.925	0.902	0.868
6	NN1	V	PPG2	0.611	0.957	0.356	0.677	0.714	0.667	0.842	0.677	0.333	0.786	0.500	0.677	0.455
7	NN2	B	PPG2	0.725	0.859	0.800	0.639	0.638	0.838	0.619	0.707	0.741	0.848	0.698	0.671	0.686
8	NN2	B	PPG2	0.787	0.868	0.654	0.778	0.839	0.846	0.750	0.667	0.876	0.857	0.699	0.718	0.857
9	NN2	S	PPG1	0.841	0.842	0.843	0.844	0.845	0.846	0.847	0.848	0.849	0.850	0.851	0.852	0.853
10	NN2	S	PPG1	0.513	0.688	0.480	0.441	0.526	0.688	0.500	0.682	0.392	0.688	0.490	0.536	0.449
11	NN2	V	PPG1	0.832	0.969	0.733	0.828	0.773	0.939	0.880	0.727	0.773	0.954	0.800	0.774	0.773
12	NN2	V	PPG2	0.513	0.842	0.377	0.615	0.429	0.571	0.885	0.571	0.097	0.681	0.529	0.593	0.158
13	ML	B	PPG1	0.265	0.214	0.241	0.167	0.400	0.375	0.250	0.095	0.300	0.273	0.246	0.121	0.343
14	ML	B	PPG2	0.667	0.556	0.846	0.843	0.637	0.878	0.452	0.589	0.730	0.681	0.589	0.694	0.681
15	ML	S	PPG1	0.894	0.903	0.926	0.885	0.862	0.903	0.962	0.793	0.926	0.903	0.943	0.836	0.893
16	ML	S	PPG2	0.531	0.743	0.522	0.000	0.710	0.520	0.400	0.000	0.667	0.612	0.453	0.000	0.688
17	ML	V	PPG1	0.265	0.273	0.222	0.077	0.360	0.667	0.071	0.037	0.290	0.387	0.108	0.050	0.321
18	ML	V	PPG2	0.265	0.214	0.241	0.167	0.400	0.375	0.250	0.095	0.300	0.273	0.246	0.121	0.343
19	NN1 (128)	B	PPG1	0.805	0.883	0.750	0.792	0.778	0.912	0.808	0.803	0.671	0.897	0.778	0.797	0.721
20	NN1 (128)	B	PPG2	0.849	0.924	0.842	0.770	0.863	0.880	0.716	0.882	0.891	0.901	0.774	0.822	0.877
21	NN1(128)	S	PPG1	0.832	0.895	0.818	0.840	0.806	0.850	0.818	0.808	0.853	0.872	0.818	0.824	0.829
22	NN1 (128)	S	PPG2	0.611	0.733	0.719	0.375	0.556	0.815	0.622	0.643	0.429	0.772	0.667	0.474	0.484
23	NN1 (128)	V	PPG1	0.894	0.920	0.839	0.933	0.889	0.958	1.000	0.848	0.800	0.939	0.912	0.889	0.842
24	NN1 (128)	V	PPG2	0.558	0.704	0.425	0.567	0.625	0.731	0.708	0.567	0.303	0.717	0.531	0.567	0.408
25	NN2 (128)	B	PPG1	0.714	0.756	0.776	0.597	0.732	0.827	0.634	0.708	0.698	0.790	0.698	0.648	0.714
26	NN2 (128)	B	PPG2	0.739	0.695	0.658	0.750	0.850	0.864	0.588	0.718	0.810	0.770	0.621	0.734	0.829
27	NN2 (128)	S	PPG1	0.894	0.935	0.826	0.926	0.875	1.000	0.826	0.806	0.933	0.967	0.826	0.862	0.903
28	NN2 (128)	S	PPG2	0.487	0.607	0.435	0.409	0.475	0.895	0.556	0.273	0.442	0.723	0.488	0.327	0.458
29	NN2 (128)	V	PPG1	0.903	0.966	0.889	0.786	0.966	0.966	0.960	0.957	0.778	0.966	0.923	0.863	0.862
30	NN2 (128)	V	PPG2	0.558	0.917	0.333	0.632	0.769	0.667	0.792	0.429	0.357	0.772	0.469	0.511	0.488
31	NN1 (64)	B	PPG1	0.848	0.892	0.829	0.894	0.789	0.949	0.797	0.756	0.887	0.919	0.813	0.819	0.835
32	NN1 (64)	B	PPG2	0.851	0.897	0.843	0.811	0.849	0.875	0.776	0.882	0.868	0.886	0.808	0.845	0.859
33	NN1 (64)	S	PPG1	0.903	0.792	0.917	0.879	1.000	1.000	0.786	0.967	0.889	0.884	0.846	0.921	0.941
34	NN1 (64)	S	PPG2	0.584	0.774	0.679	0.360	0.483	0.727	0.475	0.474	0.667	0.750	0.559	0.409	0.560
35	NN1 (64)	V	PPG1	0.903	0.963	0.792	1.000	0.853	0.929	0.864	0.875	0.935	0.945	0.826	0.933	0.892
36	NN1 (64)	V	PPG2	0.478	0.773	0.264	0.632	0.579	0.500	0.875	0.522	0.275	0.607	0.406	0.571	0.373
37	NN2 (64)	B	PPG1	0.753	0.873	0.730	0.704	0.734	0.797	0.667	0.781	0.775	0.833	0.697	0.740	0.754
38	NN2 (64)	B	PPG2	0.796	0.852	0.707	0.742	0.881	0.885	0.833	0.754	0.725	0.868	0.765	0.748	0.796
39	NN2 (64)	S	PPG1	0.903	0.885	0.964	0.885	0.879	1.000	0.844	0.920	0.879	0.939	0.900	0.902	0.879
40	NN2 (64)	S	PPG2	0.584	0.810	0.593	0.576	0.438	0.531	0.640	0.594	0.583	0.642	0.615	0.585	0.500
41	NN2 (64)	V	PPG1	0.894	1.000	0.737	0.926	0.857	0.970	0.778	0.862	0.909	0.985	0.757	0.893	0.882
42	NN2 (64)	V	PPG2	0.637	0.917	0.472	0.759	0.429	0.667	0.893	0.733	0.136	0.772	0.617	0.746	0.207
43	NN1 (512)	B	PPG1	0.772	0.829	0.861	0.671	0.744	0.863	0.785	0.724	0.728	0.846	0.821	0.696	0.736
44	NN1 (512)	B	PPG2	0.864	0.908	0.833	0.815	0.892	0.885	0.855	0.815	0.892	0.896	0.844	0.815	0.892
45	NN1 (512)	S	PPG1	0.885	0.731	1.000	0.862	0.938	0.950	0.765	0.893	0.968	0.826	0.867	0.877	0.952
46	NN1 (512)	S	PPG2	0.584	0.690	0.696	0.760	0.306	0.909	0.457	0.528	0.550	0.784	0.552	0.623	0.393
47	NN1 (512)	V	PPG1	0.920	1.000	0.815	1.000	0.882	0.960	0.917	0.824	1.000	0.980	0.863	0.903	0.938
48	NN1 (512)	V	PPG2	0.566	0.829	0.435	0.423	0.667	0.906	0.714	0.579	0.118	0.866	0.541	0.489	0.200
49	NN2 (512)	B	PPG1	0.738	0.860	0.592	0.731	0.753	0.914	0.714	0.695	0.632	0.886	0.647	0.713	0.688
50	NN2 (512)	B	PPG2	0.791	0.863	0.734	0.716	0.852	0.851	0.653	0.851	0.802	0.857	0.691	0.778	0.826

(continued)

Table 6. (*continued*)

Ex.	Group	En.	Prep. Group	Acc.	Prec. (T)	Prec. (F)	Prec. (U)	Prec. (NR)	Rec. (T)	Rec. (F)	Rec. (U)	Rec. (NR)	F1 (T)	F1 (F)	F1 (U)	F1 (NR)
51	NN2 (512)	S	PPG1	0.858	0.844	0.893	0.731	0.963	0.964	0.926	0.950	0.684	0.900	0.909	0.826	0.800
52	NN2 (512)	S	PPG2	0.513	0.583	0.615	0.818	0.103	0.955	0.552	0.305	1.000	0.724	0.582	0.444	0.188
53	NN2 (512)	V	PPG1	0.876	0.929	0.857	0.708	0.970	0.500	0.742	0.680	0.552	0.836	0.842	0.791	0.914
54	NN2 (512)	V	PPG2	0.549	0.958	0.370	0.516	0.500	0.742	0.680	0.552	0.214	0.836	0.479	0.533	0.300
55	NN1 (256)	B	PPG1	0.797	0.876	0.883	0.732	0.714	0.886	0.688	0.822	0.779	0.881	0.774	0.774	0.745
56	NN1 (256)	B	PPG2	0.815	0.881	0.766	0.821	0.795	0.868	0.671	0.821	0.897	0.874	0.715	0.821	0.843
57	NN1 (256)	S	PPG1	0.903	0.975	0.889	0.778	0.947	0.951	0.889	0.955	0.783	0.963	0.889	0.857	0.857
58	NN1 (256)	S	PPG2	0.619	0.645	0.625	0.652	0.571	0.870	0.600	0.577	0.513	0.741	0.612	0.612	0.541
59	NN1 (256)	V	PPG1	0.885	0.889	0.900	0.941	0.829	0.800	0.857	0.889	0.944	0.842	0.878	0.914	0.883
60	NN1 (256)	V	PPG2	0.681	0.808	0.641	0.605	0.800	0.875	0.862	0.852	0.242	0.840	0.735	0.708	0.372
61	NN2 (256)	B	PPG1	0.761	0.783	0.768	0.756	0.737	0.890	0.697	0.756	0.709	0.833	0.731	0.756	0.723
62	NN2 (256)	B	PPG2	0.765	0.798	0.697	0.735	0.802	0.835	0.630	0.694	0.862	0.816	0.662	0.714	0.831
63	NN2 (256)	S	PPG1	0.867	0.938	0.867	0.880	0.769	0.968	0.963	0.733	0.800	0.952	0.912	0.800	0.784
64	NN2(256)	S	PPG2	0.496	0.533	0.400	0.538	0.519	1.000	0.522	0.636	0.269	0.696	0.453	0.583	0.354
65	**NN2 (256)**	**V**	**PPG1**	**0.912**	**1.000**	**0.889**	**0.800**	**0.938**	**0.906**	**0.857**	**0.952**	**0.938**	**0.951**	**0.873**	**0.870**	**0.938**
66	NN2 (256)	V	PPG2	0.628	0.857	0.604	0.500	1.000	0.667	0.784	0.741	0.182	0.750	0.682	0.597	0.308

In Table 6, the rows with bold fonts are the experiments with the best performance for each parameter. These results show that:

- Experiment 3, in red, is our best result.
- Experiment 15 shows the best result for ML-Group3.
- Experiment 44 shows the best result for the Bagging ensemble and Processing-Group2.
- Experiment 47 shows the best result for the Simple Soft Voting ensemble.
- Experiment 65 shows the best result for NN-Group2.

6   Conclusions and Discussion

As we presented, the results of this study show that a Stacking ensemble composed of 4 neural networks (LSTM, Stacked LSTM, RCNN, and Bidirectional GRU), using the preprocessing techniques of removing URLs and mentions of the text in a Twitter post, produces the best results when trying to detect fake news in the Twitter16 dataset. This ensemble showed an accuracy of 0.93 and an average precision of 0.97, being the "non-rumor" label the one with lowest precision with a value of 0.885. The second-best result was produced by the Soft Voting ensemble, which showed an accuracy of 0.92. The last ensemble we tested was a Bagging ensemble which produces the worst results of all the 3 ensembles, with its best accuracy result of 0.86.

We also tested different configurations of neural networks, changing the units of dimensionality and the dropout value, but the results show that varying these parameters does not affect the performance of the neural networks significantly. Finally, we tested different machine learning models, using 2 other groups apart from the previously mentioned, which were, one composed of the neural networks Bidirectional GRU and Stacked Bidirectional GRU, and the other, a group composed of classic machine learning algorithms, Naive Bayes, SVM, and Random Forest. The groups composed of neural networks always performed better than the one composed of classical algorithms, and moreover, the best group (LSTM, Stacked LSTM, RCNN, and Bidirectional GRU) always performed better than the other ones.

From these results, it is of interest to be able to explore the behavior using datasets in other languages, using different types of embedding for the text, as well as incorporating other features of the data.

Acknowledgements. Mr. Mendoza acknowledges funding from the Millennium Institute for Foundational Research on Data. Mr. Mendoza was also funded by grants ANID BASAL FB210017 and ANID FONDECYT 1200211.

References

1. Aggarwal, A., Dixit, A.: Fake news detection: an ensemble learning approach. In: Proceedings of the International Conference on Intelligent Computing and Control Systems, ICICCS 2020, pp. 1178–1183 (2020)
2. Ahmad, I., Yousaf, M., Yousaf, S., Ahmad, M.: Fake news detection using machine learning ensemble methods. Complexity **2020**, 8885861:1–8885861:11 (2020)
3. Akhter, M., Zheng, J., Afzal, F., Lin, H., Riaz, A., Mehmood, S.: Supervised ensemble learning methods towards automatically filtering Urdu fake news within social media. PeerJ Comput. Sci. **7**, 1–24 (2021)
4. Al-Rakhami, M., Al-Amri, A.: Lies kill, facts save: detecting COVID-19 misinformation in Twitter. IEEE Access **8**, 155961–155970 (2020)
5. Breiman, L.: Bagging predictors. Mach. Learn. **24**(2), 123–140 (1996)
6. Castillo, C., Mendoza, M., Poblete, B.: Information credibility on Twitter. In: Proceedings of the 20th International Conference on World Wide Web, WWW 2011, Hyderabad, India, pp. 675–684. ACM (2011)
7. Dong, S., Qian, Z., Li, P., Zhu, X., Zhu, Q.: Rumor detection on hierarchical attention network with user and sentiment information. In: Zhu, X., Zhang, M., Hong, Yu., He, R. (eds.) NLPCC 2020. LNCS (LNAI), vol. 12431, pp. 366–377. Springer, Cham (2020). https://doi.org/10.1007/978-3-030-60457-8_30
8. Hunt, K., Agarwal, P., Zhuang, J.: Monitoring misinformation on Twitter during crisis events: a machine learning approach. Risk Anal. **2020**, (Early View) (2020)
9. Kaur, S., Kumar, P., Kumaraguru, P.: Automating fake news detection system using multi-level voting model. Soft. Comput. **24**(12), 9049–9069 (2019). https://doi.org/10.1007/s00500-019-04436-y
10. Kim, Y., Kim, H., Kim, H., Hong, J.: Do many models make light work? Evaluating ensemble solutions for improved rumor detection. IEEE Access **8**, 150709–150724 (2020)
11. Kotteti, C., Dong, X., Qian, L.: Ensemble deep learning on time-series representation of tweets for rumor detection in social media. Appl. Sci. (Switzerland) **10**(21), 1–21 (2020)
12. Kumar, S., Asthana, R., Upadhyay, S., Upreti, N., Akbar, M.: Fake news detection using deep learning models: a novel approach. Trans. Emerg. Telecommun. Technol. **31**(2), e3767 (2020)
13. Lin, H., Zhang, X., Fu, X.: A graph convolutional encoder and decoder model for rumor detection. In: 2020 IEEE 7th International Conference on Data Science and Advanced Analytics (DSAA), pp. 300–306 (2020)

14. Ma, J., Gao, W., Wong, K.-F.: Detect rumors in microblog posts using propagation structure via kernel learning. In: Proceedings of the 55th Annual Meeting of the Association for Computational Linguistics (Volume 1: Long Papers), pp. 708–717, July 2017
15. Malhotra, B., Vishwakarma, D.: Classification of propagation path and tweets for rumor detection using graphical convolutional networks and transformer based encodings. In: Proceedings - 2020 IEEE 6th International Conference on Multimedia Big Data, BigMM 2020, pp. 183–190 (2020)
16. Mendoza, M.: A new term-weighting scheme for Naïve Bayes text categorization. Int. J. Web Inf. Syst. **8**(1), 55–72 (2012)
17. Mikolov, T., Sutskever, I., Chen, K., Corrado, G.S., Dean, J.: Distributed representations of words and phrases and their compositionality. In: Advances in Neural Information Processing Systems 26: 27th Annual Conference on Neural Information Processing Systems 2013. Proceedings of a Meeting held Lake Tahoe, Nevada, United States, 5–8 December 2013, pp. 3111–3119 (2013)
18. Providel, E., Mendoza, M.: Using deep learning to detect rumors in Twitter. In: Meiselwitz, G. (ed.) HCII 2020. LNCS, vol. 12194, pp. 321–334. Springer, Cham (2020). https://doi.org/10.1007/978-3-030-49570-1_22
19. Providel, E., Mendoza, M.: Misleading information in Spanish: a survey. Soc. Netw. Anal. Min. **11**(1), 36 (2021)
20. Tian, L., Zhang, X., Wang, Y., Liu, H.: Early detection of rumours on Twitter via stance transfer learning. In: Jose, J.M., et al. (eds.) ECIR 2020. LNCS, vol. 12035, pp. 575–588. Springer, Cham (2020). https://doi.org/10.1007/978-3-030-45439-5_38
21. Ting, K., Witten, I.: Stacking bagged and dagged models. In: Proceedings of the Fourteenth International Conference on Machine Learning, ICML 1997, San Francisco, CA, USA, pp. 367–375. Morgan Kaufmann Publishers Inc (1997)
22. Vijeev, A., Mahapatra, A., Shyamkrishna, A., Murthy, S.: A hybrid approach to rumour detection in microblogging platforms. In: 2018 International Conference on Advances in Computing, Communications and Informatics, ICACCI 2018, pp. 337–342 (2018)
23. Vu, D.T., Jung, J.J.: Rumor detection by propagation embedding based on graph convolutional network. Int. J. Comput. Intell. Syst. **14**, 1053–1065 (2021)
24. Zhang, X., Ghorbani, A.: An overview of online fake news: characterization, detection, and discussion. Inf. Process. Manag. **57**(2), 102025 (2019)
25. Zhou, Z.-H.: Ensemble Methods: Foundations and Algorithms, 1st edn. Chapman & Hall/CRC (2012)
26. Zubiaga, A., Aker, A., Bontcheva, K., Liakata, M., Procter, R.: Detection and resolution of rumours in social media: a survey. ACM Comput. Surv. **51**(2), 1–36 (2018)
27. Zubiaga, A., Liakata, M., Procter, R., Hoi, G.W.S., Tolmie, P.: Analysing how people orient to and spread rumours in social media by looking at conversational threads. PLoS ONE **11**(3), e0150989 (2016)

A Methodological Framework
for Facilitating Explainable AI Design

Jiehuang Zhang[1,2] and Han Yu[1(✉)]

[1] School of Computer Science and Engineering,
Nanyang Technological University, Singapore, Singapore
`han.yu@ntu.edu.sg`
[2] Alibaba-NTU Singapore Joint Research Institute, Singapore, Singapore

Abstract. The advancement of artificial intelligence (AI) technologies
has enabled many feats that were previously thought to be impossible.
However as AI systems become prevalent, there is an increasing need to
interpret how they work in order to build trust with the users. In important fields like medical diagnosis and self driving cars, there are high
stakes that require the AI system to perform well and provide reasonable explanations for their decisions. While there are many techniques
developed to make black box algorithms more transparent, there is a
lack of design methodologies to allow design teams to systematically
surface and address explainability issues during the AI solution design
process. To bridge this gap, we propose the Explainability in Design
(EID) methodology to address this gap in the literature. EID enables AI
software designers to surface and explore complex explainability issues
systematically, which might potentially be overlooked otherwise. With
our methodological tool, we aim to decrease the barrier of entry, the
time and experience needed to effectively make well informed decisions
for integrating explainability into their AI solutions.

Keywords: Explainable AI · Design method · Design tool · Values
sensitive design · Ethics · Design methodology

1 Introduction

In the fourth industrial revolution [16], one of the many enabling technologies is artificial intelligence (AI). AI systems are responsible for many advances
in diverse fields, such as healthcare [18], algorithmic crowdsourcing [20] and
autonomous vehicles [21]. AI systems, especially deep learning and neural
network-based systems, have allowed us to perform feats of incredible scale and
finesse, achievements that were believed to be impossible before. While automation enabled by AI has resulted in a paradigm shift in the way we live and work
[13], there are ethical issues present in these systems that we must pay attention
to. This is particularly important as more decision support systems previously
powered by humans are being transferred to the hands of AI. Due to this transfer

G. Meiselwitz (Ed.): HCII 2022, LNCS 13315, pp. 437–446, 2022.
https://doi.org/10.1007/978-3-031-05061-9_31

of autonomy and responsibility to algorithmic systems, there is a higher likelihood of mistakes or improper handling being overlooked. Thus as a society, we must consider the ramifications of such a shift and guide the advancement of AI towards a trajectory with proper oversight that is beneficial to humanity as a whole [4].

At present, most of the scientists and engineers in the AI community measure the performance of AI systems based on their accuracy and efficiency. Although these metrics are useful, there is also evidence that they may not provide the full picture on the behaviour of the AI solution. While state-of-the-art AI systems can produce useful results, they tend to lack explainability and are not easily understood by humans [9]. Even though algorithms are trained on logical data and make decisions based on the representative dataset, it does not mean that they are invulnerable to mistakes of misjudgements. As a result, we are unable to understand how they arrive at their decisions and find the problems that lie in that decision making process. Using spectral heatmaps, Lapuschkin et al. discovered that standard performance evaluation metrics can possibly be oblivious to certain types of problems in the algorithm [11]. The black box nature of modern AI algorithms can be a problem in certain fields where explainability is necessary or even crucial (e.g., medical diagnosis). Hence it is hugely important that the AI systems in such fields provide a satisfactory explanation to accompany the decisions that they make.

Explainability is a complex topic in the context of AI as it is multifaceted and defined differently depending on the type of AI systems. The field of explainable artificial intelligence (XAI) [5] aims to create a suite of machine learning techniques that enable users to understand, trust and effectively manage the emerging generation of AI systems [8]. These goals are achievable when system developers intentionally design AI algorithms to have effective explanation features, that are comprehensible for the end users. Despite the difficulty of XAI, there has been much progress made in this field [19] (e.g., Layer-wise Relevance Propagation [14]). These advancements in the XAI field are encouraging as we gain deeper insight into how complex AI models such as neural networks operate. However, we also must consider the need of integrating XAI design considerations early in the design and conception stage of the AI life cycle. Currently, the field lacks a methodological framework for AI development teams to incorporate XAI considerations into their AI solution designs.

In this paper, we bridge this gap by proposing the Explainability in Design (EID) methodological framework. The goal of EID is to assist software design teams to systematically consider XAI in AI systems, by reducing the barrier of entry and also inducing critical thinking during all stages of the design process. We believe that explainability is an important aspect of ethical AI and it allows the AI community as a whole to understand how algorithms work, then in turn to improve them. However due to the complexity of explainability, it can be difficult for lay people to grasp the insights necessary to work on it. Furthermore, including explainability can sometimes lead to trade offs in performance or accuracy as computing resources or human effort is diverted away from the core objectives. EID aims to address this gap by enabling teams to brainstorm and explore potential explainability issues for their AI products. This is achieved by enabling software design teams to analyse explainability needs and criteria specific to their application domain and stimulate thinking from the perspective of different groups of stakeholders.

2   Related Work

XAI can be broadly classified into two main categories, namely Post Hoc and Integrated approaches [5]. The difference between the two categories lies in the stage in which explainability is applied in the AI algorithm. Integrated XAI refers to the in building of explainability features during the design and construction of the algorithm, while post hoc XAI means that the explainability of an algorithm is only investigated after the output has been produced. Both types of explainability have respective advantages and disadvantages. However, the slight advantage of post hoc explainability is that it is unlikely to interfere with the performance of the AI system. As a result of XAI being an active research field, new and improved methods of attaining explainability, such as Shapley values [1] and LIME [15], have emerged.

Currently, the existing ethical AI design methodology is acquired from Value Sensitive Design (VSD). It originates from the field of information systems design and human computer interaction (HCI) to address design issues by emphasizing on the ethical values of direct and indirect stakeholders. By considering diverse methods to engage with values depending on the application scenario, VSD is able to help designers gain more insights and become adaptable to integration with other methodologies. One main workflow of VSD is to stimulate perspectives of direct and indirect stakeholders, then explore how their values are impacted. Direct stakeholders are people that use the AI product directly and are impacted by its use, while indirect stakeholders are people who do not use the AI product but are still impacted by its use. We aim to focus more on the indirect stakeholders in our methodological framework as they tend to be overlook by software developers. VSD has produced two methodological card games, Envisioning Cards [6] and Judgement Call [3]. Envisioning Cards help to stimulate critical thinking and considers four factors: stakeholders, time, values and pervasiveness. They allow designers to ponder the long term and possible systemic problems in AI design. Meanwhile, Judgement Call is a card game that AI developer teams can play to discover ethical problems in a AI product. It consists of cards that focuses on four factors: ethical value, the stakeholder, the number of stars in a review and wild cards that encourage exploratory thinking.

Liao et al. [12] recognises that while AI systems require explainability features, there is still a need to address real world user needs before we understand AI. The authors created a question bank in which user needs for explainability are represented as prototypical questions users might ask about the AI and use it as a study probe. Then, they interviewed Usability Experience (UX) and design experts on the current gaps between XAI algorithmic work and practices.

[10] showed that there is a lack of a principled framework that can provide the basis for the development of a XAI framework. Then the authors came up with four foundational components that can assist to create a simple framework to guide the design of XAI systems.

Nevertheless, there is no software engineering design methodology to guide an AI solution development team to discuss and determine what XAI notions should be incorporated into a given AI system design. The proposed EID methodological framework aims to bridge this gap.

3 Preliminaries

In this paper, we have narrowed the definitions of explainability to the following three main types as shown in Fig. 1: 1) transparency, 2) interpretability and 3) explainability [5]. This allows novice participants to focus on the most important metrics of XAI, and decide on their relevance to their application domains.

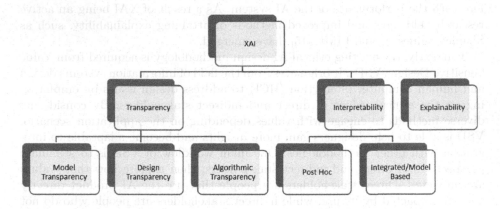

Fig. 1. Metrics of explainability in AI categorised into 6 types

1. Transparency: a model is transparent if by itself it is understandable. A model can feature different degrees of understandability
   (a) Model Transparency: Degree of human understandability of how the components (e.g. filters used, layers of a NN) of the trained model contribute to an output.
   (b) Design Transparency: Degree of human understandability of design decisions made to construct the ML model.
   (c) Algorithmic Transparency: Degree of human understandability of the training process that resulted in the trained ML model.
   We differentiate the model and design transparency by the focus area of the transparency effort: model transparency pertains to the arrangement of the components individually, and how they are selected to contribute to the entire model. On the other hand, the design transparency is the deliberate decision

of the model designer in order to promote the understandability of the model as a whole. While the two can be viewed by lay people as largely similar, we aim to use the classification system to distinguish between the minute differences.

2. Interpretability: the ability to explain or to provide the meaning in understandable terms to a human
   (a) Integrated Interpretability: Design of a ML model that involves specific design choices for better understandability.
   (b) Post Hoc Interpretability: Ability to analyse information pertaining to how the output of a trained ML model is obtained from the input.

3. Explainability: associated with the notion of explanation as an interface between humans and a decision maker, the focus is on the human and how the human can understand the mechanics of a algorithm.

The main difference between interpretability and explainability is such that, interpretability is how understandable the AI model is, while explainability concerns an output explanation for the decision made by the AI model. In the EID methodology, we focus on the two types of stakeholders, namely direct and indirect stakeholders. Direct stakeholders are people who directly use the AI system and are impact directly by its use, while indirect stakeholders do not use the AI system, but are impacted indirectly by its use. In the context of our XAI metrics classification, interpretability is closely related to the direct stakeholders, since it is in their best interests to understand how the model works. On the other hand, explainability is more related to the indirect stakeholders, since they are not interested in the inner workings of the algorithm, but just wish to understand how the model explains its decisions making process (Fig. 2).

Fig. 2. Comparison between direct and indirect stakeholders

4 The Explainability in Design Methodology

The Explainability in Design (EID) Methodology is in the form of a physical card game. It can be used by both lay people and experienced users alike,

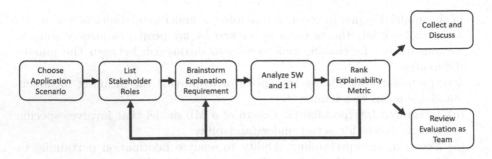

Fig. 3. The explainability in design workflow

enabling them to brainstorm and discover issues around XAI. We intend for the methodology to be application agnostic and allow the design team to specify the properties of their envisioned target application.

The workflow for a team to use EID to facilitate team discussions around XAI issues is shown in Fig. 3. We discuss the details of the step by step guide to using EID workflow:

1. First, AI design team members must choose an application domain that sets the context for the user study. The application domain can be real or fictitious, but mostly importantly it is preferably in a field where explainability is valued. If possible, the team members should be familiar with the domain such that as many details are included in the process of the user study. The reason is that sometimes in specific domains, there are possible considerations and trade offs that can affect the decision making process.
2. In Step 2, team members must choose the type of application card that accurate depicts their application domain. We have drawn from the work from Shneiderman's classification for usability motivation in the Human Computer Interaction literature [17]. They are 1) life-critical systems, 2) industrial and commercial uses, 3) office, home and entertainment, 4) exploratory, creative, collaborative applications, and 5) socio-technical applications.
3. For Step 3, the team must brainstorm and identify the list of direct and indirect stakeholders that are crucial to the analysis. Direct stakeholders are people that use the AI product directly and are impacted by the use, while indirect stakeholders are the other group of people that do not use the AI product directly but are impacted by its use [7]. Once the list of stakeholders is completed, each team member will take on the perspective of a stakeholder and conduct a deep analysis of the explanation details pertaining to that stakeholder. The EID methodological framework provides a number of guiding questions to facilitate the thinking process in this step. The guiding questions involve asking questions pertaining to the who, what, when, where, why and hows of the explanation parameters. Some examples include:

(a) Besides the direct stakeholders, who else is the explanation generated meant for?
(b) When is the best timing to show the explanation?
(c) What is the level of trust in the user and how frequently do they use the system?
(d) How do the emotional state of the user impact the explanation required?
(e) Why did the algorithm make this specific decision?
(f) What is the scope of the explanation and what depth is required

Ultimately, team members should strive to understand deeply the explanation details from the perspective of their respective stakeholder. This is especially important when the impact of the system on indirect stakeholders are not so obvious and require deep analysis to discover.

4. In this step, based on the analysis done, the team members must rank the most appropriate explainability metric pertaining to the application scenario. They must also give reasons for their ranking and how they arrive at the metric that they chose. This step is especially important as it identifies the most important explainability metrics pertaining to the participants' application scenario.

5. For the final step, the team collects all the responses of the members and shuffle them for anonymity before reading it aloud. With anonymity, this encourages team members to be truthful and list all their thought processes. The team can then evaluate the responses to determine if they are valid and worth addressing.

6. When the whole process of the methodology is completed, the team can then repeat it by going back to Step 3 and start a new stakeholder analysis, or simply conclude the session. Alternatively, the team members can decide to focus on an area of the user study that they believe warrants more attention. By asking exploratory questions, it is likely that the team can deepen their understanding of the specific area of focus.

Once the entire workflow of the methodology is completed, the team can gain deeper insights into how explainability issues are related to their application scenario. As a team, they can then decide to delve deeper into a specific topic of discussion, allowing every team member to discuss and brainstorm around that specific topic. This deep dive process will allow comprehensive exploration as well as surfacing potential complex explainability issues which otherwise might be overlooked.

The outputs of the methodology includes the following deliverables:

1. Identifying the explainability metrics that is appropriate for the application scenario.
2. Determine priorities of specialised requirements for the format, scope and type of explanations.
3. Ascertain the usage and purpose of the output explanation.
4. Quantifying improvements in the explainability knowledge levels of methodology users.

5. Enable team members to decipher the specific part of their algorithm that requires attention.

With these deliverables, the team can then make decisions for the improvement of their AI product or system.

5   Conclusions and Future Work

In this paper, we have discussed the gaps in the ethical AI design methodologies and highlighted the need for a framework that facilitates team design processes. After exploring current literature in AI and Machine Learning, we engaged the theory of VSD and various recent studies to create a methodological framework, Explainability in Design, to help design teams to navigate complex ethical choices around explainability. By designing EID to be low barrier to entry and only taking a short amount of time, it is effective and allow team members to make better decisions for explainability in their AI products and pipelines.

In subsequent research, we plan to conduct user studies to evaluate the effectiveness our EID, and see if our objectives are attainable. Currently, the format of our users studies will be conducted in teams of 2–4 members that are working or have worked on AI products in the past. As the new normal of working from home has been in effect for the past years, the online form of EID will be taking priority in our planned user studies. Then the team members can collaborate and use the EID methodology online accordingly. We aim to include project management functions for all team members to manage work allocation and timeline management. In addition to the above, we plan to identify specific application domains such as autonomous vehicles [2] and medical healthcare diagnosis [18] to apply our methodology and investigate deeper on how the different complexities interact.

Acknowledgements. This research is supported, in part, by Nanyang Technological University, Nanyang Assistant Professorship (NAP); Alibaba Group through Alibaba Innovative Research (AIR) Program and Alibaba-NTU Singapore Joint Research Institute (JRI) (Alibaba-NTU-AIR2019B1), Nanyang Technological University, Singapore; the RIE 2020 Advanced Manufacturing and Engineering (AME) Programmatic Fund (No. A20G8b0102), Singapore; and the Joint SDU-NTU Centre for Artificial Intelligence Research (C-FAIR). Any opinions, findings and conclusions or recommendations expressed in this material are those of the author(s) and do not reflect the views of National Research Foundation, Singapore.

References

1. Ancona, M., Oztireli, C., Gross, M.: Explaining deep neural networks with a polynomial time algorithm for shapley value approximation. In: International Conference on Machine Learning, pp. 272–281. PMLR (2019)

2. Atakishiyev, S., Salameh, M., Yao, H., Goebel, R.: Explainable artificial intelligence for autonomous driving: a comprehensive overview and field guide for future research directions. arXiv preprint arXiv:2112.11561 (2021)
3. Ballard, S., Chappell, K.M., Kennedy, K.: Judgment call the game: using value sensitive design and design fiction to surface ethical concerns related to technology. In: Proceedings of the 2019 on Designing Interactive Systems Conference, pp. 421–433 (2019)
4. Croeser, S., Eckersley, P.: Theories of parenting and their application to artificial intelligence. In: Proceedings of the 2019 AAAI/ACM Conference on AI, Ethics, and Society, pp. 423–428 (2019)
5. Došilović, F.K., Brčić, M., Hlupić, N.: Explainable artificial intelligence: a survey. In: 2018 41st International Convention on Information and Communication Technology, Electronics and Microelectronics (MIPRO), pp. 0210–0215. IEEE (2018)
6. Friedman, B., Hendry, D.: The envisioning cards: a toolkit for catalyzing humanistic and technical imaginations. In: Proceedings of the SIGCHI Conference on Human Factors in Computing Systems, pp. 1145–1148 (2012)
7. Friedman, B., Hendry, D.G., Borning, A.: A survey of value sensitive design methods. Found. Trends Hum.-Comput. Interact. 11(2), 63–125 (2017)
8. Gunning, D., Aha, D.: Darpa's explainable artificial intelligence (XAI) program. AI Mag. 40(2), 44–58 (2019)
9. Heinert, M.: Artificial neural networks-how to open the black boxes. In: Application of Artificial Intelligence in Engineering Geodesy (AIEG 2008), pp. 42–62 (2008)
10. Kim, M.Y., et al.: A multi-component framework for the analysis and design of explainable artificial intelligence. Mach. Learn. Knowl. Extract. 3(4), 900–921 (2021)
11. Lapuschkin, S., Wäldchen, S., Binder, A., Montavon, G., Samek, W., Müller, K.R.: Unmasking clever Hans predictors and assessing what machines really learn. Nat. Commun. 10(1), 1–8 (2019)
12. Liao, Q.V., Gruen, D., Miller, S.: Questioning the AI: informing design practices for explainable AI user experiences. In: Proceedings of the 2020 CHI Conference on Human Factors in Computing Systems, pp. 1–15 (2020)
13. Makridakis, S.: The forthcoming artificial intelligence (AI) revolution: its impact on society and firms. Futures 90, 46–60 (2017)
14. Montavon, G., Binder, A., Lapuschkin, S., Samek, W., Müller, K.-R.: Layer-wise relevance propagation: an overview. In: Samek, W., Montavon, G., Vedaldi, A., Hansen, L.K., Müller, K.-R. (eds.) Explainable AI: Interpreting, Explaining and Visualizing Deep Learning. LNCS (LNAI), vol. 11700, pp. 193–209. Springer, Cham (2019). https://doi.org/10.1007/978-3-030-28954-6_10
15. Ribeiro, M.T., Singh, S., Guestrin, C.: "why should i trust you?" Explaining the predictions of any classifier. In: Proceedings of the 22nd ACM SIGKDD International Conference on Knowledge Discovery and Data Mining, pp. 1135–1144 (2016)
16. Schwab, K.: The Fourth Industrial Revolution. Currency (2017)
17. Shneiderman, B., Hochheiser, H.: Universal usability as a stimulus to advanced interface design. Behav. Inf. Technol. 20(5), 367–376 (2001)
18. Tjoa, E., Guan, C.: A survey on explainable artificial intelligence (XAI): toward medical XAI. IEEE Trans. Neural Netw. Learn. Syst. (2020)
19. Xu, F., Uszkoreit, H., Du, Y., Fan, W., Zhao, D., Zhu, J.: Explainable AI: a brief survey on history, research areas, approaches and challenges. In: Tang, J., Kan, M.-Y., Zhao, D., Li, S., Zan, H. (eds.) NLPCC 2019. LNCS (LNAI), vol. 11839, pp. 563–574. Springer, Cham (2019). https://doi.org/10.1007/978-3-030-32236-6_51

20. Yu, H., Miao, C., Chen, Y., Fauvel, S., Li, X., Lesser, V.R.: Algorithmic management for improving collective productivity in crowdsourcing. Sci. Rep. **7**(1), 1–11 (2017)
21. Zhang, J., Shu, Y., Yu, H.: Human-machine interaction for autonomous vehicles: a review. In: Meiselwitz, G. (ed.) HCII 2021. LNCS, vol. 12774, pp. 190–201. Springer, Cham (2021). https://doi.org/10.1007/978-3-030-77626-8_13

Social Media Impact on Society and Business

The Role of Moral Receptors and Moral Disengagement in the Conduct of Unethical Behaviors Against Whistleblowers on Social Media

Stefan Becker[1]([✉]) and Christian W. Scheiner[1,2]

[1] Universität zu Lübeck, Lübeck, Germany
{stefan.becker,christian.scheiner}uni-luebeck.de
[2] Christian-Albrechts-Universität zu Kiel, Kiel, Germany

Abstract. Within the last years, whistleblowing has received considerable attention, as calls for ethical behavior in the workplace have grown louder and more forceful. Despite this fact, research has shown that individuals who blow the whistle are often frowned upon and treated poorly. In this context, while social media allows for offering support to the whistleblower, people may also hide behind online anonymity and engage in unethical behavior towards the whistleblower. Previous research already tried to explain unethical behavior on social media by linking moral receptors with moral disengagement in the context of unethical behavior. We built upon this approach in developing a set of propositions linking the moral receptors "harm/care", "fairness/reciprocity", "ingroup/loyalty", "authority/respect", and "purity/sanctity" with moral disengagement in order to explain unethical behavior against whistleblowers on social media. Furthermore, we discuss some contextual boundary conditions that may aggravate the negative behavior towards whistleblowers.

Keywords: Moral receptors · Moral foundations · Moral foundation theory · Unethical behavior · Moral disengagement · Whistleblower · Social media

1 Introduction

In recent years, whistleblowing has received greater attention, as calls for ethical behavior in the workplace have grown louder [74]. Instead of tributing whistleblowers for their engagement in uncovering illicit activities, they are blamed and often become victims of retaliation [44,54,60]. The ubiquity of social media in daily life not only allows whistleblowers to inform the public about illegal or immoral activities within an organization, but people can also easily offer support for whistleblowers. On the other hand, the perception of social media has dramatically changed within the last years. Despite the possibility of bringing

G. Meiselwitz (Ed.): HCII 2022, LNCS 13315, pp. 449–467, 2022.
https://doi.org/10.1007/978-3-031-05061-9_32

people together, social media can also help to separate people from each other and increase differences between groups [69].

Previous research has already identified the role of moral disengagement as a possible explanation, why decent people conduct malign behavior in general [7,11,46] and in social media in specific [62,64]. Hence, researchers have started to explore the mechanisms which enable people to disengage from their own moral standards without feeling any pain or regret.

In his conceptional paper, Scheiner [62] proposed to use moral receptors as a starting point to trigger unethical behavior. Building upon this idea, we argue that this framework is also capable in describing the boundary conditions and mechanisms for when and why retaliation against whistleblowers on social media occurs. We further argue that there are also different contextual factors influencing the unethical behavior towards whistleblowers positively as well as negatively. Specifically, in organizations with a highly unethical climate or in groups with a high level of group cohesion, retaliation against whistleblowers is proposed to be more likely.

2   Theoretical Background

2.1   Whistleblowing and Retaliation

Whistleblowing has brought several scandals to light in healthcare [26], finance [1], and sport [73,79]. A common definition of whistleblowing is the "deliberate disclosure of information about illegal, immoral, or illegitimate activities of an organization and its members to persons or organizations that may be able to effect action, generally by current or former organizational members" [48]. In an era of corporate fraud causing severe damages, whistleblowing is found to be a major source of fraud detection. It is also socially significant as it impacts employees, patients, organizations, and society in general [20]. As a result, it is unsurprising that whistleblowing has received considerable attention, as calls for ethical behavior in the workplace have grown louder and more forceful [20,74]. Moreover, by reporting misconduct, whistleblowers can help organizations to avoid massive reputational and financial losses, lawsuits, and reputational damages [47,50]. Within the EU, corruption is estimated to cost €120 billion per year [23]. According to the U.S. Department of Justice, in Fiscal Year 2020 alone, the government recovered over $2.2 billion in FCA settlements and judgments. Over $1.6 billion of that can be attributed to whistleblower-initiated cases [76]. Given its relevance, understanding the factors that encourage or prevent individuals from reporting observed unethical conduct is of critical importance to help organizations promote this type of behavior, especially given the consequences that some whistleblowers may encounter.

In fact, despite the pursuit of greater transparency in the workplace and the public's growing dissatisfaction with unethical behavior in modern organizations, research has shown that individuals who report unethical behavior are often frowned upon and treated poorly [22,44,45,54,59,60]. Instead of being applauded for exposing illegal actions, whistleblowers are far too often castigated for trying to challenge and overturn the status quo. In their survey with 87 US whistleblowers, Soeken and Soeken [65] found that only one of them had not experienced retaliation and harassment. Another study by Rothschild et al. [60] reports that approximately two thirds of the whistleblowers had experienced different forms of retaliation ranging from job loss (69%) to being blacklisted from getting another job in their field (64%). Similar numbers can also be found within the study of Cortina and Magley [19]. Here, 30% of the whistleblowers experienced only social retaliation (i.e., informal), while another 36% experienced both social as well as work-related retaliation (i.e., formal). Organizations and its members can use very different forms of retaliation, ranging from more formal undesirable actions (such as termination, demotion, involuntary transfer, assignment of unmanageable tasks, and professional blacklisting) to more informal ones (e.g., social ostracism, and bullying) [19,60]. Retaliation can also include tactics aimed at stigmatizing the individual to the general public, such as character assassinations or accusations of being a disgruntled employee, spy, or traitor [36], which can further be supported by social media.

So, instead of tributing whistleblowers for their engagement in uncovering illicit activities, they are blamed, defamed, and retaliated. By this, the question arises: Why these "tragic heroes battling the system" [28] are more likely to be seen as traitor in reality?

Immense research has already been done on individual and contextual determinants to report unethical conduct. In terms of individual factors, studies have examined the role of demographic characteristics (e.g., gender, age, and education) [15,44,45], personality traits [20], and value orientations [55]. Studies on situational factors have focused on characteristics of misbehavior [44,49] or organizational context, such as organizational culture and climate [38,45], threat of retaliation [44], as well as support from supervisors and coworkers [42,45]. Both individual and situational factors are determinants in whistleblowing, but studies also show that situational variables tend to be more influential than individual ones [15,44,49]. For instance, Mesmer-Magnus and Viswesvaran [44] have found that individuals blowing the whistle on serious or frequently occurring transgressions are likely to be met with retaliation. The same holds for a lack of supervisor support [44] as well as the use of external channels to report the misconduct [44,45].

Despite this current research on reporting unethical behavior, little work has been done examining the specific mechanisms for when and why retaliation against whistleblowers occurs. For instance, Miceli et al. [45] argue that organizations which are highly dependent on misconduct to obtain valuable resources and rewards, are more likely to resist paying attention to whistleblowers and limiting unethical behavior. As a result, whistleblowers who attempt to disrupt an unethical but organizationally beneficial activity are more likely to face retaliation. Literature also suggests that individuals may experience strong emotions, such as anger, fear, or shame, when accused of wrongdoing [24,71,80] and cognitively disengage from ethically questionable acts in order to justify unethical behavior [4,6,18]. Furthermore, these cognitive and emotive forces may be strongly influenced by the extent to which individuals perceive the whistleblower as a significant threat [72]. In combination, cognitive (i.e., moral disengagement [6]) and affective (i.e., moral emotions [33,71]) mechanisms can be seen as the underlying drivers on retaliation against whistleblowers.

2.2   Social Media and Unethical Behavior

Following Kaplan and Haenlein [37], social media can be defined as "a group of Internet-based applications that build on the ideological and technological foundations of Web 2.0, and that allow the creation and exchange of User Generated Content" (p. 61). According to the Organisation for Economic Cooperation and Development (OECD) [52], user-generated content needs to fulfill three basic characteristics: First, the work is published (e.g., on a publicly accessible online medium). Second, it needs to show a certain amount of creative effort. At this, a person has to create content or adapt existing work to create something new. Finally, the content is created outside of professional practices and routines.

The rapid technological developments in communication technologies in the last decades results in a tremendous growth of social media use. In January 2021, 4.2 billion people were using social networks [70]. The most popular social network in 2021 was Facebook, with more than 2.895 billion active users, followed by YouTube with 2.291 billion active users, WhatsApp 2 billion active users, Instagram 1.393 billion active users, and Facebook Messenger with 1.3 billion active users [68]. The penetration rate in January 2021 was 53.6% on average globally and ranged from 79% in Western and Northern Europe to 8% in Middle Africa [67]. The average daily social media usage of users worldwide amounted to 144 min per day [66].

Given these numbers, social media is a global phenomenon that has become an important element of the daily lives of more than half of the total global population. Individuals, communities as well as organizations can create, share, and consume information regardless of gender, race, social status, educational level, age, or other characteristics [29,53]. The ubiquity of social media in daily life provides whistleblowers a new channel to inform the public about illegal or immoral activities within an organization. On the other hand, the public, too, now has a platform for offering support to whistleblowers [56]. But, besides this "bright side" of social media, e.g., in increasing the freedom of expression as well as transparency in government, and in offering a better access to information [69], there is also a "dark side" that attempts people to engage in cyberbullying, trolling, online witch hunts, fake news, and privacy abuse [3]. Instead of offering support to the whistleblower, people may hide behind online anonymity and engage in unethical behavior towards the whistleblower, for example, by destroying whistleblowers' reputation. Hence, social media combines both good as well as bad elements.

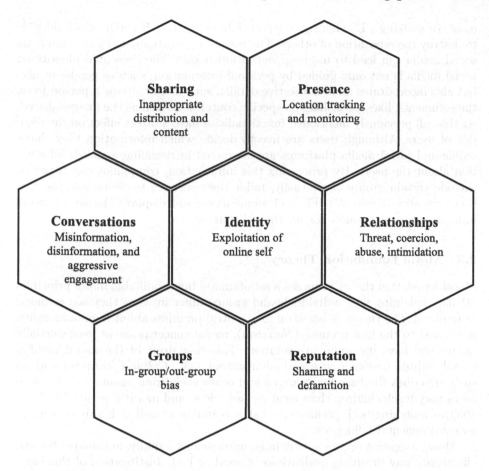

Fig. 1. The dark side of social media. [3]

In order to describe the dark and morally problematic side of social media, Baccarella et al. [3] developed a framework based on the honeycomb framework of Kietzmann et al. [39]. In both frameworks, social media functionalities are unpacked into seven building blocks (1) conversations, (2) sharing, (3) presence, (4) relationships, (5) reputation, (6) groups, and (7) identity to describe different features of the social media user experience. While Kietzmann et al. [39] focused on the bright side of social media, Baccarella et al. [3] show evidence that each of these building blocks also possesses a negative flip side.

Within the conversation building block, users can show behavior that is aggressive. In conversations, false information can intentionally or unintentionally be spread. Inappropriate content can be shared or distributed. The location and availability of users can be tracked without their awareness and used without their consent. Social media help to establish and reveal relationships. At this, they can also be used in a detrimental way (e.g., cyberbullying, online harassment, or stalking). By sharing inappropriate content, it is furthermore possible to destroy the reputation of others. The possibility to create and join a group on social media can lead to in-group/out-group biases. The presented identity on social media is not only guided by personal information, such as gender or age, but also incorporates more subjective details, such as the groups a person joins, the comments, likes, or dislikes on specific contents, as well as the images shared. At this, all previously mentioned functionalities have a direct effect on the identity of users. Although users are free to decide which information they share, online and social media platforms are interested in revealing as much information about the users. By processing this information, companies can target or exclude certain groups more easily, tailor their services to users, and increase their attention levels [25]. Figure 1 summarizes and displays the social media functionalities as a source for unethical behavior.

2.3   Moral Foundation Theory

Moral foundation theory proposes a set of innate but modifiable moral principles [31, 33] explaining why individuals and groups differ in what they see as moral or immoral. The theory is based on four central premises about human morality. According to the first premise (*Nativism*), moral concerns are at least partially innate and have its origin in evolution. This first draft of the moral mind is heavily edited by experience and environment [33] (*Cultural learning*) and, as such, describes the base for "parents and other socializing agents [...] to build on as they teach children their local virtues, vices, and moral practices" ([27], p. 1030). Combining both premises, morality is innate as well as highly dependent on environmental influences.

Moral judgment occurs mostly in an unconscious, rapidly, automatically, and effortlessly way involving evaluations of good or bad. Justification of this judgment is done afterward recruiting deliberate and rational reasons [32]. At this, the third key premise states that moral intuitions precede moral reasoning. Lastly, as there were numerous adaptive social challenges throughout evolutionary history, there are also many different moral foundations that emerged in response to those challenges. In order to identify these roots, Haidt and Joseph were searching for universal, cultural variable building blocks of morality as well as blocks that could be found in other primates [32]. They used a scoring system and identified five related but distinct moral foundations: "harm/care", "fairness/reciprocity", "in-group/loyalty", "authority/respect", and "purity/sanctity" [30, 31, 33]. Table 1 summarizes these moral foundations.

The first foundation, "harm/care", roots from our ancestors' experience of attending their own offspring. Those who possess the ability to sense others' pain and the willingness to help others in trouble would be more sensitive to their children's need improving the likelihood of survival. In addition to compassion for a suffering person, anger is often expressed towards the person causing the pain and suffering. Reciprocity as second moral foundation was essential to establish cooperation and collaboration with others; especially with non-kin, which

Table 1. The five foundations of intuitive ethics [33].

	Harm/Care	Fairness/ Reciprocity	In-group/ Loyalty	Authority/ Respect	Purity/ Sanctity
Adaptive challenge	Protect and care for young, vulnerable, or injured kin	Reap benefits of dyadic cooperation with non-kin	Reap benefits of group cooperation	Negotiate hierarchy, defer selectivity	Avoid microbes and parasites
Proper domain (Adaptive triggers)	Suffering, distress, or threat to one's kin	Cheating, cooperation, deception	Threat or challenge to the group	Signs of dominance and submission	Waste products, diseased people
Actual domain (Examples of modern triggers)	Baby seals, cartoon characters	Marital fidelity, broken vending machines	Sports teams one roots for	Bosses, respected professional	Taboo ideas (Communism, racism)
Characteristic emotions	Compassion	Anger, gratitude, guilt	Group pride, belongingness, rage at traitors	Respect, fear	Disgust
Relevant virtues (and vices)	Caring, kindness, (cruelty)	Fairness, justice, honesty, trust-worthiness, (dishonesty)	Loyalty, patriotism, self-sacrifice, (treason, cowardice)	Obedience, deference, (disobedience, uppitiness)	Temperance, chastity, piety, cleanliness, (lust, intemperance)

are indispensable for our ancestors to fight against their hostile living environment. Since cheating can cause distrust and harm cooperation, it is necessary to encourage fairness and discourage cheating. At this, the fairness foundation is activated by perceiving violations of reciprocity, equality, justice, and individual rights. "In-group/loyalty" as a third moral foundation derives from our sense of connection and obligation to groups with which we identify (e.g., family, or sports teams) resulting in a sense of belonging to those who contribute to or sacrifice themselves for the group. At the same time, the importance of in-group/out-group effects are highlighted. "Authority/respect" as the fourth moral foundation is based on our tendency to create hierarchically structured societies of dominance and subordination. In order to accept hierarchies and derive advantages from them, it was necessary to distinguish those of higher rank from those of lower rank. On the one hand, obedient and considerate behavior toward those higher in rank is particularly important. On the other hand, subordinates expect superiors to protect them from external threats.

Finally, the "purity/sanctity" foundation roots from our ancestors' instinct to avoid diseases. Behaviors signifying degradation, such as incest, sexual indulgence, and being unhygienic are likely to cause the spread of diseases. At this, avoiding dangerous microbes and parasites improved the likelihood of survival and acceptance within social groups. This foundation can further be generalized to the aversion of violating social norms and virtues.

Moral foundation theory further states that certain moral transgressions trigger certain, "characteristic" moral emotions [32,33]. Care violations, for instance, should elicit compassion, while fairness or purity violations should end up with

anger or disgust, respectively. However, the moral foundations act not independent from each other, but are often triggered simultaneously, causing specific
responses mediated by certain emotions.

2.4 Moral Disengagement

According to Bandura's social cognitive theory [5], people internalize behavioral
standards via socialization guiding behavior. Behaving in accordance with these
standards is supported by positive self-sanctions (e.g., self-respect), while violations, such as unethical behavior, activate negative self-regulatory mechanisms
(e.g., guilt) subsequently prevent the individual from engaging in that behavior.

However, this self-regulatory is not always successful, as it does not create a fixed control mechanism. Consequently, the same behavior can sometimes be followed by self-reward or self-punishment [5]. In an extension of his
social cognitive theory, Bandura's moral disengagement theory suggests that
self-regulatory processes can be deactivated using moral disengagement techniques [6, 46]. These techniques help to disengage self-regulatory processes, thus
preventing self-censure or guilt and rendering the unethical behavior unproblem-
atic for one's conscience [8].

Bandura [5] distinguishes eight psychological processes which can disengage
morality from conduct and are located at different points in the self-regulatory
process.

In rationalizing the immoral conduct by using different mechanisms, the reprehensible conduct is not evaluated as such. These mechanisms include moral justification, palliative comparison, and euphemistic labeling [6]. Moral justification
legitimizes the immoral conduct. The harmful action is justified by attributing
to it a moral or social purpose that is considered to be more significant than
the harm caused. By using euphemistic labeling, immoral conduct is verbally
sanitized by the perpetrator by using a language, making it seem respectable.
Finally, by comparing the reprehensible conduct against worse conduct, it makes
the immoral conduct more acceptable.

A necessity for moral control is the acknowledgment of one's own wrongdoing.
When perpetrators stress that they are not responsible for immoral conduct as
they acted, for instance, due to social pressure or following the instructions of a
superior, displacement of responsibility occurs. On the other hand, diffusion of
responsibility allows perpetrators to reject responsibility for the immoral conduct
of a group, arguing that the perpetrator does not feel personally liable for that
behavior.

For self-censure and self-sanctions to take place, not only the action itself
and responsibility for the action have to be accepted, but also the detrimental
effects of immoral conduct have to be acknowledged. Consequently, individuals
can morally disengage by ignoring or verbally minimizing the negativity of the
outcome.

The degree to which self-regulation processes will be triggered also depends
on how perpetrators view the victims of mistreat. If the victim is not considered
as a human being, people act less empathically as they are not seen as equal. At

this, the victim of immoral conduct is worthy of being harmed. Finally, attribution of blame refers to the situation in which perpetrators seek to blame the victim for the immoral conduct in order to relieve themselves of responsibility. Figure 2 summarizes the moral disengagement mechanisms.

3 Propositions and Conceptual Model Development

Moral foundations are innate psychological mechanisms for evaluating particular problems or situations. The first draft of the innate moral mind is highly edited by experience and environment. Haidt [30] use the analogy of "taste buds" on the tongue to illustrate the functioning of moral foundations. While the taste buds of the tongue collect perceptual stimuli with respect to food and drinks, moral taste buds detect the "flavor" of moral stimuli (e.g., kindness or cruelty, fairness or dishonesty, loyalty or cowardice, deference or disobedience, chastity or lust).

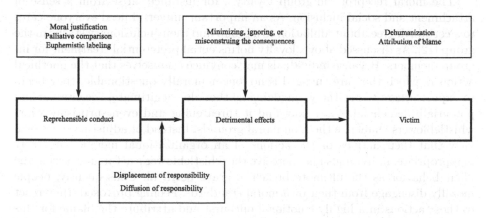

Fig. 2. Moral disengagement mechanisms [5].

As soon as our moral receptors sense a stimulus, a rapidly, near-unconsciously, and automatically experience is made guiding subsequent decisions about how to react. If this control mechanism leads to a positive experience (like), the triggering person or object is approached. On the other hand, a negative experience (dislike) will result in avoidance [33]. In addition, a positive or negative reaction can be triggered against the source of the stimulus. If the reaction is positive, it is highly unlikely that people deactivate their self-regulation system to disengage from their own moral standards and engage in retaliation against whistleblowers.

Regarding the whistleblower, in deciding whether or not to expose another person's misconduct, individuals face what is known as the "whistleblower's dilemma" [78]. Here, two important moral foundations, namely fairness and loyalty, may be in conflict. While fairness norms require treating everyone equally, loyalty norms lead to preferential treatment of members of one's own group.

Depending on which foundation is preferred, individuals are more likely to blow the whistle (fairness) or not (loyalty) [78]. But not only the whistleblower faces this tradeoff between fairness and loyalty but also the potential retaliator. As such, it can be assumed that the preference for fairness over loyalty is positively related to giving support to the whistleblower. Hence, it can be assumed that moral receptors are negatively related to moral disengagement.

Proposition 1. *Moral receptors are negatively related to moral disengagement.*

If the stimulus is, however, in contradiction to the individuals' ethical standards and beliefs, a negative reaction can occur against the source of the stimulus. In case of retaliation against whistleblowers, whistleblowing often triggers emotions, such as anger, fear, and shame, that retaliators must resolve. Individuals often work hard to minimize their experience of unpleasant and negative emotions, especially when they fail to do the right thing. In response to this emotional distress, retaliation against the whistleblower may be seen as justified behavior [51].

The moral receptor "in-group/loyalty", for instance, arise from a sense of attachment and social inclusion. As an important universal need everyone cares to a certain degree about affiliation to a group and their inclusionary status in the group [12]. As discussed above, loyalty norms entail preferential treatment for in-group members. Because individuals may convince themselves that the unethical action in which they are engaged is no longer morally questionable, they begin to expect others to see the wrongdoing as they do—legitimate, acceptable, and potentially beneficial. They may find it threatening and even surprising when whistleblowers challenge them on moral grounds. Instead of adjusting one's own view that their actions or the actions of an organizational member are truly inappropriate, individuals may perceive the whistleblower's act in uncovering the illicit behavior as the ultimate breach, a grave betrayal. Consequently, people morally disengage from their own moral standards. Feeling provoked, they react to these actions in a highly emotional outburst and attribute the blame for this unethical conduct to the victim. Studies also show that people respond with even stronger negative emotionally charged reactions against in-group members violating shared norms than out-group members who exhibit the same behavior (known as black sheep effect, [43,75]). Additionally, Parmerlee et al. [58] found evidence that older whistleblowers are more likely to be retaliated against than younger ones. One possible explanation is that older individuals are expected to be more loyal to the organization. Therefore, if they choose to speak up, other members may feel more betrayed. Studies of Near et al. [49] also show that retaliation against whistleblowers is more likely and more severe when external rather than internal channels are used to report the wrongdoing. The results can be explained by the fact that the use of external channels not only triggers the moral receptor "in-group/loyalty" but is also perceived as a violation of the authority structures present within the organization.

Considering these arguments, each moral foundation can be a source of unethical behavior. As such, it can be assumed that moral receptors are also positively related to moral disengagement.

Proposition 2. *Moral receptors are positively related to moral disengagement.*

According to Bandura, individuals using moral disengagement to set aside–guilt, shame, self-condemnation–that come from violating their personal moral standards. Once these sanctions are deactivated, people are free to participate in immoral activities, such as theft, financial fraud, the manufacturing and sale of harmful products, and corruption. Previous research already examined the role of moral disengagement in order to explain why "good" people go "bad". By using moral disengagement mechanisms, these people rationalize catastrophic corporate disasters [9], unethical behavior in idea competitions [63], killing characters in violent video games [34], unethical work behaviors [46], or violating legal and moral rules in the production process [14]. Using these mechanisms, people not only participate in unethical behavior, but also tend to challenge anyone who may disagree with their unethical, but no longer morally questionable actions, such as whistleblowers, increasing the likelihood of retaliation.

The rapid technological developments in communication technologies within the last decades and the growing popularity of social media gave also rise for new opportunities for unethical behavior, such as cyberbullying [57], trolling, online witch hunts, fake news, and privacy abuse [3]. Past research on social media has been able to show that social media possesses specific characteristics, i.e., anonymity, asynchrony, and dissociative imagination, that make it easier for people to act in an unethical way.

Anonymity on social media makes it much easier for people to minimize, ignore, or misconstrue the detrimental effects of their behavior. Santana et al. [61] stated that asynchrony can lead to reduced availability of social cues. As a result, people feel less inhibited with others and are more inclined to engage in confrontation. By simply turning off the computer or closing the app, people split or dissociate online fiction from offline fact allowing them to easily disengage from their moral standards [61]. This dissociative imagination results in a belief that the online world follows a different set of rules than the offline world [61].

Recent research also draw attention to the "shallowing hypothesis" [2,3], where the development of communication technologies has led to a dramatic decline in ordinary daily reflective thought. According to this hypothesis, certain social media activities such as sharing and conversing promote rapid, shallow thought that can result in cognitive and moral triviality, again, making it easier for people to act unethically.

Based on these findings, it can be assumed that moral disengagement is positively related to the tendency to show unethical behavior on social media, in particular, against whistleblowers.

Proposition 3. *Moral disengagement is positively related to the tendency to show unethical behavior against whistleblowers on social media.*

Together, Propositions 1, 2 and 3 suggest an influence of moral receptors on the relationship between moral disengagement and the tendency to show unethical behavior against whistleblower on social media.

Proposition 4. *Moral disengagement mediates the relationship between moral receptors and the tendency to unethical behavior against whistleblowers on social media.*

Having outlined the relationship between moral receptors, moral disengagement, and unethical behavior, there might also be some contextual boundary conditions that may exacerbate the negative behavior towards whistleblowers.

The ethical climate of an organization includes organizational procedures, policies, and practices that have moral consequences. When people are faced with a moral dilemma, expectations of (un)ethical behavior but also consequences of such behavior are provided helping them to distinguish clearly between right and wrong at work and to develop appropriate approaches to moral evaluation and problem solving [10, 21, 41]. Thus, ethical climate influences the process of self-regulation, which reduces or reinforces negative behaviors and enables individuals to maintain a positive image of themselves even when they engage in deviant actions [11, 17, 74]. To draw attention towards this interaction between organizational climate and (un)ethical behavior, Victor and Cullen [77] proposed a theoretical typology of ethical climates distinguishing three types of ethical climates: *egoistic, principled,* and *benevolent.*

Organizations with a largely *benevolent* ethical climate endorse the principle of reciprocity and the welfare of others. When judging on ethical decision, the impact on relevant groups is always considered [10], where relevant groups might include employees, customers, other stakeholders, or society at-large. Within a benevolent climate, there is a (shared) perception that caring for others is valued by the organization being an important part of the social structure of the organization. Studies have shown that a benevolent climate leads people to avoid unethical activities. Because this ethical climate emphasizes the importance of the welfare of others, people are more likely to formulate their intentions in accordance with their own moral judgments [10]. On the other hand, in a *principled* organizational climate, decisions are based on formal guidelines such as laws, policies, rules, and codes of appropriate conduct [21]. Adhering these guidelines is of particular importance and decisions are considered as ethical if they comply with the provided rules [10]. Accordingly, through a focus on welfare of others (benevolent) or emphasis on rule-abiding behavior (principled), both the benevolent and principled climates are likely to encourage fewer unethical choices.

Finally, in organizations with an *egoistic* climate, individuals perceive the organizational environment in such a way that self-interest is prioritized over all other benefits [21]. The starting point of individual decision making is the intention to realize personal interests. At this, individuals tend to evaluate the costs and benefits of an action before considering ethical concerns. The associated impact on others or on the organization is not considered. Although such an organization does not intend to harm the interests of others, some of its members pursue self-interested actions at the expense of others. Such a climate therefore encourages self-serving and often immoral decisions without considering the social consequences of one's actions [10, 40, 41].

Combining above considerations, when individuals are confronted with whistleblowers, ethical climates may influence perception about the appropriateness of retaliation. It is proposed that ethics-oriented climates can reinforce ethical considerations based on rules or laws, for instance, laws for whistleblower protection (principled) and the welfare of others (benevolent). Together, these factors act as environmental cues enhancing the likelihood of giving support to whistleblowers and discouraging individuals in participating in unethical behavior, for instance, retaliation against whistleblowers [16,40]. On the other hand, in organizations where unethical behavior is widespread and considered as the "best" way to behave, individuals who are highly sensitive to the thread posed by whistleblowers are more likely to disapprove with the whistleblower's actions. As a result, whistleblowers revealing illegal behavior are likely to be punished and face negative consequences. Furthermore, Barnett and Vaicys [10] found evidence that ethical climate moderates the mechanism by which moral judgment affects behavior intention. As such, the ethical climate may play a moderating role in controlling or altering moral biases that may lead to unethical behavior. Therefore, it can be assumed that the organizational ethical climate moderates the relationship between moral disengagement and retaliation against whistleblowers.

Proposition 5. *A highly unethical organizational climate will moderate the relationship between moral disengagement and the tendency for retaliation against whistleblowers on social media.*

Studies have already examined the positive effects of group cohesion on group performance (e.g., [13]). Despite this positive effect, there is also a downside, best illustrated by the steroid scandal that shook Major League Baseball in the 1990s and 2000s. Although many players, managers, and doctors knew about and witnessed the illegal use of steroids, the tight bond among former and current baseball players helped keep the scandal from becoming public for years. Nothing was supposed to disrupt the fraternity's strong shared identity, and anyone who threatened to do so was demoted in the status hierarchy, ignored, or ostracized [73]. In such cohesive groups, individuals who fit the prototype of one's group are more likely to be viewed favorably by other group members, while those who are viewed as "negative outliers" may be ostracized and rejected by the group [35].

Whistleblowers who threaten the status quo trigger the "in-group/loyalty" moral receptor, such that would-be retaliators can easily disengage from their moral standards. In groups characterized by high group cohesion, even higher standards are expected from group members, such that deviants are likely to be treated more harshly and are more likely to be subject to retaliation. It can therefore be assumed that group cohesion will moderate the relationship between the moral receptor "in-group/loyalty" and moral disengagement.

Proposition 6. *High levels of group cohesion will moderate the relationship between the moral receptor "in-group/loyalty" and moral disengagement.*

Figure 3 summarizes the propositions and the conceptual model.

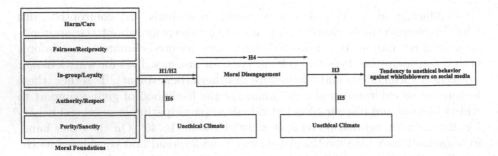

Fig. 3. Propositions and conceptual model

4   Conclusion

Unethical behavior in organizations is an all-too-common phenomenon. Although whistleblowers have brought numerous scandals to light in recent years, individuals who decide to blow the whistle are often seen as rats instead of being praised for uncovering immoral conduct. At this, they are frequently subject to retaliatory actions [44,45]. The main goal of this study was to provide a conceptual model that give insights into the interplay between moral receptors and moral disengagement in the context of unethical behavior against whistleblowers on social media. Using the model of Scheiner, a set of propositions have been provided linking the moral foundations "harm/care", "fairness/reciprocity", "ingroup/loyalty", "authority/respect", and "purity/sanctity" with moral disengagement as mechanism deactivating the self-regulatory system in order to explain immoral conduct on social media. In addition to Scheiner, two different contextual factors influencing the unethical behavior towards whistleblowers were integrated into the provided model, namely the influence of organizational ethical climate as well as group cohesion.

The conceptual model is rooted in existing research on the moral foundation theory, moral disengagement as well as unethical behavior on social media. Avenues for research in near future include a further empirical testing and theoretical discussion of the provided model to better understand the underlying motives for retaliation against whistleblowers. A better understanding of the precursors of retaliation is of particular importance as it allows for developing interventions to stop or at least reduce retaliation.

References

1. Andon, P., Free, C., Jidin, R., Monroe, G.S., Turner, M.J.: The impact of financial incentives and perceptions of seriousness on whistleblowing intention. J. Bus. Ethics **151**(1), 165–178 (2016). https://doi.org/10.1007/s10551-016-3215-6
2. Annisette, L.E., Lafreniere, K.D.: Social media, texting, and personality: a test of the shallowing hypothesis. Personal. Individ. Differ. **115**, 154–158 (2017). https://doi.org/10.1016/j.paid.2016.02.043

3. Baccarella, C.V., Wagner, T.F., Kietzmann, J.H., McCarthy, I.P.: Social media? It's serious! Understanding the dark side of social media. Eur. Manag. J. **36**(4), 431–438 (2018). https://doi.org/10.1016/j.emj.2018.07.002

4. Banaji, M.R., Bazerman, M.H., Chugh, D.: How (un)ethical are you? Harv. Bus. Rev. **81**(12), 56–64 (2003)

5. Bandura, A.: Social Foundations of Thought & Action – A Social Cognitive Theory. Prentice Hall, New Jersey (1986)

6. Bandura, A.: Moral disengagement in the perpetration of inhumanities. Personal. Soc. Psychol. Rev. **3**(3), 193–209 (1999). https://doi.org/10.1207/s15327957pspr0303_3

7. Bandura, A.: Impeding ecological sustainability through selective moral disengagement. Int. J. Innov. Sustain. Dev. **2**(1), 8 (2007). https://doi.org/10.1504/IJISD.2007.016056

8. Bandura, A., Barbaranelli, C., Caprara, G.V., Pastorelli, C.: Mechanisms of moral disengagement in the exercise of moral agency. J. Personal. Soc. Psychol. **71**(2), 364–374 (1996). https://doi.org/10.1037/0022-3514.71.2.364

9. Bandura, A., Caprara, G.V., Zsolnai, L.: Corporate transgressions through moral disengagement. J. Hum. Values **6**(1), 57–64 (2000). https://doi.org/10.1177/097168580000600106

10. Barnett, T., Vaicys, C.: The moderating effect of individuals' perceptions of ethical work climate on ethical judgments and behavioral intentions. J. Bus. Ethics **27**(4), 351–362 (2000). https://doi.org/10.1023/A:1006382407821

11. Baron, R.A., Zhao, H., Miao, Q.: Personal motives, moral disengagement, and unethical decisions by entrepreneurs: cognitive mechanisms on the "slippery slope". J. Bus. Ethics **128**(1), 107–118 (2014). https://doi.org/10.1007/s10551-014-2078-y

12. Baumeister, R.F., Leary, M.R.: The need to belong: desire for interpersonal attachments as a fundamental human motivation. Psychol. Bull. **117**(3), 497–529 (1995). https://doi.org/10.1037/0033-2909.117.3.497

13. Beal, D.J., Cohen, R.R., Burke, M.J., Mclendon, C.L.: Cohesion and performance in groups?: a meta-analytic clarification of construct relations. J. Appl. Psychol. **88**(6), 989–1004 (2003). https://doi.org/10.1037/0021-9010.88.6.989

14. Brief, A.P., Buttram, R.T., Dukerich, J.M.: Collective corruption in the corporate world: toward a process model. In: Turner, M. (ed.) Groups at Work, pp. 471–499. Erlbaum, Mahwah (2001)

15. Cassematis, P.G., Wortley, R.: Prediction of whistleblowing or non-reporting observation: the role of personal and situational factors. J. Bus. Ethics **117**(3), 615–634 (2012). https://doi.org/10.1007/s10551-012-1548-3

16. Chordiya, R., Sabharwal, M., Relly, J.E., Berman, E.M.: Organizational protection for whistleblowers: a cross-national study. Public Manag. Rev. **22**(4), 527–552 (2020). https://doi.org/10.1080/14719037.2019.1599058

17. Christian, J.S., Ellis, A.P.J.: The crucial role of turnover intentions in transforming moral disengagement into deviant behavior at work. J. Bus. Ethics **119**(2), 193–208 (2013). https://doi.org/10.1007/s10551-013-1631-4

18. Chugh, D., Bazerman, M.H., Banaji, M.R.: Bounded ethicality as a psychological barrier to recognizing conflicts of interest. In: Conflicts of Interest, pp. 74–95. Cambridge University Press, Cambridge (2005). https://doi.org/10.1017/CBO9780511610332.006

19. Cortina, L.M., Magley, V.J.: Raising voice, risking retaliation: events following interpersonal mistreatment in the workplace. J. Occup. Health Psychol. **8**(4), 247–265 (2003). https://doi.org/10.1037/1076-8998.8.4.247

20. Culiberg, B., Mihelič, K.K.: The evolution of whistleblowing studies: a critical review and research agenda. J. Bus. Ethics **146**(4), 787–803 (2016). https://doi.org/10.1007/s10551-016-3237-0

21. Cullen, J.B., Victor, B., Bronson, J.W.: The ethical climate questionnaire: an assessment of its development and validity. Psychol. Rep. **73**(2), 667–674 (1993). https://doi.org/10.2466/pr0.1993.73.2.667, http://journals.sagepub.com/doi/10.2466/pr0.1993.73.2.667

22. Dyck, A., Morse, A., Zingales, L.: Who blows the whistle on corporate fraud? J. Financ. **65**(6), 2213–2253 (2010). https://doi.org/10.1111/j.1540-6261.2010.01614.x

23. European Commission: Report from the Commission to the Council and the European Parliament - EU Anti-Corruption Report. Tech. rep. (2014)

24. Fitness, J.: Anger in the workplace: an emotion script approach to anger episodes between workers and their superiors, co-workers and subordinates. J. Organ. Behav. **21**(2), 147–162 (2000). https://doi.org/10.1002/(SICI)1099-1379(200003)21:2⟨147::AID-JOB35⟩3.0.CO;2-T

25. Fosch-Villaronga, E., Poulsen, A., Søraa, R.A., Custers, B.H.: A little bird told me your gender: gender inferences in social media. Inf. Process. Manag. **58**(3), 102541 (2021). https://doi.org/10.1016/j.ipm.2021.102541

26. Francis, R.: Freedom to Speak Up - A Review of Whistleblowing in the NHS. Tech. rep, London (2015)

27. Graham, J., Haidt, J., Nosek, B.A.: Liberals and conservatives rely on different sets of moral foundations. J. Personal. Soc. Psychol. **96**(5), 1029–1046 (2009). https://doi.org/10.1037/a0015141

28. Grant, C.: Whistleblowers: saints of secular culture. J. Bus. Ethics **39**, 391–399 (2002). https://doi.org/10.1023/A:1019771212846

29. Groshek, J., Cutino, C.: Meaner on mobile: incivility and impoliteness in communicating contentious politics on sociotechnical networks. Soc. Media + Soc. **2**(4), 205630511667713 (2016). https://doi.org/10.1177/2056305116677137

30. Haidt, J.: The Righteous Mind: Why Good People are Divided by Politics and Religion. Penguin, London (2013)

31. Haidt, J., Graham, J.: When morality opposes justice: conservatives have moral intuitions that liberals may not recognize. Soc. Justice Res. **20**(1), 98–116 (2007). https://doi.org/10.1007/s11211-007-0034-z

32. Haidt, J., Joseph, C.: Intuitive ethics: how innately prepared intuitions generate culturally variable virtues. Daedalus **133**(4), 55–66 (2004). https://doi.org/10.1162/0011526042365555

33. Haidt, J., Joseph, C.: The moral mind: how five sets of innate intuitions guide the development of many culture-specific virtues, and perhaps even modules. In: The Innate Mind, Volume 3, vol. 15, pp. 367–392. Oxford University Press (2008). https://doi.org/10.1093/acprof:oso/9780195332834.003.0019

34. Hartmann, T., Vorderer, P.: It's okay to shoot a character: moral disengagement in violent video games. J. Commun. **60**(1), 94–119 (2010). https://doi.org/10.1111/j.1460-2466.2009.01459.x

35. Hogg, M.A., Terry, D.J.: Social identity and self-categorization processes in organizational contexts. Acad. Manag. Rev. **25**(1), 121–140 (2000). https://doi.org/10.2307/259266

36. Jubb, P.B.: Whistleblowing: a restrictive definition and interpretation. J. Bus. Ethics **21**(1), 77–94 (1999). https://doi.org/10.1023/A:1005922701763

37. Kaplan, A.M., Haenlein, M.: Users of the world, unite! The challenges and opportunities of social media. Bus. Horiz. **53**(1), 59–68 (2010). https://doi.org/10.1016/j.bushor.2009.09.003

38. Kaptein, M.: From inaction to external whistleblowing: the influence of the ethical culture of organizations on employee responses to observed wrongdoing. J. Bus. Ethics **98**(3), 513–530 (2011). https://doi.org/10.1007/s10551-010-0591-1

39. Kietzmann, J.H., Hermkens, K., McCarthy, I.P., Silvestre, B.S.: Social media? Get serious! Understanding the functional building blocks of social media. Bus. Horiz. **54**(3), 241–251 (2011). https://doi.org/10.1016/j.bushor.2011.01.005

40. Kish-Gephart, J.J., Harrison, D.A., Treviño, L.K.: Bad apples, bad cases, and bad barrels: meta-analytic evidence about sources of unethical decisions at work. J. Appl. Psychol. **95**(1), 1–31 (2010). https://doi.org/10.1037/a0017103

41. Martin, K.D., Cullen, J.B.: Continuities and extensions of ethical climate theory: a meta-analytic review. J. Bus. Ethics **69**(2), 175–194 (2006). https://doi.org/10.1007/s10551-006-9084-7

42. Mayer, D.M., Nurmohamed, S., Treviño, L.K., Shapiro, D.L., Schminke, M.: Encouraging employees to report unethical conduct internally: it takes a village. Organ. Behav. Hum. Decis. Process. **121**(1), 89–103 (2013). https://doi.org/10.1016/j.obhdp.2013.01.002

43. Mendoza, S.A., Lane, S.P., Amodio, D.M.: For members only: ingroup punishment of fairness norm violations in the ultimatum game. Soc. Psychol. Personal. Sci. **5**(6), 662–670 (2014). https://doi.org/10.1177/1948550614527115

44. Mesmer-Magnus, J.R., Viswesvaran, C.: Whistleblowing in organizations: an examination of correlates of whistleblowing intentions, actions, and retaliation. J. Bus. Ethics **62**(3), 277–297 (2005). https://doi.org/10.1007/s10551-005-0849-1

45. Miceli, M.P., Near, J.P., Dworkin, T.M.: Whistle-Blowing in Organizations. Routledge, New York (2008)

46. Moore, C., Detert, J.R., Treviño, L.K., Baker, V.L., Mayer, D.M.: Why employees do bad things: moral disengagement and unethical organizational behavior. Pers. Psychol. **65**(1), 1–48 (2012). https://doi.org/10.1111/j.1744-6570.2011.01237.x

47. Morrison, E.W., Milliken, F.J.: Speaking up, remaining silent: the dynamics of voice and silence in organizations. J. Manag. Stud. **40**(6), 1353–1358 (2003). https://doi.org/10.1111/1467-6486.00383

48. Near, J.P., Miceli, M.P.: Organizational dissidence: the case of whistle-blowing. J. Bus. Ethics **4**, 1–16 (1985). https://doi.org/10.1007/978-94-007-4126-3_8

49. Near, J.P., Miceli, M.P.: Whistle-blowing: myth and reality. J. Manag. **22**(3), 507–526 (1996). https://doi.org/10.1177/014920639602200306

50. Near, J.P., Miceli, M.P.: After the wrongdoing: what managers should know about whistleblowing. Bus. Horiz. **59**(1), 105–114 (2016). https://doi.org/10.1016/j.bushor.2015.09.007

51. Norgaard, K.M.: "People want to protect themselves a little bit": emotions, denial, and social movement nonparticipation. Sociol. Inq. **76**(3), 372–396 (2006). https://doi.org/10.1111/j.1475-682X.2006.00160.x

52. OECD: Participative Web and User-Created Content. Organisation for Economic Co-operation and Development, Paris (2007). https://doi.org/10.1787/9789264037472-en

53. Papacharissi, Z.: Democracy online: civility, politeness, and the democratic potential of online political discussion groups. New Media Soc. **6**(2), 259–283 (2004). https://doi.org/10.1177/1461444804041444

54. Park, H., Bjørkelo, B., Blenkinsopp, J.: External whistleblowers' experiences of workplace bullying by superiors and colleagues. J. Bus. Ethics **161**(3), 591–601 (2018). https://doi.org/10.1007/s10551-018-3936-9

55. Park, H., Blenkinsopp, J., Park, M.: The influence of an observer's value orientation and personality type on attitudes toward whistleblowing. J. Bus. Ethics **120**(1), 121–129 (2013). https://doi.org/10.1007/s10551-013-1908-7

56. Park, Y.J., Jang, S.M.: Public attention, social media, and the Edward Snowden saga. First Monday **22**(8) (2017). https://doi.org/10.5210/fm.v22i8.7818

57. Parlangeli, O., Marchigiani, E., Bracci, M., Duguid, A.M., Palmitesta, P., Marti, P.: Offensive acts and helping behavior on the Internet: an analysis of the relationships between moral disengagement, empathy and use of social media in a sample of Italian students. Work **63**(3), 469–477 (2019). https://doi.org/10.3233/WOR-192935

58. Parmerlee, M.A., Near, J.P., Jensen, T.C.: Correlates of whistle-blowers' perceptions of organizational retaliation. Adm. Sci. Q. **27**(1), 17 (1982). https://doi.org/10.2307/2392544

59. Richardson, B.K., McGlynn, J.: Rabid fans, death threats, and dysfunctional stakeholders: the influence of organizational and industry contexts on whistleblowing cases. Manag. Commun. Q. **25**(1), 121–150 (2011). https://doi.org/10.1177/0893318910380344

60. Rothschild, J., Miethe, T.D.: Whistle-blower disclosures and management retaliation. Work Occup. **26**(1), 107–128 (1999). https://doi.org/10.1177/0730888499026001006

61. Santana, A.D.: Virtuous or vitriolic: the effect of anonymity on civility in online newspaper reader comment boards. Journal. Pract. **8**(1), 18–33 (2014). https://doi.org/10.1080/17512786.2013.813194

62. Scheiner, C.W.: The role of moral receptors and moral disengagement in the conduct of unethical behaviors on social media. In: Meiselwitz, G. (ed.) HCII 2020, Part I. LNCS, vol. 12194, pp. 335–348. Springer, Cham (2020). https://doi.org/10.1007/978-3-030-49570-1_23

63. Scheiner, C.W., Baccarella, C.V., Bessant, J., Voigt, K.I.: Participation motives, moral disengagement, and unethical behaviour in idea competitions. Int. J. Innov. Manag. **22**(06), 1850043 (2018). https://doi.org/10.1142/S1363919618500433

64. Scheiner, C.W., Krämer, K., Baccarella, C.V.: Cruel intentions? – The role of moral awareness, moral disengagement, and regulatory focus in the unethical use of social media by entrepreneurs. In: Meiselwitz, G. (ed.) SCSM 2016. LNCS, vol. 9742, pp. 437–448. Springer, Cham (2016). https://doi.org/10.1007/978-3-319-39910-2_41

65. Soeken, K., Soeken, D.: A survey of whistleblowers: their stressors and coping strategies. Proc. Hear. HR **25**(1), 156–166 (1987)

66. Statista: Daily Time Spent on Social Networking by Internet Users Worldwide from 2012 to 2020. https://www.statista.com/statistics/433871/daily-social-media-usage-worldwide/

67. Statista: Global Social Network Penetration Rate as of January 2021, by Region. https://www.statista.com/statistics/269615/social-network-penetration-by-region/

68. Statista: Most Popular Social Networks Worldwide as of October 2021, Ranked by Number of Active Users. https://www.statista.com/statistics/272014/global-social-networks-ranked-by-number-of-users/

69. Statista: Share of Internet Users Worldwide who Believe that Social Media Platforms have had an Impact on Selected Aspects of Daily Life as of February 2019. https://www.statista.com/statistics/1015131/impact-of-social-media-on-daily-life-worldwide/

70. Statista: Social Media - Statistics & Facts. https://www.statista.com/topics/1164/social-networks

71. Tangney, J.P., Stuewig, J., Mashek, D.J.: Moral emotions and moral behavior. Ann. Rev. Psychol. **58**(1), 345–372 (2007). https://doi.org/10.1146/annurev.psych.56.091103.070145

72. Tomaka, J., Blascovich, J.: Effects of justice beliefs on cognitive appraisal of and subjective physiological, and behavioral responses to potential stress. J. Personal. Soc. Psychol. **67**(4), 732–740 (1994). https://doi.org/10.1037/0022-3514.67.4.732

73. Torre, J., Verducci, T.: The Yankee Years. Doubleday, New York (2009)

74. Treviño, L.K., Weaver, G.R., Reynolds, S.J.: Behavioral ethics in organizations: a review. J. Manag. **32**(6), 951–990 (2006). https://doi.org/10.1177/0149206306294258

75. Ufkes, E.G., Otten, S., van der Zee, K.I., Giebels, E.: Neighborhood conflicts: the role of social categorization. Int. J. Confl. Manag. **23**(3), 290–306 (2012). https://doi.org/10.1108/10444061211248985

76. U.S. Department of Justice: Justice Department Recovers over $2.2 Billion from False Claims Act Cases in Fiscal Year 2020 (2020)

77. Victor, B., Cullen, J.B.: The organizational bases of ethical work climates. Adm. Sci. Q. **33**(1), 101 (1988). https://doi.org/10.2307/2392857

78. Waytz, A., Dungan, J., Young, L.: The whistleblower's dilemma and the fairness-loyalty tradeoff. J. Exp. Soc. Psychol. **49**(6), 1027–1033 (2013). https://doi.org/10.1016/j.jesp.2013.07.002

79. Whitaker, L., Backhouse, S.H., Long, J.: Reporting doping in sport: national level athletes' perceptions of their role in doping prevention. Scand. J. Med. Sci. Sports **24**(6), e515-521 (2014). https://doi.org/10.1111/sms.12222

80. Zhong, C.B., Liljenquist, K.: Washing away your sins: threatened morality and physical cleansing. Science **313**(5792), 1451–1452 (2006). https://doi.org/10.1126/science.1130726

Dynamics of Distrust, Aggression, and Conspiracy Thinking in the Anti-vaccination Discourse on Russian Telegram

Svetlana S. Bodrunova(✉) 🆔 and Dmitry Nepiyuschikh

St. Petersburg State University, 7-9 Universitetskaya nab., St. Petersburg 199034, Russia
s.bodrunova@spbu.ru

Abstract. The COVID-19 pandemic has brought along an unprecedented amount of social fear and uncertainty. The infodemic has spurred the spread of distrust to elites and their rationality, as well as an outburst of conspiracy theories, around the world. Most studies that investigate the relations between trust to social institutions and public perception of the COVID-related threats employ self-reporting, which may distort the results. This is why it is crucial to also explore discussions on social media. Despite the already existing abundance of datasets collected for misinformation, anti-vaccination, and COVID-19-related conspiracy theories, several research gaps may be identified. First, anti-vaxxer communities are rarely studied beyond the English-language context. Second, directions and main attractors of popular distrust are rarely mapped. Third, the dynamics of distrust and conspiracist thinking towards various actors of the pandemic is not explored. To address these gaps, we assess the 282,000+ comments in the largest antivaxxer community on Russian Telegram, namely *anti_covid21* (January to July 2021), including 12,200+ comments being coded manually. We find that 'the discourse of distrust' is highly politicized, where distrust to national and global actors may be a mediator to vaccine distrust. We show that conspiracies may be a mechanism of secondary coping not only for a person but also within aggressive discussions, as dynamics of their appearance depends on discussion outbursts and aggression in them. We identify a 'spiral of distrust' as a cumulative effect of interaction between distrust, aggression, and intensity of commenting, and show that mechanisms of trust building in the antivaxxer community are tribal, unlike the media-like ones in more rational pro-vaccination channels.

Keywords: Trust · Aggression · Conspiracy theories · Anti-vaccination · Telegram · COVID-19 · Cumulative deliberation · Discussion dynamics

1 Introduction

The COVID-19 pandemic has brought along an unprecedented amount of social fear and uncertainty. Among other social effects, it has spurred the spread of distrust to elites and their rationality, as well as an outburst of conspiracy theories, around the world.

G. Meiselwitz (Ed.): HCII 2022, LNCS 13315, pp. 468–484, 2022.
https://doi.org/10.1007/978-3-031-05061-9_33

Since nearly the beginning of the pandemic, both traditional and social media were convicted of creating an infodemic – a storm of information and communication on the coronavirus-related issues that, instead of proper informing the world public, gave rise to conspiracy thinking and created immense pools of disconnected information bits without proper contextualization, thus reinforcing the uncertainty by letting people get lost in information rather proper guidance to survival. The infodemic has been declared one of the 10 biggest dangers of the pandemic [1]. WHO's anxiety echoed earlier concerns on spread of anti-vaccination moods and vaccine hesitancy [2, 3].

One of the effects evident on social media has been a high level of polarization along a new line: rationality vs. mythological mind. Inability of rationally proving a position is usually accompanied by aggression. It also makes people stray into packs and defend their positions. Political polarization and educational gaps only foster and deepen social suspicions [Hamilton]. In non-democracies, during the pandemic, the situation has even further been aggravated by the pre-formed 'triangle of mistrust' be-tween political elites, mainstream media, and publics [4].

Most studies that investigate the relations between trust to social institutions and public perception of the COVID-related threats employ self-reporting surveys or experi-mental designs that allow for respondents' self-reflection, which may distort the results. This is why it is crucial to also explore discussions on social media which di-minishes reflexive self-reporting, even if creates distortions of another sort, including management of the so-called context collapse [5] and selective posting in environments of varying opinions. Polarization of views makes users search for comfort zones where their views will be accepted and they will not undergo social exclusion [6]. This may form echo chambers of same views [7] where mis-beliefs may intensify, as their carriers may gain social approval from co-believers.

While echo chambers on social networking platforms like Facebook or microblog-ging services like Twitter tend to be fuzzy and permeable [8, 9], they may be more solid within technically bound online milieus like Tele-gram channels. The role of online com-munities in reinforcing medical dissident views has been previously explored for differ-ent cultures including Russia [10, 11]. The most explored case of a COVID-conspiracist community is, undoubtedly, QAnon [12] which has been shown to have considerably boosted the conspiracist narratives. Nonetheless, despite the already existing abundance of datasets collected for misinformation, anti-vaccination, and COVID-19-related con-spiracy theories (e.g., for a review of Twitter datasets in various languages, see [13]), anti-vaxxer communities are rarely studied beyond the English-language context. Their com-parisons with rational community-based discourse are even more rare (for excep-tions, see a Harvard-based study [14] and a Twitter study [13]), and they, again, focus on the English-language context.

The second research gap is the direction of popular distrust. In particular, the con-figuration of distrust to various institutions relating to COVID-19 has been explored rarely enough, overall. Also, it is expected to vary in democracies and non-democracies [4, 15], and the patterns of public distrust during the pan-demic remain highly under-explained for countries like Russia.

The third gap is a missing link between the two previous ones – that is, under-exploration of the linkage between the patterns of social and institutional (dis)trust, on one hand, and discursive patterns of irrational speech, on the other. In particular, it is important to know what institutions or objects within the COVID-related discourse evoke people's worst outbursts of aggression and conspiracy thinking.

The fourth gap is a low number of studies that relate platform affordances to the nature of anti-vaccination discourse. While there is a substantial amount of works that focus on platform affordances in spreading misinformation and denialist views, a much smaller segment of research has looked at how platforms shape the irrational COVID-related discourse.

As some authors have argued, addressing these gaps in complex demands a more holistic approach to the infodemic on the whole [16] and relations between distrust and denial in particular. We aim at partial covering these research gaps by in-depth studying of three Russian-language Telegram communities, two of which created by medical professionals and one by anti-vaccination activists. Telegram, unlike Twitter [13] or Facebook, remains under-researched in terms of spread of COVID-related misinformation and discussions on trust. Our overall sample includes 283,000+ comments collected in 2020–2021. We assess the patterns of (dis)trust in the communities, linkages between distrust to institutions, aggression, and conspiracy thinking, and the construction of trust. We discuss the role of Telegram and its affordances for formation of divergent 'communities of (dis)trust' and how Telegram allows for creating an architecture of (dis)trust in a country with no long democratic tradition. We show that the patterns of (dis)trust growth are cumulative, on both micro-level (within commenting on one post) and macro-level (within larger time spans).

The remainder of the paper is organized as follows. Section 2 is dedicated to literature review, and Sect. 3 is where we pose the research questions. Section 4 describes our methods in short. Section 5 demonstrates the results. We conclude by putting them into a wider country-level context.

2 Literature Review: Online Media and Vaccine Denialism

2.1 Politics and Media as Factors in Growth of Anti-vaccine Movements

Suspicion to vaccination, at least partly, is supported by weighting populational benefits (even if substantial) against personal risks (even if very or extremely rare). In addition, awareness on possible negative impact, of unclear origin and uncertain probability, contributes to vaccine hesitancy. This hesitancy naturally rockets when the vaccines are created within emergency circumstances and are approved much quicker than usual.

However, it is not only the breach of conventional vaccine approval procedures that has caused anti-vaccination movements around the world, but the general levels of situational uncertainty [17] and eternal insecurity that have found an exit hole, as well as deep-lying mistrust to the ruling elites, medical systems, and pharmaceutical giants. As one study has shown, users who expressed distrust in authorities 'employed discursive practices that both mirrored authoritative discourses and subverted official advice, by appealing to scientific language and "alternative" evidence' [18: 1161]. The same study revealed that authoritative institutional sources may be a site of both trust and mistrust, as the institutions that are expected to be the bearers of ultimate truth and knowledge (and are described that way), while, at the same time, evoke doubt and resistance to their advice when their decisions do not go in line with individual expectations.

In general, there is mixed evidence on how trust to scientific knowledge, including medical expertise, is linked to social media consumption. On one hand, consuming news via social networking platforms has been shown to boost trust in science [19]. On the other hand, the aforementioned works [20, 21] demonstrated de-cline in trust to experts under the impact of social media use. One may suggest that it is the difference in practices of consumption that matters: exposure to verified news, whether via social media or not, may foster trust into rational messages, while consumption of user talk and belonging to denialist communities may decrease trust and reinforce irrational and anti-scientific beliefs. Thus, the Harvard study [22] has shown that the use of social media fosters and intensifies wrong beliefs into negative effects of anti-COVID-19 vaccination, while the use of traditional media tends to correct misinformed beliefs. For the American context, negative trust to medical experts has been shown to be a predictor for misbeliefs stronger than traditional predictors like 'education, income, age, religiosity, and conservative news media consumption' [22]. However, this study was not specific about how exactly people who believed in false vaccination used social media; in particular, the role of dissident and rational communities on social media was not examined. A Norwegian study has also shown that destructive beliefs about COVID-19, including beliefs in misinformation and an opinion that the threat of COVID-19 is overrated, correlated with distrust to traditional media and authorities [23]. The destructive beliefs, the same study shows, are linked to modes of protective behavior during a health crisis. A Canadian study of 2015, though, states that there is mixed evidence on whether informing via traditional media and governmental websites is efficient in terms of behavioral change: thus, parents demanded more information on child vaccination to be placed on them, but other studies had shown that such informing may be either ineffective or even counterproductive [24].

Even if so, taken together, distrust to experts, traditional media, and authorities rises as a crucial factor for formation of open or latent individual/group denial-based behavior during the pandemic [4].

2.2 Online Community Structure as a Factor Shaping Anti-vaxxer Discourses

As stated in the introduction, on social media, people tend to divide into communities that are bias-related [9]. As the authors state, this is consistent with both the 'confirmation bias' hypothesis [25] and the social judgment theory which both state that people tend to choose information that corresponds to the beliefs core to their identity and reject the opposite views [26].

Previous research has suggested that COVID-19 misinformed communities are denser and more organized than communities of rational and informed users [13]. A correlation between misinformed online populace and anti-vaxxer communities has also been suggested; however, there is still scarce knowledge on how the communities based on rational and irrational discourses differ in their structure, whether they can be detected via social network analysis, and whether the conspiracy theories can be a detectable sign of anti-vaxxer groupings online. We suppose that anti-vaxxer communities might be denser in structure, due to their discourse not allowing for diverse opinions, while rational communities may look like more open and sparce discussion spaces. In open-structure social media like Facebook or Twitter, one might expect echo chambers to form, distinguishable even if permeable; in messengers like Telegram, the borders of rational and irrational communities will be more rigid, as they are technical.

It is also crucial to know how various negative or destructive discursive elements relate to the users and user constellations – that is, whether the same users bear uncondi-tioned distrust, aggression, conspiracy thinking and other attributes of non-enlightened mind. A small-sample study of vaccine denialism on Facebook has shown that dis-trust and conspiracy tended to belong to different users [27], which is counter-intuitive and needs further investigation. Another Facebook study on Polio [28] also divided the vaccine-skeptical comments into groups, thus making distrust and denial diverge in user comments. However, we see that, on the Russian-speaking Telegram communities, users constantly express many aspects of COVID-related skepticism intertwined within the comments. This intertwining also goes against established views on causal relations between social media use and decline in institutional trust [20]. Other works, though, show that, unlike in the traditional 'linear' relations between policymakers, traditional media, experts, and publics during a health crisis, networked communication is signif-icantly more complex and may demonstrate 'capricious flows' of information 'where nonexperts and citizens, whose voices, amplified by social media, [gain] traction in unexpected and inexplicable ways' [21: 9].

2.3   Contextual Factors in COVID-Denialism

Social, political, and technological context of the COVID-19 crisis is another dimension that is to be taken into consideration. Only very rare works contextualize conspiracies and denialist beliefs. Thus, the authors [29] have shown that COVID-19 conspiracy theories in Africa were related to deeply-rooted suspicions towards the West and its paternalism, corporations, and local political elites. The lack of such studies creates a significant gap in understanding of the role of context in configuration of distrust across countries and in relations between salience of conspiracy theories and political cleavages.

In our earlier works [30, 31], we have suggested several dimensions of *context behind the discussion* – that is, the contextual background that does not constitute itself during and within the discourse but shapes it from outside. Among them, the combination of the recent past on a certain territory ('historic context') combines with factors related to acceptance of innovation, including use of various platforms for social networking by different populational strata. For example, if '[d]istrust [is] a strategy to minimize the risks that come with the engagement with an untrustworthy other' [32: 2671], active distrust to authorities in post-Soviet countries creates conditions when the decisions by them are perceived as pressure, not as search for ways out of crises. In the Russian context, distrust shapes most relations between the state-affiliated actors and citizens [33], which, during the pandemic, has resulted into circa 30% of people simply not believing that the pandemic was there, according to several polls.

Conspiracy Theories as Extremal Priming of Contextual Distrust. 'In the context of the COVID-19 pandemic, the sudden lack of control and increased uncertainty may have made people particularly vulnerable to conspiracy theories', which are 'explanations for events that posit powerful actors are working together in secret to achieve self-serving or malicious goals' [34]. In linkage to the pandemic, they unite suspicions towards the 'secret cliques ruling the world' and misinformation on the origin of disease, its treatment, and surrounding issues.

In general, the reasons for believing to conspiracy theories may be epistemic (satisfying curiosity, uncertainty avoidance, priming nuanced explanations), existential (restoring the sense of security and control), and social (privileging an allegedly underestimated or discriminated social group, including the speaker's own one) [35]. During major crises, however, people navigate the complex information environments that cannot eliminate the high uncertainty. This, in contradiction to the appraisal theory that implies the people engage in rational problem-focused coping, e.g., via active information seeking, users adopt maladaptive behaviors in which they engage with non-scientific explanations. As a consequence, they may learn less from their information environment and become susceptible to conspiracy theories [17].

If conspiracy theories were mere single-statement publications, they would be much easier to counter; however, they are narratives of a specific sort. The narrative frameworks fueling conspiracy stories 'rely on the alignment of otherwise disparate domains of knowledge' [36: 279] and try to prime interpretations of complicated realities in a simplistic way that has principal gaps in logic filled in by conspiracy thinking. 'You can't just factcheck, label, or remove a narrative', as it has been wisely noted [37].

Conspiracy theories as a special type of distrustful discourse have been shown to affect behavior during the pandemic [38, 39], including anti-vaccination inclinations [40], mostly based on distrust. They tend to affect perception of governmental anti-COVID measures as too strict [41]. Growth of conspiracy thinking has also been linked to general high perception of threat [17] as well as to several distinct fears and needs also linked to distrust, such as, e.g., refusal to trust science in general or the concept of biomedicine [39] and, e.g., a need for uniqueness (which is also linked to lower compliance with social norms) [42]. However, it is yet unclear how conspiracy thinking relates to aggressive behavior. It is logical to suggest that aggression must be linked to conspiracy theories spread, as both aggression and conspiracy thinking are types of defensive behavior. Thus, we see that both aggression and distrust may have heavy linkage to conspiracy thinking, but this remains under-explored.

As the authors [40] note, in their study, none of the conspiracy theories discussed dangers of the vaccines but rather addressed political or general medical issues. This shows that, first, conspiracy mind is shaped by factors external to the very virus and, second, that conspiracy theories, despite the global character of some of them, are contextual, just as the fears and insecurities that provoke them are context-bound. However, the extant research mostly views the conspiracies as more or less universal, describing them case-by-case without relating them much to contextual factors.

Conspiracy theories might be contested when posted on 'openspace' social media; there, their perception by other users is shaped by discourse features like friend/enemy scheme, verbal attacks, violent threats, or satire and humor [43]. At the same time, we do not know whether and how conspiracy theories can be (and whether they are at all) contested in close-up communities like those on Telegram.

Telegram as a Milieu for (Semi-)public Discussions: Platform Affordances and the Russian Context. As already stated above, Telegram differs from other 'openspace' social media like Facebook or Twitter. Open groups on Telegram stand in between open-talk social interfaces like Facebook and messenger-like close-up groups in, e.g., WhatsApp. Joining open groups on Telegram does not require invitations. Such groups differ both from groups by invitation and from channels where comments might be enabled or disabled. Telegram's affordances like 'the functions of communication one-to-many, one-to-one, and… encrypted chatrooms' have allowed it to shift from information dissemination via Facebook and Twitter to 'empowering supporters to consistently and in a more immediate way collaborate as content producers', while keeping the network of participatory information disseminators [44: 1889].

Created first for the Russian market, Telegram has become popular in many countries. Due to its cyphering system, it is less easy for the security services to collect data from it. For less democratic contexts, this has been a liberation tool; however, the platform has also been criticized for being home for far-right extremism after its ban on other social media [45], as well as for terrorist groups like those of ISIS. Attempts to ban Telegram in response to non-provision of encryption keys to the security services [46] have, in effect, failed. Critics said it was not only due to technical complications of banning but also because, at the moment of banning, a lot of anonymous political channels were deployed by both the pro-establishment and pro-oppositional sources in the Russian political and media elites, and, thus, Telegram was a necessary element of distorted political communication in the country. With the growth of restrictive regulation on media and Internet in Russia [47], Telegram has gradually grown into a milieu where free speech has faded less than elsewhere in Runet. Thus, both free political discussion and expression of non-conventional and radical views have flourished there. Unlike on Facebook or other social networks, there is no chance to label misinformation or conspiracy theories as such on Telegram (which would be a step towards countering anti-vaccination views [48]), both due to the platform affordances and Telegram's creator Pavel Durov's libertarian position on freedom of speech. Due to the two features – freedom of expression and relatively easy echo chambering in groups that do nor overlap with other communities, Telegram has gathered the largest echo-chamber-like community of COVID-denialists and anti-vaxxers in Runet, called *anti_covid21*.

3 Research Questions

Based on what is stated above, we have posed the following research questions:

RQ1. Mapping user denialism:

RQ1a. How does distrust distribute among its addressees?
RQ1b. To what addressees of distrust are conspiracy theories linked to?

RQ2. Connection of distrust to aggression and conspiracy:

RQ2a. What is the connection between aggression and distrust?

RQ2b. What is the connection between aggression and conspiracy thinking?

RQ3. Cumulative dynamics of the denialist discourse:

RQ3a. Do distrust, aggression, and conspiracy thinking grow in time?
RQ3b. Do they spur each other?
RQ3c. Do they spur the discussion, and on which level?

RQ4. What are the means of construction of user trust, if any, in *anti_covid19*?

We do not pose any strict hypotheses, as our research is exploratory. However, we will use the methods of quantitative nature to demonstrate dependencies in our data.

4   Methods

4.1   Data Collection

We have collected the data from the community *anti_covid21* of January to July 2021. This comprises 282,000+ comments with their respective posts, as well as metadata (the date, authors, and links of posting/commenting). To contrast this dataset to a clearly different discourse, we have also downloaded the full data from the Telegram communities created by doctors (*reddoctors* and *ratsionalno_o_koronaviruse*) and dedicated to rational discussion on COVID-19 and debunking false information. Their volume of comments was much smaller, circa 1,000 comments altogether; however, this content has allowed for qualitative assessment of mechanisms of rational proof versus irrational construction of arguments in *anti_covid21*.

The datasets were collected with the help of a web crawler with changeable modules patented by the authors.

4.2   Data Processing and Analysis

Data Processing. To diminish the dimensionality of data and subject it to manual analysis, we have read over 20,000 comments to create a dictionary of distrust, aggression, and conspiracy theory that included 620 tokens, including stems and stem bigrams. We have applied it to the *anti_covid21* dataset to reduce the number of comments and detect those that might contain open distrust, aggression, or conspiracy theories. As a result, the dataset has been reduced to 82,000+ comments.

Data Coding. Then, it was coded for five variables (presence of distrust, addressee of distrust, presence of aggression, and presence of conspiracy theories, and presence of active trust). A team of 30 coders have been tested for inter-coder reliability, reaching minimum average Randolph kappa reliability of 0.6 (of −1 to 1). After that, each sixth comment was coded via randomized coding, resulting in more than 13,600 coded comments. After post-coding cleaning, 12,188 comments remained. Additional groups of comments related to one post were coded for one-post-level Granger tests.

Data Analysis. For RQ1 to RQ3, we have used descriptive statistics, including Spearman and Pearson correlations, and Granger causality testing. For RQ4, we have employed interpretive reading and discourse analysis.

5   Results

RQ1a: mapping of distrust. The discourse of the *anti_covid21* community may be called the discourse of distrust. Of all randomized-coded sample of 12,188 comments, 59% contained identified distrust. Extrapolating this to the sub-dataset of 82,000+ comments, we receive 48,380 comments; of 282,000+, this makes ~17,2%. That is, approximately each sixth post in the original dataset contains active distrust.

The distribution of distrust demonstrates its high politicization. Figure 1 shows the actual distribution of distrust by addressee, and Fig. 2 reveals the aggregated distrust percentages by domain (health, elites, people, and culture).

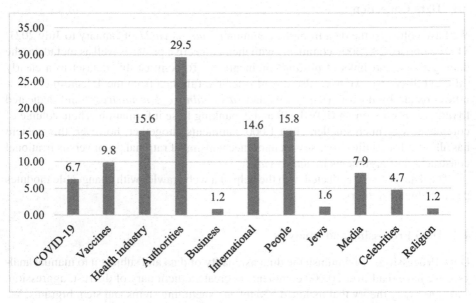

Fig. 1. Distribution of distrust by addressee, in %

Figures 1 and 2 demonstrate that it is neither the virus itself nor the vaccines that is gaining most distrust in the anti-vaccination community; thus, distrust to vaccines may be mediated via distrust to other entities or even be activated by other addressees whom the users do not trust. In other words, distrust to vaccination looks like being secondary, while distrust to other actors is primary, rather than vice versa. This definitely needs further investigation.

Neither is it local business, unlike in other countries. However, international players, mainly businessmen like Bill Gates and abstract 'globalists', are second in frequency of mentioning. This links the Russian anti-vaccination discourse with the global one.

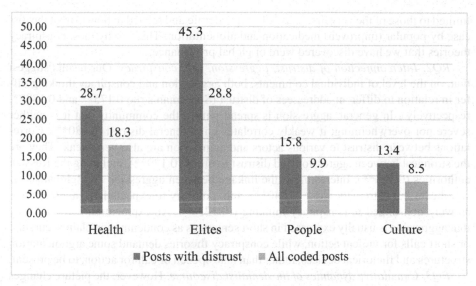

Fig. 2. Distribution of distrust by domain, in %

Nearly one third of all comments had local political actors, including the country leaders, as addressees of open distrust. The constellation of distrust to vaccination and to the authorities contradicts the widespread perception of less educated illiberal Russians as supporting the establishment and authorities. The antivaxxer movement in Russia has revealed a deeply lying massive distrust to and distancing from the president, government, and regional authorities, and this was especially evident among the irrational vaccine-denialist communities. Their attitudes resemble those of 'angry citizens' [8] or 'angry patriots' [49] of Moscow and other big cities who, in the 2010s, directed their discontent towards both the groups that irritated them (like immigrants or liberals) and the authorities. With the advent of COVID-19, this 'sleeping' trend has intensified.

We expected more interlinkages between different addressees of distrust. However, we have revealed that COVID-19 is provenly mentioned together with vaccines and people, and *not* with authorities or elites. Just as well, the health industry is provenly mentioned separately from elites and cultural agents, and vaccines are *not* mentioned with international actors. Thus, different issues within the anti-vaccination discourse are attached to different agents of distrust.

RQ1b: distrust and conspiracy thinking. Our results suggest that, understandably, conspiracies are actively *not* linked to distrust to people ($-0.195**$), media ($-0.179**$), or aggregated cultural entities including media, celebrities, and religion ($-0.243**$). It neither relates to internal politics (the correlation is insignificant), which is less expected. We see the only strong linkage of discussing conspiracy theories is to discussing international bodies ($0.320**$), which links the Russian conspiracy thinking with the global arena. We also suppose that, due to high suspicion whether the pandemic was at all there, local conspiracy theories were not that attractive for the Russian COVID-denialists, being limited to those of the very first phase of the pandemic and related to how to treat the disease by popular (unproven) medication and alimentation. This is why most conspiracy theories that we have discovered were of global provenance.

RQ2. Interconnection of distrust, aggression, and conspiracy. Our results suggest that, on the level of individual comments, both aggression and conspiracy thinking differ in relation to different addressees of distrust (Spearman's rho 0.175** and 0.188**, respectively). In general, aggression is spread within the community, but it is neither severe nor overwhelming; it weakly correlates with general distrust (0.180**). Correlations between distrust to various actors and aggression are also very weak. They are the strongest between aggression and distrust to elites (0.129**), including the Russian authorities (0.125**). Interestingly, the linkage between aggression and 'globalists' is significant but extremely weak (0.040**), and, contrary to expectations, aggression is actively non-linked to conspiracy thinking (-0.123**). An explanation here might be that aggression is usually expressed in short sentences as condemnation claims, cursing, or short calls for violent action, while conspiracy theories demand some argumentation structures and rhetorical devices, rather than hate speech or call for action, to be present.

RQ3. Cumulative dynamics of the denialist discourse. However, the picture changes when one looks at higher-level dynamics of the discussion. To assess it, we have first counted the daily percentages of distrust (in general and to the main addressees), aggression, and conspiracy thinking. Here, our findings are the following.

First of all, one needs to note that the daily dynamics of the discussions depends on the micro-issues that are discussed – and, thus, is highly unpredictable, as it depends on what the community administrators decide to post upon. Second, we have not detected any weekly or monthly cyclic dynamics in how distrust (see Fig. 3), aggression, or conspiracy thinking (see Fig. 4) developed. Third, we definitely see growth in percentages of general distrust and distrust to the health industry (including the existence of the virus itself, the vaccines, and the medical industry) with time.

Fourth, as the Pearson correlations show, distrust is not directly linked to conspiracist thinking, but aggression mediates their connection, as it correlates both with the aggregated distrust (0.294**) and conspiracy theories (0.342**). Thus, on the daily level, there are signs of cumulative effects that appear in the form of tighter connection between aggressive discourse and conspiracist / suspicious thinking in group speaking. Also, interestingly, conspiracy thinking correlates with expressions of trust (0.307**), as users tend to describe the 'trusted' sources of a given conspiracy theory. However, the construction of trust uses mechanisms that differ from those in rational communities (see below).

Figure 4 also shows that, most probably, the peaking distrustful discourse evokes aggression peaks, not vice versa (see black arrows). Also, the middle part of the graph shows how an outbreak of distrust (conditioned by the growth of distrust mostly to the health industry and vaccines themselves, as seen on Fig. 3) leads to peaking conspiracist discussion, and then to a simultaneous rise of aggression and expressions of trust, thus creating very peculiar deliberative conditions. In them, both aggression and attempts to bring on some argumentation to 'trusted' sources grew till the moment when trust rapidly dropped, and both aggression and conspiracies peaked (see light-blue arrow). Here, we see how the spiral of discussion that gradually heated up by conspiracy theories

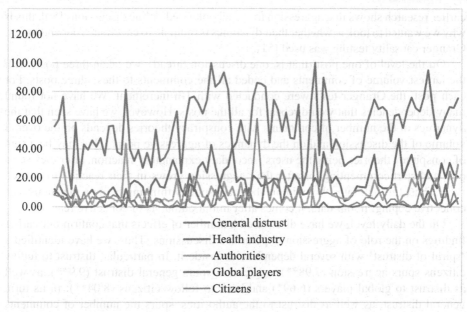

Fig. 3. Daily dynamics of distrust by addressee, January to July 2021, in % of comments

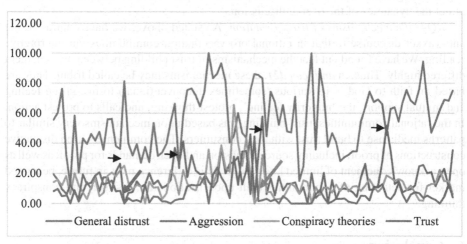

Fig. 4. Daily dynamics of distrust, aggression, conspiracy thinking, and trust, January to July 2021, in % of comments (Color figure online)

became aggressive and broke the attempts to bring on trust (however justified) into the anti-vaccination discourse. We also see the same pattern on an earlier stage of discussion, not that evident (see green arrow). This cumulative effect of drop of trust also deserved closer studying in future.

We have also checked whether, on different levels of discussion, distrust, aggression, and conspiracy theories spur each other and fuel the intensity of the discussion. Our earlier research shows that aggression may fuel polarized online discussions [50], this is why we wanted to look at whether their dynamics is mutual. To check such dependencies, Granger causality testing was used [51].

On the level of one post (that is, one discussion thread), we taken three posts with the largest volume of comments and coded all the comments to these three posts. For each post, the Granger tests were conducted with 1-h increment. We have not found many dependencies that would repeat for all the posts. However, we have seen that the dynamics of the number of comments with conspiracy theories depends on the overall volume of the discussion and on the dynamics of aggressive posts. This may be a sign of conspiracy theories being the users' secondary explanatory reaction, after they see a post and start commenting, or after their aggression grows in their reaction to a given post. Thus, this validates the opinion about conspiracy thinking being a mechanism of collective coping; in our data, it comes after initial outbursts of collective reaction.

On the daily level, we have discovered a number of effects that confirm our earlier findings on the role of aggression in polarized discussions. Thus, we have identified a 'spiral of distrust' with several dependencies inside it. In particular, distrust to fellow citizens spurs aggression (7.98**); aggression spurs general distrust (9.9**), as well as distrust to global players (6.69*) and back to fellow citizens (8.04**); in its turn, general distrust, as well as distrust to the authorities, spurs the number of comments (5.32* and 10.5**, respectively). Thus, starting from distrust to fellow commenters, aggression grows, with distrust to many actors expressed, and the levels of distrust affect the dynamics of the overall discussion.

RQ4: the mechanisms of trust formation. As stated above, we have contrasted the anti-vaxxer discourse to that in rational pro-vaccination communities via interpretive reading. We have found out that the mechanisms of trust building between users in them differed highly. Thus, in *anti_covid21*, these mechanisms may be called tribal. They are based on faith to God in its various (sometimes unconventional) forms, group feeling and mutual support, the 'minority feeling', protest thinking, and calls to protest action. In the rational communities, trust building was based upon mechanisms very similar to patterns media use to build trust with their consumers. The commenters used discursive constructions of proof, including source provision and open demands for proof, as well as open acknowledgement of limited knowledge. This difference may, in future, be a good marker for detecting of communities built upon irrationality, distrust, and conspiracy thinking.

6   Conclusion

In this paper, we have looked at how distrust to various sources, aggression, and conspiracy theories are interlinked in the anti-vaccination discourse of the Russian Telegram. We have found out that the anti-vaxxer talk is highly politicized, and it is not the vaccines themselves that cause the major distrust, but distrust to vaccines may be mediated by that to elites, including national authorities and global players, as well as to fellow citizens. The interlinkage between distrust, aggression, and conspiracy is more evident on the aggregated level, rather than within individual comments or comment threads. We have hinted to conspiracy theories being a mechanism of secondary reaction to posts that cause outbursts of comments, including aggressive ones, and have stated that attempts of bringing on trusted (even if flawed) sources may rapidly drop under the pressure of aggression. We have identified several dependencies that, taken together, may form a 'spiral of distrust' that starts from distrusting fellow citizens and results in the discussion being shaped by aggression and distrust to political actors. We have also shown that trust building between users employs either tribal or media-like mechanisms, depending on the channel's orientation to either mythology or rationality.

Acknowledgements. This research has been supported in full by Russian Science Foundation, project 21-18-00454 (2021–2023).

References

1. World Health Organization: Ten threats to global health in 2019. https://www.who.int/news-room/feature-stories/ten-threats-to-global-health-in-2019. Assessed 22 Jan 2022
2. Black, S., Rappuoli, R.: A crisis of public confidence in vaccines (2010). https://www.science.org/doi/abs/10.1126/scitranslmed.3001738. Assessed 22 Jan 2022
3. President Jean-Claude Juncker's State of the Union Address 2017. Speech. Brussels: European Commission, 12 September 2017. http://europa.eu/rapid/press-release_SPEECH-17-3165_en.htm. Assessed 22 Jan 2022
4. Bodrunova, S.S.: Russia: a glass wall. In: Lilleker, D., et al. (eds.) Political Communication and COVID-19, pp. 188–200. Routledge, London (2021)
5. Davis, J.L., Jurgenson, N.: Context collapse: theorizing context collusions and collisions. Inf. Commun. Soc. **17**(4), 476–485 (2014)
6. Noelle-Neumann, E., Petersen, T.: The spiral of silence and the social nature of man. In: Handbook of Political Communication Research, pp. 357–374. Routledge, London (2004)
7. Garrett, R.K.: Echo chambers online? Politically motivated selective exposure among internet news users. J. Comput.-Mediat. Commun. **14**(2), 265–285 (2009)
8. Bodrunova, S.S., Blekanov, I., Smoliarova, A., Litvinenko, A.: Beyond left and right: real-world political polarization in Twitter discussions on inter-ethnic conflicts. Media Commun. **7**, 119–132 (2019)
9. Bruns, A.: It's not the technology, stupid: how the 'Echo Chamber' and 'Filter Bubble' metaphors have failed us. International Association for Media and Communication Research speech (2019). http://eprints.qut.edu.au/131675/7/It%E2%80%99s%20Not%20the%20Technology%2C%20Stupid%20%28paper%2019771%29.pdf. Assessed 22 Jan 2022
10. Meylakhs, P., Rykov, Y., Koltsova, O., Koltsov, S.: An AIDS-denialist online community on a Russian social networking service: patterns of interactions with newcomers and rhetorical strategies of persuasion. J. Med. Internet Res. **16**(11), e261 (2014)
11. Dudina, V., Tsareva, A.: Studying stigmatization and status disclosure among people living with HIV/AIDS in Russia through online health communities. In: Bodrunova, S.S. (ed.) INSCI 2018. LNCS, vol. 11193, pp. 15–24. Springer, Cham (2018). https://doi.org/10.1007/978-3-030-01437-7_2
12. Thomas, E., Zhang, A.: ID2020, Bill Gates and the Mark of the Beast: how Covid-19 catalyses existing online conspiracy movements. Australian Strategic Policy Institute (2020). http://ad-aspi.s3.amazonaws.com/2020-06/ID2020%2C%20Bill%20Gates%20and%20the%20Mark%20of%20the%20Beast_%20how%20Covid-19%20catalyses%20existing%20online%20conspiracy%20movements.pdf. Assessed 22 Jan 2022

13. Memon, S.A., Carley, K.M.: Characterizing covid-19 misinformation communities using a novel Twitter dataset. arXiv preprint arXiv:2008.00791 (2020)
14. Jamison, A.M., Broniatowski, D.A., Dredze, M., Sangraula, A., Smith, M.C., Quinn, S.C.: Not just conspiracy theories: vaccine opponents and proponents add to the COVID-19 'infodemic' on Twitter. Harvard Kennedy Sch. Misinf. Rev. 1(3) (2020). https://misinforeview. hks.harvard.edu/article/not-just-conspiracy-theories-vaccine-opponents-and-pro-ponents-add-to-the-covid-19-infodemic-on-twitter/
15. [Edelman] 2020 Edelman trust barometer (2020). http://www.edelman.com/trustbarometer. Assessed 22 Jan 2022
16. Alam, F., et al.: Fighting the COVID-19 infodemic in social media: a holistic perspective and a call to arms. arXiv preprint arXiv:2007.07996 (2020)
17. Heiss, R., Gell, S., Röthlingshöfer, E., Zoller, C.: How threat perceptions relate to learning and conspiracy beliefs about COVID-19: evidence from a panel study. Pers. Individ. Differ. 175, 110672 (2021)
18. Bergman, K., Eli, K., Osowski, C.P., Lövestam, E., Nowicka, P.: Public expressions of trust and distrust in governmental dietary advice in Sweden. Qual. Health Res. 29(8), 1161–1173 (2019)
19. Huber, B., Barnidge, M., Gil de Zúñiga, H., Liu, J.: Fostering public trust in science: the role of social media. Publ. Underst. Sci. 28(7), 759–777 (2019)
20. Mari, S., et al.: Conspiracy theories and institutional trust: examining the role of uncertainty avoidance and active social media use. Polit. Psychol. (2021). https://doi.org/10.1111/pops. 12754
21. van Dijck, J., Alinejad, D.: Social media and trust in scientific expertise: debating the Covid-19 pandemic in the Netherlands. Soc. Media Soc. 6(4), 2056305120981057 (2020)
22. Stecula, D.A., Kuru, O., Jamieson, K.H.: How trust in experts and media use affect acceptance of common anti-vaccination claims. Harvard Kennedy Sch. Misinf. Rev. 1(1) (2020). https://misinforeview.hks.harvard.edu/article/users-of-social-media-more-likely-to-be-misinformed-about-vaccines/. Assessed 22 Jan 2022
23. Filkuková, P., Ayton, P., Rand, K., Langguth, J.: What should I trust? Individual differences in attitudes to conflicting information and misinformation on COVID-19. Front. Psychol. 12, 588478 (2021)
24. Greenberg, J., Dubé, E., Driedger, M.: Vaccine hesitancy: in search of the risk communication comfort zone. PLoSCurr. 9, (2017). https://doi.org/10.1371/currents.outbreaks.0561a0 11117a1d1f9596e24949e8690b
25. Oswald, M.E., Grosjean, S.: Confirmation bias. In: Pohl, R.F. (ed.) Cognitive Illusions: A Handbook on Fallacies and Biases in Thinking, Judgement and Memory, pp. 79–96. Psychology Press, Hove (2004)
26. Nyhan, B., Reifler, J.: When corrections fail: the persistence of political misperceptions. Polit. Behav. 32(2), 303–330 (2010)
27. Hoffman, B.L., et al.: It's not all about autism: the emerging landscape of anti-vaccination sentiment on Facebook. Vaccine 37(16), 2216–2223 (2019)
28. Orr, D., Baram-Tsabari, A., Landsman, K.: Social media as a platform for health-related public debates and discussions: the Polio vaccine on Facebook. Israel J. Health Policy Res. 5(1), 1–11 (2016)
29. Gagliardone, I., Diepeveen, S., Findlay, K., Olaniran, S., Pohjonen, M., Tallam, E.: Demystifying the COVID-19 infodemic: conspiracies, context, and the agency of users. Soc. Media Soc. 7(3), 20563051211044233 (2021)
30. Bodrunova, S.S.: When context matters: analyzing conflicts with the use of big textual corpora from Russian and international social media. Partecipazione e conflitto 11(2), 497–510 (2018)

31. Bodrunova, S.S.: The boundaries of context: contextual knowledge in research on networked discussions. In: Antonyuk, A., Basov, N. (eds.) NetGloW 2020. LNNS, vol. 181, pp. 165–179. Springer, Cham (2021). https://doi.org/10.1007/978-3-030-64877-0_11

32. Bodó, B.: Mediated trust: a theoretical framework to address the trustworthiness of technological trust mediators. New Media Soc. **23**(9), 2668–2690 (2021)

33. Ledeneva, A.V.: Can Russia Modernise? Sistema, Power Networks and Informal Governance. Cambridge University Press, Cambridge (2013)

34. Dow, B.J., Johnson, A.L., Wang, C.S., Whitson, J., Menon, T.: The COVID-19 pandemic and the search for structure: social media and conspiracy theories. Soc. Pers. Psychol. Compass **15**(9), e12636 (2021)

35. Douglas, K.M.: COVID-19 conspiracy theories. Group Process. Intergroup Relat. **24**(2), 270–275 (2021)

36. Shahsavari, S., Holur, P., Wang, T., Tangherlini, T.R., Roychowdhury, V.: Conspiracy in the time of corona: automatic detection of emerging COVID-19 conspiracy theories in social media and the news. J. Comput. Soc. Sci. **3**(2), 279–317 (2020). https://doi.org/10.1007/s42 001-020-00086-5

37. Wardle, C., Singerman, E.: Too little, too late: social media companies' failure to tackle vaccine misinformation poses a real threat. BMJ, **372** (2021). https://web.archive.org/web/202102130 83638id_/https://www.bmj.com/content/bmj/372/bmj.n26.full.pdf. Accessed 22 Jan 2022

38. Bierwiaczonek, K., Kunst, J.R., Pich, O.: Belief in COVID-19 conspiracy theories reduces social distancing over time. Appl. Psychol. Health Well Being **12**(4), 1270–1285 (2020)

39. Imhoff, R., Lamberty, P.: A bioweapon or a hoax? The link between distinct conspiracy beliefs about the Coronavirus disease (COVID-19) outbreak and pandemic behavior. Soc. Psychol. Pers. Sci. **11**(8), 1110–1118 (2020)

40. Bertin, P., Nera, K., Delouvée, S.: Conspiracy beliefs, rejection of vaccination, and support for hydroxychloroquine: a conceptual replication-extension in the COVID-19 pandemic context. Front. Psychol. **11**, 2471 (2020)

41. Rieger, M.O., Wang, M.: Trust in government actions during the COVID-19 crisis. Soc. Indic. Res. 1–23 (2021)

42. Imhoff, R., Lamberty, P.: How paranoid are conspiracy believers? Toward a more fine-grained understanding of the connect and disconnect between paranoia and belief in conspiracy theories. Eur. J. Soc. Psychol. **48**, 909–926 (2018)

43. Fuchs, C.: Users' reactions to COVID-19 conspiracy theories on social media. In: Fuchs, C. (ed.) Communicating COVID-19: Everyday Life, Digital Capitalism, and Conspiracy Theories in Pandemic Times, pp. 145–189. Emerald Publishing (2021)

44. Krona, M.: Collaborative media practices and interconnected digital strategies of Islamic State (IS) and pro-IS supporter networks on Telegram. Int. J. Commun. **14**, 1888–1910 (2020)

45. Urman, A., Katz, S.: What they do in the shadows: examining the far-right networks on Telegram. Inf. Commun. Soc. 1–20 (2020)

46. Wijermars, M.: Selling internet control: the framing of the Russian ban of messaging app Telegram. Inf. Commun. Soc. 1–17 (2021)

47. VendilPallin, C.: Internet control through ownership: the case of Russia. Post-Soviet Aff. **33**(1), 16–33 (2017)

48. Zhang, J., Featherstone, J.D., Calabrese, C., Wojcieszak, M.: Effects of fact-checking social media vaccine misinformation on attitudes toward vaccines. Prev. Med. **145**, 106408 (2021)

49. Bodrunova, S.S., Litvinenko, A.A., Blekanov, I.S.: Comparing influencers: activity vs. connectivity measures in defining key actors in Twitter ad hoc discussions on migrants in Germany and Russia. In: Ciampaglia, G.L., Mashhadi, A., Yasseri, T. (eds.) SocInfo 2017. LNCS, vol. 10539, pp. 360–376. Springer, Cham (2017). https://doi.org/10.1007/978-3-319-67217-5_22

50. Bodrunova, S.S., Litvinenko, A., Blekanov, I., Nepiyushchikh, D.: Constructive aggression? Multiple roles of aggressive content in political discourse on Russian YouTube. Media Commun. **9**, 181–194 (2021)
51. Wessa, P.: Bivariate Granger Causality (v1.0.4) in Free Statistics Software (v1.2.1), Office for Research Development and Education (2016). http://www.wessa.net/rwasp_grangercausa lity.wasp/. Assessed 06 Mar 2022

Gender-Sensitive Materials and Tools: The Development of a Gender-Sensitive Toolbox Through National Stakeholder Consultations

Eirini Christou[1] , Antigoni Parmaxi[1]([⊠]) , Maria Perifanou[2] ,
and Anastasios A. Economides[2]

[1] Cyprus University of Technology, Limassol, Cyprus
eirini.christou@cyprusinteractionlab.com,
antigoni.parmaxi@gmail.com
[2] University of Macedonia, Thessaloniki, Greece

Abstract. The values of gender equality are being promoted worldwide. The importance of gender equality for sustainable development is well highlighted by the United Nations Sustainable Goal 5 which notes that the need to end all discrimination against women and girls. Nowadays, most modern scholars argue that the world has made great progress towards gender balance, however, it is far from perfect. For encouraging and empowering women to remain active in every field, it is important to raise awareness about their rights, with emphasis on the vital role of girls and women in the workforce. This is especially important for the Science, Technology, Engineering and Mathematics (STEM) field where women are still underrepresented. This study aims to report on the materials and tools (digital and traditional) that can be used for sensitizing and raising awareness on issues related to gender-equality and women's empowerment. On this endeavor, we collected information on existing materials used in different contexts through national consultations in Greece, Cyprus, Italy, Spain and Slovenia. The tools and materials collected uncover the various levels of gender equality material available - digital and traditional - taking into account the various facets of gender-equality and provide a comprehensive view to the wider academic and industrial community. This study is expected to provide structured knowledge on a new and rapidly developing topic and add more information to existing contour of knowledge regarding available gender-sensitive materials and tools.

Keywords: Gender-sensitive materials · Gender-sensitive tools

1 Introduction

The Council of Europe (COE) stated in a 2002 report that "gender balance is not only about getting more of the under-represented sex into all areas of decision-making, but also about making balanced groups work more effectively, and maintaining gender balance over time" [1]. The Statistics Division of the United Nations describes Gender balance as "commonly used in reference to human resources and the equal participation of

G. Meiselwitz (Ed.): HCII 2022, LNCS 13315, pp. 485–502, 2022.
https://doi.org/10.1007/978-3-031-05061-9_34

women and men in all areas of work, projects or programmes. In a scenario of gender equality, women and men are expected to participate in proportion to their shares in the population" [2].

In recent years, gender balance theories have progressed beyond the classic scope, due to the rise in understanding of the inherent biological gender differences and societal expectations, such as women's role in child-bearing responsibilities. Thus, the main thrust is that while differences exist between the genders, it shall not be utilized to discriminate against women, but rather should contribute to an equal power sharing economically, socially, and politically and in other spheres of social influence [3, 4]. Therefore, modern scholars have emphasized that gender balance theories must strike on the following issues as necessary conditions for evaluating gender balance: First, they must appreciate that differences between the genders exist, and second, they must accept the right to be different. These rights must be respected and equally reflected in all structures of the society, in power relations, policy formulations, and elimination of all hierarchies or imbalances based on gender [3]. Nowadays, most modern scholars argue that the world has made great progress towards gender balance, however, it is far from perfect in all aspects of societal life, and especially in the developing world [5–7].

Even though it is almost impossible to refer to equality without referring to gender, the way gender equality is being treated is often superficial (see for example, [8–13]. For encouraging and empowering women to remain active in every field, it is important to raise awareness about their rights, with emphasis on the vital role of girls and women in the workforce. This is especially important for the Science, Technology, Engineering and Mathematics (STEM) field where women are still underrepresented [14–21]. Academic publications and the web are considered an influential means in this regard as it has the potential to widely distribute gender-sensitive information that can mold public opinion.

This study aims to report on the materials and tools (digital and traditional) that can be used for sensitizing and raising awareness on issues related to gender-equality and women's empowerment. On this endeavor, we collected information on existing materials used in different contexts through national consultations in Greece, Cyprus, Italy, Spain and Slovenia. The tools and materials collected uncover the various levels of gender equality material available - digital and traditional - taking into account the various facets of gender-equality and provide a comprehensive view to the wider academic and industrial community. This study is expected to provide structured knowledge on a new and rapidly developing topic and add more information to existing contour of knowledge regarding available gender-sensitive materials and tools.

2 Methodology

In order to collect different types of gender-sensitive materials, national stakeholders' consultation was conducted in five countries (Cyprus, Greece, Italy, Spain and Slovenia) through comprehensive review of tools and materials in a cyclical process as it is demonstrated in Fig. 1.

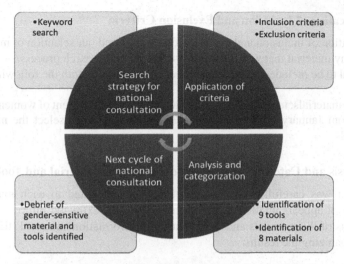

Fig. 1. Process adopted for the national stakeholder consultations.

2.1 Collection of Gender-Sensitive Materials and Tools

Six organisations (Cyprus University of Technology, Cyprus; University of Macedonia, Greece; CESIE, Italy; Magenta, Spain and Geoss, Slovenia; ARIS, Cyprus), conducted a thorough research to identify relevant digital or traditional gender sensitive materials and tools completed the information specified below (see Table 1). Through national stakeholder consultations, we gathered two types of data:

(a) gender sensitive materials, i.e. any material with gender-sensitive content (aim to raise awareness on issues related to gender-equality and women's empowerment). These materials could range from an infographic or poster demonstrating the successes of women rights to inspire and energise other women to virtual reality tours or any relevant gender-related source available on the web.

(b) gender sensitive tools, i.e. any tools that have been employed under a gender-sensitive scope.

Both gender sensitive materials and tools were documented and included the information that deemed necessary for the classification of the material (see Table 1).

2.2 Search Strategy

Appropriate materials for inclusion was selected via keyword search on the web. The search employed is as follows: ((women OR gender OR females) AND science) OR ((women OR gender OR females) AND engineering) OR ((women OR gender OR females) AND mathematics) OR ((women OR gender OR females) AND (informatics OR computing)). The search extended in Google, Google Scholar as well as on the Erasmus + project results database.

2.3   Application of Inclusion and Exclusion Criteria

The application of inclusion and exclusion criteria refined our selection of material and excluded any material that was incorrectly selected in the search process.

Material to be included in our dataset needed to conform with the following criteria:

1. Include materials, tools or practices related to the empowerment of women in STEM.
2. Date from January 2017 to March 2022 as we sought to select the most recent materials.

2.4   Analysis and Categorization of Gender-Sensitive Material and Tools

All material was carefully screened and information related to each material was described as it appears in Table 1. For each material the extracted information guided the synthesis of the materials and tools. Thus, the information extracted (IE) was then used for organizing the results.

Table 1. The four categories of the information extracted (IE) for classification of the material collected.

Group 1. Material identification	Group 2. Activities reported in the material
IE1. Material title IE2. Material ID IE3. Year of publication IE4. Authors' name(s) IE5. Institution IE6. Source of the material (Uniform Resource Locator (URL)	IE7. Objective IE8. Short description of the material IE9. Classification of material IE10. Context
Group 3. Basis of the publication	**Group 4. Evaluation of material**
IE11. Tool/software used IE12. Main technological features	IE13. Pricing IE14. Primary evaluation IE15. Future modifications/suggestions

3   Summary of Findings

Through national stakeholder consultations, we collected 448 gender-sensitive materials and tools. Based on our findings, the materials and tools can be categorized in terms of their type into digital and physical and based on their aim as factual, enlightening, practical/training and mentoring-provision. A thorough description of the tools and materials collected appears in the following sections, together with indicative examples of each category.

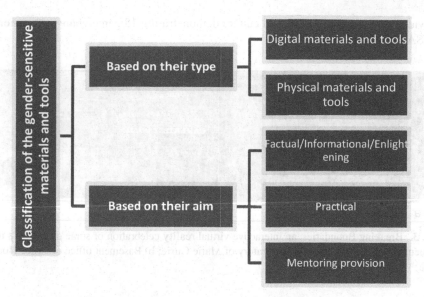

Fig. 2. Classification of the gender-sensitive materials and tools.

3.1 Classification of the Gender-Sensitive Materials and Tools Based on Their Type

Figure 2 demonstrates the types of tools collected through the national stakeholder in the partner countries. The material varies from face-to-face and online interventions to tangible materials such as Lego playset.

Digital Materials and Tools. Digital materials and tools included different multimedia (e.g., online photo galleries, videos, podcasts, talks or movies), websites with rich source of gender-sensitive content (e.g. inspirational quotes for girls and women or advice on how to overcome challenges in the area of STEM) and emerging technologies. Emerging technologies included different types of technologies such as virtual tours and augmented reality tours with gender-sensitive content with an eye to inform, empower and/or raise awareness on gender sensitive issues. For example, 'Breaking Boundaries in Science' is an interactive virtual reality celebration of some of history's most influential women scientists. Their legacies are not only of groundbreaking discoveries, but also of overcoming obstacles, following passions, and breaking the mold [22]. The specific virtual reality was designed and is available exclusively for the Sam-sung Gear, and immerses players in the life and times of famous figures like Jane Goodall, Marie Curie, and Grace Hopper. Players in the specific tour explore each scientist's real-life work environments and gain an intimate knowledge of their lives and achievements through fully voice-acted vignettes that are steeped in historical context, including narration from Jane Goodall herself. Relevant examples appear in Fig. 3. Similar virtual tours are also available through Google Arts and Culture which features from around the world on different of artworks and cultural artifacts. Available gender-sensitive tours in Google Arts and Culture include 8 Virtual Tours of Where Powerful Women Lived or the Online exhibit provided through Google arts and culture demonstrating 15 game-changing women of NASA (See Fig. 4a).

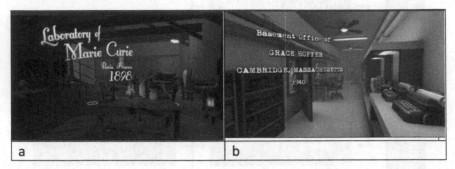

Fig. 3. Breaking Boundaries, an interactive virtual reality celebration of some of history's most influential women scientists. a) Laboratory of Marie Curie; b) Basement office of Grace Hopper [22].

This category also includes different communities, associations, organizations, foundations, initiatives, movements, campaigns or even platforms that have been developed with an eye to support and advocate for women in STEM. Such examples include ACMW (https://acmweurope.acm.org/) which supports, celebrates, and advocates internationally for the full engagement of women in all aspects of the computing field, providing a wide range of programs and services to ACM members and working in the larger community to advance the contributions of technical women, the She Can STEM campaign (https://shecanstem.com/) which shows girls how interesting STEM can be. Moreover, this category involves online mentoring platforms, for example the Womenpower platform, a community platform which aims to bring together women mentors and mentees from different fields for promoting women's empowerment, equality, and social coherence [23]. Online interventions were also noted, with an eye to empower girls and women to remain active in STEM. An example of such intervention is the CareerWISE online resilience training which aims at addressing the shortage of effective mentors and role models who have been shown to increase the persistence of women in STEM fields (see more in [24]).

Finally, this category includes mobile applications with gender sensitive content. One indicative example is the SheBoard application (see Fig. 4b) which was created by Plan International to change the narrative around gender. After downloading the app, it uses predictive text to suggest empowering and less sexist language in context to girls and women [25]. For example, if you were to type "Girls are," the suggested word might be "bold."

Physical Materials and Tools. This category includes materials and tools aiming to appear in the physical world. Such materials included posters, factsheets, books or plays. This material was observed to be more colorful, whilst its development did not encompass high cost of software or application (see for example Fig. 5a). In the same category, we also identified trainings, camps or workshops that took place face to face.

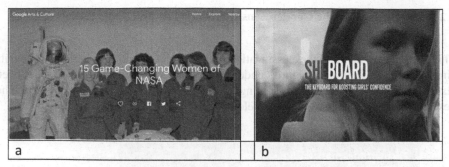

Fig. 4. a) Online exhibit provided through Google arts and culture demonstrating 15 game-changing women of NASA (Credits: Nasa). b) Example of mobile application with gender-sensitive content. Sheboard is a mobile app which uses predictive text to suggest empowering and less sexist language in context to girls and women (Credits: SheBoard).

Physical materials also included tangible materials with gender sensitive content. Such material include a new playset from LEGO aiming to honour four key women in NASA history—astronomer Nancy Grace Roman, computer scientist Margaret Hamilton, and pioneering astronauts Sally Ride and Mae Jemison (see for example Fig. 5b).

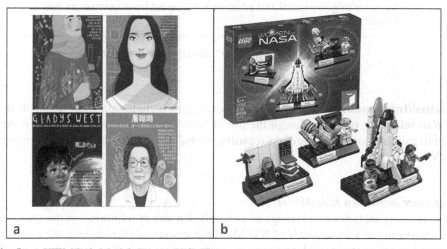

Fig. 5. a) STEM Role Models Posters [26]. The posters were created by a podcast which celebrates women transforming teaching and learning through technology. The posters are available for download for schools/workplace.; b) Example of a tangible tool with gender-sensitive content [27].

3.2   Classification of the Gender Sensitive Materials and Tools Based on Their Aim

This section demonstrates the classification of the gender sensitive material based on its aim (see Fig. 6). The four categories under which the gender sensitive material has been classified are: a) Factual; b) Enlightening; c) Practical and d) Mentoring provision. The description of the categories appears below.

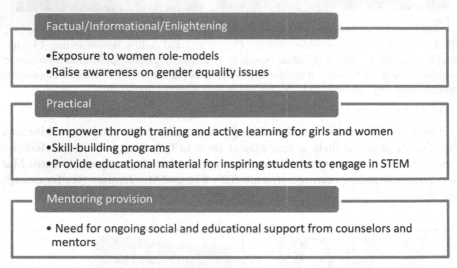

Fig. 6. Classification of the material based on its aim.

Factual/Informational/Enlightening. This category includes materials and tools that aim to inform or persuade though the use of facts or information. Additionally, they aim to enlighten or elicit awareness regarding women in STEM. Specific subcategories follow.

Exposure to Women Role-Models
This category includes a wide range of materials that aim at informing on the successes of women in the various fields of STEM. These resources include lists of prominent women and their stories, videos or movies of prominent women or even virtual reality tours of the successes (and often failures) of prominent women in the area of STEM (see for example [28]). As noted by [29], students studying all kinds of majors seemed to benefit from the role model exposure, since it is increasing the academic sense of belonging, self-efficacy, expectations, and educational degree intention, compared to students who were not exposed to role models [29].

International Women in Medical Physics and Biomedical Engineering Task Group proposes that identifying and promoting female role models that have achieved work-life balance, is a strategy that can attain gender balance in STEM [30]. Having someone to look up to is a vital empowering factor for females in challenging and male-dominated sectors [31, 32]. However, women in STEM careers have often expressed the lack of role models and real-life examples, to motivate them and to identify with [33, 34]. [35] research experiment showed that an inclusivity statement with a gender-inclusive course image depicting females as role models, at the enrolment page, raised the proportion of women enrolling in that specific STEM course. Nevertheless, [36] experiment's findings showed that female STEM students were influenced only by familiar, successful and known female leader role models, and not from someone non-famous. Moreover, [37] demonstrated that when individuals consider engaging in a potentially beneficial activity (additive behavior), they are more likely to choose positive role models as a means of harnessing their motivation, but when individuals consider reducing a potentially damaging activity (subtractive behavior), they are more likely to choose negative role models as a means of sustaining their motivation. A person's own regulatory orientations influence his role model preferences [37]. From this perspective, students who are eventually positively affected by role models, are the ones that were anyway amenable to choosing STEM as a field of study, or career.

Raise Awareness on Gender Equality Issues

This category includes materials and tools that aim at raising awareness on gender equality issues, by highlighting the difficulties that women encounter in STEM, strategies that all genders need to adhere to in order to adopt a gender-sensitive approach. These resources include lists of prominent women and their stories, videos or movies of prominent women or even virtual reality tours of the successes (and failures) of prominent women in the area of STEM (see [38]).

Video interventions with expert interviews and narratives illustrating gender bias have proven to increase awareness about the gender factor in male-dominated environments. This kind of intervention manages to enrich knowledge about gender inequity, and shape more positive attitudes towards women in STEM [39, 40].

Practical Materials and Tools. This category involves materials, training and tools that allow for hands on activities and/or training.

Empowerment Through Training and Active Learning for Girls and Women

This category involves the development and implementation of training programs for supporting, and advocating for women to remain active in STEM. These programs may include technical training for girls and women, soft skills training (see for example [36]) or training on raising awareness on gender equality issues (see for example [41]). This category also encompasses activities that engage women in a co-design adventure that allows them to gain STEM knowledge and to be empowered to be active in the field.

Emerging research surrounding informal education provides evidence that after-school and summer programs can be utilized as an effective strategy for increasing female interest in the STEM-related areas [42]. One approach that many institutions have taken towards increasing interest is to offer camps [43] which have proved to be an extremely efficient practice for developing and increasing STEM knowledge and interest among young women. The focus of the existing paradigms is on single-gender camps, where female participants interact only with other females. Such camps, mentioned in studies by [44–46], contributed to girls coming closer to STEM disciplines and values, by making science and technology enjoyable. Camps' activities strengthened their confidence and helped them see what women are capable of doing in these challenging sectors. Girls Tech Camp, described by [45] introduced middle-school girls to a range of technologies. For seven days, participants had the opportunity to acquaint themselves with 3D modelling, 3D printing, arduino microcontrollers, augmented reality, light sensors, digital video production, computer coding and conductive crafts. The fact that all the girls were a vital part of the experience and the learning process, enriched their knowledge and increased their confidence in using such tools. This camp had been held for three years in a row (2016–2018), and each year the curriculum of the camp was updated by incorporating new creative technologies so it could meet every era's specific needs. For example, in 2018, 3D modelling, 3D printing and augmented reality were embedded in group projects with the objective to design and improve a children's book for visually impaired people. Other activities of the Girls Tech Camp included field trips, children's prosthetics designing, and research for female scientists. Another noteworthy aspect of the camp is the free time that was given to the participants in order to resolve their own technological problems that emerged from their projects. This practice enhanced their persistence and resilience and as a result, their confidence was increased and their knowledge was enriched. During another similar camp, the Tech Trek, middle-school female students attended app development courses [44]. Its duration was 7 days as well, and during the week students attended 12 h of app development lessons. Participants learned to use the App Inventor, and were guided towards the exploration and the creation of applications for smartphones and tablets. Engaging with app development offered them new knowledge regarding computer science and programming, increased their confidence and helped them be more aware of the existing opportunities in STEM. These three elements contributed to a positive reinforcement towards a possible occupation in STEM fields. An additional example is the GIRLS camp, mentioned by [46] where the girls-only aspect of the program helped middle-school female participants see what girls can do with science. As a consequence, after the end of the camp girls reported that they were feeling confident and good enough to get involved with STEM. The goal of the camp was to introduce girls to STEM career opportunities through hands-on activities that can be associated with real life situations. The schedule was focused on group work and cooperation and included interaction with women professionals in STEM, and tours in laboratories, animal shelters, local waterways, organic farms, and collection and analysis of data. These activities help girls to become acquainted with science and technology, and therefore, their self-confidence was boosted. Contacts with professionals who already work in the field, worked as a supportive factor for the girls' ambitions to follow a similar path. The study of [47] advocate that strong self-efficacy expectations and assertiveness were shown to be associated with increased willingness to engage in the career-related activities of non-traditional occupations. According to them, some examples of traditional occupations are nurse, elementary school teacher, and secretary. On the contrary, two paradigms of non-traditional occupations are the engineer and the mathematician. Considering the previous points, the confidence that young girls can build by participating in camp's activities and courses, can play an extremely important role in the future, during the profession selective procedure. Table 2 summarises the camps' features collected from the literature. The features are considered as the strong point of the specific activity and are highlighted as the elements that can encourage girls and women to remain active in STEM-related professions.

Table 2. Summary of camps' features collected from the literature.

hands-on activities
engagement in the learning process
group work & projects
tools designing
encouragement to create
familiarization with modern tools and technologies
association of scientific concepts with real life situations
interaction with professionals (women)
tours & field trips
collection & analysis of data
collaborative learning
authentic learning
networking with peers and/or ex-perts/mentors

Skill-Building Programs

Another well-established practice for empowering young women to follow STEM domains by developing essential skills that are needed during their challenging journey, are the skill-building programs. Skill-building programs include activities usually guided by a teacher with support and scaffolding for helping students learn. Skills developed in skill-building programs include self-efficacy, self-awareness, self-promotion, emotional intelligence, bias recognition, coaching and teaching capability, problem-solving skills, communication skill and resilience. Findings from skill-building programs demonstrate that engagement in educational procedures contributes apart from the development of the specific skills but also to the development of STEM identity. Indicative examples of such activities follow:

Leadership Lab, a leadership development program, was designed to provide a supporting learning environment for female professionals working in STEM, to succeed in male-dominated and technology driven organizations, by helping them develop self-efficacy, self-awareness, emotional intelligence, coaching capability, prepare them to recognize bias, and be able to mitigate it [48]. The program lasted three months, and was led by experienced women in STEM professions. Its purpose was to assist participants recognize their worth in a STEM workplace by understanding bias, barriers and opportunities, skill development, and leadership. During the program they had the chance to experience guidance by an executive coach, and practice peer coaching. Outcomes showed increased self-awareness, self-efficacy, and ability to persist and succeed in their occupation.

A second example is the STEAMpunk Girls Co-Design, 11-day program. Its intention was firstly to empower high school females by positioning them as experts on how they wanted to engage with STEM and STEAM, and secondly, to design an educational pilot through a platform for their peers, using co-design methodology [49]. The program started with a workshop, which was followed by nine days of interviews (from their friends) and was completed with a second workshop and a group project. After their participation, girls reported that the program helped them comprehend STEAM values and disciplines, recognize the importance of studying STEM subjects, and expressed increased confidence in STEM.

During InSTEP, an inquiry-based science and technology enrichment program with one-week duration, middle-school female students learned more about problem-solving procedures, and were practicing with hands-on activities [50]. These student-centered learning methods encouraged girls that participated to see themselves as scientists, developed their conceptual understanding of science, and made them able to interact with others in discussions about science concepts. Consequentially, their attitudes towards sciences in general were improved, and they started to think of the idea of a career in STEM, as a possible and attractive scenario.

Another effort respecting the skill-building concept is CareerWise, an online resilience training mentoring program, designed as an alternative to traditional in-vivo mentoring [24]. CareerWise was established to develop self-talk, self-promotion, problem solving, communication skills, resilience and copying efficacy among women with STEM interests. It was designed to provide the necessary psychological support for overcoming possible obstacles women may face when studying STEM or pursuing a STEM career. The website is featuring 50 educational modules and self-tests. Some indicatives are: Build on your strengths, Self-talk, Your personality and preferences, Consider other perspectives, Stereotype threat, Recognise Sexism, Family-friendly policies, HerStories (videos of interviews from successful women in STEM occupations). All modules are based on psychological research, with emphasis on the support and encouragement. The personalised guidance material, practices, and role models as inspiration that can be found in the website, can help girls face or mitigate discouraging factors, handle possible obstacles and achieve personal and professional ambitions and dreams.

Provision of Educational Material for Inspiring Students to Engage in STEM.
This category involves resources that provide educational material (e.g. curriculum activities) for supporting and inspiring girls and women to remain active in STEM. Examples of such activities appear in Girl Scouts of the United States of America (2016-2021). More specifically, this website offers: 1) a "fun with purpose" K–12 curriculum to inspire girls to embrace and celebrate scientific discovery in their lives; 2) Leadership Journeys in order to explore a wide range of interests along their Journeys —everything from the arts to the outdoors and, of course, STEM; 3) It's Your Planet—Love It! series, girls can explore the natural world by learning about the water cycle, completing energy audits, assessing air quality, calculating their "food print" and learning kitchen science, and figuring out how much trash is created and how to reduce it. Several inspiring videos are included; 4) Badges; 5) Partners in Inspiration; 6) research Reports and much more.

Mentoring Provision. Mentoring is defined as a close, intense, mutually beneficial relationship between someone who is older and more experienced, with someone younger or with less experiences [51]. This category involves the provision of mentoring to young girls and women. Mentoring can be provided either face-to-face or online and can be supported either by peers or by prominent professionals from the industry. Examples of mentoring activities are available in [23] and [52]. Social support during women's studentship, particularly from mentors, is proven to be helpful and advantageous, by enabling students to cope with negative experiences and envision their future within the STEM sector without any limiting factors and hesitation [34]. One such in-formal mentoring program, PROGRESS, was described by [53], and included work-shops, an online peer support community, and a large network of female mentors that participants were connected with. It lasted for an academic semester and all participants were first-year and second-year female undergraduates. Guidance from faculty members was proved to be definitive, since it strengthened the scientific identity of participants and increased their interest in following a career in environmental sciences.

Women who work in STEM, have repeatedly confessed in interviews that mentorship worked as an important motivation for them during the demanding years of university. It encouraged them to keep trying to achieve their goals [34, 54]. On the other side, [55] found that negative experiences may arise between mentor and protégé, like in any other relationship. Their results showed three types of damaging experiences: Protégé performance problems, interpersonal problems, and destructive relational patterns. However, findings by [56] and [31], reveal that underrepresented students, like for example females in STEM courses and degrees, need ongoing social and educational support from counselors and mentors. The study of [57] adds that there is a need for peer mentoring programs that will connect self-doubting achievers, who do not believe in their abilities, in STEM courses, with high achievers. Interventions including a sense of social belongingness, such as the exposure to female peers who are excelling in STEM, can help women who fit in the self-doubting achiever profile [57].

In [58] study, mentoring was found to be a common strategy for retaining women scientists, since many of them express a need for it, and suggest that it is unavailable, or even missing. Psychosocial mentoring appears to be the most beneficial form of mentoring to women since it can build self-esteem, enable women to both promote themselves academically and within the research arena and make their voices and choices heard [59].

4 Discussion - Conclusion

Numerous factors can cause women's disadvantage regarding their engagement with STEM and their persistence in a stable and successful career. Deterring circumstances could emerge from the women's social environment, starting from the family, the academic or the professional sphere. The collection of materials and tools provided in this study can contribute to the confrontation of the above-mentioned stumbling blocks and work as an inspiring, encouraging and supportive stimulus for every girl who is keen on sciences and technologies. Towards this direction, many researchers have studied and suggested specific practices that can be used within the context of formal or informal education and provide beneficial and constructive outcomes for the emboldening of young female aspiring STEM achievers.

From the collection of materials and tools we have noticed a distinction of the database in two categories. First, materials and tools that are digital and are available (either free or on a paid basis) through the web and secondly physical materials and tools. Moreover, a classification of the materials and tools collected reveals their primary aims: to expose participants to role models, raise awareness, encourage train, mentor or empower.

Women's under-representation and underestimation in STEM domains does not belong in the past [60]. Instead, it has been revealed that it constitutes a contemporary problematic situation, with significant negative impact on womens' life, career choices, productivity and effectiveness. Because of this unwelcoming environment, young female students are reluctant to choose to pursue further education in STEM subjects or follow a STEM-related career path [14–21].

This happens not only due to the large number of challenges females have to face within these sectors, but also, due to ingrained gender stereotypes. Gender stereotypes are not only adopted by the females' social surrounding, but most importantly, these types of prejudices do not allow girls to perceive themselves as potential scientists. The qualities, attributes, and general characteristics of femininity are usually associated with social sciences and humanities, and contrariwise, masculinity has always been more related to science and engineering. Therefore, STEM classrooms and workplaces ended up being male dominated, while at the same time, females who are bold enough to take this challenge and enter these fields, have to bear with dismissive and unwelcoming treatment that often has repercussions on their confidence, resilience, and aspirations.

The international academic community has acknowledged this pressing issue and proceeded with various researches and projects towards the confrontation of the problem, and the mitigation of the gender gap in STEM. The focus of the studies is geared towards the motivation of young women through their engagement in the learning procedure and in STEM related activities. Trainings, camps and skill-building programs are related to active learning methods. In this case the learner is more included, engaged and energetic during the educational procedure. The student becomes the creator and the generator of new ideas and learns science through action-oriented activities. The new skills students will gain will make them more prepared and more self- confident for their demanding journey. Mentoring provides support, encouragement and motivation for the female students, and helps them develop their confidence within the academic and professional environments. The exposure to role models is also a motivation for young girls in STEM, since they get the opportunity to interact with women who faced the similar, if not the same, difficulties as they do, and they managed to succeed against all odds. Shining examples of women who already work in STEM can inspire youngsters and make them more determined and persistent to overcome barriers in order to accomplish their goals. Moreover, audio-visual material has the potential to function as a knowledge enrichment tool, for topics related to gender-sensitivity, in order to positively affect students' attitudes. Role models and video interventions do not require high levels of engagement of the student but are significantly helpful since they can provide important support through knowledge and raise awareness for specific issues.

Women tend to be more vulnerable than men when it comes to choosing a STEM degree and persist in a field-related career path, there is a persistent need for effective measures that should be taken in order to make STEM welcoming for both genders equally. The tools provided in this study can help teachers in higher education institutes to develop an informed conceptualization of inclusion and contribute to the establishment of contemporary and suitable teaching methods that will take into account gender-sensitivity issues. This study has also identified a number of issues that should be taken into consideration when using or designing gender-sensitive materials tools. Based on the analysis of the present situation and circumstances, the following components should have a fundamental role when designing gender-sensitive approaches:

- Provide support for the teachers in the understanding and application of the principles of a gender-sensitive teaching methodology.
- Help teachers become familiar with the technological and pedagogical resources
- Provide the teachers the means to assist students in making shareable exhibits and promoting gender-equality issues.
- Provide support for the students in the understanding of the gender-equality issues.
- Motivate female students to remain active in STEM.
- Engage students in the learning procedure as this will enable the better understanding of the material and help them to be able to remember most of it.
- Mentoring is considered as an important motivation for women to remain active in STEM.
- Identifying and promoting female role models is a strategy that can attain gender balance in STEM.

Acknowledgment. This work has been funded by the European Union's Erasmus Plus programme, grant agreement: 2019–1-CY01-KA203–058407 (Project: FeSTEM). This publication reflects the views only of the authors, and the European Commission cannot be held responsible for any use which may be made of the information contained there.

References

1. Woodward, A.E.: Going for gender balance (2002)
2. Unistats, United Nations Stat. Div. (2020). https://unstats.un.org/home/
3. Omotosho, B.J.: Gender balance. In: Idowu, S.O., Capaldi, N., Zu, L., Gupta, A.D (Eds.) Encyclopedia of Corporate Social Responsibility. Springer, Berlin, Heidelberg, pp. 1195–1204 (2013). https://doi.org/10.1007/978-3-642-28036-8_624
4. UNECA, Proceedings of Expert Review of the African Women's Report, in: Addis Ababa (2009)
5. Kouta, C., Parmaxi, A., Smoleski, I.: Gender equality in academia, business, technology and health care: a womenpower view in Cyprus. Int. J. Caring Sci. **10**, 1224–1231 (2017). www.internationaljournalofcaringsciences.org
6. Woetzel, J., et al.: The power of parity: advancing women's equality in Asia pacific, Shanghai (2018). www.mckinsey.com/mgi. Accessed 26 Mar 2021
7. Gerson, K.: The unfinished revolution: how a new generation is reshaping family, work, and gender in America (2010)
8. Djerf-Pierre, M.: The difference engine: gender equality, journalism and the good society. Fem. Media Stud. **11**, 43–51 (2011). https://doi.org/10.1080/14680777.2011.537026

9. Djerf-Pierre, M., Edstrom, M.: Comparing Gender and Media Equality across the Globe, GEM (2020). https://www.gu.se/en/research/comparing-gender-and-media-equality

10. Grizzle, A.: Enlisting media and informational literacy for gender equality and women's empowerment. Media Gend. A Sch. Agenda Glob. Alliance Media Gender. 79–91 (2014)

11. Morna, C.: Promoting Gender Equality In And Through The Media (2002). https://caluniv.ac.in/global-mdia-journal/Documents/D.4. SOUTH AFRICA 2002.pdf

12. Padovani, C.: Media gender-equality regimes: exploring media organisations' policy adoption across nations (2020)

13. Ross, K., Padovani, C. (eds.) Gender Equality and the Media: A Challenge for Europe - Google Books, Routledge (2016). https://books.google.com.cy/books?hl=en&lr=&id=tqauDAAAQBAJ&oi=fnd&pg=PP1&dq=Ross,+K.,+%26+Padovani,+C.(2016).+Gender+equality+and+the+media:+A+challenge+for+Europe.+Routledge.&ots=loxtdQnbsr&sig=sXpBGe6MqJKWB5JiTb8kQZEd8t0&redir_esc=y#v=onepage&q=Ross%2CK.%2C%26Padovani%2CC.&f=false. Accessed 26 March 2021

14. Badaloni, S., Brondi, S., Contarello, A., Manganelli, A.: The threatened excellence. Reasoning about young women's scientific and technological careers in Padua University, Italy. Sci. Technol. Careers Women Men. 119–136 (2011). http://scholar.google.com/scholar?hl=en&btnG=Search&q=intitle:Threatened+excellence:+Reasoning+about+young+women+'+s+scientific+and+technological+careers+in+Padua+Italy#0

15. Caroni, C.: Graduation and attrition of engineering students in Greece. Eur. J. Eng. Educ. **36**, 63–74 (2011). https://doi.org/10.1080/03043797.2010.539676

16. Eurostat, (2018). https://ec.europa.eu/eurostat/. Accessed 26 March 2021

17. OECD, Women in scientific production (2013). http://www.oecd.org/gender/data/women-in-scientific-production.htm. Accessed 26 March 2021

18. OECD, Education at a Glance 2018. Indicator B5: Who is expected to graduate from tertiary education? 206–216 (2018). https://doi.org/10.1787/eag-2018-18-en

19. OECD, Key charts on education (2018). http://www.oecd.org/gender/data/education/#d.en.387789. Accessed 26 March 2021

20. OECD, Mathematics performance (PISA) (indicator) (2021). https://doi.org/10.1787/04711c74-en

21. GSGE, General Secretariat for Gender Equality (2019). https://www.isotita.gr. Accessed 3 March 2020

22. Oculus, Breaking Boundaries in Science (2018). https://www.oculus.com/experiences/go/1973697659322414/

23. Parmaxi, A., Vasiliou, C.: Communities of interest for enhancing social creativity: the case of Womenpower platform (2015)

24. Dawson, A.E., Bernstein, B.L., Bekki, J.M.: Providing the psychosocial benefits of mentoring to women in STEM: career WISE as an online solution. New Dir. High. Educ. **2015**, 53–62 (2015). https://doi.org/10.1002/he.20142

25. P. International, Shashing gender stereotypes with She board app (2017). https://plan-international.org/smashing-gender-stereotypes-sheboard-app

26. Nevertheless, STEM role models posters (2018). https://medium.com/nevertheless-podcast/stem-role-models-posters-2404424b37dd

27. Lego, Women of NASA (2019). https://www.lego.com/en-gb/product/women-of-nasa-21312

28. Google India, Ritu Karidhal - The Rocket Woman of India | Women in STEM - YouTube (2018). https://www.youtube.com/watch?v=GpMWetdGS_Q&list=PL-kIBfSqQg3upn0g68pKuLBApO59lRqz5&index=12. Accessed 17 Feb 2022

29. Shin, J.E.L., Levy, S.R., London, B.: Effects of role model exposure on STEM and non-STEM student engagement (2016)

30. Barabino, G., et al.: Solutions to gender balance in stem fields through support, training, education and mentoring: report of the international women in medical physics and biomedical engineering task group. Sci. Eng. Ethics. **26**, 275–292 (2020). https://doi.org/10.1007/S11 948-019-00097-0/TABLES/1

31. Bystydzienski, J.M., Eisenhart, M., Bruning, M.: High school is not too late: developing girls' interest and engagement in engineering careers. Career Dev. Q. **63**, 88–95 (2015). https://doi. org/10.1002/j.2161-0045.2015.00097.x

32. Nehmeh, G., Kelly, A.: Women physicists and sociocognitive considerations in career choice and persistence. J. Women Minor. Sci. Eng. **24**, 95–119 (2018). https://doi.org/10.1615/JWo menMinorScienEng.2017019867

33. Cabay, M., Bernstein, B.L., Rivers, M., Fabert, N.: Chilly climates, balancing acts, and shifting pathways: what happens to women in STEM doctoral programs. Soc. Sci. **7**(2), 23 (2018). https://doi.org/10.3390/socsci7020023

34. Amon, M.J.: Looking through the glass ceiling: a qualitative study of STEM women's career narratives. Front. Psychol. **8**, 236 (2017). https://doi.org/10.3389/fpsyg.2017.00236

35. Kizilcec, R.F., Saltarelli, A.J.: Psychologically inclusive design cues impact women's participation in STEM education. In: Proceedings of the 2019 CHI Conference on Human Factors in Computing Systems, pp. 1–10 (2019). https://doi.org/10.1145/3290605.3300704

36. Latu, I.M., Mast, M.S., Bombari, D., Lammers, J., Hoyt, C.L.: Empowering mimicry: female leader role models empower women in leadership tasks through body posture mimicry. Sex Roles **80**, 11–24 (2019). https://doi.org/10.1007/S11199-018-0911-Y/TABLES/3

37. Lockwood, P., Sadler, P., Fyman, K., Tuck, S.: To do or not to do: using positive and negative role models to harness motivation. Soc. Cogn. **22**, 422–450 (2005). https://doi.org/10.1521/ Soco.22.4.422.38297

38. Nobel Prize Outreach, Women who changed the world (2022). https://www.nobelprize.org/ prizes/lists/nobel-prize-awarded-women/

39. Moss-Racusin, C.A., et al.: Reducing STEM gender bias with VIDS (Video Interventions for Diversity in STEM). J. Exp. Psychol. **24**, 236–260 (2018)

40. Pietri, E.S., et al.: Using video to increase gender bias literacy toward women in science. Psychol. Women Q. **41**, 175–196 (2017). https://doi.org/10.1177/0361684316674721

41. Canali, C., Moumtzi, V.: Digital girls summer camp: bridging the gender ICT divide, Institutional Chang. Gend. Equal. Res. (2019). https://edizionicafoscari.unive.it/it/edizioni/collane/ scienza-e-societa/

42. Weber, K.: Gender differences in interest, perceived personal capacity, and participation in STEM-related activities. J. Technol. Educ. **24**, 18–33 (2012). https://doi.org/10.21061/jte.v24 i1.a.2

43. Burge, J.E., Gannod, G.C., Doyle, M., Davis, K.C.: Girls on the go: a CS summer camp to attract and inspire female high school students. In: Proceeding of the 44th ACM technical symposium on Computer science education, pp. 615–620 (2013). www.appcelerator.com. Accessed 17 Feb 2022

44. Banister, S., Ross, C.: Creating an engaging app development course for girls: catalyzing young women's interest and abilities in STEM. In: Society for Information Technology & Teacher Education International Conference (2017)

45. Stapleton, S.C., et al.: Girls tech camp: librarians inspire adolescents to consider STEM careers. Issues Sci. Technol. Librariansh. (2019). https://doi.org/10.29173/istl22

46. Hughes, R.: An investigation into the longitudinal identity trajectories of women in science, technology, engineering, and mathematics. J. Women Minor. Sci. Eng. **21**, 181–213 (2015). https://doi.org/10.1615/JWomenMinorScienEng.2015013035

47. Nevill, D.D., Schlecker, D.I.: The relation of selfefficacy and assertiveness to willingness to engage in traditional/nontraditional career activities. Psychol. Women Q. **12**, 91–98 (1988). https://doi.org/10.1111/j.1471-6402.1988.tb00929.x

48. Van Oosten, E.B., Buse, K., Bilimoria, D.: The leadership lab for women: advancing and retaining women in STEM through professional development. Front. Psychol. **8**, 2138 (2017). https://doi.org/10.3389/FPSYG.2017.02138/BIBTEX

49. Saddiqui, S., Marcus, M.: STEAMpunk girls co-design: exploring a more integrated approach to STEM engagement for young women. In: 28th Annual Conference of the Australasian Association for Engineering Education (AAEE 2017), Australasian Association for Engineering Education (2017). https://search.informit.org/doi/abs/https://doi.org/10.3316/INFORMIT.395738434588372. Accessed 17 Feb 2022

50. Kim, H.: Inquiry-based science and technology enrichment program for middle school-aged female students. J. Sci. Educ. Technol. **25**, 174–186 (2016). https://doi.org/10.1007/S10956-015-9584-2/TABLES/4

51. Bullough, R.V., Draper, R.J.: Making sense of a failed triad: mentors, university supervisors, and positioning theory. J. Teach. Educ. **55**, 407–420 (2004). https://doi.org/10.1177/002248 7104269804

52. STEMconnector, Million Women Mentors (2020). https://www.millionwomenmentors.com/facts

53. Hernandez, P.R., et al.: Role modeling is a viable retention strategy for undergraduate women in the geosciences. Geosphere. **14**, 2585–2593 (2018). https://doi.org/10.1130/GES01659.1

54. Thomas, M.: Exploring the Advancement of Women in Science, Technology, Engineering, and Mathematics (STEM) Executive Management Positions in the Aerospace Industry: Strategies Identified by Women That Enable Success - ProQuest, University of La Verne (2017). https://www.proquest.com/docview/1957409367?pq-origsite=gscholar&fromopenview=true. Accessed 17 Feb 2022

55. Eby, L.T., Allen, T.D., Evans, S.C., Ng, T., DuBois, D.L.: Does mentoring matter? a multidisciplinary meta-analysis comparing mentored and non-mentored individuals. J. Vocat. Behav. **72**, 254–267 (2008). https://doi.org/10.1016/J.JVB.2007.04.005

56. Dennehy, T.C., Dasgupta, N.: Female peer mentors early in college increase women's positive academic experiences and retention in engineering. Proc. Natl. Acad. Sci. USA **114**, 5964–5969 (2017). https://doi.org/10.1073/PNAS.1613117114/-/DCSUPPLEMENTAL

57. Robnett, R.D., Thoman, S.E.: STEM success expectancies and achievement among women in STEM majors. J. Appl. Dev. Psychol. **52**, 91–100 (2017). https://doi.org/10.1016/J.APP DEV.2017.07.003

58. Thomas, N., Bystydzienski, J., Desai, A.: Changing institutional culture through peer mentoring of women STEM faculty. Innov. High. Educ. **40**, 143–157 (2015). https://doi.org/10.1007/S10755-014-9300-9/TABLES/2

59. Obers, N.: Career success for women academics in higher education: choices and challenges, South African. J. High. Educ. **28**, 1107–1122 (2014). https://doi.org/10.10520/EJC159132

60. Parmaxi, A., et al.: Understanding the challenges and expectations of women in science, technology, engineering, and mathematics: the academic and industrial perspective. In: 5th Annual International Technology Education Development Conference, pp. 8–9 (2021)

A Multidisciplinary Approach to Leadership During the COVID 19 Era. The Case of Romania

Adela Coman(✉), Mihaela Cornelia Sandu, Valentin Mihai Leoveanu,
and Ana-Maria Grigore

University of Bucharest, Bucharest, Romania
{adela.coman,mihaela.sandu,valentin.leoveanu,
ana.grigore}@faa.unibuc.ro

Abstract. The COVID 19 pandemic hit on a global scale: more than 219 countries were affected. Governments are struggling to raise awareness regarding the need to get vaccinated. Efforts have been made relentless since 2019 to prevent the infection globally. How to increase the vaccine acceptance rate proved to be a challenging task for governments since vaccination is the only efficient measure to efficiently protect people from disease. According to various studies, it is necessary to vaccinate about 90% of the population to create herd immunity and decelerate the transmission of the virus. According to the numbers published by the Romanian Health Ministry (January, 2022), 40.9% of the population got vaccinated. In terms of age segments, about 35% of people between 19 and 25 years of age had at least one dose of vaccine while the new variant of COVID (Omicron) spreads and infects about 25.000 each and every day (as of January, 2022. We all need to better understand why some people prefer to vaccinate while others oppose to vaccination. The focus is on raising the vaccination rate and therefore, public health officials and governments work on finding the strategy to impact people so that they decide to vaccinate.

While vaccination questions trust in government officials, strategies and policies, recent studies (Hui 2020) show that in times of crisis such as the COVID 19 pandemic is also a matter of leadership and leadership authenticity. Some evidence found in the literature suggests that authentic leadership is increasingly in demand (source). Therefore, we looked into the issue of authentic leaders – what are the traits, skills, and behaviors that people associate with authentic leadership in the specific context of (COVID) crisis leadership.

The research methodology is based on the construction of a questionnaire structured on types of questions related to what authentic leadership means in relation with the present pandemic. The second part of the questionnaire looked into the beliefs and behaviors of our respondents toward vaccination. Data were collected and analyzed. The model of authentic leadership that emerged and the formulated hypotheses were tested followed by a statistical analysis, interpretation and conclusions.

Keywords: Authentic leadership · Pandemic · Influence · Trust · Public policies

© The Author(s), under exclusive license to Springer Nature Switzerland AG 2022
G. Meiselwitz (Ed.): HCII 2022, LNCS 13315, pp. 503–516, 2022.
https://doi.org/10.1007/978-3-031-05061-9_35

1 Introduction

Corona virus disease (COVID 19) is an infectious illness caused by severe acute respiratory syndrome coronavirus 2 (SARS-CoV-2). As the World Health Organization warned, the whole world was affected by this syndrome, and because of the magnitude of the disease and the speed of propagation, it is considered to be a pandemic. Governments have tried to warn citizens about the dangers associated with the pandemic and adopt the necessary measures to protect people and economies. And yet, it seems that in many countries, governments have failed, therefore distrust in their ability of efficiently handle the pandemic is on the rise. (OECD). As a direct effect of distrust, the citizens became skeptical regarding the usefulness of the imposed new rules and regulations, questioning the competence of the public health and officials and the possibility of a quick return sooner rather than later to a normal economic and social life. Moreover, the pandemic has triggered widespread disinformation that has undermined both understanding and acceptance of science and public policies (de Figueredo et al. 2020), and this extends to the issue of vaccine acceptance. Despite widespread recognition that COVID 19 is a critical issue to people all around the globe, many remain unwilling to be vaccinated, However, in February 2021, an average of 76% of the population across 11 OECD countries indicated willingness to be vaccinated, an increase from only 66% in December, 2020 (Ipsos 2021). The international consulting company Kantar shows in a study that "recent data from seven OECD countries showed that a quarter of the population in France, Germany and the United States may refuse COVID 19 vaccination, and an even higher proportion among younger population cohorts. More than 50% of French 25- to 34-year-old and one third of Dutch 25- to 34-year-old said they would probably or definitely not get vaccinated" (Kantar 2021).

Authentic leadership and trust in government seem to be missing in the Romanian public space (source). In spite of the fact that government officials (doctors and medical experts) repeatedly tried to raise public awareness of people and communities, the vaccination rate is still low in Romania: approximately 50% of the population got a vaccine (by February 2022), placing this country at the bottom of the hierarchy in the EU (right next to Bulgaria). According to NIS (s), the age cohort of 19- to 25-year-old is that expressed interest in vaccination (35%) is very low among the young people from the rural areas (less than 10%).

The intent of our study is to research the significance and the meaning of authentic leadership for the younger generation considering the COVID 19 pandemic crisis. Our research questions are: (1) What are the features, skills and behaviors that young people appreciate most in authentic leaders? (2) Who are the personalities they think of as authentic (people they respect and identify with) while constantly following them on social media networks? Given the fact that the decision to vaccinate is influenced by psychological factors/personal and social factors (Catoiu and Teodorescu 2001), we advanced the idea that authentic leaders may be key to successful vaccination campaigns.

This investigation is structured in three parts: the first part considers the literature review related to the concept of authentic leadership and to the COVID leadership crisis; the second part is dedicated to the methodology and statistical analysis of data. A number of questions from the MLQ (Walumba, Avolio et al. 2008) and ALQ (Bass, Avolio) were used and adapted and based on our respondents' answers, we designed a model of authentic leadership in the context of the COVID 19 pandemic. Both model and hypotheses were tested, followed by discussion and interpretation of results.

2 Literature Review

According to Walumba, Avolio et al. (2008, p.7), authentic leadership is "a pattern of leader behavior that draws upon and promotes both positive psychological capacities and a positive ethical climate to foster greater self-awareness, an internalized moral perspective, balanced processing of information and relational transparency on the part of leaders working with followers, fostering positive self-development". It reflects an interactive and authentic relationship that develops between leaders and followers (Gardner, Avolio, Lufthans et al. 2008). In other words, authentic leaders show to others what they genuinely desire to understand their own leadership to serve others more effectively (George 2003). This author says "They act in accordance with deep personal values and convictions to build credibility and win the respect and trust of followers". By encouraging diverse viewpoints and building networks of collaborative relationships with followers, they lead in a manner that followers perceive and describe as authentic (Avolio et al. 2004).

Studies (Avolio 1999; Bass 1998) show that authentic leadership is placed at the intersection of authentic and transformational leadership. Therefore, there are some conceptual overlaps between the two. While transformational leadership is about charisma, idealized influence, motivation, intellectual stimulation and individual consideration, authentic leadership is *a process* that draws from both positive psychological capacities of the leader *and* a highly developed organizational context which result in both greater self-awareness and self-regulated positive behaviors on the part of leaders and associates (Mumford et al. 2012)).

What does it mean to have an idealized influence? A leader who manifests as such think that their followers' needs are more important than their own; risk-sharing is guided by accepted principles; and moral values are the foundation of each and every decision. Such leaders are role models for followers to emulate; can be counted on to do the right thing; and display high standards of ethical and moral conduct (Avolio 1999; p. 43). Finding meaning in work and carefully listening and understanding their followers' problems and challenges inspires followers and motivates them. Meaning-making and inspiration are thus part of inspirational motivation. Employees who are open to find new and creative solutions to old and new problems are intellectually stimulated by their leaders and actively contribute to a stronger organizational culture. Individual consideration means at least two things: leaders act as coaches/mentors for their followers: if employees feel supported in their personal and professional growth, if they are encouraged to learn, they will be more productive and willing to contribute to both their own and the company's growth.

Authentic leaders lead by example (Avolio et al. 2004). Honest leaders inspire adherents to drive their actions based on integrity and honesty. And because of that, followers perceive them as models to be followed so much so that they come to respect and embrace the same values and standards as the leaders. Beliefs about integrity and fair-play become self-defining and incorporated into followers' system of beliefs. In this context, followers end up by perceiving themselves as being moral persons.

The context of the pandemic challenges authentic leadership. Previous studies (Deitchman 2013) sought to answer the question of which leader attributes are needed for successful crisis leadership within the context of public health emergencies. The author highlights that "The attributes identified for leaders in public health to successfully navigate a crisis include knowledge of public health science, situational awareness, decisiveness, communication and the ability to build trust and coordinate diverse groups of people".

Du Brin (2013) explores "the specific traits, attributes and behaviors of leaders that proved to be effective in crisis". The author says "Effectiveness in this context includes both bottom-line factors, such as productivity, and employee factors such as morale. Through an examination of theory and research, the author proposed a set of attributes that define successful crisis leadership: crisis leaders are strategic, charismatic and have high emotional intelligence, meaning they are able to both express and understand emotion. As a result, necessary behaviors for effective leaders in crisis emerge (source). These behaviors include strong communication and a directive style of leadership. Decisiveness, compassion and agility were also found to be critical, personal attributes for leaders to possess in order to be effective in crisis".

Bennis and Thomas (2002) suggest that "a bi-directional relationship exists between leaders and crisis: leaders have an impact on crisis and crisis can make leaders". Moreover, crisis "opens up the possibility of self-discovery and growth for leaders".

Beyond who leaders are as individuals and what they do during difficult times, Stern (2013) asserts that a leader needs to carefully prepare for navigating through challenging events such as crises. To be prepared and to be ready to successfully lead your team toward your desired goal and vision is about sense-making and decision-making. It is about being a student and learning along the way. Preparedness is part of crisis leadership in action. Preparedness refers to sense-making, decision-making, meaning-making and learning along the way. How the manifests himself in a crisis defines who the leader is. According to Gigliotti's research (2020), a crisis shapes and develops a leader's identity. This observation is consistent with the findings of Bennis and Thompson (2002).

3 Methodology

The research methodology is based on the construction of a questionnaire structured on types of questions related to what authentic leadership means in correspondence with the present pandemic. A second part of the questionnaire looked into the beliefs and behaviors of our respondents toward vaccination. Data were collected and analyzed. The model of authentic leadership that emerged and the formulated hypotheses were tested followed by a statistical analysis and interpretation.

Our research questions were: (1) What are the features, skills and behaviors that young people appreciate most in authentic leaders? (2) Who are the personalities they think of as authentic (respect and identify with) while constantly following them on social media networks? Given the fact that the decision to vaccinate is influenced by membership groups (e.g., family and friends) and reference groups (Catoiu and Teodorescu 2004), we advanced the idea that influencers such as the authentic leaders identified by our respondents are best fit to have an influence in the official campaigns pro-vaccination in Romania that proved unsuccessful so far.

The trait approach in leadership has generated much interest among researchers for its explanation of how traits influence leadership (Bryman 1992). For example, Kirkpatrick and Locke (1991) went so far as to claim that effective leaders are distinct types of people. Researchers also looked into the social intelligence of leaders – as an ability to understand behaviors, feelings and thoughts- their own thoughts and the followers' and then act accordingly and appropriately (Zaccarro 2002). Lately, Zaccaro et al. (2018) found that personality traits and capacities contribute to make the difference between WHO emerges as a leader and to a better understanding that a leader's traits and abilities are essential for a leaders' effectiveness. Some of the traits that are consistent with the majority of studies regarding leadership are: intelligence, self-confidence, determination, integrity and sociability (Northouse 2018). Leaders needs certain skills/competencies that are put to work in the leadership process, such as the ability to communicate and manage employees and the ability to adapt to changing conditions (Boin et al. 2013). As a consequence, a leader's authority derives from his competencies/skills that he manifests as the company faces uncertainty and risks in difficult situations (crises).

The trait approach does not clearly define a body of principles regarding the type of leader needed in certain situations or the actions needed to be taken by him; in contrast, the trait approach highlights the essential feature of effective leadership, namely that of having a leader with a kit of traits and skills in this regard. It is the leader and the leader's traits and skills that are central to the leadership process (Northouse 2018).

Therefore, we advance the hypothesis that a leader's authority/skills correlate with a leader's traits (characteristics) and:

H1: There is a positive two-way relationship between an authentic leader's skills (authority) and traits.

The skills approach in leadership – generally emphasizes the skills and abilities that a leader can learn and develop, skills that are needed for effective leadership. There are numerous researches that highlight the relevance of individual attributes in leadership. For example, Mumford and colleagues developed a skill model of leadership based on 5 components: competencies, individual attributes, career experiences, environmental influences and leadership outcome (Mumford et al. 2000).

A leader's authority is derived largely from his skills and competencies accumulated as a result of his past experiences. Individual attributes such as creativity may stimulate the leaders to work and acquire new skills or make better the ones that he already possesses. According to Mumford et al. (2012), problem-solving skills, social judgment skills and knowledge are the key-factors that account for effective performance, although individual attributes, career experiences and environmental influences have also an impact on leader competencies. Challenging situations may improve the leaders' problem-solving skills and by thus, raising their effectiveness as authentic leaders.

Since skills and competencies drive leaders to manifest specific behaviors, we ascertain that a leader's skills trigger behaviors, with a bidirectional, positive correlation between the two. Therefore:

H2: There is a bidirectional relationship between the authentic leaders' authority (skills) and behavior.

Catastrophic events such as floods and hurricanes, earthquakes and global pandemics require specific and prompt responses from leaders. In this respect, the devastating events of hurricane Katrina in 2005 serve as a case of crisis leadership. Admiral Thad Allen was one of the leaders who took charge and based his strategy on "direct communication, quick thinking and willingness to try new and untested approaches" (Kearns et al. 2019). Kearns concluded that trust-building, empathy and accountability together with a directive style of leadership translated into more effectiveness in crises response (Kearns et al. 2019). Challenging situations drive leaders towards more flexibility in their behavior (s) while helping to successfully navigate the crises.

Therefore, we propose that:

H3: There is a two-way relationship between situations and the behavior of leaders.

Integrity and accountability (leadership traits) act as drivers of leadership behavior in the COVID 19 pandemic (source). A leader's traits influence and transforms behaviors in many ways. Everyday decisions imply risk-taking to a certain degree. A recent study found that this crisis determined leaders to strengthen relationships with employees and other stakeholders, to express emotions, empathy and willingness to help people in need (OECD 2020). The behavior of leaders induces the followers to identify with the leader and his vision by transferring higher level values, such as altruism. As a result, they work harder to improve their performance at work.

On the other hand, a directive leadership style was found to be prevalent in the present pandemic since new norms and new rules need to be followed in order to protect public health (source). Uhl-Bien (2014) highlight a model in which leaders influence their followers in two ways. First, the system of values that the leader embodies tend to motivate followers to act according to those values. The second way is about building a collective identity based on the unique identities of the followers: How they view themselves is triggered by the leader's actions. Therefore, values and self-identity mediate the linkage between the leader's traits, and the followers' behavior.

In this respect, we propose that:

H4: A leader's traits directly influences behaviors (the leader himself and followers).

The COVID 19 epidemic raised many challenges for leaders and followers alike. Work-from-home became a "must" in 2021 in particular, when the spread of the virus took its toll in peoples' lives. Government imposed restrictions and sanctions were administered against those who did not conform to the rules.

Against the backdrop of COVID 19 and the struggle the companies faced in making decisions to adapt through reinventing business models or shutting down either temporarily or permanently, Fox et al. (2020) explored the relationship between authentic leadership, corporate social responsibility (CSR) and flexibility models adopted by businesses in crisis. Their findings suggest that authentic leaders are successful in steering their firms through crises such as COVID 19, because they are more engaged with stakeholders (Fox et al. 2020). The degree to which a firm enjoys a flexible business model also was reflective of agility and stakeholder engagement (Fox et al. 2020). Considering COVID-19 pandemic crises, Fox et al. reveal several roles and skills had to be uncovered or learned by leaders such as: assessing risks, building relationships, actively communicating, role – modeling a proactive culture and adapting new standards and processes while working-from-home.

Therefore, we propose that:

H5: Situations (crises) directly influence a leader's skills.

4 Statistical Analysis

We used the MLQ questionnaire (Bass, Avolio 2004) and ALQ questionnaire ((Avolio et al. 2008) as a starting point for the construction of our model. We then tested the relationships between the components of an authentic leader - behavior, skills, traits and situations. Situations in this case refers to the specific crisis context created by the COVID 19 pandemic. Therefore, we considered the following model (Fig. 1):

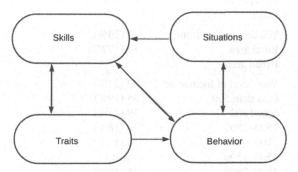

Fig. 1. Authentic leadership model to be tested

A summary of our hypotheses is presented below:

H1: There is a two-way relationship between the skills and the traits of leaders.

H2: There is a two-way relationship between the skills and behavior of leaders.

H3: There is a two-way relationship between situations and the behavior of leaders.

H4: Leaders' traits directly influence leader's and followers' behavior.

H5: Situations directly influence leaders' skills.

5 Data

The sample consists in 185 respondents aged between 18 and 54 years, 59% undergraduate or master's degree students within the University of Bucharest; 148 are women and 37 are men (Table 1).

Table 1. Descriptive statistics

Variable	Statistics
Age:	min = 18, max = 54
Gender:	148 (80%)
Female	37 (20%)
Male	
Your occupation is:	117 (63%)
Student	59 (32%)
Employee	2 (1%)
Student & employee	1 (0.5%)
Entrepreneur	6 (1.5%)
Other	
You are coming from:	42 (23%)
Rural area	143 (77%)
Urban area	
Your level of income is:	39 (21%)
Less than 500	35 (19%)
500–1500	39 (21%)
1500–2500	30 (16%)
2500–3500	15 (8%)
3500–4500	9 (5%)
4500–5500	18 (10%)
More than 5500	

Source: our elaboration using R Studio based on questionnaire answers

6 Statistical Method Used

Walumba, Avolio et al. (2008) recommend the model with structural equations in order to test hypotheses. When testing a predefined model, at the beginning of the analysis we will perform the confirmatory analysis (CFA) and then the model itself, respectively path analysis.

7 The Results of the Statistical Analysis

To test whether the data is suitable for factor analysis and implicitly for the construction of a model with structural equations, we will calculate the Kaiser-Mayer-Olkin coefficient (KMO). We find that the value of this coefficient is 0.9 therefore we can say that based on this test, a factor analysis can be built.

Another test that helps us determine if we can build a factor analysis based on our sample is the Bartlett Sphericity Test. The corresponding p-value is $1.809323e-249 < 0.05$ which indicates that factorial analysis can be performed.

The determinant of the correlation matrix can also be calculated. Its value $2.755388e-05$ is positive which also indicates that the factorial analysis can be performed.

The model we will test has 4 latent variables obtained from the 22 questions of the questionnaire related to the 4 components of our model - behavior, skills, traits and situations. We will first perform confirmatory factor analysis (CFA) to see if our data support the model being tested (Table 2).

Table 2. CFA fitting values for baseline model

Indicator	Expected value	Value in the model
Convergence & number of iterations		Yes, 46 iterations
Observations	As big as possible	185
p-value, Chi-square	<0.05	0.000
CFI	>0.95	0.832
TLI	>0.95	0.808
RMSEA	<0.07	0.018
90% Confident Interval	(0; 1)	(0.078; 0.097)
SRMR	<0.08	0.064
AIC	As small as possible	7642.017

Source: our elaboration using R Studio based on questionnaire answers

8 Adequacy Test

The root means the square of residuals (RMSR) is 0.04; this is acceptable as this value should be closer to 0.

The Tucker-Lewis Index (TLI) is 0.808 an acceptable value considering that this value is close to 0.9.

Next, the authors check the root mean square error of approximation index which has the value of 0.018 showing that this is a good model fit as it is indicating a value below 0.05.

In order to see if the established hypothesis is accepted or rejected, the authors consider that for a positive estimate value in the table above, there is a direct or positive relation between the variables. To determine if the relation between variables is statistically significant, the authors compare p-value for each hypothesis with the critical value 0.05 (for p-value it is less than 0.05, the relation is statistically significant) (Table 3).

We see that all 5 hypotheses are statistically accepted. Based on this observation, we then built the tested model with structural equations – presented in Fig. 2.

Table 3. SEM indices and decision for hypothesis tested

Hypothesis	Estimate	Standard error	z-value	P	Decision
H1	0.984	0.057	26.543	0.000	Accepted
H2	1.047	0.035	30.230	0.000	Accepted
H3	0.976	0.025	39.816	0.000	Accepted
H4	0.939	0.044	21.575	0.000	Accepted
H5	1.038	0.039	26.543	0.000	Accepted

Source: our elaboration using R Studio based on questionnaire responses

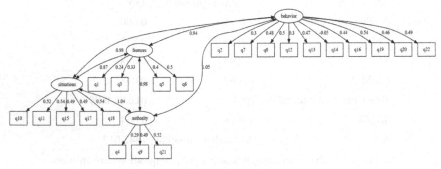

Fig. 2. Structural equation model

9 Interpretation of Results and Discussion

Our questionnaire was built on the concept of authentic leadership as developed in prior research and from the MLQ (Bass, Avolio 2004) and ALQ (Authentic Leadership Questionnaire) of Avolio et al. (2008).

Based on these two instruments, we designed a number of questions in order to research what authentic leadership means in times of crisis for our respondents. Our results show that 67% of our respondents associate authentic leaders with personality traits such as; (inspiring) trust, integrity, and strong moral and ethical values, while 74% think that skills and abilities such as effective communication, energy and an entrepreneurial spirit in challenging times are essential for a leader. Roughly over 80% of them say that leadership authenticity manifests in the leader's behavior: encouraging creativity and innovation, taking responsibility for one's decisions, efficiency in resource administering (material and human) while sharing risks and caring for the others - are seen as meeting expectations of the followers.

A number of definitive practices in the OECD countries (source) show that leaders are challenged to offer an adequate response to the crisis, to prepare for change, to adapt to change and take decisions that help people navigate through difficult situations. Context matters: this is why trust and effective communication are decisive in successfully navigating the present pandemic (Hu et al. 2020). A good example in this respect is Jacinda Ardern, New Zealand's Prime Minister who has been profiled as an authentic and successful leader, particularly in her ability *to communicate effectively and transparently through the crisis* (McGuire et al. 2020). Trust in leaders and trust in government officials were also central in the New Zealand 's pandemic response (Wilson 2020).

Studies show that people need to be adequately informed about what the COVID crisis means and how it can affect their health: open communication and competent public *communication strategies* can contribute to increasing vaccination rate (source). Governments are important regarding the access in real time to accurate information from trusted sources – institutions and people. For example, in Belgium it is the experts and scientists the ones that were asked to deliver information to the public related to health measures and rules. In Spain, the public officials began to monitor disinformation campaigns. Therefore, actions were initiated and laws were adopted so that the government made sure that the population was properly informed and protected. A different approach was adopted in UK and Finland social media influencers and leaders participated in delivering messages through their social networks. As a result, awareness and acceptance of vaccination increased and so did the number of people who got a vaccine (WHO).

10   The Case of Romania

Romania was severely hit by the first 3 waves of pandemic, the number of people who lost their lives being very high: 62.624 (February, 2022). In 2021, the Romanian government has led an aggressive campaign of informing the population regarding the importance of vaccination and the benefits that vaccination entails – since it improves immunity and protects public health. However, the percentage of persons who took a vaccine is still low of about 50% as of February 2022. Most people who refused to vaccinate live in the rural area. There are also important differences in terms of age cohorts. For example, young people between 20 and 29 years of age register a low rate of vaccination of only 10% while those who range between 40 and 49 years old register a significantly higher number: 19% (February, 2022. Statistics also highlight the fact that Romania ranked first in the world – with 16.6 deaths per million of inhabitants, a terrible signal for authorities regarding the efficiency and effectiveness of health policies adopted.

When asked about their hesitancy to take a vaccine, some of our respondents mentioned the lack of trust in vaccines, misguiding or incomplete information related to the vaccines and future effects on health, and lacking trust in public authorities. Indeed, public officials were not able to raise awareness of the public concerning the emergency and the need for the citizens to take initiative and decide to vaccinate.

One of our respondents explicitly said: "We are overwhelmed with all kind of information about the virus every day. In a TV campaign sponsored by the Romanian government, I heard a doctor saying that each and every one of us will get COVID in the end, but if you have a vaccine, then the manifestations of the illness will be less severe. My next thought was: if we still get COVID, even we have the vaccine, why should I do it? They did not convince me to do anything. I don't think that I want to do it...".

Our research shows what respondents think about authentic leadership and how do they perceive the authentic leader' traits, skills and behaviors in a crisis context. They were asked to name some of the authentic leaders they already follow on their social media networks. Several names were advanced – famous personalities in the Romanian business world: Dan Sucu, Cristian Onetiu, Denisa Mandache and Elena Lasconi seem to be the favorites on the students' list. They are all successful entrepreneurs, except Elena Lasconi, a mayor (but she roots her celebrity in the business environment as well). These people are appreciated for the way they respond to their followers' needs, for empathy and transparency, for honesty and openness in communication, and for their entrepreneurial spirit and success in their businesses.

This is one of the reasons that why we believe that new approaches need to be undertaken in the COVID 19 leadership crisis. A change of perspective on the pandemic translates into a change is strategy (Hu et al. 2020). Therefore, if raising awareness regarding the pandemic is essential for protecting public health, we suggest that the Romanian public officials initiate partnerships with authentic leaders that inspire trust (models for the young in particular) such as the ones nominated by our respondents. May be a change of perspective (a closer look into who is best fit to influence the public) would help the government adopt better communication strategies. Also, it worth testing the idea that with this type of persons acting as influencers and "messengers" while leading a vaccination campaign, could contribute to more engagement, a higher /accelerated vaccination rate among the young and, by consequence, to a healthier and more resilient population.

11 Conclusions

This paper took a multidisciplinary approach to explore authentic leadership in times of crisis: it discusses authentic leadership and crisis leadership; it looks into raising public health awareness and improving public health. It also deals with the role of public authorities in raising the public trust in COVID-19 vaccination.

Our research questions were: (1) What are the traits, skills and behaviors that young people appreciate most in authentic leaders? (2) Who are the personalities they think of as authentic (respect and identify with) while constantly following them on social media networks? Given the fact that the decision to vaccinate is influenced by membership groups (e.g., family and friends) and reference groups (Catoiu and Teodorescu 2004), we advanced the idea that influencers such as the authentic leaders identified by our respondents are best fit to have an influence in the official campaigns pro-vaccination in Romania that proved unsuccessful so far.

We have also proposed a model of authentic leadership in times of crisis – namely the COVID Based on the respondents' perspective on authenticity and what an authentic leader is associated with, a model emerged that reflects the intersection and interdependencies between 4 leadership components: traits, skills, behavior and context/situation, we proposed our own model of crisis leadership: what authenticity means for leaders and how they act in the specific context of COVID 19 We proposed 5 hypotheses and tested their validity (all 5 were valid). Based on a questionnaire adapted from the MLQ and ALQ questionnaires, we found that our respondents associate authentic leadership with traits, skills and behaviors such as: inspiring trust, empathetic, effective communicator, responsibility and integrity, effective and resourceful with an entrepreneurial spirit, creative, caring for the others' needs, a role model for the community. Our respondents identified a number of public personalities as authentic leaders, most of them successful in their businesses, others - suggested by the respondents themselves.

Limitations and future directions of research.

As a major limitation of our research is the small number of our respondents, most of them students. In order to look deeper into what determines people to vaccinate (or to positively respond to a public health issue) needs a broader perspective on the beliefs and motivations of people. Vaccination is a personal act and implies a personal decision. It has a psychological dimension and a social dimension as well. Therefore, more responses are needed from diverse categories of people, different cohorts of age and varied occupations.

This paper contributes to the leadership literature in that it adds new insights into how authenticity is challenged considering the pandemic crisis. We believe that a future direction of research could be related to how to enhance public trust in crisis leadership. Acknowledging that to navigate through a crisis is first and foremost a leadership issue may contribute to understand better what needs to be done. Raising awareness by leading convincing and persuasive official health campaigns by the right people is a must. To what extent authentic leaders may contribute and help the public health officials with successfully reaching their objectives and impacting larger audiences is yet another direction that needs to be researched.

References

Avolio, B.J., Bass, B.M.: Multifactor leadership questionnaire. Manual and sampler set. Redwood: Mindgarden (2004)

Avolio, B.J., Gardner, W.L., Walumba, F.O.: Authentic leadership questionnaire. Mind Garden, Inc. (2007)

Avolio, B.J., Wernsing, T., Gardner, W.L.: Revisiting the development and validation of the authentic leadership questionnaire: analytical clarifications. J. Manag. **44**(2), 399–411 (2018)

Bennis, W.G., Thomas, R.J.: Crucibles of leadership. Harvard Bus. Rev. **80**(9), 39–45 (2002)

Boin, A., Kuipers, S., Overdijk, W.: Leadership in times of crisis: a framework for assessment. Int. Rev. Public Adm. **18**(1), 79–91 (2013)

Comfort, L.K., Kapucu, N., Menoni, S., Siciliano, M.: Crisis-decision making on a global scale: transition from cognition to collective action under threat of COVID 19. Public Adm. Rev. **80**(4), 616–622 (2020)

DuBrin, A.J.: Handbook of Research on Crisis Leadership in Organizations. Edward, Elgar (2013)

De Figueredo, A., et al.: Mapping global trends in vaccine confidence and investigating barriers to vaccine uptake: a large-scale retrospective temporal modelling study. Lancelot **396**(10255), 898–908 (2020)

Eldridge, C.C.: Communication during crisis: the importance of leadership, messaging and overcoming barriers. Nurs. Manage. **51**(8), 50–53 (2020)

Fox, C., Davis, P., Baucus, M.: Corporate social responsibility during unprecedented crises: the role of authentic leadership and business model flexibility. Manag. Decis. **58**(10), 2213–2233 (2020)

Gardner, W.L.: Authentic leadership theory and practice: origins, effects and development. JAI (2009)

Gigliotti, R.A.: Leader as performer; leader as human: a discursive and retrospective construction of crisis leadership. Atlantic J. Commun. **24**(4), 185–200 (2016)

Hannah, S.T., Uhl, M., Avolio, B.J., Cavarretta, F.L.: A framework for examining leadership in extreme contexts. Leadersh. Q. **20**(6), 897–919 (2009)

Hu, J., He, W., Zhou, K.: The mind, the heart, and the leader in times of crisis: how and when COVID 19 triggered mortality salience relates to state anxiety, job engagement and prosocial behavior. J. Appl. Psychol. **105**(11), 1218–1233 (2020)

Ipsos. Global attitudes on a COVID 19 vaccine: Ipsos survey for the World economic Forum (2021), http://www.ipsos.com

Kantar. COVID-19 vaccine faces an increasingly hesitant public (2021), https://www.kantar.com/inspiration/coronavirus/covid-19-vaccine-faces-an-increasingly-hesitant-public

Kearns, K., Alexander, C., Duane, M., Gardner, E., Morse, E., McShane, L.: Leadership in a crisis. J. Public Affairs Educ. **25**(4), 542–557 (2019)

Kirkpatrick, S.A., Locke, E.A.: Leadership: do traits matter? Executive **5**(2), 48–60 (1991)

Lufthans, S., Avolio, B.J.: Authentic leadership: a positive development approach. In: Cameron, K.S., Dutton, J.E., Quinn, R.F. (Eds). Positive Organizational Scholarship, pp. 241–261. Barrett-Koehler (2003)

Maher, P.: Leadership and government in a crisis: some reflections on COVID 19. J. Account. Organ. Chang. **16**(4), 579–585 (2020)

McGuire, D., Cunningham, J.E., Reynolds, K., Matthew-Smith, G.: Beating the virus: an examination of the crisis communication approach by New Zealand prime minister Jacinda Ardern during the COVID 19 pandemic. Hum. Resour. Dev. Int. **23**(4), 361–379 (2020)

Mumford, M.D.: Pathways to outstanding leadership: a comparative analysis of charismatic, ideological and pragmatic leaders Lawrence Erlbaum Associates (2006)

Northouse, P.J.: Leadership: Theory and Practice. Sage (2018)

OECD. Enhancing public trust in COVID 19 vaccination: The role of governments (2021). https://www.oecd.org/coronavirus

OECD. The COVID 19 Crisis: A Catalyst for Government Transformation? OECD Policy Responses to Coronavirus. OECD Publishing, Paris (2020)

Reiter, P.M., Katz, M.: Acceptability of a COVID 19 vaccine among adults in the US. How many people would get vaccinated? Vaccine, **38**(42), 6500---6507 (2020)

Stern, E.: Preparing: the sixth task of crisis leadership. J. Leadersh. Stud. **7**(3), 51–56 (2013)

Stern, E.K.: Crisis, leadership and extreme contexts. In: Holenweger, M., Jager, M., Kernic, F. (eds) Leadership in Extreme Situations. Advanced Sciences and Technologies for Security Applications. Springer, Cham. https://doi.org/10.1007/978-3-319-55059-6_3

Stoller, J.K.: Reflections on leadership in the time of COVID 109. BMJ Leader **4**(2), 77–79 (2020)

Zaccaro, S.J.: Organizational leadership and social intelligence. In: Riggio, R.E., Murphy, S.E., Pirrozzolo, F.J. (Eds) Multiple Intelligences and Leadership, pp. 29–54 (2002)

Alumba, F.O., Avolio, B.J., Gardner, W.L., Wernsing, T.S., Peterson, S.J.: Authentic leadership. Development and validation of a theory-based measure. J. Manage. **34**(1), 89–126 (2008)

Wilson, S.: Pandemic leadership: lessons from New Zealand's approach to COVID 19. Leadership **16**(3), 279–293 (2020)

World Health Organization (2021). WHO Coronavirus (COVID 19) Dashboard

Young, M., Camp, K.M., Bushardt, S.C.: Leadership development: a hierarchy of followership skills during a crisis. J. Leadership Accountability Ethics **17**(5), 127–133 (2020)

Twitter and the Dissemination of Information Related to the Access to Credit for Cancer Survivors

The Case of the "Right to Be Forgotten" in France

Renaud Debailly[1,4]([⊠]) [iD], Hugo Jeaningros[1,4] [iD], and Gaël Lejeune[2,3,4] [iD]

[1] GEMASS, Groupe d'Etude des Méthodes
de L'Analyse Sociologique de la Sorbonne, Paris, France
[2] STIH, Sens Texte Informatique Histoire, Paris, France
[3] CERES, Expérimentation en méthodes numériques pour les Recherches en SHS,
Paris, France
[4] Sorbonne Université, Paris, France
renaud.debailly@sorbonne-universite.fr

Abstract. Patient expertise does not lie solely in the production of knowledge about diseases or about daily medical experiences of illness (e.g. post-treatment, pain and side effects, psychological distress, etc.). Patients also develop sets of knowledge regarding everyday life with and after illness, especially with regards to socio-economic harm. Internet and social media can be mobilized by patients and patient organizations not only to learn or communicate about an illness and its consequences, but also to promote political measures aiming to address these. In this paper, our aim is to examine the way in which information about life after an experience of illness is produced and circulates through social media, and the role of patients and patient organizations in this process. To do so, we focus on the case of a recent French policy regulating loan insurance practices with regards to former cancer patients (Convention AERAS).

Keywords: Cancer survivorship · Loan insurance · Social media · Patient expertise

1 Introduction

1.1 Access to Loans After Cancer: The Emergence in France of the Right to Be Forgotten

Over the last decades, medical progress in the treatment of cancer has radically increased patient's survival rates. In developed countries, the survival rate 5 years after diagnosis currently exceeds 80% for childhood cancers [15] and 90% for breast cancer. These improvements have led to a growing population of

G. Meiselwitz (Ed.): HCII 2022, LNCS 13315, pp. 517–528, 2022.
https://doi.org/10.1007/978-3-031-05061-9_36

living former cancer patients[1], also raising new issues regarding the burdens they bear. Life after cancer literature shows that in addition to medical issues, cancer survivors carry non-medical burdens such as lower education levels (for former childhood cancer patients), lower employment rates, and lower incomes than the rest of the population [5,6]. This "double penalty" affecting former patients is moreover not limited to these most obvious indicators, for the inequalities they experience are found in many areas, such as mortgages and access to professional loans.

This paper is part of a larger study focusing on the difficulties of access to loan insurance for former patients, as well as the effects and limits of regulatory measures designed to reduce them (ELOCAN - Effects of the right to be forgotten in accessing LOan-related insurance after CAncer)[2]. Loan insurance is the main (and almost mandatory) way to secure a loan in France. Perceived by insurers as being high risk, former cancer patients encounter problems when applying for loans and mortgages. They may face several situations: (1) a refusal of the loan insurance, (2) a higher premium, or (3) an exclusion of guarantees such as death or illness. Access to loan insurance and to a loan itself is thus often compromised for these former patients [14]. In 2015 the French national cancer institute conducted the first large study in France (VICAN 5) to shed light on former cancer patients' quality of life 5 years after the diagnosis. Aside from items related to health status, or private and professional trajectories, the study included items related to access to loans and loan insurance. While the average refusal rate for a loan is around 10% among the general population, among former cancer patients included in the study, 29.5% of professional loans and mortgage requests had been refused, and 30% of the former patients indicated they obtained their loan "with difficulties". The study also shows that among former cancer patients who did not ask for a loan, about 10% did not do so because they were anticipating a refusal [7].

To reduce these inequalities, patient organizations campaigned for new regulations, on the grounds that not only did refusals and higher premiums constitute a double penalty, but also that former cancer patients were treated as higher risks than they actually were. Following this mobilization, the Convention AERAS was signed in 2016. Insurance actuarial practices now have to comply with a reference framework (AERAS framework or "Grille de reference AERAS") framing the price and guarantees of loan insurance contracts for former cancer patients on the basis of available epidemiological knowledge. In addition to this, 5 to 10 years after the end of treatments, former patients no longer have to share information regarding their cancer with their insurer (Right to be forgotten, or "Droit à l'oubli", RTF). Belgium and the Netherlands recently implemented the same regulations, and the issue of access to credit after cancer has been included as a priority in the most recent European Beating Cancer Plan communicated by the European Commission (COM(2021)44).

---

[1] The European Union has about 7 million people living at least ten years after a cancer diagnosis, according to the European Cancer Patient Coalition.

[2] ELOCAN is funded by the French national cancer institute (Inca) and the Integrated cancer research site (SiRIC) CURAMUS. https://elocan.sorbonne-universite.fr.

1.2 The Role of Patient-Expertise and Social Media to Promote the RtF and the AERAS framework

The dissemination of information about the RtF and the AERAS framework is crucial for many actors due to the relative lack of knowledge of RtF among former patients, and to the complexity of the AERAS framework which is constantly evolving. More generally, the arcane nature of insurance practices and especially of loan insurance means that a significant number of borrowers are unaware of its mechanisms and of their rights. In this situation, social media may be used by various stakeholders to inform former patients about the measures from which they can benefit. Within social media, the expertise held by former patients [2] who have benefited from these policy measures can play an important role alongside traditional information campaigns led by patient organizations, banks or insurance companies.

Several authors have investigated the role of the Internet and social media in patients' search for information online and in the reconfiguration of the caregiver-patient relationship [13]. The study of Internet uses also provides a relevant angle of analysis of the modalities of production and dissemination of the expertise of patients and patient organizations. In this respect, social media such as Twitter provide a specific framework for mediating patients' experience, in which a series of uses are deployed. Testimonies and exchanges between patients on social media can participate in the redirection of research or in the emergence of medical categories, a recent example of which was the term "long covid", analyzed by Callard and Perego [4] and by Roth and Gadebusch-Bondio [11]. These shared experiences can also contribute to the constitution of informal groups or emotional communities [10] that can help patients and former patients to break the social isolation or psychological distress associated with the disease. Finally, social media can be mobilized by patients and patient groups to disseminate information about rights, claims, or events [1,9,12]. We chose to investigate Twitter instead of forums or patient organizations' websites for two main reasons. First, Twitter is widely used by both individuals and organizations, and is likely to contain various kinds of messages (appeals, information, testimonials, etc.). The population is also more diverse, ranging from experts, to patients and the general public. Second, the format of message production and relay (tweet and retweet) allows for a directed tracking of information dissemination.

In this paper, using the case of the RtF and the AERAS framework, our aim is to examine the way in which information about life after and experience of illness are produced and circulated through social media, and the role of patients and patient organizations in this process. To do so, we build an analysis from a dataset of Tweets related to the RtF and the AERAS framework, and show the specific role that former cancer patient expertise holds in the production and dissemination of information. Two main hypotheses are tested: (1) Twitter is mainly used to publicize action in the defense of RtF and the AERAS framework; and (2) former patients' expertise is mediated by organizations, and individual expertise (such as testimonies, and individual experiences) is rather peripheral compared to the rest of the network.

2   Methods and Dataset

2.1   Dataset Collection and Annotation

We collected a Twitter dataset through the Twitter API, and restricted our research to tweets written in French. The tweets were selected using keywords and Twitter accounts. We used very generic keywords like *droit à l'oubli* (Right to be forgotten), *cancer AND (assurance OR emprunt)* (cancer AND insurance OR loan). This dataset (hereafter `keywords_dataset`) comprises tweets collected from 01/11/2021 to 31/01/2022. The dataset is relatively small because the API limits access to tweets published in the last two weeks. We then created a list of 36 Twitter accounts (hereafter `users_dataset`) that we thought would publish tweets potentially relevant to our research, for instance people campaigning for the right to be forgotten, and cancer organizations . . . [3]). This dataset contains 60,740 tweets (plus 3957 tweets collected in january 2022).

Table 1. Number of tweets and retweets in the `users_dataset`

	2010	2011	2012	2013	2014	2015	2016	2017	2018	2019	2020	2021	2022
Tweets	21	65	655	1125	1673	1795	3243	2989	8396	9882	14571	16325	3957
Retweets	0	46	284	523	938	704	1449	1474	4655	5627	7926	8712	1823

There were of course many false positives. We therefore decided to build a classifier to reduce this noise. For this purpose we manually annotated a subset of 420 tweets to get a ground-truth, in which each tweet was annotated by three annotators (the three authors of this study) in two classes: positive (relevant for our task) and negative (irrelevant for our task). We needed to have a reasonable number of annotated positive tweets in the subset to build a reasonably efficient classifier. Since `users_dataset` seemed to be less noisy, we selected 420 tweets from it:

1. one third of tweets including one of our top keywords (*droit AND oubli, emprunt, assurance*)
2. one third of tweets obtained with less specific keywords (*droit, oubli*)
3. one third of tweets randomly taken from the rest of the data

We first annotated a set of 200 tweets (evenly spread across the three subsets mentioned above), to reach an inter-annotator agreement (Krippendorff's alpha [8]) of 0.75. We analyzed the disagreement cases to obtain more precise annotation guidelines. One of the issues was to determine whether tweets about the right to be forgotten for other diseases were relevant or not. When we annotated the remainder of the data (220 tweets), we obtained an alpha of 0.9.

We corrected the disagreements, keeping the majority annotation on each instance, to obtain a final annotated dataset that was reliable for machine learning with 420 tweets: 296 negative and 124 positive (29.5%).

---

[3] The complete list can be found in a dedicated repository: https://github.com/rundimeco/right-to-be-forgotten.

2.2 Training of the Classifier

We split the dataset in train/test with a 0.7/0.3 ratio, and transformed the tweets into a matrix vector using COUNTVECTORIZER from the python scikit-learn library. We first tried a very classical representation using word unigrams (*Bag of Words* and then, like [3], explored character strings to see if the particular writing of tweets makes character-based models more efficient. We also explored different configurations (lowercase VS uppercase, with or without stopwords . . .). Since our dataset is quite unbalanced (as a reminder, 29.5% of the tweets in the ground truth dataset are positive) we chose macro F-score as a performance measure. The macro F-score is the unweighted mean of the F-score of all classes, thus increasing the importance of the minority class. The F-score is the harmonic mean of precision (proportion of True positives among all the instances deemed positives by the classifier) and recall (proportion of True Positives among all the positives instances in the dataset).

Our first finding was that removing stopwords and/or lower casing in the tweets did not improve the results. Using term weighting techniques such as Tf-Idf did not help to improve the results either. We quickly saw that sub-word representations (substrings of words) offered better performance than words or unrestricted character strings. The results are presented in Table 2. Of the classifiers that we tested[4], the Ridge Classifier presented the best results when used with a sub-word representation.

Table 2. Results (macro F-score) with different representations (char_wb and different classification algorithms, the best results for each algorithm (row) are underlined, the best results for each representation (column) is in bold

Feature type	Characters		Char_wb		Words	
Length (min:max)	(1:10)	(2:7)	(1:10)	(2:7)	(1:1)	(1:2)
Perceptron	0.9403	0.9366	0.9648	0.9460	0.9280	0.8990
Ridge Classifier	0.9270	0.9270	**0.9732**	**0.9822**	0.9357	0.8856
Multinomial Naive Bayes	0.8526	0.8916	0.9139	0.8977	0.8571	0.4981
Logistic Regression	0.9260	0.9547	0.9357	0.9357	0.9061	0.8856
Decision Tree	0.9553	0.9553	0.9724	0.9652	**0.9366**	**0.9270**

Using this classifier helped us to filter out most of the irrelevant tweets that we had in the corpus. In fact, less than 1% of the tweets were considered relevant by the classifier. Although we are not able to assess recall precisely, it is pretty clear that we got rid of a lot of noise, which facilitated the sociological analysis of the data presented in the remainder of the paper.

---

[4] SVM and deep learning classifiers are not represented in this table since they exhibited worse results, certainly due to the fact that there are too few instances for these algorithms.

3   Analysis

3.1   Time Evolution of Tweets Related to the RTF on twitter

The overall dataset contains 60,740 Twitter posts and 30,889 retweets related to the RTF and/or the AERAS framework. These posts were published between 2010 and 2021 by a total of 28 different users. Of the accounts present in the dataset, those with the highest representation in terms of number of tweets posted are related to patient organizations such as France Assos Santé (@Fr_Assos_Sante), a federation of patient organizations, RoseUP (@RoseU-pAsso) in which the leading organization is related to Breast Cancer, or to La Ligue contre le Cancer (@laliguecancer).

The increase in the number of tweets and retweets related to the RTF and/or the AERAS framework over time (Table 1) shows the growing importance of these topics over the last 10 years. The first version of the AERAS convention, with a more restricted scheme, came into force in 2007. Over the period 2011–2014, the increase in the number of tweets is not representative, considering the overall development of Twitter in the French market. Following the mobilization of patient organizations and the increased involvement of public authorities, the AERAS convention was largely amended and expanded. The RTF was adopted in 2015 and came into force in 2016. Thanks to the RTF, the time limit for not declaring a cancer to the insurer was reduced from 20 years to 10 years after the end of treatments. During this period we witness a first peak of tweets: either posts consisting of "demands" by the patient or former patient organizations, or posts disseminating information.

The amended AERAS convention of 2016 guaranteed a regular update of the AERAS framework. Several times a year, the AERAS commission, composed of representatives of patient organizations, the medical profession, the insurance industry and the public authorities, is responsible for updating the reference data and examining the files submitted by patient organizations, to include new pathologies. This has led to a strong mobilization by the patient organizations to demand both a continuous improvement of the guarantees offered by the convention (delays in the implementation of the RTF, reduction in excess premiums, etc.) and the inclusion of new diseases in the convention. This intense associative activity goes hand in hand with a significant increase in the number of tweets recorded over the 2017–2021 period, with a rise from 3,243 annual tweets related to the RTF and the AERAS framework in 2016, to more than 16,000 tweets in 2021.

3.2   A Politicization of the RTF by patient organizations in the field of health

When we look at the network obtained from users and hashtags (Fig. 1), we obtain a particularly dense network in which the accounts of certain users stand out. Those who produce the most tweets about the RTF are accounts mainly

Fig. 1. Network built from the entire dataset. (Layout: OpenOrd; Gephi)

Table 3. Description of the network

Nb of nodes	Nb of edges	Degrees	Weighted Degrees
16 312	21 289	1.54	4.85

linked to patient organizations or collectives. The only exception in this network is the account of Valérie Trillet-Lenoir who is a Member of the European Parliament. This former oncologist is a member of a special committee for the fight against cancer. The first challenge in relation to our research questions is to identify the users within this social network and then see how they use it (Table 3).

To make this first network more readable, we distinguished between different Twitter accounts to highlight the relationships between the actors within the graph. In our database, the most active tweeters are the patient organizations. We can identify those in the field of health at national level (the Ligue contre le Cancer, RoseUp, France Assos Santé, Renaloo, AIDES, etc.) and those at European level (EORTC). National organizations are sometimes organized with branches at regional level. This is the case of the Ligue contre le cancer. Finally, the last case encountered was the French Insurance Federation (FFA), the only actor in the network specialized in the field of credit. We have assigned a different color to each of these types of accounts (association, regional association, European association, and insurance). While the presence of these different actors in the network reflects the way in which we vacuumed up the data, there are several important points to note.

First of all, it is clear that the RTF and the AREAS framework are the subject of tweets mainly from organizations that are active around survival issues and are trying to put these issues on the political agenda or to extend the framework. In comparison, organizations in the field of health or other diseases such as diabetes for example remain less visible. The only exception is the case of AIDES, which played an important role in the inclusion of the RTF in the administrative arsenal, alongside organizations fighting cancer. We can also highlight the important interconnection between the users of the network. We have not found any isolated community or one that stands out from the others. Finally, the network analyzed remains almost exclusively national, as one would expect. Apart from the European association that works for the adoption of the RTF in other countries (Portugal, Netherlands), there are no other users from other countries that stand out.

Thanks to this first graph, we can therefore say that the network helps us to identify specific use of Twitter. It appears that the subject is mainly addressed by activist groups to recognize the difficulties related to life after cancer or to promote measures to address these. For example, in January 2022, RoseUP posted a call asking former patients to appeal to Members of Parliament to vote on a bill:

"mobilize and help us by contacting your MP to maintain the text voted yesterday in the Senate so that these historic advances are voted by Parliament."[5].

Twitter is then a means of disseminating information to relay the success of a long-standing struggle, to maintain attention or simply to remind its members and beyond of the existence of this device, as shown by this example of a tweet from La Ligue contre le cancer:

"Since September 1, the right to be forgotten has been reduced from 10 to 5 years (after the end of the therapeutic protocol) for loan applicants who had cancer before age 21. For more information on the #AERAS convention, contact our AIDEA service.".

The individuals and the way they contribute to these exchanges are presently difficult to identify.

3.3 The Articulation Between Politicization by Associations and Tweets from Individual Accounts

To more fully grasp the logics at a meso and micro level, we retained only those Twitter accounts with fewer than 500 followers (Fig. 2), as a source of data. This filter is used to exclude collectives that have been working on these issues for many years. We thus have access to accounts that are active without having the visibility of the large organizations mentioned above. This representation

---

[5] Exemples of tweets are presented in Annexe 1.

of the network highlights the interactions between regional branches of health-related organizations and individuals. It appears here that the network studied is not reduced to an opposition between a logic of dissemination of information at a global level and a logic of testimony at an individual level. While some organizations less known to the general public are identifiable on this graph, we also find some global branches or individual accounts attached to a large organization. The meso level, with these regional branches, is important to avoid opposing these logics (logic of dissemination vs. logic of testimony). In the same perspective, among the contributors present, we find managers who also use their personal accounts to disseminate information in their own networks (Marc Morel for FAS, and Catherine Simonin for the Ligue contre le cancer and France Assos Santé) (Fig. 2 and Table 4).

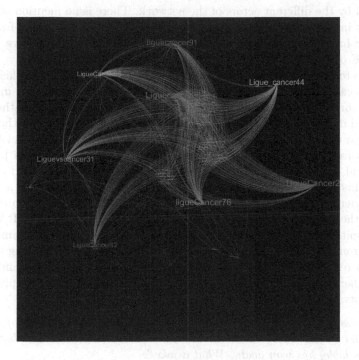

Fig. 2. Network of tweets over the entire period, accounts with < 500 followers. Number of degrees < 3. (Layout: OpenOrd; Gephi)

Table 4. Description of the network

Nb of nodes	Nb of edges	Degrees	Weighted Degrees
178	902	5.067	5.067

Before considering the tweets of individuals, it should be noted that the most used hashtags in the network are #cancer, #health. Those that refer directly to the RTF such as "AERAS", "Droit à l'oubli" also appear but are much less frequently used than the hashtags #cancer and #santé. Their use even seems to be correlated to certain actors in particular. For example, the term AERAS is often found among actors in the banking and insurance industries.

Beyond the presence of hashtags, Twitter's format obviously imposes a constraint in terms of form on the text that users can write. With a reduced number of characters, one cannot expect former patients to be particularly vocal. The search for patient expertise may therefore seem futile on this type of social network. However, there are testimonies or reactions to tweets that refer to the definitions of patient expertise mentioned in the introduction to this chapter. If we take the example of the AERAS framework, the content of the device is not discussed by the different actors of the network. There is no mention of the type of organs included in this framework or of the delays for benefiting from it. On the other hand, it is easier to find testimonies from people who have benefited from RTF or who would have liked to have benefited from it.

These testimonies have two characteristics. The first relates to their expression. We can say that this user discourse is "mediated" insofar as individuals do not express themselves directly but in reaction to a tweet. In other words, individual discourse and patient expertise are not on the surface of the network but appear as a sub-text. It is therefore by looking at how users react to official messages from the accounts of large patient organizations, or from journalists or political figures that we can capture these discourses. The second characteristic of this discourse is the fact that it targets either the RTF as a device or the experience of access to the loan. In the first case, the reactions are negative because the individuals underline the limits of the RTF or the AERAS framework despite the progress announced by this or that actor. For example, after a tweet from the Ligue contre le Cancer in October 2020 announcing that RTF had been reduced from 10 to 5 years for loan applicants who had cancer before age 21, there is a message highlighting the relative progress this represents for all patients:

"This is only for those under 21. In other words, the segment of the population least concerned by cancer or the most serious diseases. The banking and insurance lobby has won again. What a joke".

This message may raise a discussion about the understanding of the RTF because it assumes that the extension of the RTF only applies to people under the age of 21, when in reality what is targeted are cases related to pediatric cancers. Behind these denunciations or these strong reactions, it is above all a question of the aggression felt by individuals who have a real estate loan project. The second case of messages written by individuals is linked to the individual experiences and journeys of former patients who wanted to take out a loan. In this context, individuals sometimes provide explanations on the RTF, the AERAS framework or the AERAS convention. In this way, they contribute to making clearer and more understandable a system known for its complexity.

Finally, the testimonies are also an opportunity to communicate on solutions to the difficulties encountered. For example, in the message below, the user indicates the name of a banking institution that could be of interest to other people who are unable to find a solution:

"After many failures despite the AERAS convention, I was able to get a real estate loan thanks to XXX [name of a bank] which no longer asks its loyal customers for a health questionnaire. Without additional insurance costs. In case it can be useful to others."

With the latter example, we find a well-known figure in the literature on patient expertise, that of the user who has built up expertise through experience and who is able to give advice to other people likely to find themselves in the same situation.

4   Discussion

This work on the place of RTF and patient expertise using Twitter yielded several findings. First, it appeared that Twitter was a social network where issues related to access to credit were not very personalized. The massive presence of tweets from associations in the health field gives a specific image of the RTF, the AERAS framework and the AERAS convention. This is a cause promoted by certain actors who have a specific use for this social network. Another use then appeared in our work. At an individual level, the tweets generated by collective actors or authorities provoked reactions and opened up the production of individual testimonies. Patient expertise thus seemed to us to be "mediated" because it was not expressed spontaneously but in reaction to messages from collective actors who were active on these issues.

This initial research using Twitter could be extended in two ways. At the beginning of 2022, as new measures have been published, we can enquire whether the flow of tweets that was generated allows us to confirm the place and content of individual users. One observation is fairly easy to make in relation to our work. The measures announced at the beginning of the year were widely relayed on Twitter by actors in the real estate industry who had little presence in our dataset, while the consequent extension of the database is also a methodological issue. The tool that was specifically designed for this research can be tested and improved upon with new data collection.

Finally, this research must be extended by re-inscribing the use of Twitter in a sociology of the uses of digital tools. Given that we could draw conclusions from an analysis of Twitter, it would indeed be important to re-situate the use of an individual social network in relation to other social networks, and especially to examine how this network is used by actors who seek information on different media. This aspect would allow us, for example, to place our conclusions in relation to observations on the use of forums on the websites of patient organizations.

References

1. Akrich, M., Méadel, C., Rabeharisoa, V.: Se mobiliser pour la santé : Des associations témoignent. Sciences sociales, Presses des Mines, Paris, May 2013, http://books.openedition.org/pressesmines/1467
2. Arksey, H.: Expert and lay participation in the construction of medical knowledge. Soc. Health Illness **16**(4), 448–468 (1994), https://onlinelibrary.wiley.com/doi/abs/10.1111/1467-9566.ep11347516
3. Buscaldi, D., Le Roux, J., Lejeune, G.: Modèles en Caractères pour la Détection de Polarité dans les Tweets. In: Atelier DEFT 2018. Rennes, France, May 2018, https://hal.archives-ouvertes.fr/hal-01988907
4. Callard, F., Perego, E.: How and why patients made Long Covid. Soc. Sci. Med. **268**, 113426 (2021), https://www.sciencedirect.com/science/article/pii/S0277953620306456
5. Dumas, A., et al.: Educational and occupational outcomes of childhood cancer survivors 30 years after diagnosis: a French cohort study. British J. Cancer **114**(9), 1060–1068 (2016)
6. Handbook of Cancer Survivorship. Springer, Cham (2018). https://doi.org/10.1007/978-3-319-77432-9_21
7. INCa: La vie cinq ans après un diagnostic de cancer, June 2018, https://www.e-cancer.fr/Expertises-et-publications/Catalogue-des-publications/La-vie-cinq-ans-apres-un-diagnostic-de-cancer-Rapport
8. Krippendorff, K.: Content Analysis: An Introduction to its Methodology. Sage Publications (2018)
9. Majchrzak, A., Faraj, S., Kane, G.C., Azad, B.: The contradictory influence of social media affordances on online communal knowledge sharing. J. Comput.-Mediated Commun. **19**(1), 38–55 (2013)
10. Petersen, A., Schermuly, A., Anderson, A.: Feeling less alone online: patients' ambivalent engagements with digital media. Soc. Health Illness **42**(6), 1441–1455 (2020), http://onlinelibrary.wiley.com/doi/abs/10.1111/1467-9566.13117
11. Roth, P.H., Gadebusch-Bondio, M.: The contested meaning of "long COVID" - Patients, doctors, and the politics of subjective evidence. Soc. Sci. Med. **292**, 114619 (2022), https://www.sciencedirect.com/science/article/pii/S0277953621009515
12. Rueger, J., Dolfsma, W., Aalbers, R.: Perception of peer advice in online health communities: access to lay expertise. Soc. Sci Med. **277**, 113117 (2021)
13. Smailhodzic, E., Boonstra, A., Langley, D.J.: Social media enabled interactions in healthcare: towards a taxonomy. Soc. Sci. Med. **291**, 114469 (2021), https://www.sciencedirect.com/science/article/pii/S0277953621008017
14. Torregrosa, C., Gesbert, C., de Fallois, M., Mamzer, M.F.: Quel oubli pour les patients atteints de cancer ? Ethics, Med. Public Health **3**(1), 107–115 (2017)
15. Wojtyla, C., Bertuccio, P., Wojtyla, A., La Vecchia, C.: European trends in breast cancer mortality, 1980–2017 and predictions to 2025. Eur. J. Cancer (Oxford, England: 1990) **152**, 4–17 (2021)

Patient-Led Medicalisation and Demedicalisation Processes Through Social Media - An Interdisciplinary Approach

Juliette Froger-Lefebvre[1]([✉]) and Julia Tinland[2]

[1] Sorbonne Université, CNRS, Paris, France
juliette.froger-lefebvre@cnrs.fr
[2] Sorbonne Université, SiRIC CURAMUS, Paris, France
julia.tinland@sorbonne-universite.fr

Abstract. Our goal in this interdisciplinary paper is to analyse how groups of individuals have used - and still use - social media in order either to broadcast their claim to an institutional recognition of their status as being sick or, on the contrary, to rid themselves of the sick label that they feel has been unjustifiably imposed on them. « Medicalisation » is the process through which experiences and problems that can be understood as political, social or even religious come to be under the purview of medicine and « demedicalisation » points to how problems previously considered to be medical in nature cease to be so.

The increasing involvement of patients and patient groups on social media threw a wrench in the conceptualisation of medicalisation as a necessarily harmful mechanism from which patients remain the passive victims. Several groups are indeed very active, involved and vocal today on social media regarding the medical status of their own experiences. Though control ultimately remains (mostly) in the hands of medical institutions, their online presence and organisation has had profound social and medical ramifications.

The cases of endometriosis, eating disorders, chronic fatigue and a few others can highlight how the « sick role » is indeed tied to many material, social and symbolic benefits to which several online groups feel entitled. But the « sick role » is also tied with significant disadvantages as it may sometimes become an unfair and stigmatising burden. Implicit or explicit demands for medicalisation and demedicalisation are numerous online: social media have been instrumentalised widely to promote claims for both medicalisation/demedicalisation and to recruit followers internationally. Patients have invested forums, websites, groups and personal profiles on Facebook, Instagram or WhatsApp quite proficiently in order to raise public awareness of these issues and, ultimately, to institutionalise their movements.

In order to explore patient-led medicalization and demedicalisation through social media, we start by introducing sociohistorical and philosophical perspectives on both processes as well as on related movements using social media. Secondly, we present the more detailed case of eating disorders with the French support

---

While not all individuals involved in such online movements are considered to be « patients » by medical professionals and institutions, access to or negation of the status and role of « patient » is what is at stake for many of them.

© The Author(s), under exclusive license to Springer Nature Switzerland AG 2022
G. Meiselwitz (Ed.): HCII 2022, LNCS 13315, pp. 529–544, 2022.
https://doi.org/10.1007/978-3-031-05061-9_37

group Outremangeurs Anonymes (Overeaters Anonymous). Lastly, we analyse the ethical implications of people's online involvement in medicalisation and demedicalisation processes, especially in light of issues regarding over-medicalisation and under-medicalisation.

Keywords: Medicalisation · Demedicalisation · Patient expertise · Social media · Institutionalisation of patient groups

1 Introduction

In this paper, our goal is to analyse how groups of individuals have used - and still use - social media in order either to broadcast their claim to an institutional recognition of their status as « sick » or, on the contrary, to rid themselves of the « sick » label that they feel has been unjustifiably imposed on them.

« Medicalisation » as a concept was initially used as a criticism of medicine's supposed overreach into parts of human and social life that ought to have remained free from it [1]. Here, medicalisation is an insidious mechanism through which, for example, sociopolitical problems such as alcoholism are reinterpreted as medical ones [2]; biomedical control is exerted over what was previously criminal, religious or simply unusual phenomena, such as when the sainthood associated with one's privation of nourishment disappeared in favour of the diagnosis of anorexia [3]; or through which individual and collective responsibilities and/or suffering can be fled [4]. In contrast, « demedicalisation » is the process whereby problems previously considered to be medical in nature - or the answer to which was medical - cease to be so.

The increasing involvement of individuals, patients and patient groups on social media threw a wrench in the conceptualisation of medicalisation as a harmful mechanism from which patients are necessarily the passive victims. Several groups are indeed very active, involved and vocal today on social media regarding the medical status of their own experiences. Though medicalisation may ultimately be driven mostly by professional and commercial interests [5] or controlled by medical institutions, patients' online presence and organisation has had profound social and medical ramifications. The cases of eating disorders, chronic fatigue, breast cancer predisposition and a few others will be presented more thoroughly and used to highlight how the « sick role » - as a way to comply with medical diagnosis and care [6] - is tied to many material, social and symbolic benefits to which several online groups feel they ought to have access. However, the « sick role » is also tied with significant disadvantages as it may sometimes become an unfair and stigmatising burden. Implicit or explicit demands for demedicalisation are thus also numerous online: pro-ana forums, for example, denounce medicalisation as an attempt to control supposedly socially deviant behaviour [7].

In both cases, social media platforms have been instrumentalised in order to promote widely claims for medicalisation or demedicalisation, to criticise over-medicalisation or under-medicalisation, and to recruit followers internationally [8]. Patients have invested forums, websites, groups and personal profiles on Facebook, Instagram or WhatsApp quite proficiently in order to raise public awareness of these issues and, ultimately, to institutionalise their movements.

Our aim is to detail how various online mobilisations can occur on social media in order to help shape medical nosographies and to outline the ethical questions that arise with such online involvement through both sociological and philosophical approaches. This work is profoundly interdisciplinary and benefits from diverse methodological and research perspectives. As a consequence, it relies on different levels of analysis: a macro-level theoretical approach which offers critical and conceptual perspectives and another more micro-level approach based on ethnographic field work investigating some talking groups by means of observations, interviews and the analysis of their official litterature, and which stems from a thesis project. Though finding the balance between philosophy and sociology is no easy task as the need remains to acknowledge and respect disciplinary boundaries [9], it brought both depth and insight to the analysis of a topic like this one. This paper also proposes a short literature review regarding the concepts of medicalisation and demedicalisation as well as an overview of some of the ethical issues stemming from individuals' involvement in (de)medicalisation processes.

Firstly, we introduce sociohistorical and philosophical perspectives on medicalisation and demedicalisation processes as well as on related movements using social media. Secondly, we present the more detailed case of eating disorders with the French support group Outremangeurs Anonymes (Overeaters Anonymous) [10]. Lastly, we analyse the ethical implications of people's online involvement in medicalisation and demedicalisation processes, especially in light of issues regarding over-medicalisation and under-medicalisation.

2   Medicalisation and Demedicalisation: Labels and Legitimacy at Stake

Medicalisation is a process that has been unfolding from the very birth of medicine as a field of knowledge and practice regarding the diagnosis, treatment, and prevention of human ills. Demedicalisation, on the other hand, is a slightly more recent phenomenon. A historical approach both of the concepts themselves and of the processes they designate can bring light to the high stakes that are attached to them, especially with regards to labelling and legitimacy.

2.1   History of the Concepts - Medicalisation, Demedicalisation, Over-Medicalisation and Under-Medicalisation

Though the term « medicalisation » itself does not appear to entail necessarily negative connotations, its emergence in the late 1960's through to the 1980's was typically associated with a strongly critical outlook, thus conflating it with what we might call « over-medicalisation » today. Though such criticisms originated from sometimes vastly different perspectives, they all participated in framing the process of medicalisation as medicine's illegitimate overreach into areas of human life that ought to have remained free from it.

Many of these early criticisms were tied to the anti-psychiatry movement. They denounced psychiatry's lack of legitimacy as a medical field, its attempts to distinguish pathological behaviour from normal behaviour as well as the harms incurred by patients wrongfully diagnosed and treated. In psychiatry perhaps more than in somatic medicine, the potential overextension of medical authority underlying medicalisation processes could be made readily apparent. Without a legible anatomopathological substrate upon which to base its diagnostic classifications, psychiatry's reliance on value-laden parameters was seen to epitomise medicine's tendency to abuse its power over society's downtrodden, (over-)medicalising antisocial behaviours. According to Thomas Szasz [11], the political state allied itself with the medical establishment, subverting the process of diagnosis and transforming non-disease into disease, thus implementing a form of social control. To Szasz, psychiatry did not deal with actual diagnoses, illnesses or treatments but with the metaphorical (though effective) uses of these terms.

Sometimes tied to such criticisms of psychiatry were the identification and incrimination of a wider form of «biopower», meaning the state' regulation of social practises, individual bodies and populations. Michel Foucault [12] famously delineated how the challenges of administering a territory gave way to attempts to govern the lives of the populations residing there, creating a link between power and knowledge. In supporting the hypothesis that, contrary to what might be thought, capitalism gave rise to a form of collective medicine rather than a private one, Foucault points out that the body was socialised according to a certain idea of the productive labour force [13]. Society's control over individuals is not only implemented through ideology but also through - and with - the body as a biopolitical entity, using medicine as a biopolitical strategy to compel compliance with political imperatives [12]. To some, the growing social dominance of medicine is underlied by the decreased social dominance of religion, and medicine has slowly replaced religion as the dominant moral ideology of advanced capitalist societies [14].

Some critical perspectives of (over-)medicalisation see it less as the product of professional « imperialism » through the willful, premeditated construction of a medical empire than as a social process rooted in western societies' complex technological and bureaucratic systems. Phenomena such as ageing, drug addiction, alcoholism and pregnancy, once regarded either as normal natural processes or as human foibles within specific social contexts, have progressively become problems to be solved, leading to the emergence of new medical specialties [2]. Encouraged to place our trust in experts, we turn to them to guide us through these problems, expecting the solution to come from them. A larger concern here points to the rampant « depoliticisation » of many politico-social issues, which sees them labelled as « diseases » as a result of their medicalisation [15]. This encourages a (politically convenient) shift in responsibility from the state (or from society more generally) to the medical establishment and patients themselves.

Others went further in the denunciation of our growing dependence on medical experts, seeing in it a refusal to confront ourselves to difficulty or pain. Illich [4] disputed the idea that modern medicine leads to an overall reduction in human suffering and argued instead that medicine actually causes a great many ills by promising cures for all problems, including those that were not considered pathological by previous generations. In doing so, medicine creates false hopes that all suffering can be avoided and saps the individual and collective resources that allow us to cope with the inevitable difficulties of life, thus turning us into passive consumers of medical services [4].

All in all, many of these early critiques often do not distinguish between medicalisation and over-medicalisation, weaving into the concept of medicalisation itself the idea of its illegitimacy. However, the need to detach medicalisation as a morally neutral process from its critical and normative counterpart: « over-medicalisation » has since been made clearer [16].

Interestingly, criticisms surrounding the under-medicalisation of certain conditions highlight how the medicalisation process can rather become the most appropriate answer to address certain issues. Claims of under-medicalisation are prominently made by suffering individuals and groups themselves [17], which runs counter to the often prevailing idea that patients must be the passive victims of (over-)medicalisation. Indeed, more attention has started to be paid to the very active role that individuals and patients, through their criticisms of under-medicalisation, can take in the process of medicalisation itself, but also (when they take it upon themselves to voice their own concerns regarding over-medicalisation) of demedicalisation, whereby specific conditions and phenomena either cease to be defined as medical problems or to be treated medically.

It is therefore important to clarify the closely-related but distinctive concepts of medicalisation and demedicalisation as morally neutral processes designating either specific phenomena's introduction into or exclusion from the medical field, and the ethical (normative) issues of over-medicalisation and under-medicalisation. Their conflation is a significant problem when one aims to hone in on and analyse patients' own normative discourses regarding these ethical issues as well as their role in the processes themselves.

2.2 History of the Phenomena - Examples of Medicalisation and De-medicalisation Through History

Many examples of medicalisation and demedicalisation processes can be given. We chose two significant cases in order to illustrate how these processes profoundly change our perceptions of specific situations and social practises.

According to Jacques Maître [18], anorexia and bulimia are ancient phenomena historically linked to religious extreme fastings. During the Middle Ages, it represented a ritualised and institutionalised practise used by women, « the Holy Anorexics », to reach unprecedented social status within an eclesiastical hierarchy [3]. Some of them, like the famous figure of Catherine de Sienne (who counselled the Pope), were sanctified. During the nineteenth century, this practise, which used to be given religious meaning, became a medical pathology through the first clinical descriptions of bulimia and anorexia[1]. These medical works transformed the social sense given to and the role played by these kinds of practises, which became the symptoms of feminine madness. This example of medicalisation shows how the qualification of social practises can change, be turned into medical diagnoses and thus contribute to the secularisation of society.

On the other side of that coin, homosexuality remains to this day one of the foremost examples of demedicalisation throughout history (though some argue that its remedicalisation remains a possibility [21]). The medicalisation of male homosexuality can be

---

[1] We can refer to Gull [19] and Lasègue's [20] descriptions of anorexia and bulimia in the late 19th century.

traced back to the mid-nineteenth century and offers a striking example of how medicalisation itself can sometimes be presented as an alternative to the criminalisation of certain phenomena. It was officially included in the Diagnostic and Statistical Manual of Mental Disorders (DSM-II) much later, however, in 1968 [22]. The « gay liberation » movement of the late 1960s, especially in the United States, saw social activists and their allies set out to highlight the stigma associated with « homosexuality » when categorised as a psychiatric diagnosis. They played an integral part in the progressive demedicalisation of homosexuality from 1973 onward, as the American Psychological Association ceased to consider it a mental illness [21].

Medicalisation and demedicalisation processes have long been influenced by the social context within which they took place, but also through the growing involvement of individuals and groups defending their direct and indirect interests, which was compounded by the advent of online movements.

2.3   Patient Mobilisations and Social Media

The involvement of patients and other citizens regarding concerns of over-medicalisation and under-medicalisation, especially online, has gotten stronger in recent years. Taking over social media, online medical platforms, patient associations' webpages and forums, their voices have gained prominence and have had on occasion a notable impact on discussions surrounding the medicalisation or demedicalisation of certain conditions.

Ethical issues surrounding under-medicalisation, more particularly, have sometimes come to the forefront of these discussions, whereas over-medicalisation used to be the more prominent area of interest. The case of medically unexplained symptoms is an especially interesting foray into the stakes attached to under-medicalisation, as many people suffering from them have decried the lack of legitimacy attached to their experiences due to the absence of clear medical recognition. This can have a deleterious impact on their sense of self, their relationships with their general practitioners or with their loved ones [23]. Chronic fatigue syndrome and fibromyalgia are said to affect between 1 and 3% of the adult population, but both remain controversial medical entities, their boundaries still too loose and overlapping to fit within a clear physical or psychological disease category [24]. Many patients report experiencing delegitimisation and stigma as a common occurrence, be it in relation to their families, their colleagues and other acquaintances, or to the medical professionals they encounter [25]. Interestingly, such delegitimisation or culpabilisation is said to occur also through the tendency of health professionals to focus on socio-cultural and psychiatric explanations of these experiences [25], as it fuels the perception that people who suffer from chronic fatigue syndrome or fibromyalgia are simply trying to escape their responsibilities and burdens. In response, many people suffering from medically unexplained symptoms or controversial medical entities such as chronic fatigue syndrome or fibromyalgia have mobilised online in order to shine a light on their experiences and their commonality, on the resistance they face and on the lack of medical research and support which they endure [26].

Furthermore, nowadays social media represents a very important medium for patient groups who struggle to gain recognition for their cause. The existence of some of these groups is strongly dependent on the internet, which allows them to maintain a form of mobilisation despite the geographical and social distance separating their members.

Created in 2016, the French association Geneticancer gathers women - the great majority of this association's members are women - affected with cancer predisposition genes (such as BRCA 1 or BRCA 2) on a national scale[2]. While they are not considered to be « sick » as such (because the genes do not systematically trigger a cancer), these women are often subjected to prophylactic surgery in order to prevent cancer. This association promotes preventive measures and encourages women to share their experiences of illness, of the predisposition announcement and of preventative surgery. Geneticancer is very active on Instagram, is present on Facebook and Whatsapp, and employs social media in two ways. First, it allows the organisation to showcase itself, to present the events organised by its members and to promote their products. In this sense, it provides a way to collect money from benefactors. Secondly, it allows members to share their personal experiences and makes it so that visitors (and potential new participants) can identify with them.

Though patients can be very active in their mobilisations against under-medicalisation, they are also significantly invested against what they perceive to be the over-medicalisation of their own experiences and actively advocate for their demedicalisation. On one side of the coin are thus online discourses against under-medicalisation, highlighting how delegitimising, limiting and stigmatising the absence of a medical label can be in one's attempts to gain access to the « patient role »; while on the other side of that coin are patients' discourses against over-medicalisation which point to the stigmatisation, passivation and subordination that can arise from the overuse of such labels.

Psychiatry is once again one of the areas in which grievances against medicalisation and demands for demedicalisation have been loudest in patient circles, especially online. The voice hearers' movements in the United States and United Kingdom, for example, aim to re-frame auditory hallucinations as voice hearing in a bid to change how such experiences can be understood, especially outside of a purely biomedical approach. The movements encourage a more psychological understanding of these experiences, informing the ways in which support can be offered in coping with voice hearing rather than in trying to eliminate such experiences altogether through medication, leading to the creation of a network of voice hearing groups throughout the world [27]. The Hearing Voices Network (HVN) is a global initiative which spreads over 20 nationally-based networks. Its presence online is growing and aims to raise general awareness regarding the diversity of experiences that can exist in relation to hearing voices. It also aims to challenge negative stereotypes, stigma and discrimination, to help create more spaces for people to talk freely about voice-hearing and to encourage a more positive response within healthcare settings and in wider society [28]. HVN groups' stance against the over-medicalisation of these experiences can be seen in their deliberate efforts to use ordinary language in order to describe voice hearing without referring to clinical, medical or technical terms which would preclude other (non-medical) understandings of these phenomena [29]. They actively call into question the traditional psychiatric

---

[2] This association is the current subject of a sociological study led by Juliette Froger-Lefebvre (GEMASS - UMR 8598) and Marie Mathieu (Cermes3 - UMR 8211)) within the framework of their research on the sociology of cancer.

view of auditory hallucinations as psychotic symptoms and challenge clinicians to work differently with persons who hear voices.

Online mobilisations can also gather collective support in their reach for autonomy against medical domination. After it appeared in 2001 in the United States, the « pro-ana » phenomenon (for pro-anorexia) has been criticised by the media and dubbed a dangerous internet movement. If some identifiably pro-ana blogs or websites showcase self- performances featuring thin bodies [30], others insist on fostering collective support and control their content. In a way, bloggers can both play the role of content autoregulators and, at the same time, of content creators partaking in what is considered to be « dangerous » behaviours [7]. Online groups can participate in the de-stigmatisation of anorexia and in the autonomisation of patients, sharing their experiences and offering advice so as to limit bodily damages and lessen hospitalisations. Meanwhile, they promote preventive messages against eating disorders and they create a supportive community. While they freely spread medical information on eating disorders, they do not encourage medical diagnosis or hospitalisation. Ana-mia (another name for pro-ana) websites and blogs epitomise the autonomisation of an online support community when faced with medical institutions. In this configuration, social media constitute a way for patients to escape from medical domination or surveillance and to create their own spaces, dedicated to sharing both personal experiences and support.

The idea that patients and individuals can only be the passive victims of medicalisation (thus understood as over-medicalisation) has been called into question: not only must the processes of medicalisation and demedicalisation themselves be distinguished from normative judgements regarding over-medicalisation and under-medicalisation, but patients, individuals and groups have also been vocal - especially online - about the preservation and promotion of their interests, should they take the form of medicalisation or demedicalisation. We now propose to delve deeper into the analysis of a specific example: the case of Overeaters Anonymous support groups.

3   The Complexity of Label's Use. The Case of Overeaters Anonymous Support Groups in Their Struggle for Recognition

In order to analyse how (de)medicalisation processes can be striven for by lay groups, we will now present one example of mobilisation: the Overeaters Anonymous, also named OA [10]. As a talking group meant for people who suffer from eating disorders, the OA meet together every week in a non-medical environment. They evolved in the context of a strong therapeutic market dominated by mental health specialists and legitimised by governmental institutions and public policies meant to deal with bulimia and anorexia, but even more so with binge eating and obesity [31]. Far from medical institutions, support groups offer a caring model based on trust and collective care. The OA community was created in the United States in the 1960's. After two decades of expansion all over the United States, they spread in Europe and were introduced in France in 1983 thanks to the rise of charismatic leaders who ensured the preservation of Tradition.

Observations of these groups and interviews with OA members in France and in Canada enabled us to follow how discourses regarding both medicalisation and demedicalisation were mobilised by activist groups. On one hand, the process of medicalisation (and thus the recognition of their sickness) is supported in claims for a better acknowledgement of their suffering and is used as a way to legitimise the existence of the group. On the other hand, these groups gain a lot by keeping a certain distance from - and by criticising - medical institutions and their nosography of eating disorders. Sickness (understood medically) is used as a label guaranteeing institutional recognition (medical and administrative) but also recognition from potential future members of an association and/or of a therapeutic group.

3.1 The Recognition of Sickness

Open to all, even without a medical diagnosis, OA groups attract residual populations whose care was not taken on by institutions: if some of the participants do benefit from medical care, others have not ever been diagnosed for the eating disorder from which they claim to suffer. As such, the entrance fee to be part of a group is very low compared with institutional talking groups excluding non-diagnosed individuals. Without any clear therapeutic pathway, some of the non-diagnosed can find a welcoming environment in which their suffering is taken into account. As a consequence, the OA can address a larger number of people for the recruitment of their members. Additionally, they don't have to take on a « negative role » of control [32, 33] in having to distinguish what is normal from what is pathological - a role usually assumed by physicians or subalterne medical professions more dedicated to « doing the dirty work ». At the same time, the first step for persons who want to gain membership and take part in OA meetings is to recognise their sickness. Whether they are diagnosed or not, OA members start their career [34] in talking groups by becoming aware that they are sick and that they need to work on their deviance. As such, sickness represents the common mark of their affiliation. « Sickness » is actually a keyword for OA groups, pronounced by Rozanne S., the charismatic founder of these meeting groups:

> *"Overwhelmed, I thought: I'm not alone anymore*
>
> *That night, I learned that I wasn't wicked or sinful; I was sick.*
>
> *I had an illness, which I was later to call compulsive overeating.*
>
> *It was all a revelation to me".*[3]

Encouraged to mimic the founder's conduct, every OA member must introduce him- or herself during the meeting, by means of an acknowledgement of his or her own sickness. This way, before speaking at all, the participant is prompted to say: « My name is… I'm a compulsive overeater » (or anorexic, or bulimic, etc.).

If the use of the « sickness » label does not need medical approval, its general understanding refers to medical definitions. For the OA, overeating problems pervade a person's body and mind (and even their soul). Dealing with them requires external help (from medical institutions but also from people who are suffering from similar issues) and, sometimes, the very careful planning of meals. These disorders cannot be easily

---

[3] Overeaters Anonymous, *Beyond our Wildest Dreams. A History of Overeaters Anonymous as Seen by the Founder*, 2nd Edition, Rio Rancho, Overeaters Anonymous Inc., 2005, p.41.

handled and we frequently hear during the meetings « I can't help it ». In this sense, medical references are central during and after OA groups' meetings.

In the end, « sickness » symbolises three social functions for these groups. First of all, it creates a strong bond between members of the talking groups: they all recognise themselves in this appellation that they use to qualify their suffering. Secondly, the framework of « sickness » reinforces OA members in their « sick role » [6]: they acknowledge the fact that they are sick and they wish to be taken care of. Finally, the term « sickness » refers to medical definitions and provides talking groups with a legitimate cause in the pursuit of their efforts, as its symbolic value becomes commonly acknowledged and approved by society.

3.2 Social Media and Recruitment

The main goal of these meeting groups is to promote their expansion and to insure their survival within the competitive eating disorders' space, which is flooded with various therapeutic offers from numerous patients associations, medical institutions, political actors or even dieting industries. Consequently, the first difficulty for OA groups is to oversee the smooth running of their meetings. The recruitment of new members represents a real preoccupation for OA groups. Their influence on eating disorders' medical care and on public policy depends on it. If their influence in France is very limited, they have powerful supporters (medical, political or religious) at their disposal in the United States.

In a way, reference to sickness allows the groups to gain legitimacy in the eyes of institutions but also in the eyes of potential future members. First of all, OA groups use the claim that they are proper patients' talking groups to get close to medical and administrative institutions. In France, they acquired the associative status (based on the 1901 Law [35]) in the beginning of the 2000's. As a consequence, they earned the possibility to occupy freely a room for organisational purposes within the Associative House. They also gain credibility through their « public campaigns »: in order to draw physicians' attention to their cause, OA representatives produce publicity documents which they distribute in medical institutions and in clinical settings. As a patients' association, they benefit from medicalisation processes which see eating disorders gain in public legitimacy.

In this bid for recruitment, OA groups use social media to lure newcomers to meetings. On the internet, the OA website is very well optimised and appears on search engines' first page results for a search on eating disorders. Their website lists every OA meeting, including location details, so that joining a group or calling its members is made easier. Ever since the Covid-19 pandemic of 2020, meetings have been taking place online. Simultaneously, Founders' publications (literature, leaflets or public presentations) are distributed through various social media, such as facebook pages, OA websites, forums, blogs…

In conclusion, OA groups' need to fulfil one of their main goals, which is to spread the message and to recruit new members within a particularly competitive market, requires them to rely on social media as an efficient tool for recruitment, especially when targeting marginalised social groups struggling for legitimacy.

3.3   Distance and Criticism of the Medical Field. A Religious Idea of Recovery

If OA groups need the approval of medical institutions to legitimise their existence, they still foster and preserve their autonomy, especially when it comes to defining their own therapeutic program. This program is developed from the Alcoholic Anonymous Tradition and uses its own definition of recovery based on abstinence [36]. This definition allows them to attract at the same time people who « failed » in following prevailing medical therapies and people who refuse medical classifications and therapeutic care: people going through « therapeutic aimlessness ». Sickness is associated with human weakness and full recovery is not an option. For the Twelve Steps Movement, people who suffer from alcoholic or eating disorders have to learn to live *with* their deviance. The only way to get better is to abstain from alcohol or compulsive eating and to go frequently to meetings, which creates a dependency on these groups.

This model of recovery finds its roots in the evangelical tradition and in the American Temperance movement of the early nineteenth century [37]. Meant to impose a new lifestyle, morally-bound and respectable, this movement advocates abstinence when it comes to alcohol consumption (though this approach can be applied to other domains of existence). It also contributes to reinforcing therapeutic individualism - a very specific conception of health which places individuals as actors of their own health.

Selling salvation and health goods, OA groups offer a religious solution to eating disorders, which they see as a sickness and a disease of the soul. Through this religious proposition of recovery, this therapeutic offer engages with the demedicalisation process of eating disorders. The social frontiers of sickness and « normal » behaviour are scrambled, widening the population potentially concerned by deviance.

4   Political and an Ethical Questions Regarding Patient-Led Medicalisation and Demedicalisation: Health Democracy at Stake

Patients' increasingly vocal criticisms of over-medicalisation and under-medicalisation online, and their involvement in the processes of medicalisation and demedicalisation themselves, have political as well as ethical dimensions which need to be carefully delineated, considered and balanced.

4.1   Claim for Autonomy

What is mostly at stake here is the development of health democracy and patient expertise based on experiential knowledge. Health Democracy as a claim for equal access to healthcare was used in the 1960's by a movement of physician activists in France [38]. Nowadays, the political and social meanings of this notion have been significantly reduced through its use by healthcare institutions, such as Regional Health Agencies (ARS), as they reintroduced the concept of health democracy in a bid to engage patients' individual responsibility.

However, patients and individuals, as they develop situated knowledge and everyday skills drawn from having to deal with their symptoms or out of ordinary experiences, with their pathology (when it has been diagnosed), and with therapy or coping mechanisms as well as with the healthcare system more generally, also reappropriated the concept. As they form networks of cooperation through patient associations or through the use of information and communication technologies (forums, webpages, support groups, etc.), their voice gains traction and influence. Asserting their status as « knowers » and crucial stakeholders in the healthcare system, they demand a seat at the table when it comes to decisions regarding medical care, practises and research.

The development of health democracy thus results from a moral claim to having autonomous power over the ways in which medical care, practises and research may impact one's life. Disregarding the concerns, the voices and the preferences of the people whose life is most directly affected by such decisions can be seen as an overly paternalistic expression of medical power and dominance. In this sense, collective and emancipatory social practises emerge which aim to promote the autonomy of dominated groups through their empowerment, meaning through individuals' efforts to implement their collective projects and interests. The aim is thus to encourage destigmatisation as well as a shift in the current balance of power [39]. While autonomy and the obtention of informed consent has been a cornerstone of biomedical ethics when it comes to individual medical decisions [40], the role that individuals and groups can play on a wider scale in medicine has been put to the forefront of discourses surrounding the development of health democracy [41].

Though these new developments might appear ethically straightforward, it remains important to identify and address some of the potentially problematic implications they may have.

4.2 Representativity and Invisibilisation

First and foremost, questions can be raised regarding the extent of the representativity of online movements. Some people have sufficient access to online resources, the ability to develop the explicit, taught skills as well as the implicit skills (public speaking, self-confidence, etc.) needed in representative roles. These untaught skills are essential in order to pursue a representative career, whether it is to get involved in political parties or in informal movements [42]. This implies a form of tacit social selection and exclusion based on feelings of illegitimacy and on self-censorship which are linked to social origins. In a way, some people are more likely to get involved in such movements and may not be able to fully represent the concerns and the interests of many people who are less involved and yet have just as much at stake - if not more - in healthcare decisions and practises. Studies on the determinants of volunteering which have examined the socio-demographic characteristics, motivations, attitudes and values of volunteers highlight that middle-aged, middle-class, married women with higher education degrees and with dependent school-age children were more likely to be involved than any other social groups, including when volunteering as expert patients [43]. As such, even online, the social recruitment of these movements tends to exclude disadvantaged social classes and thus the representativity of these groups can be questioned. From an ethical standpoint, this means that the interests, the needs and the experiences of those who are more vulnerable than others when facing medical authority and domination [44] may continue to be disregarded, misrepresented or misunderstood.

4.3 Demedicalisation, Personal Responsibility and Public Health

The progressive rise of health democracy through individuals' empowerment and claims for autonomy - including in regards to questions related to over-medicalisation or under-medicalisation - might appear to offer a much needed counter-balance to the ethical issue of medical dominance and authority, but it does bring about new concerns. Through its insistence that patients (and people more generally) ought to have more say in their experiences of the healthcare system and to gain in autonomy, this movement entails the idea that individuals must become active participants (moral agents), rather than simply (moral) patients, in the preservation of their health. However, this does put a larger and larger onus on patients and individuals themselves to take responsibility for their own health. This may lead to a form of disregard towards the role that social inequalities may play in one's state of health and acccss to healthcare [45]. Where society used to recognise and address such inequalities more directly, personal responsibility and autonomy are now centre-stage in public health discourses [46].

Demedicalisation processes can be another instantiation of this tendency: the right to define for oneself what is and especially what is not a medical problem can sometimes overshadow how inequalities are at play regarding how people may be more or less able to cope with - or adapt to - specific issues. Where the process of medicalisation allows for the provision of medical attention and support for these problems (though this is also where ethical issues of medical dominance, paternalism and authority may arise), the process of demedicalisation sometimes sees individuals themselves claim the freedom to deal with such experiences perhaps more autonomously, but also on their own. This may compound inequalities and lead to a lack of access to healthcare resources with regards to what may sometimes be distressing experiences, going against the stated goals of public health to address justly health-related issues in the best interests of all. Additionally, it may reduce collective issues to personal problems and invisibilise the political aspect of these matters.

5 Conclusion

To conclude, we started by showing that medicalisation and demedicalisation processes are a way either to (dis)qualify social practises from a medical standpoint or, on the contrary, to discontinue the medical interpretation of what was considered to be a medical problem (like homosexuality). The use of « sickness » as a label is now commonly shared by online movements who may want to acquire it or to reject it. In order to be able to reflect on such movements with sufficient nuance, one must distinguish demedicalisation and medicalisation as processes from over-medicalisation and under-medicalisation as normative judgements present in online discourses.

The example of OA talking groups shows that the use of the « sickness » label may be more complex than it may appear at first sight. First of all, the affirmation of their « sickness » is a way to justify their own existence as a group. Also, the use of medical definitions of eating disorders legitimises their revendications as therapeutic groups. In this case, medicalisation offers talking groups a legitimate cause to pursue their goals. It guarantees the possibility to recruit new members and to institutionalise the movement. However, OA groups offer at the same time a religious solution to eating disorders. If the « sickness » label is used, it is then mainly understood as a disease of the soul. Through this redeeming proposition of recovery, this therapeutic implicitly imposes a non-medical definition of eating disorders. OA groups mobilise both under-medicalisation and over-medicalisation critiques in their bid to take more control over the perception and care of their own experiences.

While online movements' claims regarding over-medicalisation or under-medicalisation are an example of individuals' growing empowerment within the development of health democracy, which sees their experiential knowledge and their autonomy more openly valued in response to medical paternalism and dominance, they also give rise to other ethical concerns. They may participate in the invisibilisation of the experiences, needs and interests of many as they tend to emanate from more privileged social groups and as they sometimes emphasise personal responsibility over collective support. Though these possibly problematic implications must be recognised and acknowledged, such direct involvement from online communities has given a much needed shake-up to discourses surrounding over-medicalisation and under-medicalisation, possibly impacting medicalisation and demedicalisation processes themselves.

References

1. Conrad, P.: Medicalization and social control. Ann. Rev. Sociol. **18**, 209–232 (1992). https://doi.org/10.1146/annurev.so.18.080192.001233
2. Zola, I.K.: Medicine as an institution of social control. Sociol. Rev. **20**, 487–504 (1972). https://doi.org/10.1111/j.1467-954X.1972.tb00220.x
3. Maître, J.: Anorexies religieuses, anorexie mentale: essai de psychanalyse sociohistorique. de Marie de l'Incarnation à Simone Weil. Éd. du Cerf, Paris (2000)
4. Illich, I.: The medicalization of life. J. Med. Ethics **1**, 73–77 (1975). https://doi.org/10.1136/jme.1.2.73
5. Conrad, P.: The Shifting Engines of Medicalization. J. Health Soc. Behav. **46**, 3–14. Chicago University Press (2005)
6. Parsons, T.: The sick role and the role of the physician reconsidered. The Milbank memorial fund quarterly. Health Soc. **53**, 257–278 (1975). https://doi.org/10.2307/3349493
7. Casilli, A.A., Tubaro, P., Méadel, C.: Le phénomène pro ana: troubles alimentaires et réseaux sociaux. Presses des Mines, Paris (2016)
8. Mercklé, P.: La sociologie des réseaux sociaux. la Découverte, Paris (2016)
9. Louvel, S.: The impact of interdisciplinarity on disciplines: a study of nanomedicine in France and California. Rev. Fr. Sociol. **56**, 75–103 (2015)
10. Froger-Lefebvre, J.: « Les voix du rétablissement » : sociologie politique des groupes de parole, le cas des Outremangeurs Anonymes. http://www.theses.fr/2020PA100070, (2020)
11. Szasz, T.S.: The myth of mental illness. American Psychological Association, United States, North America (1992)
12. Foucault, M.: Histoire de la sexualité I. Gallimard (1994)
13. Foucault, M.: Histoire de la médicalisation. Hermès, La Revue. **2**, 11–29 (1988)
14. Turner, B.S.: The body and society: explorations in social theory. B. Blackwell, Oxford, [Oxfordshire] , New York, NY, USA (1984)
15. Busfield, J.: The concept of medicalisation reassessed. Sociol. Health Illn. **39**, 759–774 (2017). https://doi.org/10.1111/1467-9566.12538
16. Rose, N.: Beyond medicalisation. Lancet **369**, 700–702 (2007). https://doi.org/10.1016/S0140-6736(07)60319-5

17. Sholl, J.: The muddle of medicalization: pathologizing or medicalizing? Theor. Med. Bioeth. **38**(4), 265–278 (2017). https://doi.org/10.1007/s11017-017-9414-z
18. Maître, J.: Mystique et féminité: essai de psychanalyse sociohistorique. Cerf, Paris (1997)
19. Gull, W.W.: Anorexia Nervosa (Apepsia Hysterica, Anorexia Hysterica). Trans. Clin. Soc. London **7**, 8–22 (1874)
20. Lasègue, E.-C.: De l'anorexie hystérique. J. français de psychiatrie. **32**, 3–8 (1873)
21. Conrad, P., Angell, A.: Homosexuality and remedicalization. Society **41**, 32–39 (2004). https://doi.org/10.1007/BF02688215
22. American Psychiatric Association: Committee on Nomenclature and Statistics: Diagnostic and statistical manual of mental disorders: DSM-II. American Psychiatric Association, Washington, D.C. (1968)
23. Nettleton, S.: 'I just want permission to be ill': towards a sociology of medically unexplained symptoms. Soc. Sci. Med. **62**, 1167–1178 (2006). https://doi.org/10.1016/j.socscimed.2005.07.030
24. Fan, P.T.: Fibromyalgia and chronic fatigue syndrome. APLAR J. Rheumatol. **7**, 14 (2004)
25. Dickson, A., Knussen, C., Flowers, P.: Stigma and the delegitimation experience: an interpretative phenomenological analysis of people living with chronic fatigue syndrome. Psychol. Health **22**, 851–867 (2007). https://doi.org/10.1080/14768320600976224
26. Fibromyalgia Forum. https://www.fibromyalgiaforums.org/community/
27. Sapey, B., Bullimore, P.: Listening to voice hearers. J. Soc. Work. **13**, 616–632 (2013). https://doi.org/10.1177/1468017312475278
28. Hearing Voices Network: About HVN. https://www.hearing-voices.org/about-us/
29. Styron, T., Utter, L., Davidson, L.: The hearing voices network: initial lessons and future directions for mental health professionals and systems of care. Psychiatr. Q. **88**(4), 769–785 (2017). https://doi.org/10.1007/s11126-017-9491-1
30. Casilli, A.: Resumen. Communications **92**, 111–123 (2013)
31. Saguy, A.C.: What's wrong with fat? s.n., S.l. (2014)
32. Hughes, E.C., Riesman, D., Becker, H.S.: Social Role and the Division of Labor. In: The Sociological Eye. Routledge, Milton Park (1984)
33. Avril, C., Vacca, I.R.: Se salir les mains pour les autres. Métiers de femme et division morale du travail. Travail, genre et societes. **43**, 85–102 (2020)
34. Becker, H.S.: Outsiders. Studies in the Sociology of Deviance. Free Press of Glencoe, Chicago (1963)
35. Loi du 1er juillet 1901 relative au contrat d'association - Légifrance
36. Suissa, J.: Le monde des AA: alcooliques, gamblers, narcomanes. Presses de l'Université du Québec, Québec, Québec (2009)
37. Gusfield, J.R.: Symbolic Crusade: Status Politics and the American Temperance Movement. University of Illinois Press, Champaign (1963)
38. Mariette, A., Pitti, L.: Médecin de première ligne dans un quartier populaire. Agone **58**, 51–72 (2016)
39. Bacqué, M.-H., Biewener, C.: L'empowerment, une pratique émancipatrice ? La Découverte (2015)
40. Beauchamp, T.L., Childress, J.F.: Principles of Biomedical Ethics. Oxford University Press, New York (1989)
41. Gross, O., Gagnayre, R.: Caractéristiques des savoirs des patients et liens avec leurs pouvoirs d'action : implication pour la formation médicale. Revue française de pédagogie. Recherches en éducation, pp. 71–82 (2017). https://doi.org/10.4000/rfp.7266
42. Bargel, L.: Learning a craft that can't be learned. Careers within party youth organizations. Sociologie **5**, 171–187 (2014)

43. Macdonald, W., Kontopantelis, E., Bower, P., Kennedy, A., Rogers, A., Reeves, D.: What makes a successful volunteer expert patients programme tutor? Factors predicting satisfaction, productivity and intention to continue tutoring of a new public health workforce in the United Kingdom. Patient Educ. Couns. **75**, 128–134 (2009). https://doi.org/10.1016/j.pec.2008.09.024
44. Wolff, J., de-Shalit, A.: Disadvantage. Oxford University Press, Oxford (2007)
45. Gelly, M., Pitti, L.: Une médecine de classe ? Inégalités sociales, système de santé et pratiques de soins. Agone **58**, 7–18 (2016)
46. Giroux, É.: La "santé publique de précision" : un changement de paradigme pour la santé publique ou la perte de son âme? Actualité et dossier en santé publique (2021)

Social Intelligence Design for Social Computing

Renate Fruchter[1], Toyoaki Nishida[2]([⊠]), and Duska Rosenberg[3]

[1] PBL Lab, CIFE, Stanford University, California, USA
[2] University of Fukuchiyama, Kyoto, Japan
toyoakinishida@gmail.com
[3] University of London, London, UK

Abstract. Social intelligence Design (SID) is about the impact and significance of technology in our lives, work, home, and on the move. Social Intelligence is defined as the ability of people to relate to, understand and interact effectively with others. The central question is how it can be empowered using emerging technologies.

In the information society, new technologies have huge impact on the way people work, interact, and collaborate. They influence the ways they develop personal relationships, as well as enhancing interpersonal communication and professional performance.

At the same time, these technologies might amplify miscommunication and bring about new threats and fears. A notorious example is a flaming war, a barrage of postings containing abusive personal attacks, insulting, or chastising replies to other people, which has not been so disastrous before the networked society. Moreover, they provide effective channels for spreading misinformation, catfishing or grooming potential victims, thus amplifying the damaging effects of social misdemeanours.

In this paper we address the issues of both beneficial and damaging impacts of emerging technologies on social intelligence and suggest ways of addressing them in the context of social intelligence design (SID). SID is focused not only on the technology design but also on the cognitive, social and organizational context of its use. In this paper we take a holistic approach to emerging technologies inspired by Artificial Intelligence research, bearing in mind their real-life significance [1].

Keywords: Social intelligence design · Community support systems · Collaboration technologies

1 About the Authors

Toyoaki Nishida is a computer scientist working on Conversational Informatics within Artificial Intelligence and Human-Computer Interaction. He studies intelligent actors, both natural and artificial, interacting with each other. This research considers the complexity of such interactions from a computational point of view and examines design principles for intelligent human-agent systems.

Renate Fruchter is a designer and engineer whose research is focused on the transformative impact of media channels in remote collaboration on design tasks, how technology can facilitate the communication, collaboration, and coordination concerning task,

© The Author(s), under exclusive license to Springer Nature Switzerland AG 2022
G. Meiselwitz (Ed.): HCII 2022, LNCS 13315, pp. 545–558, 2022.
https://doi.org/10.1007/978-3-031-05061-9_38

team, organisation and the relation between technology and people-place-process in the emerging world where the physical and virtual merge.

Duska Rosenberg is a linguist and social scientist, specialising in informatics, whose main research is in human communication and individual, social and organisational impact of technology. In particular, she studies how intelligent actors, both natural and artificial, organise joint activities to create mutual understanding, trust, social cohesion and engagement with technologies, as well as interpersonal communication mediated by technologies.

The three have worked together over many years. Renate is using scientific results from Duska and Toyoaki and develops engineering solutions. She then helps us evaluate A—how our science is implemented and B—how users will use it.

The SID concept has motivated our collaboration, as we are concerned with what technology does for people, what it does to people and what the costs are, not only in terms of money but also considering the changes technology introduces into their lives—whether they are in the workplace at home, or on the move.

We do not individually have the answer—but together we could find one. The task is to integrate our individual perspectives and then creatively address the gaps. From the perspective of this parallel session, the aim of this paper is to propose a unifying theoretical basis for addressing the issues relevant to emerging social technologies that the other speakers will present.

2   Background

The concept of Social Intelligence has been defined by a team of multi-disciplinary researchers working together on the design of computer systems that enhance the opportunities for human action and co-action in computer-mediated environments. The term 'social intelligence' thus appears in the inter-disciplinary context that includes studies within social science focused on communication, collaboration, and teamwork. It also includes social awareness of political and public constraints of human action and interaction. From the social point of view, SID approach [1] is crucially involved with impacts of technology on people, on the social and cognitive processes in their interactions, as well as emotion, awareness, trust and in fact, all phenomena that we understand to be profoundly human in nature.

Studies of computer-based, artificial intelligence as carried out within computer and cognitive sciences are particularly relevant to the notion of social intelligence. Today's research in Artificial Intelligence and its applications relies substantially on the scientific work that aims to adopt the results of scientific investigations to inform the design of systems that support everyday human activities.

SID is also an eco-system as the relationships created by way of action and interaction result in beliefs, norms of appropriate behaviour, affection and loyalty. These in turn colour future action, leading to growth and emergence of new beliefs and emotions. In this paper we assemble a glossary of concepts, ideas, terms related to SID so that we can show its scope from theoretical, methodological and implementational perspectives.

The main research themes within SID have been defined as integral parts of a unified framework for the design of socially aware systems [2, 2] that include:

- Multidisciplinary perspectives—exploring social intelligence at the intersection of different disciplines that bring technology, social behaviour and action aspects together, such as, investigating the relationship between people who interact with one another, environments where this interaction takes place and the communicative processes that they engage in. These then provide the social context where technologies are implemented and used.
- Natural Interactions—covering theory, modelling and analytical frameworks that have been developed with SID in mind, to generate precise definitions of the concepts and relationships that influence technology development and application in the social context of use. SID approach is informed by the studies of situated computation, embodied conversational agents, sociable artifacts, socially intelligent robots as well as emerging forms of social computing.
- Communities—one of the most significant outcomes of natural interactions in real-life is the formation of social groups, collaborating teams, that is, communities that are characterised by shared aims, values and rules of appropriate behaviour. Technologies that shape community creation include the media that support communication in online, networked and anonymous communities.
- Collaboration Technologies and tools—represent innovations to support interactions within communities, covering a range from knowledge-sharing systems, multi-agent systems and interactive systems. Increasingly, emerging technologies make use of artificial intelligence, big data, mixed and virtual reality applications that augment and enhance people's experience of the world they inhabit.
- Application Domains—including technology design, smart workspaces, education, e-commerce, entertainment, digital democracy, smart cities, policy and business that all have particular requirements for technologies to fit the social contexts of their use.

3   Social Intelligence Design Workshops

A series of SID workshops explored various ways of designing socially aware systems based on our scientific understanding of their impact on the lives of their users in various application domains.

Traditionally, social intelligence is the ability of an actor/agent to relate to other actors/agents in a community, to understand them, and interact effectively with them. This is contrasted with other kinds of intelligence such as problem-solving intelligence (ability to solve logically complex problems) or emotional intelligence (ability to monitor one's own and others' emotions) and more generally, the ability to use the information from the social and physical world outside to guide one's thinking and actions.

In addition, social intelligence might be attributed to a collection of actors/agents and defined as an ability to manage complexity of interpersonal relationships and learn from experience as a function of social structure. This view emphasizes the role of social conventions that constrain the way individual agents interact with each other with reference to the social structures that emerge in the course of such interactions.

SID thus integrates our understanding of social intelligence with the principles of social computing in a uniform framework for the design of technologies that support action and interaction in socially significant situations. This can be accomplished by bringing together multi-disciplinary scientific approaches that address real-life cognitive and social phenomena.

Scientific understanding of these phenomena then provides the theoretical basis for predicting and evaluating the effect of a particular communication medium on natural interactions and community building. It also underpins the engineering approaches to the development and implementation of systems—ranging from collaboration support systems that facilitate joint, goal-oriented actions to community support systems used in large-scale online discussions.

The analytical and the engineering approaches are thus complementary to each other as good systems cannot be built without good understanding and vice versa.

4   Research Issues Addressed in SID Workshops

The main research question is how new technologies mediate human communication and collaboration across geographical and cultural divides.

New technologies such as socially intelligent agents, including cognitive robots, collaborative information environments, personalized information tools and interactive community media, are rapidly changing our language, our work and our lifestyle in general.

Typical applications of SID are group/team support systems and community support systems. Community support systems provide rather long-range, bottom-up communicative functions in the background of daily life. Major research tasks are to understand:

1. how people create shared awareness with other members of their social group,
2. how we as researchers study networks of shared knowledge,
3. how shared awareness and shared knowledge characterise communities.

In contrast, group/team support systems focus on facilitating closer collaboration among members and emphasize more task-driven, short-range collaboration, although awareness the social process in such teamwork is equally emphasized.

From this point of view, SID is also concerned with design and implementation of novel communication means for interaction among people and agents. The scope ranges from preliminary and preparatory interactions among people such as knowing who's who, to more focused interaction aimed at creating conditions for mutual understanding. People jointly negotiate the common ground that provides the basis for complex social processes, such as group formation, collaboration, negotiation, public discussion or social learning [3]. Theoretical aspects, as well as pragmatic aspects, should be considered in designing, deploying, and evaluating social intelligence support tools. The scope of SID as a discipline is summarized in Table 1.

Table 1. Horizon of social intelligence design.

- methods of establishing the social context
  - awareness of connectedness
  - circulating personal views
  - sharing stories
- embodied conversational agents and social intelligence
  - knowledge exchange by virtualized egos
  - conversational agents for mediating discussions
  - a virtual world habited by autonomous conversational agents
  - social learning with a conversational interface
  - conversations as a principle of designing complex systems
  - artifacts capable of making embodied communication
- collaboration design
  - integrating the physical space, electronic content, and interaction
  - using multi agent system to help people in a complex situation
  - evaluating communication infrastructure in terms of collaboration support
- public discourse
  - visualization
  - social awareness support
  - integrating Surveys, Delphis and Mediation for democratic participation
- theoretical aspects of social intelligence design
  - understanding group dynamics of knowledge creation
  - understanding consensus formation process
  - theory of common ground in language use
  - attachment-based learning for social learning
- evaluation of social intelligence
  - network analysis
  - hybrid method

4.1 Theoretical Aspects of Social Intelligence Design

At the theoretical level, consideration of the nature of social intelligence has been made from various perspectives, as illustrated in Fig. 1, and presented in the literature on SID by a number of authors. (cf. https://sites.google.com/view/toyoakinishida/sid).

From the theoretical base, mathematical modelling of real-life reasoning [4] emphasizes the role of mathematical models in cognitive domains. He addresses modelling real-life human reasoning to gain greater understanding that might serve as a basis for teaching reasoning skills to technology-naive people. To this end, he presents a framework of reflective and evidence-based reasoning.

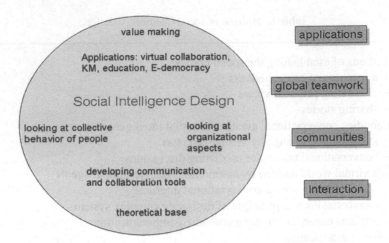

Fig. 1. Nature of social intelligence from various perspectives.

From the cognitive psychological perspective, Miura et al. [5] investigate communications in online chat to find out the optimal interval for interactive communication and to see how chat skills affect human communication behaviour. Through psychological experiments based on the two-phase human information processing model consisting of the information acquisition phase and the information output phase, they obtained several interesting findings such as that experienced users tend to post shorter messages more frequently than the inexperienced users when no congestion takes place or that the existence of message congestion might increase the number of messages posted by experienced users.

Sociologists point out that introduction of technologies has produced profound changes in the structure of our societies in a very short time [6]. Indeed, there was a warning against the unqualified acceptance of rapid technological change before a knowledge base of experiences of individuals is formed.

Likewise, the interpretive analysis approaches also suggest that the technologizing of society is establishing a prevailing knowledge base that provides a range of new opportunities for not only communicating at work, but also increasing mobility within and without the workplace, raising employment prospects, and nurturing a greater sense of self-sufficiency, accountability and responsibility. K. Gill [7] analyses community communication from philosophical perspectives, characterizing communities in terms of the structure of relationship among their members. Notsu and Katai [8] propose a framework of sympathetic caring interactions, by combining the Roy adaptation model (a nursing theory) and the notion of perceptual balance in Heider's Naive Psychology. They present a computational method for diagnosing a self-narrative of a client. If an unbalanced substructure is found in the client's network for self-narrative, their method can suggest how to amend it by introducing two kinds of caring interactions called "know new relations" and "incorporating new relations from another balanced network".

Beyond interpersonal interactions, the advanced technology brings about cognitive and sociable artifacts that may cohabit with people. Based on the experiences with several emotional conversational agents (ECAs) with varying degrees of verbal and nonverbal communication skills, Nijholt et. al [9] study the formation of long-term relationships with their human conversational partner, such as friendship, a voluntary interdependence between two persons over time, that is intended to facilitate social emotional goals of the participants, and may involve varying types and degrees of companionship, intimacy, affection, and mutual assistance. He attempts to formulate the discussion using theories of social psychology such as reinforcement theory, social exchange theory, or equity theory, then explores the possibility of adopting the main aspects of human-human friendship to human-ECA friendship, discussing how the findings can be incorporated in the ECA-design process, using a scenario-based design. Cardon and Campagne [10] discuss artificial emotion and sensation for cognitive artifacts in search for better relationship with human users or other agents.

4.2 Community Support Systems

Community is an important subject for SID both from practical and academic points of view. Significant portion of SID is concerned with a new approach to understanding communities that can inform the development of community support systems.

Fukuhara et. al [11] report on a community support system called Public Opinion Channel (POC), which is a participatory automatic broadcasting system that collects messages from community members and allows edited messages to be disseminated to the community. POC was designed not only to help the community knowledge creation process by facilitating information circulation in a community, but also help researchers make field work and psychological experiments to obtain better understanding by collecting and analyzing log and message data. Horita [12] presents a couple of methods for soliciting interesting "stirring" arguments from an archive of online discussions. The algorithm is implemented an incorporated in CRANES, an argumentation support system that has the capability of visualizing the structure of arguments, facilitating a dialogue among a large number of users and evaluating various characteristics (e.g., strength, coherence, etc.) of arguments. Murayama reports an integrated community communication called S-POC. S-POC extends POC in a number of ways. First, it allows the user to create vivid dynamic contents containing video information and dynamic image using digital camera work. Second, it incorporates an interface agent with a high-level communication skills. Third, it is designed to allow novice users to create, modify, and manage dynamic contents. In particular, it employs what is called the card-to-presentation, which releases the user from a complex agent programming. Fourth, it is designed to integrate external communication components such as CRANES as its integral part so that the user can receive the benefits from multiple assets in a uniform fashion.

Social psychological theories and their implementation help us design and evaluate community support systems. Azechi [13] proposes an information humidity model in which he classifies communities into three types, wet, anonymous, and dry communities, according to how much meta information is included in the communication medium for the given community. Azechi [13] discusses how meta information affects the nature of communications in a community. Based on experiments, he argues that lack of personal characterization and identification in the dry community tends to encourage message posting but give a biased view that the number of people who have the opposite opinions is smaller than the fact. The former may encourage the increase of shared knowledge but the latter may prevent healthy discussion. Miura et. al [5] investigate the features that may facilitate community communication. Based on psychological experimentations, he suggests both newcomers and accumulation of messages may activate message postings, replies to other messages in particular. They describe a model that explains the dynamism of 2channel, which is the most popular online forum in Japan. The model is obtained by applying the structural equation modelling to log data in 5748 message threads taken from 2channel with eight indices, suggesting that the dynamism may be caused by three latent factors called specific expression, discussion type, and chitchat type.

4.3 Collaboration Technologies

Collaboration technologies address goal-oriented interactions within communities, in search for better support of virtual teamwork and other collaborative work. Fruchter [14] presents a new perspective of the impact of collaboration technology on the degrees of engagement and specific interaction zones in interactive workspaces. She models the collaboration in terms of bricks (physical aspects of workspaces), bits (rich electronic content such as video, audio, sketching, CAD), and interaction (the ways people behave in communicative events). She presents an innovative multimodal collaboration technology, called RECALL™, that supports the seamless, real time capture of concept generation during project brainstorming and project review sessions. RECALL™ is deployed in an interactive workspace that supports real project review sessions called Fishbowl. Based on the analysis of interactions in Fishbowl sessions, she suggests that the boundary of action, reflection, and observation zones, emerges depending on the degree of participants' engagement. Fruchter also presents a study of the use of an innovative videoconference system called the Virtual Auditorium (VA), and describes how it was tested and deployed in a cross-disciplinary, geographically distributed teamwork course offered at Stanford in collaboration with university partners in Europe and US. She suggested that a videoconference platform such as VA brings about "the big win" as all remote participants were visible and created a sense of togetherness. This is currently the standard UI in all videoconference platforms. The RECALL™ tool served the dual purpose as a communication channel and collaboration platform, as well as a data collection instrument that enabled participants to revisit the rich multi-modal and multi-media session. This approach was at that time innovative and supported context-based SID R&D. Today most of the online applications and web and social media platforms collect data seamlessly as input to sophisticated machine learning algorithms to customize user feeds. This in turn raises awareness of data privacy and confidentiality that needs to be addressed.

S. Gill [15] sheds light on skilled cooperative action such as being able to understand the communicative situation and knowing how and when to respond appropriately for the purpose at hand. Skilled cooperative action is dynamic representation of knowledge, characterized as a form of social intelligence for sustainable interaction where knowledge in co-action occurs. Creating mediating interfaces that can become invisible to us as an extension of ourselves is a challenge of social intelligence design. She reports a study of the way people use surfaces that afford graphical interaction, in collaborative design tasks, in order to inform the design of intelligent user interfaces. It is a descriptive study rather than a usability study, to explore how size, orientation, horizontal and vertical positioning influence the functionality of the surface in a collaborative setting. In search for systematic study of architectural practice, Martin et al. [16] consider storytelling as a vehicle for studying projects that are in the process of being designed and built, for stories provide a dense, compact way to deal with and communicate the complex reality of a real-world project, while respecting the interrelated nature of events, people and circumstances that shape its conception. They investigate how stories can be stored, organized and accessed so as to turn the growing story repository into a convenient instrument for students, educators and practitioners.

4.4 Application Domains

The potential application of the SID approach is broad. The work presented at SID-2003 spans education, e-commerce, and chance discovery. Fukuhara et. al [11] discuss how POC can be applied to the learning environment. He reports an educational practice at a junior high school and analyzes the result. Based on the experience, he proposes the TEC (Think-Express-Communicate) model as a model of student-centered media-based learning. Shoji et. al [17] address the customer support problem in online-shopping. They propose the "purchasing as concept articulation" view and presents a system called S-Conart that facilitates purchasing as concept articulation through support for customer's conception with spatial-arrangement style information presentation and for their conviction with scene information presentation.

Cornillon et. al present a conceptual framework for the study of social intelligence in a real-life environment, focused on dialogue organisation in argumentation, in particular how our understanding of dialogue phenomena in mediated communication may help us to support natural interaction in classroom debates and how this can be used in the design, implementation and evaluation of a feedback adviser tool to automatically search for topics in debates.

5 SID Research Today

The idea of SID served as a basis for further development. This section portrays several new directions.

5.1 Conversational Informatics and Common Ground Technology

At the micro-level, social intelligence manifests itself as joint activities of various forms in conversation, as suggested in [3]. People are proficient in collaboratively forming and maintaining gatherings thereby shaping and cultivating collective thoughts through fluent conversational interactions. The critical value of human intelligence centers on bearing and cultivating new thoughts until they become full-fledged. Understanding conversation is critical to uncover the mechanism of inter-personal intelligence that help people develop shared thoughts in collaboration. Technological enhancement of conversation is no less important. Although new thoughts need to be discussed and extended from different angles in their infancy, rarely is everyone a skillful author being able to well express thoughts in progress including incompleteness, ambiguity, vagueness, and even inconsistency.

Conversational informatics [18] is a field of research that focuses on investigating conversational interactions and designing intelligent artifacts that can augment conversational interactions. The field draws on a foundation provided by artificial intelligence, natural language processing, speech and image processing, cognitive science, and conversation analysis. Platforms need to be built to support a broad range of conversation augmentation systems among people, their avatars and artificial conversational agents across cyber and physical spaces. Methods for conversation measurement and analysis are necessary to understand conversations in a quantitative fashion. Each method should be able to help developers specify the behaviour of the target system in detail. Methods for model building are needed to understand conversations in a structured fashion and determine the communicative behaviours of conversational agents. A suite of tools for content production needs to be developed so that content producers can easily create content for augmented conversation systems without much technical knowledge. Applications need to be built so that people can benefit from augmented conversational environment. Evaluation is needed to understand the achievement and limitation of individual projects.

We assume common ground—a shared cognition among people—underlies conversation and characterize conversation as a continuous update of common ground in a collaborative fashion by bringing together pieces of knowledge, belief, and experience. Common ground is versatile, uncertain, complex, and ambiguous, as we can learn from literary works, movies, their reviews, etc. People normally want to stabilize the common ground process and capture it for deeper and more solid understanding of our daily communication. The process of design and analysis of the common ground process should be bootstrapped to overcome its versatility, uncertainty, complexity, and ambiguity. Meanwhile, people often take adventurous development of common ground as a source of entertainment.

Conversation envisioning is basic to design and analysis of the common ground process. We would like to portray in a computational fashion how common ground is updated in conversation. On the engineering side, we believe it will help us build a conversational AI that can participate in our conversation and help us establish and maintain common ground. On the scientific side, it will lead us to better understanding of conversation, and eventually our mind. There are numerous challenges in conversation envisioning, such as building AI actor and producer.

5.2 SID and Emerging Technologies

We are in the midst of the 4th industrial revolution.

- What role can SID in general, and conversational informatics and emerging technologies specifically, play as enablers towards a more ecologically, economically, and socially sustainable future environment?

- How can they become an integral part of innovative holistic approaches to address challenges such as climate change, globalization, digitalization, and skilled workforce shortage?

The 4th industrial revolution — scale, speed, interoperability, and intelligence—drive digital transformations. It is a time when the physical and digital worlds merge; where the world becomes a big data set as everyone and everything that is connected to the Internet becomes data source; where AI, machine learning, VR/AR/MR/XR, robotics, autonomous mobility will continuously reshape how and where we live, work, and learn. Convergence of these emergent technologies can act as exponential accelerators as well as raise new challenges.

Leveraging principles from conversational informatics, building common ground theory, and conversational agents is critical as we see the emergence of digital twin solutions that give agency to inanimate objects that are part of our daily life such as airplanes, cars, construction equipment, buildings, and infrastructure. For instance, Obayashi Corporation provides an innovative real-world example on their fully assisted and autonomous construction equipment[1] on the construction site of a dam in Japan. Fruchter's PBL Lab research team recently focused on developing, prototyping, and testing a computational framework to harmonize occupant well-being and sustainable building performance by giving both the human occupant and the building spaces agency towards building common ground and joint decision making [19, 20].

As digital twin local and national efforts are being launched, it will be critical to keep the human in the loop through humanistic conversational informatics and common ground principles embedded in future digital solutions.

5.3 SID and New Application Domains

Research into the relationship between people and technology — its cognitive, social and organisational impacts — currently involves the application of SID approach to the design of smart technologies and their application in digital inclusion. The theoretical basis originates from the Common Ground [3, 18] and was extended to the development of the people-place-process model for interaction in physical and Augmented Reality enhanced interaction spaces [21].

The role of space in the wider contexts for interaction both face-to-face and remote was studied in the EU-funded project SANE (Sustainable Accommodation in the New Economy) https://cordis.europa.eu/project/id/IST-2000-25257. Since emerging technologies, such as, Augmented Reality for example, opened up interaction spaces beyond their physical boundaries, new issues had to be addressed for both theoretical and implementational reasons. These are explored in a number of research projects[2] and discussed

---

[1] https://www.obayashi.co.jp/thinking/detail/project61.html.

[2] ACROSSING — (http://www.acrossing-itn.eu), 2016 – 2020; Caring Homes as Learning Environments, 2017; SWAN Innovation Project, 2009 – 2011; Ifdentity project, funded by UNESCO, 2010 – 2011; IS-VIT (Interaction Space of the Virtual IT Workplace), 2009 – 2010; Digital Evidence in Legal Practice (UK, Croatia, Saudi Arabia) 2020–2022 - details available from D. Rosenberg research@icomict.org.

in joint publications [22–26] aimed at answering three main research questions about emerging technologies and their end users in smart living environments, with special emphasis on the present research in Digital Inclusion of the Elderly and Vulnerable members of a local community[3]:

• How to keep them healthy?

We are developing usability frameworks to evaluate monitoring technologies in real-life living spaces with emphasis on gathering health-related data from patients, distributing patient data to medical and care personnel and representing the health data in the form suitable for professionals, non-professional carers and patients themselves.

• How to keep them safe?

We are developing usability frameworks to evaluate privacy-invading technologies such as monitors, sensors, video surveillance. The focus is on gradual introduction and refinement of already available alert systems, such as Apple Watch that detects falls, and the representation of information from various sources on user interfaces. This also includes application of Augmented and Mixed Reality in preserving patient privacy whilst providing necessary information for services such as doctors, nurses, firemen, etc.

• How to keep them happy?

By far the most common problem faced by vulnerable groups is loneliness, isolation and lack of opportunities to participate in the lives of their communities. We are exploring the potential of visualisation applications (Virtual, Augmented, Mixed Reality), emphatic and other cognitive robots, Artificial Intelligence in Art and similar emerging technologies. These may—when sensitively applied—provide not only entertainment, but also smart interactive services to combat both physical and social isolation, thus preventing or delaying the onset of mental illness, insofar as this may be due to impoverished living conditions.

References

1. Nishida, T.: Social intelligence design — an overview. In: Terano, T., Ohsawa, Y., Nishida, T., Namatame, A., Tsumoto, S., Washio, T. (eds.) JSAI 2001. LNCS (LNAI), vol. 2253, pp. 3–10. Springer, Heidelberg (2001). https://doi.org/10.1007/3-540-45548-5_1
2. Fruchter, R., Nishida, T., Rosenberg, D.: Understanding mediated communication: the social intelligence design (SID) approach. AI Soc. **19**(1), 1–7 (2005). https://doi.org/10.1007/s00146-004-0297-y
3. Clark, H.: Common ground. In: Using Language (Using Linguistic Books), pp. 92–122. Cambridge University Press, Cambridge (1996). https://doi.org/10.1017/CBO9780511620539.005

---

[3] Digital Inclusion project funded by the local authority awarded to Age UK and Multicultural Richmond Charities 2022 – 2024.

4. Devlin, K.: A Framework for modelling evidence-based, context-influenced reasoning (2003). https://web.stanford.edu/~kdevlin/Papers/ContextLogic.pdf
5. Miura, A., Matsumura, N.: Social intelligence design: a junction between engineering and social sciences. AI Soc. **23**, 139–145 (2009). https://doi.org/10.1007/s00146-007-0139-9
6. Nishida, T.: Social intelligence design for the web. Comput. **35**(11), 37–41 (2002)
7. Gill, K.S. (ed): Human-Machine Symbiosis: the Foundations of Human-Centred Systems Design. Springer (2012) ISBN-13:978–3–540–76024–5
8. Notsu, A., Ichihashi, H., Honda, K., Katai, O.: Visualization of balancing systems based on naïve psychological approaches. AI Soc. **23**(2), 281–296 (2009). https://philpapers.org/rec/NOTVOB
9. Nijholt, A., Stock, O., Nishida, T.: Social intelligence design in ambient intelligence. AI Soc. **24**, 1–3 (2009). https://doi.org/10.1007/s00146-009-0192-7
10. Cardon, J.C., Cardon, A.: Artificial emotions for robots using massive multi-agent systems; Social Intelligence Design International Conference (SID 2003), London (2003)
11. Fukuhara, T., Fujihara, N., Azechi, S., Kubota, H., Nishida, T.: Public opinion channel: a network-based interactive broadcasting system for supporting a knowledge-creating community. In: Howlett, R., Ichalkaranje, N., Jain, L., Tonfoni, G. (eds.) Internet-Based Intelligent Information Processing Systems, chapter 7, World Scientific Publishing (2003)
12. Horita, M.: Folding arguments: a method for representing conflicting views of a conflict. Group Decis. Negot. **9**(1), 63–83 (2000)
13. Azechi, S.: Informational humidity model: explanation of dual modes of community for social intelligence design. AI & Soc **19**, 110–122 (2005). https://doi.org/10.1007/s00146-004-0304-3
14. Fruchter, R.: Degrees of engagement in interactive workspaces. AI Soc. **19**(1), 8–21 (2005)
15. Gill, S.: Designing for Knowledge Transfer; Springer (2012). ISBN-13:978–3–540–76024–5
16. Martin, M., Heylighen, A., Cavallin, H.: The right story at the right time. AI Soc. **19**(1), 34-47 (2005)
17. Shoji, H., Hori, K.: S-Conart: an interaction method that facilitates concept articulation in shopping online. AI Soc. **19**(1), 65–83 (2005)
18. Nishida, T., Nakazawa, A., Ohmoto, Y., Mohammad, Y.: Conversational informatics: In: A Data-Intensive Approach with Emphasis on Nonverbal Communication, Springer (2014)
19. Fruchter, R.: When 21st Century Technologies Meet the Oldest Engineering Discipline, Keynote at the CIB W78 International Conference, Luxemburg (2021). https://youtu.be/jHFrv8iIC5Q
20. Grey, F.: Space-mate: a framework to harmonize occupant well-being and building sustainability, Ph.D. Thesis, Civil and Environmental Engineering Department, Stanford University (2019) https://purl.stanford.edu/ks134sx2073
21. Walkowski, S., Doerner, R., Lievonen, M., Rosenberg, D.: Using game controller for relaying deictic gestures in computer mediated communication. Int. J. Hum.-Comput. Stud. **69**(6), 362–374 (2011). ISSN 1071–5819, https://doi.org/10.1016/j.ijhcs.2011.01.002
22. Stipancic, T., Rosenberg, D.: PLEAA: a social robot with teaching and interacting capabilities. J. Pacific Rim Psychol. Spec. Issue Equity Qual. Learn. Glob. Digit. World. Forthcoming **15**(18344909211037019) (2021)
23. Stipancic, T., Rosenberg, D., Jerbic, B.: Computation Approach for Realisation of Context-Aware Robots, Conference on Information and Graphic Arts Technology (2018)
24. Stipancic, T., Nishida, T., Rosenberg, D., Jerbic, B.: Context driven model for simulating human perception — a design perspective. In: DCC 2016 (Design, Computing and Communication (2016) http://dccconferences.org/dcc16/

25. Villena, S., et al.: Image super-resolution for outdoor digital forensics. Usability Legal Aspects, Elsevier Comput. Ind. **98**, 34–47 (2018)
26. Wagner, I., He, Y., Rosenberg, D., Janicke, H.: User interface design for privacy awareness in eHealth technologies. In: 1st International Workshop on Ambient Assisted Living (AALEH 2016) in Proceedings of IEEE CCNC 2016 (2016) http://ccnc2016.ieee-ccnc.org/content/wor kshops

Influence Vaccination Policy, Through Social Media Promotion (Study: West Java, East Java, and Central Java)

Ekklesia Hulahi[✉], Achmad Nurmandi, Isnaini Muallidin, Mohammad Jafar Loilatu, and Danang Kurniawan

Departement of Government Affairs and Administration, Universitas Muhammadiyah Yogyakarta, Bantul, Indonesia

ekklesia.hulahi@gmail.com, nurmandi_achmad@umy.ac.id

Abstract. The number of Covid-19 cases in Indonesia is still high. One of the efforts made by the government is vaccination. West Java, East Java, and Central Java have also implemented vaccination policies. The priority targets for vaccination in these three provinces are the general public. This study looks at the effect of vaccination policies in West Java, East Java, and Central Java through promotions on social media. The rapid growth of cases globally has caused panic, fear, and anxiety during this time. This study uses a qualitative descriptive approach and Qualitative Data Analysis Software (QDAS) data taken through Twitter accounts (@dinkesJabar, @DinkesJatim, and @DinkesJateng). Furthermore, the data is encoded using NVivo 12 plus. Researchers will describe the results of the NVivo process to find out promotions related to vaccination in the three accounts. And the concepts used are Policy Influence Vaccination and Social Media. The results show that data is one of the critical factors in the current digitalization era, which can be an essential source of information. The government also uses big data to create work efficiency and effectiveness in policymaking as a form of effort. The effect of vaccination policy through social media promotion is not very effective because it can be seen from the rise of the @Dinkesjabar @Dinkesjatim and @Dinkesjateng Twitter accounts that the hashtag related to vaccination is still very minimal. Then the data obtained shows the success of immunization in West Java, East Java. Central Java is the vaccination policy through the door-to-door, mobile vaccination mobilization, mobile vaccine bus, and cooperation with the TNI-Polri, academics, and the community in general. This success fulfills what is targeted by the provincial government.

Keywords: Vaccination · Policy · Social media · Promotion

1 Introduction

The global pandemic caused by COVID-19 has affected more than 129 million people worldwide [1]. Coronavirus disease 2019 (COVID-19) is not the only primary concern from a medical and health point of view [1]. However, the COVID-19 coronavirus

G. Meiselwitz (Ed.): HCII 2022, LNCS 13315, pp. 559–567, 2022.
https://doi.org/10.1007/978-3-031-05061-9_39

pandemic has presented new challenges to be overcome and brought to the attention of nation-states. In particular, that is about how countries respond and work to prevent and stop the much wider spread of the virus.

Many countries carry out policies implemented within their territories, such as lockdown policies or policies to maintain social distance or social distancing from the community. Since its emergence, the coronavirus pandemic has continued to develop in various parts of the world, reaching 220 countries and regions on December 9, 2020. The government has tried to contain the outbreak by considering a series of actions, not all by public opinion. So far, the rapid growth in the number of cases globally has caused panic, fear, and anxiety among the public [2].

Some countries show success, but some show failure from this policy. These two policies are examples of social vaccines being undertaken, becoming an essential element in solving a public health crisis. Many vaccines are already mass-produced and supplied in many countries [1]. Vaccines are biological preparations that increase immunity against certain diseases. Immunization programs will be effective only if vaccines are distributed equitably and maintained quality [3]. The rollout of COVID-19 vaccinations continues to grow at a significant pace worldwide [4]. An already occurring pandemic in Indonesia since March 2020 began to show a bright spot with the discovery of a vaccine [5].

The Indonesian government has begun to prove its statement made at the end of 2020 that the Indonesian state has a target in early 2021. The Indonesian people will start getting the COVID-19 vaccine in stages and divided into several waves of vaccination. Vaccination efforts by stipulating Regulation of the Minister of Health Number 10 of 2021 concerning Implementation of Vaccination in the Context of Combating the Covid-19 Pandemic [1]. The current situation in Indonesia is the implementation of vaccination. For this reason, the use of social media has increased. This is due to connecting people from geographically different places and exchanging ideas.

What's more, many people seem to rely on Social Media for more information. People show different views, opinions, and emotions on various events due to the pandemic and vaccination policies through this social media [2].

The topic of vaccination is one of the themes that raises a series of questions on social media, most of which are related to the safety of the whole process [2] Consequently, many studies have analyzed the impact of different social media campaigns on vaccination doubts on general public sentiment regarding the vaccination process. Vaccine indecision (VH) is one of the most critical threats to global health, especially in low- and middle-income countries [6]. relatively short time required for vaccine development [2] by using the public Media as a source of information about COVID-19, particularly considering COVID-19 as a health threat. A better understanding of the factors that contribute to vaccine doubt is urgently needed to dispel doubts, increase vaccination rates, and create herd immunity, to stop further transmission of the virus [7].

The success of the National Vaccination can be seen from several regions that have shown success, namely the Provinces of West Java, East Java, and Central Java. Socialization is one method to convey information about the benefits of vaccination [8] and support the vaccination mobilization program. federal involvement of the military, academics, and service users to improve welfare services to help better manage their difficulties [9]. Service through social media socialization is a strategy to mobilize vaccinations in the community and understand national vaccination through social media [10]. Social Media is also a communication using the internet and can be accessed by many people who have Internet access via smartphone or laptop. With social media, people can access information, one of which is information about national vaccinations. Therefore, this study can describe how vaccination mobilization from West, East, and Central Java areas through social Media informs national vaccination facilitation policies or programs and shows the success of vaccination in Indonesia, especially in West Java, East Java.

2 Literatur Review

2.1 Policy Influence Vaccination

Policy Influence Vaccination or the influence of vaccination policy[22] to planning, making and formulating decisions, implementing decisions, and evaluating the impact of implementing these decisions on the people who are the target of the policy (target group). Vaccination Policy is an Indonesian government program; government policies can affect compliance with containment, delay, and mitigation policies such as physical distancing, hygienic practices, use of physical barriers, implementation of testing, contact tracing, and vaccination programs. [21] to systematically implicate the public policy process to improve vaccine positioning, messaging, and administration. These insights are expected to contribute to policies that can accelerate vaccination programs and bring the nation closer to the goal of immunity [11].

2.2 Social Media

The influence of social media [6] is critical because Social Media is a mode of communication that uses the internet and can be accessed by everyone who has an internet network via smartphones and laptops. Social Media is used to access information, express opinions, participate in surveys, and share information. There are five social media functions, including (a) using media for various real-time information services with community consultation. (b) using social media to provide the public with information about services, tariffs, updates about ongoing and upcoming projects. And precisely, information related to services. (c) using social media to engage with the public based on their feedback and sentiments through social media analysis. (d) using social media to promote public transport services and increase passengers, (e) collecting data from social media. Media advocates for organizational goals and regulates them [20]. Therefore, it is essential to distribute information about the use of vaccines in combating the spread of the coronavirus in society through digital advertising, social media, and Media. It is hoped that in the future, it can be used as a basis for government agencies to invite these communities to collaborate in disseminating the vaccination program [19].

3   Research Method

To answer the problem formulation above, this research uses a qualitative descriptive approach and Qualitative Data Analysis Software (QDAS) data taken through Twitter accounts (@dinkesJabar, @DinkesJatim, and @DinkesJateng). Furthermore, the data is encoded using NVivo 12 plus. Researchers will describe the results of the NVivo process to find out promotions related to vaccination in the three accounts using the hashtags used. [16].

4   Results and Discussion

4.1   Vaccination Goals and Targets

Government plans a five-group priority list of recipients of coronavirus vaccines. Goal Vaccines are one of the most critical aspects of health care [12]. The world's first proven and effective vaccine for [18]. To prevent safety and the need for vaccines, the government provides vaccinations to target the elderly, the general public and vulnerable people, public officers, human health resources, and children aged 12–17 years. This effort was taken to achieve the government's target of bringing up herd immunity or group immunity against coronavirus transmission.

The targets and targets for the first dose of vaccination can be seen in the diagram below

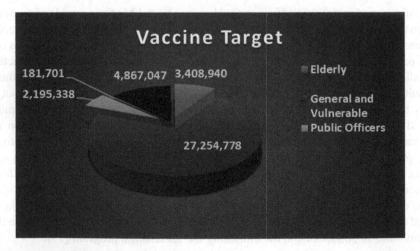

According to the diagram above, it can be seen that the government is targeting the first dose of vaccination at the elderly target of (3,408,940) General and Vulnerable

People (27,254,788,) Public Officers (2,195,338), Health Human Resources (181,701), and Age 12–17 Years (4,867,047).

Diagram 2. Target Coverage and Vaccination Targets for the 2nd Dose in West Java, East Java, and Central Java

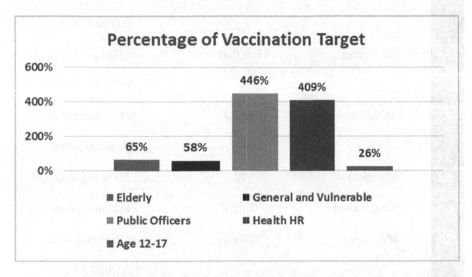

Due to the higher risk of complications associated with vaccine-preventable infections. (Boucher et al. 2021) The targets and targets for vaccination are the Elderly, General and Vulnerable Community, Public Officers, Health Human Resources, and Age 12–17 years. The success of a vaccination program depends on undeniable scientific safety data combined with a high level of public acceptance and population coverage [17].

4.2 Vaccination Promotion Strategy in West Java, East Java, and Central Java Through Social Media

To maximize the absorption of the COVID-19 vaccine, health authorities must promote the vaccine's effectiveness, proactively communicate the presence or absence of vaccine side effects, and ensure prompt and extensive media communication about local vaccine coverage [13].

No.	Hastag	Amoun and Percentage	Akun Twitter
1.	#Sobatsehat	65	@Dinkesjabar
2.	#Germas	42 & 0,7%	@Dinkesjabar & @Dinkesjateng
3.	#Covid19	11 & 0,4%	@Dinkesjabar & @Dinkesjateng
4.	#Dirumahaja	10 & 0,4%	@Dinkesjabar & @Dinkesjateng
5.	#Jatimsehat	12	@Dinkesjatim
6.	#Cegahcovid	0.10%	@Dinkesjatim
7.	#Vaksinasi	0,40%	@Dinkesjatim & @Dinkesjateng
8.	#Lawancovid	0,40%	@Dinkesjateng
9.	#Jagajarak	0,10%	@Dinkesjateng
10.	#Pakaimasker	0,20%	@Dinkesjateng

Based on the table above, the Twitter accounts @DinkesJabar @DinKesJatim, and @DinKesJateng with the highest number of hashtags are #SobatSehat with 65, #Germas 42 and 0.7%, #Covid19 11 and 0.4%, #Dirumahaja 10 and 0.4% hashtags. #JatimSehat 12, #CegahCovid 0.1%, #Vaccination 0.4%, #LawanCovid 0.4%, #JagaJarak 0.1%, and #Use Mask 0.2%. This information inspired a new idea to use Twitter as the primary social media tool to collect and inform [14]With social media, citizen participation in government is increasingly moving online, and initiatives are becoming passive. Social Media provides three essential functions for developing society:

- Providing information.
- Reducing the costs of political participation.
- Increasing the ability of opposition forces to mobilize.

Media can generate public political awareness because of its features that facilitate interactivity, proximity, and easy access. The position of the Media as a new social movement strategy plays a role in facilitating reasoning and thinking in gathering support [23]. Then, to answer how Twitter functions in promoting vaccination-related. Twitter carries out the promotional model accounts @DinkesJabar, @DinKesJatim, and @DinKesJateng. With the above facts, the promotion is not direct hashtag vaccination (Covid and Policies, n.d.), but the increase in immunization in these three provinces has met the target. Because in addition to promoting on social media, the provincial government carries out the door to door vaccinations, mobile vaccine mobilization, and Mobile Vaccine Buses to the community and involves the TNI-Polri and Academics to make the Vaccination Program a success so that there is an increase in vaccination policies in the West Java, East Java and East Java areas. Central Java.

5 Conclusion

Data is one of the critical factors in the current digitalization era, which can be an essential source of information. The government also uses big data to create work efficiency and effectiveness in policymaking as a form of effort [15]. The effect of vaccination policy through social media promotion is not very effective. It can be seen from the rise of the @Dinkesjabar @Dinkesjatim, and @Dinkesjateng Twitter accounts that the hashtag related to vaccination is still very minimal. Then the data obtained shows the success of immunization in West Java, East Java. Central Java is the vaccination policy through the door-to-door, mobile vaccination mobilization, mobile vaccine bus, and cooperation with the TNI-Polri, academics, and society in general. This success fulfills what is targeted by the provincial government.

References

1. Ai, Y., et al.: Wastewater SARS-CoV-2 monitoring as a community-level COVID-19 trend tracker and variants in Ohio, United States. Sci. Total Environ. **801**, 149757 (2021). https://doi.org/10.1016/j.scitotenv.2021.149757
2. Yulita, W., et al.: Analisis Sentimen Terhadap Opini Masyarakat Tentang Vaksin Covid-19 Menggunakan Algoritma Naïve Bayes Classifier. JDMSI **2**(2), 1–9 (2021)
3. Susyanty, A.L., Sasanti, R., Syaripuddin, M., Yuniar, Y.: Sistem Manajemen Dan Persediaan Vaksin Di Dua Provinsi Indonesia. Bul. Penelit. Kesehat. **42**(2), 108–121 (2014)
4. Acharya, A., Lam, K., Danielli, S., Ashrafian, H., Darzi, A.: COVID-19 vaccinations among Black Asian and Minority Ethnic (BAME) groups: learning the lessons from influenza. Int. J. Clin. Pract. **75**(10), 8 (2021). https://doi.org/10.1111/ijcp.14641
5. Panjaitan, A., Priyowidodo, G., Budianad, D.: Taktik self-presentation Joko Widodo dalam Menangani COVID-19 di Instagram. Publication.Petra.Ac.Id, no. 1959 (2021). http://publication.petra.ac.id/index.php/ilmu-komunikasi/article/view/11544
6. LandichoGuevarra, J., et al.: Scared, powerless, insulted and embarrassed: hesitancy towards vaccines among caregivers in Cavite Province, the Philippines. BMJ Glob. Heal. **6**(9), 1–11 (2021). https://doi.org/10.1136/bmjgh-2021-006529
7. El-Far Cardo, A., Kraus, T., Kaifie, A.: Factors that shape people's attitudes towards the Covid-19 pandemic in Germany—the influence of media, politics and personal characteristics. Int. J. Environ. Res. Publ. Health **18**(15) (2021). https://doi.org/10.3390/ijerph18157772

8. Stepita, M.E., Bain, M.J., Kass, P.H.: Frequency of CPV infection in vaccinated puppies that attended puppy socialization classes. J. Am. Anim. Hosp. Assoc. **49**(2), 95–100 (2013). https://doi.org/10.5326/JAAHA-MS-5825

9. Eriksson, E.: Incorporation and individualization of collective voices: public service user involvement and the user movement's mobilization for change. Voluntas **29**(4), 832–843 (2018). https://doi.org/10.1007/s11266-018-9971-4

10. Nurlaela, N., Hudrasyah, H.: Communication Strategy of the Importance of Vaccination Using Social Media & Public Relations a Case Study at BFM, pp. 2165–2178 (2013)

11. Ali, G.G.M.N., et al.: Public perceptions about COVID-19 vaccines: policy implications from US spatiotemporal sentiment analytics. SSRN Electron. J. **2019**(July), 1–32 (2021). https://doi.org/10.2139/ssrn.3849138

12. McClure, C.C., Cataldi, J.R., O'Leary, S.T.: Vaccine hesitancy: where we are and where we are going. Clin. Ther. **39**(8), 1550–1562 (2017). https://doi.org/10.1016/j.clinthera.2017.07.003

13. Leng, A., Maitland, E., Wang, S., Nicholas, S., Liu, R., Wang, J.: Individual preferences for COVID-19 vaccination in China. Vaccine **39**(2), 247–254 (2021). https://doi.org/10.1016/j.vaccine.2020.12.009

14. Aeschbacher, C.: Analysis of Switzerland's public health communication during the COVID-19 pandemic. (July) (2021). https://doi.org/10.13140/RG.2.2.22801.15205

15. Nurmandi, A., Kurniawan, D.: A meta-analysis of big data security: how the government formulates a model of public information and security assurance into big data. In: Stephanidis, C., Antona, M., Ntoa, S. (eds.) HCII 2021. CCIS, vol. 1499, pp. 1–8. Springer, Cham (2021). https://doi.org/10.1007/978-3-030-90179-0_60

16. Setiawan, R.E.B., Nurmandi, A., Muallidin, I., Kurniawan, D., Salahudin: Technology for governance: comparison of disaster information mitigation of COVID-19 in Jakarta and West Java. In: Ahram, T., Taiar, R. (eds) IHIET 2021. LNNS, vol. 319, pp. 130–137. Springer, Cham (2022). https://doi.org/10.1007/978-3-030-85540-6_17

17. Yusuf, A.M., Saputro, M.R.G., Maharani, W.: Identifying influencers on Twitter for Covid-19 education and vaccination using social network analysis. In: 7th International Conference on Software Engineering and Computer Systems and 4th International Conference on Computational Science and Information Management, ICSECS-ICOCSIM 2021, pp. 488–492 (2021). https://doi.org/10.1109/ICSECS52883.2021.00095

18. Finney Rutten, L.J., et al.: Evidence-based strategies for clinical organizations to address COVID-19 vaccine hesitancy. Mayo Clin. Proc. **96**(3), 699–707 (2021). https://doi.org/10.1016/j.mayocp.2020.12.024

19. Phillips, R., et al.: Cohort profile: the UK COVID-19 public experiences (COPE) prospective longitudinal mixed-methods study of health and well-being during the SARSCoV2 coronavirus pandemic. PLoS ONE **16**(10 October) (2021). https://doi.org/10.1371/journal.pone.0258484

20. Finley, C., et al.: A peer-based strategy to overcome HPV vaccination inequities in rural communities: a physical distancing-compliant approach. Crit. Rev. Eukaryot. Gene Expr. **31**(1), 61–69 (2021). https://doi.org/10.1615/CritRevEukaryotGeneExpr.2021036945

21. Boucher, V.G., Pelaez, S., Gemme, C., Labbe, S., Lavoie, K.L.: Understanding factors associated with vaccine uptake and vaccine hesitancy in patients with rheumatoid arthritis: a scoping literature review. Clin. Rheumatol. **40**(2), 477–489 (2020). https://doi.org/10.1007/s10067-020-05059-7

22. Court, J., Carter, S.M., Attwell, K., Leask, J., Wiley, K.: Labels matter: use and non-use of 'anti-vax' framing in Australian media discourse 2008–2018. Soc. Sci. Med. **291**, 114502 (2021). https://doi.org/10.1016/j.socscimed.2021.114502

23. Pratama, I., Nurmandi, A., Muallidin, I., Kurniawan, D.: Social media as a tool for social protest movement related to alcohol investments. In: Ahram, T., Taiar, R. (eds.) IHIET 2021. LNNS, vol. 319, pp. 138–146. Springer, Cham (2022). https://doi.org/10.1007/978-3-030-85540-6_18

The Role of the Financial Services Authority (OJK) in Preventing Illegal Fintech Landing in the COVID-19 Pandemic in Indonesia

Bella Kharisma[✉], Achmad Nurmandi, Isnaini Muallidin, Danang Kurniawan, and Mohammad Jafar Loilatu

Department of Government Affairs and Administration, Jusuf Kalla School of Government, University of Muhammadiyah Yogyakarta, Yogyakarta, Indonesia
bellakharisma92@gmail.com, nurmandi_achmad@umy.ac.id

Abstract. This study seeks to examine the role of the Financial Services Authority and its efforts in overcoming public problems in the form of fintech lending or online loans in the era of the COVID-19 pandemic through the application of the e-commerce concept. The increase in the use of e-commerce makes online loans also increase during the covid-19 pandemic. Based on the Financial Services Authority Regulation Number 77/POJK.01/2016 and the Financial Services Authority Regulation Number 13/POJK.02/2018 it becomes a legal reference in the implementation of financial service activities carried out by providers fintech lending. The state institution authorized to overcome the above problems is a state institution in the form of the Financial Services Authority together with the Ministry that is authorized and involved in its implementation. This study uses a descriptive qualitative approach by analyzing reports based on reports from financial service authority institutions related to online loans, both legal and non-legal. This approach is used to see the government's efforts in dealing with problems related to legal and illegal online-based money loans in Indonesia. The research data that will be used are secondary data obtained through books, scientific journals, social media, official government websites, and news portals. The results of the study show that currently illegal online loans have a higher graph than legal online loans, the government's efforts to reduce illegal online loans are by blocking the sites used. However, technology is so sophisticated that it is difficult for the government to eradicate these crimes. Another effort made by the government is asking the public to be more vigilant and careful before making transactions using online loan services by finding out in advance whether the site already has a permit in managing the online loan site.

Keywords: E-commerce · Online loans · Illegal · Covid-19 · Indonesia

1 Introduction

Activities of modern society today cannot be separated from the use of information technology. The development of technology 4.0 provides many changes in the behavior of traditional society into a modern society that cannot be separated from technology

G. Meiselwitz (Ed.): HCII 2022, LNCS 13315, pp. 568–577, 2022.
https://doi.org/10.1007/978-3-031-05061-9_40

and the internet (Shofiyah and Susilowati 2019). The development of ICT in various aspects, such as in the digital economy, is able to facilitate processes carried out such as marketing goods/services effectively (Bahtiar 2020). The application of ICT through the application of e-commerce according to says that e-commerce is able to have a positive impact on the economy through its ability and flexibility in building market access. In addition to having an impact on the economy, technological developments also affect the Indonesian financial industry with the emergence of financial technology which is an innovation in financial services in a more efficient and modern direction (Supriyanto and Ismawati 2019).

Financial technology or fintech is the use of technology in the financial system that will produce a product, service, technology so that it has an impact on monetary stability, financial system, efficiency and fluency to security in the payment system (Bank Indonesia 2018). Implementation in utilizing financial technology is useful for improving banking and financial services which are often carried out by startup companies (P. A, 2017). The concept used is an adaptation of technological developments in the financial sector so as to provide modern innovation. According to Bank Indonesia, the basic forms used by fintech lending are payments, investment, financing, insurance, cross-processing, and infrastructure (Priliasari 2019).

In the era of covid-19, many people are experiencing a difficult economy, so that the interest of the public to borrow money through online loan service providers is getting higher (Kompas 2021). The ease of accessing loans provided by companies through applications is an attraction to attract prospective borrowers to use their services. However, currently there are still many illegal online loan companies that have not received permission from the financial services authority (Kompas 2021). So that many people are affected by illegal online loans which actually make financial conditions worse.

Therefore, to overcome illegal online loans, the government established regulations related to online loans regulated in the Financial Services Authority Regulation Number 77/POJK.01/2016 concerning Information Technology-Based Lending and Borrowing Services as a legal reference for providers that must be obeyed regarding everything that moves regarding online loans (OJK 2014). Through state institutions, the government urges the public to use the services of providers fintech lending that have been registered and have received permission from the Financial Services Authority. Based on data as 8 september 2021 to fintech lending that is already registered as many as 107 organizers and fintech lending unregisteredas much as 182 organizers (OJK 2021). Although there are institutions in charge of dealing with online loans and regulations that must be set by the organizers, in fact there are still many illegal online loan sites that are found to be detrimental to the community. The public is advised to always be careful and thorough before making transactions so that unwanted events do not occur (Kominfo 2021).

This study aims to see how the role of the Financial Services Authority is related to regulations that were formed in dealing with the handling of illegal online loans in financial transaction activities in the Covid-19 era.

2   Literature Review

2.1   The Role of Financial Services Authorities Regarding Criminal Law Regulation

Role is a pattern of behavior, attitudes, and goals expected by a person based on his position in society (Widayatun 1999). In addition, roles show the same nature without any deviation from someone in certain situations (Frideman 1998) in (Putri 2019). The Financial Services Authority is an independent institution that has the functions, duties, and authorities of regulation, supervision, examination, and investigation and has a vision of becoming a trusted supervisory agency for the financial services industry, protecting the interests of consumers and the public, and being able to make the financial services industry a pillar of the national economy. that is globally competitive and realizes general welfare (OJK 2017). Where, regulation is an instrument in realizing a state policy in order to achieve state goals. Instruments as regulations must be able to be formed in a good way so that the regulations created are able to encourage the implementation of social dynamics and are able to encourage performance for state administrators (Djalil 2015). The function of regulation is as a means to order or guide behavior for both formal and informal activities, as an instrument of development and as an integration factor. The authority to make regulations lies with the central government.

2.2   The Concept of E-Commerce Related Online Loans in the Era of Covid-19

According to the WTO (2013) e-commerce or e-business is the sale of goods and services through a network of computers that have been designed with a method for receiving and ordering goods and services conducted between companies, individuals and households, government agencies, public and private organizations.. According to E-commerce is the process of buying and selling goods or services using technology in the form of the internet. Activities that occur in e-commerce are in the form of marketing, advertising, security, payments, shipping and other activities that are able to support trading activities. According to Kalakota and Whinston in Sijabat (2016) the perspective in e-commerce is classified into 4 namely communication, business, information services or online. Fintech lending or online lending is the provision of financial services to help bring together lenders and loan recipients in order to enter into loan agreements directly using an electronic system based on the rupiah currency (OJK 2017). The COVID-19 pandemic is a non-natural disaster for the entire world in the form of the disease coronavirus 2019 that attacks the human respiratory system. The spread of the Covid-19 virus is very easy to be contaminated by touching the surface of objects or touching the surface of the face. Symptoms of being exposed to COVID-19 are fever, runny nose and cough (Ayu and Lahmi 2020).

3 Research Method

This research conducts a study on how to make efforts to overcome the increase in illegal online loans by using descriptive analysis methods as a technique in managing research data. The research data that will be used are secondary data obtained through books, scientific journals, social media, official government websites, and news portals. Based on the literature review and previous findings, using logic and discussing and analyzing the potential related to the role of the government, namely the Financial Services Authority regarding the role of financial services authorities regarding the increase in illegal online loans. This data can be seen based on the official government website during the covid-19 era in Indonesia, the government evaluates monthly related online loan sites that are legal and those that are still illegal.

4 Discussion

Indonesia as a state of law, Philipus M. Hadjon said that the authority possessed by the government is aimed at the existence of the right to govern or act. The financial services authority is an independent state institution. The institution is authorized to form regulations and legal regulations that are useful for overseeing the course of activities in fintech landing or financial technology. Attributable to the financial services authority through the division of power by the laws and regulations derived from law number 21 of 2011 concerning the financial services authority.

The following is data regarding the comparison of legal online loans, potentially related to ethnic crimes and illegal in 2020 to 2021, seen in Fig. 1 regarding the online loan graph:

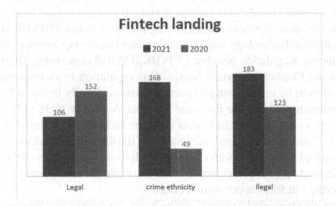

Fig. 1. (source: https://www.ojk.go.id/)

Based on the picture above, online loans that are used as a form of utilizing financial technology are divided into three parts, namely legal and illegal online loans, but illegal online loans are used as ethnic crimes. Legal online loans are loans that come from fintech companies that are safe, have low interest and have been registered on the official website of the financial services authority. Furthermore, illegal online loans do not come from the fintech lending section, illegal online loans are usually not registered and have permission from the OJK, usually illegal online loan offers come from whatsapp and sms with high enough interest to ask for access to personal data such as contacts, photos, videos and so forth. This illegal online loan is included in the category of crime in the form of fraud under the guise of information technology.

Furthermore, data regarding servers related to illegal online loan sites from Indonesia, abroad and their origin are unknown. The following is picture 2 related to the graphic of the illegal online loan server:

Fig. 2. (source: https://www.ojk.go.id/)

Illegal online loans that continue to increase from year to year become PR for state institutions authorized to solve these problems. The Financial Services Authority is authorized to establish regulations and regulations regarding online loans. This can be seen from the Financial Services Authority Regulation Number 77/POJK.01/2016 concerning Information Technology-Based Lending and Borrowing Services and Financial Services Authority Regulation Number 13/POJK.02/2018 concerning Digital Financial Innovation in the Financial Services Sector. This regulation is an important reference that must be obeyed by online loan providers so that it is legally licensed from the OJK.

The government through the Financial Services Authority (OJK) along with other authorized parties such as Bank Indonesia (BI), the Indonesian National Police (Polri), the Ministry of Communication and Information of the Republic of Indonesia (Kominfo), and the Ministry of Cooperatives and Small and Medium Enterprises of the Republic of Indonesia (Kemenkop UKM) together strive to eradicate the emergence of illegal online loans by carrying out various prevention efforts.

Prevention efforts related to illegal online loans are a form of concrete action taken by the government with the authority it has, so that people can be protected from illegal online loans and are not harmed anymore. The policy actor assigned to run and eradicate illegal online loans within the OJK is through the Investment Alert Task Force (SWI).

Through the results obtained above, it can be explained that the role of the government can be said to be still unable to deal with illegal online loans in Indonesia, this is because the data obtained by the organizers of legal online loan sites in 2021 has actually

Table 1. Legal Framework for

Regulatory	Substance	Solutions
Law Number 21 of 2011 concerning the Financial Services Authority	Explains the definition, understanding, and rules and provisions stipulated in the Law on the Financial Services Authority Explains the legal basis for the establishment, independent status, and position of the OJK	As a legal basis for strengthening the financial services authority
POJK Number 77/POJK.01/2016	Regarding information technology-based lending and borrowing services	As a guideline for the service procedures for obtaining legal online loans in accordance with applicable regulations
Financial Services Authority Circular Letter Number 18/SEOJK.02/2017	SEOJK concerning Governance and Risk Management of Information Technology in Information Technology-Based Lending and Borrowing Services	As a legal basis explaining the impact of the risks of using technology-based lending and borrowing services

decreased compared to 2020. illegal investment entities with the potential for crime in 2021 experienced a very high increase compared to 2020, and online loan services that were still not licensed or illegal in 2021 experienced an increase compared to 2020. Based on data from the OJK in 2020 on the number of loans channeled through legal online loan sites amounting to Rp. 155.9 trillion, up 91.29% from the previous period of Rp. 81.5 trillion, namely in 2019.

Through the data above, the government has tried to reduce the high level of illegal online loans that are still common in society. Efforts that have been formed and implemented by the government through regulations related to illegal online loans are still widely ignored by the perpetrators of implementing online loan providers.

In addition, according to the Financial Services Authority, this illegal online loan is a form of cross-border crime practice, where most of the illegal online lending servers are located abroad. According to the OJK Investment Alert Task Force, servers regarding online loans in Indonesia are as much as 22 percent, overseas as much as 34 percent and 44 percent of unknown origin. These criminal practices are not only Indonesian citizens but also the involvement of foreign nationals who take advantage of people in Indonesia as targets in making illegal online loans.

The Financial Services Authority is an independent state institution. The financial services authority aims to ensure that all activities in the financial services sector. OJK as the organizer functions to implement the regulatory and supervisory system that is related to all activities in the financial services sector. The task of the OJK is to regulate and supervise financial service activities from various sectors such as banking, capital markets and IKNB (Ojk 2021).

In ensuring legal and illegal online loan sites through a mechanism established by the financial services authority through the official website and application Whatsapp. Online loan service users can check through the website www.ojk.go.id by opening the page and then clicking the IKNB menu then selecting the menu fintech at the bottom right, then selecting the link for a provider of fintech lending registered and licensed based on the latest period (Finance 2021). Through this website, service users can also contact OJK contact number 157 and send messages via OJK email at Consumer@ojk.go.id. Then use the application Whatsapp by saving the OJK contact number, then opening the contact and sending the name of the online loan site you want to check, then sending the message there will be a reply regarding the answer whether the site has been registered with OJK or not (Kompas 2021).

Several efforts have been made by OJK through the Investment Alert Task Force (SWI), including providing education to the public regarding the use of online loan services that have been licensed so that they are legal and have been registered with the OJK institution. The Investment Alert Task Force has also carried out efforts cyber patrol, blocked sites and applications related to illegal online loans, offered online loans to savings and loan cooperatives, committed violations related to payment gateways to provide penalties for criminals in illegal online loans. In addition, the government also requested cooperation from Google to help provide requirements for online loan applications. This is done because there is a lot of abuse by people who are not responsible for the emergence of various illegal online loan sites. Therefore, Google added another condition in the form of eligibility for a personal loan application in the form of a license document that has been registered and has been given permission from the OJK.

The government and institutions involved in the implementation of online loans hope to reduce the high rate of illegal online loans, in addition to preventive and curative efforts the Investment Alert Task Force must be able to build an integrated system to fight illegal online loans in Indonesia that can harm the public. OJK stated that in creating an online lending ecosystem that is conducive and able to provide benefits to the community, it is necessary to collaborate with government actors, stakeholders and even the private sector so that they can encourage the national economy in Indonesia.

Shared losses felt by the community related to the rise of illegal online loans during the COVID-19 pandemic. Covid-19 has made it difficult for the community's economy, so that people can easily make online loan transactions but do not seek first whether the site or application used is registered and licensed or is still illegal. Most Indonesians rarely check the sites they use, so that after the transaction has been made, they will feel the impact of their inaccuracies. Based on the involvement of various government actors through joint commitments in discussions about online loans, steps taken such as prevention, handling, complaints, and law enforcement to eradicate illegal online loans.

Some forms of prevention are providing education programs to the public to remain careful in using an existing platform in making online loans and always maintaining personal data, strengthening communication programs to the public as a whole to stay alert regarding offers given in illegal online loans, the government need to strengthen cooperation with authorities and application developers in order to prevent the spread of illegal online loans which are usually done via cellular phones in order to spread awareness information to the public about online loan offers and prohibit banks, payment service providers *non-bank*, *aggregators*, and cooperatives from collaborating and facilitating online loans are illegal and are required to comply with the principle of using services in accordance with the applicable rules and regulations.

Complaints made by the public regarding illegal online loans can be made by opening access to public complaints and following up on public complaints in accordance with the authority of each Ministry/Agency and/or reporting to the Indonesian National Police for legal proceedings. Meanwhile, law enforcement regarding online loans can be carried out through legal processes against illegal online loan actors according to the authority of each Ministry/Agency and carry out international cooperation in the context of eradicating illegal cross-border online loan operations, which is also common in Indonesia.

The following are the efforts that the government can take to stop the proliferation of illegal online loans, namely the collaboration between Kominfo, OJK, and the Police in supervising online loan services through the government's official website through steps in preventing illegal online loans through the announcement of a list of legal online loan sites. request for blocking so that the government can cut off access to finance and report it to the police. To improve the digital literacy of the community regarding technology-based loans, the public must know the regulations or laws regarding the protection of online loan transactions such as provisions, impacts and so on so that people are always vigilant before carrying out these lending and borrowing activities. Furthermore, the importance of regulation on protection for consumers who use illegal online loan services, even though there have been several regulations implemented, but officially OJK can only close illegal online loan companies, even though OJK has closed but still these individuals will open new companies again so it is very much needed regulations for users of illegal online loan services. Finally, an evaluation is needed regarding the mechanism for obtaining permits and registration of online loan companies at OJK. The authority possessed by the FSA is to supervise the online loan company that has been registered, so the emergence of illegal online loans as a result of difficult licensing mechanism in the OJK (Budiyanti 2019).

Based on the explanation above regarding legal and illegal online loans, the following is a list on the official OJK website as of November 17, 2021 which was *uploaded* on December 10, 2021 regarding *fintech lending* legal in Table 2:

Table 2. The List of legal fintech lending 2021

Number of legal fintech lending	Types business			Operating system		
	Conventional	Sharia	Conventionaland sharia	Android	Android and IOS	-
104	96	7	1	53	19	32

5 Conclusion

Based on the research that has been done above it can be seen that the increase in illegal online loans in Indonesia is still high and continues to increase compared to the previous year. This can be caused by the difficulty of obtaining permits for loan service companies safe online which is still illegal. Efforts made by the government in eradicating it can still be said to be ineffective because the number of illegal online loans is still found by many people. One of the government's other efforts is by blocking sites or applications used in illegal online loans, but due to the rapid development of ICT, it is difficult for the government to eradicate them, even though they are blocked, other illegal sites will appear. In addition, the public also needs to educate themselves so as not to get involved in illegal online loans that are very detrimental, and always be aware of offers regarding online loan services by checking the previous company's site through the OJK official website so as not to experience cases of information technology-based fraud. In stopping the development of illegal online loans, there is a need for legal regulations that are in accordance with the needs of people who experience losses to illegal online loan services.

References

Agung, A.A., Erlina: Legal protection for consumers using online loan services. Alauddin Law Dev. J. (ALDEV) **2**(3), 432–444 (2020)

Annas, M., Ansori, M.A.: Problems in determining interest in peer-to-peer lending in Indonesia. J. Legal Media **28**(1), 102–116 (2021)

Asti, N.P.: Legal efforts by the financial services authority in overcoming illegal online loan services. ACTA COMITAS J. Notary Law **5**(1), 111–122 (2020)

Ayu, S., Lahmi, A.: The role of e-commerce in the Indonesian economy during the Covid-19 pandemic. JKMB J. Bus. Manag. Stud. **9**(2), 114–123 (2020)

Bahtiar, R.A.: Potential, Government's Role, and Challenges in E-commerce Development in Indonesia. Research Center, Secretariat General of DPR RI, 13–25 (2020)

Bank Indonesia: Bank Indonesia, 1 December 2018. Retrieved from www.bi.go.id: https://www.bi.go.id/id/edukasi/Pages/menkenal-Financial-Teknologi.aspx

BF, AR, Wisudawan, I.G., Setiawan, Y.: Online or Fintech Business Arrangements According to Positive Law in Indonesia. Faculty of Law, University of Mataram, pp. 464–475 (2020)

Budiyati, E.: Effort to overcome online loan services. Field Econ. Public Policy Brief Info **XI**(04), 19–24 (2019)

Finance: detikfinance, 4 September 2021. Retrieved from finance.detik.com/fintech/d-5709320/3-cepat-pinjol-illegal-itu-jahat-banget: https://finance.detik.com/fintech/d-5709320/3-cepat-pinjol-illegal-that's-so-evil

Dei, M.F.: Online loan transactions reviewed from the law concerning electronic information and transactions. In: National Conference on Law Studies: Legal Development Towards a Digital Society Era, Jakarta, pp. 126–149 (2020)

Firanda, G.A., Prananingtyas, P., Lestari, S.N.: Nagih debt (debt collector) online loans based on financial technology. Diponorogo Law J. **8**(4), 2523–2538 (2019)

Hendarsyah, D.: E-commerce in the era of industry 4.0 and society 5.0. IQTISHADUNA Our Sci. J. Econ. **8**(2), 171–184 (2019)

Kompas: Be careful, Illegal Online Loans are Increasing in the Middle of the Corona Virus, 29 April 2021. Retrieved from antarnews.com: https://www.antaranews.com/berita/2223006/6-tips-ajukan-pinjaman-online-dengan-minimrisiko-di-situasi-darurat-apapun

Indriani, I., Nurhayati, Utaminingsih, S.: Analysis of legal risk and impacts on online loan practices in pandemic times. J. Legal Stud. **4**(1), 95–107 (2021)

Krisnadi, I., Setiawan, H.: The Role of ICT in Accelerating Digital Transformation to Improve the Economy During the Covid-19 Pandemic 1–9 (t. yr.)

Lisnawati: The Importance of Public Education About Illegal Online Loans. Issues of the Economy and Public Policy (2021)

Lukito, I.: Legal challenges and government's role in e-commerce development. Center for Research and Policy Development, Research and Development Agency for Law and Human Rights, Ministry of Law and Human RightsRepublic of Indonesia, pp. 349–367 (2017)

Mufallihah, M.: Supervision of Financial Services Authority on Online Loan Services with Unlimited Cooperatives in Financial Services Authorities (Perspective of Law Number 21 of 2011 concerning Financial Services Authority and Islamic Law). Maulana Malik Ibrahim Malang Islamic University, 99 (2021)

Nugroho, H.: Legal protection for parties in online loan transactions. Justitia J. Law Humanit. **7**(2), 328–334 (2020)

Nurmandi, A., Almarez, D., Roengtam, S., Salahudin, Jovita, H.D., Dewi, D.S.: To what extent is social media used in city government policy making? Case studies in three asean cities. Viešoji Politika Ir Administravimas Public Policy Adm. **17**(4), 600–618 (2018)

Nurwahridya, M.M., Hartlwiningsih: The role of the police in managing cyber crime by the online loan collector desk. J. Crim. Law Crime Prev. **9**(1), 43–49 (2020)

OJK: Fintech Lending Operator Registered and Licensed at OJK as of October 25, 2021, 4 November 2021. Retrieved from ojk.go.id: https://www.ojk.go.id/id/kanal/iknb/financialtec hnology/Pages/Penyelenggara-Fintech-Lending-Registered-and-Permissioned-at-OJKper-25-October-2021.aspx

Pardosi, R.O., Primawardani, Y.: Protection of the rights of online loan service users in human rights perspective. J. Hum. Rights **11**(3), 353–367 (2020)

Priliasari, E.: The importance of personal data protection in peer to peer lending transactions (the urgency of personal protection in peer to peer lending). Natl. Law Mag. **49**(2), 1–27 (2019)

Putri, E.K.: The role of mothers in stimulating language development for toddler age children. Muhammadiyah Univ. Ponorogo, 1–55 (2019)

Reongtam, S., Nurmandi, A., Almarez, D.N., Kholid, A.: Does social media transform city government? A case study of three ASEAN cities: Bandung, Indonesia, Iligan, Philippines and Pukhet, Thailand. Transforming Gov. People Process Policy 343–376 (2017)

Shofiyah, E.N., Susilowati, I.F.: abuse of loan receiver's personal data in peer to peer lending. NOVUM J. Law **9**(2), 100–107 (2019)

Sofhian, S.: Overview of the role and functions of state institutions in Indonesia. J. Religious Educ. Train. **12**(33), 159–168 (2018)

Sudarmanto, H.L.: Legal approach to overcoming e-commerce problems in Indonesia. Surya Kencana Satu J. Dyn. Legal Justice Probl. **11**(1), 37–52 (2020)

Suharini, Hastasari, R.: The role of financial services authority against illegal fintech in Indonesia as consumer protection measures. J. Akrab Juara **5**(3), 25–38 (2020)

Sungangga, R., Sentoso, E.H.: Legal protection against illegal online loan (Pinjol) users. PAJOUL (Pakuan Justice J. Law) **1**(1), 47–61 (2020)

Supriyanto, E., Ismawati, N.: Online loan fintach information system. J. Inf. Syst. Inf. Technol. Comput. 100–107 (2019)

Suswanta, Kurniawan, D., Nurmandi, A., Salahudin: Analysis of the consistency policy Indonesia's capital relocation in the pandemic era. J. Soc. Polit. Stud. **5**(1), 35–48 (2021)

Sutanto, H.: The role of financial services authorities related to consumer protection in peer to peer lending fintech services reviewed in the Maqashid Sharia perspective (Study at the Lampung Financial Services Authority). State Islamic University Of Raden Intan Lampung, pp. 1–88 (2021)

Wicaksono, F.P.: Online-Based Money Loans from the Theory of Human Freedom (2019)

Engagement as Leadership-Practice for Today's Global Wicked Problems: Leadership Learning for Artificial Intelligence

Wanda Krause[1,2](✉) 🆔 and Alexandru Balasescu[1,2]

[1] Royal Roads University, 2005 Sooke Road, Victoria, BC V9B 5Y2, Canada
wanda.1krause@royalroads.ca
[2] Associate Faculty, Royal Roads University, 2005 Sooke Road, Victoria, BC V9B 5Y2, Canada

Abstract. As the world becomes more volatile, uncertain, complex, and ambiguous (VUCA), we require highly relationship focused, culturally sensitive, and contextually nuanced global leadership practices for our ever-changing new realities. Turbulent times have accelerated a need for leadership practice for our complex challenges and opportunities, including how to lead with developed technology. However, artificial intelligence (AI) has many lessons to learn from approaches that essentialize and universalize people, and their challenges, so as to do better in addressing challenges in different contexts. The future of leadership depends on how AI is integrated in the VUCA reality. The answer is not to hope that this tool will reduce any elements of this complexity (which is in fact the majority's current approach to AI), but how to understand the complexity, incertitude, and risks that AI potentially generates. Leadership will not only mean how to work with machines, but also to be aware of where in the leadership's space the machines are necessary, desirable, or outright unwelcome and potentially harmful.

Keywords: Global leadership · Engagement · Artificial intelligence · Feminism

1 Introduction

This paper provides a framework for understanding how to engage from a global leadership and leadership-as-practice perspective, that is, a holistic and democratic approach to address wicked problems of our VUCA times. In line with the works of Rost (1997) and Raelin (2016), the paper advances a novel approach that decenters our understanding of leadership as person-centered; as such, focusing on the engagement rather than the person. It argues for a shift in thinking from a person-centric mindset to a relation-centric mindset; that is, between people and as self-in-systems. Therefore, Balasescu introduces three types of relationality in leadership for the integration of AI as part of leadership practice for the 21st century: (1) individual human machine interaction; (2) cultural human machine interaction; and (3) species level symbiotic relationship with the machines. The individual level of human machine interaction refers to the myriad of ways in which particular individuals may, and will, relate to the automated or computerized systems used in their daily interactions, be them professional or personal, public or

G. Meiselwitz (Ed.): HCII 2022, LNCS 13315, pp. 578–587, 2022.
https://doi.org/10.1007/978-3-031-05061-9_41

intimate, and to the types of behaviours and attitudes they engender. This depends highly on the context and the subject position of the individual and the automated system which interact. The cultural level refers to the ways in which this interaction is, and must be, understood as culturally situated. Such includes whether we are talking about the design of automated systems - the first step of human machine interaction that is - or about their use, pre-existent cultural practices, space-time arrangements, and patterns that offer the larger systemic framework within which this interaction takes place. They both influence the interactions' dynamic and are influenced by them. Lastly, at the level of our species, we can talk about the fact that humans and technology are co-creations. Most anthropologists agree on these points. Every new technology re-created the meaning of what it means to be human by reshaping our relationships to the world, all the while as we are constantly re-technologizing this meaning, and the world we live in. However, due to the fact that today's technologies augment not only our muscle power but also our brain power, we need to further recognize the ways in which a symbiosis with technology can also be a possible way of understanding and framing humanity's relationship with the computerized machines. Using this approach, Balasescu and Krause explore how leaders and followers can be engaged collaboratively in order to address wicked problems of the 21st century through frameworks that can best support them.

Today, artificial intelligence (AI) is a term that subsumes most of the computerized system. The classical definition of the term, not undisputed, is that it refers to automated systems and/or robots that can perform with high accuracy and speed particular tasks usually associated with human intelligence, including learning. Our interest is in how our understanding of and relating to artificial intelligence can best avoid the same dangers and embrace lessons learned in the attempt to move from person centred to relationship centred leadership.

In this endeavor, our work seeks to provide lessons for leadership into the 21st century that is aware not to model approaches that universalize and essentialize people, how people interact with one another, and their engagement with macro-level issues. As such, our paper introduces the idea of leadership occurring as a practice sensitive to difference, intersectionality, cultures, and context, rather than leadership residing in the traits or behaviours of particular individuals (Raelin 2016). Each of these dimensions are reciprocally shaped by the other, and they form a web of interactions that imply stakeholders that are both human and non-human alike. Accordingly, our paper conceptualizes global leadership as a social relationship whereby leadership is an influence relationship among leaders and collaborators (Rost 1997), which takes shape according to their own contexts.

It delineates concrete ways to enhance leadership and followership practice to support the health, learning, and success of all around the globe by offering a framework and then providing an example. For the framework, it relies on the work of Wilber (2001) to explore holistically how to integrate the interpersonal or subjective, the intercultural or organizational, and the macro-level social, political, and economical and international with the practice of leadership. The paper argues that it is imperative to first understand how to align practice to context and that awareness before practice must be informed through these perspectives. The key focus in the argument of this framework, therefore, is bringing awareness to how leadership and collaborators, human and non-human or machine, are inextricably linked for leadership-in-practice in our 21st century. It further brings awareness to how one must encompass an appreciation of and attuning to different perspectives and contexts to address wicked problems in the international arena. For Raelin (2016), leadership-as-practice emphasizes more of what people may accomplish together. It is, therefore, concerned with how leadership emerges and unfolds through day-to-day experience. From this understanding of leadership, engagement becomes key because people are engaged in leadership at any given time due to their relations. From this standpoint, people do not reside outside of leadership but are rather embedded within it. Following this position on how to enact the kind of leadership needed for our VUCA times, we must look to the practice within which it is occurring.

To influence practice, it argues for a shift in thinking from a person-centric, leader-centric mindset to a relation-centric mindset; that is, also between people and as self-in-systems. Therefore, we can talk about three types of relationality in leadership for the integration of AI in leadership practice: individual human machine interaction; cultural human machine interaction; species level symbiotic relationship with the machines. In this approach, it includes an exploration of how leaders and followers both lead collaboratively in order to address wicked problems of the 21st century. To demonstrate how AI needs to align to leadership-in-practice for individual human machine interaction, cultural human machine interaction, and species level symbiotic relationship with machines, Krause provides examples of every-day-practice of women in Latin America (2004) and the Middle East (2008; 2012; 2020). Through these examples, the authors highlight how a universal feminist approach, among other approaches, miss essential information, similar to approaches that present algorithms unrefined to context and everyday practice.

2 Individual Human Machine Understanding and Interaction

What are some of the leadership lessons we can derive for enhancing our understanding of individual human machine interaction? As Jones (2018) argued, when a leader is contemplating a decision with life-changing implications affecting one or more persons, they will consider its emotional impact and, for that, use the human quality of empathy. AI, Jones further points out, is logical, and this is a result of how its algorithms are programmed. From a feminist standpoint, women's desires and subjectivity have been shaped by a power dynamic whereby their struggles for autonomy has been shaped through resistance, or where failed, compliance. It is assumed that their politics of resistance is in relation to that force, namely men, and perhaps the state, historically a space occupied by men, through its laws that have been biased towards men. Mohanty (1984) argued: "Privilege and ethnocentric universality on the one hand, and inadequate self-consciousness about the effect of Western scholarship on the "third world" in the context of a world system dominated by the West on the other, characterize a sizable extent of Western feminist work on women in the third world" (Mohanty 1984, p. 53). She further argued that "it is in the production of this "Third World Difference" that Western feminisms appropriate and "colonize" the fundamental complexities and conflicts which characterize the lives of women of different classes, religions, cultures, races and castes in these countries. It is in this process of homogenization and systematization of the oppression of women in the third world that power is exercised in much of recent Western feminist discourse..." (p. 54).

There is the bias that when women's activism in geographies outside the West are studied, strategically formulated goals that shape women's movements are given significance and practical or what Molineux calls feminine goals, such as a group of women in pursuit of better economic conditions, are seen as insignificant. As such, Krause (2012) argued that it is important to see power and agency not within a finite form as 'power over' or controlling power, which may result in compliance or resistance, but described as power to, power with, power within, and power for. By expanding our repertoire of lenses, we might be able to identify multiple forms of desire and goals, contextualized by context and what appears to be more immediate concerns and needs for the women in question. Feminists have widely cited Molyneux's (1985) distinction between strategic and feminine acts, placing the former above the latter. This has resulted in the marginalization of the very acts that matter most to many women outside the feminist framing of politically relevant, that is state focused, or patriarchy focused, action. Significantly, many geographies outside the west, where privileged most often white women have served to define the feminist agenda everywhere, are contextualized by poverty among various intersectional, political, and cultural differences. It is through Krause's investigation of women's activism in these contexts that abundant forms of activism are to be found illustrating the multifaceted nature of women's interests, desires, goals, needs, and therefore activism that matters most immediately and relevantly.

The feminist agenda is misconstrued through little understanding of the subject or, rather, the subjectification of women to a feminist subjectivity created within a Eurocentric mindset, race, history, and the concomitant desires of a class of women with a particular set of socio-political conditions, although challenging in critical ways that which gave rise to feminism still are not the same. In her study on women's activism in Chile and Argentina during the 1970s, Krause asked: "can a woman's identity as mother enable her to play an effective role in democratisation?" (2004, p. 367). She argued that not only in the feminist framework but frameworks to understand politics and change, in general, such as democratisation in Latin America and the developing world, misconstrue women's political import and impact because they tend to focus on elite institutional politics to interpret and explain political change. She pointed out that when the focus rests on men or government institutions, women, who represent a fraction of those involved in formal political sphere in many countries, appear minimal in their influence (2004, 2012). "The significance of the women's struggles can most accurately be understood within the context of political rule, the history of their countries, and the social environment in which these women live" (Krause 2004, p. 369). The epistemological challenge posed in such framework is in helping us better understand the characteristics of actors that contribute to or motivate an interface with systems that repress. We miss seeing the repertoire of characteristics that inform the choices for action and "selves" in systems, in fact among the majority of the world's women, because of the actions predicted by a framework that values some and devalues others as created using a Eurocentric mindset for a self in systems.

In her study on Argentina and Chile, Krause recognized "their strength also lay in the solace and empowerment they experienced in their meetings and arpillerista workshop" (2004, p. 371). Agosin had found, too, that "[d]rawing together, when faced with a crisis and creation of community networks, converts women into an extremely powerful political force in epochs of crisis' (1987: p. 10). In the case of Chile and Argentina, Krause put forward a bold thesis: "[T]he factor unifying the Mothers was the loss of their children. When no other group would dare to defy the military government, motherhood-an exclusively innate and instinctive nature-was the indefatigable drive to struggle against such a terrifying regime, despite enormous risks" (2004, p. 372). The military junta in Chile and Argentina exercised extreme political repression, sustained by the systematic and massive forced disappearance of people in torture centres and concentration and extermination camps, most centrally, in the 1970s and 1980s. The military used a broad definition of 'subversives'. Krause pointed out that these women had no legal voice in Chilean or Argentinean opposition, and that their being women in fact enabled them to move into public spaces, such as streets and plazas, initially virtually unnoticed. The context is that the societal belief in *marianismo* attributes to women masked their strengths and underestimated their potential; hence, they could organise despite the presence of violently repressive regimes. Further, their identities and consciousness as mothers evoked through emotional impact, as Jones (2018) refers to as a key consideration, which arguably motivated them into the streets to find the whereabouts of their disappeared sons and fight for democracy, a shaping of a subjectivity that would go unnoticed, undetected, and marginalized in an artificial intelligence that would likely use data formed around an identify of anything but a mother to begin a fight for democracy under such repressive conditions.

3   Cultural Human Machine Interaction

The understanding of current and future leadership will need to align with the understanding of relationality not only between humans and the automated decision-making processes, but through culture - machine interaction. Such entails exploring how algorithms interact with cultural spaces beyond the narrow scope of their design. The success of AI adoption depends on the web of relations between the agents that produce, develop, redistribute, and share knowledge: humans and machines alike.

Automation systems are not designed, and do not operate, in a void. They are created in places with a specific culture, and they are spread around the world more often than not with the expectation that they will work as intended. Usually though, they tend to work as designed. In the literature on feminism, it is recognized that 'the personal is political'. However, this understanding, as applied to various contexts, is still challenged. Feminist frameworks on women's participation and their effect on the political help us see that "If a woman disrupts the patriarchal, patrimonial or even economic frameworks that govern her everyday life, that creates a ripple effect on the world outside the home" (Krause 2012, p. 2). However, in the case of women in Egypt, as an example for various forms of women's participation, Krause found that still "[l]arge sectors of the population, such as women, the poor, or those who use Islamic frameworks within their strategies remain hardly represented and little understood" (p. 3). Even women although central to the study of feminism become marginalized when their activism is not seen as politically relevant or effective for change. Orientalism has remained influential in the works of some contemporary scholars whereby a Middle Eastern essentialism is applied to women studied in their various contexts in the Middle East. Middle Eastern essentialism sees people's behavior as circumscribed by culture and religion that is static and unchangeable and further not in alignment with a liberal ethos. Western essentialism typically sees deep religiosity and Arab culture as absolute impediments to the civic and institutional values of liberalism and civility. Feminist frameworks developed in the West under the historical trajectories and focus on liberalism have shaped its lens to include all that is identified as liberal and for change state or men referent, marginalizing women in historically, culturally or religiously different spaces. Feminist frameworks, thus, have much to adapt to be inclusive and useful. Similarly, automation systems need to be not only adopted, but also adapted.

The users and beneficiaries of AI come from a variety of cultural spaces with different perspectives of health, body, privacy, wellbeing, and ethics. For example, patients and health providers may experience care in the same institutional context, but in intellectually and emotionally different manners. The simple interposition of a device for recording electronic medical data, for example, sometimes generates two contradictory feelings and attitudes. On the one hand, some patients feel neglected while the health worker is absorbed behind the screen, and for all the patient knows, while she or he is in pain, the health worker may as well be playing Solitaire. On the other hand, the health worker feels both pressed by the urgency of giving care and by the necessity of spending time in front of the computer, only an outstretched arm away from the patient, but completely absorbed and unable to provide comfort. In even more dramatical ways, automation systems may be used for gathering information about, or even taking decisions about, the allocation of the limited care unit resources in ways that may literally end lives.

At the same time, the algorithms implicated in these decisions are more often than not inscrutable, not only for the health workers but even for IT specialists themselves. However, they are programmed with principles in mind that may not be as universal as we like to believe, but, instead, highly dependent on cultural biases. In Krause's research in the Middle East, she recognized that most women cringed at or disapproved of the very term itself 'feminism', viewed as synonymous with liberation from the foundations of religion and culture to allow for copying the West. There are many ways in which a dominant culture based on the designer's biases serves to impose its culture and neglect key variables within another's culture that shapes their desires, needs, and serves their inclusion or exclusion – and responses to their exclusion. From such experience, we might ask questions for AI, such as, how important is the age of the patient when decisions about resource allocation are made? We all agree that saving human life is the ultimate goal of healthcare, but when push comes to shove, which lives matter most? From whose perspective is the decision taken? There are cultures that are biased towards youth, while others favour a long-lived life. Is it more important to save the potential of a yet unlived life, or the accumulated experience of a fully lived one? The answer varies on different meridians, and it is never satisfactory for everyone, or universally accepted.

What cultural biases get encoded in the algorithmic models, and can they be culturally redesigned? These are not necessarily new questions. However, today's leadership will need to provide new answers, in order to move away from the reproduction and augmentation of the status quo towards a dynamic understanding of the context in which leadership is needed. To demonstrate what AI needs to become conscious of to align to leadership-in-practice sensitive to cultural human machine interaction, the exploration of pitfalls of a universal feminist approach, among other approaches, present critical lessons for how algorithms unrefined to context and everyday practice within those contexts, similarly miss essential information. At the higher level, the argument here is for a moving from a model-based understanding and approach to leadership (and human machine interaction), to a practice based one, to be applicable from the moment of designing systems - be them automated or policy shaping systems.

4 Species Level Symbiotic Relationship with the Machines

What does symbiosis with automated systems really mean? And how does this relate to the politics of change and leadership? One of the ways forward in understanding our relationships with computers/automated systems/machines is to reframe it in terms of symbiosis. The fear of being replaced is replaced by the certitude of being needed, as symbiosis is based on reciprocal interdependence. What resurfaces in symbiosis is not what data-based systems can and cannot do, but how their users relate to them, what they expect from automation, and how optimization and optimal decision making have transformed political decisions. Most of the time, the expectations are that the automated system will give us the best solution in the shortest time. But, do we formulate the demand correctly? The classic example of failure here is the thought experiment in which an AI destroys civilization when programmed to make metallic paper clips. Given absolute power and access, the AI powered machine would use every scrap of metal available on earth, the notion of what efficiency means being translated into making paper clips at all costs. What happens though if, under the mirage of efficiency promised and professed by automatic systems, we transfer this model to leadership in all its aspects? How can we avoid transforming all political action into paperclips?

To help us work through such questions, we might here, too, refer to lessons that can be gained through how the framework of feminism has caused similar challenges. In her further research, Krause notes how writings on the Arab uprisings tend to affirm and re-inscribe an 'a priori landscape of domination and resistance' even though this dichotomisation is not supported by empirically nuanced research. Sherry Ortner (1995) has argued that ethnography is a solution to the 'thinness' of such framing on resistance and what she calls its 'bizarre' refusal to know, write about, and engage the lived worlds of those who resist including their internal politics, cultural richness, and subjectivity. Women studied in the Middle East, for example, may not be seen to be necessarily taking to the streets in protest, except for during the 2011 Uprisings. Yet, these women are, in fact, chipping away at oppressive policies and practices of autocratic regimes. They do so by finding intelligent ways to keep contacts, building critical political infrastructure and tools for engagement, and expanding hope (Krause and Finn 2018). Krause and Finn (2018) found, for example, that Islamic women activists are injecting the political realm with knowledge, capacities, and *communities*. They argue that these are qualities which show responsibility to others, and thus, concomitantly, an expectation that the government responsibly protect in kind. As such, there exists a form of symbiosis that is practiced at the state and civil society or grassroots levels that becomes missed with frameworks that use lenses of conflict, tension, and resistance/compliance dichotomies. In the UAE, Krause researched and found numerous instances and examples of networking between women and individuals in government sharing common goals around human rights issues that required the grassroots information sharing, knowledge on the ground, and influence to change laws within the spheres of influence necessary.

It is important to underscore that such symbiosis and interweaving of action across these spheres is considering a shared recognition of, in fact, repressive forces and real threats to engagement. The significance of such networking to lift higher values and ethics into awareness, practice, and into policy and laws is the broader context of these repressive forces. As such, we acknowledge and emphasize that power and privilege feature into all systems, particularly given that all complex systems tend to be hierarchical. However, the goal of accentuating how engagement across these spheres can produce a symbiosis is to highlight the power individuals and collectives can wield in the face of such threats and, with the examples above, how these may be creatively enacted across the very spheres viewed as binary and exclusive within the frameworks we apply. Indeed, their boundaries are permeable as are all living systems, and from this view AI can develop an understanding of how symbiosis can be nurtured and led across and within systems.

We hope to decenter the centrality of leadership at the state level and prominent individual level as the most relevant when looking for change. Leadership becomes symbiotic and presents itself as an emergence out of the network of interactions between humans, machines, practices, actions, and models alike. With examples above, we explored how various groups, organizations, and networks of women activists integrate the interpersonal and subjective, the intercultural and organizational, and the macro-level social, political, and economical with their practice of leadership. Instead of a zero-sum game between those at different levels and within different spheres of influence, women empower each other, set direction for their communities, and contribute to the unfolding of civil societies in ways that support socio-political and economically relevant change. Similarly, AI can learn that a symbiosis is possible whereby engagement and collaboration is possible and that one sphere or level ought not necessarily to assimilate, create power-over conditions, or force resistance.

5 Conclusions

At every step of life, humans face the need to make decisions. Sometimes the role of decision maker was attributed to the leadership space. Now it may shift towards automated systems. The "natural" desire to make the right decision meets the impossibility to know beforehand the effects of the decision. The decision itself changes the parameters of the environment in which the decision is implemented. If only we could know better, or more, or perhaps everything, when making a decision. We all want hindsight with our foresight! As such, we hope that an exploration of frameworks, such as feminism, applied to various contexts outside the historical geographies from which the concept and its framing has been borne, allows us to understand its pitfalls. From the lessons, we hope that leadership-as-practice can move AI past existing and potential pitfalls as feedforward. The challenge ahead is recognizing how such frameworks allow goal-maintaining systems and how to lead goal-changing systems (Ackoff 1994) to be sensitive to culture and oriented to symbiosis.

For many, data-based technology and decision-making promise precisely a universal *apriori* hindsight, a complete control of the environment and an almost perfect predictability of results—if only we had enough data. The automation system also promises rapidity in decision making, and optimisation. From urbanisation to decisions enhanced by the paradigm of smart cities; from data-based decision making in healthcare, to shopping lists and eating habits, optimisation permeates all aspects of life. The desperate race to reach a desired objective in the shortest time and with the minimum expenditure of resources, while ignoring the way to get there, deludes us into optimizing the things that matter the least. How can we address this almost natural tendency? Do we need a new type of thinking to navigate the pitfalls of automation? We certainly do, particularly in the leadership realm.

What we know for sure is that none of the domains can align with a leadership model of "one size fits all works", similar to how the self-professed universalism of the western movement of liberation, as found in dominant feminist approaches, has proven to be rather problematic. The design place of the model imprints the model with the local flavour, and the only universal space that we can talk about is in fact that of the necessary symbiosis. We hope that symbiosis of both adoption and adaptation can be taken forward from these lessons as an urgent realization as the world becomes more volatile, uncertain, complex, and ambiguous (VUCA). It is certain that we require highly relationship focused, culturally sensitive, and contextually nuanced global leadership practices for our ever-changing new realities, and hope AI can embrace and enable us to lead in such way.

References

Ackoff, R.L.: Systems thinking and thinking systems. Syst. Dyn. Rev. **10**(2–3), 175–188 (1994). https://doi.org/10.1002/sdr.4260100206

Jones, W.A.: Artificial intelligence and leadership: a few thoughts, a few questions. J. Leadersh. Stud. **12**(3), 60 (2018). https://doi.org/10.1002/jls.21597

Krause, W.: How to become a resilient and adaptive leader in turbulent times: lessons from women in the Middle East. Int. Stud. J. (4) (2020)

Krause, W.: Civil Society and Women Activists in the Middle East: Islamic and Secular Organizations in Egypt. I.B. Tauris, London (2012)

Krause, W.: Women in Civil Society: The State, Islamism, and Networks in the UAE. Palgrave-Macmillan, New York (2008)

Krause, W.: The role and example of chilean and argentinian mothers in democratisation. Dev. Pract. **14**(3), 366–380 (2004). https://doi.org/10.1080/0961452042000191204

Krause, W., Finn, M.: Refusal and citizenship mobilization post-Arab revolts: an inquiry into Islamic women activists in Qatar. In: Rivetti, P., Kraetzschmar, H. (eds.) Islamists and the Politics of the Arab Uprisings: Governance, Pluralisation and Contention. Routledge, London (2018)

McGranahan, C.: Theorizing refusal: an introduction. Cult. Anthropol. **31**(3), 319–325 (2016). https://doi.org/10.14506/ca31.3.01

Molyneux, M.: Mobilisation without emancipation? Women's interests, the state, and revolution in Nicaragua. Feminist Stud. **11**(2), 227–254 (1985). Palgrave-Macmillan

Ortner, S.: Resistance and the problem of ethnographic refusal. Comp. Stud. Soc. Hist. **1**, 173–193 (1995)

Raelin, J.A. (ed.): Leadership-as-Practice: Theory and Application. Routledge, Upper Saddle River (2016)

Rost, J.: Leadership for the Twenty-First Century. Praeger, Westport (1997)

Wilber, K.: A Theory of Everything: An Integral Vision for Business, Politics, Science and Spirituality. Shambhala, Boulder (2001)

Open Innovation within Life Sciences: Industry-Specific Challenges and How to Improve Interaction with External Ecosystems

Niclas Kröger[1,2(✉)], Maximilian Rapp[3,4], and Christoph Janach[2]

[1] University of Innsbruck, Innsbruck, Austria
niclas.kroeger@hyve.net
[2] HYVE, Munich, Germany
christoph.janach@hyve.net
[3] Skolkovo Institute of Science and Technology, Moscow, Russia
[4] EY (Ernst and Young), Munich, Germany
maximilian.rapp@de.ey.com

Abstract. While Open Innovation (OI) and user integration have been applied across industries over the last two decades more professionally and digitally, the life science (LS) sector has various obstacles and regulations to overcome in order to implement and execute OI initiatives as well as entire programs. Moreover, scientific research has been scarce on analyzing data in this industry. The number of companies applying OI methodologies has risen significantly in the last couple of years. However, intellectual property (IP) and data protection make it hard to get access to good insights. However, we have accompanied ten LS companies (including 48 OI initiatives/programs) using an action research approach based on various data (e.g., interviews) over ten years, including the time of the outbreak and dissemination of the COVID-19 pandemic. In this paper, we will share insights about 1) the unique characteristics of OI in the LS sector compared to other industries, 2) the identification of used OI methodologies and their success criteria, as well as 3) the change and influence of the COVID-19 pandemic on the use of OI programs in the LS sector.

Keywords: Open innovation · Social media platforms · Community · Crowdsourcing · Life science · Healthcare · Open health · User innovation · Patient journey

1 Introduction and Overview to this Research

Open Innovation (OI) or user innovation, often leveraging digital platforms, has become an integral part of many companies' innovation processes. Also, companies from the life science (LS) industry have been using associated methods to create innovations over the past few years more intensively. While researchers have already investigated some aspects of Open Innovation in the healthcare sector, like collaborations with startups [1], there is still a significant gap regarding user innovation and integration in the research field. The reasons for the scarce research insights are manifold. Firstly, innovation and R&D processes in LS are highly shielded and secured, even compared to other protective industries. Secondly, there are fewer projects because it is complicated for companies to reach the right target groups in this environment, as regulations block the direct communication between stakeholders and patients.

Our paper aims to fill this gap by looking at how LS companies can implement OI focusing on users. Here, we look particularly at the following Open Innovation methodologies: (1) Co-creation, (2) Netnography, (3) Crowdsourcing, and (4) Online Research Communities (ORCs). We describe each of the methods and explain the opportunities and challenges as well as frontiers for the LS sector compared to other companies. Our research looks at specific requirements regarding regulations, pharmacovigilance, suitable applications for these methods, and challenges regarding the industry's customer groups, especially doctors and further HCPs (Healthcare Professionals) as well as patients. In that scope, we look to answer the following research questions:

1. What distinguishes OI in the life science sector from other industries?
2. What are the most appropriate methods in OI within life sciences, and what key success factors can be identified?
3. How will future OI projects within life sciences change after the digitization push through the COVID-19 pandemic?

This research is highly relevant for both researchers and practitioners as it discusses the adoption of new methodologies in a highly regulated space. Furthermore, our learnings help managers to understand the requirements and opportunities to apply these methods to their daily work and help them understand how they can leverage the power of OI given the specific circumstances in their industries.

To answer our research questions, we structured our paper as follows. First, we provide an overview of the current research landscape, define Open Innovation and corresponding methods for life sciences, and explain theoretical foundations. We then describe our data and research methodology comprised of in-depth case studies on more than 40 OI initiatives conducted by ten LS companies. Using the method of participatory action research, our team was actively participating in these projects [2]. We also interviewed project managers, stakeholders, and consultants with experience with OI projects in LS and other industries. We combine our insights from the action research approach and our interviews to generate insights to answer our research questions. Finally, we summarize our findings with managerial implications to successfully help practitioners implement OI tools and methods. We conclude with an outlook and further areas for research.

2 Theoretical Background

Open Innovation as a new research paradigm has been the focus of many scientific studies in recent years. However, there are still uncharted territories that seek further scientific inquiry. Our research aims to derive organizational and managerial implications for a specific sector, namely the LS sector, to successfully implement OI initiatives focusing on user innovation. The LS sector can be defined as organizations applying "biology and technology to health improvement, including biopharmaceuticals, medical technology, genomics, diagnostics, and digital health" [3]. Our research focuses on "inbound open innovation," where organizations open their innovation processes to gain insights from outside sources like external stakeholders [4]. While some research has already looked into partnerships with external organizations and start-ups [5], we want to focus on the collaboration with individuals (i.e., patients, doctors, lead users, innovators, experts, other HCPs, etc.), which is referred to as "user innovation" [6]. In his ground-laying book, Eric von Hippel explicitly mentions the medical device industry as offering a high potential to profit from user innovation. Still, we have found minimal research regarding user innovation in the LS sector, even though the literature supports the initial claim for high potential. For example, Smits & Boon derive a variety of reasons why pharmaceutical companies should involve users in their innovation processes to increase their societal and commercial rate of return [10]. Tracking and recording patient needs can improve the adoption rate by turning users into lead users who influence other stakeholders. Through the involvement of patient groups, they can increase the (cost)effectiveness of R&D processes through charity funding, lobbying for government funding, or assisting in running clinical trials. Further, early involvement of users can help mitigate societal resistance and foster technology acceptance. In case of market failure (e.g., lack of drug development for rare diseases), users can advocate funding research publicly. Even beyond their arguments, another study showed that patients/caregivers develop their own solutions to deal with rare diseases and even diffuse these innovations to other patients [7]. Another study investigating medical smartphone apps indicates that solutions developed by patients and healthcare professionals perceived better app store ratings than professionally developed apps [8]. While research suggests that user innovation in the LS sector offers excellent opportunities, the actual implementation of the associated methods seems to lack behind.

Therefore, we aim to understand how LS companies can use OI and especially user innovation in their daily practice, focusing on four common approaches:

- **Co-creation:** In co-creation approaches, the customer becomes a partner in the innovation process with whom the company stands in the continuous dialog to learn from the customer's social and experiential knowledge [9]. Co-creation can take place virtually on online platforms [10] and in workshop settings which currently, during the COVID-19 pandemic, have also moved to virtual settings [11].
- **Netnography:** This qualitative research approach takes the concept of ethnography and applies the principles to the internet and especially social media where data is freely shared [12]. The approach shows relevance within the LS sector [13], for example, to understand patients and caregivers [14]. With this approach, the users are not actively involved, but the users' knowledge, insights, and ideas are still obtained.
- **Crowdsourcing:** The term coined by Howe [15] describes the outsourcing of a task to a large (undefined) group of people. For this purpose, platforms using social media mechanisms engage users in open calls. In the LS sector, crowdsourcing can be used for different purposes, such as scientific ideation [16] or medical research [17].

- **Online Research Communities:** An Online Research Community (ORC) is a closed online community employing social media mechanisms. Users are invited to share experiences, and ideas in a moderated process [18]. In the LS sector, the concept can be adapted to include patients or other stakeholders [19].

3 Empirical Analysis

3.1 Methodological Approach

For our research, we use the qualitative Participatory Action Research (PAR) approach in which the researchers become a part of the social construct they want to study [2]. The method allows the researchers to get an in-depth understanding of the functioning of an intricate social system. We base our research on in-depth case studies of companies in the LS sector. Ten companies have been engaged in OI initiatives. We define "initiatives" as efforts of using OI in the broader sense of programs and holistic approaches, not just particular projects like a single workshop with doctors.

Through their jobs in leading consulting companies in the field of innovation (HYVE – the innovation company, based in Munich, Germany, and Ernst & Young, headquartered in London, UK), the researchers have been involved with the study objects during the evaluation, conception, and implementation of OI initiatives. Besides projects in the LS sector, the researchers draw on their experience of consulting companies in different industries in hundreds of OI projects.

3.2 Data Collection

This research study draws on a vast data set. The data used in the analysis of this paper include the following:

1. Observations and active participation in OI initiatives in the LS sector. In the projects, we observed stakeholders and understood the circumstances that make OI in this sector different from other sectors. Further, we talked to selected stakeholders from the LS sector who had not yet run an OI project to gain a better understanding of the burdens that keep actors from pursuing OI.
2. Semi-structured interviews with project managers who have been involved in these projects.
3. Semi-structured interviews with further stakeholders and consultants who have supported OI in the LS sector. Through their experience in other sectors, they can point out differences that insiders may not see as they lack the external multi-industry perspective of consultants.
4. Additional data from the projects. This data includes project reports, minutes, notes, slide decks, and other recordings created during the projects.

Table 1 shows an overview of the OI initiatives that form the basis for our research. For each project, it indicates the timeframe when the projects were conducted, company size, the primary focus of the company, format, and quantity of the OI initiative, the

Table 1. Data sample of the research study

Company	Duration OI initiatives	Industry	Workforce (approx.)	Number OI initiatives	Number and type of initiatives	Function of project lead	Included departments
A	2012-2016; 2021	Biotechnology and pharma	50.000	11	6 Crowdsourcing, 3 Co-creation, 1 Netnography, 1 ORC	Head of Innovation Management	Innovation Management & R&D
B	2016-2022	Pharma	40.000	6	2 Crowdsourcing, 3 Co-creation, 1 ORC	New Venture Lead, Project Management Professional New Business Development & Innovation Management	Innovation Management, Business Insights & Analytics
C	2019-2022	Pharma	25.000	8	1 Crowdsourcing, 7 Co-creation	Head of Strategy	Commercial
D	2021-2022	Pharma	100.000	4	3 Crowdsourcing, 1 Netnography	Head, Global Communications, Marketing Manager	Communications, Marketing
E	2021-2022	Biotechnology and pharma	10.000	3	2 Crowdsourcing, 1 Netnography	Senior Director Patient Advocacy	Patient Advocacy
F	2019-2021	Pharma	50.000	1	1 Crowdsourcing	Head of Commercial	Commercial
G	2011; 2019-2020	Medicine and health technology	10.000	2	1 Crowdsourcing, 1 Co-creation	Head of competitive intelligence	Commercial
H	2017	Medicine and health technology	450	3	2 Co-creation, 1 ORC	Head of Innovation Management	Innovation Management
I	2011-2016	Bio-chemistry	20.000	6	1 Crowdsourcing, 1 Netnography, 3 Co-creation, 1 ORC	Head of R&D	R&D
J	2015-2016	Health & chemicals	5.000	4	3 ORC, 1 Co-Creation	Director, Corporate Marketing	Innovation & Strategy
10	2011-2022	Life science	ø 31.000	48	20 Co-creation, 4 Netnography, 17 Crowdsourcing, 7 Online Research Community		

position(s) of the responsible project leader(s), and the department(s) involved in the project(s). Overall, our research is based on 48 OI initiatives across ten companies within the LS sector. These include 20 co-creation, four Netnography, 17 Crowdsourcing, and seven Online Research Community initiatives and programs between 2011 and 2022.

To better understand the kind of projects that were part of the study, we wanted to sketch out an exemplary process of one holistic innovation initiative (not just single projects) that leveraged several of the OI methods we analyzed for this research.

Company A (see Table 1) was looking to improve medication for children. They structured their project in three phases, 1) research, 2) ideation, and 3) design. In a first exploratory research phase, their goal was to identify children's needs, pains, and insights when taking medication. For this purpose, they used the Netnography approach, where they screened and analyzed online dialog in forums and social media. In the next phase of the project, they wanted to ideate to solve the issues identified in the previous research. As a result, they launched an innovation contest on their crowdsourcing platform. On the platform, they invited innovators worldwide, not necessarily related to the LS sector, to submit ideas and concepts. After the contest, they pre-selected promising concepts to further prototype and develop them in a co-creation workshop with global HCPs and potential users.

4 Discussion and Findings

4.1 Differences and Barriers for LS Compared to Other Industries When Implementing OI

As previously mentioned in the introduction, there is very little research on OI initiatives in the LS sector due to the highly protective and secured industry. For this reason, it is crucial to know the specific characteristics which apply to the LS sector compared to the private market to understand certain limitations and requirements for OI initiatives and suitable OI methods. There are mainly three aspects to consider: 1) regulations, 2) ethics, and 3) development costs:

Regulations and Compliance: Compliance is highly critical as no flexibility exists in the LS sector. The companies working in the LS sector must respect the legal framework, which lives by numerous regulations, and is constantly changing. In addition, the LS sector has a solid self-regulation system and constantly adapting requirements. Therefore, innovation and R&D processes in the life science industry are highly protected and secured. That is one reason why OI initiatives are difficult to implement.

Ethics and Communication in OI Initiatives: It must be ensured that no cure promises are implied and that the patient is not exploited. Therefore, communication with patients is very different from OI projects in the private sector. Often indirect research through doctors is the only feasible approach, and immersion in the patients' world is complicated. However, bias can occur because responses are based on guidelines, social desirability, or wishful thinking.

The target groups (i.e., the patients) are challenging to find, select and address due to the background of the disease, the course of the disease, and the sensitive data. For this reason, much more time needs to be spent on creating the different contents, as there is no margin for error.

R&D Associated Costs and Resources: In the life science sector, value creation is most often achieved in R&D; the competitive part of the innovation process is low. Long development cycles and high-profit margins due to comparatively low manufacturing costs in the LS sector combined with the cost-intensive development and long time-to-market play a significant factor in intellectual property. This also influences the corporate structure. Our research has also shown that there is usually no dedicated innovation function in the LS sector. If there is an assigned innovation manager, it is usually a single person and not a dedicated team. Another development is that many companies have outsourced their resources in the LS sector over the last ten to twenty years and rely on collaborations for joint development of innovations.

An additional intriguing point is that the challenges are often too complex, and there is often a lack of understanding of the exercise faced by non-experts (e.g., regulations are not known, or specific expert scientific knowledge is lacking for a feasible solution).

Nevertheless, the industry is under increasing pressure, as the expectations for product user-friendliness are rising through the experiences with other products or brands and technologies. Therefore, OI initiatives and user-centered development will become increasingly important.

In summary, and illustrated in Fig. 1, the legal and ethical barriers are significant challenges to introducing OI initiatives. Excellent stakeholder management throughout the entire process is crucial for success. Although LS companies try to implement OI methodologies, they still lack the confidence to put it on a broader scale or push regulatory boundaries to "open up" even further. Therefore, healthcare professionals, especially doctors, need to be considered when setting up and implementing initiatives due to the strict legal framework.

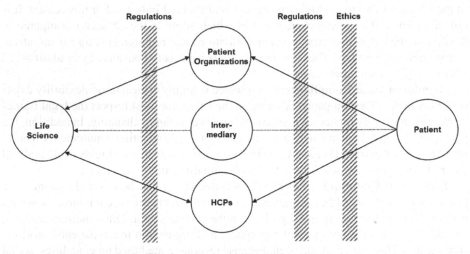

Fig. 1. Patient communication in LS

4.2 Open Innovation Methodologies in Life Sciences and Key Success Criteria

Our research has identified methods that are specifically well suited for use in the LS sector. They allow us to bridge some of the challenges pointed out in the previous section and profit from external input (mainly due to the digitization push through the COVID-19 pandemic). Depending on the phase of the innovation process (i.e., research, ideation, development), different methods may be suitable.

(Indirect) Integration of Unknown Stakeholders.
Our study identified the pattern that many LS companies prefer to involve stakeholders in their innovation process that they don't have a direct relationship with. This can be done through Netnographies, Crowdsourcing, or role-play.

The anonymity of the Internet offers unique opportunities for LS companies to conduct netnographies. Online, users tend to share uncomfortable and intimate topics very openly. Companies can analyze this publicly available online content to understand patients' wants and needs better. These topics often could not be asked in a face-to-face interview because the patients' medical conditions cause insecurity and anxiousness. Conducting netnographic research allows LS companies to gain a deeper understanding and to discover hidden needs. While the lack of personal contact makes this approach to research well suited to bridge ethical burdens, LS companies still need to consider reporting any adverse events they find online.

In contrast, **Crowdsourcing** allows involving external stakeholders unknown to the seeker, which avoids the challenges associated with involving patients/caregivers directly. In Crowdsourcing, the reporting of adverse events requires close monitoring of the community and all user-generated content, making active community management an integral part of innovating with the crowd. As possibilities for communication with the participants may also be limited, predefined and approved wording for feedback can be helpful. While not all ideas coming from an unspecialized crowd may be implementable from a regulatory standpoint, crowdsourcing offers an excellent opportunity to collect concepts and use cases from people with an outsider's perspective. Further, challenges in the LS sector also provide a great potential to attract intrinsically motivated users. The method is best used in the ideation phase.

The interface between doctors and patients is the most exciting for research. Why do problems or conflicts often arise here? The big problem is that there is often no contact with the patient, and there is only direct communication with the doctor, but the knowledge then comes second-hand. Hence, **role-play** emerged as another method of indirectly involving patients in the innovation process. When direct contact and the involvement of patients is not possible, HCPs can be asked to act out the patient perspective to allow them to empathize with patients' problems. While this method requires that the HCPs have a deep understanding of the patient, it offers the possibility to gain a deep understanding of the patients' situation, needs, and pain points in a more memoizable and engaging manner. Roleplay is exciting as inspiration in the ideation phase.

Active Integration of Known Stakeholders.
Our research shows that LS companies involve stakeholders directly, even if this is often more difficult than indirect integrating. Methods to integrate known stakeholders that we

have identified include ORCs with doctors, co-creation with doctors, co-creation with universities, co-creation with patient organizations, and co-creation with patients.

Online Research Communities with doctors and/or HCPs are platforms with a selected invitation to just a few persons who interact online over a longer time than within a one-day workshop. They are challenging to develop as the target group has limited time available and is not always motivated by monetary incentives. Therefore, building a community may be difficult. Structures need to be established to attract doctors driven by extrinsic (financial) and intrinsic motives to contribute. As doctors are very familiar with the processes and regulations, it is less likely that a doctor will post about an unreported adverse event in the online community that would have to be reported. These communities that aim to generate qualitative insights are well suited for the research phase.

Co-creation with universities also offers a great approach to integrating stakeholders who understand the specific circumstances of innovating in the LS sector through an academic view. Higher education institutions provide access to young innovators with industry knowledge and an agile mindset. Universities are often involved during ideation and development phases with their students or even dedicate whole courses or special courses to a specific challenge or topic of a company.

Co-creation with patients is very complicated to facilitate as LS players have various restrictions regarding integrating and direct contact with patients (respective not allowed due to regulations at all). Furthermore, their input may not always be implementable as they often lack knowledge about regulations. Their involvement is most helpful in the early innovation process during the research phase. It becomes applicable on special occasions. Not all organizational units may communicate with patients. As patient data must not be processed, intermediaries are required (see Fig. 1). Due to these restrictions and efforts, LS companies prefer other methods for practical and organizational reasons even when the output of direct patient interaction could be more informative.

In our research sample, companies preferred to **co-create with doctors and/or further HCPs**. While it is usually expensive to work with them (due to monetary incentives for participation), it is possible to form long-term relationships and involve them throughout more extended programs. Also, doctors are aware of regulations and may provide value and feasible insights and ideas that can have a positive impact, especially during ideation and development phases.

Co-creation with patient organizations has more similarities to working with doctors. They are aware of the regulations, and LS companies often have contracts that allow for collaboration. As they have an excellent understanding of patients' needs, they are often involved in the early stages to define potential areas for innovation and in later steps to give feedback on ideas and prototypes.

Overall Success Factors of Open Innovation in the Life Science Industry.
Through our research, we have identified general success factors for OI in the LS sector that can help managers develop an OI strategy:

- Precise **coordination and clarification**, including determination of responsibility, are essential for OI projects. Particularly in the kick-off phase, this is much more resource-intensive, as the topics are complex, and it takes more time to understand the challenge.

● Stakeholder management and **managing expectations** are critical to success. The team's commitment is crucial, especially for projects over an extended period. It is essential to involve all internal stakeholders (especially compliance/legal) from the beginning.

● It is also critical to involve all external stakeholders like HCPs and patients throughout the **entire process**. Often, certain groups are only involved in specific phases leading to a limited understanding of the stakeholders' wants and needs.

● Many LS company employees have little methodological knowledge and experience to pull the right levers for a successful OI project. Especially the **moderation** in workshops and handling sensitive and often very intimate topics need qualified personnel who can be found in or outside the organization.

● When working with patients, **data protection** is essential and highly critical. Ideally, most work can be done without collecting any patient data or indirectly getting the patients' views. Furthermore, target group-specific and sensitive communication are critical as topics are usually related to negative impulses like diseases and sickness.

● Regulations can be a burden to "open up" to the outside world. They can increase the effort for applying OI principles. However, once the regulations and limitations are understood, it is possible to develop **mechanisms and procedures** that allow LS companies to follow the regulations while ensuring project success.

4.3   Recent Developments on LS OI Initiatives Through the COVID-19 Pandemic

In accompanying projects in recent years, we could identify and pinpoint precise adjustments on OI initiatives through the influence of the COVID-19 pandemic starting to spread in the western world, especially in early 2020. While companies in unsecure and tough financial times are most likely to cut innovation budgets and focus on securing operations and the core business [21], the majority have intensified their OI engagements and intentions, as our research shows. This can be due to multiple reasons, where the most important ones are summarized in the following:

Acceleration of Digitalization through COVID-19: LS companies have vastly accelerated digitization like all other industries. Internal processes, IT landscape, client engagement, marketing campaigns, service developments, etc., got a massive push towards digital implementation or offerings, as communication and interactions in LS have been focused on face-to-face interaction and onsite relationships before. This increased digital mindset led to more understanding and reasoning for further disruptive approaches and ideas or methodologies and the need to push the organization outside its comfort zone. Through uprising competitors from the MedTech and Startup environment and the changing demand for services around products (beyond the pill concept), LS companies now feel the pressure of making more significant innovation leaps than before. OI fosters the outside-in perspectives and focuses on innovation topics around digital processes, products, services, communication, marketing, or production topics. Patients' expectations mirror those from other industries, where they experience a stronger user focus and quick improvements.

Usage of IT-Based OI Tools for Improved Experience and Flexibility: As the target group of LS OI programs and initiatives are, as stated above, dependent on rare experts with seldom knowledge, the recruitment and activation of those have always been very complicated. This is due to both the regularities (e.g., objective vs. subjective perspective in invitations) and the different expectations of participants compared to other industries (e.g., scientists and doctors expect higher incentives than brand fans in other industries). Furthermore, the availability of high-profile stakeholders is usually hard to predict, and their locations tend to be scattered around the world, which makes onsite workshops hard to execute as time windows and travel efforts must be considered and are more oriented on the participants than on the organizers. However, the rollout of tools to support OI methodologies like Mural or Miro has changed the possibilities for LS companies tremendously. According to our analysis, those tools give them the chance to engage with their target groups independent of space and with less time investment, as global traveling is not necessary. Especially co-creation - which in our sample seem to be the preferred methodology of LS companies - has profited through the new tools, as interviews are more accessible with various target groups, particularly dial-ins from experts for just 20 min instead of one-day workshops for quick evaluations or the participation of mixed groups to boost co-creation now being more feasible. Additional functions like polls or breakouts allow the initiators to get information and insights that differ from those before using the digital tools. Finally, we should add that while OI prior to COVID-19 was often conducted on an either-or basis of OI methodologies (either crowdsourcing or co-creation approach), the easier and more accessible tools led to more courage at LS stakeholders to mix and combine different methods and let them run for extended periods as approaches such as design thinking (quick feasibility checks, evaluation sessions and feedback scrums within the ecosystem) are simpler to handle and to implement.

Collaboration and Co-creation as "New Normal": The pandemic has been targeted by many pharma companies to develop vaccination, therapies, or medicines. The emergency and necessity of a global virus without boundaries (and therefore a global market scale) have led to various LS company collaborations to fight this pandemic at record speed. Co-creation and newly established alliances have been formed and developed a new mindset and understanding of how to "open up" and accelerate innovation and hence, push co-creation as well as OI with other companies or Non-Profit Organizations (e.g., foundations or governmental institutions). This completes the significant shift towards innovation, innovation management, and disrupting innovation, which started around ten years across industry and focuses on new roadmaps on changing consumer/patient interests, desires, and mindsets [22].

Missing Difference Between Innovation and Digitization: The LS sector - like the public, finance, or insurance sector- tends not to be first movers but rather late followers on using innovation methodologies [23]. In addition, our analysis shows that in many LS companies, the term and definition of digitization are often mixed up or seen as equal to innovation (especially when it comes to automatization or incremental innovation on processes through IT-based innovations). While industries like automotive or FMCG differ clearly between R&D, innovation management, and digital units, many LS companies lack a clear distinction and vision. As a result, the necessity of pushing digitization through the COVID-19 situation and the need for innovative collaboration had a positive push also on OI initiatives, as project managers and stakeholders got the financial and top-down backing to further invest in innovation/digital OI programs.

While this might lead to the conclusion that digital (using tools to include more ecosystem players regularly) OI initiatives or programs (combined methodologies for a more extended period) have set a new agenda, the quality is not necessarily better. The reason for this is mainly the missing link of personal partnerships, ad hoc discussions within onsite sessions, and the foundation of LS still being a people business. As our analysis shows, online workshops find all stakeholders in general speaking very openly and sharing ideas and comments (even if the boss is in the workshop – in many onsite workshops, the boss is a dominant participant and opinion leader). However, online tools lack the possibility to communicate objections or interpret non-verbal communication in many cases. Cases of social isolation reflected in the way of participating in crowded online workshops have been detected as well as the ongoing frustration of long sessions behind the screen without real-life interaction and consequently a lack of interest, participation, and excitement.

5 Outlook and Further Research

Our intention in this research was to answer and deliver insights to the introduced scientific questions, namely about 1) the unique characteristics of OI in the LS sector compared to other sectors, 2) the success criteria to execute various OI methodologies in LS, and 3) the potential change of the use of OI in LS through the COVID-19 epidemic. However, our research comes with certain limitations and restrictions. First of all, we have to mention that the use and definition of OI have changed over the years. So, the categorization and quantification of the analyzed data had to be balanced across the ten companies we accompanied over ten years. Moreover, this research has collected data around two years into the COVID-19 pandemic. At this point, it is hard to predict in which direction and how long the remote work (work from home, no face-to-face meetings, etc.) will last and how many new mutations might occur. In contrast, especially the effects and influence of the COVID-19 situation on OI programs are an exciting field for further research, as many adaptions and changes will occur on handling user integration over the upcoming years. Even if the future is hard to predict through the just mentioned limitations, we would like to share a selected outlook on how LS companies might proceed to use OI in the upcoming years for managerial implications:

- **Solutions "beyond the pill":** While product developments around treatments, pills, and drugs are a matter of 10 + years and multi-billion-dollar budgets, the rising pressure to think disruptive, patient and data-driven as well as digital will lead to new horizons covering more than the existing markets and IPs. Digital services, platform ecosystems, big data, or even the metaverse must be tackled and will lead to new business models. This puts up pressure for further adaption. Until now, every LS company started to explore opportunities around those new business models "beyond the pill", but the efforts have mainly been shabby. The search around this new holy grail will be a rat race for the upcoming years and must be seen separate from the core business with high margins on drugs and treatments.

- **Combination of OI methodologies and execution of holistic programs:** The time of piloting in the field of OI is long gone, and so LS companies will try to combine topics from different business units and target groups (HCPs, patients, caregivers, non-profits organizations) into a holistic program over a more extended period and with various objectives, like new digital services, prototypes, products or strategies. Here, the adaption of design thinking from trend to idea to MVP/prototype (end-2-end process) will become more and more part of the future LS companies' DNA. The status-quo shows a different face, as companies tend to stop innovation processes or OI initiatives often prematurely or don't connect the OI initiatives with other existing processes or innovation cycles.
- **Legal and compliance will adapt – again:** If we turn back time for around ten years, many of today's methodologies, IT processes, or interactions with players around the ecosystem would not have been possible due to compliance or legal constraints. We pointed out the massive restrictions within LS. However, much of today's impossible will change into being possible as acceleration, incubation, user interaction, data, and technology will speed up even further. Regulations must shift focus to secure a patient-driven and friendly environment in the future. New profiles and roles will arise in entities like legal and compliance, which must bridge those gaps and drive innovation even within those departments. Direct interaction with patients will still be complex to pull off, but there will be evolving opportunities and frameworks to do so.
- **On- & Offline combination:** More than ever will the best of two worlds (online: collaboration tools, platforms, independence of place; offline: workshops, personal relationship) go hand in hand. Companies will try to engage experts, HCPs, or patients in online feedback, evaluation, or ideation sessions to save money, travel, and avoid bureaucratic, legal, or compliance topics beforehand. However, in-depths ideation and prototyping to proceed in the design thinking cycles with the same stakeholders will still be onsite to secure quality.
- **Extension of the ecosystem and cross-industry players:** The ecosystem of LS companies has been extending over the past years through OI approaches, startup scouting, many M&A deals, and the outsourcing of various services. This trend is continuing - so goes our predictions - as a cross-industry partnership will become more attractive in topics like distribution, supply chain, technology, and service developments beyond the pill.

References

1. Wikhamn, B.R., Styhre, A.: Open innovation as a facilitator for corporate exploration. Int. J. Innov. Manag. **21**(06), 1750042 (2017)
2. Baskerville, R.L., Wood-Harper, A.T.: A critical perspective on action research as a method for information systems research. J. Inf. Technol. **11**(3), 235–246 (1996). https://doi.org/10.1080/026839696345289
3. Bell, J.: Life sciences: industrial strategy. London Off. Life Sci. **30** (2017)
4. Chesbrough, H., Crowther, A.K.: Beyond high tech: early adopters of open innovation in other industries. R&D Manag. **36**(3), 229–236 (2006)

5. Bianchi, M., Cavaliere, A., Chiaroni, D., Frattini, F., Chiesa, V.: Organisational modes for Open Innovation in the bio-pharmaceutical industry: an exploratory analysis. Technovation **31**(1), 22–33 (2011)
6. Von Hippel, E.: Democratizing Innovation. MIT Press, Cambridge (2006)
7. Oliveira, P., Zejnilovic, L., Canhão, H., von Hippel, E.: Innovation by patients with rare diseases and chronic needs. Orphanet J. Rare Dis. **10**(1), 1–9 (2015)
8. Goeldner, M., Herstatt, C.: Are patients and relatives the better innovators? The case of medical smartphone applications. Technol. Innov. Manag. Hambg. Univ. Technol. Work. Pap., no. 91 (2016)
9. Sawhney, M., Verona, G., Prandelli, E.: Collaborating to create: the Internet as a platform for customer engagement in product innovation. J. Interact. Mark. **19**(4), 4–17 (2005)
10. Füller, J., Mühlbacher, H., Matzler, K., Jawecki, G.: Consumer empowerment through internet-based co-creation. J. Manag. Inf. Syst. **26**(3), 71–102 (2009)
11. Benson, T., Pedersen, S., Tsalis, G., Futtrup, R., Dean, M., Aschemann-Witzel, J.: Virtual co-creation: a guide to conducting online co-creation workshops. Int. J. Qual. Methods **20**, 16094069211053096 (2021)
12. Kozinets, R.V.: Netnography: The Essential Guide to Qualitative Social Media Research. Sage, Thousand Oaks (2019)
13. Bartl, M., Füller, J., Mühlbacher, H., Ernst, H.: A managers perspective on virtual customer integration for new product development. J. Prod. Innov. Manag. **29**(6), 1031–1046 (2012). https://doi.org/10.1111/j.1540-5885.2012.00946.x
14. Graffigna, G., Libreri, C., Bosio, C.: Online exchanges among cancer patients and caregivers: constructing and sharing health knowledge about time. Qual. Res. Organ. Manag. Int. J. (2012)
15. Howe, J.: The rise of crowdsourcing. Wired Mag. **14**(6), 1–4 (2006)
16. Christensen, I., Karlsson, C.: Open innovation and the effects of crowdsourcing in a pharma ecosystem. J. Innov. Knowl. **4**(4), 240–247 (2019)
17. Tucker, J.D., Day, S., Tang, W., Bayus, B.: Crowdsourcing in medical research: concepts and applications. PeerJ **7**, e6762 (2019)
18. Joutsela, M., Korhonen, V.: Capturing the user mindset–using the online research community method in packaging research. Packag. Technol. Sci. **28**(4), 325–340 (2015)
19. Stones, S.R., Bull, S., Becerra, S.: OP0162-PARE An exploration of lived experiences amongst adults with rheumatoid arthritis using an online research community platform: a pilot study. BMJ Publishing Group Ltd (2017)
20. Eisenhardt, K.M.: Building theories from case study research. Acad. Manag. Rev. **14**(4), 532–550 (1989)
21. Baig, A., Hall, B., Jenkins, P., Lamarre, E., McCarthy, B.: The COVID-19 recovery will be digital: a plan for the first 90 days. McKinsey Digital, vol. 14 (2020)
22. Rapp, M., Kröger, N., Scheerer, S.: Roles on corporate and public innovation communities: understanding personas to reach new frontiers. In: Meiselwitz, G. (ed.) HCII 2020. LNCS, vol. 12194, pp. 95–109. Springer, Cham (2020). https://doi.org/10.1007/978-3-030-49570-1_8
23. Rapp, M., Kröger, N., Scheerer, S.: Inside-out: how internal social media platforms can accelerate innovation and push external crowdsourcing towards new frontiers. In: Nah, F.-H., Siau, K. (eds.) HCII 2021. LNCS, vol. 12783, pp. 500–514. Springer, Cham (2021). https://doi.org/10.1007/978-3-030-77750-0_32

Moderation of Deliberation: How Volunteer Moderators Shape Political Discussion in Facebook Groups?

Sanna Malinen(✉) 🔟

Department of Social Research, University of Turku, Turku, Finland
sanna.malinen@utu.fi

Abstract. Platform moderation policies are described as a "black box" because the reasoning behind them has remained opaque to users. This study investigates how moderation is enacted in Facebook groups, which are user-created and voluntarily administered social media communities hosted by a global platform company. By exploring one part of the complex moderation structure of Facebook, the volunteer moderators of Facebook groups, this study aims to find out how their moderation strategies shape political debate on social media. What are their strategies for dealing with problematic content and how do these strategies support deliberation and autonomy of participants? This qualitative study is based on thematic interviews for 15 volunteer moderators of Finnish Facebook groups with focus on political and societal topics. Based on the findings, this article discusses the extent to which the public debate on social media, currently shaped by the affordances and design choices made by global platforms, serves the benefit of general audience who is seeking political information from social media.

Keywords: Social media platforms · Moderation · Online communities · Political communication

1 Introduction

Social media are associated with openness and deliberation, and because they are intended to nurture participation and social connection between people, moderation is often viewed as a threat to participation and strongly against the notions of free web and online deliberation [14, 28]. Social media platforms, however, have to moderate actively in order to protect their users from exposure to harmful content. Prior studies on platform moderation have pointed out to many of its problems, most notable ones being the opacity of the governance policy and asymmetrical power relation of users and the platform [14, 25, 28]. Previous studies suggest that moderation policies of platforms and their underlying logics are intentionally being kept guarded in order to maintain a sense of openness and freedom that any form of moderation and content control would typically be strongly against [14, 28].

© The Author(s), under exclusive license to Springer Nature Switzerland AG 2022
G. Meiselwitz (Ed.): HCII 2022, LNCS 13315, pp. 602–616, 2022.
https://doi.org/10.1007/978-3-031-05061-9_43

Facebook has become a host for a great variety of user-created groups focusing on political and ideological topics, which constitute a notable public sphere for citizens' political debate. Even though Facebook was originally designed for personal networking, it has become an important enabler of civic engagement [35]. In particular, social media promote political awareness and communication, which are well presented in Facebook groups, the object of this study. As noted by Halpern and Gibbs [17], Facebook's democratic potential lies in its ability to integrate diverse sectors of the population into networked information access through its communicative features that promote the quality of argumentation. Furthermore, Facebook is known from its accessibility and huge popularity worldwide as it has set favourable conditions for low threshold participation because it is free, requiring only an internet connection and e-mail address from its users [21].

Since the early days of the internet, scholars greeted online discussions for their potential to democratize communication and expected them to increase citizen participation by providing more people with information and access to public sphere [1]. Particularly free access and the ability for anyone to contribute information have named as key features of the potential for good political discourse online [21]. Based on these expectations, the hopes were high that democracy could move on towards the deliberative ideal, allowing decision-making to be better informed [1]. Engaging in this communicative action can lead to civic activity and political participation beyond online environment in, for example, social movements [21]. Research has shown that social media have a positive effect on participation [3] but the deliberative quality of online discussions is not completely clear. Instead, social media companies have been accused for enabling the massive spread of problematic content. Particularly Facebook has been named as a key actor in distributing political misinformation to be used for undemocratic purposes [2]. The spread of problematic content is explained by features of platform infrastructure; namely, their vague policies, outsourced content moderation teams, and the algorithmic content curation that promotes content that attracts reactions in users [8, 26]. As Ekman [8] argues, currently social media platforms are major contributors to the growing anti-immigrant and racist sentiments, and anti-immigrant actors benefit from their structural affordances.

Particularly the possibility to hide non-preferred political views and consequent lack of diversity have been identified as critical factors behind polarization and radicalization of online communities, and they are named as significant obstacles for online deliberation [32]. Diversity is lacking, when there is not exchange of opinions, no reasoned debate between opponents, and therefore no common ground or shared concerns. Content selection is inherent to platforms as they have developed various personalization tools for adjusting and filtering content, in addition to algorithmic content curation by platforms themselves [14]. From the perspective of the platform, automated content reduction is easy and allows for more flexibility than creating a clear set of moderation rules and obeying them. Particularly the automated moderation tools are associated with the lack of accountability, whereas human moderators are expected to have a more sophisticated understanding of the quality of discussion. This study evaluates their role as the enablers for deliberative discussion.

Social media platforms have been described as the 'hidden custodians' who maintain order and safety by guarding user-generated content [14]. Their moderation strategies are based on fast and easy removal of visual and textual content. However, for the deliberative public debate, concealing moderation decisions from users involves problems. When users are not aware of what is deleted and why, content removal works against learning new things and development of more justified political views. Previous research indicates that moderation is an important part of platform design that has consequences for the deliberative quality of online discussion [12, 13, 34]. The critical question is how to enable constructive, rational conversations to constitute suitable environment for online deliberation [19]? This study agrees with Wright and Street [34], who state that technology is both shaped by and shaping online political discussion. In other words, websites and platforms such as Facebook, the object of present study, are the product of design choices, which influence not only in the form of interaction but also the social and political processes they make possible [34]. Moderation is another design feature of online discussion and the way how it is designed has consequences for the deliberative quality of online debates [1, 12, 13].

As more and more discussion moves online, moderation warrants more attention and research. This study explores the complex moderation structure of Facebook groups from the viewpoint of volunteer moderators, to find out how their moderation strategies shape political discussion. Through thematic interviews for 15 volunteer moderators of Finnish Facebook groups with focus on political and societal discussion, this study investigates how they draw lines for appropriate public speech, and what are the implications of their decisions for deliberation. By analyzing moderators' views on their work, this study discusses how moderation affects the deliberative quality of online participation through Facebook groups.

2 Related Work

2.1 Users as Moderators

In early online communities, moderation was mostly a responsibility of community founders. When online communities grew large, enforcing local social norms became increasingly difficult and social media platforms had to develop new, more scalable moderation strategies for monitoring the massive amount of user-generated content [14, 19]. As a result, today's popular platforms use complex, multi-layered content-moderation systems that involve different socio-technical mechanisms. According to Gillespie [14], the main strategies for handling user-generated content are editorial review, community flagging, and automated detection. These three approaches involve different stakeholders. Automated detection is performed with software tools, whereas social media users are needed for reporting problematic content to platforms by flagging it and thereafter, paid moderator teams to review the reported content. Despite the various strategies and tools, the main challenge for all moderation remains: when to intervene and where to draw the line for the acceptable behavior.

This study focuses on one part of the complex moderation system of Facebook, human moderation by volunteering users. Together with the technical moderation tools, volunteer users are still considered as the most effective moderators because they are familiar with group norms and cultural context of discussion [14, 31]. Prior research has shown that volunteer moderators tend to engage personally in the moderation process and view it as an opportunity to growth through conversation and negotiation for both themselves and their communities [22, 29, 31]. When moderation decisions are left to algorithms or outsourced moderation teams, communities and their human moderators miss opportunities for discussion and reflecting the values and norms of the group [29, 31]. As shown in a study by Ruckenstein and Turunen [29], when moderators were reduced to similar position to machine and they could only remove content without explaining their decisions to users, they felt that they could not contribute to the growth of community and became frustrated.

Moderation is generally associated with content removal and banning of those who violate the rules. However, prior studies have shown that it involves a great variety of tasks and roles: moderators consider themselves as caretakers and community builders, who nurture the community and help it to grow over time through moderation and responsive negotiation [29, 31]. Human moderators have a difficult task in balancing between two demands: ensuring the participants' freedom of expression while protecting them from harmful content. Prior studies of human moderators have highlighted the continuous interaction between moderators and members, and conceptualized moderation as an ongoing negotiation in which the meaning of moderation is continuously defined amongst the platform, community members, and fellow moderators [14, 22, 31]. Thus, community guidelines and policies are not fixed but they can evolve over time because of the negotiation process [25].

Today's social media platforms are hosts for multiple user-communities, each with their own local group norms alongside with site-wide policies. For users, this structure is often complicated and it is difficult to keep track of and obey all the norms. In addition, local groups may not have explicit rules, or their rules are vague and vary across subgroups [11]. For instance, each Reddit community has its own set of guidelines and moderators who create these guidelines [11, 19]. Most users attend to many communities and use different platforms, and for them it is particularly tedious to comply with all the rules and act as expected. Particularly in the context of commercial content moderation systems, scholars have criticized the insignificant position of users. As Myers West [25] argues, users remain absent, and they are only given the role of laborers who can report content they deem objectionable.

Prior research has named fairness and transparency as the main elements of good moderation: when users understand the policy and rules of moderation, they are more likely to consider content removals as fair [19]. Particularly silent removals have caused anger and frustration in users, leading to protesting opaque moderation policies and developing their own explanations for silent moderation [14, 25]. In addition, the human moderation has raised concerns about subjectivity and biases. According to Kalsnes and Ihlbaek [20], moderation can be framed as political activity; while moderators give visibility to some views, they permit others. Due to their power position in relation to other participants, their choices have implications to communication rights of citizens and shaping the discussion on public sphere [13, 14, 19].

2.2 Role of Moderation in Online Deliberation

In his famous definition, Habermas [16] views deliberation as an interchange of rational–critical arguments in discussion to seek a common understanding between participants. Ideally, participants of the reasoned discussion are open to ideas expressed by others [6]. Even disagreements are useful for deliberative debate, if arguments are presented in a civilized and constructive manner [23]. However, hateful or disrespectful tones can weaken the deliberative quality of discussion. There is evidence that the lack of civil discourse and fear of conflicts can restrain people from discussing political topics on social media [21]. Scholars' views on the deliberative quality of social media debate are controversial. The key function of social media, connecting users through online networks and communities of interest, has a huge democratic potential as it facilitates information flow and diversity of opinion leading to more informed and democratic decisions [17, 27]. However, online discussions are often characterized by incivility and polarization rather mutual respect and consensus. The ideal notion of public sphere seems to collide with the reality in which the social media platforms have become enablers for the spreading of hate speech and disinformation.

Findings on Facebook's features that promote deliberative quality are controversial. Although Facebook scores high in two main predictors of online deliberation, the affordances of identifiability and networked information access, cross-platform comparison shows that Facebook discussion has lower deliberative quality in comparison to other types of discussion sites, e.g. news media forum and government-run websites [17]. Then again, when compared to more anonymous and deindividuated YouTube, discussions in Facebook present a more egalitarian distribution of comments between discussants and higher level of politeness in their messages [17]. In general, social media platforms promote users' general engagement but their discussions have the lowest level of deliberative quality [9, 10, 17].

Theory of deliberative democracy emphasizes exposure to information that challenges an individual's pre-existing political beliefs and attitudes as a prerequisite for informed political discussion [24]. In the consumption of political information, selective exposure is known to increase polarization of political attitudes, which is an undesirable outcome from the perspective of deliberation. Social media platforms customize information to improve user-experience by offering people content they prefer. However, hidden content removal and reduction work against deliberation, learning and developing more justified political views. Hiding contradicting views from the political discussion is usually associated with algorithmic content moderation but exists in moderation by humans as well [20].

Scholars have sought the favorable conditions for rational-critical debate online. Deliberation is a demanding type of communication characterized by rationality, reciprocity, constructiveness, and mutual respect between participants [9, 10]. Kruse and colleagues [21] highlight the possibility for citing sources, critical thinking, examination of data and facts, and searching for the truth for enhancing deliberation. According to Jenkins [18], unlimited access to information, including diversity of opinions are required. In addition, Esau and colleagues [9] found that moderation, asynchronous discussion, a well-defined topic, and the availability of information enhance the level of deliberative quality of user comments. This study looks particularly into the role of moderation in the deliberative quality.

Moderators have an important role in maintaining the conversational coherence. When the topic is well defined, users' contributions are more likely to be constructive and rational [9]. Different technological choices regarding moderation have consequences for deliberation. In Facebook groups, moderators can both pre-moderate and post-moderate content. In pre-moderation, the moderators read each message prior to its publication. This is an effective way prevent offensive content but it disrupts the natural rhythm of the discussion by causing delay [29]. Another problem of pre-moderation lies in its invisibility; when messages are post-moderated, offensive messages are visible for participants and they can estimate whether the moderation was justified. The benefit of pre-moderation is that in asynchronous discussion participants can take more time to elaborate their arguments and justify their views, which is likely to increase rationality of their arguments [9]. This way, pre-moderation is expected to contribute to respectful and factual discussion [9].

In particular, political groups are at a risk of evolving into echo chambers if dissenting opinions are not allowed. Research recommends that to support the deliberative process, moderators should act as facilitators of discussion who ensure the existence of competing voices [9]. However, different moderation strategies lead to different outcomes in terms of diversity. Research suggests that exposure to diverse opinions is encouraged by providing "safe spaces" where people feel comfortable sharing their views. Safe space policies aim to prevent the marginalization of those who are in a more vulnerable position [12, 13]. However, safe spaces tends to involve more moderator intervention, deleting and self-censoring [13]. The downside is, that in the free speech strategy with low moderator intervention, discussion contains negative and aggressive tones, which may cause retaining from sharing one's views [12, 13]. In general, online political discussion often remains superficial and its deliberative potential is not realized. One main reason for this is that users are hesitant to express their political leanings for fear of harassment and the potential effect it has on their employment or social relationships [21]. Thus, social media networks involve surveillance elements, which can cause intimidation and work against sharing one's views and participating.

3   Research Goals and Materials

3.1   Research Questions

In order to promote deliberation, public discussions should work as catalysts and provide their participants with new ideas and diverse views. As moderators have the power to determine the topics of discussion and silence unwanted voices in Facebook groups, this study looks into their views and actions. By looking into the views and strategies of volunteer moderators, this study explores (RQ1): *Who is protected by moderation and from what?* The chosen moderation strategy affects the degree that the deliberative potential is realized through discussion. Particularly invisible modes of moderation and unclear rules and policies are harmful for participants' learning and personal development [14, 19, 20]. If people do not know why their messages were moderated, they cannot learn from their mistakes and continue to post content of higher quality [19]. Therefore, the moderators' emphasis should be on education, rather than mere removal, and by helping users to understand the community norms and encouraging them to participate constructively, they can improve the quality of discussion [19]. Continuing the research on online deliberation, this study explores to what extent online political discussions live up to the standards of deliberation, and asks (RQ2): *How does the moderation on Facebook groups support the deliberative quality of discussion?*

3.2 Data Collection

This qualitative study uses data obtained from 15 semi-structured interviews with Facebook group moderators as research data. The face-to-face interviews were conducted between December 2019 and February 2020 in Finland. The informants were selected first by searching for active Finnish Facebook discussion groups labeled as political or societal. Then persons named as moderators or administrators of these groups were identified through the each group's public page and contacted in person via Messenger. Initially, interview requests was sent to 20 individuals, of whom five either declined or did not see the invitation. The interviews were recorded and transcribed, and the duration of the voice files varied from 58 to 170 min.

This study uses semi-structured interviews because of the flexibility of this format. In addition to predetermined research themes, it allows other relevant themes to develop throughout the interviews [4]. Therefore, it can bring out new and unexpected results and allow the study to take new directions. Data analysis followed the principles of thematic analysis, which is a process of identifying patterns and themes within the data [5].

4 Findings

4.1 Moderation for Diversity or Harmony?

Moderation is usually understood as mere content removal or banning of those who violate the rules. However, the moderators interviewed for this study reported possessing a wide range of tools for monitoring and controlling participation in the group. Some of these moderation strategies were visible and some invisible to participants. Explicit moderation activities contain post-moderation and public interventions in discussions, so that all the members can see the inappropriate content, how the moderators deal with it, and participate in the public discussion about moderation if they want to. In addition to this, moderators have many ways of guiding and controlling participation silently without leaving any visible traces into the group. They mentioned using the following tools for invisible intervention: applicant-screening, pre-moderation, hiding content without letting the participant know, and private discussions with members who cause problems and with other moderators.

Pre-moderation was commonly used in groups that discussed more sensitive topics, such as racism or gender equality, to ensure that offensive content would not surface in the discussion even temporarily. In these groups, moderators paid special attention to careful member selection to avoid trolls and troublemakers in advance. The policy of these groups draws from the idea of safe spaces, which aim to protect those who are in a weaker position in the society and ensure that their voices are heard in discussion. However, there was also a more practical reason for pre-moderation: problematic content usually occurs during nights and weekends when there is no one watching. Although pre-moderation receives criticism for disturbing the natural flow of discussion, the interviews of this study showed that pre-moderation is contributing to the quality of discussion and can be used to avoid thoughtless commenting when low quality posts are not published.

The most significant benefit of pre-moderation is that it ensures that trolls and trouble-makers do not appear when moderators are absent. Reviewing comments before publishing is the most effective way to repel hateful content but it leads to slowing down the rhythm of discussion. As noted by Towne and Herbsleb [33], making users' contributions immediately visible lowers perceived entry barriers especially for newcomers. However, sometimes controlling the amount of posts and the hectic rhythm of discussion can improve its overall quality. As one moderator describes, in a very active group, controlling the reported content was impossible without pre-moderation, because group members saw offensive posts and started to report them to moderators.

At first, we did not moderate before publishing, but then we noticed that it became a chaos. People just post loads of content and we have no time to react to it. Then they evolve into fights and people start blaming moderation and the group members for the chaos. And we don't even know what's going on. (Moderator 10).

Facebook enables applying of different moderation strategies for different people and some moderators admitted using that. Posts by those who have posted controversial comments earlier, were pre-screened while valued members who were known from their good quality contributions were not screened. Pre-moderation was perceived as effective especially in those groups that had zero tolerance to negative and aggressive language or had members who posted messages at night, often while they were drunk. Interestingly, moderators emphasized, that pre-screening is used not only to protect members from offensive content but also to protect members who have a tendency to write too personal, emotional or messy comments, and this way, are at risk of losing their reputation in the group.

I often feel that we need to protect people from themselves, like prevent them for causing harm to themselves. --- Even famous politicians may say something stupid and then other people get mad at them. And then I'll just close the discussion quickly to avoid more damage. Because you never know if someone is under the influence of alcohol. People just don't think clearly all the time, especially at night, and then they might become ridiculed for that afterwards. (Moderator 6).

My ethical rule is that if someone writes a post at 2am at weekend, it is very likely that s/he would not want it to be published next morning, when s/he is sober. So, I will either delete it or send a private message and ask if you still want to publish this. (Moderator 14).

This type of protective behavior can be interpreted as an attempt to save face of those who fail in obeying the codes of conduct in the group. Denoting to Erving Goffman's [15] classic concept of face-work, Gibson [12] argues that moderation policies are explicitly defined rules for maintaining the face of all users. In Goffman's view, during social interaction, participants are motivated to maintain both their own and their discussion partner's self-worth and autonomy [15]. Moderators' role here is to help maintaining the participants' faces and guide them in navigating between various rules and norms. Silent moderation often involves discretion, which lacks from harder moderation strategies: many moderators reported that they are frequently giving users feedback and personal advice on how to write better comments. However, as one moderator described, personal guidance became demanding when the group grew in popularity.

In the beginning, I was very active and welcomed everyone personally. I tried to govern the group by being nice to everyone and reminded them about the rules. For example, if someone called names, I sent them private message asking to behave kindly to avoid punishment. Nevertheless, that became very laborious and when there were thousands of members, I could not read all comments anymore and the nasty messages stayed. (Moderator 11).

Moderators emphasized that they wanted to avoid permanent discharging from the group because of its visibility. Removing someone from the group permanently may cause negative reactions within the group because very often, those who were banned, ended up questioning the group norms and its' moderation policy in public. Furthermore, the discharged members sometimes decided to create their own competing Facebook group as a protest for the previous group and its moderators. Therefore, to avoid public disputes and negative discussion about moderation policy following from someone's dismissal, moderators rather preferred invisible strategies. If unwanted behavior occurred, they rather contacted these persons privately to reprimand them for their messages or to give a warning, or negotiated about moderation decisions with their fellow moderators. Preferring silent moderation strategies shows that moderators did not want to disturb the harmony within the group members by attracting their attention to conflicts. Because restricting someone's access to the group and its discussions is the most tangible form of moderation, moderators wanted to avoid that by giving shorter bans instead of a permanent removal from the group and by cherry-picking preferred applicants based on their profile information, interests and other group memberships.

Pre-screening of applicants was commonly used in the Facebook groups of this study and moderators had developed their personal checklists and rules for estimating people and their suitability to the group. However, this selection involves some problems for deliberation, the most notable being that the process is invisible and based on moderators' subjective views. In addition, too exclusive selection may suppress critical discussion in the group. As stated by one moderator, a good discussion requires diverse viewpoints and critical comments:

It is difficult because also the critical viewpoints are good for the development of the group, especially if there is a gray area within the rules. Public debate promotes discussion and keeps the group active. Although members often report dissenting comments, I don't think that they should be forbidden unless they are clearly against rules. Actually, it is nice to see how people become more active when someone challenges them. Unpleasant persons can thus be very important for the group and help its members to better justify their views. (Moderator 12).

The most effective ways of preventing unwanted content, namely screening the member applications and moderating content before accepting it, are also the hardest forms of moderation as they prevent some users' participation. Therefore, they have implications to successful discussion. One fundamental question about moderation is to what extent it contributes to diversity or conformity of a group and its members. From the viewpoint of deliberation, invisible moderation practices have two sides. They maintain the norms and harmony within the group by mitigating risks that dissenting views and attacking members may cause. The downside of invisibility is that when participants are not aware of the criteria of member selection and they do not see the private negotiations behind moderating decisions, they cannot estimate the legitimacy of policy nor participate in its development. Previous research suggests that online communities are capable of regulating norm-deviant behavior effectively themselves if they are given the opportunity to do so [7]. This was not realized in most of the groups as moderators did not usually involve group members in the moderation process and avoided public discussion about problems. As suggested by Ditrich and Sassenberg [7], when Facebook group members are setting their rules and exerting control over other members, it is likely to strengthen the group identity and its impact on its members.

4.2 Supporting the Deliberative Quality with Moderation

Deliberation is a reciprocal process and arguments should be not only articulated but also heard and responded. For online deliberation, it is crucial to investigate the communicative nature of written comments: to which extent individual comments are interactive and do they elicit more comments. A well-defined topic has a positive impact on the number of constructive contributions [9, 10], and on Facebook, posts are usually brief news titles or links to stories, and thus they provide information for only a low level of reasoning. Technically, the form of Facebook discussion was perceived as hard to follow, especially in the groups that were very active. Moderators found it difficult to keep track of new posts and conversation topics, and often users posted similar news to the group and discussed them under different posts at the same time. They criticized Facebook's group discussion for the lack of coherence. Discussions tend to become chaotic and without a common ground, particularly if there were many comments. One effective way to improve the quality of user comments was to require framing of the posts by adding some context to the news story instead of just sharing links to news sites. In some groups, link-sharing without explanation was forbidden:

It annoys me when someone shares a link to a 30-min long Youtube video without any explanation. Nobody opens it. We just reject it and ask the person to add an accompanying note to it. (Moderator 14).

We have this rule that if you share a story from media, you'll have to write your own views about it. Just sharing the link is not leading the discussion anywhere. We expect you to have an opinion about the topic. (Moderator 1).

Moderators had different views on how well-suited Facebook as a platform is for political discussion. The benefit of Facebook groups is that they bring people easily together because they are easy to create and invite members to. Grass-root political activity is often realized as spontaneous of social movements and it is relatively easy to set up a group to promote a certain political goal on Facebook. Facebook groups were acknowledged for supporting group identity as members are identifiable and they can deepen their connections via mutual conversation. However, moderators were unanimous in their conclusion that Facebook's non-anonymous profiles do not reduce hate speech or inappropriate behavior.

There's been a lot of talk about hate speech recently but it isn't going anywhere. There were hopes that when people speak with their own face it would be less aggressive. That is not true, people can say horrible things with their own identity and they do not see anything wrong with it. There really is no difference to the old discussion boards. Non-anonymity does not prevent even trolling. (Moderator 6).

In some groups, moderators noticed that there were attempts to use the group for spreading disinformation. Especially, in groups with polarizing topics, fake profiles emerged regularly and they were usually recognized and denied an access to the group. Discussion about climate change was mentioned to attract systematic spreading of disinformation. As one moderator explains, it is sometimes difficult to judge whether to allow the discussion about fake news stories or to remove the content. If someone truly believes in fake news content, it may be better to have a critical discussion than just delete the content.

There is a clear agenda of denialism regarding the climate change in our group. There are some users who are specialized in posting that climate change is not true and it is a conspiracy. It is clear because they always post massive amounts of links to different sites. They are usually fake news sites which are disguised as scientific sources. They are easy to remove but we have to think about these persons motives. Do they believe in the news or are they just trolling. (Moderator 9).

Engagement in rational-critical debate requires good rhetoric skills and knowledge about the topic of discussion. Some moderators pointed out that particularly in political discussions, the quality of writing skills impacts a person's reputation within the group. Some members may even become silenced in the group if they have not adequate skills for expressing their opinions. Facebook is generally considered as an easy to join and widely accessible social media site but political discussion requires a wider set of resources than an internet connection. Accessibility of political debate is thus connected rather with the right use of terminology and language than material resources. Moderators admitted that knowledge of correct terminology could divide people so that those who do not understand the more academic terms, may be closed from the group. This leads to a question: Is the requirement of rational-critical debate exclusive to a wider audience of political discussants and this way against participation for citizens coming from different socio-economic backgrounds? Previous research confirms that the participants of political discussion and content creation are more likely coming from elite backgrounds in comparison to other forms of online content creation [30]. One moderator explained having tried to teach grammar for the discussants to improve the quality of their comments with little or no success and emphasized the meaning of good grammar in getting one's message through in social media:

In social media, you may have excellent ideas but if you cannot express them properly and with good grammar, you end up losing your credibility. Because some of us master the language and we don't trust in the views of those who write with bad grammar. (Moderator 7).

I think that it is problematic if you do not possess certain linguistic skills. It distances people from the actual idea of discussion and we don't want that. --- We cannot use too academic language here because if people do not understand the discussion it alienates them from the original purpose of the discussion. (Moderator 5).

Facebook groups have become popular and accessible public spheres for people from diverse backgrounds. Do these digital spaces allow participation for people with diverse backgrounds and skills? As argued by Jhaver and colleagues [19], platform moderation needs to support not just those who are already familiar with these sites and how they operate, but also to those who may be unaware of the normative practices of online communities. Do current social media platforms allow participation for those users who have failed in their attempts to express their views and discuss them with others but are motivated to learn more? As research highlights the opportunity to personal growth as one main criteria for successful online community moderation [19], moderation systems should give opportunities to participate and growth also for the less-skilled newcomers.

5 Discussion

The findings show that the main goals of group moderation are protecting its borders and members by avoiding disturbances in the discussion. Hence, moderators are committed to enforcing the group norms and uniformity in discussion. In this sense, political Facebook groups include elements that may nurture the homogeneity of views. The recent echo chamber debate suggests that particularly private groups that are strictly moderated are at the risk of becoming polarized if they do not tolerate dissenting opinions. In this study, most groups are ideologically biased as they have been created for promoting certain political goals or views, and moderators want the discussion to obey the preferred political line. However, a clear and well-defined topic of discussion contributes to deliberation as it increases rationality and coherence of discussion. This way, group moderators have a crucial role in determining whether the groups develop into deliberative spaces or into echo chambers. From the viewpoint of deliberation, it is essential that group members are aware of what is hidden from them and on which grounds does the selection made by moderators lie.

This study has some limitations and the most obvious one is that it only investigates viewpoints of one stakeholder group, the group moderators. To analyze moderation process as a whole and to better grasp the effects of different moderation strategies, future work should be extended to cover the objects of moderation, that is, the group members. Particularly, the views of those who have been excluded from the group would add value to the existing work on moderation. All in all, we need more research on what kind of participation social media platforms produce and enforce through their complex moderation systems. Do they invite users to encounter diverse viewpoints and learn through discussion or enforce echo chamber effect? Furthermore, for civilized debate, is it required that users agree about everything?

This study has shed light on the conditions under which one popular form of political debate occurs online, Facebook groups. From the perspective of deliberative discussion, it is clear that we need more versatile strategies to cope with content instead of mere removal. Perhaps the current ideal of the exchange of rational–critical arguments can be too narrow and exclusive: also those individuals who have not much prior experience of participating in public debate should be encouraged to express their emotions and frustrations with politics in public spheres.

6 Conclusion

Moderation of political discussion is often a tradeoff between limiting users' opportunities for participation and allowing conflicting voices in the discussion. The findings suggest that it is a demanding but not an impossible task. Limiting someone's right to participate can promote deliberation if moderator justifies the reasons for being moderated and encourages participation within the limits of rules. When creating the groups, most moderators had ideal views of guiding participants personally to ensure the good quality of their comments. When the groups grew, personal guidance became difficult and they had to move on to stronger forms of moderation. The main problem for the deliberation became that the critical voices are easily suppressed in these groups.

In this study, moderators possessed various tools for improving the deliberative quality of discussion. They were active in private guidance and face-saving behavior, which helped in participants' personal growth, autonomy and maintaining their reputation in the eyes of the others. However, moderation mostly involved strategies that remained invisible for the rest of the group. These silent moderation strategies do not disturb the cohesion of the group but because of their privacy, they leave other group members unaware of moderation and do not give them an opportunity to have their say in moderation policy. From the deliberative viewpoint, this type of moderation limits users' opportunities for participation in problem solving and community development. As a result, possibilities for strengthening their group identity will be missed.

Acknowledgments. The author wishes to thank Emmi Lehtinen for her valuable help in the analysis of research data. This study was funded by Academy of Finland, decision number 321608.

References

1. Albrecht, S.: Whose voice is heard in online deliberation?: A study of participation and representation in political debates on the internet. Inf. Commun. Soc. **9**(1), 62–82 (2006). https://doi.org/10.1080/13691180500519548
2. Bessi, A., et al.: Homophily and polarization in the age of misinformation. Eur. Phys. J. Spec. Top. **225**(10), 2047–2059 (2016). https://doi.org/10.1140/epjst/e2015-50319-0
3. Boulianne, S.: Social media use and participation: a meta-analysis of current research. Inf. Commun. Soc. **18**(5), 524–538 (2015)
4. Choak, C.: Asking questions: Interviews and evaluations. In: Bradford, S., Cullen, F. (eds.) Research and Research Methods for Youth Practitioners, pp. 90–112. Routledge (2012)
5. Creswell, J.W.: Qualitative Inquiry and Research Design: Choosing Among Five Traditions. SAGE, Thousand Oaks (2013)
6. Carpini, M.D., Cook, F.L., Jacobs, L.R.: Public deliberation, discursive participation, and citizen engagement: a review of the empirical literature. Annu. Rev. Polit. Sci. **7**(1), 315–344 (2004)
7. Ditrich, L., Sassenberg, K.: Kicking out the trolls – antecedents of social exclusion intentions in Facebook groups. Comp. in Hum. Beh. **75**, 32–41 (2017). https://doi.org/10.1016/j.chb. 2017.04.049
8. Ekman, M.: Anti-immigration and racist discourse in social media. Eur. J. Commun. **34**(6), 606–618 (2019)

9. Esau, K., Friess, D., Eilders, C.: Design matters! an empirical analysis of online deliberation on different news platforms. Policy Internet **9**(3), 321–342 (2017)

10. Esau, K., Fleuss, D., Nienhaus, S.M.: Different arenas, different deliberative quality? Using a systemic framework to evaluate online deliberation on immigration policy in Germany. Policy Internet **13**(1), 86–112 (2021)

11. Fiesler, C., Feuston, J.L., Bruckman, A.S.: Understanding copyright law in online creative communities. In: Proceedings of the 18th ACM Conference on Computer Supported Cooperative Work & Social Computing, pp. 116–129 (2015)

12. Gibson, A.: Safe spaces & Free speech: Effects of moderation policy on structures of online forum discussions. In: Proceedings of the 50th Hawaii International Conference on System Sciences (2017)

13. Gibson, A.: Free speech and safe spaces: how moderation policies shape online discussion spaces. Social Media+ Society 5 **1**, 2056305119832588 (2019)

14. Gillespie, T.: Custodians of the Internet: Platforms, Content Moderation, and the Hidden Decisions that Shape Social Media. Yale University Press, New Haven (2018)

15. Goffman, E.: On face-work: an analysis of ritual elements in social interaction. Psychiatry **18**(3), 213–231 (1955)

16. Habermas, J. The Structural Transformation of the Public Sphere: An Inquiry into a Category of Bourgeois Society. MIT Press, Cambridge, MA (1989)

17. Halpern, D., Gibbs, J.: Social media as a catalyst for online deliberation? Exploring the affordances of Facebook and YouTube for political expression. Comp. in Hum. Beh. **29**(3), 1159–1168 (2013). https://doi.org/10.1016/j.chb.2012.10.008

18. Jenkins, H.: Convergence Culture: Where Old and New Media Collide. New York University Press, New York (2006)

19. Jhaver, S., Appling, D.S., Gilbert, E., Bruckman, A.:"Did you suspect the post would be removed?" Understanding user reactions to content removals on Reddit. In: Proceedings of the ACM on Human-Computer Interaction, vol. 3(CSCW), pp. 1–33 (2019)

20. Kalsnes, B., Ihlebaek, K.A.: Hiding hate speech: political moderation on Facebook. Media Cult. Soc. **43**(2), 326–342 (2021)

21. Kruse, L.M., Norris, D.R., Flinchum, J.R.: Social media as a public sphere? Politics on social media. Sociol. Q. **59**(1), 62–84 (2018). https://doi.org/10.1080/00380253.2017.1383143

22. Matias, J.N.: The civic labor of volunteer moderators online. Social Media + Society, **5**(2), 2056305119836778 (2019)

23. Mouffe, C.: The democratic paradox. Verso, London, England (2000)

24. Mutz, D.C.: Hearing the other side: deliberative versus participatory democracy. Cambridge University Press, New York (2006)

25. West, S.M.: Raging against the machine: Network gatekeeping and collective action on social media platforms. Media Commun. **5**(3), 28–36 (2017)

26. Nikunen, K.: From irony to solidarity: affective practice and social media activism. Stud. Trans. States Soc. **10**(2), 10–21 (2018)

27. Noveck, B.S.: Wiki government: How technology can make government better, democracy stronger, and citizens more powerful. Brookings Institution Press, Washington, DC (2009)

28. Roberts, S.T.: Commercial content moderation: Digital laborers' dirty work. Media Studies Publications 12 (2016). https://ir.lib.uwo.ca/commpub/12

29. Ruckenstein, M., Turunen, L.L.M.: Re-humanizing the platform: content moderators and the logic of care. New Media Soc. **22**(6), 1026–1042 (2020). https://doi.org/10.1177/146144481 9875990

30. Schradie J.: The digital production gap: the digital divide and Web 2.0 collide. Poetics **39**(2), 145–168 (2011)

31. Seering, J., Wang, T., Yoon, J., Kaufman, G.: Moderator engagement and community development in the age of algorithms. New Media Soc. **21**(7), 1417–1443 (2019)

32. Terren, L., Borge, R.: Echo chambers on social media: a systematic review of the literature. Rev. Commun. Res. **9**, 99–118 (2021). https://doi.org/10.12840/ISSN.2255-4165.028
33. Towne, W.B., Herbsleb, J.D.: Design considerations for online deliberation systems. J. Inform. Tech. Polit. **9**(1), 97–115 (2012). https://doi.org/10.1080/19331681.2011.637711
34. Wright, S., Street, J.: Democracy, deliberation and design: the case of online discussion forums. New Media Soc. **9**(5), 849–869 (2007)
35. Warren, A.M., Sulaiman, A., Jaafar, N.I.: Facebook: The enabler of online civic engagement for activists. Comput. Hum. Behav. **32**, 284–289 (2014)

The Platform-of-Platforms Business Model: Conceptualizing a Way to Maximize Valuable User Interactions on Social Media Platforms

Jürgen Rösch[1] and Christian V. Baccarella[2]([⊠])

[1] Faculty of Media, Bauhaus-Universität Weimar, Weimar, Germany
[2] School of Business and Economics, Friedrich-Alexander-Universität Erlangen-Nürnberg, Nürnberg, Germany
christian.baccarella@fau.de

Abstract. Influencer marketing has become big business. The growth of influencer marketing is not only relevant for firms' marketing departments but is also interesting from an entrepreneurial value creation perspective. The advent of social media influencers has spawned a new generation of entrepreneurs who are active on social media platforms and eagerly create their network of followers. The simultaneous presence of many smaller networks results in a specific type of business model for digital platforms: the Platform-of-platforms (PoP) business model. The PoP business model describes a digital platform that allows other platforms to run on its infrastructure and to use its network effects to pursue entrepreneurial activities. To increase value-creating activities, social media platforms need to provide an adequate infrastructure and governance to support the creation of a constant stream of original content. In this paper, we conceptualize the advantages of this PoP business model for social media platforms as whole and for the user groups on the platforms. In that way, we offer interesting insights for theory and practice.

Keywords: Digital platforms · Social media · Business model · Network effects · Weak ties · Autonomy · Entrepreneurship

1 Introduction

Influencer marketing has become big business. This development is illustrated by the fact that spending on influencer marketing has increased from 1.7 billion US dollars in 2016 to over 13 billion US dollars in 2021 [1]. The growth of influencer marketing is not only relevant for firms' marketing departments [2] but is also interesting from an entrepreneurial value creation perspective. People who were formerly regular users on a social media platform can now make a living by inviting others to participate in their daily lives. This opportunity to earn money on digital platforms has encouraged social media users to provide content on diverse topics across a variety of different domains.

The advent of online influencers has spawned a new generation of entrepreneurs who are exclusively active on social media platforms. These entrepreneurs would be deprived of their economic grounds if they were no longer allowed to be active on a specific platform. Therefore, their entrepreneurial activities are heavily dependent on the digital infrastructures that social media platforms provide. We refer to these entrepreneurs as on-platform entrepreneurs (OPEs) and further define them in the next section.

This new breed of entrepreneurs with their corresponding follower bases has serious implications in relation to hosting digital platforms. The simultaneous presence of many smaller OPE networks results in a Platform-of-Platforms[1] (PoP) business model. The PoP business model describes a digital platform that allows other independent platforms to run on its infrastructure and to use its network effects to pursue entrepreneurial activities. For users, this means that they have access to many connected small platforms without having to switch between providers. Compared to off-platform entrepreneurs, OPEs benefit from the existing network effects and infrastructure of the social media platform when creating their own advertisement-financed network of followers.

To increase value-creating activities, social media platforms therefore need to provide an adequate infrastructure and governance to support OPEs in the creation of a constant stream of original content because this content is needed to attract the attention of potential advertisers that themselves generate revenue for the digital platform.

In this paper, we conceptualize the advantages of this PoP business model for social media platforms, for OPEs, and for users and advertisers. Our argumentation builds strongly on the importance of network effects and especially on the fact that when indirect network effects are strong, one global network can be more efficient than many independent small networks (e.g., [3–5]).

The paper is organized as follows. We first describe the underlying rationale for the PoP business model. We then outline five propositions that describe the underlying mechanisms of the PoP business model for OPEs, users of the social media platform, and the digital ecosystem as a whole. We discuss the implications of the PoP business model for social media platforms and describe the risks that might emerge from this approach. We conclude this article with some final remarks.

2 The Emergence of On-Platform Entrepreneurs and the Platform-of-Platforms Business Model

Social media platforms connect people and allow them to create, share, and exchange information [6]. To enable interactions between users, most social media platforms allow multiple user groups on their platforms that are connected via indirect network effects (e.g., regular users and advertisers). However, what makes social media platforms unique is that they combine direct network effects between users with indirect network effects between other user groups (e.g., advertisers). Direct network effects mean that every user joining the social media platform increases the value for existing users of the same

---

[1] To our knowledge, the term platform-of-platforms has not been used in this context; however, the term has appeared in a blog before: https://stratechery.com/2020/stripe-platform-of-platforms/.

group on the platform. For instance, the more friends you have on Facebook, the more value you receive by using the platform. Users benefit from membership externalities (i.e., the sheer number of people who are on the social media platform). However, they mainly benefit from usage externalities, i.e., they benefit from interacting with others on the platform [7].

The interactions of users on a digital platform generate the context for advertisers to place targeted advertisements. By adding advertisers to a social media platform as an additional user group, the platform benefits from indirect network effects. This means the more people using a social media platform, the more valuable it becomes for advertisers. Advertisers benefit from a high number of users and corresponding interactions (i.e., from strong indirect network effects from users to advertisers). In addition, targeted ads better fit users' current needs, which in turn increases their perceived value. Users then also benefit from useful and informative advertisements. If ads are a valuable addition to the content, there is also a positive indirect network effect from advertisers to users (e.g., [8, 9]).

Connecting different user groups has changed the original idea of social media (i.e., "connecting people") to a viable business model for digital platforms. In this context, we have recently seen the emergence of so-called social media influencers. Social media influencers are defined as "individuals, groups of individuals, or even virtual avatars who have built a network of followers on social media and are regarded as digital opinion leaders with significant social influence on their network of followers" [10, p. 28]. We argue that there is a certain sub-group of influencers that join a digital platform with the intention to monetize their networks by creating content and sharing it with other users. Based on this notion, we define these on-platform entrepreneurs (OPEs) as individuals who build their networks of followers on social media platforms with the intention to monetize it. OPEs thus use the platform and their networks to proactively identify and pursue entrepreneurial activities.

Based on the interconnection of OPE networks, we propose that a Platform-of-Platforms (PoP) business model aims at supporting OPEs to create their follower networks and allowing them to run them on their platform infrastructure. The purpose of the PoP business model is to maximize the value of user interactions within and across these "sub-networks" by providing necessary support for OPEs. A digital platform that follows the PoP strategy integrates and manages the common infrastructure and guarantees a global user experience independent of the many different and autonomous networks. It is comparable to a marketplace, such as Amazon, where the offerings consist of many independent pipeline businesses. Similarly, it also comparable to an Appstore which allows apps with many different business models to run on the underlying platform. Although competing platforms can be a threat to the core platform (see for example [11–13]), we argue that the advantages of the PoP business model for a social media platform outweigh the potential shortcomings (see Fig. 1).

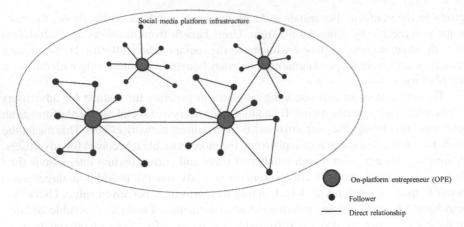

Fig. 1. The Platform-of-Platforms concept

The PoP business model is related to the platform envelop strategy [11, 14], a multi-platform strategy [15], organization-of-organization [16], and the idea of ego networks [17]. However, the PoP business model focuses on the perspective of the "global digital platform," thereby aiming to explain how the efficiency and effectiveness of hosted OPE networks can be maximized (without the intention to own the on-platform platforms). Thus, this paper explains the concept of finding an adequate market organization for platforms that want to adopt a PoP business model.

In the following, we present the underlying mechanisms that highlight the advantages of a PoP business model for a digital platform.

3   Underlying Mechanisms in the Platform-of-Platforms Business Model

3.1   Lower Market Entry Barriers and Implications for the Chicken-Egg Problem

Large user networks on social media platforms create business opportunities that are hardly feasible or non-existent outside the boundaries of a platform. For example, starting a travel or fitness blog without a social media platform is relatively easy and inexpensive. However, building up and nurturing a user network and finding the right advertisement partners requires considerable effort. On the contrary, joining a social media platform, creating a profile, and starting to build an audience within the existing network of the social media platform requires almost zero costs for the infrastructure and very low costs for setting up the profile. Therefore, using a digital platform can significantly lower the barriers for market entry.

The main advantage, however, is that OPEs can freeride on the existing network effects of the social media platform to solve the so-called chicken-egg problem [18]. They can piggyback on top of the global network effects of the social media platform to create their own audience. While piggybacking is typically used without the permission of the underlying platform [19], OPEs are allowed and encouraged to tap into the existing network of the social media platform to attract new followers until they have reached the critical mass. Once they have acquired sufficient followers, they can begin offering advertisement inventory to firms. Platforms like Starngage[2] or Upfluence show that OPEs can also receive help finding the right advertisement partners when they run on a PoP. This means that OPEs benefit through existing followers and advertisers that are already present on a social media platform.

Another advantage is that OPEs do not depend on allowances from the social media platform or any kind of funding in the first place. This makes it attractive to apply different content creation strategies or to find unforeseen new opportunities while experimenting on the social media platform. Combined with the low entry barriers, this leads to an increased market entry by many different OPEs, which all create content for the social media platform and make the global platform attractive for more users to join and for advertisers to use it to promote their products.

Proposition 1: The PoP business model supports OPEs in monetizing their network of followers by lowering the barriers of entry and by helping them to solve the chicken-egg problem.

3.2 The Benefits of On-Platform Competition

As more and more OPEs gain traction and learn how to refine their own business models on a digital platform, the attention of users becomes a tightly contested currency. This on-platform competition for attention leads to a diversification of provided content offerings. As the number of OPEs on the platform grows, OPEs strive to define their own position to attract users and advertisers [20, 21]. This leads to a differentiation and specialization of the content by OPEs ranging from, for example, educational content to tutorials of all sorts to food and lifestyle influencers to people showing how they produce and then sell their canned farts. Popular topics cumulate competitive OPEs, while other OPEs explore the long tail of user preferences (for information about the long tail, see, [22, 23]).

For the social media platform, this competition has a further benefit: each OPE not only creates content choices for users but also creates new advertisement opportunities for firms. The various content offerings of competitive and diversified OPEs combined with the option to use in-content placement of ads creates, for example, opportunities for small and regional brands to use the social media platform to reach their specific target groups. The already mentioned Starngage and Upfluence are examples of platforms that connect influencers with specific brands and their target audience. With the help of these platforms, advertisers can directly work with OPEs that are specialized in their product category, target group, or target region.

The intra-competition for attention and advertisers on the platform challenges OPEs to continuously create content to stay competitive, which increases the pressure for each OPE. However, an increased density of OPEs on social media platforms also comes with advantages. The aggregation of networks on a digital platform protects each local platform and the global social media platform from inter-competition from other digital platforms. Each OPE strengthens its own network effects on the social media platform by

---

[2] https://starngage.com/ or https://get.upfluence.com/.

trying to attract new users and advertisers to its network. By combining the joint efforts of all OPEs, the social media platform forms a global network "on top" of the many local networks. This in turn leads to an interlocking of the individual network effects, which then increases the resilience of the entire platform. To challenge this position, new entrants need to either join the social media platform or compete with it on a global scale [24].

Proposition 2: A social media platform that adopts a PoP business model creates a protected on-platform economy with diverse offerings and manifold advertisement inventory that enhances its competitive position.

3.3 The Importance of On-Platform Multihoming

Users that consume or share content online only for private purposes benefit from having access to several on-platform platforms. A multi-platform bundle enables a digital platform to leverage a shared user experience, a connection between the local networks, and a common technological infrastructure (e.g., [11] or [24]). Therefore, users have access to multiple different content platforms that are not even a click away. Having access to a comparable amount and variety of content would otherwise be associated with comparable high transaction costs. Users would need to browse online and search for specific content on, for example, food, do-it-yourself, lifestyle, travel, or entertainment blogs and check for new updates of each provider individually. A digital platform with a PoP business model can provide diversified content frictionless without any effort to search, choose, or evaluate its quality.

Providing access to diversified networks has various advantages for users and platforms alike. First, users do not have switching costs (e.g., [25]) as they do not have to go to a new webpage with a different navigation or different UX-design to create a new profile on an unknown platform. Second, algorithms and curation mechanisms of the social media platform significantly reduce search costs for finding content and identifying high-quality offerings that fit preferences. For example, algorithms learn clicking and consuming behavior, while likes, comments, and recommendations by other users curate between high- and low-value content. Third, browsing across multiple local platforms also reveals the taste and preferences of users to the digital platform (e.g., [14]), which increases the targeting opportunities for digital platforms, making them even more attractive to new advertisers.

By allowing for multihoming, a digital platform also solves a coordination problem (e.g., [26, 27]) that users, OPEs, and advertisers would otherwise face outside the platform. To interact with each other the same way as on the social media platform, they would need to coordinate themselves to meet on a specific webpage, a blog, or any other medium. Having access to multiple networks solves this coordination problem not only for users but also for OPEs and their advertisers. Without having the option to multihome different networks, users would need to decide which other platforms they should use and where to find different content that fits their needs. Moreover, advertisers would not benefit as much from a more fragmented market that would increase the costs for finding and addressing their target groups.

Proposition 3: Frictionless multihoming between all on-platform platforms reduces transactions costs for users as well as for advertisers.

3.4 The Strength of Weak Ties for OPEs on Digital Platforms

As discussed, social media platforms incorporate many different "sub-networks" that are curated and nurtured by OPEs. From a network effect perspective, the direct relationships between OPEs and their followers determine the benefits that OPEs can leverage from their network. The more users they can attract, the more valuable their networks become (e.g., for potential advertisers); however, there are different types of relationships, such as direct relationships between different OPEs and indirect relationships between followers, that are not directly connected. On the one hand, the more OPEs who join a digital platform, the more crowded and the more competitive a social media platform becomes. On the other hand, the more users other OPEs bring to the platform, the more likely it becomes that new users will also join other OPEs' networks.

The importance of platform users that are part of other OPEs' networks can be explained by the differentiation between weak and strong ties on digital platforms. This helps to understand the underlying mechanisms regarding why a social media platform benefits from hosting and connecting several OPEs, even when users are not directly connected with each other or when networks only partially overlap. In his seminal paper "The Strength of Weak Ties," Granovetter [28] proposes that weak interpersonal ties (e.g., a friend of a friend) allow for tapping into other networks to gain access to valuable resources that would otherwise not have been accessible (e.g., novel information). Individuals with many weak ties will allow information to move more freely in contrast to tightly meshed networks that consist only of strong ties (e.g., a clique of close friends who keep to themselves) [29]. Research has found evidence that weak ties also play an important role in the diffusion of information on online social networks [30], and it has been shown that social influence is a key driver in this context [31, 32].

Weak ties thus promote the speed of information dissemination on social media platforms. Individuals may not only have faster access to novel information, but they may also find shared information more relevant to them because being connected through a weak tie increases the chances that individuals are similar to one another with respect to interests and characteristics [28].

As OPEs act independently on the platform but are still connected through the social media platform, they may also benefit from weak ties that exist with other sub-networks. Because individuals base their own behavior on the behavior of other individuals in a social environment [33], we argue that when users who are part of the network of a specific OPE are exposed to a network of another OPE, it is likely that they will ultimately also join and "spillover" to the new network. Therefore, it not only makes sense for OPEs to gather as many followers as possible to increase their own network, but it also becomes important to leverage as many opportunities as possible to create touchpoints with other networks (e.g., by sharing content of other OPEs through their own account). By creating these contact points between independent networks, OPEs can mutually benefit from each other by increasing the chances of these user spillovers. The greater the complementarity of two networks, the greater the chances that social influence will prompt participants to also join the other network. Of course, this makes particular sense when OPEs mutually share their content because then the network becomes permeable in both directions.

A digital platform can actively promote the interconnection of networks. For example, with "Instagram Stories,"[3] OPEs can not only create and share their own content but can also share content (e.g., videos or images) of other OPEs. Currently, there are over 500 million active daily Instagram Stories users [34], which emphasizes the reach this tool has. By sharing other users' content with Instagram Stories, OPEs allow their own followers to access other networks that they are not yet connected to but that may interest them. The smaller the overlap of users that already follow both OPEs, the greater the benefit in terms of network growth potential.[4]

Taken together, weak ties between OPEs' networks are valuable in terms of network growth. If the social media platform manages to leverage this potential, it creates direct network effects among OPEs, which again strengthen the overall network effects of the social media platform. In addition, this weakens the competition between OPEs and creates a cooperative and competitive environment. This means that OPEs need to create as many touch points as possible between networks, allowing weak network ties to become strong network ties. The better a digital platform enables these touch points, the more likely it is that networks will attract additional participants.

Proposition 4: Weak and strong ties between the sub-network of OPEs create direct network effects among OPEs, which strengthens the position of each OPE as well as the position of the social media platform as a whole.

3.5 The Importance of Granting OPEs the Autonomy to Manage and Adjust Their On-Platform Business Models

To help OPEs engage in entrepreneurial value creation, a digital platform needs to provide adequate circumstances. In this context, entrepreneurial autonomy is another important factor of a PoP business model [35]. Autonomy allows entrepreneurs to "remain free to act independently, to make key decisions, and to proceed" [36, p. 140]. Therefore, OPEs need to have an environment that promotes autonomy without unnecessary constraints [37]. This is supported by studies that have shown that firms that grant autonomy to their members have higher levels of competitiveness and effectiveness compared to firms that constrain or complicate their decision-making freedom [38]. Consequently, an environment that gives individuals the freedom to pursue their entrepreneurial activities fosters renewal, innovation, and growth.

Social media platforms have a direct influence on how much autonomy OPEs have on their platforms. On the one hand, digital platforms need to provide an adequate regulatory framework based on the technological architecture, the purpose of the platform, or on legal requirements [39]. This regulatory framework ensures that users navigate in a protected space that specifies allowed or prohibited online behaviors (e.g., hate speech). By setting the rules, a platform acts as the "choice architect" [40] in the sense that it

---

[3] https://about.instagram.com/features/stories.

[4] Of course, a platform may also allow users to "indirectly" tap into other networks through, for example, recommendation algorithms. The better the fit of a suggested OPE, the more likely a user will connect to a suggested network.

creates the decision space for OPEs that they can use to design und operate their own business models.

As mentioned, the on-platform entrepreneurial environment is characterized by competition between OPEs on the one hand and by advantages that arise due to cooperation between OPEs on the other hand. To help OPEs thrive in such a co-optive environment, platforms need to grant OPEs a considerable amount of autonomy to allow for entrepreneurial behaviors. By providing a platform governance that supports OPEs in their entrepreneurial endeavors, OPEs' networks are more likely to grow (in terms of increasing revenues and attracting new users). Social media platforms need to allow OPEs to individually adjust their business models depending on the respective strategic approach. For example, because OPEs' personal backgrounds differ, a one-size-fits-all solution or unnecessary platform-induced regulations would limit the OPEs' decision-making and action space, hindering them from adequately reacting to changing circumstances.

OnlyFans is a good example of a platform that connects different OPEs while allowing them to individually monetize their network of followers.[5] On OnlyFans, OPEs can create content (i.e., photos, texts, or videos) that is accessible to their followers. They can monetize their content by charging a monthly subscription fee that ranges from $4.99 to $50 (or they can decide to offer their content for free). Moreover, OPEs can charge exclusive commissioned content and can ask for tips. According to their corporate website, OnlyFans has paid out over $5 billion to their creators and has currently over 150 million registered users. One major success factor of OnlyFans is the autonomy they grant their creators, who have the possibility to decide which mode of monetization best fits their needs. Therefore, OPEs can strategically and proactively use this freedom to their advantage. By having the opportunity to adjust their business models according to changing circumstances, OnlyFans not only accounts for the diversity of content creators but also increases its chances that other OPEs will also join the platform. Because registered users can multihome and follow as many creators as they want with almost no switching costs, OnlyFans can leverage indirect network effects because the more creators who join the platform, the more users who will register to follow their content.

Instagram is another good example of a platform that allows OPEs to monetize their networks in various ways. OPEs can negotiate "off platform" with potential advertisers that like to see their products or services in the OPEs' Instagram accounts. Along with sponsored content, OPEs can also generate income with affiliated links or sell merchandise [41]. Recently, Instagram began directly supporting creators to monetize their networks. For example, Instagram now shares revenue with creators when ads appear in their standalone video application IGTV [42]. In addition, Instagram introduces various features to support OPEs in monetizing their networks (for example, users can purchase products from shops they follow, or they can save them to their wish list and shop later). The unique aspect of this approach is that Instagram constantly tries to support OPEs in their entrepreneurial activities. In doing so, they give them a large amount of freedom to decide which strategy is most appropriate for their needs.

---

[5] https://onlyfans.com.

Proposition 5: Social media platforms can create a vivid on-platform economy by granting entrepreneurial autonomy and autonomous decision making within the governance framework of their platforms.

4   Implications for Social Media Platform Business Models

The core idea of the PoP business model is that social media platforms benefit from many autonomously managed platforms that emerge on it. Global network effects of the social media platform grow as the number of OPEs and the number of interactions between users, OPEs, and advertisers increase. Consequently, the more successfully individual OPEs operate on the platform, the more value they create for the digital platform as a whole. To achieve this, a digital platform needs to find ways to support its OPEs to create frequent high-value content and to find users who benefit most from the content.

Therefore, we propose a new category of social media business models. To better illustrate this, we have categorized existing social media and software-as-a-service (SaaS) platform business models in Fig. 2 along two dimensions: (1) on-platform autonomy and the existence of (2) indirect and between-network effects. Compared to communities and traditional advertisement-financed social media platforms, a PoP business model grants more autonomy to content creators. Compared to communities or SaaS, a PoP business model capitalizes on different user-based networks integrated in one global network.

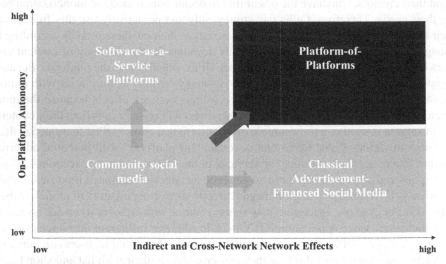

Fig. 2. Platform-of-Platforms compared to other digital platform business models

This categorization offers useful insights for social media scholars and practitioners. Let us consider the example of a social media community that has gained some traction and now intends to monetize the interactions on the platform (lower left quadrant in Fig. 2). Typically, it is difficult to introduce a price in the form of, for example, a subscription fee for users when the community was free of charge before. A direct monetization of users might be possible with freemium offers when new features are introduced. This pricing decision, however, can be detrimental to the network effect: some users will refrain from using the platform as they have to pay for it, or they will not use the platform to its full extent as they do not want to pay for extra features. In either case, network effects will stagnate or even suffer from such pricing schemes.

Introducing advertisers as an additional user group on the platform (right quadrant in Fig. 2) can help monetize some of the interactions by allowing the placement of ads. However, to generate revenue with advertisements the community needs to reach a critical mass and needs to sustain a constant stream of interactions to create the context to allow targeting the ads. Content creation and user engagement rely, in this case, on implicit incentives, such as social recognition or peer acknowledgement. Furthermore, community platforms typically offer only very limited means to create more advanced content. This can be solved by moving upwards to the first upper quadrant of Fig. 2. SaaS or creator platforms provide their users with tools and services that helps them create their own services, such as newsletters, podcasts, or videos. They give creatives and potential entrepreneurs the means to bring their ideas to life. A SaaS-platform approach helps in understanding what is necessary for OPEs to create and manage their own business models. Substack is a good example for such a platform. It allows for creating and distributing newsletters and offers services, such as a homepage, mailing list, payments, analytics, and different subscriptions models[6]. Substack thereby supports its users in managing their work more professionally without the need to build their own infrastructure. Users, however, must build their own audience without the help of the existing underlying network of Substack.

A PoP business model combines the logic of SaaS platforms with advertisement financed social media platforms. It gives users the tools to create content while giving them the opportunity to monetize their network of followers. The main implication for social media platforms is therefore that to further develop their business models, they need to support OPEs in acquiring easy access to the platform, allowing them to attract existing platform users and advertisers. To achieve this, a social media platform needs to provide the right tools and environment for OPEs to generate, curate, and cultivate their content. Social media platforms could, for example, also support OPEs by providing tools for billing or tax statements. Furthermore, social media platforms could offer OPEs dynamic pricing schemes like OnlyFans or could integrate shop functionalities to sell products directly over a social media platform like Instagram. Social media platforms then combine ecommerce functions similar to the ecommerce platform Shopify with creator tools like Substack that provide tools for subscription-based business. However, the main task of social media platforms remains orchestrating the interactions on the platform and between the on-platform platforms.

A PoP business model also bears risks that should be kept in mind. With increased autonomy, misbehavior of OPEs might fall back to the social media platform itself. The musician Neil Young, for example, withdrew its music from Spotify as it allowed a podcast that contained misinformation about the COVID-19 pandemic. Thus, a social media platform needs mechanisms to guarantee the quality of provided content [39].

---

[6] https://substack.com/.

OPEs face the risk of being dependent on a "gatekeeper." Their whole business model is based on being active on certain digital platforms. Nevertheless, they have no influence on the strategic choices the platform makes. For example, research shows that digital platforms can divert users to unwanted or least preferred content to increase the profit for the platform [43]. The recommendation algorithms may also intensify competition between OPEs on the platform to maximize profits for the digital platform [44, 45]. Finally, the social media platform needs to set rules to decide which content is appropriate, and the rules can also change. OPEs have no influence on the decisions of the social media platform. Therefore, created content may be censored or banned from a platform.

5   Conclusion

In this paper, we have conceptualized the PoP business model as a distinct competitive strategy for social media platforms. From an economic point of view, the PoP business model is closely related to the question regarding the ideal market structure of digital markets. Tighter market structures might be beneficial to all users of the platform and the platform itself. They might also lead to "tipping," where in the sense of a "winner-takes-it-all" market, only one platform with significant market power will remain. However, this monopoly can then lead to disadvantages for users and for the economy as it opens the door for anti-competitive behaviors, such as the foreclosure or self-preferencing of the dominant platform. This is subject to for example the Digital Market Act and the Digital Service Act and many national antitrust laws (see for example [46, 47]). In the case of social media, platforms can behave as gatekeepers and can decide who is allowed to participate and under which conditions. However, a platform needs to protect its network effects as they can be challenged by other platforms. An exact understanding of the success factors of the platform combined with more intense antitrust efforts, data portability, and interoperability schemes can open an innovative space for new entrepreneurs to create more diverse digital offerings. Although we are confident that the PoP business model idea offers a fresh perspective of digital platforms, we acknowledge that our postulated propositions require empirical testing, and we hope that our paper encourages future research in this area.

References

1. Statista: Influencer marketing market size worldwide from 2016 to 2021. https://www.statista.com/statistics/1092819/global-influencer-market-size
2. Goldenberg, J., Lanz, A., Shapira, D., Stahl, F.: The Research Behind Influencer Marketing. Impact at JMR (2021). https://www.ama.org/2022/02/16/the-research-behind-influencer-marketing
3. Tirole, J.: Competition and the industrial challenge for the digital age. Paper for IFS Deaton Review on Inequalities in the Twenty-First Century (2020)
4. Evans, D. S., Schmalensee, R.: Markets with two-sided platforms. Issues Compet. Law Policy 20, 667–693 (2008)
5. Dewenter, R., Rösch, J.: Market entry into emerging two-sided markets. Econ. Bull. 32(3), 2343–2352 (2012)

6. Wagner, T.F., Baccarella, C.V., Voigt, K.I.: Framing social media communication: investigating the effects of brand post appeals on user interaction. Eur. Manag. J. **35**(5), 606–616 (2017)
7. Rochet, J.C., Tirole, J.: Two-sided markets: a progress report. Rand J. Econ. **37**(3), 645–667 (2006)
8. Anderson, S.P., Gabszewicz, J.J.: The media and advertising: a tale of two-sided markets. Handb. Econom. Art Cult. **1**, 567–614 (2006)
9. Kaiser, U., Wright, J.: Price structure in two-sided markets: evidence from the magazine industry. Int. J. Ind. Organ. **24**(1), 1–28 (2006)
10. Leung, F.F., Gu, F.F., Palmatier, R.W.: Online influencer marketing. J. Acad. Mark. Sci. **50**, 226–251 (2022)
11. Eisenmann, T., Parker, G., Van Alstyne, M.: Platform envelopment. Strateg. Manag. J. **32**(12), 1270–1285 (2011)
12. Hagiu, A., Teh, T.H., Wright, J.: Should platforms be allowed to sell on their own marketplaces?. Available at SSRN 3606055 (2020)
13. Padilla, J., Perkins, J., Piccolo, S.: Self-preferencing in markets with vertically-integrated gatekeeper platforms. Available at SSRN 3701250 (2020)
14. Condorelli, D., Padilla, J.: Harnessing platform envelopment in the digital world. J. Compet. Law Econ. **16**(2), 143–187 (2020)
15. Klimmek, M., Hermes, S., Schreieck, M., Krcmar, H.: Platform Roles in a Multi-Platform Strategy: Insights from Google (2021)
16. Kretschmer, T., Leiponen, A., Schilling, M., Vasudeva, G.: Platform ecosystems as meta-organizations: implications for platform strategies. Strategic Manage. J. **43**(3), 405–424 (2022)
17. Everett, M., Borgatti, S.P.: Ego network betweenness. Soc. Netw. **27**(1), 31–38 (2005)
18. Eisenmann, T., Parker, G., Van Alstyne, M.: Strategies for two-sided markets. Harv. Bus. Rev. **84**(10), 92 (2006)
19. Parker, G., Van Alstyne, M., Choudary, S.P.: Platform revolution: how networked markets are transforming the economy and how to make them work for you. WW Norton & Company (2016)
20. Lee, J.A., Eastin, M.S.: I like what she's endorsing: the impact of female social media influencers' perceived sincerity, consumer envy, and product type. J. Interact. Advert. **20**(1), 76–91 (2020)
21. Leung, F.F., Gu, F.F., Palmatier, R.W.: Online influencer marketing. J. Acad. Mark. Sci. 1–26 (2022). https://doi.org/10.1007/s11747-021-00829-4
22. Brynjolfsson, E., Hu, Y.J., Smith, M.D.: From niches to riches: anatomy of the long tail. Sloan Manage. Rev. **47**(4), 67–71 (2006)
23. Donnelly, R., Kanodia, A. Morozov, I.: The Long Tail Effect of Personalized Rankings. Available at SSRN 3649342, 1–56 (2021)
24. Zhu, F., Iansiti, M.: Why some platforms thrives and others don't what Alibaba, Tencent, and Uber teach us about networks that flourish. The five characteristics that make the difference. Harvard Bus. Rev. **97**(1), 118–125 (2019)
25. Klemperer, P.: Markets with consumer switching costs. Q. J. Econ. **102**(2), 375–394 (1987)
26. Roson, R.: Two-sided markets: A tentative survey. Rev. Netw. Econom. **4**(2), 142–160 (2005)
27. Hagiu, A., Spulber, D.: First-party content and coordination in two-sided markets. Manage. Sci. **59**(4), 933–949 (2013)
28. Granovetter, M.S.: The strength of weak ties. Am. J. Sociol. **78**(6), 1360–1380 (1973)
29. Gilbert, E., Karahalios, K: Predicting tie strength with social media. In: Proceedings of the SIGCHI Conference on Human Factors in Computing Systems, pp. 211–220 (2009)

30. Bakshy, E., Rosenn, I., Marlow, C., Adamic, L.: The role of social networks in information diffusion. In: Proceedings of the 21st International Conference on WorldWide Web, pp. 519–28. Association for Computing Machinery, New York (2012)
31. Aral, S., Walker, D.: Tie strength, embeddedness, and social influence: a large-scale networked experiment. Manage. Sci. **60**(6), 1352–1370 (2014)
32. Bapna, R., Umyarov, A.: Do your online friends make you pay? A randomized field experiment on peer influence in online social networks. Manage. Sci. **61**(8), 1902–1920 (2015)
33. Asch, S. E.: Studies of independence and conformity: I. A minority of one against a unanimous majority. Psychol. Monograph. General Appl. **70**(9), 1 (1956)
34. Chernev, B.: 33+ Incredible Instagram Stories Statistics You Need to Know in 2022. https://techjury.net/blog/instagram-stories-statistics/#gref
35. Dess, G.G., Lumpkin, G.T.: The role of entrepreneurial orientation in stimulating effective corporate entrepreneurship. Acad. Manag. Perspect. **19**(1), 147–156 (2005)
36. Lumpkin, G.T., Dess, G.G.: Clarifying the entrepreneurial orientation construct and linking it to performance. Acad. Manag. Rev. **21**(1), 135–172 (1996)
37. Lee, S.M., Peterson, S.J.: Culture, entrepreneurial orientation, and global competitiveness. J. World Bus. **35**(4), 401–416 (2000)
38. Lumpkin, G.T., Cogliser, C.C., Schneider, D.R.: Understanding and measuring autonomy: an entrepreneurial orientation perspective. Entrep. Theory Pract. **33**(1), 47–69 (2009)
39. Choudary, S.P., Parker G., Van Alstyne, M.: Platform scale: how an emerging business model helps startups build large empires with minimum investment. Platform Thinking Labs (2015)
40. Thaler R., Sunstein C.: Nudge: Improving Decisions About Health, Wealth, and Happiness. Penguin Books, London (2009)
41. Bradley, S.: How much money Instagram influencers make. https://www.businessinsider.com/how-much-money-instagram-influencers-earn-examples-2021-6
42. Carman, A.: Instagram will share revenue with creators for the first time through ads in IGTV. https://www.theverge.com/2020/5/27/21271009/instagram-ads-igtv-live-badges-test-update-creators
43. Hagiu, A., Jullien, B.: Why do intermediaries divert search? Rand J. Econ. **42**(2), 337–362 (2011)
44. Bourreau, M., Gaudin, G.: Streaming platform and strategic recommendation bias. J. Econom. Manage. Strategy **31**, 25–47 (2018)
45. Lee, K.H., Musolff, L.: Entry into two-sided markets shaped by platform-guided search. mimeo, Princeton University (2021)
46. Cabral, L., Haucap, J., Parker, G., Petropoulos, G., Valletti, T.M., Van Alstyne, M.: The EU Digital Markets Act: A Report from a Panel of Economic Experts (2021)
47. Bourreau, M., De Streel, A.: Digital conglomerates and EU competition policy. Available at SSRN 3350512 (2019)

The Impact of Tweets, Mandates, Hesitancy and Partisanship on Vaccination Rates

Cheng Lock Donny Soh(✉)ⓘ and Indriyati Atmosukartoⓘ

Singapore Institute of Technology, 10 Dover Drive, Singapore 138683, Singapore
{donny.soh,indriyati}@singaporetech.edu.sg

Abstract. This study aims to carry out a cross sectional study of how a state's majority political affiliation, vaccine hesitancy and exposure to misinformed news on social media affects overall vaccination rates. The target country in this study is the United States of America and the granularity of data used is on a state level. Analysis is conducted using a fusion of information from various data sources such as CDC, US Census, Twitter API. Specifically, vaccination rates are taken from CDC, hesitancy rates from US Census, the state's political affiliation from past elections, and social media misinformation from twitter feeds.

The three main study research findings are summarised as follows. First, the findings show the strong correlation between political party affiliation and vaccination rates/vaccine hesitancy rates. As a corollary, there is also a strong negative correlation between vaccination hesitancy and vaccination rates. Second state-wide vaccination mandates typically increased the vaccination rates. Interestingly, states without a state-wide mandate had similar vaccination/vaccine hesitancy rates as states who had banned the state-wide mandates. Finally, the paper shows there will always be a baseline of misinformation in social media articles. However when either (1) the average retweet or (2) the average number of followers reached per tweet crosses a certain critical mass, it will adversely affect the vaccination rate of a state. This suggests that eradicating all misinformation from social media will be counterproductive, rather it is more critical to curtail misinformation before they reach a critical mass.

Keywords: Covid-19 · Vaccine · Twitter · Partisanship

1 Introduction

Although it is intuitive that one's political affiliation, vaccine hesitancy and exposure to misinformed news on social media affects vaccination rates, there are few large scale studies which explore these factors. Most studies approach this research topic by conducting various longitudinal surveys, typically analyzing in the range of hundreds or at most thousands of data points.

Supported by Singapore Institute of Technology.

For instance, Wilson et al. [12] evaluated the effect of social media on vaccine hesitancy globally. Their study looked at the use of social media from two different aspects, including level of negatively oriented discourse about vaccines on social media based on geocoded tweets in the world, and the level of foreign-sourced coordinated disinformation operations. Results showed that there is significant relationship between organization on social media and public doubts of vaccine safety.

Brunson [13] looked at the impact of parents' social networks on the parents' vaccination-decision making. The study showed that most predictive of parents' vaccination decisions was the percent of parents' people networks recommending nonconformity to recommended vaccination schedule. This strongly suggests that parents' social network play an important role in parents' vaccination decision making.

Alfateas et al. [14] conducted a cross-sectional study in Saudi Arabia to look at the impact of social media on acceptance of the COVID-19 vaccine. The study looked at survey results from population over the age of 18 and who were eligible for COVID-19 vaccine. The study confirms that majority of the surveyed population (74.6%) agreed that the vaccine is highly misrepresented via social media and as such it is imperative for the government to share up-to-date scientific information about vaccination that can be helpful in assisting the population in making decisions regarding vaccines.

This study aims to carry out a cross sectional study of data sources from Twitter and various public datasets from the CDC and US government census data. The objective of the study is to determine the extent to which various factors, such as social media, may play in both vaccine hesitancy and vaccination rates in different states. The target geographical region for this study will be the United States of America. Data on vaccination rates from December 2020 to December 2021 and vaccine hesitancy (survey done in July 2021) will be used to study how vaccine hesitancy affects vaccination rates within states. As a context, vaccines were available in USA only on 14th December 2020 [8]. Data from the social media platform Twitter will also be used to determine if there is a correlation between the prevalence of vaccine misinformation versus that of vaccine hesitancy and vaccination rates. Lastly, other factors such as political slant of a state in 2020 is also analyzed and correlated to vaccination rates.

The contribution of our work can be summarised as follows:

- There is a strong correlation between partisanship versus vaccine hesitancy and vaccination take up rates.
- There is a strong correlation between enacting vaccine mandates and vaccination take up rates.
- When a threshold of misinformation is reached, it will adversely affect the vaccination rates of a state.

2   Data Sources

Analysis was performed on five (5) sources of data:

1. The first source of data is the vaccination hesitancy of citizens from a government census taken in April 2020 [1]. This census estimates the percentage of people who are unsure, hesitant and strongly hesitant of the Covid-19 vaccines. These people are clustered into these three categories based on their response if they would take the COVID-19 vaccine. Strongly hesitant respondents would indicate that they would "definitely not" receive a COVID-19 vaccine. Hesitant respondents had indicated they would "probably not" or "definitely not" receive a COVID-19 vaccine when available. Unsure respondents had indicated that they would "probably not" or "unsure" or "definitely not" receive a COVID-19 vaccine when available [1]. The data source clarifies that only the figures for strongly hesitant should be used, hence this paper only makes use of the strongly-hesitant figures.

2. The second data source are vaccination rates. These vaccination rates are in the public domain and obtained from the CDC. The vaccination data is updated daily in CDC [2]. However, the data used in this study ranges only from 13th December 2020 to 2nd February 2022.

3. The third data source is the political slant of a state in 2020. This is estimated by the percentage of votes received who voted Democratic or Republican in the 2020 presidential elections. The data is obtained from the MIT Election Data and Science Lab [3]. The final US Electoral map after election 2020 is shown in Fig 1.

4. The fourth data source is the tweet database on Covid-19 misinformation [4]. The twitter API is used to hydrate the original tweets based on the tweet IDs. Certain fields from the database such as geographical location are used to determine the prevalence of Covid-19 misinformation within various states in the United States of America.

5. The final data source is the aggregated data on Covid-19 vaccine mandates [10]. This data shows the different states and clusters them into three categories: states with mandates for some workers, states who has blocked vaccine mandates and states without a statewide policy for mandates. A visual map of these states can be seen in Fig. 2.

It is noted that although county data is available for data source #1 (vaccine hesitancy) and data source #3 (political slant), this paper has chosen to use state-wise data (as opposed to county data) as the location granularity of tweets where available are typically state-wise.

The vaccination rates for states which majorly voted for the republican candidate or democratic candidate can be seen in Fig. 3. For ease of reference in this paper, the term"red state" will be used to refer to the former, while the term "blue state" will be used to refer to the latter. The top chart refers to vaccination rates for red states while the bottom chart refers to vaccination rates for blue states. As can be seen in top chart of Fig. 3, there is an anomaly in the vaccination rates for West Virginia (cyan color line) due to incorrect reporting [9]. This explains the sudden sharp increase in vaccination rates followed by a sharp decrease in vaccination rates. Some additional statistical information related to the vaccination rates of the two states can be seen in Table 1.

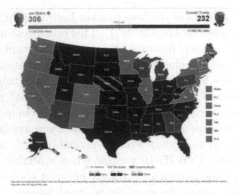

Fig. 1. US Presidential Electoral map in election 2020 [15]. The states colored in blue denote Democratic states, while those colored in red denote Republican states.

Fig. 2. US Vaccine Mandates as of Dec 2021 [10]. The states colored in yellow denote states which have some state wide mandates. The light grey states are states which have no statewide policy on mandates. The dark grey states are states which have banned/blocked statewide vaccine mandates. (Color figure online)

There are several other relevant tweet public data sources such as the covid-19 tweet dataset in kaggle [6] and the dataset from covid19misinfo [7]. The former was not used for several reasons: (1) The tweets from this dataset are not updated regularly. For instance, the last tweet from the dataset was in 30th August 2020 and much has changed since; (2) The date range of the dataset is very limited, only 37 days from 24th July 2020 to 30th August 2020; (3) The size of the dataset has only 179,108 tweets, relatively small compared to the datasets of our choice; (4) Of the 179,108 tweets, only 142,337 tweets contain geographical information. Of the 142,337 tweets, there are 26,921 unique locations with many of these locations not in the United States. All these reasons would limit carrying out a strong cross sectional study across states of the US. The dataset [7] was also not used as the dataset collected tweets only from June 2020 and the intention of the study is to include tweets from when the census was taken (April 2020).

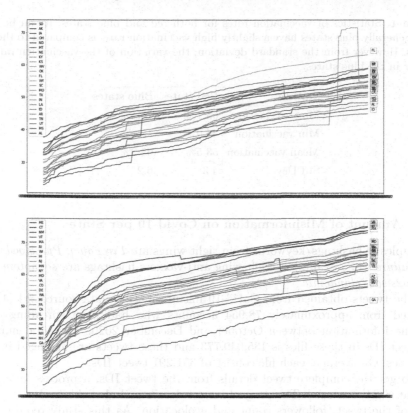

Fig. 3. Top: Vaccination rates for states which voted for the republican candidate in the 2020 presidential elections. Bottom: Vaccination rates for states which voted for the democratic candidate in the 2020 presidential elections. Note that the highest vaccination rate in the top graph is 63.1% (Florida) while that of the bottom graph is 75.1% (Vermont) .

3   Methodology

3.1   Vaccine Hesitancy Rates per State

The data source for vaccine hesitancy [1] has the vaccine hesitancy rates based on PUMAs (Public Use Microdata Areas). The PUMA data source [5] is used to first extract the population size per PUMA, after which it is merged with the data source for vaccine hesitancy. The parameter PWGTP, Pearson's weight, is used to determine the population size per PUMA. This gives the estimated population in each PUMA who is vaccine hesitant. The vaccine hesitant populations for all the PUMAs in each state are summed up and divided by the population of the state. This provides the percentage of the population per state who is vaccine hesitant. To verify that the population per PUMA is correct, the calculated population by state is compared with the census population by state and the difference was less than 1.1 percent (1.04%)

Table 1. Statistics of vaccination rates for both red and blue states. It can be seen that generally blue states have a slightly high vaccination rate as compared to the red states. However from the standard deviation, the variation of the vaccination rates is higher in the blue states.

Statistic	Red states	Blue states
Max vaccination	63.1%	75.1%
Min vaccination	44.9%	50.6%
Mean vaccination	53.5%	65.2%
Std Dev	4.3	6.2

3.2   Amount of Misinformation on Covid-19 per State

Examples of hashtags/keywords from right wings are *FireFauci, Pureblood, vaccinefailure* while examples of left wing anti-vaxxers hashtags are *saynotomasks, sayno2swab, plandemic*.

The tweets obtained from Covid-19 misinformation data source #4 [4] are curated from approximately 78,954 accounts who had been spreading anti-vaccine information between October and December 2020. The total number of tweet IDs in these files is 135,949,773 and these tweets are arranged in 387 text files. On average, each file consist of 351,291 tweet IDs.

To get the complete tweet details from the tweet IDs, a process known as hydrator [11] was used. The complete tweet details include fields such as the full text of the tweet, followers count and geolocation. As this study requires the geographical location of the user, only tweets that have the geolocation enabled are considered.

From the total of 135,949,773 tweets, about 1,155,573 or 0.85% of these tweets have geolocation enabled. About 404,450 or 35% of these geolocation enabled tweets occur in the United States of America. From these tweets, the following basic metrics were captured at the state level:

- Total number of tweets
- Total sum of all retweets
- Total number of tweet content creators

4   Results and Analysis

4.1   Effect of Party Affiliation on Vaccine Hesitancy and Vaccination Rates

Several comparisons were conducted to analyze the voting percentages seen in 2020 versus that of vaccine hesitancy and vaccination rates. The results are shown in Figs. 4 and 5. The red horizontal line in the figures indicate the 50% mark. States above the red line would have had voted more then 50% Republican in the 2020 Presidential elections. Do note that for some states (e.g. Arizona), both major parties obtained less than 50% of the votes and the y-axis plot here only refers to the percentage of Republican votes.

Figure 4 depicts a strong positive correlation between states who have majorly voted Republican in the 2020 Presidential elections versus percentage of strongly vaccine hesitant in state population. The Spearman's correlation value is 0.8. Majority of states that are below the red line, indicating that less than 50% voted Republican, have less than 10% of its state population who are strongly hesitant of Covid-19 vaccine, in contrast to the Republican states that are up to 16% strongly hesitant of Covid-19 vaccine.

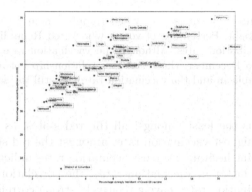

Fig. 4. Percentage of voters who voted Republican in 2020 versus percentage hesitancy. It shows the relationship where the more a state leaned towards Republican in the 2020 presidential elections, the more vaccine hesitant they were.

Figure 5 (left) shows a strong negative correlation between states who have voted Republican in the 2020 Presidential elections versus the vaccination rate in July 2021. The Spearman's correlation value is −0.85. As can be seen in the figure, the vaccination rate for states whose majority vote Republican (those above the red line in the figure) is less than 50% in July 2021. In comparison, the Democratic states (states below the red line) have up to 64% vaccination rate in July 2021. Similarly, a strong negative correlation between states who have Republican-majority vote in the 2020 Presidential elections versus vaccination rate continued to be observed in December 2021 as shown in Fig. 5 (right). A similar Spearman's correlation value of −0.85 was computed. It is noted that although the vaccination rate improved over the months from July to December, the correlative relationship between the percentage who had voted Republican and the vaccination rate remained the same.

Although it is easy to generalize states who voted Republican as having a lower vaccination rate, it is critical as well to understand the reason why. Additional analysis was conducted to look at how vaccine hesitancy may affect vaccination rates. Figure 6 shows the analysis results. Here the state of Nebraska is highlighted. Although the state voted Republican, the percentage of vaccine

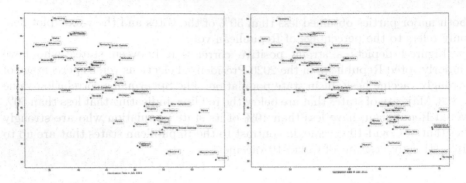

Fig. 5. Left: Percentage of voters who voted Republican in 2020 versus July 2021 Vaccination rates. Right: Percentage of voters who voted Republican in 2020 versus December 2021 Vaccination rates. Although the vaccination rate improved over the months from July to December, the correlative relationship between the percentage who had voted Republican and the vaccination rate was still the same.

hesitant residents is the least amongst all the red states, as it can be seen as having one of the highest vaccination rates amongst the red states as well.

A study of vaccine hesitancy versus vaccination rates indicates that although political slant is a factor in vaccination rates, the correlation between vaccine hesitancy and vaccination rates have a relatively strong correlation as well. The Spearman's correlation is −0.75 between vaccine hesitancy and vaccination rates for both July 2021 (Fig. 6 left) and December 2021 (Fig. 6 right). The figures show that states that have high vaccine hesitancy rate in general have less than 50% vaccination rate both in July 2021 and 6 months later in December 2021.

4.2   Effect of Vaccine Mandates on Vaccination Rates and Vaccine Hesitancy Rates

A comparison is done to determine how vaccine mandates affect both the vaccination rates and vaccine hesitancy. Results are shown in Table 2. It is interesting to note here that states with vaccine mandates have a higher vaccination rate and lower vaccine hesitancy rate than both the states with no mandates and states that ban mandates. Also worth highlighting is that the mean vaccination rates and hesitancy rates are very similar for states with no mandates versus states that ban mandates.

A graphical example can be seen in Fig. 7. The top chart compares state wide mandates with vaccination rates while the bottom chart compares state wide mandates with vaccine hesitancy rates. Each individual charts can be interpreted by the following: The top row denotes the states that have banned vaccine mandates. Note they generally have lower vaccination rates and higher vaccine hesitancy. The lowest row denotes states that have enacted vaccine mandates. Conversely they have higher vaccination numbers and lower hesitancy rates. The middle row denotes states that have chosen to do neither. Note that the outcome

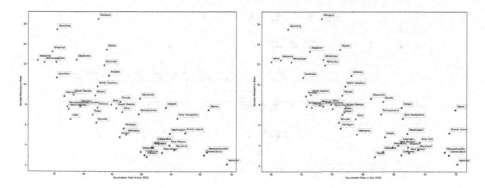

Fig. 6. Left: Vaccine hesitancy rates versus July 2021 vaccination rates. Right: Vaccine hesitancy rates versus December 2021 vaccination rates. The colors of the states denotes if they are voted Red or Blue during the 2020 Presidential elections. There is a strong trend (regardless of party affiliation) where generally when the vaccine hesitancy was higher, the vaccination rate would be lower. (Color figure online)

of these states that have chosen to do neither is similar to the states that have decided to ban the mandates.

Another point to note is that the mandates will cause vaccination rates to shift higher regardless of political affiliation. However the vaccination rates in blue states (with mandates) are generally still higher than red states (with mandates).

Table 2. Statistics of the mean vaccination rates and vaccine hesitancy rates for states who has mandates, no mandates, and had banned mandates. Note that the states that have mandates generally have a higher average vaccination rate while states with no/banned mandates have largely similar average vaccination/vaccine hesitancy rates.

Statistic	Mandates	No mandates	Ban mandates
Mean vaccination	65.3%	54.7%	54.0%
Mean hesitancy	5.44%	9.60%	9.25%

4.3 Effect of Misinformation on Vaccine Hesitancy

Study conducted by Muric et al. [4] did not show a very large correlation between the amount of twitter misinformation present in a state versus vaccine take up rate. In their results, they highlighted that only two states (Montana and New York) had a high level of misinformation. In this case, Montana has a relatively low level of vaccination, while New York has a high level of vaccination. This suggests that the misinformation is widespread and using the number of misinformation tweets alone is insufficient to determine how it will affect the vaccination take up rate. However, two other metrics from the tweet database

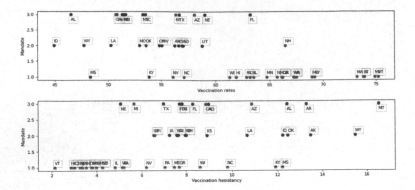

Fig. 7. Top: Vaccine Mandates versus US Vaccination Rates as of Dec 2021. Bottom: Vaccine Mandates versus US Vaccine Hesitancy Rates. The bottom markers denotes states with some state-wide mandates, the middle markers denotes states that do not have state-wide mandates and the top markers denotes states that has banned state-wide mandates. Note that the bottom markers typically have a higher vaccination rates while the both the top and middle markers have similar vaccination/hesitancy rates.

suggests an interesting relationship between misinformation found in tweets versus vaccination rates. The two metrics are:

- Average number of retweets per tweet: This metric is calculated by taking the sum of all retweets divided by the total number of tweets.
- Average follower Rate: This metric is calculated by taking the sum of all tweet followers of content creators (who had previously spread Covid-19 misinformation) divided by the number of tweets.

The first metric calculates the average popularity of the misinformation retweet, while the second metric calculates the average number of tweet followers each misinformation tweet reaches.

Figure 8 (left) shows that after the average retweet rate exceeds a certain threshold (e.g. the red horizontal line depicting 50 retweets), there will be an adverse effect on the vaccination take up rate on the citizens of the state. It can be seen that states with more than 50 retweets on average such as Texas, Tennessee, Arkansas, South Carolina, and Missouri, have lower vaccination rates compared to the other states. This implies that when there is consistently large amount of misinformation spreading in a state, there will be an adverse effect on the state vaccination rate.

Similarly, Fig. 8 (right) shows that once the average follower rate exceeds a certain threshold (e.g. the red horizontal line depicting 4000 followers), there will also be an adverse effect on the vaccination take up rate on the citizens of the state, as can be seen in the vaccination rate for states South Carolina, North Dakota, and Nevada. Intuitively, this metric shows that if the average tweet containing misinformation is able to reach out to more followers the vaccination rate for the state will be lower. Note that Rhode Island has a high vaccination rate even though it has a high average follower rate, this can be attributed to the state's vaccine mandate for most of its workers.

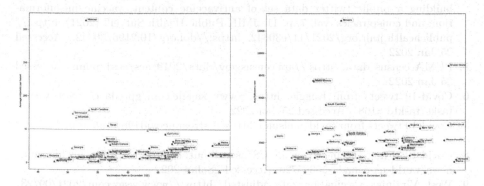

Fig. 8. Left: Average retweet rate versus vaccination take up rate in December 2021. Right: Average follower rate versus vaccination take up rate in December 2021

5   Conclusion

This paper presents a quantitative analysis investigating the correlation between the states' party affiliation, vaccine hesitancy, level of twitter misinformation, and vaccination rates. The results first establish the correlation between red states and vaccination rates/vaccine hesitancy rates. It was also shown that vaccine hesitancy had a strong negative correlation on vaccination rates. Next, the effect of state-wide mandates was analyzed and results show that while vaccination rates are generally higher for states with mandates, both vaccination rates/hesitancy rates are similar for states either with no or with banned mandates. The presence of mandates seems to indicate that regardless of political affiliation, states with vaccine mandates generally do well in terms of vaccination rates. Finally, the analysis showed that the number of misinformation tweets in a state is not sufficient to affect it's vaccination rate. However, the data suggests that when either the average number of retweets, or the average number of followers each tweet reaches a certain critical mass, it will adversely affects the vaccination rate of a state. This suggests that one should always expect the presence of a baseline of misinformation in all social media articles and the key concern is to curtail misinformation before they reach a critical mass.

References

1. Vaccine Hesitancy for COVID-19: Public Use Microdata Areas (PUMAs). https://data.cdc.gov/Vaccinations/Vaccine-Hesitancy-for-COVID-19-Public-Use-Microdat/djj9-kh3p. Accessed 25 Jan 2022
2. Covid-19 vaccination rates. https://data.cdc.gov/Vaccinations/COVID-19-Vaccination-Trends-in-the-United-States-N/rh2h-3yt2. Accessed 11 Dec 2021

3. MIT Election Data and Science Lab, Harvard Dataverse (2017). https://doi.org/ 10.7910/DVN/42MVDX. Accessed 25 Jan 2022

4. Muric, G., Wu, Y., Ferrara, E.: COVID-19 vaccine hesitancy on social media: building a public twitter data set of antivaccine content, vaccine misinformation, and conspiracies, vol. 7, p. 11. JMIR Public Health Surveill (2021). https:// publichealth.jmir.org/2021/11/e30642. https://doi.org/10.2196/30642.. Accessed 25 Jan 2022

5. PUMA census data. https://api.census.gov/data/2019/acs/acs1/pums. Accessed 25 Jan 2022

6. Covid-19 tweets from Kaggle. https://www.kaggle.com/gpreda/covid19-tweets/ tasks?taskId=1681. Accessed 25 Jan 2022

7. Covid-19 tweets from covid19misinfo. https://stream.covid19misinfo.org. Accessed 25 Jan 2022

8. Covid-19: First vaccine given in US as roll-out begins. https://www.bbc.com/ news/world-us-canada-55305720. Accessed 28 Jan 2022

9. West Virginias vaccination rate adjusted. https://www.wsaz.com/2021/09/23/ west-virginias-vaccination-rate-adjusted-after-counting-error-by-cdc-contractor/. Accessed 5 Feb 2022

10. States showing the presence/absence of vaccine mandates. https://www.nytimes. com/interactive/2021/12/18/us/vaccine-mandate-states.html. Accessed 5 Feb 2022

11. Documenting the Now (2020). Hydrator [Computer Software]. https://github.com/ docnow/hydrator

12. Wilson, S.L., Wiysonge, C.: Social media and vaccine hesitancy. BMJ Global Health (2020)

13. Brunson EK. The impact of social networks on parents' vaccination decisions. Pediatrics **131**(5), e1397–404. Epub 2013 Apr 15. PMID: 23589813. https://doi. org/10.1542/peds.2012-2452

14. Alfatease, A., Alqahtani, A.M., Orayj, K., Alshahrani, S.M.: The impact of social media on the acceptance of the COVID-19 vaccine: a cross-sectional study from Saudi Arabia. Patient Prefer Adherence **15**, 2673–2681 (2021)

15. ABC Election map 2020. https://abcnews.go.com/Politics/abc-news-crowdsourced-electoral-map-off-state. Accessed 5 Feb 2022

Exploring Public Trust on State Initiatives During the COVID-19 Pandemic

Austin Sebastien Tan(✉)(iD), Maria Regina Justina Estuar(iD), Nicole Allison Co, Hans Calvin Tan, Roland Abao, and Jelly Aureus

Ateneo de Manila University, Metro Manila, Philippines
sebastien.tan@obf.ateneo.edu
https://www.ateneo.edu/

Abstract. As most governments in the world currently face the pandemic, various policies and initiatives have been put in place in order to help control the spread of the COVID-19 outbreak. While these initiatives and interventions are taking place, a pandemic still creates a reality of risk and uncertainty. In these kinds of situations, public trust is greatly important to properly mitigate health and societal impacts of the pandemic. Social media platforms could be utilized as sources of information to gain insight on public sentiment, especially with the rise of social media utilization during the quarantine [13]. Given this, the study attempts to analyze social media sentiments particularly found in Twitter in order to not only look into the polarity of public sentiment on the government, its processes, and its policies, but particularly, to detect trust between the governed and the ones governing. Furthermore, it seeks to examine and analyze the trust narratives present in the Philippines currently. In this study, a supervised machine learning model was created using Linear SVC, utilizing TF-IDF and n-grams for feature extraction and selection in order to detect the respective trust category of a given sentiment and predict the trust category of new data points. While the results are overall negative, examining the trust categories individually demonstrates different narratives that dictate, affect, and express citizen trust towards different aspects of the government. The behavioral trust group provided narratives on certain political figures involved in a string of anomalies for the negative category, while the positive category lauded the VP for her continued service amidst the pandemic. On the other hand, narratives in the institutional trust group revolved around national and local institutions, where talks about national institutions being more prominent in the negative category, while local institutions, such as local government units, are found in the positive category. Lastly, narratives on the operational trust group focused on certain pandemic policies (lockdowns, mass testing, contact tracing) for the negative side, while vaccines and vaccinations were the focus for the positive side.

Supported by Ateneo de Manila University.

Keywords: COVID-19 · Trust sentiment · TF-IDF · N-Grams · Multi-class classification model

1   Introduction

Many governments continue to manage and mitigate effects of the pandemic brought by COVID-19. The Philippines, for instance, has been responding to mitigate the impact of surging COVID cases through the implementation of numerous interventions to varying levels and degrees of success. One such intervention is the implementation of travel restrictions and other restrictive measures such as heavy quarantine and lockdowns to ensure minimum contact and reduce the risk of transmitting the virus for both healthy and sick citizens [9]. As a result of these interventions, people spend more time on social media, a recent report on digital global insights shows that 80.7% of the population are active social media users with an average amount of four (4) h and sixteen minutes (16) a day [13].

Social networking sites have become a more important tool during the pandemic, one that allows the public access to information as well as providing them an avenue to connect with one another during a period of isolation. It has not only effectively altered the way people communicate, form, and develop relationships, but as well as opening up new possibilities for connecting with one another [3]. With its prevalence, social media can be considered as a huge source for retrieving unfiltered and unorganized data that can be used to determine the opinion of the public on various subjects [26].

However, even with interventions such as quarantines in effect, a pandemic creates a reality where uncertainty and risk are ever-looming, and it is in these kinds of situations where trust is of vital importance—trust as well as cooperation are needed to mitigate health and societal impacts [17]. An example of which could be seen through the Philippine Dengvaxia vaccine panic-where it was revealed that the dengue vaccine posed a risk when given to people who were previously unexposed to dengue, which caused vaccine refusal as well as reluctance [16]. This event shows that the effects of broken trust in critical situations such as pandemics are pertinent. The aftermath of the event revealed not only political tension and public outrage, but also revealed the implications of public trust in government health interventions and strategies. A survey made by the Vaccine Confidence Project showed that there was a dramatic drop in vaccine confidence from 93% in the 2015 survey to 32% in the more recent 2018 survey [17]. Suffice to say, trust is an important indication of the public's belief in the overall reliability, capability, and integrity of the government [7]. As ultimately, the current COVID-19 pandemic challenges government systems around the world, and while governments have varied responses, one factor that seems to dictate the degree of success is the idea of trust [7].

Given this, the study is an exploration of how social media analytics, particularly Twitter, can be used to not only identify, predict, and examine public sentiment on the government, its processes and policies, but to look into the trust between the government and its citizens. Furthermore, the study aims to provide insights on the narratives present in these trust sentiments. In order to do so, the paper first tackles the concept of trust in order to frame its definition of trust and define it. Moreover, the study also presents a trust classification model that identifies the trust category and the respective polarity attached to a sentiment. This model then allows insights as to where citizens direct their trust towards the government and its various facets.

2   Review of Related Literature

2.1   Understanding Trust and Trustworthiness

In a loose sense, to trust in someone is to believe or to have the belief that the person would do us good, whatever "good" may be [25]; however, "good" is rather subjective. One study shows that citizens may have varied criteria in evaluating what makes an institution trustworthy [8]. For example, some may believe that the trustworthiness of an institution might be based on how it is able to deliver on its promises, but for others, trustworthiness may come from sharing a common identify, or even perhaps by their moral capacity to do what is right. Some studies also have shown that trust often involves being vulnerable to a certain degree of perceived risk to oneself [2,8,24].

Trust is certainly a rich concept that covers a range of relationships between objects that change given certain contexts. The topic of trust and confidence; however, is often disputed in many studies when it comes to political institutions. While it is generally accepted that trust is extended towards other people, the dispute lies in the idea of trusting an abstract system such as the government [18]. In this case, this study will adopt the understanding that "trust in government" is synonymous with the notion of "confidence in government" which could be simply referred to as political trust.

Delving further into political trust, it can be defined as the general confidence in core political institutions. Studies often show that high levels of political trust have often been associated with higher political interest, and more involvement in civic affairs [28]. Another study suggests that the relationship between citizens and the government has an important role towards the development of trust towards the government and its services [14]. Furthermore, some studies would argue that there are certain factors in place that would determine the trustworthiness of an institution such as the government. As a matter of fact, previous studies on trust literature have characterized trust and trustworthiness in three (3) areas: Ability, integrity and benevolence [14,20,21]. This would be the definition that this study would adopt in order to determine what makes a citizen continue to trust its government. Thereby, for this study, ability represents the capacity or competency to implement policies or safeguards in the pandemic. Benevolence, on the other hand, would be the belief that the government is putting the concerns or best interest of its citizens' at the forefront, whereas integrity represents the government's honesty as well as its consistency with their promises to the citizens [14,20]. Trustworthiness could then be understood as how an object in a relationship manages to maintain the trust in the specific relationship.

2.2   Effects of Trust in Governance

Trust has varied effects on governance depending on the current level of trust. Extant literature on trust shows that lower levels of trust undermine government capacity to pursue redistributive policies; furthermore, studies also show that trust can increase law compliance [6]. In addition, other studies show that governments who face great mistrust, defined as the absence of trust, and suspicion may find its citizens ignoring and resisting its actions as well as being suspicious of its policies and declarations [10]. This is also true in terms of compliance with health policies and restrictive policies, with studies showing how greater trust in the government leads to more compliance with specific health policies such as quarantine, testing, social distancing, and other health requirements [6,27].

2.3   Trust Detection in Social Media

There are studies that have attempted trust detection in social media. One study focused on the analysis of tweets about Saudi Arabia's COVID prevention measures and found that the high positivity found in the tweets regarding implemented protective measures could be attributed towards overall trust and confidence in their government; solidified through proactive governmental approaches and swift implementation of security measures [1]. This shows that trust is therefore sensitive towards changes to current events, and examining social media sentiments in the context of COVID-19 could prove to be invaluable in today's context. Further studies were done to gain more insight on not just the polarity of Twitter sentiments during COVID-19, but the embedded emotions as well. These studies revealed that trust no longer was a prominent emotion when Twitter users discuss COVID-19 and that in the context of the Philippines, results show that Twitter users from the National Capital Region expressed much more negative sentiment along with emotions of anger, fear, and sadness [9,29]. While literature is present in trust detection in social media, most studies tend to operationalize trust as a measure of sentiment that is, utilizing sentiment as a proxy measure of trust [1,5]. Building a machine learning model capable of detecting not only the polarity but the direction of the trust towards governments could further provide the necessary information to help increase public trust by addressing specific concerns and by avoiding the consequences of low-levels of trust.

2.4   Measuring and Analyzing Trust

Drawing on trust research in government provides three trust categories that details how trust is produced and directed towards the government. Firstly, behavioral trust, defined as one's perceptions regarding the alignment between an expected behavior and the actual behavior [22,25]. This category of trust hinges on characteristics, culture, and belief. Therefore, it is trust relating to political figures, politicians, or other government personas.

Secondly, operational trust which could be defined as the trust relating towards government policies or regulations. It is the match between one's expectations and the operational attributes of the government [22]. It is the type of trust that is produced through repeated exchanges rather than based on characteristics, thus, this is the type of trust that happens over time [25].

Lastly, institutional trust or trust relating to government institutions or other extensions of the government.According to trust literature, this type of trust is produced through institutions that have become accepted as social facts and as a result of which are seldom questioned [22,25]. These three categories would serve as the categories with which the trust detection model would be trained on.

2.5   Multi-class Text Classification Model to Detect Trust

Studies on the general topic of text classification was utilized due to limited studies on trust detection models. Studies on multi-class text classification often utilize supervised learning in order to build the classification model [12,15,19]. It needs data that has been labeled and organized into human-defined classes or categories, which the model would learn how to best fit the sentiment [4].In one study, a multi-class classification of Tweets was done that utilized Support Vector Machines, Neural Networks, Naive Bayes, and Random Forests as the machine learning models [19]. The study found that SVM out-performed the other models in accuracy. Furthermore, In a more recent multi-class text classification study, a total of six machine learning models were utilized in order to assess their efficiency in predicting medical conditions of patients based on drug reviews [12]. The reviews from the dataset were first converted into feature vectors through utilizing the process of Term Frequency-Inverse Document Frequency (TF-IDF) Vectorization as feature extraction. The results of the study showed, that out of the six machine learning models, LinearSVC was found to be the most effective in predicting. LinearSVC is a variant of support vector classification that has more flexibility and utilizes a one-vs-rest multi-class reduction to fit training data. Lastly, the N-Grams feature selection technique was mentioned in another study [1]. This approach allows the model to learn not only the individual words, but the relationship between multiple words as well.

3   Methodology

3.1   Data Collection

The study would be utilizing the social media network of Twitter to acquire the necessary dataset through the scraping of tweets. Twitter was chosen as the focal point for the social media analysis because the platform has an easily accessible application programming interface. Furthermore, Twitter focuses on key words and allows posting to a wider audience when compared to other platforms like Facebook [23]. The study aims to gather tweets related to COVID-19 as well as government responses to the pandemic using key words such as COVID, COVID-19, #RESBAKUNA, #COVID19PH, #COVIDPH. In addition, other keywords that pertain to preventive measures done in the Philippines such as social distancing, quarantine, self-isolation, GCQ, ECQ, MECQ and keywords that pertain to government institutions such as @DOHgovph were used as well. Both English and Tagalog tweets were gathered and collected in the compilation of the dataset and was done through the usage of the Twitter API in Python. Aside from the tweets themselves, other features such as the date of the tweet, geo-location, tweet id, language of the tweet, like count, quote count, reply count, source of the tweet and retweet count were also gathered.

Twitter introduced a new API, Twitter API v2, which includes more features, functionalities and allows more data to be pulled and analyzed. The new Twitter API allows users to apply for the standard track or the new Academic Research track that allows for more precise, complete and unbiased data along with a higher Tweet pull capacity, and access to the complete history of archived public Tweets. For this study, only tweets from the last seven days are extracted. After initial extraction, the data needs to be properly labeled in order for the supervised machine learning model to learn and train from the dataset. This was done by two volunteers who manually tagged each data point in the dataset and assigned them a trust category. Afterwards, the output of the two volunteers was compared at each data point in order to check the inter-rater agreement. Data points that were tagged similarly by the two volunteers were retained while data points that were not similar were discarded from the dataset. Prior to the start of tagging, the volunteers were debriefed and oriented on the following trust categories. There are overall three trust categories— Behavioral, Operational, and Institutional Trust as well as the respective polarity of the sentiment—positive or negative. As a result, a total of eight categories were available for the volunteers to choose from: Behavioral-Positive; Behavioral-Negative; Operational-Positive; Operational-Negative; Institutional-Positive; Institutional-Negative; Filipino Dialect; and Irrelevant. The later two being added to label tweets that were outside the scope of the study.

3.2   Data Preprocessing and Exploration

Each tweet in the dataset and the dataset itself was preprocessed prior to applying them to the machine learning model. This is to ensure that noisy and inconsistent data would be lessened or removed altogether. First, the text and label column of the dataset was selected, all other columns were dropped. Second, all duplicate tweets in the text column were removed as having retweets of the same tweet could skew the model. Third, the tweets that were tagged as Filipino dialects and irrelevant were removed from the dataset as well. Lastly, common data preprocessing techniques were also applied, preprocessing steps that dealt with stop word removal, removing emojis. Converting the text into lower case was applied as well in order to ensure uniformity. Afterwards, punctuations were cleaned and removed to reduce the noise from the dataset. Other possible noise such as repeating characters found in social media syntax along with URLs were also removed as they provide no importance to the data. Finally, stemming and lemmatization was also performed in order to obtain higher accuracy.

After cleaning the dataset, some data exploration is performed before going into training the machine learning model. Exploring the data allows for a more thorough understanding and may provide new insights or angles in engaging with the data. For this study, the counts of each category were visualized in order to check the overall spread of the data. Furthermore, a word cloud was created for each of the categories in order to gain a deeper understanding of the data by visually representing text data in which the size of each word indicates its frequency or importance. Furthermore, n-grams was also utilized to check word relationship per trust category, as it can provide added information that individual word frequencies can not.

3.3   Feature Extraction and Selection

The machine learning model will not be able to directly process text data in as is because most of them expect numerical feature values rather than raw text data Therefore, the texts are converted to a format that the model can understand. A common approach for feature extraction in text problems is to use the bag of words model. This approach takes the presence and frequency of words in a document and takes note of it, however, this approach does not take into account the order they occur in. For this study, the text documents were used to calculate a measure called Term Frequency - Inverse Document Frequency (TF-IDF). To calculate for the tf-idf vector, the Python package sklearn.TfidVectorizer was used. Parameters such as min_df-the minimum number of documents a word must be present in to be kept was set at five, while the ngram_range was set to (1,3) to indicate the consideration of unigrams, bigrams and trigrams.

3.4   Multi-class Model Creation

After cleaning the dataset and gathering the necessary features the data would then serve as the input for the model to train on. Four machine learning models were selected and benchmarked to evaluate their respective accuracies. The four models were: 1) Logistic Regression, 2) Multinomial naïve Bayes, 3) Linear Support Vector Machine, and 4) Random Forest. These models were evaluated through the usage of cross validation and was performed at five fold cross-validation. The highest scoring model was then chosen and then further examined and evaluated. A confusion matrix was created to further look at the discrepancies between the predicted and actual labels. Furthermore, the chi-squared test was once again performed in order to find the terms most correlated with each of the categories for the model. Afterwards, the precision, recall and f1-score of each category in the model was examined. Lastly, the model was then given new unknown data points, one for each trust category in order to check if it will be able to predict the respective trust category of the newly introduced data.

4   Results

4.1   Dataset Description

The dataset was obtained by using the Twitter API Recent Search endpoint to crawl through Tweets using queries that contained specific keywords. After the collection process of Tweets using the Twitter API, a total of 9450 tweets were collected to be used for the study. These tweets were collected from the dates of September 14, 2021 up to September 25, 2021. Aside from the actual Tweets themselves other pertinent information such as author id, source of tweet, Tweet id, language, like_count, quote_count, reply_count, and retweet_count were collected as well and stored in a CSV (comma-separated values) file, which is a simple text file where information is separated by commas. The dataset consisted of 9450 Tweets by the end of the data gathering period.

The CSV file was then processed to create a separate file that only contained the Tweet itself. This file would then be provided to the two volunteers who would each manually tag the dataset, with each volunteer having a clean copy of the dataset so as to not be influenced by each other's decision. The results of both volunteers were then compiled and compared to check for their inter-rater agreement. Any data point in which the two volunteers fail to reach a consensus was dropped. The results of which can be seen in Table 1. After going through the inter-rater process, out of the original 9450 data points, the study was able to retain roughly 69% of the data.

Table 1. Trust categories

Category	Count
Irrelevant	4748
Behavioral-Negative	624
Operational-Negative	586
Operational-Positive	258
Filipino Dialects	167
Institutional-Negative	119
Behavioral-Positive	24
Institutional-Positive	22

Afterwards, Tweets that fall under the Irrelevant category and the Filipino Dialects were dropped as well. For the case of the irrelevant category was put in place to categorize Tweets that would not be useful to the studies. These include, but not limited to:

- Simple statement of facts
- Tweets that do not provide any stance
- Questions, inquiries or clarifications
- Totally unrelated topics
- Languages that are not used in the Philippines (languages that are not Filipino Dialects, Tagalog or English)

On the other hand, the category, Filipino Dialects were also removed as the researchers and the volunteers did not have the necessary knowledge in handling and deciphering the meaning behind those texts. As a result, the dataset was reduced to 1633 data points which were then pre-processed in order to remove any noise and eliminate inconsistent data when training the model.

Initial data exploration would reveal that the overall polarity of the sentiments were revealed to be negative as seen in Table 2, roughly 81% of the remaining dataset were tagged as negative. This indicates that there is an overall sense of mistrust-defined as a doubtful or cautious approach to the object of trust or even distrust-believing that the object of trust is untrustworthy [11] directed towards the government and its policies. Moreover, when looking at the overall distribution of the data points in relation to the three trust groups (Behavioral, Operational, and Institutional) as seen in Table 3, the category with the most number is the operational trust category, followed by the behavioral category, and lastly the institutional category. This means that in this initial exploratory analysis of Trust in Twitter sentiment in the Philippines the focus of trust is placed more on state initiatives and processes, figureheads and leaders rather than government institutions, and local government units.

Table 2. Trust categories

Trust sentiment	Count
Negative	1329
Positive	304

Table 3. Data distribution over trust groups

Trust sentiment	Count
Behavioral	648
Operational	844
Institutional	141

However, that is not to say that the institutional trust group is any less relevant than the other two. It was quickly found out that in assessing Twitter discourse, some categories were easier to identify than others. Twitter in particular houses massive volumes of tweets in which individuals in government are mentioned. For example, there were instances that a given tweet encompasses multiple trust groups, such as talking about the head of a particular institution (behavioral) alongside the institution in question (institutional) and even about a particular process (operational). This finding remains consistent with a prior study [22]. Therefore, to address this issue, the volunteers were advised to solely label it in the trust group in which the tweet strongly belongs to rather than allowing it to be part of all the groups that it encompassed.

Lastly, in order to get a better understanding of the narratives of each trust category a word cloud was created as a tool to visualize the text documents of each category, which can be seen in Fig. 1.

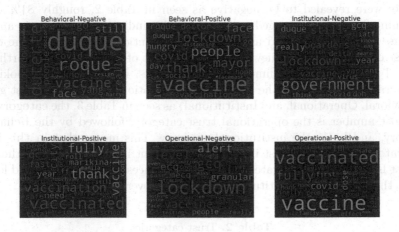

Fig. 1. Word cloud for each category

Looking at the "Behavioral-Negative" category, it mainly shows names of political figures present in the Philippines. Names like "Duterte"-the surname of the Philippine president, "Roque"- the surname of the presidential spokesperson, "Duque"-surname of the secretary of health, "Go"-surname of a senator in the Philippines. Furthermore, aside from names of political figures, it also includes names of people implicated in a particular controversy that occured in the Philippines in 2021. Labeled the "Pharmally Incident", it is one of the biggest scandals that affected the presidency of Rodrigo Duterte, it involved a string of anomalies following a multi-billion contract the Duterte administration awarded to the Pharmally Pharmaceutical Corporation. Implicated in this incident are "Yang"- a Chinese businessman, friend and former adviser of the President, senator Go, and "Lao"-former budget undersecretary who headed the procurement service of the Department of Budget and Management. Furthermore, words such as "resign", "tangina" (a Filipino slang for an offensive word), "jail", "stupid", "corruption" indicate that the general theme of this category expresses discontent towards the individuals indicated, which could then be inferred to as lack of trust or trustworthiness. Focusing on trustworthiness and referring to trust literature once again, the words indicate a lack of ability (capacity to safeguard its citizens), benevolence (to act in the best interest of its citizens), and integrity (honesty). Next, for the category of "Behavioral-Positive" results of the word cloud shows words of gratitude such as "thank", the word "hope", it also includes political figures such as "Leni"- The Vice President of the Philippines, "mayor" "rexgatchalian"- mayor of Valenzuela, other words in this category include "vaccine", "facemask". Examining the results in tandem with current events, this category seems to indicate the combined efforts of Leni and the office of the Vice President(OVP) for their COVID-19 response during the second year of the pandemic. Words such as "swab" and "cab" that is present in the word cloud indicates the mobile free COVID-19 testing initiative of the OVP. This provided free antigen testing services to the people provided that they sign up and wait for a confirmation. Furthermore, prior to this program the OVP focused on providing personal protective equipment (PPE) sets to health care workers and provide free transportation to medical frontliners, which could explain the presence of the words "facemask", " doctor", "hospital", and "frontliner".

For the Institutional trust group, the word cloud for Institutional-Negative provides words such as "government", "iatf"- meaning Inter Agency Task Force for the management of emerging infectious diseases, "duterte" and "administration", "doh"- Department of Health. This indicates that these are the main topics of negative discussion present for the institutional category in twitter. While the negative category of Institutional trust focused on country-wide institutions such as doh, iatf, etc. the positive category focuses more on local institutions. As seen in the word cloud, cities such as Marikina, "QC"- abbreviation for Quezon City, etc. are commended for feats such as "vaccination" or having its residents "vaccinated". Other words such as "fast", "process", "safe", "thank", "fully", "speed", further reinforce this insight. This could potentially mean that trust in this category, in the case of the Philippines, is geared more towards the local institutions such as local government units, rather than the national institutions in place like iatf, doh.

Finally, for the last trust group, Operational Trust, the word cloud for the negative category mainly focuses on the restrictive measures that were put in place. Words such as "lockdown", the various community quarantine policies ("gcq", "ecq", "mecq", etc.) bear the brunt of the ire of Twitter users. Aside from restrictive measures, other processes such as contact tracing, and mass testing are also topics for this category. This shows that the narrative for this category mainly deals with how hard being restricted can get especially for extended amounts of time, and with little to no improvements in critical processes such as mass testing and contact tracing creates an endless cycle of jumping from one community quarantine to another due to fluctuating cases. Inversely, the positive side focuses on being vaccinated, vaccination, vaccines, etc. The narratives present on this side are narratives on getting vaccinated, encouraging others to get vaccinated, and being thankful or happy about receiving a vaccine.

4.2   Multi-class Text Classification Model

After calculating the tf-idf for the dataset, the remaining tweets were represented by 811 features with each feature representing a tf-idf score of an n-gram. Four models were benchmarked through a 5-fold cross validation test out of the four models namely: Logistic Regression; Multinomial Naive Bayes; Linear Support Vector Machine; and Random Forests. While results were close, Linear Support Vector Machine (SVC) had the highest scoring mean accuracy as seen in Table 4.

Table 4. Model benchmark at 5-fold cross validation

Model	Mean accuracy
Linear SVC	0.737
Logistic regression	0.731
Multinomial Naive Bayes	0.728
Random forests	0.642

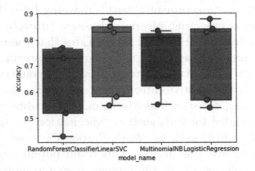

Fig. 2. Model benchmark results

The Linear SVC model was then further examined by looking at its confusion matrix. As shown in Fig. 3 while not perfect, a good majority of the predictions ended up on the diagonal where the predicted label is equal to the actual label. This is especially true for the categories of Behavioral-Negative, Operational-Negative, and Operational-Positive the three categories that had the most number of samples. Aside from the amount of samples, the misclassification rate could also be attributed to tweets that encompass more that one trust group such as tweets involving both a national institution and its head.

Looking at important classification metrics aside from accuracy-precision, recall, and f1-score, the model performs fairly well in categories that had sufficient data points. Given that the dataset is imbalanced the accuracy metric, albeit at 80% is not a very reliant measure of model reliability since it does not take into consideration the correctly classified data points of each category, only

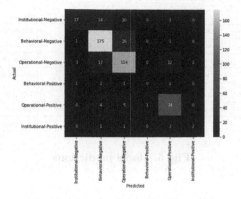

Fig. 3. Linear SVC confusion matrix

as a collective. Other metrics such as Precision, Recall and the F1 score would be a better fit in measuring model reliability. For example, a high precision would allow us to be sure that every-time the model predicts a data point to a specific class it actually belongs to that class. On the other hand, having a high recall would provide us confidence that the model is good at catching true positives. While surely each have their own merits, a good balance of the two would be the F1 score. With this, the model performs fairly well, with a weighted average f1-score of 79%. Weighted average takes into account the support of each class during calculation, thereby making the classes with less samples have less of an effect on the metric.

	precision	recall	f1-score	support
Institutional-Negative	0.77	0.40	0.53	42
Behavioral-Negative	0.82	0.86	0.84	203
Operational-Negative	0.78	0.82	0.80	188
Behavioral-Positive	0.00	0.00	0.00	5
Operational-Positive	0.80	0.88	0.84	84
Institutional-Positive	0.50	0.29	0.36	7
accuracy			0.80	529
macro avg	0.61	0.54	0.56	529
weighted avg	0.79	0.80	0.79	529

Fig. 4. Model metric report

4.3   Model Prediction

The model was then introduced to new, unknown data points in order to test its capability of predicting the trust category of a given statement. The respective statements and the model's prediction can be found in Fig. 5. The model was able to correctly predict the trust categories of the newly introduced data points, however, upon tweaking the provided statements, the prediction could change based on introduced words. For example, upon adding the words "for president", "THE BEST AND THE ONLY CHOICE FOR PRESIDENT IS VP LENI ROBREDO" to the initial statement, the model would predict it

Statement	Prediction
THE BEST AND THE ONLY CHOICE IS VP LENI ROBREDO!	Behavioral-Positive
2 years and nothing has changed! hay grabe ka duterte palpak ka talaga!!	Behavioral-Negative
Do not Total Lockdown, it's not a good policy. It stops the work of people	Operational-Negative
I just really thank god we're all fully vaccinated. My grandmother got covid Thankfully she was convinced to be vaccinated in Early Roll out of vaccines. She was infected along with the plumber that repaired the house we were in so it's important to wear a mask	Operational-Positive
our QC LGU is really good .. it never stops working to help those in need. thank you very much	Institutional-Positive
So wala po talaga kwenta ginagawa ng DOH	Institutional-Negative

Fig. 5. Model predictions

as Behavioral-Negative. This denotes that the words "for president", specifically "president" is viewed as something negative by the model and thus, consequently by the training data. On the topic of statements that encompass more than one trust category, if the last statement were changed to "So wala po talaga kwenta ginagawa ng DOH at si Duque", roughly translating to, "the actions of the DOH and Duque are fruitless", would turn the prediction of the model to Behavioral-Negative as well, showing that the model weighed Behavioral-Negative more than its previous label upon the addition of "Duque".

4.4   Word Relationship per Category

Aside from simply looking into the word frequencies given by the word cloud, examining word correlations expressed through n-grams could provide better insight on how the model decides where to categorize a statement. By grouping and examining correlated words, instead of doing so individually, more context can be derived and consequently giving more information. Looking at the top 3 correlated words in terms of unigrams, bigrams and trigrams, would give us a similar narrative as when examining the word clouds, which can be seen in the tables below.

Table 5. Behavioral-negative

Unigrams	Bigrams	Trigrams
Duterte	Government response	ecq gcq mecq
Roque	Yang duque	Duque roque duterte
Duque	Said good	Anomaly pharmally jail

With this, statements that contain these top three n-grams would more likely be predicted in the respective category that highly ranks that n-gram, which could explain the findings discussed earlier.

Table 6. Behavioral-positive

Unigrams	Bigrams	Trigrams
Leni	Hard lockdown	Im fully vaccinated
Mayor	Vaccine express	Think delta variant
Spox	Face mask	Swab quarantine boarders

Table 7. Institutional-Negative

Unigrams	Bigrams	Trigrams
Government	Lockdown solution	Alert level lockdown
iatf	Metro manila	Think delta variant
doh	Level lockdown	Delta variant attending

Table 8. Institutional-positive

Unigrams	Bigrams	Trigrams
City	Lockdown solutionspeed vaccine	Harry roque duque
qc	Pandemic response	Jail duterte duque
Vaccination	Vaccine express	Ong syndicated estafa

Table 9. Operational-negative

Unigrams	Bigrams	Trigrams
Lockdown	Vaccinated vaccine	mecq gcq mgcq
gcq	Duque roque	Alert level thought
Guidelines	Mass testing	gcq mecq ecq

Table 10. Operational-positive

Unigrams	Bigrams	Trigrams
Vaccinated	Vaccines work	Im fully vaccinated
Dose	Vaccine works	Ong syndicated estafa
Moderna	Fully vaccinated	Jail duterte duque

5 Conclusion and Future Research

Four machine learning models were benchmarked in order to test effectivity in categorizing trust based on twitter sentiments about the on-going COVID pandemic. Linear SVC was found to be the highest scoring model during the benchmarking process and was thus examined further and resulted in a weighted f1-score of 79%. Furthermore, additional unknown data points was introduced to the model in order to test its predictive capabilities, all of which the model was able to predict correctly, but it is important to note that when introducing certain words to the sentiments the prediction of the model could change. This could be attributed to Twitter discussions on the topic encompassing several trust group.

While overall sentiment was negative, it was found that certain narratives could be seen depending on the trust category examined. Behavioral-Negative provides information on the Pharmally anomaly incident that occurred in 2021 and contains narratives on the current administration. Behavioral-Positive on the other hand, contains narratives that focuses more on the Vice President and by extension, the OVP for their valued and continued service to the people through different initiatives such as the swab cab project among many others. For the Institutional trust group, negative narratives revolve around national institutions such as the department of health and the inter-agency task force, while positive narratives focuses on local institutions, such as the local government units of various cities. Lastly, the operational trust group provides negative narratives about certain processes and initiatives during the pandemic such as the various levels of lockdowns, contact tracing and mass testing. On the other hand, positive narratives for the operational trust group focus on vaccines, vaccinations and herd immunity.

To conclude, the study was able to further reinforce that social media analytics can identify, analyze, and predict trust between the government, its policies and processes as well as its citizens. Moreover, the narratives present in these were unpacked and analyze. The results that have been obtained, however, could still be improved upon. The dataset utilized was imbalanced and only included one social media network. Further studies could incorporate other social media networks in the data gathering process to further expand the diversity of the data gathered. In addition, the structure of the study focused on a multi-class approach in the classification of the trust categories. This however, is shown to be unrealistic as trust sentiments, more often than not, can encompass more than one trust group. Therefore, taking on a multi-label approach that is allowing trust sentiments inclusion to more than one trust group could be a potential angle for future research.

References

1. Alhajji, M., Al Khalifah, A., Aljubran, M., Alkhalifah, M.: Sentiment analysis of tweets in Saudi Arabia regarding governmental preventive measures to contain COVID-19 (2020)
2. Alzahrani, L., Al-Karaghouli, W., Weerakkody, V.: Analysing the critical factors influencing trust in e-government adoption from citizens' perspective: a systematic review and a conceptual framework. Int. Bus. Rev. 26(1), 164–175 (2017)
3. Arunachalam, R., Sarkar, S.: The new eye of government: citizen sentiment analysis in social media. In: Proceedings of the IJCNLP 2013 Workshop on Natural Language Processing for Social Media (SocialNLP), pp. 23–28 (2013)

4. Arusada, M.D.N., Putri, N.A.S., Alamsyah, A.: Training data optimization strategy for multiclass text classification. In: 2017 5th International Conference on Information and Communication Technology (ICoIC7), pp. 1–5. IEEE (2017)
5. Calderon, N.A., et al.: Mixed intitiative social media analytics at the world bank. In: 2015 IEEE International Conference on Big Data. IEEE (2015)
6. Devine, D., Gaskell, J., Jennings, W., Stoker, G.: Trust and the coronavirus pandemic: what are the consequences of and for trust? An early review of the literature. Polit. Stud. Rev. **19**(2), 274–285 (2021)
7. Dodsworth, S., Cheeseman, N.: Political trust: the glue that keeps democracies together. Westminster Foundation for Democracy, London (2020)
8. Fisher, J., Van Heerde, J., Tucker, A.: Does one trust judgement fit all? Linking theory and empirics. Br. J. Polit. Int. Relat. **12**(2), 161–188 (2010)
9. Garcia, M.B.: Sentiment analysis of tweets on coronavirus disease 2019 (COVID-19) pandemic from metro manila, Philippines. Cybern. Inf. Technol. **20**(4), 141–155 (2020)
10. Goldfinch, S.: Public trust in government, trust in e-government, and use of e-government. In: Encyclopedia of Cyber Behavior, pp. 987–995. IGI Global (2012)
11. Jennings, W., Stoker, G., Valgarsson, V., Devine, D., Gaskell, J.: How trust, mistrust and distrust shape the governance of the COVID-19 crisis. J. Eur. Publ. Policy **28**(8), 1174–1196 (2021)
12. Joshi, S., Abdelfattah, E.: Multi-class text classification using machine learning models for online drug reviews. In: 2021 IEEE World AI IoT Congress (AIIoT), pp. 0262–0267. IEEE (2021)
13. Kemp, S.: Digital in the Philippines: all the statistics you need in 2021 - datareportal - global digital insights, November 2021. https://datareportal.com/reports/digital-2021-philippines
14. Khan, S., Rahim, N.Z.A., Maarop, N.: Towards the development of a citizens' trust model in using social media for e-government services: the context of Pakistan. In: International Conference of Reliable Information and Communication Technology, pp. 1002–1012. Springer (2018). https://doi.org/10.1007/978-3-319-99007-1_93
15. Kurnia, R., Tangkuman, Y., Girsang, A.: Classification of user comment using word2vec and SVM classifier. Int. J. Adv. Trends Comput. Sci. Eng. **9**, 643–648 (2020)
16. Larson, H.J.: Politics and public trust shape vaccine risk perceptions. Nat. Hum. Behav. **2**(5), 316–316 (2018)
17. Larson, H.J., Hartigan-Go, K., de Figueiredo, A.: Vaccine confidence plummets in the Philippines following dengue vaccine scare: why it matters to pandemic preparedness. Hum. Vacc. Immunotherapeutics **15**(3), 625–627 (2019)
18. Luhmann, N.: Familiarity, confidence, trust: problems and alternatives. Trust: Making Break. Coop. Relat. **6**(1), 94–107 (2000)
19. Malkani, Z., Gillie, E.: Supervised multi-class classification of tweets (2012)
20. Mayer, R.C., Davis, J.H.: The effect of the performance appraisal system on trust for management: a field quasi-experiment. J. Appl. Psychol. **84**(1), 123 (1999)
21. McKnight, D.H., Choudhury, V., Kacmar, C.: Developing and validating trust measures for e-commerce: an integrative typology. Inf. Syst. Res. **13**(3), 334–359 (2002)
22. Papp, G., El-Gayar, O.F., Lovaas, P.: Citizen trust in the united states government: Twitter analytics measuring trust in government sentiments (2020)
23. Raghupathi, V., Ren, J., Raghupathi, W.: Studying public perception about vaccination: a sentiment analysis of tweets. Int. J. Environ. Res. Public Health **17**(10), 3464 (2020)

24. Ranaweera, H.: Perspective of trust towards e-government initiatives in Srilanka. SpringerPlus **5**(1), 1–11 (2016)
25. Thomas, C.W.: Maintaining and restoring public trust in government agencies and their employees. Adm. Soc. **30**(2), 166–193 (1998)
26. Umali, J.M., Miranda, J.P.P., Ferrer, A.L.: Sentiment analysis: a case study among the selected government agencies in the Philippines. Int. J. Adv. Trends Comput. Sci. Eng. **9**(3), 2739–2745 (2020)
27. Van Bavel, J.J., et al.: Using social and behavioural science to support COVID-19 pandemic response. Nat. Hum. Behav. **4**(5), 460–471 (2020)
28. Woelfert, F.S., Kunst, J.R.: How political and social trust can impact social distancing practices during COVID-19 in unexpected ways. Front. Psychol. **11**, 3552 (2020)
29. Xue, J., et al.: Twitter discussions and emotions about the COVID-19 pandemic: machine learning approach. J. Med. Internet Res. **22**(11), e20550 (2020)

Twitter as a Communication Tools for Vaccine Policy in Indonesia: An Analysis

Iradhad Taqwa Sihidi(✉), Salahudin ⓘ, Ali Roziqin ⓘ, and Danang Kurniawan ⓘ

Department of Government Studies, Faculty Social and Political Sciences,
Universitas Muhammadiyah Malang, East Java, Indonesia
iradhad@umm.ac.id

Abstract. This study aims to see Twitter as a medium for regional heads to succeed in vaccine policies. Anis Basewedan, Ridwan Kamil, and Ganjar Pranowo, as Governors, actively maximize Twitter to invite the public to support and want to be vaccinated, which is one of the keys to handling the Covid-19 crisis through the creation of herd immunity. The study used the Nvivo 12 Plus Qualitative Data Analysis Software (QDAS) approach. The data for this study used the Twitter social media of the three regional heads, namely @aniesbaswedan, @ridwankamil, and @ganjarpranowo. The results showed that the three of them were very active in informing the vaccination schedule, emphasizing the safety of vaccination, and inviting the public to want to be vaccinated. Through concise and clear messages, the three of them emphasize that one of the primary keys to getting out of the Covid-19 pandemic and recovering social and economic life is through vaccination. This research further highlights the usefulness of social media in crisis management, so it is necessary to continue to increase its use by leaders.

Keywords: Communication · Twitter · Vaccine Policy

1 Introduction

Vaccination is one of the Indonesian government's strategies to deal with Covid-19 (Pramono 2021). However, it is constrained by several things, one of which is caused by declining public trust in the government (Saechang et al. 2021) (Kriswibowo et al. 2021). The decline in public confidence in vaccination is dominantly caused by people questioning the safety of vaccines. Some of the wild issues related to the procurement of vaccinations are the composition of vaccine ingredients, side effects of vaccines, and the high number of doctors, public figures, and influencers who question the secret/conspiracy interests behind the procurement of the Covid-19 vaccine (Januraga and Harjana 2020). In addition, hoax news about vaccines has emerged, affecting public perception of the government's policy (Dwipayanti et al. 2021). The safety of vaccines regarding "halalness" has become a public discussion and side effects of vaccines such as blindness or paralysis after being vaccinated and public figures and influencers who doubt the purpose of vaccination on the grounds of conspiracy.

G. Meiselwitz (Ed.): HCII 2022, LNCS 13315, pp. 661–671, 2022.
https://doi.org/10.1007/978-3-031-05061-9_47

The role of regional heads in disseminating vaccine information and increasing public confidence about the safety and objectives of vaccines is a supporting factor for the success of vaccine policies from the central government (Wiyani et al. 2019). Several regional heads have specific strategies to convince the public about vaccine safety, including Special Capital Region (DKI) of Jakarta Jakarta Governor Anies Baswedan, West Java Governor Ridwan Kamil, and Central Java Governor Ganjar Pranowo. Various vaccination campaigns spearheaded by these three governors use the Twitter social media facility. Twitter is used as a forum for campaigning for vaccinations because the number of Indonesian active Twitter users is 15.7 million in 2021 and is the sixth country with the most Twitter users in the world after the USA, Japan, India, United Kingdom, and Brazil (Wiyani et al. 2019). Thus, the Twitter platform became one of the vaccination campaign tools that the governors could use.

The crisis communication strategy to publicize government policies and self-branding efforts is an effective way to consider the technology-savvy Indonesian people. Twitter has become an effective means to shape the public image so that this opportunity is read as a superior strategy for campaigning regarding the Covid-19 vaccination. Several previous studies related to social media as a means of policy campaigns and specific issues include Jannah and Sonni (2021), which examines the use of Makassar Terkini. id online media to disseminate policies regarding the Covid-19 pandemic. Siahaan and Adrian (2021) found that the social media used by the Palu City Government was quite effective in changing the public's perspective on the Covid-19 vaccination policy. This is also in line with the findings of Gafatia and Hadinata (2021), who see Twitter can help governments to chart acceptance rates for vaccines.

These various studies show the role of social media that focuses on official government accounts. This research is different because it focuses more on the personal accounts of the Governor of DKI Jakarta, the Governor of West Java, and Central Java in developing vaccine policy communications. This study focuses on looking further at the efforts and strategies of each Governor through the narratives delivered so that the vaccination process can run smoothly in their regions, significantly eliminating and convincing the public about the safety of the Covid-19 vaccine.

2 Literature Review

2.1 Vaccine Policy

Vaccine policy is one of the most reasonable steps to deal with Covid-19 in Indonesia (Dwipayana 2020; Heriyanto et al. 2021). Vaccines have at least four crucial benefits: stimulating immunity, reducing the risk of transmission, reducing the severe impact of the virus, and achieving herd immunity (Ministry of Health 2021). If all of this works effectively, the Indonesian government, supported by recommendations from health experts, believes that the spread of Covid-19 can be stopped so that the vaccination policy will be implemented optimally (Fuady et al. 2021).

Even though it went smoothly, marked by Indonesia's ranking as the 5th country with the largest vaccination coverage globally, this policy was met with quite a bit of public resistance at the beginning but has subsided recently. The reluctance is recorded from a survey released by the Indonesian Political India (IPI) that 41% of Indonesians refuse the vaccine (Amalia 2021; (Pradila 2021). The reasons are various, namely worried about the side effects of the vaccine produced (54.2%), the effectiveness of the vaccine (27%), feeling healthy or not in need (23.8%), and refusing to buy (17.3%).On Twitter, for example, the hashtag #TolakDivaksinSinovac had become a trending topic so that it can be concluded as a minor perception of the vaccination plan. Even though with a more localized scope, DKI Jakarta and Yogyakarta, the survey findings from CSIS also found that 63% of the proportion of youth in DKI Jakarta and 55% of young people in Yogyakarta aged 17–22 years who lacked or did not believe in vaccines (FISIPOL 2021). The belief anomaly was formed for the same reason as the findings from the IPI survey but added another factor, namely the halalness of the vaccine.

2.2 Twitter as Communication Crisis

In a crisis, social media can function, for example, providing information and understanding to the public in real-time about the current crisis and how to handle it (Charrad et al. 2019; Salahudin et al. 2020). The platform, which provides an interactive space, allows the government to understand the public's response quickly and with high accuracy (Beni-hssane 2017)so that public sentiment/managers can be appropriately managed (Singh and Verma 2020). Effective crisis communication will maintain government power amid crisis turmoil and high public panic. To be effective, social media as a means of crisis communication must be in harmony with the culture of the community (Zhu et al. 2017) so that the affection of the emphasized message does not get rejected by the public.

One of the social media that can be used is Twitter. At the time of Covid-19 9, Twitter has many benefits from the government's policy/crisis communication tool (Kaur et al. 2021; Riddell and Fenner 2021; Zeemering 2021; Sleigh et al. 2021), whose aim is to increase public (Engel-Rebitzer et al. 2021) with concise and clear information (Haupt et al. 2021). One of the essential things related to vaccine policy (Sandag et al. 2021), is vaccine policy communication (Scannell et al. 2021). In addition, Twitter can record public perceptions related to the Covid-19 handling policy (Dhar and Bose 2020; Pascual-Ferrá et al. 2020) and specifically regarding vaccine vaccination policies (Scannell et al. 2021; Jiang et al. 2021; Bonnevie et al. 2021; Park et al. 2021; Hoffman et al. 2021; Jemielniak and Krempovych 2021;Yousefinaghani et al. 2021). This information is crucial for the government, especially in making vaccination policies that are efficient and effective and reduce resistance from the public.

3   Research Method

The study used the Nvivo 12 Plus Qualitative Data Analysis Software (QDAS) approach. The data for this research uses Twitter social media through the three regional heads, namely @aniesbaswedan, @ridwankamil, and @ganjarpranowo, related to vaccine policy keywords. Data retrieved viaNCapture of NVivo 12 Plus with Chrome Web. First, the data is processed using the Cross Tab feature to calculate statistics related to the selected keywords (vaccine policy). Next, using Crosstab Query is to enter code (manual, generate, etc.), text data, and numeric data on variable and pattern data. Finally, use the Word Cloud feature, which will show words that often appear from data searches or view terms which in this case are related to vaccine policies in the three regional head accounts.

4 Results

The number of Indonesian people with 15.7 active Twitter users shows that the advancement of digital technology with all its novelties and offers of convenience is well received by the Indonesian people (Aprianita and Hidayat 2020). This opportunity is taken by public figures, politicians, officials, influencers, and social media activists to express ideas, criticism, leading issues, mini-blogs, diaries, and public policy dissemination (Candon 2019). In addition, several officials in Indonesia use the Twitter space as a means of policy communication and self-branding, including DKI Jakarta Governor Anies Baswedan, West Java Governor Ridwan Kamil, and Central Java Governor Ganjar Pranowo.

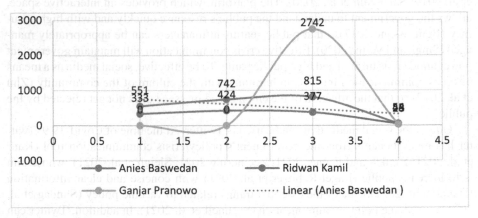

Fig. 1. Number of tweets 2019–2022, processed by Nvivo 12 Plus

The data (Fig. 1) shows that the three governors have had strong communication in Twitter for the last three years from 2019-to 2022, which is displayed based on the curve above. Figures show that the Governor of Central Java Ganjar Pranowo has 2,742 tweets in 2021, followed by the Governor of DKI Jakarta with a total of 815 tweets in the same year, and the Governor of West Java with a total of 424 tweets in 2021. The active role of these regional heads on Twitter is based on the primary need to communicate public policies both nationally as a supporter of central government policies and regional policies.

The strategy of each regional head is also different depending on the communication method created. Governor Anies Baswedan has 4.4 million followers, West Java Governor Ridwan Kamil with 4.5 million total followers, and Central Java Governor Ganjar Pranowo with 2.3 million followers have a strong influence in communicating government policies and self-branding as public figures. Central Java Governor Ganjar Pranowo, although with the least total followers among the other two governors, is active in social media, especially Twitter, becoming Governor with a commitment to government communication and self-branding with a total of 51% in terms of tweets related to the vaccine program. Meanwhile, Governor Anies Baswedan has a cumulative total of 32% of tweets related to vaccines, and West Java Governor Ridwan Kamil has a cumulative total of 17% of tweets. The cumulative figure was communicated since the central government announced the vaccination policy from 2021 to 2022 (Fig. 2). The role of government communication with vaccination in the three regions is essential because, in terms of the spread of Covid-19, it is included in Indonesia's top 3 highest regions. This form of communication aims to convince the public that vaccination is an absolute necessity to reduce the high prevalence rate in these areas.

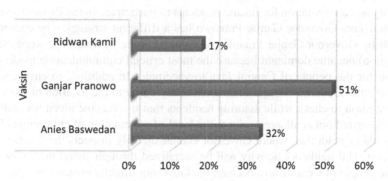

Fig. 2. The intensity of vaccine-related communication

The three regional heads have their respective strategies for communicating government policies and self-branding purposes. DKI Jakarta Governor Anies Baswedan relies on four main hashtags, namely #jakartabangkit, #vaksindulu, #jagajakarta, #kotakolaborasi (Fig. 3). The use of the hashtag has specific considerations, including DKI Jakarta as an epicenter area that has a very high rate of spread and death toll, the level of public awareness to be involved in the Covid-19 vaccination is still low, for example, the survey findings from CSIS also found 63% of the proportion of young people in Indonesia. DKI Jakarta with 17–22 years old who do not believe in vaccines. The tagline was disseminated to the public so that public fear regarding the Covid-19 vaccine would be reduced so that Governor Anies Baswedan convinced the Jakarta public that Jakarta's security from the threat with vaccination activities of Covid-19 would be reduced. Thus, the people of Jakarta can return to a more normal life. The hashtags #jakartabangkit, #vaksindulu, #jagajakarta, and #kotakolaborasi are forms of socialization and an invitation for Jakarta residents to minimize the spread of Covid-19, it is necessary to get vaccinated.

In addition, the tweets that the Governor of DKI Jakarta often tweets are "vaksin" (vaccinations) and " anak" (children). These two words are essential to convince parents that the vaccine to be given is very child-friendly, so there is no need to be afraid. The policy communication effort made by Governor Anies Baswedan is to provide a sense of security to parents so that they invite their children to participate in the vaccination program. The messages conveyed through these tweets with hashtags and several appeals

Fig. 3. World cloud Twitter of @aniesbasewedan, processed by Nvivo 12 Plus

containing invitations were a strategy by Governor Anies Baswedan to provide a sense of security and an invitation for Jakarta residents to participate in the Covid-19 vaccine.

Central Java Governor Ganjar Pranowo has a different strategy. The communication built by Governor Ganjar Pranowo tends to be more flexible. The word "vaksin" (vaccination) became dominant because the most crucial communication made was an invitation for the people of Central Java to vaccinate. In addition, to ensure that the vaccine is safe, Governor Ganjar Pranowo directly came to the location of the vaccine implementation to check while assuring residents that the vaccine given was safe. This method is carried out at all ages, even to the level of vaccination of children in elementary schools or madrasas, where Governor Ganjar directly inspects the location of the vaccination. The fieldwork activity will be socialized through social media Twitter to affect the people of Central Java because the Governor directly ensures that the "vaksinasi" given are safe (Fig. 4). The dominant word is "sehat", which encourages people to maintain their health by increasing their immune systems.

Fig. 4. World cloud twitter of @ganjarpranowo, processed by Nvivo 12 Plus

The word 'sehat' became dominant because Governor Ganjar intensively disseminated to the public to comply with health protocols. The strategy shown is unique: exercising while cycling and then walking through the streets or crowded places or markets and inviting people who still do not obey health protocols to wear masks. This method is effective because the community feels protected by Governor Ganjar's efforts to ensure a healthy Central Java. The government communication model that goes directly to the field becomes its attraction for the community because it facilitates it well. So that, building public awareness about vaccination and obeying health protocols that Governor Ganjar Pranowo often echoes through Twitter.

The strategy for using Twitter social media by the Governor of West Java, Ridwan Kamil, has its pattern. The words "vaksin" (vaccine) and Covid are still dominant because these three governors carry out activities on social media to communicate public policies and self-branding. One way of campaigning to invite the people of West Java about the convenience of the vaccine being used is to give an example of himself as the object of the vaccine. This effort makes people aware that the Governor of West Java has undergone the vaccine. The advantages of this strategy have a significant impact on influencing the perspective of the people who are still unsure, and the public's trust in the government is decreasing. The self-branding method is more effective because it is an example of inviting people. Another effort to disseminate information about Covid-19 vaccination is to use #jabartangapcovid19 (Fig. 5). This hashtag provides education to the public that to make West Java free from Covid-19, what must be done is to carry out a vaccination program.

Fig. 5. World cloud Twitter of @ridwankamil, processed by Nvivo 12 Plus

The word puskesmas is also dominant because of the efforts of the West Java government to convince its citizens that how to get the Covid-19 vaccine does not require additional efforts. Because every puskesmas in West Java has provided vaccination facilities. The invitation encourages people not to feel that the vaccine program involves a lot of money to go to the vaccine site. But what is provided by the government of Java Brat is to facilitate every puskesmas. In addition, the dominant word is "disiplin". The term became a tweet that very often appeared in the tweets of Governor Ridwan Kamil as a form of appeal to the public that by implementing a disciplined life carrying out health protocols, the effect was that West Java became a healthy area.

The word that was used the most in the tweets of the Governor of West Java was Covid-19 with 16 tweets, "saya" 8 tweets and the word "tes" (Test) had 8 tweets. Of these three dominant words, 'saya' (me) is the main differentiator of the strategy campaigned by the Governor of West Java to convince the public about vaccines. This means that efforts to make people aware must begin by setting an example through oneself. The efforts made effectively influence the public perspective with doubts and fears about vaccines. In addition, the word test became dominant because of the recommendation or invitation by the Governor of West Java. He urges his citizens always to undergo a health test at the nearest health facility if they experience symptoms like Covid-19. This effort was carried out consistently by Governor Ridwal Kamil by socializing #jabar-tanggapcovid19 to invite the community together to realize a Covid-19-free West Java. The various strategies used by each regional head show that with the development of digitalization in Indonesia increasing, the way to socialize about public policy is through Twitter. Because Twitter is an effective forum for raising issues, clarifying, and making people aware, especially for situations of information confusion, both about vaccination and Covid-19. The various strategies used by each regional head show that with the development of digitalization in Indonesia increasing, the way to socialize about public policy is through Twitter. Because Twitter is an effective forum for raising issues, clarifying, and making people aware, especially for situations of information confusion, both about vaccination and Covid-19. The various strategies used by each regional head show that with the development of digitalization in Indonesia increasing, the way to socialize about public policy is through Twitter because Twitter is an effective forum for raising issues, clarifying, and making people aware, especially for situations of information confusion, both about vaccination and Covid-19.

5 Conclusion

Vaccines are one of the best strategies to solve Covid-19 in Indonesia. However, the vaccine policy in Indonesia initially received relatively high resistance from the public because it was related to misinformation about vaccine safety which was considered a health threat and was not halal. Therefore, communication from regional heads as opinion leaders through Twitter social media is essential to convey the safety of vaccines and the benefits of vaccines for the common good. Anis Basewedan, Ganjar Pranowo, and Ridwan Kamil, even though they have dominant keywords highlighted in their respective Twitter accounts, they have the same goal of inviting the public to be willing to be vaccinated. The three of them both emphasized that the key to the revival of the Covid-19 pandemic is vaccines, so community participation is crucial. This research further confirms the critical contribution of social media in crisis communication.

References

Amalia, Y.: Indikator_41 Persen Masyarakat Enggan Disuntik Vaksin, 38,4 Persen Tolak Membeli _ merdeka. Merdeka.Com (2021)

Aprianita, D., Hidayat, D.: Analisis Pesan Kampanye #Dirumahaja Di Tengah Pandemi Covid-19. Komunikologi: Jurnal Pengembangan Ilmu Komunikasi Dan Sosial, 4(2), 78 (2020). https://doi.org/10.30829/komunikologi.v4i2.7910

Beni-hssane, A.: sciencedirect analyzing social media through big data using infosphere biginsights and apache flume. Procedia Comput. Sci. 113, 280–285 (2017). https://doi.org/10.1016/j.procs.2017.08.299

Bonnevie, E., Gallegos-Jeffrey, A., Goldbarg, J., Byrd, B., Smyser, J.: Quantifying the rise of vaccine opposition on twitter during the COVID-19 pandemic. J. Commun. Healthc. **14**(1), 12–19 (2021). https://doi.org/10.1080/17538068.2020.1858222

Candon, P.: Twitter: social communication in the twitter era. New Media Soc. **21**(7), 1656–1658 (2019). https://doi.org/10.1177/1461444819831987

Charrad, M., Bellamine, N., Saoud, B.: sciencedirect sciencedirect towards a social media-based framework for disaster towards a social media-based framework for disaster communication communication. Procedia Comput. Sci. **164**, 271–278 (2019). https://doi.org/10.1016/j.procs.2019.12.183

Dhar, S., Bose, I.: Emotions in twitter communication and stock prices of firms: the impact of Covid-19 pandemic. Decision **47**(4), 385–399 (2021). https://doi.org/10.1007/s40622-020-00264-4

Dwipayana, I.D.A.P.: Efforts in securing vaccine for Covid-19 outbreak in Indonesia. Health Notions **4**(10), 313–317 (2020). https://doi.org/10.33846/hn41003

Dwipayanti, N.M.U., Lubis, D.S., Harjana, N.P.A.: Public perception and hand hygiene behavior during COVID-19 pandemic in Indonesia. Front. Public Health **9**(May), 1–12 (2021). https://doi.org/10.3389/fpubh.2021.621800

Engel-Rebitzer, E., Stokes, D.C., Buttenheim, A., Purtle, J., Meisel, Z.F.: Changes in legislator vaccine-engagement on twitter before and after the arrival of the COVID-19 pandemic. Hum. Vaccin. Immunother. **17**(9), 2868–2872 (2021). https://doi.org/10.1080/21645515.2021.1911216

fisipol. Beragam Survei Sebut Penolakan dan Keraguan Masyarakat Terhadap Vaksin COVID-19 – Fakultas Ilmu Sosial dan Ilmu Politik. Fisipolugm.Ac.Id (2021)

Fuady, A., Nuraini, N., Sukandar, K.K., Lestari, B.W.: Targeted vaccine allocation could increase the covid-19 vaccine benefits amidst its lack of availability: a mathematical modeling study in Indonesia. Vaccines **9**(5) (2021). https://doi.org/10.3390/vaccines9050462

Gafatia, I.W.D., Hadinata, N.: Analisis Pro Kontra Vaksin Covid 19 Menggunakan Sentiment Analysis Sumber Media Sosial Twitter. Jurnal Pengembangan Sistem Informasi Dan Informatika, **2**(1), 34–42 (2021). https://doi.org/10.47747/jpsii.v2i1.544

Haupt, M.R., Jinich-Diamant, A., Li, J., Nali, M., Mackey, T.K.: Characterizing twitter user topics and communication network dynamics of the Liberate movement during COVID-19 using unsupervised machine learning and social network analysis. Online Soc. Netw. Media **21**, 100114 (2021). https://doi.org/10.1016/j.osnem.2020.100114

Heriyanto, R.S., et al.: The role of COVID-19 survivor status and gender towards neutralizing antibody titers 1, 2, 3 months after Sinovac vaccine administration on clinical-year medical students in Indonesia: role of COVID-19 survivor status and gender towards neutralizing antib. Int. J. Infect. Dis. **113**, 336–338 (2021). https://doi.org/10.1016/j.ijid.2021.10.009

Hoffman, B.L., et al.: #DoctorsSpeakUp: lessons learned from a pro-vaccine twitter event. Vaccine **39**(19), 2684–2691 (2021). https://doi.org/10.1016/j.vaccine.2021.03.061

Jannah, N., Sonni, A.F.: Konstruksi Pemberitaan Kepala Daerah di Kota Makassar Terkait COVID-19. Warta ISKI, **4**(1), 17–26 (2021). https://doi.org/10.25008/wartaiski.v4i1.100

Januraga, P.P., Harjana, N.P.A.: Improving public access to COVID-19 pandemic data in Indonesia for better public health response. Front. Public Health **8**, 8–11 (2020). https://doi.org/10.3389/fpubh.2020.563150

Jemielniak, D., Krempovych, Y.: An analysis of AstraZeneca COVID-19 vaccine misinformation and fear mongering on twitter. Public Health **200**, 4–6 (2021). https://doi.org/10.1016/j.puhe.2021.08.019

Jiang, X., et al.: Polarization over vaccination: ideological differences in twitter expression about COVID-19 vaccine favorability and specific hesitancy concerns. Soc. Media Soc. **7**(3) (2021). https://doi.org/10.1177/20563051211048413

Kaur, M., Verma, R., Otoo, F.N.K.: Emotions in leader's crisis communication: twitter sentiment analysis during COVID-19 outbreak. J. Hum. Behav. Soc. Environ. **31**(1–4), 362–372 (2021). https://doi.org/10.1080/10911359.2020.1829239

Kemenkes. 4 Manfaat Vaksin Covid-19 yang Wajib Diketahui. Kementerian Kesehatan RI (2021). http://upk.kemkes.go.id/new/4-manfaat-vaksin-covid-19-yang-wajib-diketahui

Kriswibowo, A., Prameswari, J.K.P., Baskoro, A.G.: Analisis Kepercayaan Publik Terhadap Kebijakan Vaksinasi Covid-19 Di Kota Surabaya. J. Publicuho, **4**(2), 326–344 (2021). https://doi.org/10.35817/jpu.v4i2.17912

Park, S., et al.: COVID-19 discourse on twitter in four Asian countries: case study of risk communication. J. Med. Internet Res. **23**(3), 1–17 (2021). https://doi.org/10.2196/23272

Pascual-Ferrá, P., Alperstein, N., Barnett, D.J.: Social network analysis of COVID-19 public discourse on twitter: implications for risk communication. Disaster Med. Public Health Prep. (2020). https://doi.org/10.1017/dmp.2020.347

Pradila, M.R.: Hasil Survei Sebut 41 Persen Masyarakat Tolak Vaksin Covid-19, DPR_ Masalah Serius - Pikiran-Rakyat. Pikiranrakyat.Com (2021)

Pramono, G.E.: Policing in the Covid-19 situation in Indonesia. Int. J. Soc. Sci. Hum. Res. **04**(02), 154–165 (2021). https://doi.org/10.47191/ijsshr/v4-i2-06

Riddell, H., Fenner, C.: User-generated crisis communication: exploring crisis frames on twitter during hurricane harvey. South Commun. J. **86**(1), 31–45 (2021). https://doi.org/10.1080/1041794X.2020.1853803

Saechang, O., Yu, J., Li, Y.: Public trust and policy compliance during the covid-19 pandemic: The role of professional trust. Healthcare (Switzerland) **9**(2), 1–13 (2021). https://doi.org/10.3390/healthcare9020151

Salahudin, S., Nurmandi, A., Sulistyaningsih, Tri, Taqwa, I.: Analysis of government official twitters during Covid-19 crisis in Indonesia. Talent Dev. Excellence **12**(1), 3899–3915 (2020)

Sandag, G.A., Manueke, A.M., Walean, M.: Sentiment analysis of COVID-19 vaccine tweets in Indonesia using recurrent neural network (RNN) approach. In: 2021 3rd International Conference on Cybernetics and Intelligent System (ICORIS), pp. 1–7 (2021). https://doi.org/10.1109/ICORIS52787.2021.9649648

Scannell, D., et al.: COVID-19 vaccine discourse on twitter: a content analysis of persuasion techniques, sentiment and mis/disinformation. J. Health Commun. **26**(7), 443–459 (2021). https://doi.org/10.1080/10810730.2021.1955050

Siahaan, C., Adrian, D.: Komunikasi Dalam Persepsi Masyarakat Tentang Kebijakan Pemerintah Dimasa Pandemi. Kinesik, **8**(2), 158–167 (2021). https://doi.org/10.22487/ejk.v8i2.159

Singh, R.K., Verma, H.K.: Effective parallel processing social media analytics framework. J. King Saud Univ. – Comput. Inf. Sci. (2020). https://doi.org/10.1016/j.jksuci.2020.04.019

Sleigh, J., Amann, J., Schneider, M., Vayena, E.: Qualitative analysis of visual risk communication on twitter during the Covid-19 pandemic. BMC Publ. Health **21**(1), 810 (2021). https://doi.org/10.1186/s12889-021-10851-4

Wiyani, F., Wijaya, M.E., Nawir, A.A.: Analysis study of open data implementation to improve public policy making process in jakarta provincial government based on dynamic governance. Admin. Jurnal Ilmiah Administrasi Publik Dan Pembangunan **10**(2), 93–102 (2019). https://doi.org/10.23960/administratio.v10i2.107

Yousefinaghani, S., Dara, R., Mubareka, S., Papadopoulos, A., Sharif, S.: An analysis of COVID-19 vaccine sentiments and opinions on twitter. Int. J. Infect. Dis. **108**, 256–262 (2021). https://doi.org/10.1016/j.ijid.2021.05.059

Zeemering, E.S.: Functional fragmentation in city hall and twitter communication during the COVID-19 pandemic: evidence from Atlanta, San Francisco, and Washington DC. Govern. Inf. Q. **38**(1), 101539, (2021). https://doi.org/10.1016/j.giq.2020.101539

Zhu, L., Anagondahalli, D., Zhang, A.: Social media and culture in crisis communication : McDonald ' s and KFC crises management in China. Publ. Relat. Rev. **43**(3), 487–492 (2017). https://doi.org/10.1016/j.pubrev.2017.03.006

Author Index

Abao, Roland   I-247, I-370, I-643, II-48
Acacio-Claro, Paulyn Jean   I-281
Acero, Nibaldo   I-13
Alvarez, Claudio   II-3
Anand, Pranit   II-21
Atmosukarto, Indriyati   I-631
Aubrun, Frédéric   II-268
Aureus, Jelly   I-247, I-370, I-643, II-48

Baccarella, Christian V.   I-617
Balasescu, Alexandru   I-578
Barbosa, Leticia   I-3
Bascur, Camila   II-278
Bautista, Maria Cristina   I-281
Becker, Stefan   I-449
Björkman, Mårten   I-65
Bodrunova, Svetlana S.   I-468
Bohle, Florencia   II-422
Botella, Federico   I-205
Bracci, Margherita   II-147
Brzezinska, Magdalena   II-30

Calderon, Juan Felipe   I-13
Cano, Sandra   I-134
Cao, Cong   I-170
Castro, John W.   I-28
Chen, Zichen   I-160
Christou, Eirini   I-485
Co, Nicole Allison   I-247, I-370, I-643
Coman, Adela   I-503
Cromarty, Edward   II-30

de Dios Valenzuela, Juan   II-3
de Lara-Tuprio, Elvira   II-48
de Paulo, Beatriz   I-47
Debailly, Renaud   I-517
Demir Kanik, Sumeyra U.   I-65
Dery, Taume   I-83
Dettmer, Niklas   II-210
Dey, Priyanka   I-267
Doerner, Ralf   I-97

Economides, Anastasios A.   I-485
Estuar, Maria Regina Justina   I-247, I-281, I-370, I-643, II-48
Eto, Shoki   I-110
Eugeni, Ruggero   II-223

Farouqa, Georgina   II-67, II-77
Fernández-Robin, Cristóbal   II-235, II-422
Flores, Pablo   II-235
Froger-Lefebvre, Juliette   I-529
Fruchter, Renate   I-545
Fujikake, Kazuhiro   I-146
Fukuda, Karin   I-294
Fushimi, Kengo   I-123

García, Matías   I-134
Garcia-Vaquero, Marco   II-91
Garnica, Ignacio   I-28
Ghadirzadeh, Ali   I-65
Godoy, María Paz   II-248
González, Patricia   II-331
Gordeev, Denis   I-309
Goto, Masayuki   II-344, II-388
Grigore, Ana-Maria   I-503
Guidi, Stefano   II-147
Guneysu Ozgur, Arzu   I-65

Hamam, Doaa   II-103
Han, Han   I-217
Haroon, Harshita Aini   II-115
Hulahi, Ekklesia   I-559
Hysaj, Ajrina   II-67, II-77, II-103, II-115, II-199

Inal, Yavuz   I-83
Ito, Ayaka   II-127

Janach, Christoph   I-588
Jauffret, Marie-Nathalie   II-268
Jeaningros, Hugo   I-517
Jin, Qun   I-294, I-332, I-408

Kanamori, Hitoshi   I-146
Kharisma, Bella   I-568

Kimura, Ryusei   I-146
Koren, Leon   I-318
Kragic, Danica   I-65
Krause, Wanda   I-578
Kröger, Niclas   I-588
Kurniawan, Danang   I-559, I-568, I-661

Lakhdari, Abdallah   I-358
Lejeune, Gaël   I-517
Leoveanu, Valentin Mihai   I-503
Lim, Sufang   I-389
Loh, Yuanchao   I-160
Loilatu, Mohammad Jafar   I-559, I-568

Malinen, Sanna   I-602
Marchigiani, Enrica   II-147
Márquez, Leslie   II-331
Matsuoka, Yui   II-344
Matus, Nicolás   II-127
McCoy, Scott   II-235
Mendoza, Marcelo   I-422
Miao, Chunyan   I-389
Montecinos, Catalina   II-278
Morooka, Ryo   II-292
Muallidin, Isnaini   I-559, I-568
Murata, Erina   I-332
Myers, Marie J.   II-136

Namatame, Takashi   I-110, II-292, II-359,
     II-403
Nepiyuschikh, Dmitry   I-468
Nigmatullina, Kamilla   I-345
Nishida, Toyoaki   I-545
Nurmandi, Achmad   I-559, I-568

Okada, Shogo   I-146
Opresnik, Marc Oliver   II-308
Orsag, Luka   I-318
Osop, Hamzah   I-358
Otake, Kohei   I-110, II-292, II-359, II-403

Palmitesta, Paola   II-147
Pangan, Zachary   II-48
Parlangeli, Oronzo   II-147
Parmaxi, Antigoni   I-485
Peng, Haoxuan   I-170
Pereira Neto, André   I-3
Perifanou, Maria   I-485
Piki, Andriani   II-161
Potapov, Vsevolod   I-309

Providel, Eliana   I-422
Pulmano, Christian   I-281

Qiu, Yang   I-389
Quaresma, Manuela   I-47
Quiñones, Daniela   II-317

Rapp, Maximilian   I-588
Ricko, Andrija   I-318
Rodossky, Nikolay   I-345
Rodrigueza, Rey   II-48
Rogers, Kristine M.   I-185
Rojas, Luis   II-317
Rojas, Luis A.   I-13, I-28, II-3
Rösch, Jürgen   I-617
Rosenberg, Duska   I-545
Roziqin, Ali   I-661
Rusu, Cristian   I-205, II-127, II-248, II-331
Rusu, Virginica   II-331

Sakai, Yuta   II-344
Salahudin   I-661
Sandu, Mihaela Cornelia   I-503
Sandu, Roxana   II-180
Scheiner, Christian W.   I-449
Schurz, Katharina   II-210
Soh, Cheng Lock Donny   I-631
Sorbello, Katrina   I-13
Stipancic, Tomislav   I-318
Sugon, Quirino   I-281
Suleiman, Basem   I-358
Suleymanova, Sara   II-199
Sumikawa, Yasunobu   I-123

Tago, Kiichi   I-332
Tamayo, Lenard Paulo   II-48
Tan, Austin Sebastien   I-247, I-370, I-643
Tan, Hans Calvin   I-247, I-370, I-643
Tan, Toh Hsiang Benny   I-389
Tanaka, Takahiro   I-146
Taqwa Sihidi, Iradhad   I-661
Teng, Timothy Robin   II-48
Terasawa, Shinnosuke   II-359
Thelen, Tobias   II-210
Tinland, Julia   I-529

Ueda, Masao   II-375
Ugalde, Jonathan   II-248

Valencia, Katherine   I-205
Valera, Madeleine   I-281
Villamor, Dennis Andrew   I-281

Wang, Yenjou   I-408
Weber, Felix   II-210

Yamagiwa, Ayako   II-388
Yamaguchi, Haruki   II-403
Yáñez, Diego   II-235, II-422
Yen, Neil   I-408
Yin, Wenjie   I-65

Yoshihara, Yuki   I-146
Yu, Han   I-160, I-437

Zapata, Andrés   I-422
Zhai, ChengXiang   I-267
Zhang, Hantian   I-227
Zhang, Jiehuang   I-437
Zhang, Mu   I-217
Zhang, Xuanwu   I-170
Zhao, Yansong   I-160

Printed in the United States
by Baker & Taylor Publisher Services

Printed in the United States
by Baker & Taylor Publisher Services